applied analysis for physicists and engineers

bernard j. rice
university of dayton

prindle, weber & schmidt, inc.
boston, massachusetts london sydney

Prindle, Weber & Schmidt, Inc.
53 State Street
Boston, Massachusetts 02109

Library of Congress Catalog Number 71-185189

SBN 87150-137-6

Printed in U.S.A.

preface

This book is an outgrowth of a sequence in applied mathematics that I have taught during the past five years at the University of Dayton. Enrollment consisted of undergraduate students majoring in Physics, Computer Science, and Engineering. The latter two-thirds of the material was also used in a course designed for first year graduate students from the School of Engineering.

The first post calculus course in applied mathematics is widely varied because of the diversity of backgrounds of the typical student enrolling for such a course. Some students have substantial backgrounds in differential equations, linear algebra, or vector analysis, and these same students show a distinct inability to cope with even the very elementary notions for curves and surfaces. Since no natural breakoff can be assumed, teachers of these courses out of necessity are forced to lean heavily on intuitive concepts rather than on a mathematical, rigorous development of certain topics.

To assure interest and at the same time to allow for review, this book begins by acquainting, or reacquainting, the reader with the modern notion of a vector. Hence, no linear algebra need be assumed. The treatment of most of the topics seeks to explain rather than to prove, to motivate understanding rather than mathematical precision. With so much contemporary mathematics to learn, a student in the applied fields must have his mathematical maturity developed, but hopefully not at the cost of becoming disillusioned with undue stress on detail. Neither must he consider mathematics as a purely formal manipulative skill.

In all cases, the coverage is intended only to be introductory, for I believe that excellent books exist on every topic considered. Students have often expressed the need of a readable textbook at this level to supplement these many good and readily available reference books. With this in mind, I have often been content to introduce a subject with an emphasis on the understanding of the notation and terminology. It has been my experience that this is usually sufficient to allow the physics or engineering teacher to build profitably on this foundation.

I have included a review (Appendix A) on the general topics of convergence of sequences, series, and improper integrals since many of my own students have confessed a need for such a review. Appendix A also includes a discussion of uniform convergence which must be covered in detail if one wishes a thorough coverage of Special Functions.

I have usually covered Chapters 1–5 in one semester, Chapters 1, 6–8, 10, 11, with some omissions, in a second semester, and Chapters 12–16 in another. On occasion I have substituted Chapter 9 for Chapter 16 in the third semester.

This Preface provides me the opportunity to acknowledge publicly the invaluable assistance of my colleague Martinus Esser. His recommendations significantly improved the manuscript while at the same time he improved some of my mathematical ideas. Any inadequacies of expression, awkward mathematics, or dubious pedagogy still present in the book testify to the fact that he was not completely successful in changing my mind.

Bernard J. Rice

contents

1 *vector spaces* 1

Introduction *1*. Geometric vectors *1*. Ordered triples *3*. Ordered n-tuples *5*. Piecewise continuous functions *5*. Inner product spaces *7*. Linear independence *13*. Matrices *19*.

2 *first-order ordinary differential equations* 25

Introduction *25*. Some elementary terminology *25*. Separable first-order equations *27*. Substitutions *33*. Exact differential equations *37*. Integrating factors; Linear equations of the first order *42*. Approximation by straight lines *51*. An iterative method *55*.

3 *linear differential equations* 59

Differential operators *59*. The homogeneous equation *65*. Solution to homogeneous equations with constant coefficients *68*. The nonhomogeneous linear differential equation *72*. Successive antidifferentiations *74*. Method of undetermined coefficients *76*. Method of variation of parameters *82*. Application to electric circuits *87*. Application to mechanical systems *94*. The sliding block *97*. Systems of linear differential equations *100*.

4 *the laplace transformation* 105

The general idea of a transform *105*. Existence of the Laplace transform *108*. Some general properties; The first shifting theorem *110*.

The inverse Laplace transform *119*. Solutions to ordinary differential equations *124*. A "turn-on" function; The second shifting theorem *129*. The convolution theorem; Integral equations *137*.

5 *infinite series methods* 141

Method of assuming an infinite series solution *141*. Taylor series *142*. Comments on the form of a solution *144*. Method of undetermined coefficients *147*. Singular points of a differential equation *154*. Method of Frobenius *156*.

6 *classical elementary vector analysis* 161

Introduction *161*. Vector algebra *161*. Vector functions *169*. Vector calculus *171*. Surfaces *174*. Curves *177*. Tangents to curves *185*. Curves on surfaces *190*. Dynamics of a particle *193*.

7 *differentiation of fields* 199

Introduction *199*. Scalar and vector fields *199*. Sketching scalar and vector fields *203*. Variation of a scalar field *207*. Variation of a vector field: divergence *215*. Variation of a vector field: curl *220*.

8 *integration of vector fields* 227

Introduction *227*. Line integration *227*. Line integral of a vector field *232*. Evaluation of line integrals *236*. Independence of path *240*. Surface area *246*. Surface integrals *250*. Surface integrals of vector fields *255*. Green's theorem in the plane *259*. Stokes' theorem *268*. Divergence theorem *275*. Summary of techniques of evaluation *280*.

9 *tensor analysis* 285

Introduction *285*. Notation *285*. Basis vectors and coordinate systems *287*. Reciprocal bases *293*. Covariant and contravariant components *296*. Transformation laws *299*. Relationship between covariant and contravariant components *300*. Tensors *305*. Algebra of tensors *309*. Covariant differentiation *313*.

10 fourier expansions 319

Introduction *319.* The space $PC\,[-L, L]$ *320.* Two kinds of equality *322.* Fourier coefficients *327.* Subspaces of $PC\,[-L, L]$ *333.* Properties of Fourier coefficients *336.* Convergence of a Fourier series *340.* Sine and cosine series *345.* Which Fourier expansion to use? *349.* Approximation in the mean *353.* Pointwise approximation *359.* The Fourier integral *362.* The Sturm-Liouville problem *372.*

11 partial differential equations 377

Introduction *377.* Boundary value problems *379.* Superposition *381.* Three methods *383.* The wave equation: derivation *390.* Solution by D'Alembert's method *393.* Product solution *397.* The heat equation *400.*

12 complex functions 405

The complex field *405.* The complex plane *406.* Functions and mappings *411.* Derivative of a complex function *415.* Integral and fractional powers of z *422.* The exponential function *427.* Complex trigonometric functions *431.* Inverse functions *434.* The power function *435.*

13 mapping by analytic functions 439

The point at infinity *439.* Curves in the complex plane *440.* Conformal mapping *442.* Bilinear transformations *446.* Special linear mappings *452.* Table of mappings *455.* Application to boundary value problems *456.*

14 integration in the complex plane 463

Introduction *463.* Line integral *463.* Fundamental theorem of complex integration *471.* The Cauchy-Goursat theorem *476.* Principle of deformation of path *477.* Cauchy's integral formula *481.* Derivatives of an analytic function *484.*

15 complex series 489

Introduction 489. Elementary definitions 489. Series of variable terms 493. Power series 494. Power series representation of a function 497. Laurent series 504. Poles and zeros 510. Residues 512. The residue theorem 517. Evaluation of real integrals 519.

16 special functions 531

Introduction 531. Functions defined by improper integrals 531. The gamma function 532. The beta function 539. Functions defined by indefinite integrals 542. Functions defined by differential equations 546. Bessel's functions 547. Modified Bessel functions 550. Some fundamental identities 552. Approximations to the Bessel functions 555. Generating function; Bessel's integral representation 556. Zeros of the Bessel functions 558. Fourier-Bessel series 562. The vibrating membrane 567. Phase modulation of radio waves 569.

appendix a: sequences, improper integrals, and series 572

Introduction 572. Sequences of numbers 572. Sequences of functions 576. Improper integrals 580. Improper integrals with a parameter 586. Infinite series of constants 590. Series of functions 595. Power series 599. Taylor's series 602.

appendix b: proof of stokes' theorem 604

appendix c: table of mappings 606

appendix d: table of values of $J_0(x)$ and $J_1(x)$ 610

appendix e: references 613

answers to exercises 615

index 675

vector spaces

1

1.1 introduction

In this introductory chapter we will describe what is meant by a mathematical quantity called a "vector." Much of this book could be considered a discussion of vectors, even though stress is not continuously placed on that perspective. Initially we give the general setting in which you should understand the concept of a vector, a setting which is admittedly abstract. The specialized material of later chapters should reinforce the abstraction of this chapter.

The material of this chapter will probably be only remotely necessary for the understanding of the remainder of the book. However, even a superficial understanding will give an undeniable thread of unity to many seemingly disconnected topics.

1.2 geometric vectors

In more elementary mathematics courses and in certain applications you have discussed one concept of a vector. The discussion was probably restricted to two- or three-dimensional space and a vector was probably identified with the idea of an *arrow*. We will call this a *geometric* vector to distinguish it from the generalization. A geometric vector was defined as a quantity having "direction" and "length." Since "direction" and "length" are left as undefined terms, this elementary definition has its shortcomings. However, if you accept these terms in the sense of everyday Euclidean geometry, the properties of direction and length are self-evident, and consequently arrows or geometric vectors have intuitive meaning. This intuitive meaning of a geometric vector was, and is, sufficient for many applications.

Thus a geometric vector, in the elementary sense, is usually represented by an arrow. You should realize that this vector is not the same as the *one* arrow but is the collection of *all* arrows which have the same direction and length as the representative arrow. The fact that any arrow may be replaced by a parallel arrow and still represent the same vector is expressed by saying that an arrow is "free."

Initially, two elementary operations were defined on the set of geometric vectors: addition and scalar multiplication. It probably is not a concern of the beginner, but it is precisely these two operations and certain selected properties that they have, which allow a more precise definition of the idea of a vector. This more precise definition also yields an immediate generalization of the concept for use in more diverse applications.

Addition of geometric vectors is a binary operation in which the sum of two arrows is defined by constructing the second arrow at the tail of the

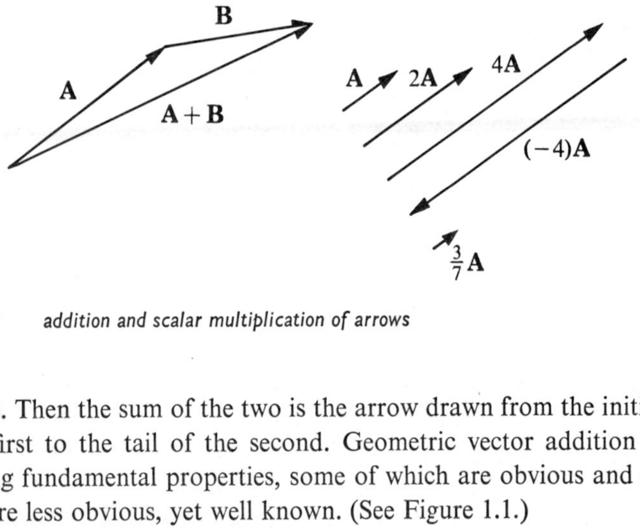

figure 1.1 *addition and scalar multiplication of arrows*

first one. Then the sum of the two is the arrow drawn from the initial point of the first to the tail of the second. Geometric vector addition has the following fundamental properties, some of which are obvious and some of which are less obvious, yet well known. (See Figure 1.1.)

1. The addition of two arrows yields another arrow. Technically we say the set of arrows is *closed* with respect to addition.
2. For any two arrows $\mathbf{A}$ and $\mathbf{B}$, $\mathbf{A}+\mathbf{B}=\mathbf{B}+\mathbf{A}$ (Commutative Property).
3. For any three arrows $\mathbf{A}$, $\mathbf{B}$, and $\mathbf{C}$, $(\mathbf{A}+\mathbf{B})+\mathbf{C}=\mathbf{A}+(\mathbf{B}+\mathbf{C})$ (Associative Property).
4. We define a special arrow, called the "zero arrow," denoted by $\mathbf{0}$ with the property that for any other arrow $\mathbf{A}$, $\mathbf{A}+\mathbf{0}=\mathbf{A}$. In elementary work

we define $\mathbf{0}$ by saying that it has zero length. We leave the direction undefined although it is useful to think of $\mathbf{0}$ having any direction you wish.

5. Corresponding to each arrow $\mathbf{A}$, there is another arrow $\mathbf{B}$, such that $\mathbf{A}+\mathbf{B}=\mathbf{0}$. $\mathbf{B}$ is called the additive inverse of $\mathbf{A}$ and is denoted by $(-\mathbf{A})$. $(-\mathbf{A})$ has the same length as $\mathbf{A}$ but is in the opposite direction.

Multiplication by a scalar of geometric vectors is an operation by a real number on an arrow. For any real number c, the quantity $c\mathbf{A}$ is an arrow $|c|$ times as long as $\mathbf{A}$, pointing in the same direction as $\mathbf{A}$ if c is positive and in the opposite direction from $\mathbf{A}$ if c is negative. The following properties are "fundamental" to this operation.

6. Multiplication by a scalar is possible and unique for every scalar and every arrow.
7. For any two real numbers b and c, $b(c\mathbf{A})=(bc)\mathbf{A}$.
8. Scalar multiplication is distributive over vector addition:

$$a(\mathbf{A}+\mathbf{B})=a\mathbf{A}+a\mathbf{B}.$$

9. $1\mathbf{A}=\mathbf{A}$; $\quad 0\mathbf{A}=\mathbf{0}$; $\quad (-1)\mathbf{A}=-\mathbf{A}$.

The space of geometric vectors is based upon the sound but intuitive notion that an arrow has direction and length. It is often called a "concrete Euclidean vector space."

The notion of a Euclidean vector space is a special case of the more general concept of an *abstract vector space*. In general, a set of objects may be called a vector space if it has two operations, addition and scalar multiplication, which satisfy properties 1 to 9, or more precisely the more abstract statement of these properties. In their abstract form, properties 1–9 are called the "vector space axioms." If a set is a vector space, its elements may be called vectors.

There are properties which are specific to Euclidean vectors but which are not assumed as vector space axioms. These properties are often meaningless in abstract vector spaces. For example, abstract vector spaces are not necessarily "three-dimensional," nor are cross products or dot products defined. There are vector spaces in which the elements look nothing like arrows.

1.3 ordered triples

By the set of ordered triples of real numbers is meant the set $\{(a_1, a_2, a_3), a_i$ a real number$\}$ where i means that this applies to every subscript

1, 2, 3. If we are to label this set correctly as a "vector space," we must tell how to add elements of the set and how to multiply each ordered triple by a real number. Then every one of the nine fundamental properties listed for geometric vectors must also be satisfied in this space.

Addition of ordered triples is defined by component addition, that is

$$(a_1, a_2, a_3) + (b_1, b_2, b_3) = (a_1 + b_1, a_2 + b_2, a_3 + b_3).$$

Likewise, multiplication by a real number is defined by

$$c(a_1, a_2, a_3) = (ca_1, ca_2, ca_3).$$

Then it is an easy task to show that the nine essential properties are satisfied and hence the elements of the space are entitled to be called vectors, even though "ordered triple" might be more popular. The terminology "vector space," or more completely and accurately, "Cartesian three-dimensional vector space," is used to describe this space if the intent is to apply the vector operations specifically to the set of ordered triples.

The zero vector in this space is the ordered triple (0, 0, 0). An ordered triple has no natural length or direction, although in some circumstances it is desirable and convenient to associate with each vector (a_1, a_2, a_3) a length defined by $(a_1^2 + a_2^2 + a_3^2)^{1/2}$.

One of the best known applications of the vector space of ordered triples is its use in describing analytically the space of geometric vectors. We give geometric meaning to an ordered triple by allowing it to represent a point in three dimensions. Then the arrow from the origin to the point can be represented by the same ordered triple as the one which represents the point. (See Figure 1.2.)

In this representation process, we have allowed the coordinate system

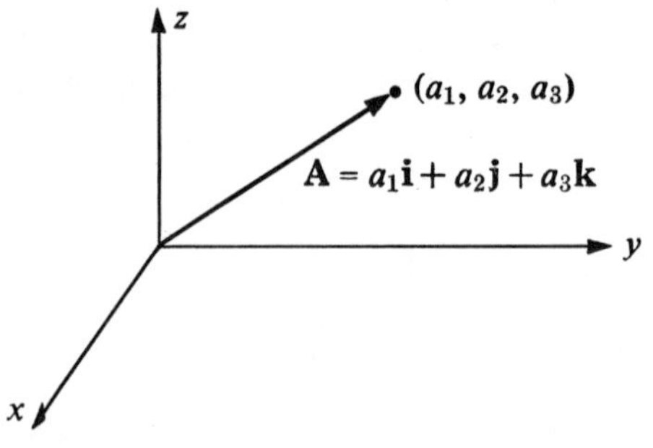

figure 1.2 *representation of arrows with ordered triples*

to influence the analytic description of the set of geometric vectors. For the Cartesian coordinate systems used in elementary calculus this creates no real difficulty and hence arrows and ordered triples may be used almost interchangeably. A re-examination of the process is required when other coordinate systems are used.

1.4 ordered n-tuples

We can describe an ordered 4-tuple, 5-tuple, or n-tuple in exactly the same way as we did ordered triples in the previous section. For example, an ordered 5-tuple looks like $(a_1, a_2, a_3, a_4, a_5)$ with vector operations

$$(a_1, a_2, a_3, a_4, a_5) + (b_1, b_2, b_3, b_4, b_5) = (a_1 + b_1, a_2 + b_2, a_3 + b_3, a_4 + b_4, a_5 + b_5),$$
$$c(a_1, a_2, a_3, a_4, a_5) = (ca_1, ca_2, ca_3, ca_4, ca_5).$$

Similarly, the set of ordered n-tuples has two operations defined on it which are natural extensions of the operations on ordered triples. As in the case of ordered triples, the operations satisfy the vector space axioms and thus the n-tuple space is justifiably called a vector space. Do not worry about whether an ordered 5-tuple, for example, has direction or length since these properties are not vector space axioms.

1.5 piecewise continuous functions

We now describe a very important space in which the elements can properly be considered to be vectors even though they in no way resemble arrows. Denote by $C(a, b)$ the set of continuous functions on the interval (a, b) and by $PC(a, b)$ the set of piecewise continuous functions on that interval. Examples of graphs of members of $C(a, b)$ are shown in Figure 1.3,

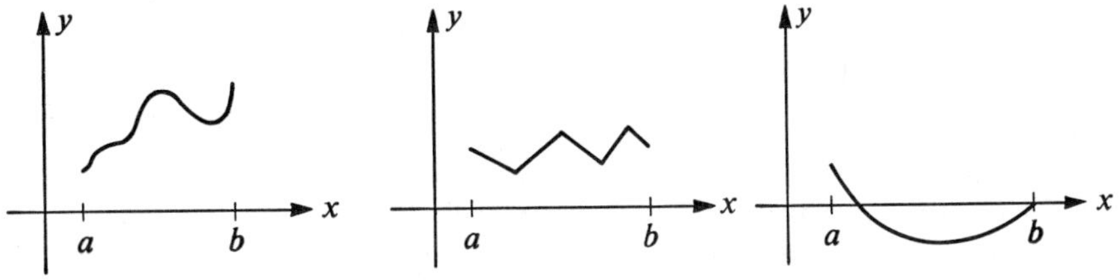

figure 1.3 members of $C(a, b)$

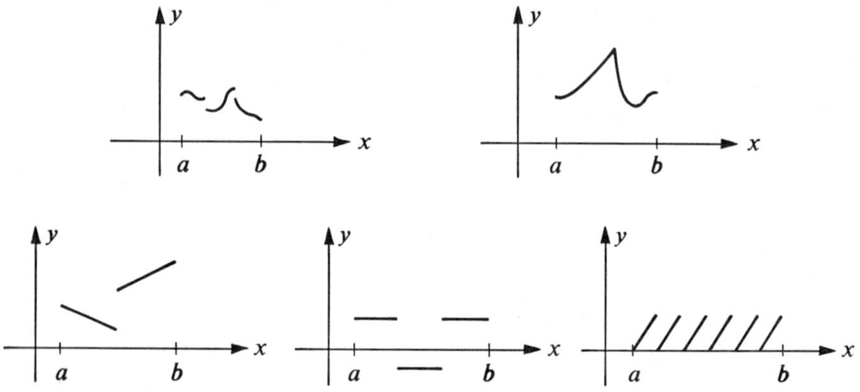

figure 1.4 *members of PC (a, b)*

and some piecewise continuous† functions are shown in Figure 1.4. Of course, all those shown in Figure 1.3 are piecewise continuous also since $C(a, b) \subset PC(a, b)$.

On $PC(a, b)$ we define operations of addition and multiplication by a scalar:

1. By $f + g$ is meant the function whose value at x is given by $f(x) + g(x)$.
2. By cf is meant the function whose value at x is given by $cf(x)$.

example 1.1: Let $f(x) = x$ and $g(x) = x^2$. Find and sketch the functions $f + g$, $\frac{1}{2}f$, and $2g$.

solution: From the definitions

$$(f + g)(x) = f(x) + g(x) = x + x^2.$$
$$(\tfrac{1}{2}f)(x) = \tfrac{1}{2}f(x) = x/2.$$
$$(2g)(x) = 2g(x) = 2x^2.$$

The graphs are sketched in Figure 1.5.

The fact that the nine fundamental properties are satisfied by the set $PC(a, b)$ with the operations of addition and scalar multiplication as defined in the preceding paragraph can be readily proved. We therefore say that

† A function f is *piecewise continuous* on the interval (a, b) if it is defined at every point on the interval and is such that the interval can be subdivided into finitely many intervals in each of which the function is continuous and the limits from the right and from the left on any subinterval are finite. Note that a function which is piecewise continuous on a closed interval is bounded on that interval.

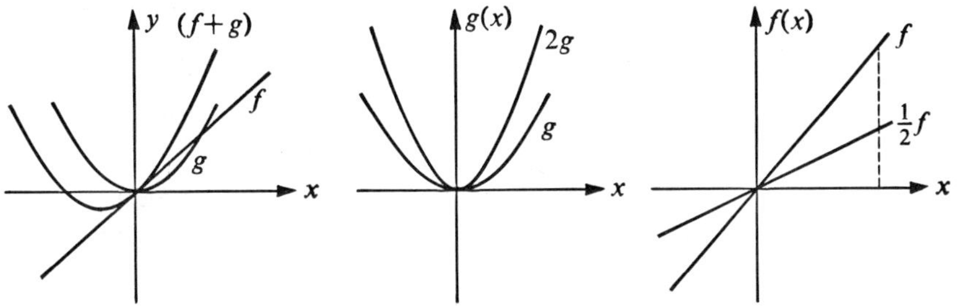

figure 1.5

$PC(a, b)$ together with the operations defined above form a vector space. Often we abbreviate the terminology and say, a bit incorrectly, that the set $PC(a, b)$ is a vector space. Each function is called a vector if we intend to emphasize the vector operations on the space. Are you surprised to be referring to the function f in the previous example as a vector? Are you wondering if there is any useful application for $PC(a, b)$ as there is for the set of arrows? Later, in Chapter 10, you will see why it is convenient to conceive of $PC(a, b)$ as a vector space. At present, you should realize that the term vector is often used in a much more general sense than arrows in three dimensional space.

In passing we note that function spaces $PC[a, b), PC(a, b]$, and $PC[a, b]$ are defined analogously to $PC(a, b)$. Further, any operations defined for $PC(a, b)$ will be appropriate for the other spaces too.

Admittedly, the function space $PC(a, b)$ and the space of arrows are completely different. But both *are* vector spaces. Elements of both spaces can be called vectors. Some vector spaces consist of elements to which we do not care to associate the notion of length and direction. Vector spaces in which a length can be assigned to each element are called *normed.* We will consider only those normed vector spaces in which there is an *inner product* defined.

1.6 inner product spaces

In the geometric vector space with which you are most familiar from elementary calculus, the four most common operations are (1) addition, (2) multiplication by a scalar, (3) dot product, and (4) cross product. Up to now we have generalized the first two operations by an abstraction, thus obtaining the notion of an abstract vector space. Now we shall consider those vector spaces on which a generalization of the dot product is defined, obtaining what is called an *inner product* space. (The notion of cross product is intrinsically restricted to three dimensions, and will never be generalized in this book.)

definition 1.1

An *inner product* is a rule which associates with every two vectors v_1, v_2 a real number, written $v_1 \cdot v_2$ or (v_1, v_2) subject to the following axioms:

1. For all v_1, v_2, v_3 in the vector space and for all real numbers a and b,

$$(av_1 + bv_2)\cdot(v_3) = a\,(v_1 \cdot v_3) + b\,(v_2 \cdot v_3).$$

   This is known as the linearity property.

2. For all v_1, v_2 in the vector space,

$$v_1 \cdot v_2 = v_2 \cdot v_1 .$$

   This property is variously called the symmetric property or commutative property.

3. For all vectors v,

$$v \cdot v \geqq 0$$

   and $v \cdot v = 0$ if and only if v is the zero vector.
   This is called the positive definiteness property.

example 1.2: Let V be the space of two-dimensional arrows. Show that the dot product is an inner product.

solution: Recall from elementary calculus (or see Section 6.2 of this book) that if **A** and **B** are two arrows, then

$$\mathbf{A}\cdot\mathbf{B} = |\mathbf{A}|\,|\mathbf{B}|\cos\theta$$

where $|\mathbf{A}|$ designates the length of **A** and θ is the angle between the two arrows when they have a common initial point.

1. $(\mathbf{A}+\mathbf{B})\cdot(\mathbf{C}) = \mathbf{A}\cdot\mathbf{C} + \mathbf{B}\cdot\mathbf{C}$. This property is called the distributive property of the dot product over addition. Also

$$(c\mathbf{A})\cdot\mathbf{B} = c\,|\mathbf{A}|\,|\mathbf{B}|\cos\theta = c\,(\mathbf{A}\cdot\mathbf{B}) \qquad \text{for } c > 0$$

   and

$$(c\mathbf{A})\cdot\mathbf{B} = |c|\,|\mathbf{A}|\,|\mathbf{B}|\cos(\pi-\theta) = c\,(\mathbf{A}\cdot\mathbf{B}) \qquad \text{for } c < 0,$$

   which proves the linearity property.

2. $\mathbf{A}\cdot\mathbf{B} = |\mathbf{A}|\,|\mathbf{B}|\cos\theta = |\mathbf{B}|\,|\mathbf{A}|\cos\theta = \mathbf{B}\cdot\mathbf{A}$.
3. $\mathbf{A}\cdot\mathbf{A} = |\mathbf{A}|^2$ which is obviously nonzero unless $\mathbf{A} = \mathbf{0}$.

example 1.3: Show how to define an inner product on the space of ordered triples.

solution: Let (a_1, a_2, a_3) and (b_1, b_2, b_3) be any two elements of the space. Then the product defined by

$$(a_1, a_2, a_3)\cdot(b_1, b_2, b_3) = a_1b_1 + a_2b_2 + a_3b_3$$

can be verified to be an inner product in a straightforward manner. (See the exercises at the end of this section.)

The inner products on the space of arrows in three-dimensional space and for the space of ordered triples are closely related. In fact when the relation is made properly, the inner product on the space of ordered triples is used to calculate the dot product of two geometric arrows.

The vector space of piecewise continuous functions is also an inner product space.

definition 1.2

Let f and g be any two elements of $PC[a, b]$. The *integral inner product*, denoted by (f, g), is defined to be

$$(f, g) = \int_a^b f(x)\, g(x)\, dx.$$

The properties of the inner product are easy to verify.

example 1.4: Calculate (f, g) in $PC[0, 1]$ and then in $PC[0, 2]$ if $f(x) = x$ and $g(x) = x^2$.

solution: In $PC[0, 1]$,

$$(f, g) = \int_0^1 (x)\,(x^2)\, dx = \tfrac{1}{4}.$$

In $PC[0, 2]$,

$$(f, g) = \int_0^2 (x)\,(x^2)\, dx = 4.$$

Once an inner product is defined on a vector space, the so-called *norm* of an element is defined automatically.

definition 1.3

The *norm* of a vector v, is denoted by $\|v\|$ and is defined by

$$\|v\| = \sqrt{v\cdot v}.$$

In the vector space of arrows, $\|\mathbf{A}\| = \sqrt{\mathbf{A}\cdot\mathbf{A}} = |\mathbf{A}|$ which is just the length of **A**. In the vector space of 3-tuples,

$$\|(a_1, a_2, a_3)\| = \sqrt{a_1^2 + a_2^2 + a_3^2}$$

which can be interpreted to be the distance from the origin to the point with coordinates (a_1, a_2, a_3). In the function space $PC(a, b)$

$$\|f\| = \left\{\int_a^b [f(x)]^2\, dx\right\}^{1/2}.$$

In the case where no confusion with absolute value arises, we may write $|v|$ instead of $\|v\|$. Thus for three-dimensional arrows we write $|\mathbf{A}|$ for its length, but for a function f, we write $\|f\|$ to distinguish it from $|f(x)|$.

If an inner product is defined, we not only can define the norm (which is the obvious generalization of the concept of length) but we can also make the important definition of orthogonality (which is the generalization of the concept of perpendicularity).

definition 1.4

Two vectors v_1 and v_2 are said to be *orthogonal* if and only if $v_1 \cdot v_2 = 0$. A set of vectors is an *orthogonal set* if every two distinct vectors of the set are orthogonal.

In the case of the space of geometric vectors, if the dot product $\mathbf{A}\cdot\mathbf{B}$ of two nonzero arrows is zero, then $|\mathbf{A}|\,|\mathbf{B}|\cos\theta = 0$ which would imply that $\theta = 90°$. Hence, in this case, orthogonality would reduce to the physical concept of perpendicularity.

In the space of ordered triples, if the dot product of (a_1, a_2, a_3) and (b_1, b_2, b_3) is zero, then $a_1b_1 + a_2b_2 + a_3b_3 = 0$. When ordered triples are used to describe analytically the geometric vectors, this is a very convenient check for perpendicularity.

In function spaces, you must be careful to specify the interval of orthogonality. For it does happen that a pair of functions may be orthogonal on one interval and not on another.

example 1.5: Show that the functions $\sin x$ and $\cos x$ are orthogonal in $PC[-\pi, \pi]$ but not in $PC[0, \pi/4]$.

solution: In $PC[-\pi, \pi]$,

$$(\sin x, \cos x) = \int_{-\pi}^{\pi} \sin x \cos x\, dx = 0.$$

In $PC[0, \pi/4]$,

$$[\sin x, \cos x] = \int_0^{\pi/4} \sin x \cos x \, dx = \tfrac{1}{4} \neq 0.$$

Do not read anything "physical" into the fact that $\sin x$ and $\cos x$ are orthogonal on $[-\pi, \pi]$. This does not mean for example that the graphs of the two functions intersect at right angles, as it is easy to show.

It is often necessary to divide an arrow, **A**, by its length to obtain a so-called unit arrow, $\mathbf{A}/|\mathbf{A}|$. In a more general setting, we will want to divide a vector, v, by its norm to obtain $v/\|v\|$. Such a vector is said to be normalized. If we normalize each vector of an orthogonal set we call the set *orthonormal.* Why this should be a desirable quality of a set of vectors will become clearer in the next section.

example 1.6: Show that the set of functions $\{\cos mx\}_{m=0}^{\infty}$ is an orthogonal set on $[0, \pi]$. Find the corresponding orthonormal set.

solution: If $m \neq n$,

$$(\cos mx, \cos nx) = \int_0^{\pi} \cos mx \cos nx \, dx$$

$$= \tfrac{1}{2} \int_0^{\pi} [\cos(m+n)x + \cos(m-n)x] \, dx$$

$$= \left[\frac{\sin(m+n)x}{2(m+n)} + \frac{\sin(m-n)x}{2(m-n)}\right]_0^{\pi} = 0,$$

which proves the orthogonality.

For $m = 0$, $\cos 0x = 1$ and

$$\|1\| = \left[\int_0^{\pi} (1)^2 \, dx\right]^{1/2} = \sqrt{\pi}.$$

For $m \neq 0$,

$$\|\cos mx\| = \left[\int_0^{\pi} \cos^2 mx \, dx\right]^{1/2} = \sqrt{\pi/2}.$$

Hence, the corresponding orthonormal set is

$$\left\{\frac{1}{\sqrt{\pi}}, \frac{\sqrt{2}}{\sqrt{\pi}} \cos x, \frac{\sqrt{2}}{\sqrt{\pi}} \cos 2x, \cdots, \frac{\sqrt{2}}{\sqrt{\pi}} \cos mx, \cdots\right\}.$$

The norm satisfies the important triangle inequality

$$\|v_1 + v_2\| \leqq \|v_1\| + \|v_2\|$$

and therefore the distance between two vectors satisfies

$$\|v_1 - v_2\| \leqq \|v_1 - v_3\| + \|v_3 - v_2\|.$$

The proof may be found in the text by Kreider, *et al.* (1966). (See Appendix E.) It is omitted here because the discussion related to it is a bit off the main topic. It is this inequality which tells us that the norm acts in much the same manner as our intuitive notion of length.

example 1.7: Find'the distance between x and x^2 first in $PC[0, 1]$ and then in $PC[0, 2]$.

solution: In $PC[0, 1]$,

$$\|x - x^2\| = \left[\int_0^1 (x - x^2)^2\, dx\right]^{1/2} = \sqrt{1/30}.$$

In $PC[0, 2]$,

$$\|x - x^2\| = \left[\int_0^2 (x - x^2)^2\, dx\right]^{1/2} = 4/\sqrt{15}.$$

In these six sections we have attempted to show how the specific concrete notions of arrows and related concepts are generalized. The following table is intended to summarize the net effect of this generalization.

Specific	*Generalization*
Arrow	Vector
Dot product	Inner product
Length	Norm
Perpendicularity	Orthogonality
Unit arrow	Normalized vector
Set of perpendicular arrows	Orthogonal set
Set of unit perpendicular arrows	Orthonormal set

You should note carefully in each case the abstraction required to implement the generalization.

exercises for sections 1.1–1.6

1. Find the norm of the following functions (with respect to the integral inner product) when considered as elements of $PC[0, 1]$:
   (a) $f(x) = x$. (b) $f(x) = x^2$. (c) $f(x) = \sin \pi x$.
   (d) $f(x) = 1$. (e) $f(x) = \cos \pi x$.
2. Find the norm of the functions of Exercise 1 when considered as elements of $PC[0, 2]$.

3. The first six Legendre polynomials are

$$P_0(x) = 1, \quad P_1(x) = x, \quad P_2(x) = \frac{3x^2 - 1}{2}, \quad P_3(x) = \frac{5x^3 - 3x}{2},$$

$$P_4(x) = \frac{35x^4 - 30x^2 + 3}{8}, \quad P_5(x) = \frac{63x^5 - 70x^3 + 15x}{8}.$$

Show that this is an orthogonal set in $PC[-1, 1]$. Is the set orthonormal?

4. Show that the trigonometric set

$$\{\cos mx\}_{m=0}^{\infty} \cup \{\sin nx\}_{n=1}^{\infty}$$

is orthogonal on the interval $[-\pi, \pi]$.

5. Under what conditions is the norm of a vector equal to zero?
6. Complete Example 1.3.
7. Find constants $a_1, b_1, b_2, c_1, c_2, c_3$ such that

$$\{a_1, b_1x + b_2, c_1x^2 + c_2x + c_3\}$$

is an orthonormal set in the space $PC[0, 1]$.

8. Find the "distance," in the sense of the norm, between the functions:
   (a) x and x^2 considered as members of $PC[-1, 0]$.
   (b) x and $\sin x$ considered as members of $PC[0, \pi]$.
   (c) x and $\sin x + \sin 3x$ considered as members of $PC[0, \pi]$.
9. Let $f(x) = 1$ on $(-1, 1)$ and $g(x) = -1$ on $(-1, 1)$. Compute:
   (a) $\|f - g\|$. (b) $\|f\| - \|g\|$.

1.7 linear independence

In a vector space the fundamental operations are addition of the elements and multiplication of the elements by real numbers. Addition is an operation between *two* vectors, but because of the associativity of the operation, the technique of adding n vectors is well defined. If we begin with a set of vectors and add together scalar multiples of members of the set, the resulting sum is called a *linear combination.*

definition 1.5

Let V be a vector space. A *linear combination* of vectors is a vector expressed in the form

$$\sum_{i=1}^{n} c_i v_i,$$

where c_i are real numbers and the v_i are vectors.

Note that every linear combination of vectors from V is another vector in V. For example, some linear combinations of the set of functions $\{e^x, e^{-x}, \sinh x\}$ are:

(a) $e^x + e^{-x} - \sinh x$, (b) $e^x - e^{-x} - 2\sinh x$,
(c) $0e^x + 0e^{-x} + 0\sinh x$, (d) $3e^{-x}$.

Sometimes a linear combination of the vectors of a set is equal to the zero vector. In fact, if all the c_i's $= 0$, then

$$\sum_{i=1}^{n} c_i v_i$$

is equal to the zero vector. Since that particular combination is so obvious, it is called the *trivial linear combination* of the vectors. In the examples of linear combinations given in the preceding paragraph, note that combination (b) is a nontrivial linear combination which is equal to the zero vector. This discussion is a prelude to the following definition and should help to motivate it.

definition 1.6

A set of vectors $\{v_i\}_{i=1}^{n}$ is said to be *linearly dependent* if and only if *some* nontrivial linear combination of the vectors in the set is equal to the zero vector; otherwise the set is *linearly independent.*

Hence the set of functions $\{e^x, e^{-x}, \sinh x\}$ is a linearly dependent set since there does exist a nontrivial linear combination, specifically $e^x - e^{-x} - 2\sinh x$, which is equal to the zero function. Note that the definition of linear dependence requires that *some* linear combination be zero, not *all* such combinations.

The definition of linear independence is often stated more positively than in Definition 1.6. We say the set

$$\{v_i\}_{i=1}^{n}$$

is linearly independent if whenever

$$\sum_{i=1}^{n} c_i v_i = 0,$$

then $c_i = 0$ for all i.

example 1.8: Under the assumption that the orthonormal set $\{\mathbf{i}, \mathbf{j}, \mathbf{k}\}$ is a linearly independent set in the space of three-dimensional arrows, show that the set $\{3\mathbf{i} + 2\mathbf{j}, \mathbf{i} + 5\mathbf{k}, 6\mathbf{i} + \mathbf{j}\}$ is linearly independent.

solution: A linear combination of the given set is:

$$c_1(3\mathbf{i}+2\mathbf{j})+c_2(\mathbf{i}+5\mathbf{k})+c_3(6\mathbf{i}+\mathbf{j}).$$

Simplifying, and equating this linear combination to the zero arrow, we obtain:

$$(3c_1+c_2+6c_3)\,\mathbf{i}+(2c_1+c_3)\,\mathbf{j}+5c_2\mathbf{k}=\mathbf{0}.$$

Since the set $\{\mathbf{i},\mathbf{j},\mathbf{k}\}$ is linearly independent, the scalar multipliers of **i**, **j**, and **k** are identically zero:

$$\begin{aligned} 3c_1 + c_2 + 6c_3 &= 0, \\ 2c_1 \quad\;\; + c_3 &= 0, \\ 5c_2 \qquad\;\; &= 0, \end{aligned}$$

The only solution for this system is $c_1=c_2=c_3=0$ and hence the set is linearly independent.

example 1.9: Determine the linear independence or linear dependence of the set of three-dimensional arrows expressed in terms of **i**, **j**, and **k** as $\{\mathbf{i}+\mathbf{j}+3\mathbf{k}, \mathbf{i}+2\mathbf{k}, \mathbf{i}-2\mathbf{j}\}$.

solution: We take an arbitrary linear combination of the three given vectors and equate it to the zero vector:

$$c_1(\mathbf{i}+\mathbf{j}+3\mathbf{k})+c_2(\mathbf{i}+2\mathbf{k})+c_3(\mathbf{i}-2\mathbf{j})=\mathbf{0},$$

from which we obtain the three equations

$$c_1+c_2+c_3=0; \qquad c_1-2c_3=0; \qquad 3c_1+2c_2=0.$$

One nontrivial solution is $c_1=2$, $c_2=-3$, $c_3=1$, which shows that the given set is linearly dependent.

One of the distinguishing factors between vector spaces is size. By this we do not mean the number of vectors in the space but rather the number of elements in a special subset of the space called a *basis*. Roughly speaking, a basis is a linearly independent set (nonunique) from which the entire space is derived by linear combinations. More precisely:

definition 1.7

A set of vectors $\{v_i\}_{i=1}^n$ is said to be a *basis* for a vector space V if:
(a) it is linearly independent.
(b) every vector in V can be expressed as a linear combination of the set. We say that the basis *generates* the vector space, V.

Even though a set of basis elements is nonunique, it is a theorem (and a very important one) that if the number of elements in a basis set is finite, then every other basis will have precisely the same number of elements. We can therefore make the following definition.

definition 1.8

The number of elements in any basis of a vector space is called the *dimension* of the space.

You may have noticed that we have unashamedly used the word "dimension" in a rather informal way up to now. Indeed, without saying so, we have assumed and will continue to assume the following important theorem.

theorem 1.1

In a vector space of dimension n, every set of n linearly independent vectors is a basis.

If the set $\{v_i\}_{i=1}^{n}$ is a basis, the symbol $V\{v_1, v_2, \cdots, v_n\}$ is often used to denote the vector space formed from the basis.

example 1.10: Show that $V\{1, x\} = V\{3, x-2\}$.

solution: We show only that $V\{1, x\} \subset V\{3, x-2\}$. The reverse inclusion is done in a similar manner. Let v be any function in $V\{1, x\}$. Then $v = a + bx$. To show v is an element in $V\{3, x-2\}$, let $a + bx = c(3) + d(x-2) = 3c - 2d + dx$. Then $d = b$ and $c = (a+2b)/3$ and hence v can be written in terms of 3 and $x - 2$.

Whereas Theorem 1.1 presupposes knowledge of the dimension of a vector space, it is nonetheless important, since it leads to a great latitude in choosing a basis. In fact, although the orthonormal set $\{\mathbf{i}, \mathbf{j}, \mathbf{k}\}$ is a convenient basis for the set of three-dimensional arrows, the linearly independent set of Example 1.8 would serve as a basis too, although far more awkwardly.

The choice of a basis for a given vector space can be very important; in effect, the choice of basis for a space is equivalent to choosing the coordinates for the space.

When a vector is written as a linear combination of the basis vectors, the scalar multipliers are called the *components* of the vector with respect to that basis. Of course, the components change if the basis is changed. For example, in Example 1.10, we say that v has components a, b with respect to the basis $\{1, x\}$ and $(a+2b)/3$, b with respect to $\{3, x-2\}$.

example 1.11: Find the components of the arrow $\mathbf{A}$ with respect to the basis $\{3\mathbf{i}+2\mathbf{j}, \mathbf{i}+5\mathbf{k}, 6\mathbf{i}+\mathbf{j}\}$ if the components of $\mathbf{A}$ with respect to the $\{\mathbf{i}, \mathbf{j}, \mathbf{k}\}$ basis are 1, 2, and -1.

solution: The components of $\mathbf{A}$ with respect to the basis are numbers a_1, a_2, a_3 such that

$$\mathbf{i}+2\mathbf{j}-\mathbf{k} = a_1(3\mathbf{i}+2\mathbf{j}) + a_2(\mathbf{i}+5\mathbf{k}) + a_3(6\mathbf{i}+\mathbf{j}).$$

The vector equation yields the following three equations:

$$\begin{aligned} 3a_1 + \ a_2 + 6a_3 &= 1 \\ 2a_1 \qquad + \ a_3 &= 2 \\ 5a_2 \qquad &= -1 \end{aligned}$$

from which we obtain the solution $a_1 = 6/5$, $a_2 = -1/5$, $a_3 = -2/5$.

In passing we note that Definitions 1.5–1.8 apply to finite linear combinations and to finite dimensional spaces. In Chapter 10, we will have need of extensions of these concepts to infinite dimensional spaces when we discuss the space $PC[-L, L]$ in more detail.

It is particularly advantageous to choose an orthonormal set (or at least one that is orthogonal) to be the basis. All the advantages are not immediately obvious but one of the reasons the basis $\{\mathbf{i}, \mathbf{j}, \mathbf{k}\}$ is used is that the formulas come out "nice." For example suppose you wish to find the components of a vector $\mathbf{A}$ with respect to an orthonormal basis $\{\mathbf{e}_1, \mathbf{e}_2, \mathbf{e}_3\}$. Then as in the previous example you must find a_1, a_2, a_3 such that

$$\mathbf{A} = a_1\mathbf{e}_1 + a_2\mathbf{e}_2 + a_3\mathbf{e}_3.$$

To find a_1 we "dot" both sides with $\mathbf{e}_1$:

$$\mathbf{e}_1\cdot\mathbf{A} = a_1\mathbf{e}_1\cdot\mathbf{e}_1 + a_2\mathbf{e}_2\cdot\mathbf{e}_1 + a_3\mathbf{e}_3\cdot\mathbf{e}_1.$$

Since the basis is orthogonal, $\mathbf{e}_2\cdot\mathbf{e}_1 = \mathbf{e}_3\cdot\mathbf{e}_1 = 0$, and since the basis elements are normalized, $\mathbf{e}_1\cdot\mathbf{e}_1 = 1$. Therefore

$$a_1 = \mathbf{e}_1\cdot\mathbf{A}.$$

Similarly, $a_2 = \mathbf{e}_2\cdot\mathbf{A}$ and $a_3 = \mathbf{e}_3\cdot\mathbf{A}$. You will readily agree that these are easy formulas to use and certainly far easier than the general procedure which we use in the case of a nonorthogonal basis as in Example 1.11.

This technique of finding the components of a vector with respect to an orthonormal basis generalizes to inner product spaces. The method is often called the "Fourier method" since it is even used for infinite dimensional spaces, where Fourier series play an important role. (See Chapter 10.)

exercises for section 1.7

1. Show that the following sets are basis sets for the set of three-dimensional "arrows":
   (a) $\mathbf{i}+\mathbf{j}, \mathbf{k}, \mathbf{i}+\mathbf{k}$. (b) $3\mathbf{i}+2\mathbf{j}+\mathbf{k}, \mathbf{i}-5\mathbf{j}+2\mathbf{k}, \mathbf{j}+\mathbf{k}$.
2. Express the vector $5\mathbf{i}+3\mathbf{j}-2\mathbf{k}$ in terms of each of the bases of Exercise 1.
3. Show that two members of $PC(a, b)$ are linearly dependent if their ratio is a constant on (a, b). Can you arrive at a test for the linear dependence of three functions?
4. Show that the following bases are orthogonal:
   (a) $\mathbf{i}+\mathbf{j}, \mathbf{i}-\mathbf{j}, \mathbf{k}$. (b) $3\mathbf{i}+2\mathbf{j}+\mathbf{k}, \mathbf{i}+\mathbf{j}-5\mathbf{k}, 11\mathbf{i}-16\mathbf{j}-\mathbf{k}$.
5. Find the corresponding orthonormal basis for each of the sets of Exercise 4.
6. Express the vector $6\mathbf{i}+2\mathbf{j}+3\mathbf{k}$ in terms of the orthonormal bases of Exercise 5.
7. Show that *any* set of vectors which contains the zero vector of the space is linearly dependent.
8. Show that the set of real numbers can be called a vector space. What is its dimension?
9. Show how the Fourier method is used in a general finite dimensional vector space to obtain the components of a vector with respect to an orthonormal basis.
10. Show that $V\{1, x, x^2\} = V\{x-1, x^2+1, x-3\}$.
11. Show that the functions $3, x-2, x+3, x^2+1, 3x^2-2$ are linearly dependent. Find a basis for the vector space of all linear combinations of these functions. What is the dimension of the space?
12. Show that $V\{e^x, e^{-x}\} = V\{\cosh x, \sinh x\}$.
13. In Chapter 12 you will learn that
    $$e^{(a\pm bi)x} = e^{ax}\cos bx \pm ie^{ax}\sin bx\,.$$
    Show that
    $$V\{e^{(a+bi)x}, e^{(a-bi)x}\} = V\{e^{ax}\cos bx, e^{ax}\sin bx\}\,.$$
    This vector space will be very important to us in Chapter 3.

If y_1, y_2, y_3 are functions having at least two derivatives on an interval, the function W given by the determinant

$$W(x) = \begin{vmatrix} y_1(x) & y_2(x) & y_3(x) \\ y_1'(x) & y_2'(x) & y_3'(x) \\ y_1''(x) & y_2''(x) & y_3''(x) \end{vmatrix}$$

is called the *Wronskian* of the functions y_1, y_2, y_3.

14. Show that if y_1, y_2, y_3 are linearly dependent on some interval $[a, b]$, then $W(x) = 0$ on the interval. Extend this result to n functions.

15. Show that the converse of Exercise 14 is false by considering the following functions:

$$y_1(x) = \begin{cases} 1 + x^3 & \text{for } x \leqq 0, \\ 1 & \text{for } x \geqq 0 \end{cases} \quad y_2(x) = \begin{cases} 1 & \text{for } x \leqq 0, \\ 1 + x^3 & \text{for } x \geqq 0 \end{cases} \quad y_3(x) = 3 + x^3 \text{ for all } x.$$

Show that the three functions are linearly independent even though their Wronskian vanishes for all x.

16. Show that x and $|x|$ are linearly dependent in $PC(0, 1)$ but linearly independent in $PC[-1, 1]$.

17. Show that the following functions are linearly independent on any interval (a, b):
(a) x and x^2. (b) x, x^2, and x^3.
(c) $\sin x$ and $\cos x$. (d) e^x and e^{-x}.

18. Show that 1, $\sin^2 x$, and $\cos^2 x$ are linearly dependent.

1.8 matrices

As a major example of how sets may be viewed as vector spaces we now consider rectangular arrays of numbers, or, as they are technically called, *matrices*. The elements which are arranged horizontally in the array are called *rows* and those arranged vertically are called *columns*. For example, a three-row by four-column matrix is given by

$$\begin{pmatrix} a_{11} & a_{12} & a_{13} & a_{14} \\ a_{21} & a_{22} & a_{23} & a_{24} \\ a_{31} & a_{32} & a_{33} & a_{34} \end{pmatrix}$$

where the a_{ij} represent real numbers. The subscript notation used here is typical. The first number of the double subscript refers to the row of the matrix and the second number to the column. In this way any element of the array can be clearly identified.

In general a matrix with m rows and n columns is called an "$m \times n$ matrix." If we wish to refer to the matrix as an entity rather than mn numbers, we write either $\mathbf{A}$ or (a_{ij}). Be sure that you understand that a matrix is not a number, unless of course the size of the array is 1×1. Note the difference between a_{ij} and (a_{ij}).

There are many applications of matrices ranging from a use as a filing system to that of solving difficult systems of equations. Hence, in some cases you will have little need of knowing more than the indexing scheme. Other uses of matrices depend upon being able to operate with them in an algebraic manner. It is in the latter case that the perspective of vector spaces becomes important.

definition 1.9

Two matrices (a_{ij}) and (b_{ij}) are said to be equal if they are of the same size and if $a_{ij} = b_{ij}$ for all i and j.

definition 1.10

By the sum of two matrices (a_{ij}) and (b_{ij}) is meant the matrix $(a_{ij} + b_{ij})$.

Note that the definition of addition of two matrices demands that the two matrices be of the same size. This definition is sometimes called "elementwise addition."

example 1.12: Let

$$\mathbf{A} = \begin{pmatrix} 7 & 3 & 4 \\ -1 & 2 & 0 \\ 2 & -4 & 4 \\ 3 & 0 & -7 \end{pmatrix} \quad \text{and} \quad \mathbf{B} = \begin{pmatrix} 4 & 2 & -1 \\ 3 & -1 & 7 \\ 2 & 0 & 8 \\ 6 & 3 & -4 \end{pmatrix}.$$

Find $\mathbf{A} + \mathbf{B}$.

solution:

$$\mathbf{A} + \mathbf{B} = \begin{pmatrix} 7+4 & 3+2 & 4-1 \\ -1+3 & 2-1 & 0+7 \\ 2+2 & -4+0 & 4+8 \\ 3+6 & 0+3 & -7-4 \end{pmatrix} = \begin{pmatrix} 11 & 5 & 3 \\ 2 & 1 & 7 \\ 4 & -4 & 12 \\ 9 & 3 & -11 \end{pmatrix}.$$

It is very easy to show that addition of $m \times n$ matrices is closed, commutative, and associative. The additive identity is the $m \times n$ matrix of all zeros. The additive inverse of the matrix (a_{ij}) is the matrix $(-a_{ij})$.

definition 1.11

By scalar multiplication of a matrix (a_{ij}) by the real number r is meant the matrix (ra_{ij}).

example 1.13: Let $\mathbf{A} = \begin{pmatrix} 2 & 5 & 3 \\ -1 & 4 & 2 \end{pmatrix}$. Find $3\mathbf{A}$.

solution: $$3\mathbf{A} = \begin{pmatrix} 3\times 2 & 3\times 5 & 3\times 3 \\ -3\times 1 & 3\times 4 & 3\times 2 \end{pmatrix} = \begin{pmatrix} 6 & 15 & 9 \\ -3 & 12 & 6 \end{pmatrix}.$$

Properties 6–9 of Section 1.2 are easily verified for this operation. Consequently the set of $m \times n$ matrices forms a vector space with the two operations as given in Definitions 1.10 and 1.11. Note that there is a different vector space for each distinct pair of positive integers m and n. However, matrices are seldom called vectors except in the case of $m \times 1$ matrices or $1 \times n$ matrices, in which case the matrices are called *column vectors* and *row vectors* respectively.

Remember that the vector space idea focuses on the taking of linear combinations of elements from the set. If this is not central to your applications of matrices, then you will probably not even bother to notice that matrices can be called vectors.

example 1.14: Determine a basis for the set of 3×3 square matrices. What is the dimension of this vector space?

solution: Let $\mathbf{E}_{ij}$ be the matrix with 1 in the ij place and 0 in the other eight places. For example,

$$\mathbf{E}_{21} = \begin{pmatrix} 0 & 0 & 0 \\ 1 & 0 & 0 \\ 0 & 0 & 0 \end{pmatrix}.$$

Then, any 3×3 matrix $\mathbf{A} = (a_{ij})$ may be written,

$$\mathbf{A} = a_{11}\mathbf{E}_{11} + a_{12}\mathbf{E}_{12} + a_{13}\mathbf{E}_{13} + a_{21}\mathbf{E}_{21} + \cdots + a_{33}\mathbf{E}_{33}.$$

Furthermore, the set of matrices $\mathbf{E}_{ij}$ is linearly independent and hence the set is a basis. Since there are nine matrices in the set, the dimension of the vector space of 3×3 square matrices is 9.

In addition to the vector space operations of addition and multiplication by a scalar, it is possible to *multiply* certain pairs of matrices. This operation may appear a bit unnatural to you if you have not seen its definition before.

definition 1.12

Let $\mathbf{A} = (a_{ij})$ and $\mathbf{B} = (b_{jk})$ where $\mathbf{A}$ is $m \times n$ and $\mathbf{B}$ is $n \times r$. By the product $\mathbf{AB}$ is meant the $m \times r$ matrix

$$(c_{ik}) = \left(\sum_{j=1}^{n} a_{ij}b_{jk}\right).$$

In words, the element in the *ik* place of the product matrix is obtained by multiplying corresponding elements from the *i*th row of **A** and the *k*th column of **B**. (See Figure 1.6.)

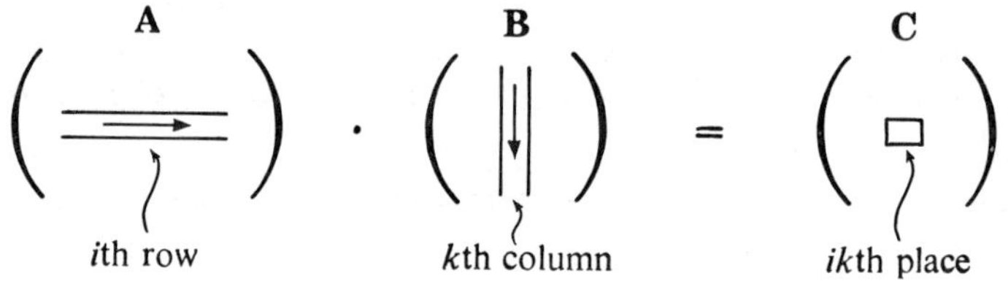

figure 1.6 *multiplication of matrices*

example 1.15: Find the product **AB** of the matrices

$$\mathbf{A}=\begin{pmatrix}1 & 2 & -3\\ 2 & 7 & 1\end{pmatrix},\quad \text{and} \quad \mathbf{B}=\begin{pmatrix}5 & 2\\ 6 & 3\\ -2 & 0\end{pmatrix}.$$

solution: The product matrix is 2×2 with elements

$$c_{11} = (1)(5) + (2)(6) + (-3)(-2);$$
$$c_{12} = (1)(2) + (2)(3) + (-3)(0);$$
$$c_{21} = (2)(5) + (7)(6) + (1)(-2);$$
$$c_{22} = (2)(2) + (7)(3) + (1)(0).$$

Hence,

$$\mathbf{AB}=\begin{pmatrix}23 & 8\\ 50 & 25\end{pmatrix}.$$

Note that the product in reverse order is

$$\mathbf{BA}=\begin{pmatrix}9 & 24 & -13\\ 12 & 33 & -15\\ -2 & -4 & 6\end{pmatrix}.$$

From the previous example we see that the product of matrices is *not* a commutative operation. In fact, the two products may be completely different types of matrices. Further, one of the products might not even exist since a product matrix is undefined if the number of columns of the first matrix is equal to the number of rows of the second.

exercises for section 1.8

Let $\mathbf{A}=\begin{pmatrix}2 & -3\\ 4 & 2\end{pmatrix}$ and $\mathbf{B}=\begin{pmatrix}-1 & 2\\ 3 & 0\end{pmatrix}.$

1. Find $\mathbf{A}+\mathbf{B}$.
2. Find $2\mathbf{A}-3\mathbf{B}$.
3. Find $\mathbf{AB}$ and $\mathbf{BA}$.
4. Are the two matrices linearly independent?
5. Determine a basis for the set of 2×3 matrices. What is the dimension of the vector space of 2×3 matrices? Generalize to the space of $m \times n$ matrices.
6. Is multiplication closed on the set of 2×3 matrices? Is it closed on the set of 2×2 matrices?

The *transpose* of a matrix $\mathbf{A}$ is denoted by $\mathbf{A}^T$ and is the matrix obtained from $\mathbf{A}$ by interchanging the rows and columns.

7. If $\mathbf{A}=\begin{pmatrix} 2 & 1 & 3 \\ 3 & 0 & 1 \end{pmatrix}$, find $\mathbf{A}^T$.
8. Let $\mathbf{A}$ be as in Exercises 1–4. Find $\mathbf{A}+\mathbf{A}^T$, $\mathbf{A}-\mathbf{A}^T$, and $\mathbf{AA}^T$.
9. Let $\mathbf{A}$ be as in Exercises 1–4. Find a matrix $\mathbf{B}$ such that $\mathbf{AB}=\mathbf{I}=\begin{pmatrix} 1 & 0 \\ 0 & 1 \end{pmatrix}$. Then find $\mathbf{BA}$. The matrix $\mathbf{B}$ is called the multiplicative inverse of $\mathbf{A}$.
10. Let $\mathbf{A}=(5 \quad -1 \quad 3)$ and $\mathbf{B}=\begin{pmatrix} 2 \\ 1 \\ -1 \end{pmatrix}$. Find $\mathbf{AB}$ and $\mathbf{BA}$.

first-order ordinary differential equations

2

2.1 introduction

In this and the next few chapters you will learn the introductory topics of the subject of ordinary differential equations. Engineers and physicists have a variety of reasons for requiring a speaking knowledge of the techniques of solving ordinary differential equations and while our discussion will not be sketchy, it will be limited to the extent that most of the practical differential equations considered will be classical. The precise nature of this limitation will be made clear in the next section and is important for you to understand.

2.2 some elementary terminology

By a differential equation is meant an equation involving an independent variable, an unknown function, and its derivatives. If the function is assumed to be dependent on one variable, the equation is called *ordinary*. Some examples of differential equations are:

(a) $y' = y$, (b) $y'' = y$, (c) $y'' = -y$, (d) $(y')^2 = y$,

(e) $yy' = 1$, (f) $xy'' = y^2 + x^3$, (g) $y'' + \sin y = 0$.

Most developments of elementary ordinary differential equations are very limited in scope and so it will be with our discussion. There are certain terms which it becomes necessary to learn in order to classify differential equations and to describe the general nature of the discussion. These terms are not important in themselves but should help you to focus your ideas.

The *order* of a differential equation is the highest order derivative occurring in the equation. Equations (a), (d), and (e) above are of the first order and the others are of the second order. Since many elementary physical laws can be formulated in terms of first- and second-order differential equations, we will usually limit ourselves to these two orders.

A differential equation is said to be *linear* when it is of the first degree in the dependent variable and the derivatives. Equations (a)–(c) are linear and the others are nonlinear. The most general linear differential equation of the first order may be put into the form

$$a_1(x)\,y' + a_0(x)\,y = f(x),$$

and a linear second-order equation may be written

$$a_2(x)\,y'' + a_1(x)\,y' + a_0(x)\,y = f(x).$$

Linear equations permit a particularly elegant theory; moreover those with *constant* coefficients of the first and second order have their analogs in important applications.

A *solution* to a differential equation is a function which when substituted into the equation reduces it to an identity. A solution is not always described by an equation or rule in the explicit form $y = f(x)$. Sometimes we will agree that a solution is defined by an implicit equation such as $x^2 + y^2 = 1$, by a graph, a table of values, or a convergent power series.

example 2.1: Show that e^x, e^{x+c}, and ce^x are solutions to $y' = y$. Show also that e^{2x} and $e^x + c$ are not solutions.

solution: The fact that e^x, e^{x+c}, and ce^x are solutions is immediately apparent from the fact that the functions and their derivatives are equal. Similarly, since $(d/dx)\,e^{2x} = 2e^{2x} \neq e^{2x}$ and $(d/dx)\,(e^x + c) = e^x \neq e^x + c$, none of these functions satisfies $y' = y$.

example 2.2: Show that $y' = 4(y/x) + x^4$ has solutions $y_1 = x^5 + x^4$ and $y_2 = x^5$.

solution: Since $y_1' = 5x^4 + 4x^3$ and $4(y_1/x) + x^4 = [(4x^5 + 4x^4)/x] + x^4 = 5x^4 + 4x^3$ it follows that $y_1(x)$ is a solution. Since $y_2' = 5x^4$ and $4(y_2/x) + x^4 = 5x^4$, $y_2(x)$ is also a solution.

We will usually not be content with finding one function which satisfies a differential equation. Rather we will want to find all solutions to the given equation. The set of all functions which satisfy a differential equation is called the *solution set*. The defining equation of the solution set is called either the *general solution* or the *complete solution*. There is a technical difference associated with the words "general" and "complete" but this technicality will not concern us.

By imposing additional conditions so that precisely one solution function can be chosen from the solution set we obtain a *particular solution*. For example, the solution set for the differential equation $y' = y$ is $\{y = ce^x \mid c$ a real constant$\}$. If we specify that when $x = 0$, $y = 1$, then the particular solution $y = e^x$ is determined. The general solution is simply $y = ce^x$.

If a (second-order) differential equation is considered over an interval (a, b) there are two fundamentally different types of auxiliary conditions:

1. If the values of the solution are specified at $x = a$ and $x = b$, the problem is generally called a *boundary value* problem.
2. If the values of the function and its derivative are specified at $x = a$, the problem is called an *initial value* problem.

The distinction between a boundary value problem and an initial value problem is not always clear-cut, nor will it be an important consideration in this book.

example 2.3: Given that $y'' + y = 0$ has the general solution

$$y = c_1 \cos x + c_2 \sin x,$$

find the particular solution which passes through (0, 1) with slope 2.

solution: The auxiliary conditions specify that $y(0) = 1$ and $y'(0) = 2$. The first condition yields $c_1 = 1$ and, since

$$y'(x) = -\sin x + c_2 \cos x,$$

the second condition gives $c_2 = 2$. Hence, the particular solution satisfying the given conditions is

$$y(x) = \cos x + 2 \sin x.$$

2.3 separable first-order equations

The most general explicit first-order differential equation is written in the form

$$y' = f(x, y)$$

or, which we will understand to be the same thing,

$$dy = f(x, y)\, dx.$$

If we write $f(x, y)$ as $-\dfrac{M(x, y)}{N(x, y)}$, then the equation has the form

$$M(x, y)\,dx + N(x, y)\,dy = 0.$$

We will use any of these general forms for a first-order equation interchangeably. For example, the equations

$$y' = \frac{-y}{x}, \quad x\,dy = -y\,dx, \quad \text{and} \quad x\,dy + y\,dx = 0$$

will mean the same thing.

Perhaps the simplest differential equation of the first order to solve is the kind you learned in elementary calculus corresponding to the case where $f(x, y)$ is independent of y. Then the differential equation has the form

$$y' = f(x)$$

and the solution set is, therefore, the family of functions

$$y = \int f(x)\,dx + c.$$

example 2.4: Solve the differential equation $x^2\,dx - dy = 0$.

solution: This equation may be put into the form

$$y' = x^2$$

and hence, the general solution, obtained by antidifferentiating, is

$$y = \frac{x^3}{3} + c.$$

Such a differential equation is only slightly less difficult than the kind you can put into the form

$$y' = \frac{f(x)}{g(y)},$$

in which case the variables are said to be separated, and the original differential equation is said to be *separable*. At least formally we can write

$$g(y)\,dy = f(x)\,dx,$$

and then the general solution is obtained by antidifferentiating both sides of this equality. The formal manipulation can of course be justified on a given interval, but we will bow to the almost universal custom of being rather cavalier with our manipulations and trust that the technique does lead to the general solution.

example 2.5: Solve the differential equation

$$y' = \frac{xy}{y+1}.$$

solution: The equation can be written

$$\frac{y+1}{y}\,dy = x\,dx.$$

When antidifferentiating both sides, we get

$$y + \ln y = \frac{x^2}{2} + c,$$

which defines the solution set implicitly. We could have written instead,

$$2(y + \ln y + \ln c_1) = x^2,$$

from which,

$$2y + \ln (c_1 y)^2 = x^2.$$

This is an alternate method of writing the general solution. There are others.

Even though the technique for solving separable equations reduces to one of finding antiderivatives, some of the applications are not trivial. For example, one of the most important differential equations is one for which the derivative of a function is directly proportional to the function. Using k as the constant of proportionality we may write,

$$y' = ky.$$

The method of separation of variables gives the family of functions

$$y = c\,e^{kx}$$

as the solution set. The constant k is fixed for the entire solution set and c is the parameter. Sufficient auxiliary conditions will allow the computation of both constants to obtain a particular solution (to a particular differential equation).

example 2.6: It has been proven empirically that radioactive elements decompose at approximately a fixed rate, called the *decay rate*. That is, if $y(t)$ is the mass of the substance present at any time, then the rate of change per unit mass, $(dy/dt)/y$, is a constant, say k. The *half-life* of a substance is the amount of time it takes for it to decay to one-half of its initial value. *Question:* Find the decay rate of a substance whose half-life is 20 years.

solution: We use the differential equation

$$\frac{dy}{dt} = -ky$$

so that the decay rate will be positive. The general solution is

$$y = c\,e^{-kt}.$$

At $t = 0$, $y = y_0$ and at $t = 20$, $y = y_0/2$. The first condition gives $c = y_0$ and the second one gives the equation

$$\tfrac{1}{2} = e^{-k20}$$

from which

$$k = \frac{\ln 2}{20} = \frac{0.693}{20} = 0.035.$$

Other physical laws besides those involving decay can be expressed in terms of separable differential equations. The situation is similar to many areas of applied mathematics: the actual mathematics is easy; the difficulty is in establishing the mathematical model or abstract description of the physical situation. It is this part of the problem which is often more demanding in terms of analytical maturity.

The following is a classical example (due to R. P. Agnew) of how simple differential equations are used. It is the kind of problem encountered in applications in that one is forced to make some fundamental assumptions to establish a good, but solvable, mathematical model.

example 2.7: (The "Snowplow Problem.") One day it started snowing at a heavy and steady rate. A snowplow started out at noon, going 2 miles the first hour and 1 mile the second hour. What time did it start snowing?

solution: Let t be the time measured in hours from noon, let h be the depth in miles of the snow at time t, and let y denote the distance moved by the plow. (See figure 2.1.) Then, if we *assume* the plow has width w and clears snow at a constant rate k, we have

$$\frac{wh\,\Delta y}{\Delta t} = k$$

or, in the limit as Δt approaches 0,

$$wh\frac{dy}{dt} = k.$$

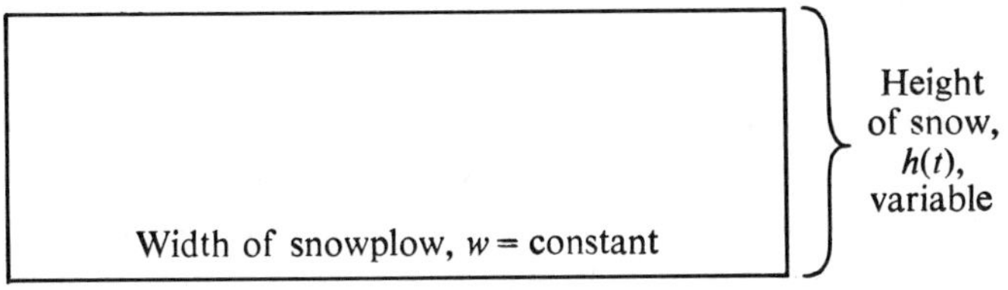

figure 2.1

The height of the snow is equal to the rate of the snowfall, a constant, say r, multiplied by the total time it has snowed,

$$h(t) = r(t + t_0),$$

which means that

$$wr(t + t_0)\frac{dy}{dt} = k$$

with the additional conditions

$$y(0) = 0, \qquad y(1) = 2, \qquad y(2) = 3.$$

Letting $k/wr = c$, and solving by the method of separation of variables,

$$dy = \frac{c}{t + t_0}\, dt.$$

Antidifferentiating,

$$y = c \ln(t + t_0) + c_1.$$

Applying the condition at $t = 0$ yields $c_1 = -c \ln t_0$ and hence

$$y = c \ln\left(\frac{t}{t_0} + 1\right).$$

The other two boundary conditions give

$$\left(1 + \frac{2}{t_0}\right)^2 = \left(1 + \frac{1}{t_0}\right)^3,$$

or, after simplifying,

$$t_0^2 + t_0 - 1 = 0.$$

Solving for t_0 and using the fact that t_0 is positive, we obtain

$$t_0 = 0.618 \text{ hours}.$$

Hence, the approximate time it started to snow was 11:23 A.M.

The next example requires you to know *Newton's law of cooling.* Experiments show that if an object is immersed in a medium and if the medium is kept at a constant temperature, then the temperature of the object will change at a rate proportional to the difference in temperature between the object and the medium.

example 2.8: A steel ball is heated to a temperature of 100° and at time $t=0$ is placed in water maintained at 10°. At $t=4$ minutes, the temperature of the ball is 60°. When is the temperature of the ball reduced to 20°?

solution: The mathematical model of the physical situation may be formulated with the aid of Newton's law of cooling as

$$\frac{dT}{dt} = -k(T-10), \qquad T(0)=100, \qquad T(4)=60,$$

where T is the temperature of the ball and k is the constant of proportionality. Separating variables and antidifferentiating,

$$T(t) = ce^{-kt} + 10.$$

The constants c and k may be obtained by applying the conditions at 0 and 4. We obtain $c=90$ and

$$k = \tfrac{1}{4}\ln\tfrac{9}{5} = \tfrac{1}{4}(2.1972 - 1.6094) = 0.147.$$

Therefore

$$T(t) = 90e^{-0.147t} + 10.$$

When $T=20°$,

$$20 = 90e^{-0.147t_0} + 10,$$

from which,

$$t_0 = \frac{\ln 9}{0.147} = \frac{2.1972}{0.147} = 14.9 \text{ minutes}.$$

exercises for sections 2.1–2.3

1. Verify that e^x, $3e^x$, and $-e^x$ are solutions to $y' - y = 0$ but that e^{-x} or e^{3x} is not a solution.
2. Verify that $\cos x$, $\sin x$, and any linear combination of $\cos x$ and $\sin x$ are solutions to $y'' + y = 0$. What can you say about the vector space of functions generated by $\sin x$ and $\cos x$?
3. Verify that e^x, e^{-x}, $\sinh x$, and $\cosh x$ are solutions to $y'' - y = 0$. Find a linearly independent subset of these functions. What can you say about the function space generated by this linearly independent set?

In Exercises 4–8, find the most general solution to the given differential equation.

4. $y' + y = 0$.

5. $(\sin x)\, y' + (\cos x)\, y = 0$.

6. $y' = \dfrac{1 - y^2}{1 - x^2}$.

7. $y' = e^{x+y}$.

8. $L \dfrac{di}{dt} + R\, i = 0$, L and R equal to constants.

In Exercises 9–12, find the particular solution having the specified auxiliary condition.

9. $y' = y$; $y(0) = 3$.

10. $y' = x^2$; $y(0) = 2$.

11. $y' - 3y = 0$; $y(1) = 7$.

12. $y' + y \sin x = 0$; $y(\pi) = 3$.

13. A steel ball is heated to a temperature of 80° and at time $t = 0$ is placed in water, maintained at 30°. At $t = 3$ minutes, the temperature of the ball is 55°. When is the temperature of the ball reduced to 40°?

14. What is the half-life of a radioactive substance of which 20 per cent disappears in 100 years?

15. After 20 years a quantity of radium has decayed to 60 grams and at the end of 50 years to 40 grams. How many grams were there in the first place?

16. Students sometimes confuse the expression $(y')^2$ with $(y^2)'$. They are not equal in general, of course, but for what functions are the two expressions the same?

17. Find all functions such that the derivative is the cube of the function.

18. Find all functions with derivative two more than the square of the function.

19. Find all functions whose derivative is equal to the reciprocal.

20. Solve Example 2.7 with the conditions changed to "the snowplow takes 1 hour to go the first mile and 2 hours to go the second mile."

2.4 substitutions

Often a substitution can be made which will reduce a differential equation to a form that you already know how to solve. In this section you should be more concerned with learning *how* to make a substitution than in memorizing particular types. It is the technique that you will have to recall; the actual substitutions used will often be suggested or self-evident.

So far, you have learned how to solve separable equations. Now we will study a few differential equations which, by use of a substitution, can be made separable. Most of the time, the substitution used will be obvious.

If a differential equation can be put into the form

$$y' = f\left(\frac{y}{x}\right),$$

then the substitution $v = y/x$ will reduce the equation to separable form. Since $y = vx$, then $y' = v + v'x$ and substituting into the equation,

$$v + v'x = f(v).$$

Separating variables,

$$\frac{v'}{f(v) - v} = \frac{1}{x}$$

which can be solved by antidifferentiating if $1/(f(v) - v)$ is relatively elementary. It is not difficult to determine when this substitution can be made profitably. Almost any book on elementary differential equations will have more details on how to make this judgment. For us, a bit of trial and error will suffice.

example 2.9: Solve the differential equation

$$y' = \frac{x - y}{x + y}.$$

solution: Note that this equation is not separable. By dividing numerator and denominator of the right-hand side by x the differential equation takes the form

$$y' = \frac{1 - (y/x)}{1 + (y/x)} = f\left(\frac{y}{x}\right).$$

Using the substitution $y = vx$, we obtain

$$v + v'x = \frac{1 - v}{1 + v}.$$

Simplifying,

$$v'x = \frac{1 - v}{1 + v} - v = \frac{1 - v - v - v^2}{1 + v}$$

or,

$$\frac{1 + v}{1 - 2v - v^2}\, v' = \frac{1}{x}.$$

The antiderivative of $\dfrac{v + 1}{v^2 + 2v - 1}$ is

$$\int \frac{v + 1}{v^2 + 2v - 1}\, dv = \int \frac{v + 1}{(v + 1)^2 - 2}\, dv = \tfrac{1}{2} \ln (v^2 + 2v - 1).$$

Using this in the differential equation,

$$\ln(v^2 + 2v - 1)^{-1/2} = \ln cx$$

and hence

$$(v^2 + 2v - 1)\, x^2 = c_1,$$

where $c_1 = 1/c^2$ and is therefore also "arbitrary." Since $v = y/x$, the solution, in implicit form, is

$$y^2 + 2xy - x^2 = c_1.$$

example 2.10: Solve the differential equation

$$(x^2 + y^2)\, dx - xy\, dy = 0.$$

solution: We may write this equation in the form

$$y' = \frac{x^2 + y^2}{xy} = \frac{1 + (y/x)^2}{y/x}.$$

The substitution $y = vx$ gives

$$v'x + v = \frac{1 + v^2}{v}$$

and hence

$$v'x = \frac{1}{v}.$$

Therefore

$$v\, dv = \frac{dx}{x}$$

whose solution is

$$\frac{v^2}{2} = \ln cx.$$

In terms of y,

$$y^2 = 2x^2 \ln cx.$$

Sometimes, two successive substitutions are required to reduce the equation to separable form. For example, there are occasions in which we express x and y linearly in terms of two new variables X and Y followed by $v = Y/X$.

example 2.11: Solve the differential equation

$$y' = \frac{x - y + 1}{x + y - 1}.$$

solution: We let $X = x - h$ and $Y = y - k$, and assume that Y is a function of X. Then

$$y' = \frac{dy}{dx} = \frac{dy}{dY}\frac{dY}{dx} = \frac{dy}{dY}\frac{dY}{dX}\frac{dX}{dx} = \frac{dY}{dX}.$$

Hence, after substitution, the differential equation is

$$\frac{dY}{dX} = \frac{X + h - Y - k + 1}{X + h + Y + k - 1}.$$

The given substitution will be completely specified when the values of h and k are fixed. We now choose h and k to eliminate the constant terms from the numerator and denominator, by simultaneously satisfying the two equations

$$h - k + 1 = 0,$$
$$h + k - 1 = 0.$$

This system is satisfied by $h = 0$, $k = 1$. The differential equation then has the form

$$\frac{dY}{dX} = \frac{X - Y}{X + Y}.$$

The solution to this was obtained in Example 2.9,

$$Y^2 + 2XY - X^2 = c.$$

Therefore, the solution to the given differential equation is

$$(y - 1)^2 + 2(y - 1)x - x^2 = c.$$

Some substitutions are suggested by the appearance of a group of terms in the differential equation.

example 2.12: Solve the differential equation

$$y' = \frac{x + y - 1}{x + y + 1}.$$

solution: The occurrence of the expression $x + y$ in both the numerator and denominator suggests that we let $u = x + y$. From $y = u - x$, we get $y' = u' - 1$ and substituting into the differential equation,

$$u' - 1 = \frac{u - 1}{u + 1}$$

$$u' = \frac{u - 1}{u + 1} + 1 = \frac{u - 1 + u + 1}{u + 1} = \frac{2u}{u + 1}$$

$$\frac{u + 1}{u}\, du = 2\, dx.$$

Antidifferentiating both sides,

$$u + \ln u = 2x + c\,.$$

The solution, in implicit form, is therefore

$$y - x + \ln(x + y) = c\,.$$

2.5 exact differential equations

In this section we consider equations which are not separable but whose solution is found by two antidifferentiations.

First recall that the *total differential* of a function of two variables $u(x, y)$ is given by

$$du = u_x\,dx + u_y\,dy\,.$$

Thus, if we know the two partial derivatives of a function, then its total differential is determined. Conversely, as the following example will show, a knowledge of the total differential is sufficient to determine the function within an arbitrary constant.

example 2.13: The total differential of a function is given by

$$du = 3x(xy - 2)\,dx + (x^3 + 2y)\,dy\,.$$

Find $u(x, y)$.

solution: The coefficients of dx and dy are u_x and u_y. Therefore

$$u_x = 3x^2y - 6x \qquad \text{and} \qquad u_y = x^3 + 2y\,.$$

Partially antidifferentiating the first equation gives

$$u(x, y) = x^3y - 3x^2 + C(y)$$

where $C(y)$ stands for an arbitrary function of y only since y is constant for the partial antidifferentiation. From this function u_y is found to be

$$u_y = x^3 + C'(y)\,.$$

Equating this to the given expression for u_y,

$$x^3 + C'(y) = x^3 + 2y\,.$$

This means that $C(y) = y^2 + c$ and therefore

$$u(x, y) = x^3y - 3x^2 + y^2 + c\,.$$

The following definition should establish the connection between the concept of a total differential and the subject of first-order ordinary differential equations.

definition 2.1

The first-order differential equation

$$M(x, y)\,dx + N(x, y)\,dy = 0$$

is said to be *exact* if the left-hand side of the equation is the total differential of some function $u(x, y)$.

Since an exact equation can be written in the form $du = 0$, its solution in implicit form may be written $u(x, y) = c$. Note that the crux of the problem in finding the solution to an exact equation is in finding the function $u(x, y)$. This function itself is not the solution but rather, when equated to c, defines implicitly a function y in terms of x.

example 2.14: Solve the differential equation

$$3x(xy - 2)\,dx + (x^3 + 2y)\,dy = 0.$$

solution: In the previous example we found a function u such that du is the left-hand side of this equation. Hence the implicit solution is

$$x^3y - 3x^2 + y^2 = c.$$

example 2.15: Solve the differential equation

$$(y\cos x - \sin y)\,dx + (\sin x - x\cos y)\,dy = 0,$$

under the assumption that the left-hand side is of the form $du = 0$.

solution: Since the left-hand side is the differential of some unknown function, the coefficient of dx must equal u_x and the coefficient of dy must equal u_y:

$$u_x = y\cos x - \sin y$$
$$u_y = \sin x - x\cos y.$$

Antidifferentiating the first of these equations,

$$u(x, y) = y\sin x - x\sin y + C(y).$$

The partial derivative of this function is

$$u_y = \sin x - x\cos y + C'(y).$$

Equating this expression for u_y to that previously given we obtain $C'(y) = 0$ and hence $C(y) = c$. Therefore the solution in implicit form is

$$y \sin x - x \sin y = k.$$

The previous example could have been solved by grouping terms. Note that the sum $y \cos x \, dx - x \cos y \, dy$ is the differential of $-x \sin y$; the sum $y \cos x dx + \sin x \, dy$ is the total differential of $y \sin x$. Hence the left-hand side is the differential of the sum $y \sin x - x \sin y$.

What would happen if you tried to apply the technique for exact equations to one which was "inexact"?

example 2.16: Attempt to solve the equation $(x + y) \, dx + (y - x) \, dy = 0$ as if it were an exact equation.

solution: (Note that this differential equation *can* be solved by use of the substitution $u = y/x$.) If you assume (erroneously) that the differential equation is exact, you obtain the two requirements on the function $u(x, y)$,

$$u_x = x + y, \qquad u_y = y - x.$$

Antidifferentiating the first of these, we get

$$u(x, y) = \frac{x^2}{2} + xy + C(y)$$

where $C(y)$ is assumed to be a function of y only. Finding u_y from this expression and equating it to the given one,

$$x + C'(y) = y - x$$

or,

$$C'(y) = y - 2x,$$

and this contradicts the assumed condition that $C(y)$ is independent of x.

The following theorem provides us with a conclusive test of the exactness of a differential equation.

theorem 2.1

Let $M(x, y)$ and $N(x, y)$ be functions with continuous first partial derivatives over some domain $D = \{(x, y) \mid a < x < b, c < y < d\}$. Then the differential equation

$$M(x, y) \, dx + N(x, y) \, dy = 0$$

is exact if and only if $M_y = N_x$.

proof: If the differential equation is exact, then there exists a function $u(x, y)$ such that the left-hand side of the equation is of the form $u_x\,dx + u_y\,dy$. Thus,

$$M = u_x \qquad \text{and} \qquad N = u_y.$$

Partially differentiating the first of these with respect to y and the second with respect to x,

$$M_x = u_{xy} \qquad \text{and} \qquad N_y = u_{yx}.$$

Since M_x and N_y are assumed to be continuous, so are u_{xy} and u_{yx}; and hence, from a theorem of elementary calculus, $u_{xy} = u_{yx}$; that is, $M_y = N_x$.
The converse of the theorem is a bit more difficult for it requires the construction of a function $u(x, y)$ from $M(x, y)$ and $N(x, y)$. The procedure parallels the examples. Define $u(x, y)$ by the equation

$$u(x, y) = \int_x M(x, y)\,dx + C(y),$$

where $C(y)$ is to be determined later. (The x under the integral sign means to antidifferentiate with respect to x, holding y constant.) From this expression, we compute u_y,

$$u_y = \frac{\partial}{\partial y}\int_x M(x, y)\,dx + C'(y)$$

and this is equal to $N(x, y)$. Therefore

$$C'(y) = N(x, y) - \frac{\partial}{\partial y}\int_x M(x, y)\,dx.$$

There will be a solution if $C'(y)$ is independent of x, and this will be the case if the derivative of the right-hand side with respect to x is zero. Carrying out this differentiation,

$$\frac{\partial}{\partial x}\left[N(x, y) - \frac{\partial}{\partial y}\int_x M(x, y)\,dx\right] = \frac{\partial N}{\partial x} - \frac{\partial}{\partial x}\frac{\partial}{\partial y}\int_x M(x, y)\,dx = \frac{\partial N}{\partial x} - \frac{\partial M}{\partial y}$$

under the assumption that the partial differentiations with respect to x and y may be interchanged. The last expression is zero because of the assumed hypothesis. This completes the proof.

example 2.17: Solve the differential equation

$$(2xy + 3y)\,dx + (4y^3 + x^2 + 3x + 4)\,dy = 0.$$

solution: Since $M_y = 2x + 3 = N_x$, the equation is exact. Therefore, the partial derivatives of the desired function $u(x, y)$ are given by

$$u_x = 2xy + 3y \qquad \text{and} \qquad u_y = 4y^3 + x^2 + 3x + 4.$$

Antidifferentiating the first of these,

$$u(x, y) = x^2y + 3xy + C(y).$$

From this expression we find u_y and equate it to the known expression for u_y,

$$x^2 + 3x + C'(y) = 4y^3 + x^2 + 3x + 4.$$

Therefore,

$$C'(y) = 4y^3 + 4,$$

and

$$C(y) = y^4 + 4y.$$

Hence, the implicit solution to the differential equation is

$$x^2y + 3xy + y^4 + 4y = c.$$

exercises for sections 2.4 and 2.5

In Exercises 1–4 solve the differential equations by first making the substitution $y = vx$.

1. $y' = \dfrac{xy}{x^2 + y^2}$.
2. $y' = \dfrac{2x + y}{x - 2y}$.
3. $y' = \dfrac{x^2 - y^2}{xy}$.
4. $y' = \dfrac{x^3 + xy^2}{yx^2 - y^3}$.

In Exercises 5–7 solve by first making linear substitutions.

5. $y' = \dfrac{x + 2y - 1}{x + 2y + 7}$.
6. $y' = \dfrac{x - y + 8}{y - 3x + 2}$.
7. $y' = \dfrac{2x + 3y - 4}{x - y}$.
8. $y' = \dfrac{y + x}{y - x}$. Solve by using a substitution.
9. Solve Exercise 8 by another method.

In Exercises 10–14, determine which of the equations are exact, and solve those which are, first by inspection and then by the more formal method of partial antidifferentiations.

10. $(x^2 + y^2)\,dy + 2xy\,dx = 0$.
11. $(y^2 - x^2)\,dx + 2xy\,dy = 0$.
12. $(2x + y)\,dx + (y - x)\,dy = 0$.
13. $(2x + y)\,dx + (x - y)\,dy = 0$.
14. $(ye^x - 2x)\,dx + e^x\,dy = 0$.

15. $(ye^{xy} + 2xy)\,dx + (xe^{xy} + x^2)\,dy = 0.$

16. Show that an equation of the form $f(x)\,dx + g(y)\,dy = 0$ is exact.

17. Under what conditions is the differential equation $(x + ay)\,dx + (y + bx)\,dy = 0$ exact?

2.6 *integrating factors; linear equations of the first order*

If a differential equation is not exact, you can sometimes "make it exact" by multiplying the equation by a quantity called an *integrating factor*. Finding an integrating factor for a first-order equation is equivalent to solving it, for as you recall from the previous section, an exact equation may be solved by finding two partial antiderivatives. Generally, we are not concerned with the uniqueness of a given integrating factor and, in fact, they usually are not unique. For example, the factors x^{-2} and y are both integrating factors of the differential equation

$$\frac{x^2}{y}\,dy + 2x\,dx = 0.$$

It is possible to reduce the study of first-order equations to one of determination of integrating factors. In this section you will learn a few important integrating factors for some rather general classes of equations, rather than analyzing the problem in its complete generality. We restrict our discussion to some special cases in which a specific *form* for an integrating factor may be assumed.

Sometimes the differential equation has an integrating factor of the form $x^p y^q$, where p and q are constants. These constants are determined from the condition for exactness.

example 2.18: Solve the differential equation

$$(x^2y^2 + y)\,dx + (2x^3y - x)\,dy = 0$$

by assuming that there is an integrating factor of the form $x^p y^q$.

solution: Multiplying the equation by $x^p y^q$, we obtain

$$(x^{p+2}y^{q+2} + x^p y^{q+1})\,dx + (2x^{p+3}y^{q+1} - x^{p+1}y^q)\,dy = 0.$$

In this equation

$$M(x, y) = x^{p+2}y^{q+2} + x^p y^{q+1},$$
$$N(x, y) = 2x^{p+3}y^{q+1} - x^{p+1}y^{q+1}.$$

Applying the condition for exactness, $M_y = N_x$, yields

$$(q+2)\,x^{p+2}y^{q+1} + (q+1)\,x^p y^q = 2(p+3)\,x^{p+2}y^{q+1} - (p+1)\,x^p y^q .$$

Equating like powers of xy gives the two equations

$$q+2 = 2p+6, \qquad q+1 = -p-1,$$

whose solutions are $p = -2$, $q = 0$.
Therefore, the integrating factor is $(1/x)^2$, and

$$M(x, y) = y^2 + x^{-2}y,$$
$$N(x, y) = 2xy - x^{-1}.$$

The function u for which $u_x = M$ and $u_y = N$ is

$$u(x, y) = y^2 x - \frac{y}{x}.$$

Therefore the solution is

$$y^2x^2 - y = cx .$$

Another kind of form natural to assume for an integrating factor is that it be a function of x only or a function of y only. To show the conditions under which $u(y)$ is an integrating factor of

$$M(x, y)\,dx + N(x, y)\,dy = 0,$$

we multiply by $u(y)$

$$u(y)\,M(x, y)\,dx + u(y)\,N(x, y)\,dy = 0$$

and then apply the condition for exactness

$$\frac{\partial}{\partial y}\{u(y)\,M(x, y)\} = \frac{\partial}{\partial x}\{u(y)\,N(x, y)\}$$
$$u'(y)\,M(x, y) + u(y)\,M_y(x, y) = u(y)\,N_x(x, y).$$

This is a differential equation in the function u, and may be put into the form

$$\frac{u'(y)}{u(y)} = \frac{N_x(x, y) - M_y(x, y)}{M(x, y)}.$$

If the right-hand side is a function of y only, say $f(y)$, then the solution to this equation is

$$u(y) = e^{\int f(y)\,dy}.$$

example 2.19: Solve the differential equation $y^2\,dx + 4xy\,dy = 0$ by first finding an integrating factor which is a function of y.

solution: To find the integrating factor, we first find

$$f(y) = \frac{N_x - M_y}{M} = \frac{4y - 2y}{y^2} = \frac{2}{y}$$

and hence the integrating factor is

$$e^{\int (2/y)\,dy} = e^{\ln y^2} = y^2 .$$

Using the integrating factor in the differential equation,

$$y^4\,dx + 4xy^3\,dy = 0$$

whose solution is (by inspection)

$$xy^4 = c .$$

Recall that all first-order linear equations may be written in the form

$$a_1(x)\,y' + a_0(x)\,y = r(x)$$

or

$$y' + p(x)\;\;y = f(x).$$

This latter form is called "standard."

We will now show that an integrating factor can always be found for this kind of differential equation and hence the entire class of first-order linear equations is solvable. To show this, assume that an integrating factor exists which is a function of x only. If the integrating factor is $v(x)$,

$$v(x)\,dy + [v(x)\,p(x)y - v(x)\,f(x)]\,dx = 0 .$$

Applying the condition for exactness

$$\frac{\partial}{\partial x}[v(x)] = \frac{\partial}{\partial y}[v(x)\,p(x)\,y - v(x) f(x)],$$

or

$$v'(x) = v(x)\,p(x) .$$

This equation can be solved for $v(x)$ to give

$$v(x) = ce^{\int p(x)\,dx}$$

where we ordinarily choose c to be 1.

Applying this integrating factor to the first-order equation in standard form we obtain

$$y'e^{\int p(x)\,dx} + p(x)\,ye^{\int p(x)\,dx} = f(x)\,e^{\int p(x)\,dx} .$$

The left-hand side of this equation is the derivative of the product $ye^{\int p(x)\,dx}$ and hence

$$ye^{\int p(x)\,dx} = \int f(x)\,e^{\int p(x)\,dx}\,dx + c .$$

Thus the solution can always be found in explicit form if the indicated anti-differentiations can be performed.

example 2.20: Solve the differential equation

$$xy' + 2y = x^2.$$

solution: First recognize that this equation is linear and of the first order. To put the equation in standard form, divide both sides of the equation by x,

$$y' + \frac{2}{x}y = x.$$

The integrating factor is

$$e^{\int (2/x)\,dx} = e^{\ln x^2} = x^2.$$

Multiplying the differential equation (in standard form) by this factor,

$$x^2y' + 2xy = x^3$$

or, in differential form,

$$x^2\,dy + 2xy\,dx = x^3\,dx.$$

The left-hand side *must* be the total differential of some function (otherwise x^2 is not an integrating factor). Thus,

$$d(yx^2) = x^3\,dx,$$

and

$$yx^2 = \frac{x^4}{4} + c.$$

In explicit form, this is

$$y = \frac{x^2}{4} + \frac{c}{x^2}.$$

Linear first-order equations arise naturally in some recognizably important applications. For example, the electric circuit with negligible capacitance, but non-negligible resistance R of an electrical circuit with

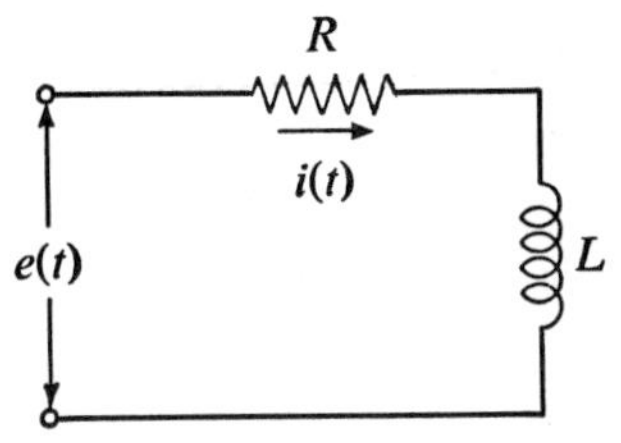

figure 2.2 *RL circuit*

dominant components of resistance R and inductance L (See Figure 2.2), is described by the first-order linear equation,

$$L\frac{di}{dt} + Ri = e(t)$$

where $e(t)$ is the driving voltage and $i(t)$ is the current in the circuit. Often L and R are considered constants, but generally there would be no actual requirement for this to be so. The solution function, which is the current in the circuit as a function of time, is sometimes called the *response*.

example 2.21: Find the current in an LR circuit where L and R are constants, $e(t)$ is a d.c. (constant) voltage, and the initial current is I_0.

solution: Let E be the input voltage. Then the differential equation of the circuit is

$$L\frac{di}{dt} + Ri = E.$$

In standard form, the equation is

$$\frac{di}{dt} + \frac{R}{L}i = \frac{E}{L}.$$

Hence, an integrating factor is $e^{Rt/L}$ which when used on the equation yields

$$d(ie^{Rt/L}) = \frac{E}{L}e^{Rt/L}\,dt.$$

Antidifferentiating and solving for i,

$$i = \frac{E}{R} + ce^{-Rt/L}.$$

The condition $i(0) = I_0$ yields $c = I_0 - (E/R)$, and therefore the current in

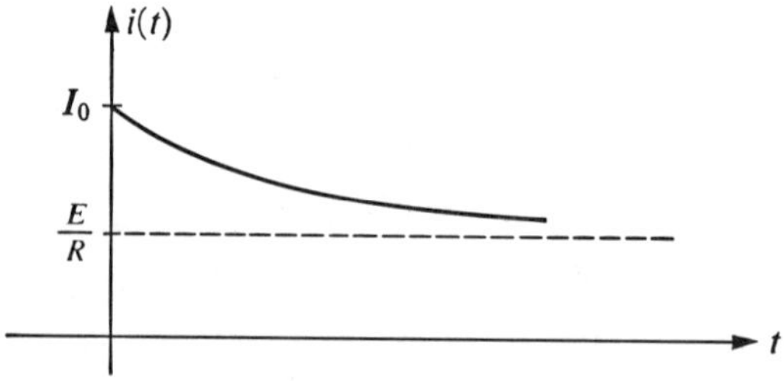

figure 2.3

the circuit is given by

$$i = \frac{E}{R} + \left(I_0 - \frac{E}{R}\right) e^{-Rt/L}.$$

See Figure 2.3.

example 2.22: Find the response of an RL circuit to a unit square wave. Assume the current is initially zero. See Figure 2.4(a).

solution: For the interval $0 \leqq t \leqq 1$, the current is the same as in the previous example, with $E = 1$ and $I_0 = 0$. Substituting these values,

$$i(t) = \frac{1}{R}(1 - e^{-Rt/L}), \qquad 0 \le t \le 1.$$

The mathematical model of the circuit for $t \geqq 1$ is given by

$$L\frac{di}{dt} + Ri = 0$$

$$i(1) = \frac{1}{R}(1 - e^{-R/L}).$$

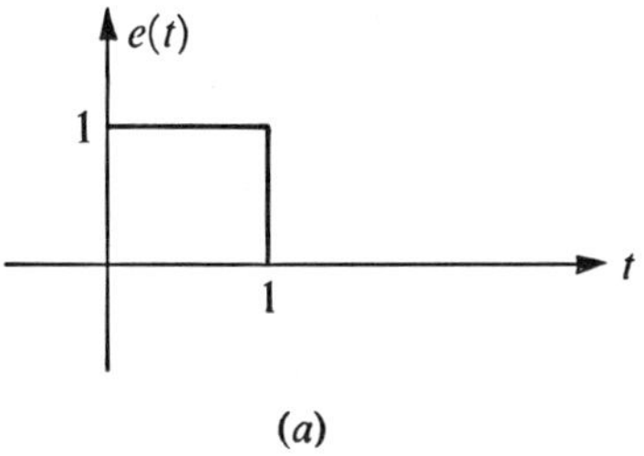

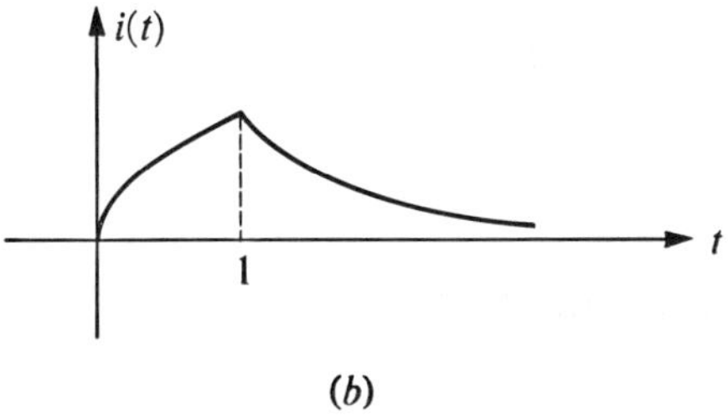

figure 2.4 (a) (b)

The differential equation is easily solved by the method of separation of variables:

$$i(t) = c_1 e^{-Rt/L}.$$

Applying the condition at $t = 1$,

$$i(1) = c_1 e^{-R/L} = \frac{1}{R}(1 - e^{-R/L}),$$

from which

$$c_1 = \frac{1}{R}(e^{R/L} - 1).$$

Therefore the current for $t \geqq 1$ (see Figure 2.4(b)) is:

$$i(t) = \frac{e^{R/L} - 1}{R} e^{-Rt/L}$$
$$= \frac{e^{-R(t-1)/L} - e^{-Rt/L}}{R}.$$

example 2.23: Find the current in an RL circuit which has zero initial current, a resistance of 2 ohms, an applied voltage of 4 volts, and an inductance which varies as

$$L(t) = 5 - t, \qquad 0 \leqq t \leqq 5$$
$$= 0, \qquad 5 \leqq t.$$

solution: The mathematical model for the circuit for $0 \leqq t \leqq 5$ is

$$(5 - t)\frac{di}{dt} + 2i = 4, \qquad i(0) = 0.$$

The solution for this time interval is obtained by first dividing by $5 - t$ to put the equation into standard form. (Note: Do not make the error of applying the result of Example 2.21 with $L = 5 - t$. That result applies only to the case in which L is a constant.)

$$\frac{di}{dt} + \frac{2}{5 - t} i = \frac{4}{5 - t}.$$

An integrating factor is

$$e^{\int (2/5 - t)\, dt} = e^{-2 \ln (5 - t)} = (5 - t)^{-2}.$$

Applying this integrating factor to the differential equation,

$$d\left(i(5 - t)^{-2}\right) = \frac{4}{(5 - t)^3}\, dt.$$

Therefore

$$i(5 - t)^{-2} = 2(5 - t)^{-2} + c$$

or

$$i = 2 + c(5 - t)^2.$$

The value of c is determined from the fact that $i = 0$ when $t = 0$,

$$0 = 2 + c(5)^2$$
$$c = -\tfrac{2}{25}.$$

Therefore,

$$i(t) = 2 - \frac{2(5 - t)^2}{25} \text{ amperes}$$
$$= \frac{2t}{5}\left(2 - \frac{t}{5}\right) \text{ amperes}.$$

For $t \geqq 5$, the differential equation becomes

$$0\frac{di}{dt} + 2i = 4,$$

whose solution is $i = 2$ amperes. (See Figure 2.5.)

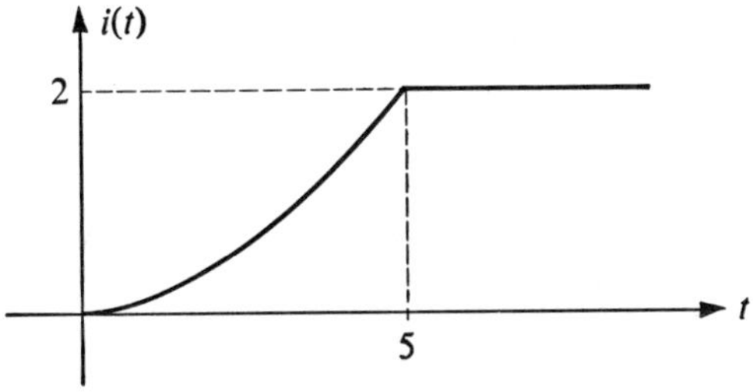

figure 2.5

Often, you may be given a rule or a formula which supposedly governs a given type of physical situation. Such a rule may not involve differentiation or integration and relates the various physical variables and their magnitudes. You may be led to believe that all you need do is to insert the values for the particular case into the rule. Such a rule is usually not general enough; it may be restricted by an assumption such as the constancy of certain components, the absence of initial current or voltage, or a specific kind of driving voltage. Hence, it is probably best always to write the mathematical model which expresses the most fundamental physical law. Then, if this model consists of implicit equations with derivatives, integrals, etc., the use of mathematical techniques will eventually give an explicit expression for the desired unknowns.

exercises for section 2.6

In Exercises 1–3, there is an integrating factor of the form $x^p y^q$. Find it and solve the equation.

1. $(xy + y^2)\,dx + (x^2 - xy)\,dy = 0.$
2. $y(x + y^2)\,dx + x(x - y^2)\,dy = 0.$
3. $(y^2 - y)\,dx + x^2\,dy = 0.$
4. Under what conditions does the differential equation

$$M(x, y)\,dx + N(x, y)\,dy = 0$$

have an integrating factor which is a function of x only?

5. Under what conditions does the differential equation

$$M(x, y)\,dx + N(x, y)\,dy = 0$$

have an integrating factor which is of the form $x^p y^q$?

In Exercises 6–8, solve the differential equations by finding an integrating factor by inspection.

6. $3x^2 y\,dx + (y^4 - x^3)\,dy = 0.$
7. $x\,dx + (\sqrt{x^2 - y^2} - y)\,dy = 0.$
8. $(\sin^2 x - y)\,dx - \tan x\,dy = 0.$

In Exercises 9–13, solve by using the standard integrating factor for linear equations of the first order.

9. $3y' + 2y + x = 0.$
10. $y' \cos x - y \sin x + e^x = 0.$
11. $y' - 4y = x - x^2.$
12. $y' + y = \cos x.$
13. $y' + 2y = e^{-2x}.$
14. Find the current in an RL circuit, initially quiescent (zero initial current), with input voltage as given for each of the following cases. R and L are assumed to be constant.
    (a) e^{-t}. (b) $\sin t$. (c) t.
15. Find the current in an RL circuit with $R = 1$ ohm, $L = 1$ henry, and input voltage as shown in Figure 2.6. Assume that the initial current is zero.

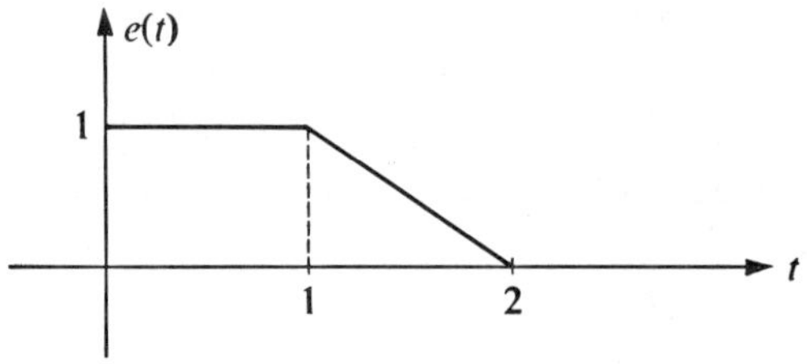

figure 2.6

16. Consider an RL circuit with $R = 10$ ohms, input voltage 20 volts, and

$$\begin{aligned} L &= 5 - t, && 0 \leq t \leq 5 \\ &= 0, && 5 \leq t. \end{aligned}$$

Find the current in the circuit if the initial current is zero.

17. Repeat Exercise 16 with the assumption that the initial current is 2 amperes.

18. Find the current in the RL circuit of Exercise 15 if the input voltage is as shown in Figure 2.7. Assume that the initial current is 1 ampere.

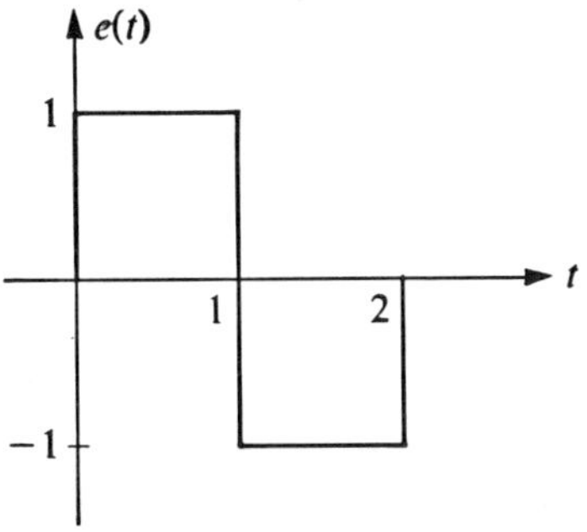

figure 2.7

19. If the input voltage of the previous exercise is extended periodically, what should the initial current be to assure a periodic output?

20. Consider the RL circuit of Exercise 15 with the periodic square wave input voltage shown in Figure 2.8. What should the initial current be to assure a periodic output?

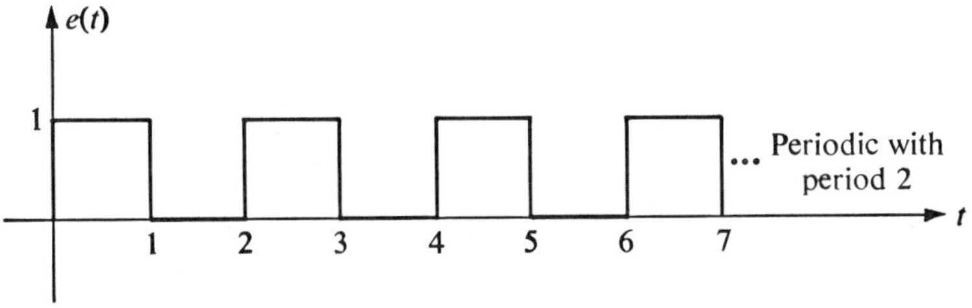

figure 2.8

2.7 approximation by straight lines

In many practical situations you will be interested only in the numerical values of the solution for one or several values of the independent variable. The description of the solution in terms of elementary functions and algebraic operations may be relatively unimportant. Thus, it may be preferable to describe the solution by a table of values such as the next one

x	2	2.5	5	3.5	4	4.5	5
y	3.72	4.62	5.19	6.94	9.11	10.01	11.23

rather than by an analytic description such as

$$y = \frac{(x^7 + \cos x)^{1/3}}{e^x + \log x}.$$

Methods for solving differential equations numerically are much more powerful (that is, they apply to a much larger collection of types of differential equations) than analytical methods. A numerical method consists in replacing the actual solution by an approximation. In practice you will have some tolerance imposed on the admissable error and this should determine (by a fairly difficult mathematical theory) which approximation should be used to calculate the solution.

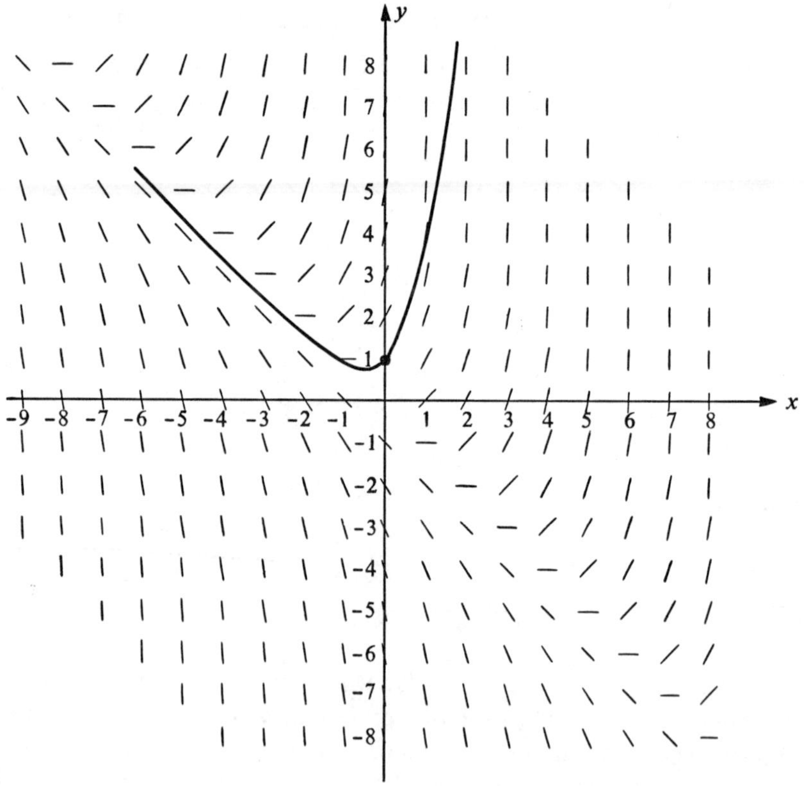

figure 2.9

In this section we consider first-order equations written in the form $y' = f(x, y)$. This differential equation essentially specifies the value of the slope of any solution at every point in the plane. We therefore attach to each point (x_0, y_0) a third number which is the value of the slope at the point. This slope is computed by evaluating $f(x_0, y_0)$. To display this graphically we sketch short line segments at a "reasonable" number of points in the plane, each line segment having slope as determined by $f(x, y)$. The complete set of line elements is known as the *direction field* of the differential equation.

The direction field specifies the direction of solution curves. Particular solutions may be sketched by beginning at a given point (a, b) and continuously modifying the slope of the curve in accordance with the direction of the line elements.

example 2.24: Sketch the direction field associated with the differential equation $y' = x + y$ and draw the particular solution curve which passes through the point $(0, 1)$.

solution: Figure 2.9 shows both the sketch of the direction field and the particular solution. Note that in sketching the set of line elements, it is helpful to consider first the curves along which the direction of the line segments is the same. Such curves are called *curves of constant slope* of the direction field, and are obtained by setting $f(x, y)$ equal to a constant. In this case we obtain the lines $x + y = c$. These lines are not solution curves but are meant to assist in the drawing of the direction field.

The value of such a graphical approach should not be underestimated. The degree of accuracy is largely dependent upon the completeness of your sketch of the direction field, and you can obtain a good qualitative view of particular solutions.

The foregoing method may be modified to give a numerical solution by using techniques learned in analytic geometry. The graphical method yields a solution curve whose tangent is continuously changing. Unless other information is given, it is not possible actually to write the equation of the curve directly from its graph. However, if the changes to the slope of the solution curve are made in discrete steps rather than continuously, then the approximation curve is a polygonal line whose analytic description can be written on every interval.

The approximating line through (a, b) will have slope $f(a, b)$ and thus its equation is

$$\frac{y - b}{x - a} = f(a, b).$$

This equation, when written in the form

$$y = f(a, b)(x - a) + b,$$

expresses y as a function which we are willing to accept as a reasonable approximation to the actual solution if x is restricted to a small interval $a \leqq x \leqq a + h$. In particular, this approximation gives the point, $x_1 = a + h$, $y_1 = f(a, b)h + b$. The point (x_1, y_1) and the slope $f(x_1, y_1)$ determine the linear approximation

$$y = f(x_1, y_1)(x - x_1) + y_1$$

which we are willing to accept over $a + h \leqq x \leqq a + 2h$. In this manner the process is continued until the desired value of x is reached. The value of the function used to describe the line segment at that value of x is considered to be the approximate value of the solution.

example 2.25: Approximate the value of the solution at $x = 3$ to the initial value system $y' = y + x^2$, $y(0) = 1$ by using line segments of width 1 along the x-axis.

solution: The first line passes through (0, 1) with slope 1. Its equation is $y = 1 + x$, from which $y = 2$ when $x = 1$. From the differential equation the slope of the direction field at (1, 2) is 3. Thus the equation of the second line is $y = 3x - 1$. Similarly, the third line passes through (2, 5) with slope 9 and its equation is $y = 9x - 13$. From this last equation we obtain the value of y when $x = 3$ to be 14. The graph is sketched in Figure 2.10.

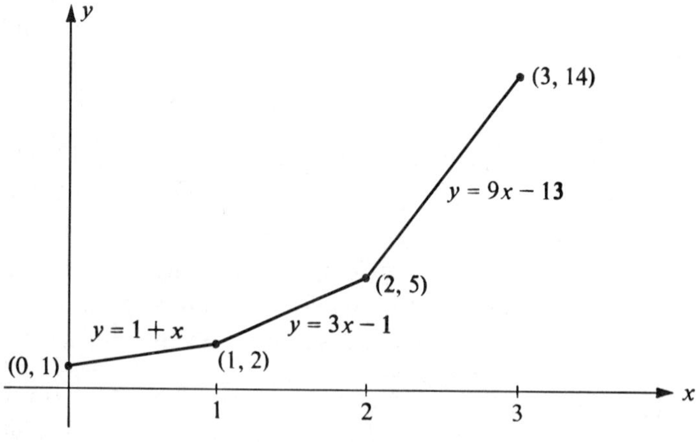

figure 2.10

The method just described yields more accurate results if you are willing to use smaller values of h. Actually, the error can be made arbitrarily small by taking h sufficiently small. Mathematically this is expressed by

$$\text{(actual solution)} = \lim_{h \to 0} \text{(approximation functions)}.$$

The length of the calculations can become quite large if we use a small value of h, but they are of a type which can be easily adapted to a computer. We do not actually use the equation of the nth line over the whole interval $x_{n-1} \leqq x \leqq x_n$ but only to calculate the ordinate y_n at the one abcissa x_n.

For the initial value system $y' = f(x, y), y(a) = b$ the successive approximations are given by the formulas

$$\begin{aligned}
x_0 &= a & y_0 &= b \\
x_1 &= a + h & y_1 &= y_0 + f(x_0, y_0)\, h \\
x_2 &= a + 2h & y_2 &= y_1 + f(x_1, y_1)\, h \\
&\vdots & &\vdots \\
x_n &= a + nh & y_n &= y_{n-1} + f(x_{n-1}, y_{n-1})\, h.
\end{aligned}$$

The points (x_n, y_n) are the vertices of the polygonal approximation to the solution curve. To "start" the computational process in obtaining these points, the point (x_0, y_0), the function $f(x, y)$, and the increment h, must be known.

2.8 an iterative method

Consider the initial value problem

$$y' = f(x, y), \quad y(a) = b,$$

where f is a continuous function over some domain in the plane. The solution to this problem can be written in indicated form in terms of an indefinite integral,

$$y = b + \int_a^x f(t, y(t))\, dt.$$

In this case, a method known as a repetitive or *iterative* technique allows the computation of a sequence of functions which under rather general conditions on the function $f(x, y)$ converges to a function $y(x)$ which is a solution to the initial value problem. The particular technique given here is based upon a classical approach due to Picard and thus the elements of the sequence are sometimes called the *Picard approximations.*

We use the given solution to define the recursive relation

$$y_n(x) = b + \int_a^x f(t, y_{n-1}(t))\, dt.$$

Thus an entire sequence is defined if y_0 is given. As a rule y_0 is chosen to be b, although this is not necessary.

The Picard iterative method is not very well suited for numerical work and is presented more for historical purposes than as a practical approach.

example 2.26: Find the Picard approximations y_1, y_2, y_3 if $y_0 = 1$ to the initial value problem

$$y' = y, \quad y(0) = 1.$$

solution: The recursive formula for the Picard approximations is given by

$$y_n = 1 + \int_0^x y_{n-1}(t)\, dt.$$

Since

$$\begin{aligned} y_0 &= 1 \\ y_1 &= 1 + \int_0^x dt = 1 + x \\ y_2 &= 1 + \int_0^x (1 + t)\, dt = 1 + x + \frac{x^2}{2} \\ y_3 &= 1 + \int_0^x \left(1 + t + \frac{t^2}{2}\right) dt = 1 + x + \frac{x^2}{2} + \frac{x^3}{6}. \end{aligned}$$

example 2.27: Find the Picard approximations y_1 and y_2 if $y_0 = 1$ to the initial value problem

$$y' = 1 + y^2, \quad y(0) = 1.$$

(Note that this is a nonlinear equation.)

solution: The Picard approximations are given by

$$y_n = 1 + \int_0^x (1 + y_{n-1}^2(t))\, dt.$$

Since $y_0 \equiv 1$, it follows that

$$\begin{aligned} y_1 &= 1 + \int_0^x (1 + (1)^2)\, dt = 2x + 1 \\ y_2 &= 1 + \int_0^x (1 + (1 + 2t)^2)\, dt \\ &= 1 + 2x + 2x^2 + \frac{4x^3}{3}. \end{aligned}$$

example 2.28: Using just one Picard approximation, approximate the solution to the initial value problem of Example 2.25 at $x = 3$.

solution: If $y_0 = 1$, then

$$y_1 = 1 + \int_0^x (1 + t^2)\, dt = 1 + x + \frac{x^3}{3}.$$

At $x = 3$, this is equal to 13.

exercises for sections 2.7 and 2.8

In Exercises 1–6 sketch the direction field associated with the given differential equation. Draw particular solution curves which pass through the points (1, 0), (0, 1), and (1, 1).

1. $y' = y$.
2. $y' = x$.
3. $y' - xy + 1 = 0$.
4. $x + 2yy' = 0$.
5. $y + xy' = 0$.
6. $y' = x^2 + y$.

In Exercises 7–10, use the method of straight line approximations to estimate the value of the solution to the given initial value problems at the indicated points. In each case, write the equations of the straight lines over the different intervals.

7. $y' = y$, $y(0) = -2$. Find $y(4)$, using increments of length 2.
8. $y' = x + y$, $y(0) = 0$. Find $y(3)$, using increments of length 1.
9. $y' = x^2$, $y(1) = 2$. Find $y(4)$, using increments of length 1.
10. $y' = 3$, $y(0) = -4$. Find $y(5)$, using increments of length $\frac{1}{2}$.
11. Compare the approximating solutions in Exercises 1–10 to the actual solutions by finding the explicit particular solutions.

In Exercises 12–15, find the first three Picard iterations, y_1, y_2, y_3.

12. $y' = y,\quad y(0) = -2$.
13. $y' = x + y,\quad y(0) = 0$.
14. $y' = x - y^2,\quad y(0) = 3$.
15. $y' = x^2 - 4,\quad y(1) = 2$.

linear differential equations

3

3.1 differential operators

In this chapter we will discuss linear differential equations of order n. These equations may be put into the form

$$a_n(x)\,y^{(n)}(x) + a_{n-1}(x)\,y^{(n-1)}(x) + \cdots + a_1(x)\,y'(x) + a_0(x)\,y(x) = f(x).$$

The $a_i(x)$ are called the coefficient functions and $f(x)$ is called the *driving* or *input* function. If $f(x) = 0$, the equation is called *homogeneous*; otherwise it is called a *nonhomogeneous* equation.

Associated with each nonhomogeneous equation is the linear differential equation

$$a_n(x)\,y^{(n)}(x) + a_{n-1}(x)\,y^{(n-1)}(x) + \cdots + a_1(x)\,y'(x) + a_0(x)\,y(x) = 0$$

which is called the homogeneous equation corresponding to the nonhomogeneous equation, or, as in this book, the *reduced* equation. For example, the corresponding reduced equation for the linear second-order equation

$$xy'' + y' + 3y = \sin x \quad \text{is} \quad xy'' + y' + 3y = 0.$$

The left member of an nth order linear differential equation is often represented by $L(y)$, where L is called a differential operator of order n. You should think of L as being a combination of coefficient functions and derivatives which operate on a function y to yield $f(x)$. The linear differential equation is thus written as

$$L(y) = f(x)$$

and the corresponding reduced equation as

$$L(y) = 0.$$

In using operator notation to represent nth order linear differential equations we seek those functions, y, which satisfy $L(y) = f(x)$ for a fixed operator L. Occasionally, we are also interested in fixing a function, say y_1, and seeking those operators for which $L(y_1) = 0$. Operators, L, for which $L(y_1) = 0$ are said to *annihilate* y_1 and are called *annihilators* of y_1. For example, the operator $\left(\frac{d^2}{dx^2} + 1\right)$ annihilates $\sin x$ since $\frac{d^2}{dx^2}(\sin x) + \sin x = 0$.

Note that there may be distinct limitations on the kind of functions for which $L(y)$ has meaning. Functions for which $L(y)$ *has* meaning are called *admissible* for that operator. Usually the criteria for admissibility will be somewhat obvious. For example, only twice differentiable functions are admissible for the operator $\left(\frac{d^2}{dx^2} + 1\right)$.

For linear differential equations, the differential operator has the two properties

$$L(y_1 + y_2) = L(y_1) + L(y_2)$$
$$L(cy) = cL(y)$$

where y, y_1, and y_2 are all admissible for the operator L. These two properties are called the linearity properties and their proof is left for the exercises. Hence, L is called a linear differential operator.

You should become familiar with at least the rudimentary algebraic laws connected with operators.

definition 3.1

Two operators L_1 and L_2 are said to be *equal* if $L_1(y) = L_2(y)$ for all admissible functions y.

definition 3.2

By the operator D^k is meant the kth derivative operator $\frac{d^k}{dx^k}$. Note that the set of k differentiable functions is admissible for D^k. By $D^0 y$ is meant y. The number k is called the *power* of the operator.

All linear differential operators may be expressed in terms of linear combinations of powers of the D operator. For example, the expression $y'' + x^2 y' - y$ may be written $(D^2 + x^2 D - 1)y$ and the expression $(D^2 - x^2)y$ represents $y'' - x^2 y$.

definition 3.3

The *sum* of two operators L_1 and L_2 is obtained by first expressing L_1 and L_2 as linear combinations of the D operator and adding coefficients of like powers of D.

For example, if $L_1 = 3D^2 + xD - 1$ and $L_1 = xD^2 + xD - x^2$, then

$$L_1 + L_2 = (x + 3)\,D^2 + 2xD - (1 + x^2).$$

definition 3.4

By the *product* of two operators L_1L_2 is meant the equivalent operator obtained by using the operator L_2 followed by L_1. If we find the product of two operators as a linear combination of the D operator, we call it the *expansion* of the product.

example 3.1: Expand the product $(xD + 1)\,(xD - 1)$.

solution: Let y be any admissible function for the product. Then

$$\begin{aligned}(xD + 1)\,(xD - 1)\,(y) &= (xD + 1)\,(xy' - y)\\ &= xD\,(xy') - xD\,(y) + xy' - y\\ &= x\,(xy'' + y') - xy' + xy' - y\\ &= x^2y'' + xy' - y\\ &= (x^2D^2 + xD - 1)\,y,\end{aligned}$$

which means that

$$(xD + 1)\,(xD - 1) = x^2D^2 + xD - 1\,.$$

Note that to expand a product of operators, you must determine its effect on any admissible function and then express this effect as a linear combination of powers of D.

It is easy to show that addition of linear differential operators is commutative and associative; that multiplication is associative; and that multiplication is distributive over addition.

Operator multiplication is generally not commutative (multiply the operators of the previous example in reverse order), but there is a very important case in which it is.

definition 3.5

Consider an operator which is a linear combination of powers of the D operator. If the coefficients are constant the operator is called a *polynomial operator*. To emphasize that an operator is of the polynomial type, we often write $P(D)$ instead of L.

A linear nth order differential equation of the type $P(D)y=f(x)$ is one in which the coefficient functions are constant.

Poylnomial operator multiplication is commutative as it is easy to show, at least for special cases. More importantly, as the following theorem shows, polynomial operator multiplication takes the same form it would if the operators were real polynomials.

theorem 3.1

Let $P_1(m)$ and $P_2(m)$ be any two real polynomials and let $P_1(D)$ and $P_2(D)$ be the corresponding polynomial operators. Then if $P_1(m)P_2(m)=P_3(m)$, it follows that

$$P_1(D)P_2(D)=P_3(D).$$

proof: (For first-degree polynomials; the general result follows with the use of mathematical induction.) Let

$$P_1(m)=a_1m+b_1 \quad \text{and} \quad P_2(m)=a_2m+b_2.$$

Then,

$$P_3(m)=a_1a_2m^2+(a_1b_2+a_2b_1)m+b_1b_2.$$

For polynomial operator multiplication, let y be any admissible function. Then,

$$\begin{aligned}P_1(D)P_2(D)y &= (a_1D+b_1)(a_2y'+b_2y)\\ &= (a_1D+b_1)(a_2y')+(a_1D+b_1)(b_2y)\\ &= a_1a_2y''+a_2b_1y'+a_1b_2y'+b_1b_2y\\ &= (a_1a_2D^2+a_2b_1D+a_1b_2D+b_1b_2)(y),\end{aligned}$$

which shows that $P_3(D)$ has the same form as $P_3(m)$ with D substituted for m. This proves the theorem.

theorem 3.2

Polynomial operator multiplication is commutative.

proof: Exercise for the special case of first-degree polynomials.

Theorems 3.1 and 3.2 along with the linearity properties of the more general linear differential operator allow multiplication or factorization of polynomial differential operators as if they were real polynomials. For example, the operators D^2+5D+4, $(D+4)(D+1)$, and $(D+1)(D+4)$ are all representatives of the same operator.

Often we must operate on the product of an exponential function and a rather arbitrary admissible function of x. Since this is equivalent to differentiating a product, the work can be quite tedious, but the following theorem will shorten the process.

theorem 3.3

Exponential Shift. If $P(D)$ is a polynomial operator, then

$$P(D)\,(e^{mx}y) = e^{mx}P(D+m)\,y.$$

proof: (i) Let $P(D) = aD + b$. Then,

$$\begin{aligned}(aD+b)\,(e^{mx}y) &= ame^{mx}y + ae^{mx}y' + be^{mx}y \\ &= e^{mx}(aD + am + b) \\ &= e^{mx}P(D+m)\,y.\end{aligned}$$

(ii) Let $P(D) = P_1(D)\,P_2(D)$, where $P_1(D)$ and $P_2(D)$ are first-order polynomial operators. Then,

$$\begin{aligned}P(D)\,e^{mx}y &= P_1(D)\,P_2(D)\,(e^{mx}y) \\ &= P_1(D)\,(e^{mx}P_2(D+m)\,y) \\ &= e^{mx}P_1(D+m)\,P_2(D+m)\,y \\ &= e^{mx}P(D+m)\,y.\end{aligned}$$

This procedure can obviously be done for any number of linear factors.

corollary: $$e^{mx}P(D)\,y = P(D-m)\,e^{mx}y.$$

proof: This is a direct application of the theorem.

example 3.2: Using the exponential shift theorem, compute

$$\frac{d^3}{dx^3}\,e^{2x}(\sin x + x^5).$$

solution: In polynomial operator form this expression may be written

$$D^3\,(e^{2x}(\sin x + x^5)).$$

Using the exponential shift, we obtain

$$\begin{aligned}\frac{d^3}{dx^3}\,e^{2x}(\sin x + x^5) &= e^{2x}(D+2)^3\,(\sin x + x^5) \\ &= e^{2x}(D^3 + 6D^2 + 12D + 8)\,(\sin x + x^5) \\ &= e^{2x}(-\cos x - 6\sin x + 12\cos x \\ &\quad + 8\sin x + 60x^2 + 120x^3 + 60x^4 + 8x^5) \\ &= e^{2x}(11\cos x + 2\sin x + 60x^2 + 120x^3 + 60x^4 + 8x^5).\end{aligned}$$

In Sections 3, 4, and 5 of this chapter you will see how polynomial operators are used to solve linear differential equations with constant coefficients. For now we will be content with some examples which will show you how some of the ideas of this section are put to good use.

example 3.3: Solve the differential equation $y''' + 6y'' + 12y' + 8y = 0$.

solution: In operator form this equation may be written

$$(D^3 + 6D^2 + 12D + 8)\,y = 0$$

or

$$(D + 2)^3\,y = 0.$$

Multiply both sides of this equation by e^{2x}:

$$e^{2x}(D + 2)^3\,y = 0.$$

Using the corollary to Theorem 3.3,

$$D^3(e^{2x}y) = 0.$$

Three antidifferentiations yield

$$e^{2x}y = c_1 + c_2x + c_3x^2,$$

from which

$$y = e^{-2x}(c_1 + c_2x + c_3x^2).$$

example 3.4: Solve the differential equation $y'' + 4y' + 4y = 0$.

solution: In polynomial operator form, this equation is

$$(D^2 + 4D + 4)\,y = 0$$

or

$$(D + 2)^2\,y = 0.$$

Multiplying by e^{2x} and applying the corollary to Theorem 3.3, this becomes

$$D^2(e^{2x}y) = 0.$$

Therefore,

$$e^{2x}y = c_1 + c_2x$$

or

$$y = e^{-2x}(c_1 + c_2x).$$

exercises for section 3.1

Perform the indicated operator multiplication in Exercises 1–7.

1. $(xD - 1)(xD + 1)$. (Compare to Example 3.1.)
2. $D(x^2D - 1)$.
3. $(x^2D - 1)D$.
4. $(D - x)(D + x)$.
5. $(D + x)(D - x)$.
6. $(D - 2)(D + 3)$.
7. $(xD^2 - D + 1)(xD + x)$.
8. Prove that $(aD + b)(cD + d) = (cD + d)(aD + b)$ (Theorem 3.2).

Factor the operators in Exercise 9–12.

9. $2D^2 + 3D - 2$. 10. $D^3 - 3D^2 + 4$.

11. $D^4 + D^3 - 2D^2 + 4D - 24$. 12. $D^2 - 6D + 9$.

In Exercises 13–15 compute, using the exponential shift theorem.

13. $D^2(e^x \sin x)$. 14. $D^3(e^{-x} \cos x)$.

15. $(D^3 + D^2 - D - 1)(e^{2x}(x^2 + x + 1))$.

16. Show that $(D - a)^n (e^{ax}x^k) = 0$ for $0 \leqq k < n$.

Using the exponential shift theorem, solve the differential equations in Exercises 17–20.

17. $y'' - 4y' + 4y = 0$.

18. $y^{(iv)} - 12y''' + 54y'' - 108y' + 81y = 0$.

19. $y^{(viii)} + 8y^{(vii)} + 28y^{(vi)} + 56y^{(v)} + 70y^{(iv)} + 56y''' + 28y'' + 8y' + y = 0$.

20. $y'' - 6y' + 9y = x$.

3.2 *the homogeneous equation*

The nth-order homogeneous linear differential equation, which we represent by $L(y) = 0$, has a simple but elegant theory associated with it. We first note some rather general vector space properties of equations of this type.

theorem 3.4

The set of solutions of an nth-order homogeneous linear differential equation is a vector space.

proof: Note that solutions, being functions, are subsets of the vector space of piecewise continuous functions and hence the only vector space properties to be verified are the so-called closure properties; that is, we must show that any linear combination of solutions is also a solution. Let y_1 and y_2 be solutions to the given differential equation. Then $L(y_1) = 0$ and $L(y_2) = 0$. Let $y_3 = c_1y_1 + c_2y_2$. Then, using the linearity properties of L,

$$\begin{aligned} L(y_3) &= L(c_1y_1 + c_2y_2) \\ &= L(c_1y_1) + L(c_2y_2) \\ &= c_1L(y_1) + c_2L(y_2) \\ &= c_1 0 + c_2 0 \\ &= 0, \end{aligned}$$

which proves the theorem.

Vector space terminology allows a characterization of the set of solutions to an nth-order homogeneous linear differential equation in a very concise statement. The following theorem will not be proved here (See Appendix E, Reference 12, page 205) but the result will be very useful.

theorem 3.5

Let $L(y)=0$ be an nth-order homogeneous linear differential equation with $a_n(x) \neq 0$. The dimension of the solution space is n.

As a result of Theorem 3.5, a basis for the solution space of $L(y)=0$ consists of n linearly independent functions, each of which satisfies the differential equation. Since the solution space is completely determined if a basis is known, we will know all solutions if we can find n linearly independent ones. If the set

$$\{y_i\}_{i=1}^{n}$$

is a basis, the general solution is

$$y(x) = \sum_{i=1}^{n} c_i y_i(x).$$

It is therefore obviously important to know when n solutions to an nth-order homogeneous linear equation are linearly independent. For the case $n=2$, we can show linear dependence on an interval by showing that the ratio of two functions is constant over that interval. (See Exercises for Section 1.7, Exercise 3.) For the general case of n arbitrary functions, direct appeal to the definition of linear independence is necessary; that is, to show that the set of functions

$$\{y_i\}_{i=1}^{n}$$

is a linearly independent set on some interval, you must show that if

$$\sum_{i=1}^{n} c_i y_i(x) = 0$$

on the interval, then $c_i = 0$ for all i.

There is a rather simple formal procedure for deciding specifically about the linear independence of solutions to homogeneous linear differential equations. This procedure is based on the nonvanishing of the Wronskian determinant.

In the Exercises for Section 1.7 the Wronskian determinant of three functions y_1, y_2, y_3 is defined to be

$$W(x) = \begin{vmatrix} y_1(x) & y_2(x) & y_3(x) \\ y_1'(x) & y_2'(x) & y_3'(x) \\ y_1''(x) & y_2''(x) & y_3''(x) \end{vmatrix}.$$

For three functions each with two derivatives on (a, b) it is easy to show that if the Wronskian of the three functions is nonzero, then the three functions are linearly independent. For example, since the Wronskian of e^x, e^{-x}, and e^{2x} is $-6e^{2x} \neq 0$, the functions e^x, e^{-x}, and e^{2x} are linearly independent.

The nonvanishing of the Wronskian on an interval is not a necessary condition for linear independence, unless the functions being considered are solutions to a homogeneous linear differential equation. We state this in the form of a theorem.

theorem 3.6

Let y_1, y_2, and y_3 be solutions to a linear third-order homogeneous differential equation with continuous coefficients. The three functions are linearly independent on (a, b) if and only if their Wronskian is nonzero on (a, b).

proof: See Appendix E, Reference 12, page 204.

This theorem is easily extended to the case of an nth-order equation.

example 3.5: Solve the differential equation $y^{(5)} = 0$.

solution: This is a linear homogeneous differential equation of order 5 and hence its solution space is 5-dimensional. Obviously, the functions 1, x, x^2, x^3, and x^4 are linearly independent solutions. The general solution is often written

$$y(x) = c_0 + c_1x + c_2x^2 + c_3x^3 + c_4x^5.$$

example 3.6: All solutions to the differential equation $y'' + 3y' + 2y = 0$ are linear combinations of terms of the form e^{mx}. Find a basis for the solution space.

solution: Substituting $y = e^{mx}$ into the differential equation we obtain $e^{mx}(m^2 + 3m + 2) = 0$, from which $m = -1$ and $m = -2$. A basis for the two-dimensional solution space is $\{e^{-x}, e^{-2x}\}$ and the general solution is $y = c_1e^{-x} + c_2e^{-2x}$.

example 3.7: All solutions to the differential equation $x^2y'' + xy' - y = 0$ are linear combinations of terms of the form x^m. Find a basis for the solution space.

solution: Substituting $y = x^m$ into the differential equation, we obtain $x^m[m(m-1) + m - 1] = 0$, from which $m^2 - 1 = 0$. Hence $m = \pm 1$ and a basis for the solution space is $\{x, x^{-1}\}$. The general solution is

$$y = c_1 x + c_2 x^{-1}.$$

exercises for section 3.2

The differential equations in Exercises 1–5 have solutions of the form e^{mx}. Find the general solution of each by finding n linearly independent solutions. Show that the functions are linearly independent by showing that their Wronskians are non-zero.

1. $y''' + y'' - 10y' + 8y = 0$.
2. $y'' + 4y' + 3y = 0$.
3. $y'' - 4y = 0$.
4. $y'' - y' = 0$.
5. $y''' - 8y'' + 7y' = 0$.

The differential equations in Exercises 6 and 7 have solutions of the form x^m. Write the general solution for each.

6. $x^2y'' + xy' - 4y = 0$.
7. $x^2y'' - xy' - 3y = 0$.
8. Show that $\{e^x, xe^x, x^2e^x\}$ are linearly independent by showing that the Wronskian of the three functions does not vanish.
9. Show that $\{e^{mx}, e^{nx}, e^{kx}\}$ are linearly independent, where m, n, and k are distinct real numbers.
10. Show that $\{\cos x, \sin x, x \sin x, \text{and } x \cos x\}$ are linearly independent.
11. Using the definition of linear independence, show that the set of functions $\{e^{rx}, xe^{rx}, x^2e^{rx}, \cdots, x^{k-1}e^{rx}\}$ is linearly independent.

3.3 solution to homogeneous equations with constant coefficients

The nth-order homogeneous linear differential equation with constant coefficients can be written in the form

$$P(D)\, y = 0$$

where $P(D)$ is a polynomial operator. The corresponding real algebraic equation $P(m) = 0$ is called the *auxiliary* equation of the differential equation. In Section 3.1 you learned that polynomial operators may be factored in the same way as real polynomials. Recall that every real polynomial can be factored into real factors of the form:

1. $(m - r)^k$ where r is a real number. The polynomial is said to have a root $m = r$ of multiplicity k. Conversely, if the polynomial has a root of

multiplicity k at $m = r$, then the polynomial has a factor of the form $(m - r)^k$.

★ 2. $[(m - a)^2 + b^2]^j$ where a and b are real numbers. The polynomial has complex conjugate roots $a \pm bi$, each of multiplicity j. If $a \pm bi$ is a root of the polynomial of multiplicity j, then the polynomial has a factor of the form $[(m - a)^2 + b^2]^j$.

The following theorem shows that to solve a homogeneous equation, you may consider separately each of the factors of the polynomial operator.

theorem 3.7

Consider the linear homogeneous differential equation with constant coefficients represented by $P(D)\,y = 0$, and suppose $P(D) = P_1(D)\,P_2(D)$. Then every solution of $P_1(D)\,y = 0$ and of $P_2(D)\,y = 0$ is also a solution of $P(D)\,y = 0$.

proof: Because of the commutativity of polynomial operator multiplication, we need prove the theorem for only one of the two factors. Suppose y_1 is a solution to $P_2(D)\,y = 0$. Then, $P_2(D)$ annihilates y_1 and hence

$$P(D)\,y_1 = P_1(D)\,P_2(D)\,y_1 = P_1(D)\,\{0\} = 0,$$

which shows that y_1 is a solution to $P(D)\,y = 0$.

Hence, the method of solving the homogeneous linear differential equation with constant coefficients essentially reduces to solving differential equations in either one of the two following forms:

1. $(D - r)^k\,y = 0$, where r is a real number; or ★
2. $[(D - a)^2 + b^2]^j\,y = 0$, where a and b are real numbers. ★

Multiplying the first of these equations by e^{rx} and applying the exponential shifting rule, we obtain

$$D^k(e^{-rx}y) = 0, \quad \text{★}$$

from which, after k antidifferentiations,

$$e^{-rx}y = c_0 + c_1 x + c_2 x^2 + \cdots + c_{k-1}x^{k-1}. \quad \text{★}$$

Therefore, the solution is

$$y = e^{rx}(c_0 + c_1 x + c_2 x^2 + \cdots + c_{k-1}x^{k-1}). \quad \text{★}$$

The set of functions $\{e^{rx}, xe^{rx}, x^2 e^{rx}, \cdots, x^{k-1}e^{rx}\}$ is linearly independent. (See the previous set of exercises, Exercise 11.) Since the equation is of order

k, the given set is a basis for the solution space and the solution as given is the general solution.

The differential equation

$$[(D-a)^2+b^2]^j\, y=0$$

may be written in the form

$$[D-(a+bi)]^j\,[D-(a-bi)]^j\, y=0\,.$$

By reasoning used in the preceding paragraph, a basis for the solution space is

$$\{e^{(a+bi)x}\, x^n\}_{n=0}^{j-1} \cup \{e^{(a-bi)x}\, x^n\}_{n=0}^{j-1}.$$

As you will learn in Chapter 12, the complex exponential may be expressed in terms of the real sine and cosine functions as

$$e^{(a\pm bi)x}=e^{ax}\cos bx \pm i\, e^{ax}\sin bx\,.$$

Further, the functions $e^{ax}\cos bx$ and $e^{ax}\sin bx$ are linearly independent. Therefore, the following set of functions is also a basis for the solution space:

$$\{x^n e^{ax}\cos bx\}_{n=0}^{j-1} \cup \{x^n e^{ax}\sin bx\}_{n=0}^{j-1}.$$

The method of solution to a homogeneous linear differential equation with constant coefficients is now clear.

1. Express the differential equation in polynomial operator form.
2. Express the polynomial operator as a product of factors of the type $(D-r)^k$ and $[(D-a)^2+\mathrm{b}^2]^j$. This is an algebraic problem of finding the roots of the auxiliary equation and may prove to be a formidable task.
3. Each of the factors $(D-r)^k$ contributes the set $\{x^n e^{rx}\}_{n=0}^{k-1}$ to the set of basis elements.
4. Each of the factors $[(D-a)^2+b^2]^j$ contributes

   $$\{x^n e^{ax}\cos bx\}_{n=0}^{j-1} \cup \{x^n e^{ax}\sin bx\}_{n=0}^{j-1}$$

   to the set of basis elements.

That the union of all these sets comprises the total set of basis elements is true because the following theorem assures their linear independence.

***theorem* 3.8**

The set of functions

$$\{e^{r_1x}x^n\}_{n=0}^{n_1} \cup \{e^{r_2x}x^n\}_{n=0}^{n_2} \cup\cdots\cup \{e^{r_jx}x^n\}_{n=0}^{n_j}$$

is linearly independent.

proof: The general proof is omitted, but for a proof in some specific cases, see Exercises 8–11 of the previous exercise set.

In many cases you will not actually have to factor the operator, but may proceed directly to the solution once the roots of the auxiliary equation have been found.

example 3.8: Solve the differential equation $y'' + 3y' + 2y = 0$.

solution: The auxiliary quadratic equation is $m^2 + 3m + 2 = 0$, whose roots are $m = -1$ and $m = -2$. Therefore any solution may be written as a linear combination of the basis elements e^{-x} and e^{-2x}. The general solution is

$$y = c_1 e^{-x} + c_2 e^{-2x}.$$

example 3.9: Solve the differential equation $y''' + 4y'' + 4y' = 0$.

solution: The auxiliary equation is $m^3 + 4m^2 + 4m = 0$. The roots are 0, -2, and -2, from which we know a basis for the solution space to be $\{1, e^{-2x}, xe^{-2x}\}$. The general solution has the form

$$y = c_1 + e^{-2x}(c_2 + c_3 x).$$

example 3.10: Solve the differential equation $y'' + 2y' + 4y = 0$.

solution: The auxiliary equation is $m^2 + 2m + 4 = 0$. The roots of this equation are

$$a \pm ib = \frac{-2 \pm \sqrt{4 - 16}}{2} = -1 \pm i\sqrt{3}$$

and hence a basis for the solution space is $\{e^{-x} \cos\sqrt{3}x, e^{-x} \sin\sqrt{3}x\}$. The general solution is

$$y = e^{-x}(c_1 \cos\sqrt{3}x + c_2 \sin\sqrt{3}x).$$

In passing we note that solutions to homogeneous linear differential equations with constant coefficients may always be written as a sum of terms of the type $x^k e^{ax}$, $x^k e^{ax} \cos bx$ and $x^k e^{ax} \sin bx$, where a and b are real numbers. It is startling that this very important class of differential equations, found in so many applications, should have a solution space which is so easy to describe.

exercises for section 3.3

In the following exercises, find a basis for the solution space.

1. $y'' + 9y = 0.$
2. $y'' - 9y = 0.$
3. $y'' + 5y' + 4y = 0.$
4. $y'' + y = 0.$
5. $y''' + y = 0.$
6. $4y^{(iv)} - 8y''' - 7y'' + 11y' + 6y = 0.$
7. $y^{(iv)} + 8y''' + 16y'' = 0.$
8. $y^{(v)} - y''' = 0.$
9. $y''' - 3y' - 2y = 0.$
10. $y'' - 2y' + 6y = 0.$
11. $y''' + y'' + 4y' + 4y = 0.$
12. $y''' + 2y'' + 5y' = 0.$
13. $y^{(vi)} + 9y^{(iv)} + 24y'' + 16y = 0.$
14. $y'' - 7y' + y = 0.$
15. Find a particular solution to the initial value problem:
$$y'' - 6y' + 9y = 0, \qquad y(0) = 1, \qquad y'(0) = 2.$$
16. Find a particular solution to the system
$$y'' - y = 0, \qquad y(0) = 1, \qquad y'(0) = 1.$$
17. Solve the system
$$y'' - (4 + \varepsilon)\, y' + (4 + 2\varepsilon)\, y = 0, \qquad y(0) = 1, \qquad y'(0) = 1$$
first when ε is zero and then when ε is a nonzero constant. Show that the solution for $\varepsilon = 0$ is the limit of the solution for $\varepsilon \neq 0$ as $\varepsilon \to 0$.
18. Solve the system
$$y'' - 4y' + ay = 0, \qquad y(0) = 0, \qquad y'(0) = 1$$
where a is a constant. Consider the three cases $a = 4$, $a = 4 - \varepsilon^2$ and $a = 4 + \varepsilon^2$, and study the limit as $\varepsilon \to 0$.
19. Find a particular solution to
$$y'' - y' - 2y = 0, \qquad y(0) = 2$$
and $y \to 0$ as $x \to \infty$.

3.4 the nonhomogeneous linear differential equation

The general technique of solving a linear nth-order differential equation with nonzero driving function is based upon the following theorem.

theorem 3.9

Let y_p be any particular solution to the nth-order linear differential equation represented by
$$L(y) = f(x)$$
and let y_c be the general solution to the corresponding reduced equation, $L(y) = 0$. Let $y(x) = y_c(x) + y_p(x)$. Then y is the general solution to $L(y) = f(x)$.

proof: Let y_{c1} be any one of the functions represented by y_c. Then,

$$\begin{aligned} L(y_{c1}+y_p) &= L(y_{c1})+L(y_p) \\ &= 0+f(x) \\ &= f(x) \end{aligned}$$

which shows that $y_{c1}+y_p$ is a solution to the nonhomogeneous equation. Conversely, let y_{p1} be any particular solution to the nth-order linear differential equation, $L(y)=f(x)$. Then

$$\begin{aligned} L(y_p-y_{p1}) &= L(y_{p1})-L(y_p) \\ &= f(x)-f(x) \\ &= 0, \end{aligned}$$

which shows that y_{p1} is of the form $y_{c2}+y_p$ where y_{c2} is one of the solutions to the reduced equàtion. Thus, any particular solution can be written as the sum of y_p and one of the solutions to the reduced equation, which is the same as saying that the general solution is

$$y(x)=y_c(x)+y_p(x).$$

Two particular solutions $y_{p1}(x)$ and $y_{p2}(x)$ may differ by a function which is a solution to the reduced equation. This accounts for the often dissimilar appearance of particular solutions to nonhomogeneous equations.

example 3.11: Reconcile the fact that $(1+\frac{1}{2}\tan x)(\sin x)$ and $\frac{1}{2}\sin x\tan x$ are both particular solutions to the differential equation $y''+y=\sec^3 x$.

solution: We simply note that the difference of the two solutions is $\sin x$ which is a solution to the corresponding reduced equation $y''+y=0$.

Theorem 3.9 says that if *any* particular solution to the differential equation is known, then finding the general solution is a matter of solving the corresponding reduced equation.

example 3.12: Show that $x-1$ is a solution to $y''-y=1-x$ and then find the general solution.

solution: Verification of the fact that $x-1$ is a solution to the differential equation is a matter of substituting $y=x-1$, $y''=0$ into the equation. Since the roots of the auxiliary equation are 1 and -1, the general solution to the reduced equation is

$$y_c(x)=c_1e^x+c_2e^{-x}.$$

Hence, the general solution to the given equation is

$$y(x) = c_1 e^x + c_2 e^{-x} + x - 1.$$

The driving function is often written as the sum of two functions, say $f(x) = f_1(x) + f_2(x)$. It is then convenient to consider the two equations $L(y) = f_1(x)$ and $L(y) = f_2(x)$ separately.

theorem 3.10

Let y_1 be a solution to $L(y) = f_1(x)$ and y_2 a solution to $L(y) = f_2(x)$. Then $y_1 + y_2$ is a solution to

$$L(y) = f_1(x) + f_2(x).$$

proof: See Exercise 1, Exercises for Sections 3.4 and 3.5.

This theorem is known as the *principle of superposition*. As a result of this principle, finding the general solution to a nonhomogeneous linear equation such as

$$3y'' - 5y' + 2y = \sin x + x^2$$

is a matter of finding

1. A general solution, y_c, to $3y'' - 5y' + 2y = 0$.
2. A particular solution, y_1, to $3y'' - 5y' + 2y = \sin x$.
3. A particular solution, y_2, to $3y'' - 5y' + 2y = x^2$.

Then, $y_c + y_1 + y_2$ is the general solution to the given differential equation.

In the following sections we will examine three well-known methods for obtaining at least *one* solution to the nonhomogeneous equation with constant coefficients. Then by Theorem 3.9 and the methods of Section 3.3 we will be able to write the general solution.

3.5 successive antidifferentiations

The method described in this section is not always the easiest but it is very general. Assume the differential equation can be written in the form

$$P(D)\,y = f(x)$$

where $P(D)$ is a polynomial operator. Find roots of the auxiliary equation $P(m) = 0$ and let r be one of these roots. Then $P(D)$ may be written in the form $(D-r)P_1(D)$ where $P_1(m)$ is of degree one less than $P(m)$. The differential equation then may be written

$$(D-a)P_1(D)\,y = f(x).$$

Let $P(D)\,y = w$. Then w is a function of x satisfying the first-order linear equation

$$(D-a)\,w = f(x),$$

or

$$w' - aw \quad = f(x).$$

This equation may be solved by the method of Section 2.6. The equation

$$P_1(D)\,y = w$$

is of lower order (by one) than the initial equation. The process of splitting off a linear factor of the polynomial operator may be repeated n times and a particular solution is thereby obtained.

Note that we have essentially replaced the problem of finding the solution to an nth-order linear differential equation (with constant coefficients) with that of finding, individually, solutions to n first-order linear equations. The process of successive antidifferentiations can be tedious and messy.

example 3.13: Find the general solution to $y'' - 3y' + 2y = e^x$.

solution: In operator form, this equation is written

$$(D^2 - 3D + 2)\,y = e^x,$$

or, in factored form,

$$(D-1)(D-2)\,y = e^x.$$

Letting $w = (D-2)\,y$, this equation becomes

$$(D-1)\,w = e^x.$$

The first-order equation $w' - w = e^x$ has e^{-x} for an integrating factor and therefore,

$$we^{-x} = \int dx = x,$$

where the constant of antidifferentiation is omitted since we seek only a particular solution. The first-order equation

$$y' - 2y = w = xe^x$$

has an integrating factor e^{-2x} and thus

$$ye^{-2x} = \int xe^{-x}\,dx$$
$$= -xe^{-x} - e^{-x}$$

from which

$$y_p = -xe^x - e^x.$$

The general solution to the reduced equation is found to be

$$y_c = c_1e^x + c_2e^{2x}$$

and thus the general solution to the given equation is

$$y = c_1e^x + c_2e^{2x} - xe^x - e^x.$$

exercises for sections 3.4 and 3.5

1. Prove Theorem 3.10.
2. The functions $x+1$, $x+e^x$, and $x-1-e^x$ are three solutions to a certain nonhomogeneous equation of the second order. Find all solutions.
3. Given that e^x is a solution to $y''+y=2e^x$, find the general solution.
4. Given that $e^{-x}-3x+\frac{3}{2}-2xe^{-x}$ is a solution to $y''-y'-2y=6x+6e^{-x}$, find the general solution.
5. Given that $e^{2x}+e^{-2x}-xe^{-x}+x+1/2$ is a solution to $y'''+y''-4y'-4y=3e^{-x}-4x-6$, find the general solution.
6. Show that if y_1 is a solution to $L(y)=f(x)$, then ay_1 is a solution to $L(y)=af(x)$.
7. Find the general solution to $y'''+y''=x+1$.

Find the general solution to the differential equations in Exercises 8–12. Find the y_p by the method of Section 3.5.

8. $y''-y=x$.
9. $y''-y=e^{3x}$.
10. $y''+4y'+4y=e^{-2x}$.
11. $y'''+4y''+4y'=1$.

3.6 method of undetermined coefficients

In this section we learn how to find a particular solution to the nonhomogeneous equation when the driving function is of a specific but very important form. The method is limited to those driving functions for which there exist polynomial operator annihilators. Recall that determining annihilators for a specific function is the reverse process of finding a solution to a homogeneous differential equation. In solving a homogeneous differential equation you find those functions which the polynomial operator annihilates; now we wish to find a polynomial annihilator for a specific function. In effect, finding an annihilator is the same as exhibiting a homogeneous linear differential equation with constant coefficients for which the given function is a solution.

Recall that the only kind of functions which satisfy homogeneous linear differential equations with constant coefficients have the form of sums of $x^k e^{ax}$, $x^k e^{ax} \sin bx$, and $x^k e^{ax} \cos bx$, where a and b are real numbers. These are precisely the kinds of functions to which the method of undetermined coefficients is limited. To use the method you must recall that

1. $(D-a)^{k+1}$ annihilates $x^k e^{ax}$.
2. $[(D-a)^2+b^2]^{k+1}$ annihilates both $x^k e^{ax} \cos bx$ and $x^k e^{ax} \sin bx$.
3. The annihilator of a sum of functions is the *product* of the individual annihilators. (Note that the annihilator of a product of functions is *not* the product of the individual annihilators.)

example 3.14: Find a homogeneous linear differential equation with constant coefficients which has a particular solution,

$$y = xe^{-3x} + x^2 + \sin x.$$

solution: We examine each term in the driving function separately.

1. The function xe^{-3x} is annihilated by $(D+3)^2$.
2. x^2 is annihilated by D^3.
3. $\sin x$ is annihilated by D^2+1.

Therefore, a polynomial operator annihilator of the function is $D^3(D+3)^2$ (D^2+1), and hence the desired differential equation is

$$y^{(\text{vii})} + 6y^{(\text{vi})} + 10y^{(\text{v})} + 6y^{(\text{iv})} + 9y''' = 0.$$

When finding an annihilator of a function you will usually (if not always) desire the polynomial annihilator of least degree. For example, it is certainly true that D^6 annihilates x^3+3x^2-x+1 but D^4 is more desirable.

The method of undetermined coefficients is based upon substituting the *general* form of the y_p into the differential equation and determining the *specific* form by forcing the assumed y_p to be a solution. To determine the general form of the y_p that should be assumed consider the nonhomogeneous equation

$$P(D)\, y = f(x)$$

where $f(x)$ is a functoin which has a polynomial differential annihilator, $P_1(D)$.

1. Find the general solution, y_c, to the reduced equation $P(D)\,y = 0$.
2. Apply the polynomial operator annihilator, $P_1(D)$, to both sides of the differential equation to obtain $P_1(D)\,P(D)\,y = 0$.
3. Solve this latter equation. This solution will consist of terms due to the factor $P(D)$; these terms are identical to the y_c. The remaining terms are the form of the y_p.
4. To find the specific form of the y_p the relation $P(D)\,y_p = f(x)$ should be used to determine the coefficients.

example 3.15: Solve the differential equation $y'' + 3y' + 2y = e^{2x}$.

solution: In operator form this equation is

$$(D+1)(D+2)\,y = e^{2x}.$$

The solution to the reduced equation, $(D+1)(D+2)\,y = 0$, is

$$y_c = c_1e^{-x} + c_2e^{-2x}.$$

An annihilator for e^{2x} is $D-2$. We apply this to both sides of the differential equation to obtain

$$(D-2)(D+1)(D+2)\,y = 0.$$

The solution to this equation is

$$y = c_1e^{-x} + c_2e^{-2x} + Ae^{2x}$$

where a different notation is used for terms which are not common to this solution and to y_c. Therefore, the form of y_p is

$$y_p = Ae^{2x}.$$

Substituting $y_p = Ae^{2x}$, $y_p' = 2Ae^{2x}$, $y_p'' = 4Ae^{2x}$ into the differential equation gives

$$4Ae^{2x} + 3(2Ae^{2x}) + 2(Ae^{2x}) = e^{2x},$$

from which $A = 1/12$, and the specific y_p is $y_p = e^{2x}/12$. Therefore the general solution is given by

$$y = c_1e^{-x} + c_2e^{-2x} + \frac{e^{2x}}{12}.$$

example 3.16: Find the general solution to the differential equation

$$y''' + 4y' = x^2 + \sin x.$$

solution: In operator form, this equation may be written

$$D(D^2+4)\,y = x^2 + \sin x\,.$$

The solution to the reduced equation, $D(D^2+4)\,y=0$, is

$$y_c(x) = c_1 + c_2\cos 2x + c_3\sin 2x\,.$$

An annihilator for the driving function is $D^3(D^2+1)$. Applying this to both sides of the differential equation we obtain the equation

$$D^4(D^2+4)(D^2+1)\,y = 0\,.$$

The general solution to this equation is

$$\begin{aligned} y = c_1 + Ax + Bx^2 + Cx^3 + c_2\cos 2x + c_3\sin 2x \\ + D\sin x + E\cos x\,, \end{aligned}$$

where, as before, alternate notation is used to distinguish the terms of y_c and y_p. The coefficients of the terms in the expression for y_p are found by substituting into the given differential equation.

$$\begin{aligned} y_p &= Ax + Bx^2 + Cx^3 + D\sin x + E\cos x \\ y_p' &= A + 2Bx + 3Cx^2 + D\cos x - E\sin x \\ y_p'' &= 2B + 6Cx - D\sin x - E\cos x \\ y_p''' &= 6C - D\cos x + E\sin x \end{aligned}$$

Therefore

$$\begin{aligned} y_p''' + 4y_p' &= 6C - D\cos x + E\sin x + 4A + 8Bx + 12Cx^2 \\ &\quad + 4D\cos x - 4E\sin x \\ &= 4A + 6C + 8Bx + 12Cx^2 + 3D\cos x - 3E\sin x \\ &= x^2 + \sin x\,. \end{aligned}$$

We equate coefficients of like terms on both sides of this equation:

Constant terms	$4A + 6C = 0$
Coefficient of x	$8B = 0$
Coefficient of x^2	$12C = 1$
Coefficient of $\cos x$	$3D = 0$
Coefficient of $\sin x$	$-3E = 1\,.$

From these equations, the coefficients are found to be $A = 1/8$, $B = 0$, $C = 1/12$, $D = 0$, and $E = (-1/3)$. Hence, the general solution is

$$y = c_1 + c_2\cos 2x + c_3\sin 2x - \frac{x}{8} + \frac{x^3}{12} - \frac{\cos x}{3}\,.$$

example 3.17: Find the general solution to the differential equation

$$y'' + y = \cos^2 x.$$

solution: At first glance you may think the method of undetermined coefficients is not applicable since $\cos^2 x$ does not have the appearance of a function for which a polynomial annihilator exists. Using the trigonometric identity

$$\cos^2 x = \frac{1 + \cos 2x}{2}$$

we see that an annihilator for $\cos^2 x$ is $D(D^2 + 4)$. From this it is easy to show that the form of the y_p is

$$y_p = A + B\cos 2x + C\sin 2x.$$

Differentiating twice,

$$y_p' = -2B\sin 2x + 2C\cos 2x$$
$$y_p'' = -4B\cos 2x - 4C\sin 2x.$$

Therefore,

$$\begin{aligned}(D^2 + 1)y_p &= -4B\cos 2x - 4C\sin 2x + A + B\cos 2x + C\sin 2x \\ &= A - 3B\cos 2x - 3C\sin 2x \\ &= \tfrac{1}{2} + \frac{\cos 2x}{2}.\end{aligned}$$

This equation gives $A = 1/2$, $B = -(1/6)$, and $C = 0$. The general solution is

$$y = c_1 \cos x + c_2 \sin x + \tfrac{1}{2} - \frac{\cos 2x}{6}.$$

The formal procedure of the method of undetermined coefficients may be shortened after you become familiar with the basic idea of selecting the form of the y_p. Then you can omit the steps involved in finding the annihilator of the driving function. As you probably have noticed, the only reason this process is included is to give you a reasonable procedure for determining the *form* of the y_p. Hopefully you will find the following two rules to be "reasonable."

1. The form of the y_p is, aside from the exceptional case outlined in statement 2. on the next page, of the same form as the driving function.
   (a) If the driving function has a term of the form $x^k e^{ax}$, the y_p should include a sum of terms, $p(x)e^{ax}$, where $p(x)$ is a polynomial of degree k.
   (b) If the driving function has a term of the form $x^k e^{ax} \cos bx$ or

$x^k e^{ax} \sin bx$, then the y_p should include a sum of terms, $p(x)e^{ax}$ $[A \cos bx + B \sin bx]$, where $p(x)$ is a polynomial of degree k.

2. If a term of the driving function is a solution to the corresponding reduced equation, then the corresponding term of the assumed form of the y_p is the same as outlined above but multiplied by a power of x. The precise power of x needed is easy to determine and in the Exercises for this section you are asked to find a rule for determining it.

example 3.18: Solve the differential equation $y'' + 3y' + 2y = e^{-x}$.

solution: Since $y_c = c_1 e^{-x} + c_2 e^{-2x}$ the driving function is a solution to the corresponding homogeneous equation. Therefore,

$$y_p = Axe^{-x}, \quad y_p' = -Axe^{-x} + Ae^{-x}, \quad y_p'' = Axe^{-x} - 2Ae^{-x}.$$

Substituting into the given differential equation, we obtain

$$Axe^{-x} - 2Ae^{-x} - 3Axe^{-x} + 3Ae^{-x} + 2Axe^{-x} = e^{-x}.$$

This yields the result, $A = 1$. Therefore the general solution is

$$y = c_1 e^{-x} + c_2 e^{-x} + xe^{-x}.$$

The method of undetermined coefficients is not only limited to specific kinds of driving functions, it is sometimes very lengthy. For example, consider the differential equation

$$y''' - 3y'' + 3y' - y = x^4 e^x.$$

In operator form, this equation is

$$(D - 1)^3 y = x^4 e^x.$$

The form of the y_p that should be used in the method of undetermined coefficients is

$$y_p = e^x (Ax^3 + Bx^4 + Cx^5 + Dx^6 + Ex^7).$$

The differentiations required and the algebraic equations which result are a bit messy. A different approach is to multiply both sides of the equation by e^{-x} and then apply the exponential shift to obtain

$$D^3 (e^{-x} y) = x^4.$$

Three antidifferentiations give (the arbitrary constants are omitted since only a particular solution is desired)

$$e^{-x} y_p = \frac{x^7}{7 \cdot 6 \cdot 5}$$

so that

$$y_p = \frac{x^7 e^x}{210}.$$

Now try to find this solution by the method of undetermined coefficients.

exercises for section 3.6

Solve the equations in Exercises 1–25.

1. $y'' - 4y' + 4y = e^x$.
2. $y'' - 4y' + 4y = e^x + 1$.
3. $y'' - 4y' + 4y = e^{2x}$.
4. $y'' - 4y' + 4y = \sin x$.
5. $y'' - 4y' + 4y = xe^{2x} + e^{2x}$.
6. $y'' - 4y' + 4y = xe^{2x}$.
7. $y'' + y = \sin 2x$.
8. $y'' + 4y = \sin 2x$.
9. $y'' + 4y' = \sin 2x$.
10. $y'' - 2y' + 5y = \sin 2x$.
11. $y'' - 2y' + 5y = e^x \cos 2x$.
12. $y'' - y = \cosh 2x$.
13. $y'' - y = \cosh x$.
14. $y'' + y = x \cos x$.
15. $y'' - 3y' + 2y = e^x + e^{2x} + e^{-x}$.
16. $y'' - y' = x^2$.
17. $y'' - y' = 3$.
18. $y'' + 3y' = x + 3$.
19. $y'' + 4y = 3$.
20. $y'' + 4y = 3x$.
21. $y''' - 3y' - 2y = \sin 2x$.
22. $y^{(iv)} - y = e^{-x}$.
23. $y''' + 4y'' + 9y' + 10y = -e^x$.
24. $y''' + y'' = 1$.
25. $y''' + y'' - 2y = x^2 + 10 \cos 2x$.
26. Consider the differential equation $y''' + 2y'' + y' = f(x)$. Determine the y_p to be used if $f(x)$ equals each of the following:
    (a) x. (b) $x + 2$. (c) $\sin x + x$.
    (d) e^{-x}. (e) xe^{-x}. (f) $\sinh x$.
    (g) $\sinh x + \cosh x$. (h) $x(1 + e^{-x})$.
27. Determine the rule for selecting the power of x to use in modifying the y_p for the exceptional case when the driving function is a solution to the reduced equation.

3.7 method of variation of parameters

In the previous section we found a particular solution to the nth-order linear equation with constant coefficients, $L(y) = f(x)$, by the method of undetermined coefficients. We saw that the method is limited to those driving functions which are solutions to some homogeneous linear equation with constant coefficients.

In this section we will consider a method which has no such restriction. In fact the method is applicable to all linear equations, including those with variable coefficients. For purposes of discussion we will study the application of the method to the second-order equation with constant coefficients,

$$a_2y'' + a_1y' + a_0y = f(x).$$

As you will see, the method requires only that the solutions to the corresponding reduced equation $L(y) = 0$ (or, in the specific second-order case, $a_2y'' + a_1y' + a_0y = 0$) be known.

Let y_1 and y_2 be any two linearly independent solutions to the corresponding reduced equation. Assume that a solution to the nonhomogeneous equation may be written in the form

$$y_p = uy_1 + vy_2$$

where u and v are unknown functions (or *parameters*) to be determined.

Since two functions are to be determined we will need two conditions to be imposed. The first condition is that $uy_1 + vy_2$ must satisfy the given differential equation, while the second condition, to be imposed for the convenience of simplifying the computations, will be that $u'y_1 + v'y_2 = 0$.

To impose the first condition, we find y_p' and y_p'' and substitute into the differential equation,

$$y_p' = u'y_1 + v'y_2 + uy_1' + vy_2'.$$

Since $u'y_1 + v'y_2 = 0$, this becomes

$$y_p' = uy_1' + vy_2'.$$

Computing y_p'',

$$y_p'' = u'y_1' + v'y_2 + uy_1'' + vy_2''.$$

Substituting into the differential equation,

$$a_2(u'y_1' + v'y_2' + uy_1'' + vy_2'') + a_1(uy_1' + vy_2') + a_0(uy_1 + vy_2) = f(x)$$

$$a_2(u'y_1' + v'y_2') + u(a_2y_1'' + a_1y_1' + a_0y_1) + v(a_2y_2'' + a_1y_2' + a_0y_2) = f(x).$$

The coefficient of u is zero since y_1 is a solution to the corresponding reduced equation. Similarly the coefficient of v is zero, so that u' and v' must satisfy the system of equations

$$u'y_1 + v'y_2 = 0$$

$$u'y_1' + v'y_2' = \frac{f(x)}{a_2}.$$

This system of equations may be solved for u' and v' (by Cramer's rule),

$$u' = \frac{\begin{vmatrix} 0 & y_2 \\ \dfrac{f(x)}{a_2} & y_2' \end{vmatrix}}{\begin{vmatrix} y_1 & y_2 \\ y_1' & y_2' \end{vmatrix}}; \quad v' = \frac{\begin{vmatrix} y_1 & 0 \\ y_1' & \dfrac{f(x)}{a_2} \end{vmatrix}}{\begin{vmatrix} y_1 & y_2 \\ y_1' & y_2' \end{vmatrix}}.$$

Note that the denominator is the Wronskian of the two functions y_1 and y_2 and since y_1 and y_2 are linearly independent solutions, by Theorem 3.6, this Wronskian is nonzero. Hence u' and v' are completely determined.

example 3.19: Solve the differential equation $y'' + y = \sec x$.

solution: Since two linearly independent solutions to the corresponding reduced equation are $\cos x$ and $\sin x$, the general solution to the given equation is

$$y = c_1 \cos x + c_2 \sin x + y_p$$

where $y_p = u \cos x + v \sin x$, and u' and v' are given by

$$u' = \frac{\begin{vmatrix} 0 & \sin x \\ \sec x & \cos x \end{vmatrix}}{\begin{vmatrix} \cos x & \sin x \\ -\sin x & \cos x \end{vmatrix}}; \quad v' = \frac{\begin{vmatrix} \cos x & 0 \\ -\sin x & \sec x \end{vmatrix}}{\begin{vmatrix} \cos x & \sin x \\ -\sin x & \cos x \end{vmatrix}}.$$

Therefore $u' = -\tan x$ and $v' = 1$, from which $u = \ln \cos x$ and $v = x$. The general solution is

$$y = c_1 \cos x + c_2 \sin x + (\ln \cos x) \cos x + x \sin x.$$

example 3.20: Solve the differential equation $y'' - y = xe^x$.

solution: Because of the nature of the driving function, the y_p could be found by the method of undetermined coefficients. In the method of variation of parameters, let $y_1 = e^x$ and $y_2 = e^{-x}$ in the expressions for u' and v' to obtain

$$u' = \frac{\begin{vmatrix} 0 & e^{-x} \\ xe^x & -e^{-x} \end{vmatrix}}{\begin{vmatrix} e^x & e^{-x} \\ e^x & -e^{-x} \end{vmatrix}}; \quad v' = \frac{\begin{vmatrix} e^x & 0 \\ e^x & xe^x \end{vmatrix}}{\begin{vmatrix} e^x & e^{-x} \\ e^x & -e^{-x} \end{vmatrix}}.$$

Therefore, $u' = x/2$ and $v' = -xe^{2x}/2$ from which $u = x^2/4$ and $v = -(xe^{2x}/4) + (e^{2x}/8)$. The general solution is

$$y = c_1 e^x + c_2 e^{-x} + \tfrac{1}{4}x^2 e^x - \tfrac{1}{4}xe^x + \tfrac{1}{8}e^x.$$

The first and last terms are usually combined to yield a new arbitrary coefficient of e^x.

In passing we note that other choices are made for the second condition for the determination of u and v instead of the computationally convenient $u'y_1 + v'y_2 = 0$. Sometimes these choices lead to usable techniques other than the method of variation of parameters. For example, the condition $v = 0$ is equivalent to making the substitution $y_p = uy_1$. Then u is chosen so that y_p satisfies the differential equation. This method is called "reduction of order" since the substitution $y_p = uy_1$ will always result in effectively reducing (by one) the order of the differential equation which must be solved.

example 3.21: Solve the differential equation in Example 3.20 by the method of "reduction of order" as outlined above.

solution: Let $y_p = ue^x$. Then $y_p' = u'e^x + ue^x$ and $y_p'' = u''e^x + 2u'e^x + ue^x$. Substituting into the differential equation,

$$u''e^x + 2u'e^x = xe^x,$$

or

$$u'' + 2u' = x.$$

Letting $u' = w$,

$$w' + 2w = x$$

which is a first-order equation in w. A particular solution to this equation is $w = (x/2) - (1/4)$, from which

$$u = \int (\tfrac{1}{2}x - \tfrac{1}{4})\, dx = \tfrac{1}{4}x^2 - \tfrac{1}{4}x.$$

The general solution is, as before,

$$y = c_1 e^x + c_2 e^{-x} + \tfrac{1}{4}x^2 e^x - \tfrac{1}{4}xe^x.$$

exercises for section 3.7

1. Find functions $u_1(x)$ and $u_2(x)$ such that $u_1(x)\,y_1(x) + u_2(x)\,y_2(x)$ is a solution to the differential equation $y'' - 3y' + 2y = x + 1$ for each of the following conditions ($y_1(x)$ and $y_2(x)$ are solutions to the homogeneous equation):
   (a) $u_1(x) = 0$. (b) $u_1(x)\,y_1(x) = u_2(x)\,y_2(x)$.
   (c) $u_1(x)$ and $u_2(x)$ satisfy the usual condition imposed in the method of variation of parameters

2. By using the method of variation of parameters show that the solution to $y'' + y = f(x)$ can be written in the form

$$y = c_1 \cos x + c_2 \sin x + \int_a^x f(t) \sin(x - t)\, dt.$$

Solve Exercises 3–12 by using the method of variation of parameters.

3. $y'' - y' - 2y = e^{2x}$.
4. $y'' + 2y' + y = \dfrac{e^{-x}}{x}$.
5. $y'' + 4y = \tan 2x$.
6. $y'' + 4y = \tan^2 2x$.
7. $y'' + 4y = \sin^2 2x$.
8. $y'' - 3y' + 2y = \cos(e^{-x})$.
9. $y'' - 4y' + 4y = \dfrac{e^{2x}}{x}$.
10. $y'' + 6y' + 9y = \dfrac{e^{-3x}}{x^2 + 1}$.
11. $y'' + 2y' + y = e^{-x} \ln x$.
12. $y'' + 2y' + y = \dfrac{e^{-x}}{x^3}$.
13. Check that x and $1/x$ are solutions to the differential equation $x^2 y'' + xy' - y = 0$. Solve the differential equation if the differential equation becomes nonhomogeneous, with driving function $x^2 \ln x$.
14. Check that x and e^x are solutions to $(1 - x)\, y'' + xy' - y = 0$. Solve, if the equation becomes nonhomogeneous with driving function $(x - 1)^2 e^{-x}$.

The second-order equation $a_2 x^2 y'' + a_1 xy' + a_0 y = f(x)$ where the a_i are constants is called the Cauchy-Euler equation.

15. Show that the substitution for the independent variable $x = e^v$ (or equivalently $v = \ln x$) reduces the equation to

$$a_2 \frac{d^2 y}{dv^2} + (a_1 - a_2) \frac{dy}{dv} + a_0 y = f(e^v)$$

which is a second-order equation with constant coefficients, and can be solved by the methods of the previous sections.

Solve the equations in Exercises 16–20.

16. $x^2 y'' + 4xy' + 2y = 0$.
17. $x^2 y'' - 2xy' + 2y = 4x + \sin(\ln x)$.
18. $x^2 y'' - xy' - 3y = x^2 \ln x$.
19. $x^2 y'' + xy' + 9y = 0$.
20. $x^2 y'' - 2xy' + 2y = (x - 1) \ln x$.
21. Classify the solutions to the homogeneous Cauchy-Euler equation by the roots of a quadratic equation.

22. Show that if y_1 is a solution to

$$a_2(x)\,y'' + a_1(x)\,y' + a_0(x)\,y = 0,$$

then the equation

$$a_2(x)\,y'' + a_1(x)\,y' + a_0(x)\,y = f(x)$$

may be solved by first order linear methods by first making the substitution $y_p = uy_1$ followed by $v = u'$. As was pointed out, this procedure is called reduction of order.

23. Solve by the method of reduction of order: $y'' - y = \cos x$.

24. Solve by the method of reduction of order: $y'' + 3y' + 2y = e^x$.

FINAL TEST

3.8 *application to electric circuits*

One of the most immediate applications of second-order linear differential equations with constant coefficients is to electric circuits. Every circuit has three passive elements which behave to impede the flow of current, change in current, or change in voltage. In elementary circuit analysis, these elements are considered as being "lumped" components called resistance, R, inductance, L, and capacitance, C. These three constants are essentially defined by the following three equations (see Figure 3.1):

$$Ri_R = v_R; \qquad L\frac{di_L}{dt} = v_L; \qquad C\frac{dv_C}{dt} = i_C,$$

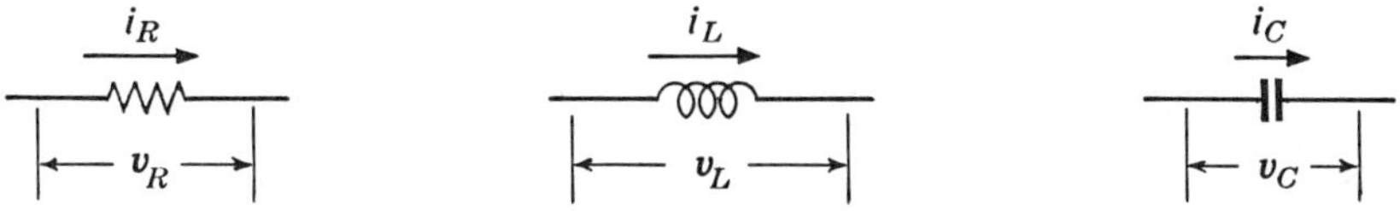

figure 3.1

where the i's and v's refer to the current through and the voltage across the lumped components. In a so-called series circuit, $i_R = i_L = i_C$, and in a parallel circuit $v_R = v_L = v_C$ (Figure 3.2). Many electric circuits can be considered as combinations of these two arrangements.

The governing physical law for an RLC series circuit is given by Kirkhoff's law which says that the driving voltage to the circuit is equal to the sum of the voltage drops. In differential equation form,

$$L\frac{di}{dt} + Ri + \frac{1}{C}\int i(t^*)\,dt^* = v$$

where v is the driving voltage. This equation relates the response, that is, the current, to the input voltage, or excitation.

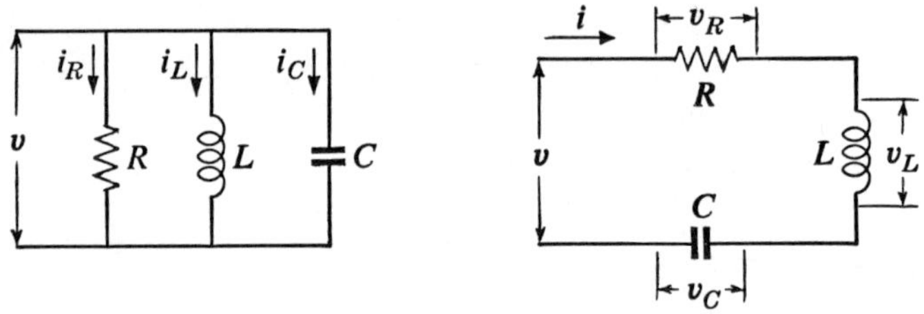

figure 3.2

It is often easier to handle the equation if both sides are differentiated with respect to t,

$$L\frac{d^2i}{dt^2} + R\frac{di}{dt} + \frac{i}{C} = \frac{dv}{dt}.$$

A very common type of input to an *RLC* circuit is some combination of sine or cosine functions. For the remainder of this section we will discuss the *RLC* circuit with a sinusoidal input voltage, $E \sin\omega t$, where E is the *amplitude* of the input and ω is the angular frequency. Since $\dfrac{dv}{dt} = E\omega \cos\omega t$, the differential equation of the circuit is

$$L\frac{d^2i}{dt^2} + R\frac{di}{dt} + \frac{i}{C} = E\omega \cos\omega t\,.$$

The corresponding reduced equation is

$$L\frac{d^2i}{dt^2} + R\frac{di}{dt} + \frac{i}{C} = 0\,.$$

The solution to the reduced equation is called the *transient current*, since, if $R \neq 0$, the current described by these terms is of relatively short duration. The auxiliary polynomial equation is

$$Lm^2 + Rm + \frac{1}{C} = 0$$

whose solution is

$$m = \frac{-R \pm \sqrt{R^2 - 4\dfrac{L}{C}}}{2L}$$

$$= -\left(\frac{R}{2L}\right) \pm \sqrt{\left(\frac{R}{2L}\right)^2 - \frac{1}{LC}}$$

$$= -a \pm b \qquad \text{if} \quad \left(\frac{R}{2L}\right)^2 - \frac{1}{LC} \geqq 0$$

$$= -a \pm i\omega_0 \quad \text{if} \quad \left(\frac{R}{2L}\right)^2 - \frac{1}{LC} < 0,$$

where

$$a = \frac{R}{2L} \quad \text{and is called the } \textit{damping factor},$$

$$b = \sqrt{\left(\frac{R}{2L}\right)^2 - \frac{1}{LC}} \quad (\text{note that } a > b), \quad \text{if} \quad \left(\frac{R}{2L}\right)^2 - \frac{1}{LC} \geqq 0$$

$$\omega_0 = \sqrt{\frac{1}{LC} - \left(\frac{R}{2L}\right)^2}, \quad \text{if} \quad \left(\frac{R}{2L}\right)^2 - \frac{1}{LC} < 0,$$

and is called the internal or *natural angular frequency.*

case 1: Overdamped. If $\left(\frac{R}{2L}\right)^2 - \frac{1}{LC} > 0$, then both roots are real and negative. The transient current has the form

$$i_c = c_1 e^{-(a+b)t} + c_2 e^{-(a-b)t}.$$

Figure 3.3 shows the current for some typical initial conditions.

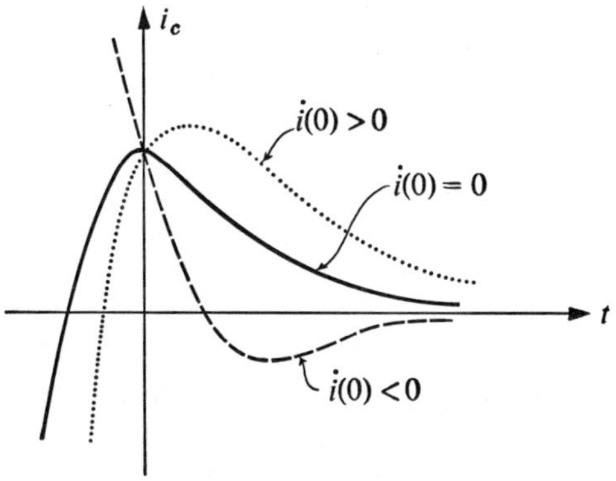

figure 3.3 *overdamped or critically damped*

case 2: *Critcally Damped.* If $(R/2L)^2 - 1/LC = 0$, then the auxiliary equation has a double root $m = -a$. Hence the solution is

$$i_c = e^{-at}(c_1 + c_2 t).$$

The transient current is of the same general form as for the overdamped case.

case 3: *Underdamped.* If $(R/2L)^2 - 1/LC < 0$, then the roots to the auxiliary equation are $m = -a \pm i\omega_0$ and the solution is

$$i_c = e^{-at}(c_1 \cos \omega_0 t + c_2 \sin \omega_0 t).$$

This is sometimes put into the form

$$i_c = e^{-at} c_4 \cos(\omega_0 t + c_4)$$

where

$$c_3 = \sqrt{c_1^2 + c_2^2} \quad \text{and} \quad c_4 = \text{Tan}^{-1}\left(\frac{c_2}{c_1}\right).$$

Figure 3.4 shows a typical underdamped case for the initial conditions $i(0) = 1, \dot{i}(0) = 0$.

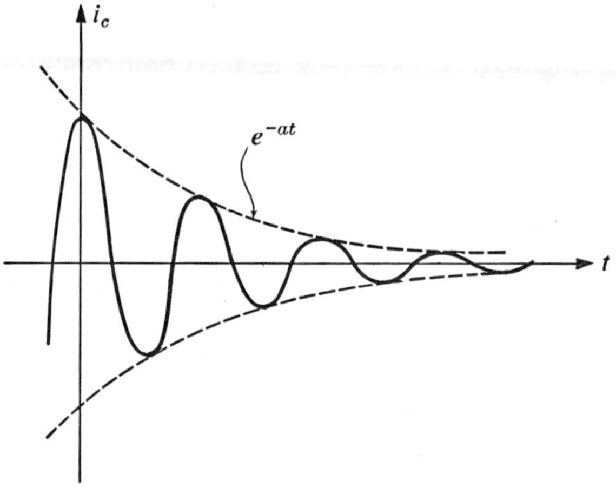

figure 3.4 *underdamped*

The graphs of the overdamped and critically damped transients cut the t-axis at most once (see Exercise 5, Exercises for Sections 3.8–3.10) while the underdamped transient is zero infinitely often.

The particular solution to the nonhomogeneous equation corresponding to $i_c = 0$ is called the *steady state* current, i_{ss}. It is easily obtained by an application of the method of undetermined coefficients. Let

$$i_{ss} = A \cos \omega t + B \sin \omega t$$

from which

$$\frac{d}{dt}(i_{ss}) = -\omega A \sin \omega t + \omega B \cos \omega t$$

$$\frac{d^2}{dt^2}(i_{ss}) = -\omega^2 A \cos \omega t - \omega^2 B \sin \omega t .$$

Substituting into the differential equation, we obtain

$$-\omega^2 AL \cos \omega t - \omega^2 BL \sin \omega t - \omega AR \sin \omega t + \omega RB \cos \omega t$$
$$+ \frac{A}{C} \cos \omega t + \frac{B}{C} \sin \omega t = E\omega \cos \omega t .$$

Equating coefficients of $\cos \omega t$ and $\sin \omega t$, we have the following two equations:

$$\left(-\omega^2 L + \frac{1}{C}\right) A + \omega RB = E\omega ,$$

$$-\omega RA + \left(-\omega^2 L + \frac{1}{C}\right) B = 0 .$$

Solving for A and B by using Cramer's rule,

$$A = \frac{\begin{vmatrix} E\omega & \omega R \\ 0 & -\omega^2 L + \frac{1}{C} \end{vmatrix}}{\begin{vmatrix} -\omega^2 L + \frac{1}{C} & \omega R \\ -\omega R & -\omega^2 L + \frac{1}{C} \end{vmatrix}}, \qquad B = \frac{\begin{vmatrix} -\omega^2 L + \frac{1}{C} & \omega E \\ -\omega R & 0 \end{vmatrix}}{\begin{vmatrix} -\omega^2 L + \frac{1}{C} & \omega R \\ -\omega R & -\omega^2 L + \frac{1}{C} \end{vmatrix}},$$

$$A = \frac{-E\omega^2 \left(\omega L - \frac{1}{\omega C}\right)}{\omega^2 \left(\omega L - \frac{1}{\omega C}\right)^2 + \omega^2 R^2}, \qquad B = \frac{E\omega^2 R}{\omega^2 \left(\omega L - \frac{1}{\omega C}\right)^2 + \omega^2 R^2}.$$

The quantity $\left(\omega L - \frac{1}{\omega C}\right)$ is called the *reactance* and is denoted by X. The quantity $\sqrt{X^2 + R^2}$ is called the *impedance* and is denoted by Z. Using these notations,

$$A = \frac{-EX}{Z^2} \quad \text{and} \quad B = \frac{ER}{Z^2} .$$

It is customary to express the steady state current in the form $\alpha \cos(\omega t + \beta)$ which displays the *amplitude*, α, of the output oscillation, and the *phase shift*, β, instead of the constants A and B. Thus,

$$\alpha = \sqrt{A^2 + B^2} = \sqrt{\frac{E^2X^2 + E^2R^2}{Z^4}} = \frac{E}{Z}; \qquad \beta = \text{Tan}^{-1}\frac{R}{X}.$$

In summary, the total output current is given by:
Output Current = Transient Current + Steady State Current.

$$i = c_1 e^{-(a+b)t} + c_2 e^{-(a-b)t} + \frac{E}{Z}\cos\left(\omega t + \text{Tan}^{-1}\frac{R}{X}\right) \qquad \textit{Overdamped}$$

$$i = e^{-at}(c_1 + c_2 t) + \frac{E}{Z}\cos\left(\omega t + \text{Tan}^{-1}\frac{R}{X}\right) \qquad \textit{Critically Damped}$$

$$i = e^{-at}c_3 \cos(\omega_0 t + c_4) + \frac{E}{Z}\cos\left(\omega t + \text{Tan}^{-1}\frac{R}{X}\right) \qquad \textit{Underdamped}$$

There is one other case that is interesting which occurs when $R = 0$:

$$i = c_3 \cos(\omega_0 t + c_4) + \frac{E}{X}\cos\omega t \qquad \textit{No Damping}$$

If we use initial conditions $i(0) = \dot{i}(0) = 0$, then $c_4 = 0$, $c_3 = -(E/X)$, and

$$i = \frac{E}{X}(\cos\omega t - \cos\omega_0 t).$$

By use of the trigonometric identity, $\cos v - \cos u = 2\sin\frac{u+v}{2}\sin\frac{u-v}{2}$, this becomes

$$i = 2\frac{E}{X}\left[\sin\left(\frac{\omega_0 - \omega}{2}\right)t \sin\left(\frac{\omega_0 + \omega}{2}\right)t\right].$$

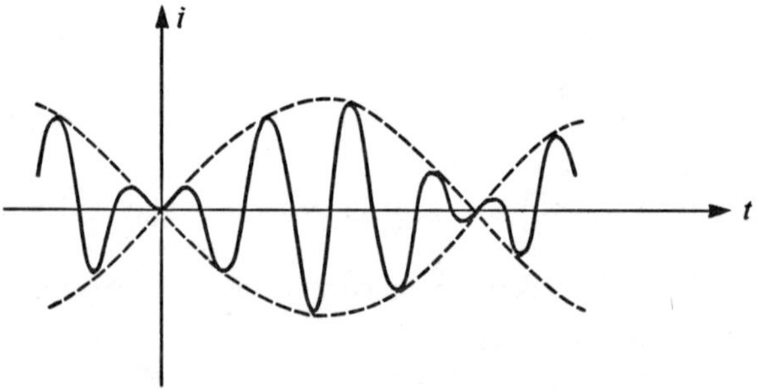

figure 3.5 beats

The point of interest is when ω is close to ω_0. This type of oscillation can be interpreted to have angular frequency close to ω (and to ω_0) with amplitude $\alpha(t)$ which fluctuates slowly with time with angular frequency $(\omega_0-\omega)/2$. Oscillations of this type are called *beats*. (See Figure 3.5.) Note that the amplitude of the beat is twice the amplitude of the steady state output when there is damping.

In the practical case there is always *some* damping so that the transient current is negligible after a period of time. The amplitude of the steady state current is dependent upon L, C, R, E, and ω, and is given by the formula

$$\alpha = \frac{E}{Z}.$$

We are interested in conditions for which this amplitude is a maximum. Conditions of this type are called *resonance*. Obviously the output is increased whenever the amplitude, E, of the input voltage is increased. For a fixed value of E, the output is maximimized by minimizing the impedance as follows:

1. For fixed L, C, and ω, by making $R=0$.
2. For fixed R, C, and ω, by making $L=1/(\omega^2C)$.
3. For fixed L, R, and ω, by making $C=1/(\omega^2L)$.
4. For fixed L, R, and C, by making $\omega=1/\sqrt{LC}$.

In summary, resonance conditions occur in the RLC circuit for a resistance as small as possible and with the condition $\omega^2LC=1$.

example 3.22: Using initial conditions $i(0)=0$, $\dot{i}(0)=0$ find the current in an RLC circuit if
(a) $L=1$, $R=0$, and $C=1$ with an input voltage, $E=\sin t$.
(b) $L=1$, $R=0$, $C=1/1.1025$, and $E=(\sin 0.95t)/0.95$.
(c) $L=1$, $R=0.2$, $C=1/1.01$, and $E=\sin t-0.05\cos t$.

solution: The three differential equations which must be solved with initial conditions $i(0)=0$, $\dot{i}(0)=0$ are
(a) $\ddot{i}+i=\cos t$.
(b) $\ddot{i}+1.1025i=\cos 0.95t$.
(c) $\ddot{i}+0.2\dot{i}+1.01i=\cos t+0.05\sin t$.

We recognize case (a) as undamped resonance,

$$i=c_1\cos t+c_2\sin t+\tfrac{1}{2}t\sin t.$$

Evaluating the constants from the initial conditions, we obtain $c_1=c_2=0$, so that $i=(t\sin t)/2$ which becomes unbounded with time.

The general solution to case (b) is easily found to be

$$i = c_1 \cos 1.05t + c_2 \sin 1.05t + 5 \cos 0.95t.$$

The initial conditions give $c_1 = -5$ and $c_2 = 0$, from which

$$i = 5(\cos 0.95t - \cos 1.05t) = 10 \sin(0.05t) \sin t.$$

This is the "beats" solution.

The general solution to case (c) is

$$i = e^{-0.1t}(c_1 \cos t + c_2 \sin t) + 5 \sin t.$$

The initial conditions give $c_1 = 0$, $c_2 = -5$, so that constants c_1 and c_2 are zero, so that

$$i = (5 - 5e^{-0.1t}) \sin t.$$

Note again that the amplitude of the "beats" solution is exactly twice that of the steady state solution of the last, damped, case.

3.9 application to mechanical systems

An elementary mechanical system may be pictured with a spring suspended vertically from a fixed support and attached to a mass, m. The mass in turn is connected to a dashpot, a device simulating viscous friction. Assume the mass is connected to some external force function, $F(t)$. Then if k is the spring constant and c is the fluid friction constant of proportionality, the differential equation which describes the system is

$$m \frac{d^2y}{dt^2} + c \frac{dy}{dt} + ky = F(t),$$

where y is the transverse motion of the center of mass. (See Figure 3.6.)

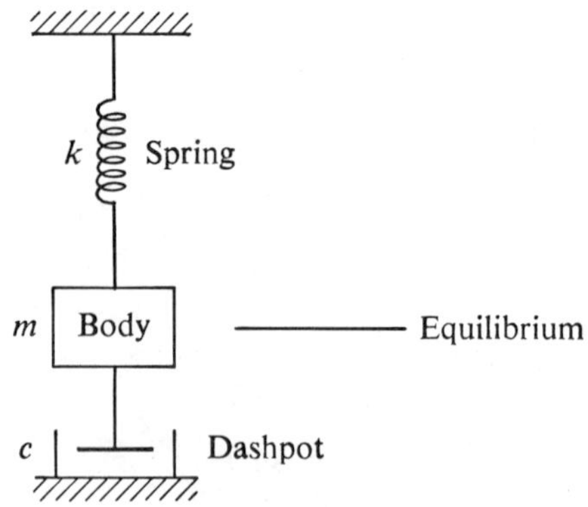

figure 3.6 *elementary mechanical system*

This is a differential equation with constant coefficients and is exactly analogous to the electric circuit discussed in the previous section. Note the correspondence between the electrical and mechanical quantities.

Mass, m	$\leftrightarrow$	Inductance, L
Fluid friction constant, c	$\leftrightarrow$	Resistance, R
Spring constant, k	$\leftrightarrow$	Reciprocal of capacitance, $1/C$
External driving force, $F(t)$	$\leftrightarrow$	Derivative of driving voltage, dv/dt
$F\cos\omega t$	$\leftrightarrow$	$E\omega\cos\omega t$
F	$\leftrightarrow$	$E\omega$
Displacement, y	$\leftrightarrow$	Current, i

With these correspondences, we can immediately write the displacement as a function of time in response to an external excitation of $F\cos\omega t$.

$$y = c_1 e^{-(a+b)t} + c_2 e^{-(a-b)t} + \frac{F}{\omega Z}\cos\left(\omega t + \operatorname{Tan}^{-1}\frac{c}{X}\right) \qquad \textit{Overdamped}$$

$$y = e^{-at}(c_1 + c_2 t) + \frac{F}{\omega Z}\cos\left(\omega t + \operatorname{Tan}^{-1}\frac{c}{X}\right) \qquad \textit{Critically Damped}$$

$$y = e^{-at}c_3\cos(\omega_0 t + c_4) + \frac{F}{\omega Z}\cos\left(\omega t + \operatorname{Tan}^{-1}\frac{c}{X}\right) \qquad \textit{Underdamped}$$

where

$$a = \frac{c}{2m}, \text{ the damping factor}$$

$$b = \sqrt{\left(\frac{c}{2m}\right)^2 - \frac{k}{m}}$$

$$\omega_0 = \sqrt{\frac{k}{m} - \left(\frac{c}{2m}\right)^2}, \quad \text{the natural angular frequency}$$

$$\omega Z = \sqrt{\omega^2 c^2 + m^2\left(\omega^2 - \frac{k}{m}\right)^2}$$

$$X = \frac{m}{\omega}\left(\omega^2 - \frac{k}{m}\right)$$

The concept of resonance is analyzed in a manner analogous to that for the case of the *RLC* circuit. Note that while resonance is often a desirable physical quality of an electric circuit, it is often undesirable for a mechanical system.

The amplitude of the steady state motion is

$$A = \frac{F}{\omega Z}.$$

The method of maximizing this output amplitude is the same as for the electrical case if F, m, c, or k is varied. If ω is varied and F, m, c, and k are fixed, the value of A is maximized by minimizing ωZ, and this quantity is a minimum if $\omega^2 Z^2$ is a minimum.

We therefore take the derivative of $\omega^2 Z^2$ with respect to ω,

$$\frac{d}{d\omega}(\omega^2 Z^2) = \frac{d}{d\omega}\left[(\omega^2 c^2) + m^2\left(\omega^2 - \frac{k}{m}\right)^2\right] = 2\omega c^2 + 4m^2\omega\left(\omega^2 - \frac{k}{m}\right).$$

Setting this quantity equal to zero, we obtain

$$c^2 = 2m^2\left(\frac{k}{m} - \omega^2\right).$$

Hence, the angular resonant frequency, ω_r, is

$$\omega_r^2 = \frac{k}{m} - \frac{c^2}{2m^2}.$$

If $c^2 > 2mk$, the equation for ω_r does not give a real value and the maximum output occurs when $\omega = 0$. If $2mk > c^2$, a positive real value of ω_r is obtained.

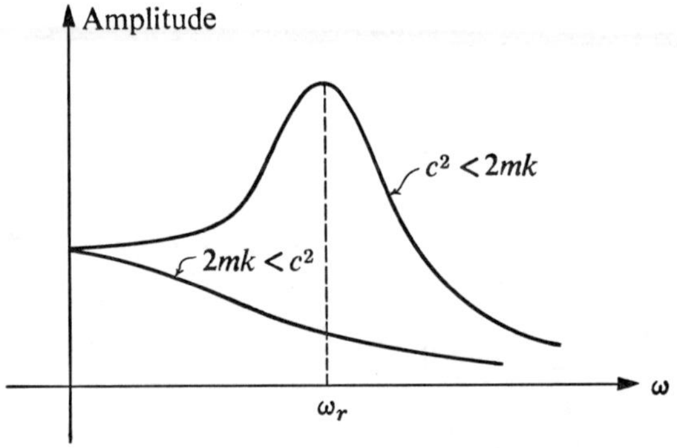

figure 3.7 *output amplitude as a function of excitation angular frequency*

Figure 3.7 shows the variation of the output amplitude with ω for the cases where the maximum output occurs for $\omega = 0$ and for the case of some positive ω_r. Note that the values of the resonant and natural angular frequencies differ by a small number if the damping is small. Hence, it is sometimes said that resonance is roughly approximated by letting the value of ω be ω_0, and this is one of the reasons that the resonant and natural angular frequencies are sometimes confused.

3.10 the sliding block

In this section we consider a variation on the mechanical system of the previous section. Consider a block of mass m on a horizontal plane with a constant frictional force which always opposes the movement of the block. A spring (with spring constant, k) is attached to a support and connected to the block. The block is displaced from equilibrium and then released with

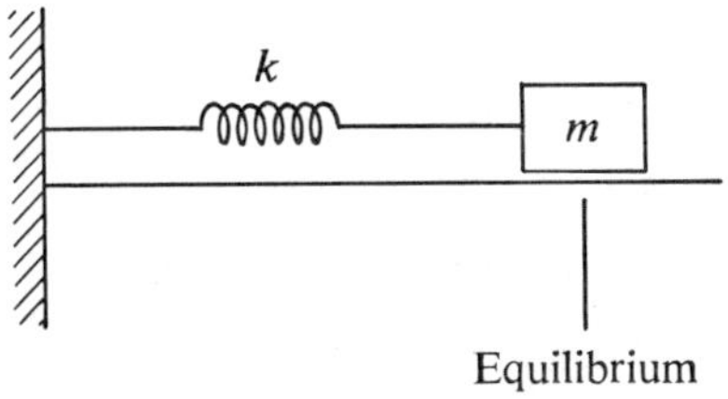

figure 3.8 *sliding block*

zero initial velocity. (See Figure 3.8.) The differential equation of the system is given by one of the two equations

$$m\frac{d^2x}{dt^2} + kx = \pm F$$

where F is a measure of the constant frictional force. The plus sign is chosen if the velocity, dx/dt, is negative and the minus sign if dx/dt is positive. The general solution to the differential equation over a specific interval will be one of the two equations

$$x(t) = c_1 \cos\omega_0 t + c_2 \sin\omega_0 t \pm \frac{F}{k}$$

where ω_0 is the natural angular frequency of the system, $\sqrt{k/m}$. The constants c_1 and c_2 are determined by the initial conditions at the time when the block changes directions. Since the initial velocity at the beginning of each time interval is 0 (the block stops instantaneously when changing directions), $c_2 = 0$ so that

$$x(t) = c_1 \cos\omega_0 t \pm \frac{F}{k}.$$

The value of c_1 is determined at the beginning of each interval of time from the value of the displacement from equilibrium.

The values of t for which the block changes direction can be predetermined from the condition that the instantaneous velocity is zero; that is, when $dx/dt = -\omega_0 c_1 \sin \omega_0 t = 0$. Thus when $\omega_0 t = n\pi$, $n = 0, 1, 2, \cdots$. The block will stop when the spring force at one of these values of time is insufficient to overcome the frictional force.

example 3.23: A block of mass 4 grams is attached to a spring with spring constant equal to 9 dynes/cm. If the frictional force is 4.5 dynes and the block is initially 3 centimeters from its position of equilibrium with zero initial velocity, determine the motion of the block and sketch the graph of the motion.

solution: For this system the natural frequency is 3/2. Hence, the block will change directions at $3t_0/2 = 0$, $3t_1/2 = \pi$, $3t_2/2 = 2\pi$, etc.

For the interval $0 \leqq t \leqq 2\pi/3$,

$$x(t) = c_1 \cos \tfrac{3}{2}t + \tfrac{1}{2}, \quad x(0) = 3,$$

which implies $c_1 = 5/2$. Hence,

$$x(t) = \tfrac{5}{2} \cos \tfrac{3}{2}t + \tfrac{1}{2}.$$

Note that the value of $x(2\pi/3) = -2$.

For the interval $2\pi/3 \leqq t \leqq 4\pi/3$,

$$x(t) = c_1 \cos \tfrac{3}{2}t - \tfrac{1}{2}, \quad x(\tfrac{2}{3}\pi) = -2,$$

which implies $c_1 = 3/2$. Hence,

$$x(t) = \tfrac{3}{2} \cos \tfrac{3}{2}t - \tfrac{1}{2}.$$

Note that $x(4\pi/3) = 1$.

For the interval $4\pi/3 \leqq t \leqq 2\pi$,

$$x(t) = c_1 \cos \tfrac{3}{2}t + \tfrac{1}{2}, \quad x(\tfrac{4}{3}\pi) = 1,$$

which implies $c_1 = 1/2$. Hence,

$$x(t) = \tfrac{1}{2} \cos \tfrac{3}{2}t + \tfrac{1}{2}.$$

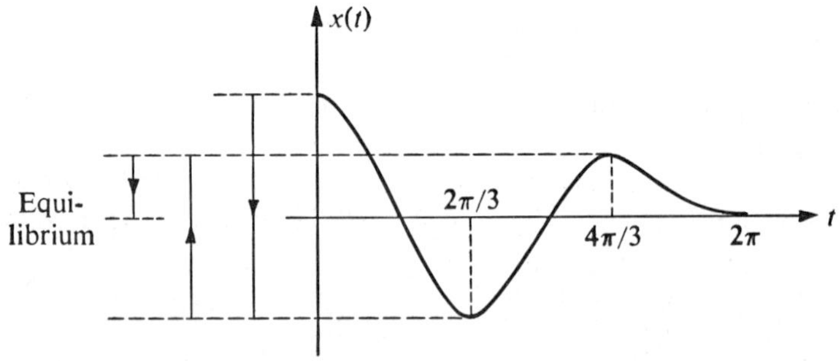

figure 3.9

Note that $x(2\pi) = 0$, which is the equilibrium point. Since the spring force is zero at that point it will obviously be unable to overcome the frictional force, and the block will stop. (See Figure 3.9.)

example 3.24: Repeat the previous example, but change the frictional force to 4 dynes.

solution: Since the points at which the block changes direction are determined by the natural frequency, they are unchanged from Example 3.23.

For the interval $0 \leqq t \leqq 2\pi/3$,

$$x(t) = c_1 \cos \tfrac{3}{2}t + \tfrac{4}{9}, \qquad x(0) = 3,$$

which implies $c_1 = 23/9$. Hence,

$$x(t) = \tfrac{23}{9} \cos \tfrac{3}{2}t + \tfrac{4}{9}.$$

Note that $x(2\pi/3) = -19/9$.

For the interval $2\pi/3 \leqq t \leqq 4\pi/3$,

$$x(t) = c_1 \cos \tfrac{3}{2}t - \tfrac{4}{9}, \qquad x(\tfrac{2}{3}\pi) = -\tfrac{19}{9},$$

which implies $c_1 = 15/9$. Hence,

$$x(t) = \tfrac{15}{9} \cos \tfrac{3}{2}t - \tfrac{4}{9}.$$

Note that $x(4\pi/3) = 11/9$.

For the interval $4\pi/3 \leqq t \leqq 2\pi$,

$$x(t) = c_1 \cos \tfrac{3}{2}t + \tfrac{4}{9}, \qquad x(\tfrac{4}{3}\pi) = \tfrac{11}{9},$$

which implies $c_1 = 7/9$. Hence,

$$x(t) = \tfrac{7}{9} \cos \tfrac{3}{2}t + \tfrac{4}{9}.$$

Note that $x(2\pi) = -1/3$. The spring force at that spot is 3 dynes and hence is insufficient to overcome the frictional force of 4 dynes. Hence, the block stops at $x = -1/3$.

exercises for sections 3.8–3.10

1. Find the transient and steady state current in an RLC circuit with $R = 10$ ohms, $L = 1$ henry, and $C = 1/9$ farad, if the input voltage is $E = 25 \sin t$ volts. Assume $i(0) = 0$ and $\dot{i}(0) = 0$.
2. Repeat Exercise 1 with $R = 1000$ ohms, $L = 10$ millihenries, and $C = 10$ microfarads.
3. Find the steady state and transient oscillations of a mechanical system

with $m=9$, $c=5$, and $k=16$ with external excitation $10\cos t$, if the initial displacement and velocity are zero.

4. Repeat Exercise 3 with $m=1$, $c=2$, $k=2$ with external force $5\sin t$.
5. Prove that the overdamped or critically damped transient current passes through zero at most once and hence is nonoscillatory.
6. How does the frequency of an underdamped mechanical vibration with no external excitation depend on the initial conditions?
7. Find the output vibration of an undamped mechanical system with $k=4$, $m=1$ if the external excitation is $\sin 2t$. Sketch the graph of the output (transverse) motion versus t. Assume $y(0)=\dot{y}(0)=0$.
8. Sketch the graph of the output amplitude as a function of input frequency for Exercise 7.
9. Sketch the graph of the damped oscillation $y=e^{-\alpha t}\cos t$. Where do the maxima and minima occur? Find the ratio of any two consecutive maxima.
10. Find the resonant angular frequency for the mechanical system of Exercise 3.
11. Find the resonant angular frequency for an RLC circuit of $R=10$ ohms, $L=10$ millihenries, $C=10$ microfarads.
12. Using the initial conditions $y(0)=0$, $y'(0)=0$, solve each of the following.
    (a) $y''+y=\sin t$.
    (b) $y''+1.1025y=\sin 0.95t$.
    (c) $y''+0.2y'+1.01y=\sin t-0.05\cos t$.
13. Determine the motion of a block of 1 gram attached to a spring with spring constant 1. The friction force is 3/8 dyne and the block is initially displaced 1 cm from equilibrium.
14. Repeat Exercise 13 with $m=4$, $k=9$, $F=4.5$, and $x(0)=1.25$.
15. Repeat Exercise 13 with $m=4$, $k=9$, $F=7$, and $x(0)=2$.

3.11 systems of linear differential equations

Most practical electrical or mechanical systems are serial arrangements of the simple systems discussed in Sections 3.8 and 3.9. The mathematical models of these more complicated systems require two or more linear differential equations. These differential equations, taken together, are called a *system* of differential equations.

A system of simultaneous linear differential equations in the unknown functions $y_1, y_2, \cdots, y_n$ is a set of linear equations in these variables, the independent variable x, and at least some of the derivatives of the unknown functions. Except for an easy example and a few exercises to the contrary, our discussion will be limited to linear systems with constant coefficients. In polynomial operator form, the equations may be written

$$P_1(D)\,y_1 + P_2(D)\,y_2 + \cdots + P_n(D)\,y_n = f_P(x),$$
$$Q_1(D)\,y_1 + Q_2(D)\,y_2 + \cdots + Q_n(D)\,y_n = f_Q(x),$$
etc.

We will not give the general theory of systems of this type but will be content with showing the more common techniques of obtaining a solution.

example 3.25: Solve the system of linear equations

$$y_1' = y_2 x\,; \qquad y_2' = x\,.$$

solution: This system has variable coefficients but is easily solved because the unknown function y_2 is found by antidifferentiating both members of the second equation. Therefore,

$$y_2 = \frac{x^2}{2} + c_1\,,$$

and hence

$$y_1' = \frac{x^3}{2} + c_1 x\,,$$

from which

$$y_1 = \frac{x^4}{8} + c_1\frac{x^2}{2} + c_2\,.$$

Systems of linear equations with constant coefficients which are expressed in terms of the D operator may be solved in a manner somewhat like that of solving algebraic systems. In the algebraic case, one successively eliminates the unknowns. In solving systems of linear differential equations, elimination of unknowns is done by forming linear combinations of the equations of the system using polynomial operators as factors.

In eliminating the unknown functions, one actually differentiates the functions, although with the operator technique you may not be completely aware of this. For this reason, the resulting system which is derived by operator multiplication may have more solutions with more arbitrary constants than the beginning system.

No simple rule exists for the determination of how many arbitrary constants to expect in the general solution. Exercises 13–16 of the next exercise set will show you how unpredictable things can be. For us, in order to eliminate any extraneous arbitrary constant, it is absolutely necessary to substitute the general solution into all the equations of the original system.

example 3.26: Solve the system of equations

$$y_1' - 2y_1 + 2y_2' = 2 - 4e^{2x}$$
$$2y_1' - 3y_1 + 3y_2' - y_2 = 0.$$

solution: In polynomial operator form, this is written

$$(D-2)\,y_1 + 2Dy_2 = 2 - 4e^{2x}$$
$$(2D-3)\,y_1 + (3D-1)\,y_2 = 0.$$

Operate on both members of the first equation with the operator $(3D-1)$ and on the second by $-2D$, to obtain

$$(3D-1)(D-2)\,y_1 + 2(3D-1)\,y_2 = (3D-1)(2-4e^{2x})$$
$$-2D(2D-3)\,y_1 - 2D(3D-1)\,y_2 = 0.$$

Adding these two equations and carrying out the indicated operation on $2-4e^{2x}$, we obtain

$$(3D^2 - 7D + 2 - 4D^2 + 6D)\,y_1 = -24e^{2x} - 2 + 4e^{2x} = -20e^{2x} - 2,$$

which after simplification is

$$(D^2 + D - 2)\,y_1 = 20e^{2x} + 2$$

or,

$$(D+2)(D-1)\,y_1 = 20e^{2x} + 2.$$

We have immediately that

$$y_{1c} = c_1e^{-2x} + c_2e^{2x}.$$

The particular solution is of the form $Ae^{2x} + B$. Using the condition $y_p'' + y_p' - 2y_p = 20e^{2x} + 2$, we obtain

$$4Ae^{2x} + 2Ae^{2x} - 2Ae^{2x} - 2B = 20e^{2x} + 2.$$

Therefore, $A = 5$ and $B = -1$, and the general solution for the unknown, y_1, is

$$y_1 = c_1e^{-2x} + c_2e^{x} + 5e^{2x} - 1.$$

We use this in the first of the two equations in the given system

$$-2c_1e^{-2x} + c_2e^{x} + 10e^{2x} - 2c_1e^{-2x} - 2c_2e^{x} - 10e^{2x} + 2 + 2y_2' = 2 - 4e^{2x}.$$

Simplifying,

$$2y_2' = 4c_1e^{-2x} + c_2e^x - 4e^{2x}$$

and hence,

$$y_2 = -c_1e^{-2x} + \tfrac{1}{2}c_2e^x - e^{2x} + c_3.$$

We now substitute the expressions for y_1 and y_2 into the second equation of the system to determine if each of the constants c_1, c_2, and c_3 is truly arbitrary.

$$-3c_1e^{-2x} - 3c_2e^x - 15e^{2x} + 3 - 4c_1e^{-2x} + 2c_2e^x + 20e^{2x} + c_1e^{-2x}$$
$$-\tfrac{1}{2}c_2e^x + e^{2x} - c_3 + 6c_1e^{-2x} + \tfrac{3}{2}c_2e^x - 6e^{2x} = 0,$$

from which c_1 and c_2 are arbitrary and $c_3 = 3$. Therefore, the solution to the system is

$$y_1 = c_1e^{-2x} + c_2e^x + 5e^{2x} - 1$$
$$y_2 = -c_1e^{-2x} + \tfrac{1}{2}c_2e^x - e^{2x} + 3.$$

In Section 4.5 you will learn to solve systems of linear equations with constant coefficients by the method of Laplace transforms.

exercises for section 3.11

Solve the systems of first-order equations in Exercises 1–6.

1. $y_1' + y_2' = 3$, $y_1' - y_2' = x$.
2. $y_1' = y_1$, $y_2' = y_2$.
3. $y_1' - y_2 = x^2$, $y_2' + 2y_1 = x$.
4. $y_1' = y_2^2y_1$, $y_2' = x$.
5. $y_1' = y_1y_2$, $y_2' = y_2^2$.
6. $y_1' - y_1 = x^2y_2$, $y_2' = x$.

Solve the systems of equations in Exercises 7–12 by first expressing them in terms of polynomial operators. Substitute your general solutions into the equations of the system to assure extraneous solutions are not included.

7. $y_1' + 2y_1 + y_2' + 2y_2 = 0$, $2y_1' - 2y_1 - 3y_2' + 3y_2 = 0$.
8. $3y_1' - 2y_2' + 5y_2 = 0$, $y_1' - 4y_1 - 4y_2' + y_2 = 0$.
9. $3y_1' - 2y_2' + 5y = e^{3x}$, $y_1' - 4y_1 - 4y_2' + y_2 = 0$.
10. $y_1'' + 6y_1 + y_2' = 0$, $y_1' + 2y_1 + y_2' - 2y_2 = 2$.
11. $y_1'' - 2y_2' + y_2 = 1$, $2y_1' + y_1 + y_2'' - 4y_2 = 0$.
12. $y_1' + 2y_1 + y_2' - y_2 = 0$, $2y_1' + 3y_1 + 3y_2' + y_2 = \sin 2x$.

Find and compare the solutions to the systems in Exercises 13–16.

13. $y_1'' - y_1 - y_2'' = -2\sin x$
$y_1'' + y_1 + y_2'' = 0.$

14. $y_1'' - y_1 + y_2'' = -2\sin x.$
$y_1'' + y_1 + y_2'' = 0.$

15. $y_1'' - y_1 + y_2'' - 2y_2 = -2\sin x$
$y_1'' + y_1 + y_2'' = 0.$

16. $y_1'' - y_1 + y_2'' - y_2' = -2\sin x$
$y_1'' + y_1' + y_2'' = 0.$

the laplace transformation

4

4.1 the general idea of a transform

In this chapter you will learn an operational tool for finding solutions to linear differential equations. The tool we introduce has many other applications, so our discussion will not always be directly concerned with its use in solving differential equations but sometimes with the tool itself. The transform concept is of rather wide interest to mathematicians since several different ideas may then be viewed from one perspective.

The general idea of a transform is that of a pairing of two functions, where to be useful, we would expect some sort of uniqueness in the pairing idea. For example, a function can be thought of as paired with its derivative. We say that the derivative is an "operator," or "transform" which we label $D\{f\}$. The derivative transform is unique in the sense that two functions with the same derivative differ at most by a constant. Similarly, another transform would be the indefinite integral of a function, $I\{f\}$, where

$$I\{f\} = \int_0^x f(t)\,dt$$

As you know, the indefinite integral transform is very important and the idea is extended a little by considering transforms of the type

$$T\{f\} = \int_0^\infty K(s, t) f(t)\,dt.$$

This is actually an entire class of transforms, since if K is changed, a new transform is defined. One of this class is called the Laplace transform.

definition 4.1

The Laplace transform of a function f, if it exists, is given by the function F defined by

$$F(s) = \int_0^\infty e^{-st} f(t)\, dt.$$

The notation $L\{f\}$ or $L\{f(t)\}$ is also used.

In Chapter 11 you will learn of another kind of integral transform, called the Fourier transform of a function.

In general, the variable s may be complex valued but in this chapter we shall restrict s to real numbers.

example 4.1: Find the Laplace transform of the function $f(t) = t$.

solution:

$$L\{t\} = \int_0^\infty te^{-st}\, dt = \lim_{B\to\infty} \int_0^B te^{-st}\, dt.$$

Integrating by parts, we obtain

$$\begin{aligned} L\{t\} &= \lim_{B\to\infty} \left(-\frac{te^{-st}}{s}\bigg|_0^B + \frac{1}{s}\int_0^B e^{-st}\, dt \right) \\ &= \lim_{B\to\infty} \left(-\frac{Be^{-sB}}{s} - \frac{e^{-sB}}{s^2} + \frac{1}{s^2} \right) \\ &= \frac{1}{s^2}, \qquad \text{for } s > 0. \end{aligned}$$

The result of Example 4.1 is typical in the sense that the function of s is defined for values of s greater than some fixed constant. It will usually be important only that there exist *some* value, $s = a$, such that the function is defined for $s > a$. The domain of the transform is always determined by assuring existence of the improper integral.

Note that the values of the function for $t < 0$ have no effect on the transform since the defining integral for $L\{f\}$ requires a knowledge of $f(t)$ for $t \geqq 0$.

example 4.2: Determine the Laplace transform of the function e^{at}.

solution:

$$\begin{aligned} L\{e^{at}\} &= \int_0^\infty e^{at}e^{-st}\,dt \\ &= \lim_{B\to\infty} \int_0^B e^{-(s-a)t}\,dt \\ &= \lim_{B\to\infty} -\frac{e^{-(s-a)t}}{s-a}\bigg|_0^B \\ &= \lim_{B\to\infty} -\frac{e^{-(s-a)B}}{s-a} + \frac{1}{s-a} \\ &= \frac{1}{s-a}, \qquad \text{for } s > a. \end{aligned}$$

example 4.3: Find the Laplace transform of $\sin at$.

solution: Recall from elementary calculus that

$$\int e^{-st}\sin at\,dt = -\frac{e^{-st}(s\sin at + a\cos at)}{s^2+a^2} + C$$

so that

$$\begin{aligned} L\{\sin at\} &= \int_0^\infty e^{-st}\sin at\,dt \\ &= \lim_{B\to\infty} -\frac{e^{-sB}(s\sin aB + a\cos aB)}{s^2+a^2} + \frac{a}{s^2+a^2} \\ &= \frac{a}{s^2+a^2}, \qquad s > 0. \end{aligned}$$

example 4.4: Find the Laplace transform of the discontinuous function

$$\begin{aligned} f(t) &= 1, \qquad 0 < t < 2 \\ &= -2, \qquad 2 < t < 5 \\ &= 0 \quad \text{otherwise}. \end{aligned}$$

solution:

$$L\{f\} = \int_0^2 e^{-st}\,dt - 2\int_2^5 e^{-st}\,dt = -\frac{e^{-st}}{s}\bigg|_0^2 + 2\frac{e^{-st}}{s}\bigg|_2^5$$

$$= \frac{1}{s}(1 - 3e^{-2s} + 2e^{-5s}).$$

The short table of Laplace transforms which follows lists most of the elementary transforms you should know. In Section 4.3 you will learn that this little table, along with some general properties, will enable you to find the transforms of surprisingly many functions.

table i

Short Table of Laplace Transforms

Original Function $f(t)$	Laplace Transform $F(s)$		
1	$1/s$	$s>0$	
t^n	$\dfrac{n!}{s^{n+1}}$	$s>0$,	n an integer
t^a	$\dfrac{\Gamma(a+1)^{\dagger}}{s^{a+1}}$	$s>0$,	a a real positive number
$\sin bt$	$\dfrac{b}{s^2+b^2}$	$s>0$	
$\cos bt$	$\dfrac{s}{s^2+b^2}$	$s>0$	
$\sinh bt$	$\dfrac{b}{s^2-b^2}$	$s>b$	
$\cosh bt$	$\dfrac{s}{s^2-b^2}$	$s>b$	
e^{at}	$\dfrac{1}{s-a}$	$s>a$	

† See Chapter 16 for a discussion of the gamma function.

4.2 existence of the laplace transform

What kind of function has a Laplace transform? In this section we shall classify the set of functions for which Laplace transforms exist. To do so we shall first have to discuss the concept of the order of a function.

definition 4.2

A function f is said to be *dominated* by a function g if

$$\lim_{t\to\infty}\frac{f(t)}{g(t)}=0.$$

A function is said to be *dominated by an exponential* if there exists a number $a \neq 0$ such that f is dominated by e^{at}.

A function which is dominated by an exponential is said to be of *exponential order.*

example 4.5: Determine which of the following functions are of exponential order: (a) e^{5t}, (b) t^3, (c) e^{t^2}, (d) $\sin e^{t^2}$.

solution: (a) e^{5t} is dominated by e^{at} for $a > 5$ and hence is of exponential order.

(b) t^3 is dominated by e^{at} for $a > 0$ and hence is of exponential order.

(c) e^{t^2} is not of exponential order, since

$$\lim_{t \to \infty} e^{t^2 - at} = \infty$$

regardless of the value of a.

(d) $\sin e^{t^2}$ is of exponential order since it has maximum value 1 and hence is dominated by e^{at} for $a > 0$.

In passing note from the example $\sin e^{t^2}$ that if a function is of exponential order, its derivative may fail to be.

The concept of exponential order enables us to describe a sufficient set of conditions for the existence of the Laplace transform.

theorem 4.1

Let f be piecewise continuous on $[0, t]$ for all $t > 0$, and of exponential order. Then $L\{f\}$ exists. (In fact the improper integral converges uniformly to the transform function on any subinterval $[c, d]$ of the domain of the function.)

proof: Since f is of exponential order there are constants a and t_0 such that for any M, $|f(t)| < Me^{at}$ when $t > t_0$. Then

$$|f(t)\, e^{-st}| < Me^{at}e^{-st} = Me^{-(s-a)t}.$$

Hence, if $s > a$,

$$\left| \int_0^\infty f(t)\, e^{-st}\, dt \right| < \int_0^\infty Me^{-(s-a)t}\, dt = \frac{M}{s-a}$$

which implies existence of the Laplace transform. (The uniform convergence follows by appeal to the "M" Test, see Theorem A.14, Appendix A.)

That the conditions of Theorem 4.1 are not necessary is seen by considering the function $1/\sqrt{t}$ which is not piecewise continuous on $[0\ \ t]$ but whose Laplace transform does exist and is equal to $\sqrt{\pi/s}$.

exercises for sections 4.1 and 4.2

1. Show conclusively that the following functions are of exponential order: (a) $\sin 3t$. (b) e^{100t}.
2. Let G be the transform defined by $G\{f\} = \int_1^5 \frac{f(x)}{sx}\,dx$.
   (a) Find $G\{x^2\}$.
   (b) How do any two transforms differ?
3. Verify the formulas for the Laplace transform of $\cos at$, $\cosh at$, and $\sinh at$.
4. Find $L\{t^2 - 2t - 5\}$.
5. Find $L\{\cos^2 at\}$.
6. Find $L\{f\}$, if
$$f(t) = 1, \quad 0 \leqq t \leqq 4$$
$$= 3, \quad 4 < t.$$
7. Find $L\{f\}$ where
$$f(t) = t, \quad 0 \leqq t \leqq 2$$
$$= 4, \quad 2 < t \leqq 5$$
$$= 0, \quad 5 < t.$$
8. Find $L\{f\}$ where
$$f(t) = \sin t, \quad 0 \leqq t \leqq \pi$$
$$= 0, \quad \pi \leqq t.$$
9. Find $L\{f\}$ where
$$f(t) = 1 \quad \text{for} \quad a \leqq t \leqq b$$
$$= 0 \quad \text{otherwise}.$$
10. Find $L\{e^t \cos t\}$.

4.3 some general properties; the first shifting theorem

At the end of Section 4.1, we listed a few functions with their Laplace transforms. There are much more extensive tables but even with such long lists of transforms the usefulness of Laplace transforms would be limited were it not for a few general properties. These properties not only make the finding of a transform much easier but also contribute to their application to ordinary linear differential equations with constant coefficients.

Theorem 4.2 shows that the Laplace transform has the linearity property. In that sense the Laplace transform behaves like the derivative, the antiderivative, and the integral.

theorem 4.2

The Laplace transform is a linear operation; that is,

$$L\{af(t) + bf(t)\} = aL\{f(t)\} + bL\{f(t)\},$$

where a and b are constants and f and g are functions whose Laplace transforms exist.

proof: By definition,

$$L\{af(t) + bg(t)\} = \int_0^{\infty} e^{-st}\{af(t) + bg(t)\}\,dt.$$

By the linearity property of the integral, this may be written

$$\begin{aligned} L\{af(t) + bg(t)\} &= a\int_0^{\infty} e^{-st}f(t)\,dt + b\int_0^{\infty} e^{-st}g(t)\,dt \\ &= aL\{f(t)\} + bL\{g(t)\} \end{aligned}$$

which proves the theorem.

In practical applications, you will often find it necessary to find Laplace transformations of functions multiplied by an exponential factor. The next theorem shows that if you know the Laplace transform of the function, then multiplying it by an exponential does not require a new computation.

theorem 4.3

Suppose the Laplace transform of a function $f(t)$ is $F(s)$, $s > \beta$. Then the Laplace transform of $e^{\alpha t}f(t)$ is $F(s - \alpha)$, $s > \alpha + \beta$. (Note: The notation $F(s - \alpha)$ denotes the function $F(s)$ "shifted" α units to the right. Consequently, the theorem is called the "first shifting theorem," implying of course that there is a second to come! See Figure 4.1.)

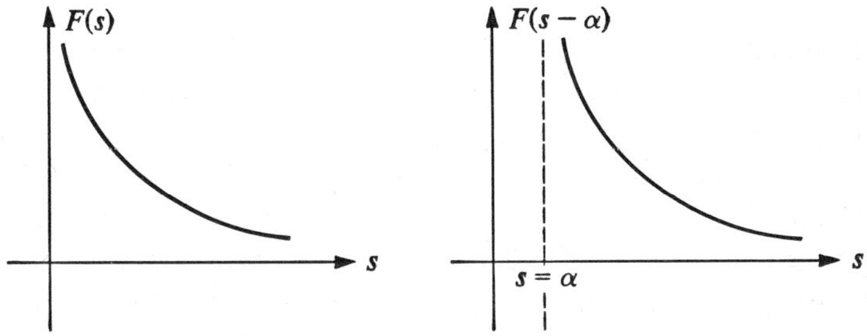

figure 4.1 *first shifting theorem*

proof:

$$L\{e^{\alpha t}f(t)\} = \int_0^\infty e^{-st}e^{\alpha t}f(t)\,dt$$
$$= \int_0^\infty e^{-(s-\alpha)t}f(t)\,dt$$
$$= F(s-\alpha), \qquad s > \alpha + \beta.$$

example 4.6: Find the Laplace transform of te^t.

solution: The Laplace transform of t, from Table 1, is $1/s^2$, $s > 0$. Hence, by the first shifting theorem,

$$L\{te^t\} = \frac{1}{(s-1)^2}, \qquad s > 1.$$

example 4.7: Find the Laplace transform of $e^t \sin t$.

solution: The Laplace transform of $\sin t$ is given in Table 1 as $1/(s^2+1)$, $s > 0$. Hence, by the first shifting theorem,

$$L\{e^t \sin t\} = \frac{1}{(s-1)^2+1}.$$

example 4.8: Suppose the Laplace transform of a function $f(t)$ is known to be $e^{-2s}/(s+2)$, $s > -2$. What is the Laplace transform of $e^{3t}f(t)$?

solution: The Laplace transform of the function $e^{3t}f(t)$ is obtained by letting s be $s-3$ in the formula for the Laplace transform of $f(t)$. Therefore,

$$L\{f(t)\} = \frac{e^{-2(s-3)}}{s-1}, \qquad s > 1.$$

For applications to problems involving differentiation and indefinite integration, the next two theorems are of obvious importance. Roughly speaking, they tell us that differentiation of the function of t corresponds to the algebraic operation of multiplication of the function of s by s, and indefinite integration corresponds to the algebraic operation of division of the function of s by s.

theorem 4.4

Let f be continuous and of exponential order and let f' be piecewise continuous and of exponential order on every finite interval $0 \leqq t \leqq t^*$. Then for $s > a$, $L\{f'(t)\} = sL\{f(t)\} - f(0^+)$, where by $f(0^+)$ is meant $\lim_{t \to 0^+} f(t)$.†

† See the footnote to Definition A.7, Appendix A, for a more precise definition.

proof: By definition,

$$L\{f'(t)\} = \lim_{B\to\infty} \int_0^B f'(t)\, e^{-st}\, dt.$$

Subdivide the interval $[0, B]$ at the points t_j where $f'(t)$ has discontinuities. Then

$$L\{f'(t)\} = \int_0^{t_1} f'(t)\, e^{-st}\, dt + \int_{t_1}^{t_2} f'(t)\, e^{-st}\, dt + \cdots + \lim_{B\to\infty} \int_{t_n}^{B} f'(t)\, e^{-st}\, dt.$$

Each of these can be integrated by parts to give

$$\begin{aligned} L\{f'(t)\} &= f(t)\, e^{-st}\Big|_0^{t_1} + s\int_0^{t_1} f(t)\, e^{-st}\, dt \\ &+ f(t)\, e^{-st}\Big|_{t_1}^{t_2} + s\int_{t_1}^{t_2} f(t)\, e^{-st}\, dt + \cdots \\ &+ \lim_{B\to\infty} f(t)\, e^{-st}\Big|_{t_n}^{B} + \lim_{B\to\infty} s\int_{t_n}^{B} f(t)\, e^{-st}\, dt. \end{aligned}$$

Since f is continuous, $f(t_k - 0) = f(t_k + 0)$ for $k = 1, 2, 3 \cdots n$. Hence,

$$L\{f'(t)\} = -f(0^+) + \lim_{B\to\infty} f(B)\, e^{-sB} + s\int_0^{\infty} f(t)\, e^{-st}\, dt.$$

Because f is of exponential order, the second term vanishes. The third term is recognized as $sL\{f(t)\}$, and the theorem is proved.

With suitable conditions on f, f', and f'', Theorem 4.4 may be successively applied to obtain the formula

$$L\{f''(t)\} = -f'(0^+) - sf(0^+) + s^2 L\{f(t)\}.$$

The formulas for the Laplace transform of f' and f'' in terms of the Laplace transform of f are most important for solving second-order ordinary differential equations with constant coefficients. For the moment, we will be content with using the formula for f'' to find the transform of f.

example 4.9: Using the formula for $L\{f''(t)\}$, find $L\{\sin 3t\}$.

solution: If we let $f(t) = \sin 3t$, then $f'(t) = 3\cos 3t$ and $f''(t) = -9\sin 3t$. Also $f(0) = 0$ and $f'(0) = 3$. Then

$$L\{-9\sin 3t\} = -3 + s^2 L\{\sin 3t\}.$$

Solving for $L\{\sin 3t\}$,

$$L\{\sin 3t\} = \frac{3}{s^2+9}.$$

If the functions which we are considering are continuous, then the formulas for the transform of f' and f'' need not contain a reference to $f(0^+)$ and $f'(0^+)$ but instead may be written $f(0)$ and $f'(0)$.

Further, if the condition on the continuity of f is relaxed so that jump discontinuities are permitted, the formula for $L\{f'(t)\}$ would have to be modified. In one of the exercises you will be asked to show that if there is a jump discontinuity at $t = t_0$, then the term

$$e^{-st_0}[f(t_0-0) - f(t_0+0)]$$

would be added to the formula for $L\{f'(t)\}$ where the term in the brackets is the jump in the value of the function at $t = t_0$. Similar changes are required for the formula for the transform of f''.

Theorem 4.4 may be used to give a formula for the transform of the indefinite integral.

theorem 4.5

Let f be piecewise continuous and of exponential order, dominated by, say, e^{at}. Then for $s > a$,

$$L\left\{\int_{t_0}^{t} f(t^*)\,dt^*\right\} = \frac{1}{s}L\{f(t)\} + \frac{1}{s}\int_{t_0}^{0} f(t^*)\,dt^*.$$

proof: Let $g(t) = \int_{t_0}^{t} f(t^*)\,dt^*$. Then $g'(t) = f(t)$ and $g(0) = \int_{t_0}^{0} f(t^*)\,dt^*$. Further, $g(t)$ may be shown to be of exponential order and hence the result of theorem 4.4 may be applied to the case of continuous functions:

$$L\{g'(t)\} = sL\{g(t)\} - g(0)$$

or

$$L\{f(t)\} = sL\left\{\int_{t_0}^{t} f(t^*)\,dt^*\right\} - \int_{t_0}^{0} f(t^*)\,dt^*.$$

Solving for $L\left\{\int_{t_0}^{t} f(t^*)\,dt^*\right\}$,

$$L\left\{\int_{t_0}^{t} f(t^*)\,dt^*\right\} = \frac{1}{s}L\{f(t)\} + \frac{1}{s}\int_{t_0}^{0} f(t^*)\,dt^*$$

which proves the theorem.

If the value of t_0 is 0, then we can say that indefinite integration corresponds exactly to division of the transform by s.

example 4.10: If $L\{f(t)\} = \dfrac{1}{s(s^2+4)}$, find $f(t)$.

solution: Note that $L\left\{\dfrac{\sin 2t}{2}\right\} = \dfrac{1}{s^2+4}$.

Hence, since $\displaystyle\int_0^t \frac{\sin 2t^*}{2}\,dt^* = -\frac{\cos 2t^*}{4}\bigg|_0^t = \frac{1}{4} - \frac{\cos 2t}{4}$, it follows that $f(t) = \dfrac{1-\cos 2t}{4}$.

Table ii summarizes the general properties you should know about the Laplace transform. Note how many transforms are known once you know that the transform of f is F.

table ii

General Properties

(for continuous functions)

***t* space**	***s* space**
$f(t)$	$F(s) = \int_0^\infty e^{-st} f(t)\,dt$
$g(t)$	$G(s)$
$f(t)+g(t)$	$F(s)+G(s)$
$af(t)$, a a constant	$aF(s)$
$f'(t)$	$sF(s)-f(0)$
$f''(t)$	$s^2F(s)-sf(0)-f'(0)$
$\int_{t_0}^t f(t^*)\,dt^*$	$\frac{1}{s}F(s)+\frac{1}{s}\int_{t_0}^0 f(t^*)\,dt^*$
$e^{at}f(t)$	$F(s-a)$

We close this section with a derivation of the formula for a general periodic function of period L. A function with period L has the property that

$$f(t) = f(t+L) = f(t+2L) = f(t+3L) \cdots = f(t+nL).$$

The transform of such functions can be computed by integrating over just one of the periods.

theorem 4.6

Let f be piecewise continuous with period L. Then,

$$L\{f(t)\} = \frac{1}{1-e^{-Ls}} \int_0^L e^{-st} f(t)\, dt.$$

Proof:

$$L\{f(t)\} = \int_0^\infty e^{-st} f(t)\, dt.$$

We write this integral as the sum of integrals over intervals of length L,

$$L\{f(t)\} = \int_0^L e^{-st} f(t)\, dt + \int_L^{2L} e^{-st} f(t)\, dt + \int_{2L}^{3L} e^{-st} f(t)\, dt + \cdots + \int_{(n-1)L}^{nL} e^{-st} f(t)\, dt + \cdots.$$

In the second integral, make the substitution $u = t - L$; in the third integral, $u = t - 2L$; in the fourth, $u = t - 3L$; in the nth integral, $u = t - (n-1)L$.

$$\begin{aligned} L\{f(t)\} &= \int_0^L e^{-st} f(t)\, dt + e^{-sL} \int_0^L e^{-su} f(u+L)\, du \\ &\quad + e^{-2sL} \int_0^L e^{-su} f(u+2L)\, du \\ &\quad + \cdots + e^{-s(n-1)L} \int_0^L e^{-su} f(u+(n-1)L)\, du + \cdots. \end{aligned}$$

Using the fact that $f(u+nL) = f(u)$ and factoring the integral $\int_0^L e^{-st} f(t)\, dt$ (recall that $\int_0^L e^{-st} f(t)\, dt = \int_0^L e^{-su} f(u)\, du$),

$$L\{f(t)\} = \left(\int_0^L e^{-st} f(t)\, dt\right)(1 + e^{-sL} + e^{-2sL} + \cdots + e^{-(n-1)sL} + \cdots).$$

The quantity in the second parentheses is an infinite geometric series with common ratio e^{-sL} (see Appendix A.6). Hence, the "sum" is $1/(1-e^{-sL})$, and the series converges to its "sum" because $0 < e^{-sL} < 1$. Therefore,

$$L\{f(t)\} = \frac{1}{1-e^{-sL}} \int_0^L e^{-st} f(t)\, dt.$$

example 4.11: Find the Laplace transform of the function shown in Figure 4.2.

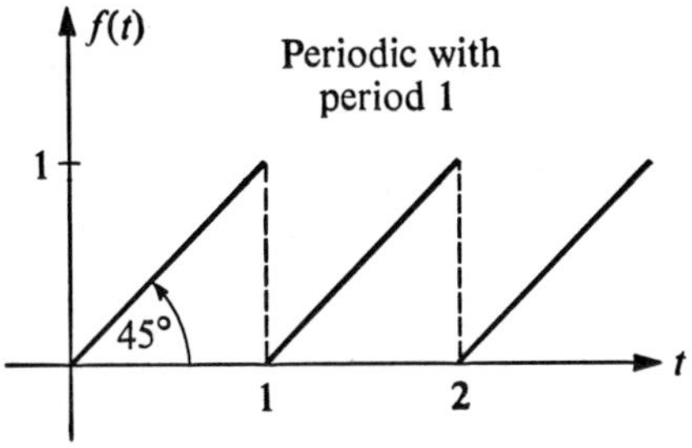

figure 4.2

solution: Over the interval $0 < t < 1$, the function is defined by the rule $f(t) = t$. Hence,

$$\begin{aligned} L\{f(t)\} &= \frac{1}{1-e^{-s}} \int_0^1 e^{-st} t\, dt \\ &= \frac{1}{1-e^{-s}} \left(-\frac{t}{s} e^{-st} \Big|_0^1 + \frac{1}{s} \int_0^1 e^{-st}\, dt \right) \\ &= \frac{1}{1-e^{-s}} \left(-\frac{e^{-s}}{s} - \frac{1}{s^2} (e^{-s} - 1) \right) \\ &= \frac{1}{s^2} - \frac{e^{-s}}{s(1-e^{-s})}. \end{aligned}$$

example 4.12: Find the transform of the full-wave rectification of $\sin t$ given that the transform of the half-wave rectification of $\sin t$ is

$$\frac{1}{(s^2+1)(1-e^{-\pi s})}.$$

(See Figure 4.3, and Exercise 15 of the next exercise set.)

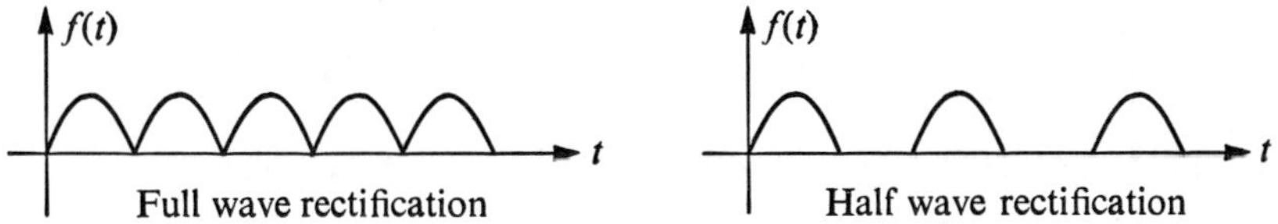

figure 4.3

solution: The definition of the half-wave rectified wave is

$$f(t) = \tfrac{1}{2}(|\sin t| + \sin t).$$

Hence,

$$L\{|\sin t| + \sin t\} = \frac{2}{(s^2+1)(1-e^{-\pi s})}.$$

Since $L\{\sin t\} = 1/(s^2+1)$ (from Table ii), it follows that

$$\begin{aligned} L\{|\sin t|\} &= \frac{2}{(s^2+1)(1-e^{-\pi s})} - \frac{1}{s^2+1} \\ &= \frac{1}{s^2+1}\,\frac{1+e^{-\pi s}}{1-e^{-\pi s}} \\ &= \frac{1}{s^2+1}\coth\frac{\pi s}{2}. \end{aligned}$$

exercises for section 4.3

1. Find $L\{\cos 2t\}$ using only the formula for $L\{(\cos 2t)''\}$.
2. Find $L\{e^t \cos 4t\}$.
3. Find $L\{e^{-t}t^2\}$.
4. Find $L\{e^{4t}(t^3+2t^2+t\}$.
5. Find $L\{e^{2t}f(t)\}$ where $f(t)$ is the function of Exercise 6 of the previous exercise set.
6. Find $L\{e^{-t}f(t)\}$ where $f(t)$ is the function of Exercise 7 of the previous exercise set.
7. Find $L\{e^{-2t}\sin 3t\}$.
8. Find $L\left\{\int_0^t \cos x\,dx\right\}$.
9. Find $L\left\{\int_1^t \cos x\,dx\right\}$.
10. Find $L\left\{\int_1^t x\,dx\right\}$.
11. Find $L\left\{e^{at}\int_1^t x\,dx\right\}$.
12. If $L\{f\} = \dfrac{1}{s(s+3)}$, find $f(t)$.
13. If $L\{f\} = \dfrac{1}{s(s^2-9)}$, find $f(t)$.
14. Show $L\{f(at)\} = \dfrac{1}{a}F\left(\dfrac{s}{a}\right)$, where $F(s)$ is the $L\{f\}$.
15. Find the Laplace transform of the half-wave rectification of $\sin t$. (See Example 4.12.)
16. Using the results of Exercises 14 and 15, find the Laplace transform of the half-wave rectification of $\sin at$.
17. Find the Laplace transform of the function periodic with period π whose definition on $0 < t < \pi$ is $f(t) = t^2$.

18. Find the Laplace transform of the function periodic with period 2 whose definition on $0<t<2$ is

$$\begin{aligned} f(t) &= t, & 0<t<1 \\ &= 2-t, & 1<t<2. \end{aligned}$$

19. Find the Laplace transform of the function periodic with period L whose definition on $0<t<L$ is

$$\begin{aligned} f(t) &= 2, & 0<t<\frac{L}{2} \\ &= -2, & \frac{L}{2}<t<L. \end{aligned}$$

20. Find the Laplace transform of the function periodic with period 3 whose definition on $0<t<3$ is

$$\begin{aligned} f(t) &= 2, & 0<t<1 \\ &= 0, & 1<t<2 \\ &= -2, & 2<t<3. \end{aligned}$$

21. Show that if f has a jump discontinuity at $t=t_0$, then

$$L[f'(t)] = -f(0^+) + e^{-st_0}[f(t_0-0) - f(t_0+0)] + sL[f(t)].$$

4.4 the inverse laplace transform

The process of finding the transform of a function is greatly facilitated by the general properties discussed in the previous section, and in that sense is like finding the formula for the derivative of a function. From just a few specific formulas many transforms are found. But, like antidifferentiation, the determination of a function whose transform is given is a less direct process.

definition 4.3

If there exists a function of t, $f(t)$, such that $L\{f(t)\}=F(s)$, then $f(t)$ is called the *inverse Laplace transform* of $F(s)$· The notation commonly used is

$$L^{-1}\{F(s)\} = f(t).$$

Not all functions of s have inverse Laplace transforms. Simple functions such as s, s^2, or $\sin s$, for example, are such that there is no corresponding function of t. In fact, the Laplace transform of any function of t which you will have to discuss has the property of having a limit of zero as $s\to\infty$.

theorem 4.7

If f is piecewise continuous on $0 \leqq t \leqq t_1$ and dominated by e^{at}, and if F is the Laplace transform of f, then

$$\lim_{s \to \infty} F(s) = 0.$$

proof: Since f is of exponential order, there exist constants M, a, and t_0 such that $|f(t)| < Me^{at}$ when $t > t_0$. Hence,

$$|F(s)| = \left| \int_0^\infty f(t)\, e^{-st}\, dt \right| < \int_0^\infty Me^{-(s-a)t}\, dt = \frac{M}{s-a},$$

and since this last expression has a limit of zero as $s \to \infty$, the theorem is proven.

The study of finding inverse Laplace transforms is extensive. Specific formulas have been developed, especially for the case in which F is a rational function of s. However, in this shortened treatment of the subject we shall not usually develop specific formulas but rather investigate the general pattern of analysis. If you work through the following examples, a general approach will become clear.

example 4.13: Find

$$L^{-1}\left\{\frac{1}{s^2 - 5s + 6}\right\}.$$

solution: *Method 1:* Factor the denominator and express the function in terms of its partial fractions expansion,

$$\frac{1}{s^2 - 5s + 6} = \frac{1}{(s-2)(s-3)} = \frac{A}{s-2} + \frac{B}{s-3},$$

where A and B are constants to be determined. There are several elementary ways to obtain these constants, but perhaps the easiest way to find the value of A is first to multiply both sides by $s-2$, simplify, and then let $s = 2$. We find that $A = -1$. Similarly, $B = 1$. Since $L^{-1}\{1/(s-2)\} = e^{2t}$ and $L^{-1}\{1/(s-3)\} = e^{3t}$, it follows that

$$L^{-1}\left\{\frac{1}{s^2 - 5s + 6}\right\} = -e^{2t} + e^{3t}.$$

Method 2: Complete the square on the quadratic in the denominator to obtain

$$\frac{1}{s^2 - 5s + 6} = \frac{1}{s^2 - 5s + (\frac{5}{2})^2 - \frac{1}{4}} = \frac{1}{(s - \frac{5}{2})^2 - (\frac{1}{2})^2}.$$

From Table i, Section 4.1, the Laplace transform of $\sinh\frac{t}{2}$ is

$$\frac{\frac{1}{2}}{s^2 - (\frac{1}{2})^2}.$$

Hence,

$$L^{-1}\left\{\frac{1}{s^2 - (\frac{1}{2})^2}\right\} = 2\sinh\frac{t}{2}.$$

The given function of s is the same as this with s replaced by $(s - \frac{5}{2})$. Hence, from the first shifting theorem,

$$L^{-1}\left\{\frac{1}{s^2 - 5s + 6}\right\} = 2e^{5t/2}\sinh(t/2).$$

The correspondence of the results of Methods 1 and 2 is easily verified.

example 4.14: Find the inverse Laplace transform of $\dfrac{1}{s^2 - 5s + \frac{13}{2}}$.

solution: The denominator of this function cannot be factored into a product of real linear factors. We complete the square on the quadratic factor,

$$\frac{1}{s^2 - 5s + \frac{13}{2}} = \frac{1}{s^2 - 5s + (\frac{5}{2})^2 + \frac{1}{4}} = \frac{1}{(s - \frac{5}{2})^2 + \frac{1}{4}}.$$

Using the fact that

$$L^{-1}\left\{\frac{1}{s^2 + \frac{1}{4}}\right\} = 2\sin\frac{t}{2}$$

and the first shifting theorem, we obtain

$$L^{-1}\left\{\frac{1}{s^2 - 5s + \frac{13}{2}}\right\} = 2e^{5t/2}\sin\frac{t}{2}.$$

The preceding two examples are representative of the techniques used to find the inverse transform of functions whose denominators are linear or quadratic factors. The first shifting theorem plays an important role.

example 4.15: Find the inverse transform of the function $\dfrac{s}{s^2 - 5s + \frac{13}{2}}$.

solution: Since, from the previous example,

$$L^{-1}\left\{\frac{1}{s^2 - 5s + \frac{13}{2}}\right\} = 2e^{5t/2}\sin\frac{t}{2}$$

and since $sF(s) = f(0) + L\{f'(t)\}$, it follows that

$$L^{-1}\left\{\frac{s}{s^2 - 5s + \frac{13}{2}}\right\} = \frac{d}{dt} 2e^{5t/2} \sin\frac{t}{2}$$
$$= e^{5t/2}\left(5 \sin\frac{t}{2} + \cos\frac{t}{2}\right).$$

example 4.16: Find the inverse transform of the function $\dfrac{1}{s(s^2 - 5s + 6)}$.

solution: Recall, from Theorem 4.5, that division by s corresponds to indefinite integration from 0 to t. Since, from Example 4.13,

$$L^{-1}\left\{\frac{1}{s^2 - 5s + 6}\right\} = -e^{2t} + e^{3t},$$

it follows that

$$L^{-1}\left\{\frac{1}{s(s^2 - 5s + 6)}\right\} = -\int_0^t e^{2x}\,dx + \int_0^t e^{3x}\,dx$$
$$= -\frac{e^{2t} - 1}{2} + \frac{e^{3t} - 1}{3}$$
$$= \frac{1}{6} - \frac{e^{2t}}{2} + \frac{e^{3t}}{3}.$$

We now give a summary of some of the fundamental techniques you should know for finding an inverse transform.

1. If $F(s)$ is of the form $sF_1(s)$, find $L^{-1}\{F_1(s)\}$ and apply Theorem 4.4. If $F(s)$ is of the form $s^2F_1(s)$, two differentiations will be required.
2. If $F(s) = F_1(s)/s$, find $L^{-1}\{F_1(s)\}$ and perform an indefinite integration from 0 to t. (See Theorem 4.5.) If $F(s) = F_1(s)/s^2$, two such indefinite integrations will be required.
3. If $F(s)$ is of the form $F_1(s - a)$, find $L^{-1}\{F_1(s)\}$ and multiply by e^{at}. (See Theorem 4.2.)
4. Suppose $F(s)$ is of the form of a rational function of s, say

$$F(s) = \frac{N(s)}{D(s)},$$

where the degree of $D(s)$ is greater than the degree of $N(s)$. Factor $D(s)$ into its real quadratic and linear factors. We accept the fact that every real polynomial can be so reduced. Then $F(s)$ can be expanded in terms of its partial fractions. Each of the unrepeated linear factors leads to an exponential. Each of the unrepeated quadratic factors leads to an exponential times a trigonometric function. Repeated factors lead to multiplication by t.

In more extensive discussions of the Laplace transformation, you will find a systematic development of a set of formulas by which the inverse transform of a rational function of s may be found quickly. These formulas are obtained by following the techniques already given to the general cases, and extending the method to the case of repeated quadratic factors.

Note that there are other functions of s which have inverse transforms, but are not of the form of a rational function of s. In Section 4.6 we will prove the second shifting theorem. This will considerably widen our ability to find inverse transforms.

example 4.17: Find the inverse transform of the function

$$F(s) = \frac{1}{(s+1)(s-3)^2(s-1)^3}.$$

solution: Expanding $F(s)$,

$$\frac{1}{(s+1)(s-3)^2(s-1)^3} = \frac{A}{s+1} + \frac{B}{s-3} + \frac{C}{(s-3)^2} + \frac{D}{s-1} + \frac{E}{(s-1)^2} + \frac{F}{(s-1)^3}.$$

It is an easy, but tedious, task to show that

$$A = -\tfrac{1}{128}, \quad B = -\tfrac{7}{128}, \quad C = \tfrac{1}{32}, \quad D = \tfrac{1}{16}, \quad E = \tfrac{1}{16}, \quad F = \tfrac{1}{8}.$$

Since $L^{-1}\{1/s^2\} = t$, and $L^{-1}\{1/s^3\} = t^2/2$, the use of the first shifting theorem allows us to write

$$L^{-1}\{F(s)\} = -\frac{e^{-t}}{128} - \frac{7e^{3t}}{128} + \frac{te^{3t}}{32} + \frac{e^t}{16} + \frac{te^t}{16} + \frac{t^2e^t}{16}.$$

example 4.18: Find the inverse transform of the function

$$F(s) = \frac{1}{(s-1)(s^2+4)}.$$

solution: The denominator is already factored into irreducible real and quadratic factors. (The factor s^2+4 can be factored, but not into *real* linear factors.) Decomposing the rational function into its partial fractions expansion,

$$\frac{1}{(s-1)(s^2+4)} = \frac{A}{s-1} + \frac{Bs+C}{s^2+4},$$

from which $A = 1/5$ and $B = C = -1/5$. Hence

$$L^{-1}\{F(s)\} = \frac{e^t}{5} - \frac{\cos 2t}{5} - \frac{\sin 2t}{10}.$$

exercises for section 4.4

Find the inverse transform of the following functions.

1. $\dfrac{1}{s^2 + 3s + 2}$.
2. $\dfrac{1}{s(s^2 + 3s + 2)}$.
3. $\dfrac{1}{s^2 + 4s + 4}$.
4. $\dfrac{1}{s^2 + 4s + 2}$.
5. $\dfrac{s}{s^2 + 4s + 2}$.
6. $\dfrac{s^2}{(s + 2)^3}$.
7. $\dfrac{s}{(s + a)^2 + b^2}$.
8. $\dfrac{b}{(s + a)^2 + b^2}$.
9. $\dfrac{s^2}{(s + 1)(s + 2)(s + 3)}$.
10. $\dfrac{2s + 1}{(s + 2)^4}$.
11. $\dfrac{1}{(s - 1)^2 (s + 2)}$.
12. $\dfrac{1}{(s - 3)^2 (s^2 + 3s + 6)}$.
13. $\dfrac{s^2 - 3s}{(s - 2)(s - 1)^2}$.
14. $\dfrac{4s - 3}{s^3 + s^2}$.
15. $\dfrac{3s^2 + 4}{s^4 + s^2}$.

4.5 solutions to ordinary differential equations

For our purposes, the most immediate and important application of the Laplace transformation is to the solution of ordinary linear differential equations with constant coefficients. Consider the equation

$$a_2 \ddot{y} + a_1 \dot{y} + a_0 y = f(t)$$

where a_2, a_1, and a_0 are constants and $f(t)$ is the driving function or "input." Taking the Laplace transformation of both sides and denoting the Laplace transform of y by Y and that of f by F, we obtain, using Theorem 4.4,

$$a_2 [s^2 Y - sy(0) - \dot{y}(0)] + a_1 [sY - y(0)] + a_0 Y = F(s).$$

Solving for Y,

$$Y(s) = \frac{F(s)}{a_2s^2 + a_1s + a_0} + \frac{(a_2s + a_1)\,y(0) + a_2\dot{y}(0)}{a_2s^2 + a_1s + a_0},$$

which is called the *Laplace transform of the response.* The first term in the expression for Y is the transform of the part of the solution due to the driving function, while the second term is due to nonzero initial conditions. In some applications it is unnecessary to find the actual response; the transform of the response is sufficient.

Note how the initial conditions are considered automatically as opposed to our previous method where the "arbitrary" constants of the general solution were found in a separate calculation. We see very clearly and unmistakeably the role of the initial conditions in affecting the total solution.

We may summarize the technique of Laplace transforms as follows:

1. Take the Laplace transform of both sides of the given differential equation. The transform of the driving function may prove difficult to find and might require the use of tables.
2. Solve the transformed equation for the transform of the solution. This is a purely algebraic step and should cause no difficulty.
3. Find the inverse transform of $Y(s)$. This may be (and usually is) the most difficult of the steps involved. It is the one limitation on the method requiring the most experience to overcome.

example 4.19: By using the method of Laplace transformations, solve the initial value problem

$$\ddot{y} + 2\dot{y} + 2y = \sin t, \qquad y(0) = 1, \qquad \dot{y}(0) = 1.$$

solution: Taking the transform of both sides, we obtain

$$s^2Y - s - 1 + 2sY - 2 + 2Y = \frac{1}{s^2 + 1}.$$

Solving for Y,

$$Y(s) = \frac{1}{(s^2 + 1)(s^2 + 2s + 2)} + \frac{s + 3}{s^2 + 2s + 2}.$$

The partial fractions expansion of the first term is

$$\frac{1}{(s^2 + 1)(s^2 + 2s + 2)} = \frac{As + B}{s^2 + 1} + \frac{Cs + D}{s^2 + 2s + 2},$$

from which $A = -(2/5)$, $B = 1/5$, $C = 2/5$ and $D = 3/5$. Hence

$$L^{-1}\left(\frac{1}{(s^2+1)(s^2+2s+2)}\right) = L^{-1}\left(\frac{1}{5}\frac{1-2s}{s^2+1}\right) + L^{-1}\left(\frac{1}{5}\frac{2s+3}{s^2+2s+2}\right)$$

$$= \frac{1}{5}L^{-1}\left(\frac{1}{s^2+1}\right) - \frac{2}{5}L^{-1}\left(\frac{s}{s^2+1}\right)$$

$$+ \frac{2}{5}L^{-1}\frac{s+1}{(s+1)^2+1} + \frac{1}{5}L^{-1}\frac{1}{(s+1)^2+1}$$

$$= \frac{1}{5}(\sin t - 2\cos t) + \frac{e^{-t}}{5}(2\cos t + 1\sin t).$$

In like manner,

$$L^{-1}\left(\frac{s+3}{s^2+2s+2}\right) = e^{-t}(\cos t + 2\sin t),$$

so that

$$y(t) = \tfrac{1}{5}e^{-t}(7\cos t + 11\sin t) + \tfrac{1}{5}(\sin t - 2\cos t).$$

Note the similarity between the application of Laplace transformations to solving ordinary differential equations and the elementary application of logarithms to perform multiplication. With logarithms we replace the supposedly more difficult task of multiplication with that of addition; it is of course necessary that a table of logarithms be available and that you have ability to find the inverse logarithm. In using Laplace transforms to solve differential equations, it is usually a matter of a little algebra to find the Laplace transform of the solution. Finding the inverse transform can be a tricky task.

example 4.20: The driving voltage to an RC circuit is e^{-t} for $t \geqq 0$. Find the current in the circuit if the initial current is zero. The circuit diagram is shown in Figure 4.4.

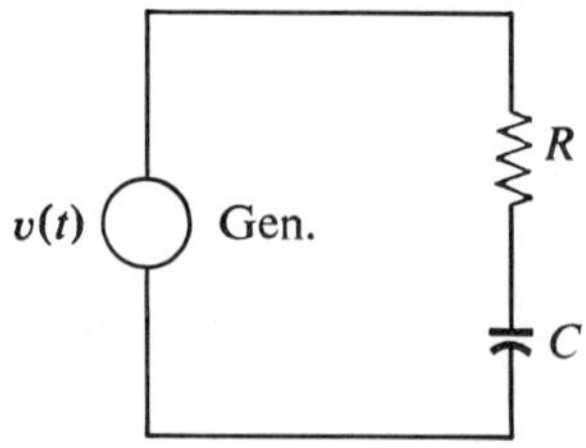

figure 4.4

solution: The equation governing the circuit is

$$Ri + \frac{1}{C}\int_0^t i(x)\,dx = e^{-t}.$$

Taking the Laplace transform of both sides and letting the transform of the current be $I(s)$,

$$RI(s) + \frac{1}{Cs}I(s) = \frac{1}{s+1}.$$

Solving for $I(s)$,

$$I(s) = \frac{sC}{(RCs+1)(s+1)} = \frac{1}{R}\frac{s}{\left(s+\frac{1}{RC}\right)(s+1)}.$$

Since

$$\frac{s}{\left(s+\frac{1}{RC}\right)(s+1)} = -\frac{1}{s+\frac{1}{RC}}\frac{1}{RC-1} + \frac{1}{s+1}\frac{RC}{RC-1},$$

it follows that

$$i(t) = \frac{-1}{R(RC-1)}e^{-t/RC} + \frac{C}{RC-1}e^{-t}.$$

With the use of Laplace transforms, systems of differential equations become systems of algebraic equations.

example 4.5: Solve the system

$$\dot{y}_1 + y_1 + 3\dot{y}_2 = 1$$
$$3y_1 + \dot{y}_2 + 2y_2 = t$$
$$y_1(0) = 0, \quad y_2(0) = 0.$$

solution: Taking the Laplace transform of both sides of both equations, we obtain

$$sY_1 + Y_1 + 3sY_2 = \frac{1}{s}$$

$$3Y_1 + sY_2 + 2Y_2 = \frac{1}{s^2}.$$

Using Cramer's rule to solve for Y_1 and Y_2,

$$Y_1(s) = \frac{\begin{vmatrix} 1/s & 3s \\ 1/s^2 & s+2 \end{vmatrix}}{\begin{vmatrix} s+1 & 3s \\ 3 & s+2 \end{vmatrix}} = \frac{s-1}{s(s^2-6s+2)},$$

$$Y_2(s) = \frac{\begin{vmatrix} s+1 & 1/s \\ 3 & 1/s^2 \end{vmatrix}}{\begin{vmatrix} s+1 & 3s \\ 3 & s+2 \end{vmatrix}} = \frac{1-2s}{s^2(s^2-6s+2)},$$

$$Y_1(s) = -\frac{1}{2s} + \frac{1}{2}\frac{s-3}{(s-3)^2-7} - \frac{1}{2}\frac{1}{(s-3)^2-7},$$

$$Y_2(s) = \frac{1}{2s} + \frac{1}{2s^2} - \frac{1}{2}\frac{s-3}{(s-3)^2-7} + \frac{1}{(s-3)^2-7}.$$

Therefore,

$$y_1(t) = -\tfrac{1}{2} + \tfrac{1}{2}e^{3t}\cosh\sqrt{7}\,t - \frac{1}{2\sqrt{7}}e^{3t}\sinh\sqrt{7}\,t$$

$$y_2(t) = \tfrac{1}{2} + \tfrac{1}{2}t - \tfrac{3}{2}e^{3t}\cosh\sqrt{7}\,t + \frac{1}{\sqrt{7}}e^{3t}\sinh\sqrt{7}\,t.$$

exercises for section 4.5

Solve the following initial value problems using the Laplace transform method.

1. $y'' + y' - 2y = 0, \quad y(0) = 0, y'(0) = 2.$
2. $y'' + 4y = 0, \quad y(0) = 1, \quad y'(0) = 1.$
3. $y'' - 16y = 0, \quad y(0) = 1, \quad y'(0) = 0.$
4. $4y'' + 4y' + y = 0, \quad y(0) = 2, \quad y'(0) = -1.$
5. $y'' + 2y' - 3y = 0, \quad y(0) = 0, y'(0) = 1.$
6. $y'' - 3y' = 2, \quad y(0) = 2, \quad y'(0) = 3.$
7. $y' = y, \quad y(0) = 1.$
8. $y'' + 2y' + 4y = x, \quad y(0) = 2, \quad y'(0) = 1.$
9. $y'' + 3y' + 2y = x, \quad y(0) = -1, \quad y'(0) = 1.$
10. $y'' + y = \sin 2x, \quad y(0) = 1, \quad y'(0) = 1.$
11. $y'' + a^2y = 0, \quad y(0) = 1, \quad y'(0) = 0.$
12. $y'' + a^2y = 0, \quad y(0) = 0, \quad y'(0) = 1.$

13. An RLC circuit has a driving voltage of t^2, with $R = 3$ ohms, $L = 1$ henry, and $C = \frac{1}{2}$ farad. If $i(0) = 1$ and $\dot{i}(0) = -1$, find an expression for the current.

14. An RLC circuit has a driving voltage of $\sin t$ volts, with $R = 4$ ohms, $C = \frac{1}{4}$ farad, $L = 1$ henry. If $i(0) = 1$ and $\dot{i}(0) = -1$, find an expression for the current.

15. A mechanical system with spring constant 5, fluid friction constant 6, and a mass of 1 gram is subjected to an external excitation of $\cos t$. If the initial deflection is 3 cm and the initial velocity is zero, find the equation for the deflection at any time t.

16. Solve the system

$$\begin{aligned} &y_1'' - 3y_1' - y_2' + 2y_2 = 14t + 3 \\ &y_1' - 3y_1 + 2y_2' = 1 \\ &y_1(0) = 0, \quad y_1'(0) = 0, \quad y_2(0) = 2. \end{aligned}$$

17. Solve the system

$$\begin{aligned} &y_1' + y_2' = 0 \\ &y_1 - y_2' = 0 \\ &y_1(0) = 0, \quad y_2(0) = 1. \end{aligned}$$

18. Solve the system

$$\begin{aligned} &y_1' + y_2 = t^2 + 1 \\ &y_1 - y_2' = -t - 1 \\ &y_1(0) = y_2(0) = 0. \end{aligned}$$

19. Solve the system

$$\begin{aligned} &y_1'' + y_2'' + y_1 + y_2 = 2 - 6t + t^2 - t^3 \\ &y_1'' - y_2' = 2 + 3t^2 \\ &y_1(0) = y_2(0) = 0, \quad y_1'(0) = 3, \quad y_2'(0) = 0. \end{aligned}$$

20. Solve the system

$$\begin{aligned} y_1'' + y_2' - y_2 &= 0 \\ 2y_1' - y_1 + y_3' - y_3 &= 0 \\ y_1' + 3y_1 + y_2' - 4y_2 + 3y_3 &= 0 \\ y_1(0) = 0, \quad y_1'(0) = 1, &\quad y_2(0) = y_3(0) = 0. \end{aligned}$$

4.6 a "turn-on" function; the second shifting theorem

There are many important driving functions which are not combinations of any well known elementary functions. Figure 4.5 shows several of these more practical functions.

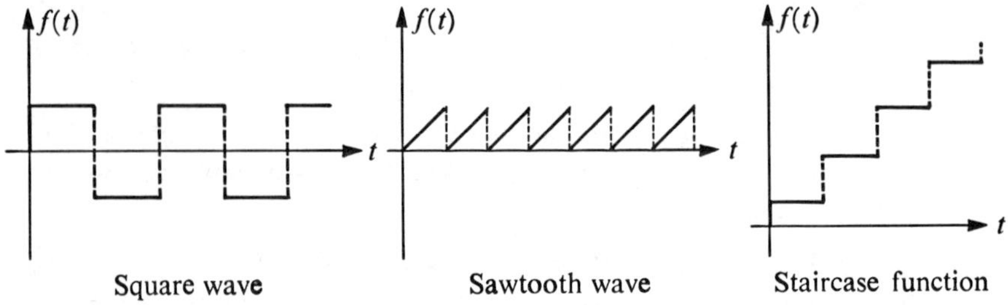

figure 4.5 *some important driving functions*

Ordinarily, to describe analytically one of these functions, you would give its "rule" over each subinterval. Sometimes though, and especially when using Laplace transforms to solve differential equations, you will find it convenient to give one rule which is valid over the entire interval of interest. To do this, we will introduce a function whose only property is that it "turns on" other functions at a specific value of t. This function is called the *unit step function*, $u_a(t)$, and is defined by

$$u_a(t) = \left.\begin{aligned} &= 0, \quad t < a \\ &= 1, \quad t \geqq a \end{aligned}\right\}, \qquad a > 0.$$

(See Figure 4.6.)

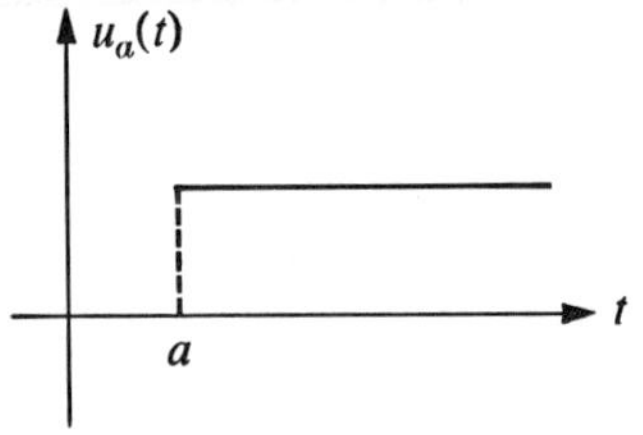

figure 4.6 *unit step function*

A related function, which mathematicians would call a "characteristic function of the interval $[a, b]$" and engineers might call a "pulse" function, is the function whose graph is shown in Figure 4.7. In terms of unit step functions this may be written $u_a(t) - u_b(t)$. It is the pulse function more specifically than the unit step function which allows a compact analytic description of many important driving functions.

example 4.22: Using unit step functions, describe analytically the function which is 0 except for the line segment connecting (0, 0) to (2, 6). (See Figure 4.8.)

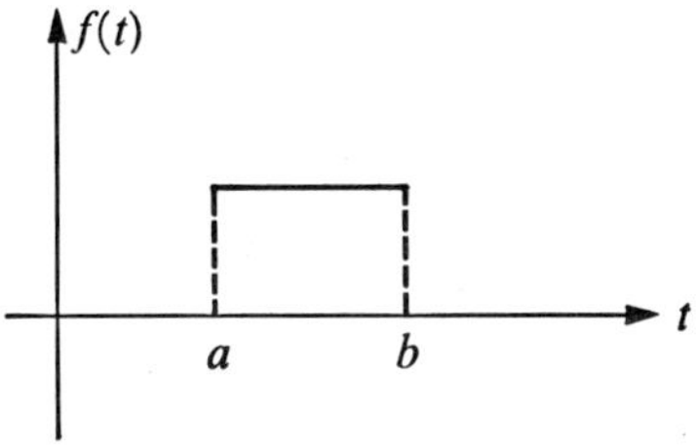

figure 4.7 *characteristic function of* [a, *b*]

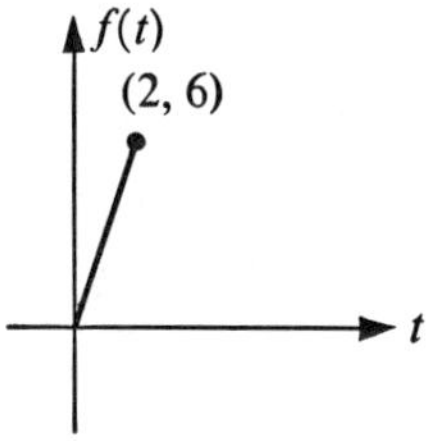

figure 4.8

solution: Note that on the interval [0, 2] the defining rule for this function is $3t$. Since the function is zero elsewhere, we may describe this by multiplying $3t$ by the pulse function which is "on" from 0 to 2,

$$f(t) = (u_0(t) - u_2(t))\,3t.$$

example 4.23: Using combinations of unit step functions, write the rule for the periodic square wave shown in Figure 4.9.

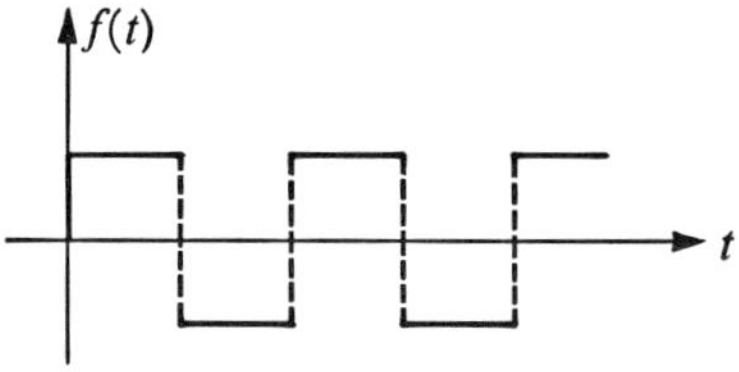

figure 4.9

solution: The square wave may be thought of as a repeated pulse function alternating between positive and negative values. Hence, the function may be written

$$\begin{aligned}[u_0(t) - u_1(t)] - u_1[(t) - u_2(t)] + [u_2(t) - u_3(t)] - \cdots \\ = u_0(t) - 2u_1(t) + 2u_2(t) - 2u_3(t) + \cdots \\ = u_0(t) + 2\sum_{n=1}^{\infty} (-1)^n u_n(t).\end{aligned}$$

The Laplace transform of the unit step function is of obvious interest and is found by direct appeal to the definition of the transform:

$$
\begin{aligned}
L\{u_a(t)\} &= \int_0^\infty u_a(t)\,e^{-st}\,dt \\
&= \int_a^\infty e^{-st}\,dt = \lim_{B\to\infty} \left.\frac{e^{-st}}{-s}\right|_a^B \\
&= \lim_{B\to\infty} \frac{e^{-sB}}{-s} - \frac{e^{-as}}{-s} \\
&= \frac{e^{-as}}{s}.
\end{aligned}
$$

example 4.24: Find the Laplace transform of the square wave discussed in the previous example.

solution: It would be possible to find the transform by using the formula for the Laplace transform of a periodic function developed in Theorem 4.6. Instead we work directly with the function as written in Example 4.23.

$$
\begin{aligned}
L\{\text{sq. wave}\} &= L\{u_0(t)\} - 2L\{u_1(t)\} + 2L\{u_2(t)\} - 2L\{u_3(t)\} + \cdots \\
&= \frac{1}{s} - 2\frac{e^{-s}}{s} + 2\frac{e^{-2s}}{s} - 2\frac{e^{-3s}}{s} + 2\frac{e^{-4s}}{s} - \cdots.
\end{aligned}
$$

After the first term, this is a geometric series with common ratio equal to $-e^{-s}$. Therefore,

$$
\begin{aligned}
L\{\text{sq. wave}\} &= \frac{1}{s} - 2\frac{e^{-s}}{s}\left(\frac{1}{1+e^{-s}}\right) \\
&= \frac{1}{s}\left(\frac{1+e^{-s}-2e^{-s}}{1+e^{-s}}\right) \\
&= \frac{1-e^{-s}}{s(1+e^{-s})} \\
&= \frac{1}{s}\tanh e^{-s/2}.
\end{aligned}
$$

Of more general interest than the Laplace transform of linear combinations of unit step functions is the transform of some well known functions multiplied by a unit step function. The following theorem does not answer this question directly but it does yield the ability to perform the necessary manipulations so that such a transform may be found.

theorem 4.8

The Second Shifting Theorem. Let $F(s) = L\{f(t)\}$. Then

$$L\{u_a(t)\, f(t-a)\} = e^{-as}F(s)$$

or, which is the same thing,

$$L^{-1}\{e^{-as}F(s)\} = u_a(t)\, f(t-a).$$

proof: By definition

$$e^{-as}F(s) = \int_0^\infty e^{-s(x+a)} f(x)\, dx.$$

In the integral, we make the substitution $t = x + a$

$$\begin{aligned} e^{-as}F(s) &= \int_a^\infty e^{-st} f(t-a)\, dt \\ &= \int_0^\infty e^{-st} u_a(t) f(t-a)\, dt \\ &= L\{u_a(t) f(t-a)\}. \end{aligned}$$

Theorem 4.8 may be viewed in two ways:

1. When finding the transform of a function times a unit step function, $u_a(t)$, the function must first be expressed in terms of $t - a$. Then the result of the theorem may be applied.
2. The multiplication of the transform by the exponential, e^{-as}, causes a shift of the inverse function to the right and a "turn-on" at $t = a$. Recall that the first shifting theorem showed how a shift of the transform function occurs when the function of t is multiplied by e^{at}. The first and second shifting theorems are analogous, but the "turn-on" is unique to the second theorem.

example 4.25: Find the Laplace transform of $t^2 u_3(t)$.

solution: We express the function t^2 in terms of $t - 3$,

$$t^2 = (t-3)^2 + 6t - 9 = (t-3)^2 + 6(t-3) + 9.$$

Hence,

$$\begin{aligned} L\{t^2 u_3(t)\} &= L\{(t-3)^2 u_3(t)\} + L\{6(t-3)\, u_3(t)\} + L\{9 u_3(t)\} \\ &= \frac{2e^{-3s}}{s^3} + 6\frac{e^{-3s}}{s^2} + 9\frac{e^{-3s}}{s}. \end{aligned}$$

example 4.26: Find the inverse of the transform function e^{-4s}/s^2.

solution: Since $L^{-1}\{1/s^2\}=t$, the factor e^{-4s} causes a shift to $t-4$ and a turn-on at that point,

$$L^{-1}\left\{\frac{e^{-4s}}{s^2}\right\}=u_4(t)\,(t-4).$$

example 4.27: Find the response of an RL circuit to a single square wave, $u_a(t)-u_b(t)$. Assume that the initial current is zero.

solution: The differential equation of the circuit is

$$L\frac{di}{dt}+Ri=u_a(t)-u_b(t).$$

Taking the Laplace transform of both sides and letting $L\{i(t)\}=I(s)$,

$$LsI(s)+RI(s)=\frac{e^{-as}-e^{-bs}}{s},$$

from which

$$I(s)=\frac{e^{-as}}{Ls\left(s+\dfrac{R}{L}\right)}-\frac{e^{-bs}}{Ls\left(s+\dfrac{R}{L}\right)}$$

$$=\frac{e^{-as}}{R}\left(\frac{1}{s}-\frac{1}{s+\dfrac{R}{L}}\right)-\frac{e^{-bs}}{R}\left(\frac{1}{s}-\frac{1}{s+\dfrac{R}{L}}\right).$$

Hence,

$$i(t)=\frac{1}{R}\left(1-e^{-R(t-a)/L}\right)u_a(t)-\frac{1}{R}\left(1-e^{-R(t-b)/L}\right)u_b(t).$$

(See Figure 4.10.)

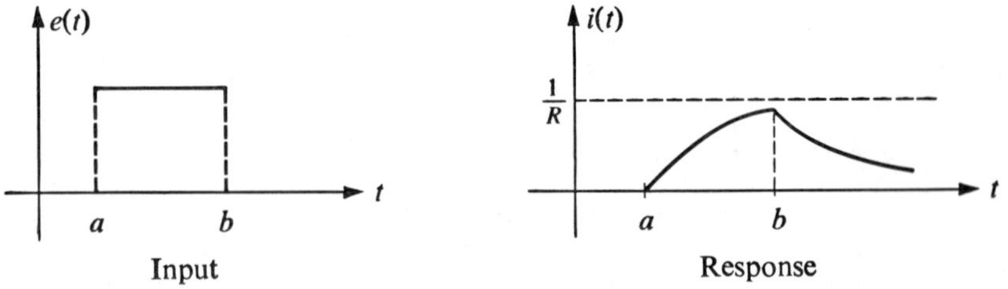

figure 4.10

exercises for section 4.6

In Exercises 1–10, sketch the graph of the given function for $t \geqq 0$.

1. tu_1.
2. $t(u_1 - u_2)$.
3. $u_1 - 3u_4 - 4u_5$.
4. $(t-3)\,u_3$.
5. $(t^2 - 4)\,u_4$.
6. $u_\pi \sin t$.
7. $u_\pi \sin(t - \pi)$.
8. $t(u_0 - u_1) + (u_1 - u_3) + (u_3 - u_4)\,(4 - t)$.
9. $u_1 + u_2 + u_3 + u_4 + \cdots$.
10. $u_1 - 2u_2 + 2u_3 - 2u_4 \cdots$.

In Exercises 10–20, represent the functions in terms of unit step functions and find their Laplace Transforms.

11. $f(t) = 2,\ 0 < t < 1$
 $= t,\ 1 < t.$
12. $f(t) = 3, \quad 0 < t < 2$
 $= t + 1,\ 2 < t.$
13. $f(t) = e^{-t},\ 0 < t < 3$
 $= 0, \quad 3 < t.$
14. $f(t) = \sin 3t,\ 0 < t < \pi$
 $= 0, \quad \pi < t.$
15. $f(t) = t^2,\ 0 < t < 3$
 $= 9,\ 3 < t < 5$
 $= 0,\ 5 < t.$
16. See Figure 4.11.
17. See Figure 4.12.
18. See Figure 4.13.
19. See Figure 4.14.
20. See Figure 4.15.

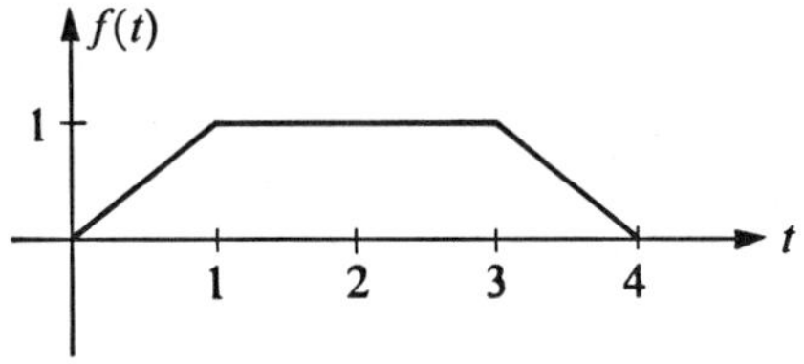

figure 4.11

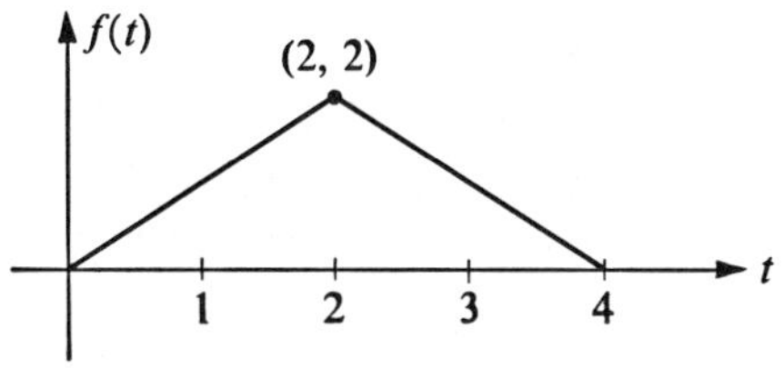

figure 4.12

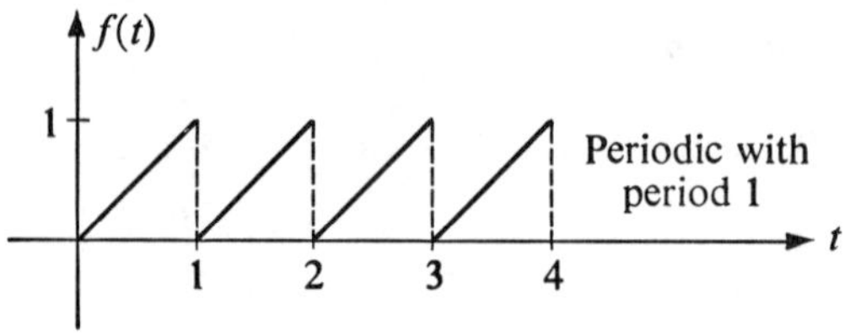

figure 4.13

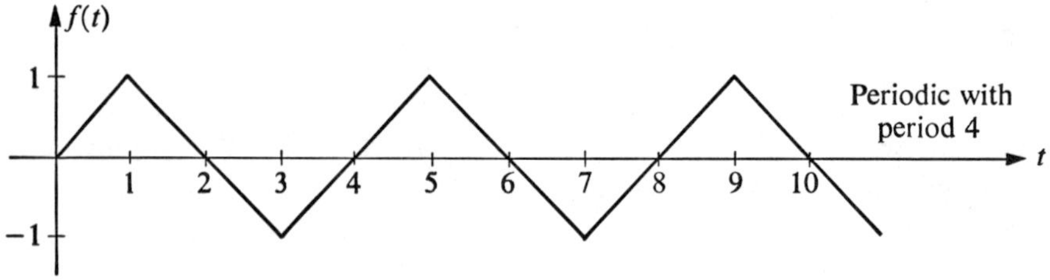

figure 4.14

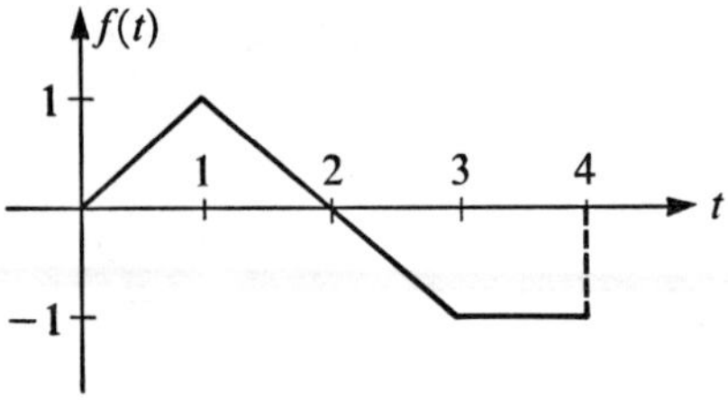

figure 4.15

In Exercises 21–25, find the inverse Laplace transform of the given function. Sketch the graph of the function of t.

21. $\dfrac{e^{-3s}}{s^2}$.   22. $\dfrac{e^{-4s}-e^{-s}}{s^3}$.   23. $\dfrac{e^{-s}}{s^2+4}$.

24. $\dfrac{e^{-s}}{(s-1)(s-2)}$.   25. $\dfrac{se^{-2s}}{s^2+4}$.

In Exercises 26–30, find the Laplace transform of the given function.

26. tu_2.   27. t^2u_3.   28. $u_\pi \cos t$.   29. $u_2(t-3)^2$.   30. e^tu_3.

31. Find the current in an RL circuit whose voltage is a constant, V_0. Assume no initial current and use the method of Laplace transforms.

32. Find the response of an RC circuit to a single square wave, of height h, between $t=a$ and $t=b$. Assume zero initial charge and use the method of Laplace transforms.

4.7 the convolution theorem; integral equations

In this section you will learn about a different kind of binary operation between two functions. The operation arises quite naturally in discussing the inverse of the product of two transform functions.

definition 4.4

Let f and g be piecewise continuous functions. Then the function of t defined by the integral

$$\int_0^t f(x)\,g(t-x)\,dx$$

is called the *convolution* of f and g, and is denoted by $f*g$.

example 4.28: Find the convolution of the sine and cosine functions.

solution:

$$\begin{aligned}(\sin t)*(\cos t) &= \int_0^t \sin(t-x)\cos x\,dx\\ &= \tfrac{1}{2}\int_0^t [\sin t + \sin(t-2x)]\,dx\\ &= \frac{x\sin t}{2}\Bigg|_0^t + \frac{\cos(t-2x)}{4}\Bigg|_0^t\\ &= \tfrac{1}{2}t\sin t.\end{aligned}$$

The fundamental theorem of this section is concerned with the transform of a convolution function.

theorem 4.9

Let f and g be piecewise continuous and of exponential order. Then $f*g$ has a Laplace transform which is given by the product of the transform of f and the transform of g.

proof:

$$\begin{aligned}L\{f*g\} &= \int_0^\infty e^{-st}\left[\int_0^t f(t-x)\,g(x)\,dx\right]dt\\ &= \lim_{B\to\infty}\int_0^B e^{-st}\left[\int_0^t f(t-x)\,g(x)\,dx\right]dt.\end{aligned}$$

The iterated integral is equal to the double integral of the function $e^{-st}f(x)\,g(t-x)$ over the region R in the tx-plane between the t-axis and the line $x=t$. (See Figure 4.16.) The change of variables $t^*=t-x$, $x^*=x$,

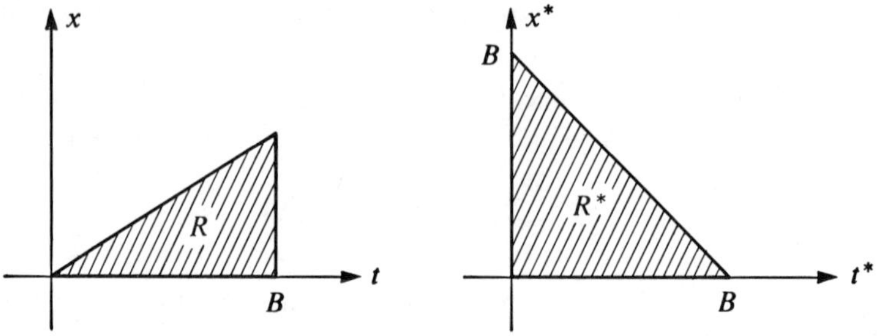

figure 4.16

changes the region R into the region R^* in the (t^*, x^*)-plane. Hence,

$$\begin{aligned}L\{f * g\} &= \lim_{B\to\infty} \int_0^B \int_0^{B-t^*} e^{-s(t^*+x^*)} f(t^*)\, g(x^*)\, dx^*\, dt^* \\ &= \int_0^\infty \left[\int_0^\infty e^{-st^*} f(t^*)\, dt^*\right] e^{-sx^*} g(x^*)\, dx^* \\ &= F(s)\, G(s).\end{aligned}$$

example 4.28: Use the convolution theorem to find

$$L^{-1}\left\{\frac{1}{s(s^2+1)}\right\}.$$

solution: Note that

$$\frac{1}{s(s^2+1)} = \frac{1}{s}\frac{1}{s^2+1} = F(s)\, G(s).$$

Since

$$L^{-1}\{F(s)\} = u_0(t),$$

and

$$L^{-1}\{G(s)\} = \sin t,$$

it follows that

$$\begin{aligned}L^{-1}\left\{\frac{1}{s(s^2+1)}\right\} &= u_0(t) * \sin t \\ &= \int_0^t u_0(t-x) \sin x\, dx \\ &= 1 - \cos t.\end{aligned}$$

example 4.29: Find

$$L^{-1}\frac{s^2}{(s^2+1)^2}.$$

solution: Note that the function is the square of $s/(s^2+1)$, whose inverse transform is $\cos t$. Hence

$$\begin{aligned} L^{-1}\left\{\frac{s^2}{(s^2+1)^2}\right\} &= (\cos t) * (\cos t) \\ &= \int_0^t \cos(t-x)\cos x\,dx \\ &= \frac{\sin t + t\cos t}{2}. \end{aligned}$$

The convolution theorem may be used to great advantage in solving equations with an unknown function under an indefinite integral sign. Equations of this type are called "integral equations." In order to use Laplace transformations to solve such equations, the indefinite integrals occurring must be either of the form of (1) a convolution integral, or, (2) an indefinite integral of the unknown function.

example 4.30: Find the function y, if

$$y = t + \int_0^t y(x)\sin(t-x)\,dx.$$

solution: We take the Laplace transform of both sides and denote the transform of y by Y,

$$Y(s) = \frac{1}{s^2} + Y(s)\frac{1}{s^2+1}.$$

Solving for Y we obtain

$$\begin{aligned} Y(s) &= \frac{s^2+1}{s^4} \\ &= \frac{1}{s^2} + \frac{1}{s^4}. \end{aligned}$$

The inverse transform of $Y(s)$ is

$$y(t) = t + \frac{t^3}{6}.$$

example 4.31: Solve the integral equation

$$y = t + \int_0^t y(x)\,dx + \int_0^t (t-x)\,y(x)\,dx.$$

solution: Taking the transform of both sides we obtain

$$Y(s) = \frac{1}{s^2} + \frac{Y(s)}{s} + \frac{Y(s)}{s^2}.$$

Solving for $Y(s)$,

$$Y(s) = \frac{1}{s^2 - s - 1}$$
$$= \frac{2}{\sqrt{5}} \frac{\sqrt{5}/2}{(s^2 - s + \frac{1}{4}) - \frac{5}{4}}.$$

Hence, the unknown function is

$$f(t) = \frac{2}{\sqrt{5}} e^{t/2} \sin \frac{\sqrt{5}}{2} t.$$

exercises for section 4.7

1. Find the Laplace transform of $\int_0^t (t-u) \sin 2u \, du$.

2. Find the Laplace transform of $\int_0^t e^{-(t-u)} \cos u \, du$.

3. Find the inverse transform of $\frac{1}{s(s^2+1)}$ by using the convolution theorem.

4. Find the inverse of $\frac{1}{s^2(s-1)}$ by using the convolution theorem.

5. Solve the equation $y(t) = t + \int_0^t \sin(t-u)\, y(u)\, du$.

6. Solve the equation $y(t) = \sin t + \int_0^t \sin(t-u)\, y(u)\, du$.

7. Solve the equation $\dot{y} + 3y + 2\int_0^t y(u)\, du = u_0(t)$, $y(0) = 0$.

8. Solve the equation $y(t) = t^2 + \int_0^t y(u) \sin(t-u)\, du$.

9. Solve the equation $y(t) = t + 4\int_0^t (u-t)^2\, y(u)\, du$.

10. Solve the equation $y(t) = t^2 - 2\int_0^t y(t-u) \sinh 2u \, du$.

11. Solve the equation $y(t) = t + \int_0^t y(t-u)\, e^{-u}\, du$.

12. Solve the equation $y(t) = 1 + \int_0^t y(u) \sin(t-u)\, du$.

infinite series methods

5

5.1 method of assuming an infinite series solution

In this chapter you will learn how to find solutions to ordinary differential equations by assuming the form of the solution to be an infinite series. Generally, the form will be a power series (that is, a Taylor series), but there are important exceptions to this general rule. The method of infinite series is in many ways the most basic method of finding solutions to differential equations, since such a wide set of equations may be solved this way. You may find the method a bit unsatisfying since ordinarily you will not be able to write the solution in terms of elementary functions. In fact, you may be able to write only the first few terms of the series; however, if the convergence of the series to the solution is already established, the first few terms will serve as an approximation to the solution. Further, the first few terms of a series often indicate a similarity between the solution and some better known elementary function.

In Sections 5.2 and 5.4 we will see two different techniques used with the one fundamental assumption that the solution has the form of a Taylor Series. The first of these methods emphasizes that the form of the coefficients in a Taylor series is

$$\frac{f^{(n)}(a)}{n!}$$

and hence the coefficients may be found by determining $f^{(n)}(a)$. Thus, the solution function is essentially known if the values of the derivatives are known at $x = a$. The second method does not take advantage of the property that the coefficients are a function of the values of the derivatives. Instead, the coefficients are found directly from the fact that the assumed power series must satisfy the differential equation.

5.2 taylor series

When using the method of Taylor series, you should assume the solution to be explicitly of the form

$$y(x) = \sum_{n=0}^{\infty} \frac{y^{(n)}(a)}{n!}(x-a)^n .$$

Hence, to write the solution, you need to know the values of the unknown function and its derivatives at $x = a$. Since we will probably use the solution for values of x near to $x = a$, a few terms will usually be sufficient. The first few terms yield an *approximating polynomial* to the infinite series solution.

In Section 5.5 we will address ourselves briefly to the rather important question of how to know when an assumption of this type is valid. For now, if the assumption turns out to be invalid, it will become obvious in the procedure of the technique. (See Exercise 16 in the next exercise set.)

The values of the derivatives are found directly from the initial conditions and the differential equation. Determining the various derivatives is time consuming and tedious, and is the major limitation to this method. However, the method is general enough to apply to all types of differential equations, both linear and nonlinear. One only has to have the stamina to continue the process of differentiation.

example 5.1: Find the first five nonzero terms of the Taylor series expression for the solution to the initial value problem

$$y'' = y^2 + x, \qquad y(1) = 2, \qquad y'(1) = -1 .$$

solution: The unknown function and the derivative are given at $x = 1$. Consequently, we want to write a Taylor series in powers of $(x - 1)$. The value of $y''(1)$ is determined from the differential equation to be

$$y''(1) = [y(1)]^2 + 1 = 4 + 1 = 5 .$$

Differentiating both sides of the differential equation leads to the following expressions for $y'''(x)$ and $y^{(iv)}(x)$:

$$y'''(x) = 2yy' + 1 ,$$
$$y^{(iv)}(x) = 2(y')^2 + 2yy'' ,$$

from which

$$y'''(1) = 2(2)(-1) + 1 = -3,$$
$$y^{(iv)}(1) = 2(-1)^2 + 2(2)(5) = 22$$

The first few terms are

$$y(x) = 2 - (x-1) + \tfrac{5}{2}(x-1)^2 - \tfrac{1}{2}(x-1)^3 + \tfrac{11}{12}(x-1)^4 + \cdots.$$

example 5.2: Find the Taylor series expression about $x = 0$ of the solution to the differential equation $y'' + y = 0$. Initial conditions on the solution and its derivative are to be left arbitrary.

solution: The solution is assumed to be of the form

$$y(x) = \sum_{n=0}^{\infty} \frac{y^{(n)}(0)}{n!} x^n.$$

To determine $y^{(n)}(0)$, we let $y(0)$ and $y'(0)$ be the constants a_0 and a_1, and then the differential equation yields

$$y''(0) = -y(0) = -a_0, \quad y'''(0) = -y'(0) = -a_1, \quad y^{(iv)}(0) = -y''(0) = a_0,$$
$$y^{(v)}(0) = a_1, \quad y^{(vi)}(0) = -a_0, \quad \cdots y^{(n)} = -y^{(n-2)}.$$

By a simple inductive proof we can show

$$y^{(2n)}(0) = (-1)^n a_0 \quad \text{and} \quad y^{(2n+1)}(0) = (-1)^n a_1$$

and hence, the solution can be written in the form

$$y = a_0 \sum_{n=0}^{\infty} \frac{(-1)^n}{(2n)!} x^{2n} + a_1 \sum_{n=0}^{\infty} \frac{(-1)^n}{(2n+1)!} x^{2n+1}.$$

You probably recognize the two infinite series as being the Taylor series for the elementary functions $\cos x$ and $\sin x$. Hence,

$$y = a_0 \cos x + a_1 \sin x.$$

example 5.3: Find the first three terms of the Taylor series expansion about $x = 2$ of the solution to the differential equation

$$y'' + y' - \ln y = e^x.$$

solution: Since $y'' = e^x + \ln y - y'$, we have that $y(2) = a_0$, $y'(2) = a_1$, $y''(2) = e^2 + \ln a_0 - a_1$. Hence, the first three terms are

$$y(x) = a_0 + a_1(x-2) + \frac{e^2 + \ln a_0 - a_1}{2}(x-1)^2 + \cdots.$$

This approximation is very rough, but you will have to agree that it was easy to obtain!

example 5.4: Solve the differential equation $y' - y + x^2 + x + 1 = 0$ by assuming a Taylor series solution valid near $x = 0$. Assume $y(0) = 1$.

solution: Let

$$y = \sum_{n=0}^{\infty} \frac{y^{(n)}(0)}{n!} x^n.$$

The given differential equation gives

$$y' = y - x^2 - x - 1, \qquad y'' = y' - 2x - 1,$$
$$y''' = y'' - 2, \qquad y^{(\text{iv})} = y''',$$

and for $n \geqq 4$,

$$y^{(n)} = y^{(n-1)}.$$

Therefore, $y(0) = 1$, $y'(0) = 0$, $y''(0) = -1$, $y'''(0) = -3$, and for $n \geqq 3$, $y^{(n)}(0) = -3$. Consequently, the solution is

$$\begin{aligned} y &= 1 - \frac{x^2}{2} - 3 \sum_{n=3}^{\infty} \frac{x^n}{n!} \\ &= 1 - \frac{x^2}{2} - 3 \sum_{n=0}^{\infty} \frac{x^n}{n!} + 3 \sum_{n=0}^{2} \frac{x^n}{n!} \\ &= 1 - \frac{x^2}{2} - 3 \sum_{n=0}^{\infty} \frac{x^n}{n!} + 3 + 3x + \frac{3x^2}{2}. \end{aligned}$$

The infinite series can be recognized as the Taylor series for e^x about $x = 0$. Hence

$$y = 4 + 3x + x^2 - 3e^x.$$

5.3 comments on the form of a solution

When assuming an infinite series form for the solution, you will inevitably be faced with deciding the appropriate way of stating the answer. There are basically four levels of achievement which you can reach and each one yields a satisfactory method of stating the solution under certain circumstances.

1. A few of the nonzero terms of the series can be obtained. This is the minimum, in the way of an answer, that you should give since this much requires at most the computation of a few derivatives. (See Examples 5.1 and 5.2.)
2. A recursion (or inductive) formula can be obtained which expresses the nth coefficient or nth derivative in terms of previous derivatives and coefficients. This is more information than in level 1 since if the recursion formula is known, one can often determine the interval of convergence.
3. You may be able to use the recursion formula to find a general formula for the coefficient of x^n, or for $f^{(n)}(a)$. This will enable you to write explicitly the infinite series form of the solution function. (See Examples 5.2 and 5.4.)
4. Finally, the infinite series form may be expressible in terms of elementary functions. (See Examples 5.2 and 5.4.)

From a theoretical standpoint the ability to express the answer in terms of elementary functions rather than in terms of an infinite power series is of little consequence. Even from a practical consideration such an attainment is overplayed. In many cases the expression in infinite series form is precisely the best form for performing evaluations and computations.

example 5.5: Write the power series centered at 0 whose coefficients are related by the inductive formula

$$a_0 = 1, \quad a_1 = 0, \quad a_n = -\frac{2}{n}\, a_{n-2} \text{ for } n \geqq 2.$$

solution: Such a power series has the form $\sum_{n=0}^{\infty} a_n x^n$. Because the recursion formula for a_n is in terms of the coefficient of the term two previous, you can see that if n is even, a_n will be in terms of a_0, and if n is odd, a_n will be in terms terms of a_1. Such a series is often written in rearranged form as

$$\sum_{n=0}^{\infty} a_n x^n = \sum_{k=0}^{\infty} a_{2k} x^{2k} + \sum_{k=0}^{\infty} a_{2k+1} x^{2k+1}.$$

The first series on the right is a series of even terms and the second is a series of odd terms.

Since a_1 vanishes, so do all the odd coefficients and the second series vanishes.

The recursion formula for the even coefficients is

$$a_{2k} = -\frac{1}{k}\, a_{2k-2}.$$

Repeatedly applying this formula to $a_{2k-2}, a_{2k-4}, \cdots, a_8, a_6, a_4$, and a_2,

$$a_{2k} = \left(-\frac{1}{k}\right)\left(\frac{1}{k-1}\right)\left(\frac{1}{k-2}\right)\cdots\left(-\tfrac{1}{4}\right)\left(-\tfrac{1}{3}\right)\left(-\tfrac{1}{2}\right)\left(-\tfrac{1}{1}\right)a_0$$
$$= \frac{(-1)^k}{k!}a_0.$$

Hence, the power series is

$$\sum_{k=0}^{\infty} \frac{(-1)^k}{k!}x^{2k}.$$

In many cases the recursion formula is much too complicated to obtain a general term. The preceding example shows how one approaches the problem of using the recursion formula to find the general coefficient.

exercises for sections 5.1–5.3

Use the method of Taylor Series to solve the following differential equations. Obtain a recursion formula for the derivatives if you can, but if you can not, write at least five nonzero terms of the series. If possible, obtain the general coefficient and try to write the solution in terms of elementary functions.

1. $y' = x + y, \quad y(0) = 1.$
2. $y' = y, \quad y(0) = 1.$
3. $y'' = y, \quad y(0) = 1, \quad y'(0) = 1.$
4. $y' + 2y = x^2, \quad y(0) = 1.$
5. $y'' + 4y' + 3y = 0, \quad y(0) = 1, \quad y'(0) = 1.$
6. $y' = y^2, \quad y(0) = 2.$
7. $y'' = xy, \quad y(1) = 2, \quad y'(1) = -2.$
8. $y'' = y^3, \quad y(1) = 1, \quad y'(1) = -1.$
9. $y'' = \ln y, \quad y(1) = 1, \quad y'(1) = 1.$
10. $y' = e^y, \quad y(0) = 1.$

In each of the following exercises you are given a recursion formula for the power series solution, $\sum_{n=0}^{\infty} a_n x^n$, to a differential equation. Find the general term, and write the solution.

11. $a_n = -\dfrac{1}{n-2}a_{n-1}, \quad n \geqq 4, \quad a_0 = 1, \quad a_1 = 0, \quad a_2 = 1, \quad a_3 = -1.$

12. $a_n = \dfrac{n+2}{n}a_{n-2}, \quad n \geqq 2, \quad a_0 = a_1 = 1.$

13. $a_n = \dfrac{(n-5)(n-6)}{n^2} a_{n-2}, \quad n \geqq 2, \quad a_0 = a_1 = 1.$

14. $a_n = \dfrac{n-5}{n(n-1)} a_{n-2}, \quad n \geqq 2, \quad a_0 = a_1 = 1.$

15. $a_n = \dfrac{3}{4n} a_{n-2}, \quad n \geqq 2, \quad a_0 = a_1 = 1.$

16. Show that the method of Taylor series cannot be used to solve the differential equation

$$(x^2 - 3x + 2)\, y'' + 3y' - y = 0, \qquad y(1) = y'(1) = 1.$$

(*Hint:* Try it!)

5.4 *method of undetermined coefficients*

The fundamental idea of assuming an infinite series solution can be easily adapted to another approach to obtain the coefficients of the power series. This technique is particularly appropriate for linear differential equations.

We assume that the power series can be differentiated term by term as often as necessary and that the series of differentiated terms converges to the derivative of the function. Implicit is the even more fundamental assumption that two power series are equal if and only if they have identical coefficients.

The coefficients are found by substituting the series into the differential equation. Then the information on each of the coefficients is essentially known. The infinite series expansion of the coefficient functions and the driving function must also be known. Hence, the case where the coefficient functions are polynomials is probably the most important application of this technique.

example 5.6: Solve the differential equation $y' + xy = x^2 - 2x$ by assuming the solution may be written in the form of a power series. Use the method of undetermined coefficients.

solution: Since $y = a_0 + a_1x + a_2x^2 + a_3x^3 + a_4x^4 \cdots$, then $y' = a_1 + 2a_2x + 3a_3x^2 + \cdots$ and upon substituting into the differential equation, we obtain

$$a_1 + 2a_2x + 3a_3x^2 + 4a_4x^3 + \cdots + a_0x + a_1x^2 + a_2x^3 + a_3x^4 + \cdots = x^2 - 2x.$$

The coefficient a_0 is obtained from a knowledge of $y(0)$ and is left as arbitrary. We now proceed by equating coefficients of like powers of x on both sides of this equation.

$$\begin{array}{ll}
\text{Coefficients of } x^0: & a_1 = 0 \\
\text{Coefficients of } x^1: & 2a_2 + a_0 = -2 \\
\text{Coefficients of } x^2: & 3a_3 + a_1 = 1 \\
\text{Coefficients of } x^3: & 4a_4 + a_2 = 0 \\
\text{Coefficients of } x^4: & 5a_5 + a_3 = 0 \\
\vdots & \vdots \\
\text{Coefficient of } x^{n-1} \text{ for } n \geqq 4: & na_n + a_{n-2} = 0, \quad n \geqq 4
\end{array}$$

From this the necessary coefficient information is

$$a_0 \text{ arbitrary}, \quad a_1 = 0, \quad a_2 = -\frac{a_0 + 2}{2},$$

$$a_3 = \tfrac{1}{3}, \quad a_n = -\frac{1}{n} a_{n-2} \text{ for } n \geqq 4.$$

This recursion formula may also be obtained with the formal use of summation notation. Let $y = \sum_{n=0}^{\infty} a_n x^n$, then $y' = \sum_{n=1}^{\infty} na_n x^{n-1}$. Upon substituting into the differential equation, we obtain

$$\sum_{n=1}^{\infty} na_n x^{n-1} + \sum_{n=0}^{\infty} a_n x^{n+1} = x^2 - 2x.$$

In the second series, shift the index back by 2,

$$\sum_{n=1}^{\infty} na_n x^{n-1} + \sum_{n=2}^{\infty} a_{n-2} x^{n-1} = x^2 - 2x.$$

Writing the first term of the first series separately and then combining into one series, we obtain

$$1a_1 x^0 + \sum_{n=2}^{\infty} x^{n-1}(na_n + a_{n-2}) = x^2 - 2x.$$

From this point, the procedure is the same as before.

This recursion formula is simple enough that we would expect to be able to obtain the general term. By the nature of the recursion formula, we can write the solution in the rearranged form

$$\sum_{k=0}^{\infty} a_{2k} x^{2k} + \sum_{k=0}^{\infty} a_{2k+1} x^{2k+1}.$$

The coefficient information for the first series is

$$a_0 \text{ arbitrary}, \quad a_2 = -\frac{a_0 + 2}{2}, \quad a_{2k} = -\frac{a_{2k-2}}{2k} \text{ for } k \geqq 2$$

and for the second series is

$$a_1 = 0, \quad a_3 = \tfrac{1}{3}, \quad \text{and} \quad a_{2k+1} = -\frac{a_{2k-1}}{2k+1} \text{ for } k \geqq 2.$$

For the first series

$$\begin{aligned} a_{2k} &= \left(-\frac{1}{2k}\right)\left(-\frac{1}{2k-2}\right)\left(-\frac{1}{2k-4}\right)\cdots\left(-\frac{1}{6}\right)\left(-\frac{1}{4}\right)a_2 \\ &= \frac{(-1)^{k-1}}{2^{k-1}(k!)}\left(-\frac{a_0+2}{2}\right). \end{aligned}$$

For the second series

$$\begin{aligned} a_{2k+1} &= \left(-\frac{1}{2k+1}\right)\left(-\frac{1}{2k-1}\right)\left(-\frac{1}{2k-3}\right)\cdots\left(-\frac{1}{7}\right)\left(-\frac{1}{5}\right)a_3 \\ &= \frac{(-1)^k\{(2k)(2k-2)(2k-4)\cdots(6)(4)(2)\}\,a_3}{\dfrac{(2k+1)!}{3}} \\ &= \frac{(-1)^k 2^k k!\,3}{(2k+1)!}\,\frac{1}{3} = \frac{(-1)^k 2^k k!}{(2k+1)!}. \end{aligned}$$

Hence, the solution is

$$\begin{aligned} y = a_0 - \frac{a_0+2}{2}x^2 + \frac{a_0+2}{2}\sum_{k=2}^{\infty}\frac{(-1)^k}{2^{k-1}k!}x^{2k} \\ + \frac{1}{3} + \sum_{k=2}^{\infty}\frac{(-1)^k 2^k k!}{(2k+1)!}x^{2k+1}. \end{aligned}$$

example 5.7: Solve the differential equation $y'' + y = e^x$ by assuming a power series solution about the origin.

solution: Letting $y = \sum_{n=0}^{\infty} a_n x^n$, then $y' = \sum_{n=1}^{\infty} a_n n x^{n-1}$ and $y'' = \sum_{n=2}^{\infty} a_n n(n-1)x^{n-2}$. The infinite series expansion for e^x is $e^x = \sum_{n=0}^{\infty}\frac{x^n}{n!}$. Substituting into the differential equation, we obtain

$$\sum_{n=2}^{\infty} n(n-1)a_n x^{n-2} + \sum_{n=0}^{\infty} a_n x^n = \sum_{n=0}^{\infty}\frac{x^n}{n!}.$$

Simplifying slightly,

$$\sum_{n=2}^{\infty} n(n-1)\,a_n x^{n-2} + \sum_{n=0}^{\infty}\left(a_n - \frac{1}{n!}\right)x^n = 0.$$

We shift the index in the second series to make the exponent the same as in the first series,

$$\sum_{n=2}^{\infty} n(n-1)\,a_n x^{n-2} + \sum_{n=2}^{\infty}\left(a_{n-2} - \frac{1}{(n-2)!}\right)x^{n-2} = 0.$$

We combine the two series to obtain

$$\sum_{n=2}^{\infty} x^{n-2}\left[n(n-1)\,a_n + a_{n-2} - \frac{1}{(n-2)!}\right] = 0.$$

Since the right-hand side is identically zero, the coefficient of every power of x on the left-hand side must be zero:

(The coefficients a_0 and a_1 are arbitrary and are obtained from the initial conditions.)

Coefficient of x^{n-2} for $n \geqq 2$:

$$n(n-1)\,a_n + a_{n-2} - \frac{1}{(n-2)!} = 0$$

or

$$a_n = \frac{1}{n!} - \frac{1}{n(n-1)}\,a_{n-2}.$$

If we let $a_n = a_n^* + \dfrac{1}{2(n!)}$, then the solution may be written in the form

$$y = \sum_{k=0}^{\infty} a_{2k}^* x^{2k} + \sum_{k=0}^{\infty} a_{2k+1}^* x^{2k+1} + \frac{1}{2}\sum_{k=0}^{\infty}\frac{x^k}{k!}$$

where

$$a_{2k}^* = \frac{(-1)^k}{(2k)!}\,a_0^*,\quad a_{2k+1}^* = \frac{(-1)^k}{(2k+1)!}\,a_1^*.$$

In this form, the solution may be recognized in terms of elementary functions,

$$y = a_0^* \cos x + a_1^* \sin x + \frac{e^x}{2}.$$

example 5.8: Solve the differential equation

$$(1-x^2)\,y'' - 2xy' + 12y = 0$$

by assuming a solution in power series form valid near the origin. This differential equation is called Legendre's equation of order three.

solution: Assuming $y = \sum_{n=0}^{\infty} a_n x^n$, then $y' = \sum_{n=1}^{\infty} a_n n x^{n-1}$, $y'' = \sum_{n=2}^{\infty} a_n n(n-1) x^{n-2}$. Substituting into the given differential equation:

$$(1-x^2) \sum_{n=2}^{\infty} a_n n(n-1) x^{n-2} - 2x \sum_{n=1}^{\infty} a_n n x^{n-1} + 12 \sum_{n=0}^{\infty} a_n x^n = 0.$$

Carrying out the indicated multiplications,

$$\sum_{n=2}^{\infty} a_n n(n-1) x^{n-2} - \sum_{n=1}^{\infty} a_n n(n-1) x^n - 2 \sum_{n=0}^{\infty} a_n n x^n + 12 \sum_{n=0}^{\infty} a_n x^n = 0.$$

Combining the last three series,

$$\sum_{n=2}^{\infty} n(n-1) a_n x^{n-2} - \sum_{n=0}^{\infty} a_n x^n \{n^2 - n + 2n - 12\} = 0,$$

or

$$\sum_{n=2}^{\infty} n(n-1) a_n x^{n-2} - \sum_{n=0}^{\infty} a_n x^n (n+4)(n-3) = 0.$$

Shifting index in the second of these summations,

$$\sum_{n=2}^{\infty} n(n-1) a_n x^{n-2} - \sum_{n=2}^{\infty} a_{n-2} x^{n-2} (n+2)(n-5) = 0.$$

This equation implies that a_0 and a_1 are both arbitrary and that for $n \geqq 2$,

$$a_n = \frac{(n+2)(n-5)}{n(n-1)} a_{n-2}.$$

Therefore, $a_3 = -5a_1/3$, and $a_{2k+1} = 0$ for $k \geqq 2$. Therefore, one of the solutions to the differential equation is the Legendre polynomial of degree three,

$$y_1 = c_1 (x - \tfrac{5}{3} x^3).$$

All Legendre polynomials have graphs which pass through the point $(1, 1)$ so that $c_1 = -3/2$ and then $y_1 = 5x^3/2 - 3x/2$.

A second solution to the equation is of the form $\sum_{k=0}^{\infty} a_{2k} x^{2k}$, where a_0 is arbitrary and $a_{2k} = \dfrac{(2k+2)(2k-5)}{2k(2k-1)} a_{2k-2}$. Then,

$$a_{2k} = \frac{(2k+2)(2k-5)}{2k(2k-1)} \cdot \frac{(2k)(2k-7)}{(2k-2)(2k-3)} \cdot \frac{(2k-2)(2k-9)}{(2k-4)(2k-5)} \cdots \frac{10 \cdot 3}{8 \cdot 7} \cdot \frac{8 \cdot 1}{6 \cdot 5} \cdot \frac{6(-1)}{4 \cdot 3} \cdot \frac{4(-3)}{2 \cdot 1} a_0.$$

Cancelling as many terms as possible in the numerator and denominator,

$$a_{2k} = \frac{(2k+2)(-1)(-3)}{(2k-1)(2k-3)(2)} a_0 = \frac{3(k+1)}{(2k-1)(2k-3)} a_0 .$$

Hence, a second solution to Legendre's differential equation of order three is

$$y = 3a_0 \sum_{k=0}^{\infty} \frac{k+1}{(2k-1)(2k-3)} x^{2k}.$$

You can easily show, by using the ratio test, that the series converges for $|x| < 1$.

The method of undetermined coefficients may also be used to find power series solutions centered at points other than $x = 0$. In the following example we will assume a power series solution about $x = 1$. The example also exhibits how "multi-term" recursion formulas arise. This kind of recursion formula makes it almost impossible to determine a general term.

example 5.9: Find a solution in power series form, centered at $x = 1$, to the differential equation $xy'' + y' + xy = 0$.

solution: The given differential equation may be written in the form

$$(x-1)\,y'' + y'' + y' + (x-1)\,y + y = 0 .$$

Assume the solution is of the form $y = \sum_{n=0}^{\infty} a_n (x-1)^n$ from which

$$y' = \sum_{n=1}^{\infty} a_n n (x-1)^{n-1} \quad \text{and} \quad y'' = \sum_{n=2}^{\infty} a_n n (n-1)(x-1)^{n-2} .$$

Substituting into the differential equation

$$\begin{aligned}(x-1) &\sum_{n=2}^{\infty} a_n n (n-1)(x-1)^{n-2} \\ &+ \sum_{n=2}^{\infty} a_n n (n-1)(x-1)^{n-2} + \sum_{n=1}^{\infty} a_n n (x-1)^{n-1} \\ &+ (x-1) \sum_{n=0}^{\infty} a_n (x-1)^n + \sum_{n=0}^{\infty} a_n (x-1)^n = 0 .\end{aligned}$$

Simplifying,

$$\begin{aligned}&\sum_{n=1}^{\infty} a_n (n)(n-1)(x-1)^{n-1} + \sum_{n=2}^{\infty} a_n (n)(n-1)(x-1)^{n-2} \\ &\quad + \sum_{n=1}^{\infty} a_n (n)(x-1)^{n-1} + \sum_{n=0}^{\infty} a_n (x-1)^{n+1} + \sum_{n=0}^{\infty} a_n (x-1)^n = 0 .\end{aligned}$$

Combining terms with like exponents,

$$\sum_{n=1}^{\infty} a_n n^2 (x-1)^{n-1} + \sum_{n=2}^{\infty} a_n n(n-1)(x-1)^{n-2} + \sum_{n=0}^{\infty} a_n (x-1)^{n+1} + \sum_{n=0}^{\infty} a_n (x-1)^n = 0.$$

Shifting indexes in the various series so that the exponents are all $n-2$,

$$\sum_{n=2}^{\infty} a_{n-1}(n-1)^2 (x-1)^{n-2} + \sum_{n=2}^{\infty} a_n n(n-1)(x-1)^{n-2} + \sum_{n=3}^{\infty} a_{n-3}(x-1)^{n-2} + \sum_{n=2}^{\infty} a_{n-2}(x-1)^{n-2} = 0.$$

We write the first term separately, and combine the other series,

$$(2a_2 + a_1 + a_0)x^0 + \sum_{n=3}^{\infty} \{a_n n(n-1) + (n-1)^2 a_{n-1} + a_{n-2} + a_{n-3}\}(x-1)^{n-2} = 0.$$

From this equation, we obtain the following coefficient information: a_0 and a_1 are arbitrary;

$$a_2 = \frac{-a_0 - a_1}{2}$$

and for $n \geqq 3$,

$$a_n = -\frac{(n-1)^2 a_{n-1} + a_{n-2} + a_{n-3}}{n(n-1)}.$$

Assuming that $a_0 = a_1 = 1$, the next few coefficients are $a_2 = -1$, $a_3 = 1/3$, $a_4 = -1/4$. The first few terms of the solution are

$$y = 1 + (x-1) - (x-1)^2 + \frac{(x-1)^3}{3} - \frac{(x-1)^4}{4} + \cdots.$$

The differential equation of the previous example is called *Bessel's equation of order zero*. The technique of solution used in the example is not standard. A solution valid near the origin is much more popular but requires a slight variation on the technique of assuming a power series solution. In Chapter 16 we shall discuss Bessel's equations and functions extensively.

A power series assumption does not always lead to a recursion formula, as is shown in the next example.

example 5.10: By use of power series find the general solution to the differential equation $xy' = x^2e^x + y$.

solution: Assuming $y=\sum_{n=0}^{\infty} a_n x^n$, then $y'=\sum_{n=1}^{\infty} a_n n x^{n-1}$. The power series for e^x is $\sum_{n=0}^{\infty} (x^n/n!)$. Substituting into the differential equation,

$$\sum_{n=1}^{\infty} a_n n x^n = \sum_{n=0}^{\infty} \frac{x^{n+2}}{n!} + \sum_{n=0}^{\infty} a_n x^n .$$

Simplifying,

$$\sum_{n=0}^{\infty} a_n x^n (n-1) = \sum_{n=0}^{\infty} \frac{x^{n+2}}{n!} = \sum_{n=2}^{\infty} \frac{x^n}{(n-2)!} .$$

Therefore, $a_0 = 0$ and a_1 is arbitrary. For $n \geqq 2$,

$$a_n = \frac{1}{(n-1)!}$$

and hence the solution is

$$\begin{aligned} y &= a_1 x + \sum_{n=2}^{\infty} \frac{x^n}{(n-1)!} \\ &= a_1 x + x e^x - x . \end{aligned}$$

exercises for section 5.4

By assuming an infinite power series solution around 0 and using the method of undetermined coefficients, solve the following differential equations.

1. $y' + y = 0.$
2. $y' + y = e^x.$
3. $y' + y = e^{-x}.$
4. $y' + y = \sin x.$
5. $y' - y = \sin x.$
6. $y' + xy = 1.$
7. $y' + x^2 y = 0.$
8. $y'' + y = 0.$
9. $y'' + xy = 0.$
10. $y'' + x^2 y = 0.$
11. $y'' + x^3 y = 0.$
12. $(1 - x^2)\, y'' + 3xy' + 5y = 0.$
13. $y'' + y = e^x.$
14. $(1 - x^2)\, y'' + xy' - y = 0.$
15. $(1 - x^2)\, y'' + xy' - y = x.$
16. $(x^2 + 1)\, y'' - 2xy' - 10y = 0.$
17. $y'' + xy' - 4y = 0.$
18. $y'' + xy' - 4y = x.$
19. $(1 + x^2)\, y'' + 3xy' - 3y = 0.$

5.5 singular points of a differential equation

The second-order linear differential equation may be written

$$a_2(x)\, y'' + a_1(x)\, y' + a_0(x)\, y = f(x)$$

or, dividing by $a_2(x)$,

$$y'' + b_1(x)\, y' + b_0(x)\, y = g(x).$$

definition 5.1

The point $x = a$ is called an *ordinary* point of the differential equation if both $b_1(x)$ and $b_0(x)$ have Taylor series representations about $x = a$. At ordinary points $a_2(x) \neq 0$.

definition 5.2

If a point is not an ordinary point it is called a *singular* point. If $(x - a)\, b_1(x)$ and $(x - a)^2\, b_0(x)$ have Taylor series representations, the singular points are said to be *regular*; otherwise they are *irregular*.

example 5.11: Locate and classify the singular points of the differential equation

$$(x - 1)^3 x^2 (x - 2)\, y'' + 3x^3 (x - 1)\, y' + (x - 1)\, y = 0.$$

solution: Since the coefficients are all polynomials, the singular points are those values of x for which the coefficient of y'' is 0, hence at $x = 0$, 1, and 2. Dividing by the coefficient of y'', we obtain

$$b_1(x) = \frac{3x}{(x - 1)^2 (x - 2)}, \qquad b_0(x) = \frac{1}{(x - 1)^2 x^2 (x - 2)}.$$

Since

$$(x - 1)\, b_1(x) = \frac{3x}{(x - 1)(x - 2)}$$

which is undefined at $x = 1$, the singular point at $x = 1$ is irregular. The other two singular points are regular.

It can be shown that if $x = a$ is an ordinary point of a differential equation, then every solution to the equation has a Taylor series representation about $x = a$. If $x = a$ is a regular singular point, then there will be at least one solution of the form of a product of x^r times a power series. If $x = a$ is an irregular singular point, the situation is more complicated and will not be considered here.

Note that the existence of a power series solution about x_0 is a necessary condition for x_0 to be an ordinary point of the differential equation. Thus you may *assume* a power series solution only around ordinary points.

5.6 method of frobenius

A series of the type

$$\sum_{n=0}^{\infty} a_n (x-a)^{n+r}$$

is sometimes called a *Frobenius series*, which means that a Taylor series is a special case corresponding to $r=0$. If we proceed to solve a differential equation by assuming a Frobenius series and find the coefficients by the method of undetermined coefficients, the technique is called the *method of Frobenius*. For a second-order equation you will always obtain at least one solution about a regular singular point which is a Frobenius series. To determine the precise form of a Frobenius series, first substitute the series into the differential equation; second, in the resulting algebraic expression equate to zero the coefficient of the lowest power of x; third, find the values of r which allow a_0 to be chosen arbitrarily.

example 5.12: Use the method of Frobenius with $a=0$ to solve the differential equation $2xy'' + 3y' - y = 0$.

solution: We assume the solution to be of the form

$$\sum_{n=0}^{\infty} a_n x^{n+r}.$$

Differentiating twice,

$$y' = \sum_{n=0}^{\infty} a_n (n+r)\, x^{n+r-1}$$

and

$$y'' = \sum_{n=0}^{\infty} a_n (n+r)(n+r-1)\, x^{n+r-2}.$$

Substituting into the differential equation,

$$2x \sum_{n=0}^{\infty} a_n (n+r)(n+r-1)\, x^{n+r-2} + 3 \sum_{n=0}^{\infty} a_n (n+r)\, x^{n+r-1} - \sum_{n=0}^{\infty} a_n x^{n+r} = 0.$$

Carrying out the indicated multiplications,

$$2 \sum_{n=0}^{\infty} a_n (n+r)(n+r-1)\, x^{n+r-1} + 3 \sum_{n=0}^{\infty} a_n (n+r)\, x^{n+r-1} - \sum_{n=0}^{\infty} a_n x^{n+r} = 0.$$

Shifting the index in the third series and combining the first two,

$$\sum_{n=0}^{\infty} a_n x^{n+r-1}(n+r)(2n+2r+1) - \sum_{n=1}^{\infty} a_{n-1} x^{n+r-1} = 0.$$

Writing the term corresponding to $n=0$ and combining those terms for $n \geqq 1$ into one series,

$$a_0 x^{r-1} r(2r-1) + \sum_{n=1}^{\infty} x^{n+r-1}\{a_n(n+r)(2n+2r+1) - a_{n-1}\} = 0.$$

Equating the coefficient of x^{r-1} to zero yields the equation

$$a_0 r(2r-1) = 0.$$

The value of r for which this coefficient is zero is $r=0$ or $r=1/2$. Hence, two linearly independent solutions to the given differential equation have the form

$$y_1 = \sum_{n=0}^{\infty} a_n x^n \quad \text{and} \quad y_2 = x^{1/2} \sum_{n=0}^{\infty} a_n^* x^n.$$

Since $a_n(n+r)(2n+2r+1) - a_{n-1} = 0$ for all n, we have the following information on the coefficients for the two series:

$$a_0 \text{ is arbitrary and for } n \geqq 1, \quad a_n = \frac{1}{n(2n+1)} a_{n-1};$$

$$a_0^* \text{ is arbitrary and for } n \geqq 1, \quad a_n^* = \frac{1}{n(2n-1)} a_{n-1}^*.$$

Finding the general terms of the two series is a matter of a little algebra,

$$a_n = \frac{2^n a_0}{(2n+1)!} \quad \text{and} \quad a_n^* = \frac{2^n a_0^*}{(2n)!}.$$

The two solutions are

$$y_1 = a_0 \sum_{n=0}^{\infty} \frac{2^n}{(2n+1)!} x^n, \quad y_2 = a_0^* x^{1/2} \sum_{n=0}^{\infty} \frac{2^n}{(2n)!} x^n.$$

The second of these is not a power series.

example 5.13: Use the method of Frobenius to solve the differential equation $4x^2 y'' + (3x+1)y = 0$ near the origin.

solution: Since the origin is a regular singular point, we assume the solution to be of the form of a Frobenius series

$$y = \sum_{n=0}^{\infty} a_n x^{n+r}.$$

Differentiating twice and substituting into the given equation,

$$4x^2 \sum_{n=0}^{\infty} a_n (n+r)(n+r-1) x^{n+r-2} + (3x+1) \sum_{n=0}^{\infty} a_n x^{n+r} = 0.$$

Simplifying and expanding,

$$\sum_{n=0}^{\infty} 4a_n (n+r)(n+r-1) x^{n+r} + \sum_{n=0}^{\infty} 3a_n x^{n+r+1} + \sum_{n=0}^{\infty} a_n x^{n+r} = 0.$$

Shifting the index in the second series and combining the first and the third series,

$$\sum_{n=0}^{\infty} a_n x^{n+r} \{4(n+r)(n+r-1)+1\} + \sum_{n=1}^{\infty} 3a_{n-1} x^{n+r} = 0.$$

Writing the first term of the first series and then combining,

$$a_0 x^r \{4r(r-1)+1\} + \sum_{n=1}^{\infty} x^{n+r} \{a_n (2n+2r-1)^2 + 3a_{n-1}\} = 0.$$

To assure that a_0 is left arbitrary, we choose r so that

$$4r(r-1)+1 = 0, \qquad \text{or,} \qquad r = \tfrac{1}{2}.$$

Hence, the form of one solution is

$$y = x^{1/2} \sum_{n=0}^{\infty} a_n x^n,$$

where the coefficient information in recursive form is:

$$a_0 \text{ is arbitrary and for } n \geqq 1, \qquad a_n = -\frac{3}{4n^2} a_{n-1}.$$

In this case, the method of Frobenius yields only one solution, since setting the coefficient for $n = 0$ equal to zero yields only one value of r. Hence the other solution is not a Frobenius series.

exercises for sections 5.5 and 5.6

Which points are not ordinary points of the following equations? Of the singular points, which are regular and which are irregular?

1. $x^2 y'' + xy' + y = e^x$.
2. $(x^2-1)^2 xy'' + (x+1) y' + xy = 0$.
3. $x(x-1)^3 y'' + xy' + (x-1)^2 y = 0$.
4. $xy'' + \dfrac{1}{x} y' + y = 7$.

5. $(2-x)\,y'' + xy' + \dfrac{y}{(x-2)^2} = 0.$

In each of the following equations, show that the origin is a regular singular point. Use the method of Frobenius to find at least one solution in powers of x.

6. $4xy'' + 2y' + y = 0.$
7. $x(1-x)\,y'' - 3y' + 2y = 0.$
8. $4x^2y'' - 2x(x+2)\,y' + (x+3)\,y = 0.$
9. $x(x+1)\,y'' + (5x+1)\,y' + 3y = 0.$
10. $(x^2+x)\,y'' + (x+1)\,y' - y = 0.$
11. $x^2y'' + xy' + (x^2-1)\,y = 0.$
12. $x^2y'' + xy' + (x^2-4)\,y = 0.$
13. $2x^2y'' + x(1-x)\,y' - y = 0.$
14. $xy'' + y' + xy = 0.$
15. $xy'' + y = 0.$
16. $4x^2y'' - 2x(x-2)\,y' - (3x+1)\,y = 0.$

For each of the following equations, find the form of at least one solution. Express the coefficients in terms of a recurrence relation.

17. $x(2x+1)\,y'' + 3y' - xy = 0.$
18. $3x^2y'' + (5x-x^2)\,y' + (2x^2-1)\,y = 0.$
19. $2xy'' - (1+x^3)\,y' + y = 0.$
20. $x^2(3+x^2)\,y'' + 5xy' - (1+x)\,y = 0.$

classical elementary vector analysis

6

6.1 introduction

In this chapter we shall be concerned with the kind of vector analysis that you can apply more immediately to physical problems. Specifically, we restrict ourselves either to the space of Euclidean geometric vectors with basis set $\{\mathbf{i}, \mathbf{j}, \mathbf{k}\}$ or to the space of Cartesian ordered triples with basis $\{(1, 0, 0), (0, 1, 0), (0, 0, 1)\}$. As you learned in Chapter 1, these two spaces may be used almost interchangeably if all the representative arrows are considered with their initial points at the origin of a Cartesian coordinate system. One arrow with its initial point at the origin represents all other arrows having the same direction and length; the totality of such arrows with the same direction and length is considered to be one vector.

Rather than treat the space of geometric vectors and the space of ordered triples independently, we will give the definitions in juxtaposition. We assume that most of the material is not new to you so that the parallel structuring might help if you have already seen some of the results as isolated facts.

6.2 vector algebra

In Chapter 1 you saw how addition of vectors and multiplication of vectors by a real number are defined in the spaces of geometric vectors and of ordered triples. These operations satisfy the fundamental properties which qualify each space to be called a vector space. Now we will give two other operations which each of these spaces has: the dot and the cross product. In fact, in Chapter 1 we assumed that you had at least a passing acquaintance with the dot product.

definition 6.1

1. In the space of Euclidean geometric vectors, by $\mathbf{A}\cdot\mathbf{B}$ is meant the real number $|\mathbf{A}|\,|\mathbf{B}|\cos\theta$ where θ is the angle between the vectors **A** and **B**.
2. In the space of Cartesian ordered triples by $(a_1, a_2, a_3)\cdot(b_1, b_2, b_3)$ is meant the real number $a_1b_1 + a_2b_2 + a_3b_3$. (See Figure 6.1.)

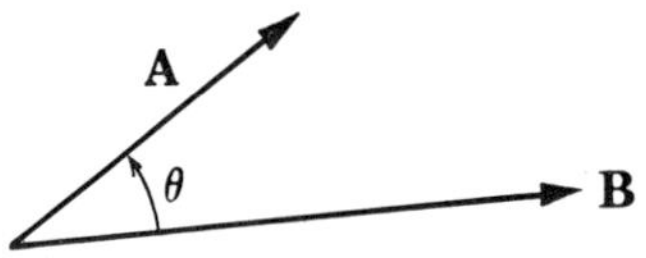

figure 6.1

The set of vectors under consideration is certainly not closed with respect to the dot product since the dot product yields a real number and not another vector. The dot product obviously has no associative property since a scalar cannot be "dotted with" a vector. However, it does qualify as an inner product, and specifically it satisfies the linearity property of Definition 1.1. (Sometimes the linearity property is referred to in this case as the distributive property over addition $\mathbf{A}\cdot(\mathbf{B}+\mathbf{C}) = \mathbf{A}\cdot\mathbf{B} + \mathbf{A}\cdot\mathbf{C}$ and the associative property $(a\mathbf{A})\cdot\mathbf{B} = a(\mathbf{A}\cdot\mathbf{B})$.) With this property we can show how to write an expression for the dot product in the space of arrows if the basis used is *not* the set $\{\mathbf{i}, \mathbf{j}, \mathbf{k}\}$.

example 6.1: Write the general formula for the dot product of two arrows if the basis used is $\{\mathbf{e}_1, \mathbf{e}_2, \mathbf{e}_3\}$.

solution: Let $\mathbf{A} = a_1\mathbf{e}_1 + a_2\mathbf{e}_2 + a_3\mathbf{e}_3$ and $\mathbf{B} = b_1\mathbf{e}_1 + b_2\mathbf{e}_2 + b_3\mathbf{e}_3$. Then

$$\mathbf{A}\cdot\mathbf{B} = (a_1\mathbf{e}_1 + a_2\mathbf{e}_2 + a_3\mathbf{e}_3)\cdot(b_1\mathbf{e}_1 + b_2\mathbf{e}_2 + b_3\mathbf{e}_3).$$

A use of the left distributive property gives

$$\mathbf{A}\cdot\mathbf{B} = \mathbf{A}\cdot(b_1\mathbf{e}_1) + \mathbf{A}\cdot(b_2\mathbf{e}_2) + \mathbf{A}\cdot(b_3\mathbf{e}_3)$$

followed by a use of the right distributive property,

$$\begin{aligned}\mathbf{A}\cdot\mathbf{B} = (a_1\mathbf{e}_1)\cdot(b_1\mathbf{e}_1) + (a_2\mathbf{e}_2)\cdot(b_1\mathbf{e}_1) + (a_3\mathbf{e}_3)\cdot(b_1\mathbf{e}_1)\\ + (a_1\mathbf{e}_1)\cdot(b_2\mathbf{e}_2) + (a_2\mathbf{e}_2)\cdot(b_2\mathbf{e}_2) + (a_3\mathbf{e}_3)\cdot(b_2\mathbf{e}_2)\\ + (a_1\mathbf{e}_1)\cdot(b_3\mathbf{e}_3) + (a_2\mathbf{e}_2)\cdot(b_3\mathbf{e}_3) + (a_3\mathbf{e}_3)\cdot(b_3\mathbf{e}_3).\end{aligned}$$

The application of the associative property to each of the nine dot products yields $(a_i\mathbf{e}_i)\cdot(b_j\mathbf{e}_j) = (a_ib_j)\,(\mathbf{e}_i\cdot\mathbf{e}_j)$, and therefore,

$$\begin{aligned}\mathbf{A}\cdot\mathbf{B} = a_1b_1(\mathbf{e}_1\cdot\mathbf{e}_1) + a_1b_2(\mathbf{e}_1\cdot\mathbf{e}_2) + a_1b_3(\mathbf{e}_1\cdot\mathbf{e}_3)\\ + a_2b_1(\mathbf{e}_2\cdot\mathbf{e}_1) + a_2b_2(\mathbf{e}_2\cdot\mathbf{e}_2) + a_2b_3(\mathbf{e}_2\cdot\mathbf{e}_3)\\ + a_3b_1(\mathbf{e}_3\cdot\mathbf{e}_1) + a_3b_2(\mathbf{e}_3\cdot\mathbf{e}_2) + a_3b_3(\mathbf{e}_3\cdot\mathbf{e}_3).\end{aligned}$$

This formula shows the interesting and useful fact that the dot product of any two vectors is known if the dot products of the elements of the basis set are known. In particular, *if the basis is orthonormal*, then

$$\mathbf{A}\cdot\mathbf{B} = a_1b_1 + a_2b_2 + a_3b_3.$$

The last formula of Example 6.1 is very important in that it shows that the relationship, described in Section 1.3, between the space of arrows and the space of ordered triples may be extended to dot products if the basis for the arrows is orthonormal. The example should also show you what influence the basis has upon the formulas of vector analysis and specifically how convenient it is to choose the basis to be orthonormal.

When we say that arrows are "useful" in physical applications we mean that statements are much more conveniently stated in vector terminology. We rarely mean that the law itself is dependent on the language of vectors. The following examples, showing the application of the concept of the dot product, may help you to understand this.

example 6.2: The concept of *scalar projection* is exemplified in Figure 6.2. Show how this may be expressed using the dot product.

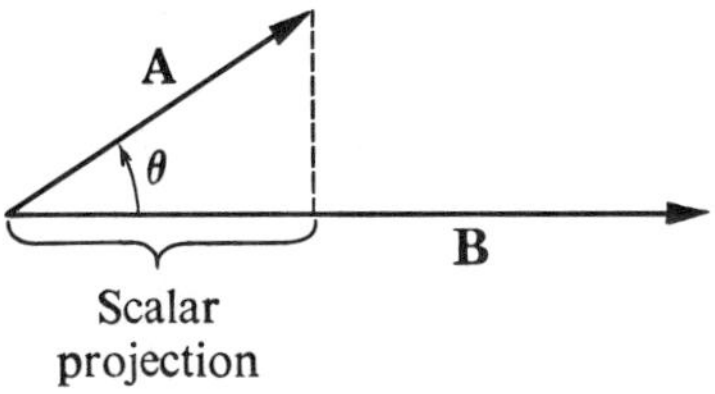

figure 6.2

solution: The scalar projection of **A** onto **B** is given by the number $|\mathbf{A}|\cos\theta$. $|\mathbf{A}|\cos\theta = |\mathbf{A}|\,(|\mathbf{B}|/|\mathbf{B}|)\cos\theta = \mathbf{A}\cdot\mathbf{u}_B$ where $\mathbf{u}_B$ is a unit vector in the direction of **B**. The number $\mathbf{A}\cdot\mathbf{u}_B$ is called the numerical component (or scalar projection) of **A** in the direction of **B**. The vector quantity $(\mathbf{A}\cdot\mathbf{u}_B)\,\mathbf{u}_B$ is called the *vector component* of **A** in the direction of **B**.

In mechanics the concept of work is defined to be the product of a force, **F**, with a displacement, **D**, multiplied by the cosine of the angle between the line of direction of **F** and of **D**. Hence, as:

$$\text{work} = |\mathbf{F}|\,|\mathbf{D}|\cos\theta = \mathbf{F}\cdot\mathbf{D}.$$

In three-dimensional analytic geometry, a plane may be defined as a surface passing through a given point, such that the direction of its normal does not vary. The use of the dot product is a nice application to describe this. Let (a_1, a_2, a_3) be a fixed point through which the plane passes and let the vector $\mathbf{N} = n_1\mathbf{i} + n_2\mathbf{j} + n_3\mathbf{k}$ be normal to the plane. Let **A** be a "position vector" to (a_1, a_2, a_3), that is, a vector from $(0, 0, 0)$ to (a_1, a_2, a_3), and let **R** be a position vector to any other point (x, y, z) on the plane. Then, the vector $\mathbf{R} - \mathbf{A}$ is in the plane and hence the angle between $\mathbf{R} - \mathbf{A}$ and **N** is 90 degrees. Therefore, $(\mathbf{R} - \mathbf{A})\cdot\mathbf{N} = 0$ is the equation of the plane.

example 6.3: Write the equation of the plane passing through $(3, -1, 4)$ to which the vector $-\mathbf{i} + \mathbf{j} + 2\mathbf{k}$ is normal.

solution: The vector **R** is given by $x\mathbf{i} + y\mathbf{j} + z\mathbf{k}$ and hence, the equation of the plane is $[(x-3)\,\mathbf{i} + (y+1)\,\mathbf{j} + (z-4)\,\mathbf{k}]\cdot(-\mathbf{i} + \mathbf{j} + 2\mathbf{k}) = 0$. Simplifying,

$$-x + y + 2z = 4.$$

The other operation which we find useful in the space of arrows and of ordered triples is the cross product.

definition 6.2

1. In the space of Euclidean geometric vectors, by $\mathbf{A} \times \mathbf{B}$ is meant a vector with magnitude $|\mathbf{A}|\,|\mathbf{B}|\sin\theta$, where θ is the angle between **A** and **B**, and with direction that of a right-hand screw in turning **A** into **B**. The cross product has no unique direction associated with the cases $\mathbf{A} = \mathbf{0}$, $\mathbf{B} = \mathbf{0}$, $\theta = 0$.

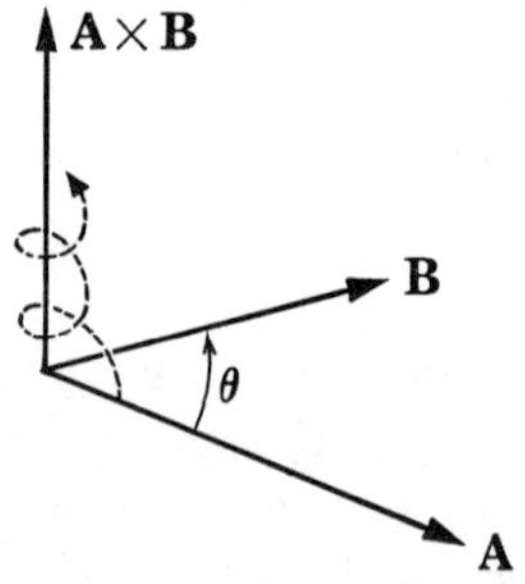

figure 6.3

2. In the space of Cartesian ordered triples,

$$(a_1, a_2, a_3) \times (b_1, b_2, b_3) = (a_2b_3 - a_3b_2, a_3b_1 - a_1b_3, a_1b_2 - a_2b_1).$$

(See Figure 6.3.)

The cross product is not an associative operation (consider $\mathbf{i} \times (\mathbf{i} \times \mathbf{j})$ and $(\mathbf{i} \times \mathbf{i}) \times \mathbf{j}$) and is anticommutative, that is, $\mathbf{A} \times \mathbf{B} = -\mathbf{B} \times \mathbf{A}$. Further, the cross product is distributive over addition, a property which, as in the case of the dot product, allows us to derive a formula for the cross product of any two vectors in terms of the components of the vectors and the cross products of the basis vectors. (See Exercise 3.) If the basis $\{\mathbf{e}_1, \mathbf{e}_2, \mathbf{e}_3\}$ is orthonormal, and if $\mathbf{e}_1 \times \mathbf{e}_2 = \mathbf{e}_3$, then

$$\begin{aligned}\mathbf{A} \times \mathbf{B} &= (a_1\mathbf{e}_1 + a_2\mathbf{e}_2 + a_3\mathbf{e}_3) \times (b_1\mathbf{e}_1 + b_2\mathbf{e}_2 + b_3\mathbf{e}_3) \\ &= (a_2b_3 - a_3b_2)\,\mathbf{e}_1 + (a_3b_1 - a_1b_3)\,\mathbf{e}_2 + (a_1b_2 - a_2b_1)\,\mathbf{e}_3.\end{aligned}$$

Once again the relationship between the space of arrows and the space of ordered triples is extended to the cross product *if an orthonormal basis is used.*

The formula for the cross product is often written in "determinant" form as

$$\mathbf{A} \times \mathbf{B} = \begin{vmatrix} \mathbf{e}_1 & \mathbf{e}_2 & \mathbf{e}_3 \\ a_1 & a_2 & a_3 \\ b_1 & b_2 & b_3 \end{vmatrix}.$$

This is not a determinant in the ordinary sense because of the vector entries in the first row. However, if we apply the rule for evaluating a third-order determinant, the result is the same as the one derived above. Hence, the determinant form for writing a cross product is strictly a device to assist you in remembering the formula for the cross product *when an orthonormal basis is used* (if $\mathbf{e}_1 \times \mathbf{e}_2 = \mathbf{e}_3$).

In calculating the magnitude of the cross product, you may find it convenient to use only dot products.

example 6.4: Express $|\mathbf{A} \times \mathbf{B}|$ in terms of dot products.

solution:

$$\begin{aligned}|\mathbf{A} \times \mathbf{B}|^2 &= |\mathbf{A}|^2\,|\mathbf{B}|^2 \sin^2\theta \\ &= |\mathbf{A}|^2\,|\mathbf{B}|^2 (1 - \cos^2\theta) \\ &= |\mathbf{A}|^2\,|\mathbf{B}|^2 - |\mathbf{A}|^2\,|\mathbf{B}|^2 \cos^2\theta \\ &= (\mathbf{A}\cdot\mathbf{A})\,(\mathbf{B}\cdot\mathbf{B}) - (\mathbf{A}\cdot\mathbf{B})^2.\end{aligned}$$

As with the dot product, the cross product may be used in some elementary physical definitions. For example it sometimes is convenient to speak of a *directed area*, that is, an area directed by a vector. The area of a parallelogram with sides **A** and **B** is equal to $|\mathbf{A}|\,|\mathbf{B}|\sin\theta$, which is $|\mathbf{A}\times\mathbf{B}|$. Hence, the directed area $\mathbf{A}\times\mathbf{B}$ is defined to be the area of the parallelogram with direction $\mathbf{A}\times\mathbf{B}$. (See Figure 6.4.)

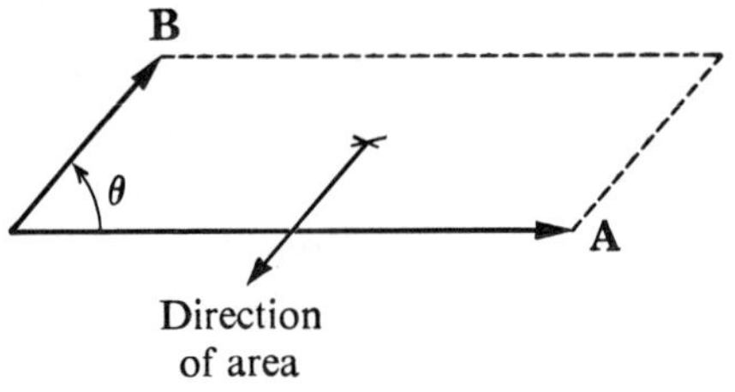

figure 6.4 *directed area*

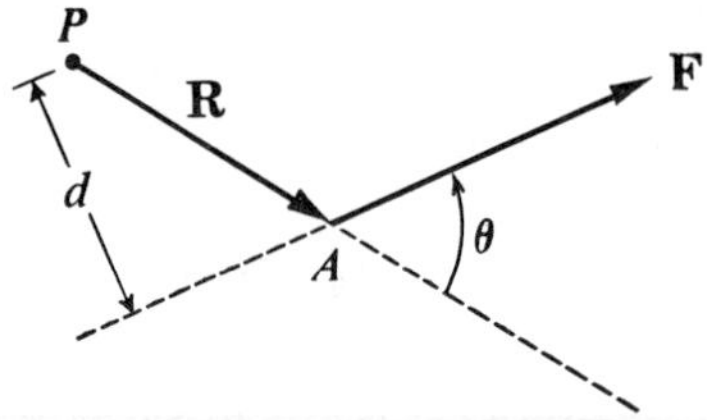

figure 6.5 *vector moment*

In mechanics, the moment of a force **F** about a point P is defined to be $m=|\mathbf{F}|\,d$ where d is the distance of P from the line of action of the force. If A is any point on the line of force and **R** is a vector from P to A, (see Figure 6.5) then $d=|\mathbf{R}|\sin\theta$ and therefore

$$m=|\mathbf{F}|\,|\mathbf{R}|\sin\theta,$$

or in terms of the cross product,

$$m=|\mathbf{F}\times\mathbf{R}|=|\mathbf{R}\times\mathbf{F}|.$$

The so-called *vector moment* is thus defined to be

$$\mathbf{M}=\mathbf{R}\times\mathbf{F}.$$

The dot and cross products may be combined in essentially two ways to give other meaningful quantities. (An expression such as $(\mathbf{A}\cdot\mathbf{B})\times\mathbf{C}$ has no meaning since $\mathbf{A}\cdot\mathbf{B}$ is a real number and the cross product is defined only for two vectors.) Products such as $\mathbf{A}\cdot(\mathbf{B}\times\mathbf{C})$ and $(\mathbf{A}\times\mathbf{B})\times\mathbf{C}$ are called *triple products*. The first is called a *scalar triple product* since from it is obtained a real number, and the second is called a *vector triple product* since it yields a vector quantity.

It is merely a matter of verifying the computations to note that if (a_1, a_2, a_3), (b_1, b_2, b_3), and (c_1, c_2, c_3) are ordered triples, then their scalar triple product is given by the value of the determinant

$$\begin{vmatrix} a_1 & a_2 & a_3 \\ b_1 & b_2 & b_3 \\ c_1 & c_2 & c_3 \end{vmatrix}.$$

Further, if **A** and **B** are arrows with components (with respect to an orthonormal basis) equal to the three ordered triples then the magnitude of $\mathbf{A} \cdot \mathbf{B} \times \mathbf{C}$ is the same as that given by the above determinant. Note that the formula would be significantly different for arrows *if the basis used were not an orthonormal set.*

The scalar triple product is important because of the following theorem.

theorem 6.1

Three vectors (either ordered triples or arrows with the components given with respect to an orthonormal basis) are linearly independent if and only if their scalar triple product is nonzero.

proof: (For arrows.) In one of the exercises you will be asked to show that the scalar triple product may be interpreted as the volume of a parallelepiped with **A**, **B**, and **C** as adjacent edges. Hence, $\mathbf{A} \cdot \mathbf{B} \times \mathbf{C}$ is zero if and only if **A**, **B**, and **C** are coplanar. (See Figure 6.6.) But a plane is a two-dimensional subspace and hence any three vectors lying in a plane are linearly

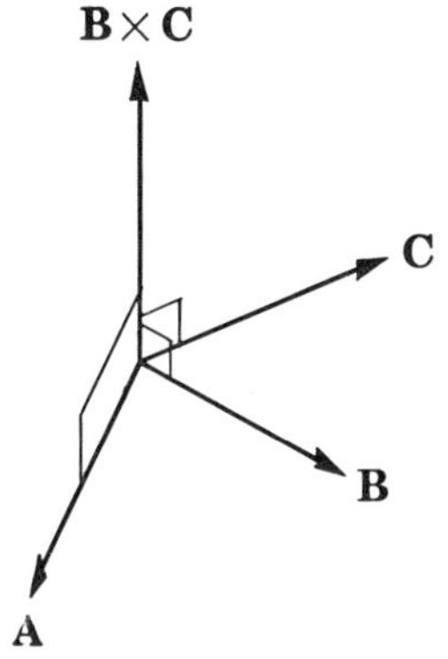

figure 6.6

dependent. Conversely, any three linearly dependent arrows must lie in the same plane. This completes the proof.

The vector triple product $(\mathbf{A} \times \mathbf{B}) \times \mathbf{C}$ can be shown to be a vector which lies in the plane of **A** and **B**. (All initial points are assumed to be at the origin.) Hence, if **A** and **B** do not point in the same direction, the quantity $(\mathbf{A} \times \mathbf{B}) \times \mathbf{C}$ can be expressed as a linear combination of the vectors **A** and **B**. This linear combination is given by the following formula:

$$(\mathbf{A} \times \mathbf{B}) \times \mathbf{C} = -(\mathbf{B} \cdot \mathbf{C})\, \mathbf{A} + (\mathbf{A} \cdot \mathbf{C})\, \mathbf{B}.$$

exercises for sections 6.1 and 6.2

1. Show that the dot product of two arrows or of two ordered triples is a commutative operation.
2. Find the dot product of $5\mathbf{i} + 3\mathbf{j} - 2\mathbf{k}$ and $-3\mathbf{i} + \mathbf{j} - \mathbf{k}$ by first expressing the two vectors in terms of the basis $\{\mathbf{i} + \mathbf{j}, \mathbf{k}, \mathbf{i} + \mathbf{k}\}$ and then using the result of Example 6.1. Repeat the process for the basis $\{3\mathbf{i} + 2\mathbf{j} + \mathbf{k}, \mathbf{i} - 5\mathbf{j} + 2\mathbf{k}, \mathbf{j} + \mathbf{k}\}$.
3. Find the general formula for the cross product of two arrows if the basis used is $\{\mathbf{e}_1, \mathbf{e}_2, \mathbf{e}_3\}$.
4. Find the equation of the plane passing through $(5, -1, 3)$ if a vector normal to the plane is $2\mathbf{i} + \mathbf{j} - \mathbf{k}$.
5. Find the work done by a constant force $\mathbf{F} = \mathbf{i} + \mathbf{j}$ when a particle moves from $(0, 0)$ to $(1, 0)$.
6. Find the distance of the plane $x + 3y - 4z = 12$ from the point $(1, 0, 3)$.
7. Find the equation of the plane which contains the point $(1, 0, 1)$ and is parallel to the plane determined by the vectors $\mathbf{i} + \mathbf{j} - \mathbf{k}$ and $2\mathbf{i} + 3\mathbf{j} + 2\mathbf{k}$.
8. Find the vector component of the vector $\mathbf{A} = \mathbf{i} + 3\mathbf{j} - 7\mathbf{k}$ in the direction of the vector $\mathbf{B} = 2\mathbf{i} - \mathbf{j} + \mathbf{k}$.
9. Find the area of the parallelogram with coordinates of vertices $(0, 0)$, $(3, 0)$, $(1, 2)$, and $(4, 2)$.
10. Find the vector moment of the force $\mathbf{F} = \mathbf{i} + 3\mathbf{j} + \mathbf{k}$ about the origin when applied at the point $(2, -1, 1)$.
11. Find the work done by a constant force $\mathbf{F} = 3\mathbf{i} - 2\mathbf{j} + \mathbf{k}$ in moving a particle from $(-1, 0, 2)$ to $(3, -2, 1)$.
12. Find $\mathbf{A} \times \mathbf{B}$ if $\mathbf{A} = 2\mathbf{e}_1 + 5\mathbf{e}_2 - \mathbf{e}_3$ and $\mathbf{B} = \mathbf{e}_1 - 2\mathbf{e}_2 - 4\mathbf{e}_3$ where $\mathbf{e}_1 \times \mathbf{e}_2 = \mathbf{i} - \mathbf{j}$, $\mathbf{e}_1 \times \mathbf{e}_3 = \mathbf{j} + \mathbf{k}$, $\mathbf{e}_2 \times \mathbf{e}_3 = \mathbf{i} + \mathbf{k}$.

13. Find $\mathbf{A}\cdot\mathbf{B}$ if $\mathbf{A} = 3\mathbf{e}_1 + 5\mathbf{e}_2 - \mathbf{e}_3$ and $\mathbf{B} = \mathbf{e}_1 - \mathbf{e}_2 + \mathbf{e}_3$ if $\mathbf{e}_1\cdot\mathbf{e}_1 = 2$, $\mathbf{e}_2\cdot\mathbf{e}_2 = 3$, $\mathbf{e}_3\cdot\mathbf{e}_3 = 5$, $\mathbf{e}_1\cdot\mathbf{e}_2 = -1$, $\mathbf{e}_1\cdot\mathbf{e}_3 = -3$, $\mathbf{e}_2\cdot\mathbf{e}_3 = 2$.
14. Determine if the vectors $\mathbf{i} - \mathbf{j} + \mathbf{k}$, $\mathbf{i} + \mathbf{j} - \mathbf{k}$, and $\mathbf{k}$ are linearly dependent or independent by using Theorem 6.1.
15. Repeat Exercise 14 with $\mathbf{i} - \mathbf{j} + 2\mathbf{k}$, $2\mathbf{i} + \mathbf{j} + \mathbf{k}$, $5\mathbf{i} - 2\mathbf{j} + 7\mathbf{k}$.
16. What is the scalar triple product if two of the vectors are equal?
17. Show that the scalar triple product may be interpreted as the volume of a parallelepiped with **A**, **B**, and **C** as adjacent edges.
18. How are $\mathbf{A}\cdot\mathbf{B}\times\mathbf{C}$ and $\mathbf{A}\cdot\mathbf{C}\times\mathbf{B}$ related? How are $\mathbf{A}\cdot\mathbf{B}\times\mathbf{C}$ and $\mathbf{B}\cdot\mathbf{C}\times\mathbf{A}$ related? Generalize.
19. Prove Theorem 6.1 for the case of ordered triples.

6.3 vector functions

Recall that a function, also called a mapping or a correspondence, from one set to another is a pairing such that for any given element X of the first set, there is at most one element of the second set which is paired with it. The second element is denoted variously by $f(X)$, $F(X)$, etc. In the study of real functions, both the first and second sets are sets of real numbers and are called the domain and range, respectively, of the function. We say that a function is "from the domain to the range."

More generally, either the domain or range, or both, may be sets consisting of elements other than real numbers. For example, the first set may be a set of arrows with initial points at the origin and the second set may be a set of real numbers, in which case the function is called a "real valued function of a vector variable." Or, the domain may be a set of real numbers and the range a set of vectors with initial points at the origin, and the function is then called a "vector valued function of one real variable."

A (geometric) vector function **F** of one real variable is usually specified by an equation of the type

$$\mathbf{F}(t) = x(t)\,\mathbf{i} + y(t)\,\mathbf{j} + z(t)\,\mathbf{k}.$$

This one vector equation is equivalent to three *parametric* equations, $x = x(t)$, $y = y(t)$, $z = z(t)$. One geometric interpretation that we will give to vector functions of one real variable is that of a mapping from the real line (in this case called the *parameter line*) to arrows whose initial points are the origin and whose terminal points are $(x(t), y(t), z(t))$.

example 6.5: If $\mathbf{F}(t) = t\mathbf{i} + t^2\mathbf{j} + t^3\mathbf{k}$, what parametric equations are equivalent to this function? What is the image of $t = 1$? Of $t = 2$?

solution: The parametric equations are $x(t) = t$, $y(t) = t^2$, $z(t) = t^3$. $\mathbf{F}(1) = \mathbf{i} + \mathbf{j} + \mathbf{k}$ and $\mathbf{F}(2) = 2\mathbf{i} + 4\mathbf{j} + 8\mathbf{k}$. (See Figure 6.7.)

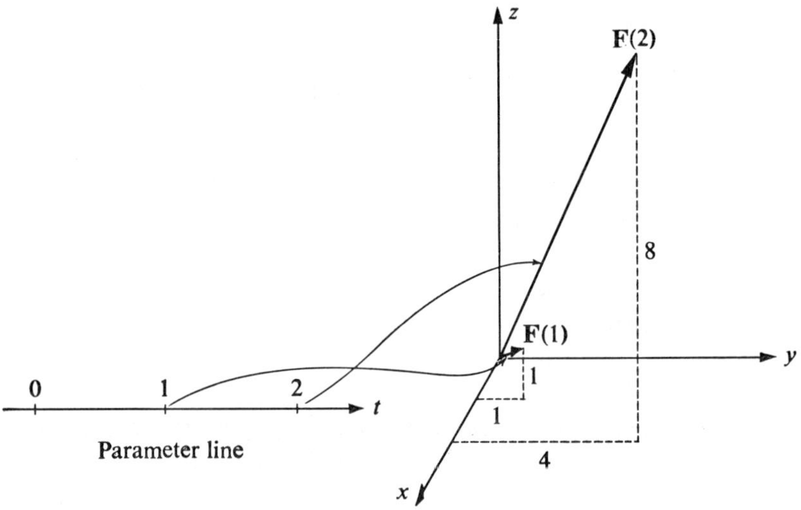

figure 6.7

A (geometric) vector function of two real variables is a mapping from a set of ordered pairs of real numbers to the set of vectors and is specified by an equation of the type

$$\mathbf{F}(u, v) = x(u, v)\,\mathbf{i} + y(u, v)\,\mathbf{j} + z(u, v)\,\mathbf{k}.$$

If we let $\mathbf{P} = (u, v)$ and $\mathbf{Q} = (x, y, z)$, then the one equation $\mathbf{Q} = \mathbf{F}(\mathbf{P})$ is equivalent to the three equations $x = x(u, v)$, $y = y(u, v)$, and $z = z(u, v)$. The three real functions $x(u, v)$, $y(u, v)$, and $z(u, v)$ are completely determined by the vector function and conversely the three real functions are sufficient to define a vector function of two real variables. As with vector functions of one variable, the geometric interpretation which will most interest us is that of a mapping from the *uv*-plane (called the *parameter plane*) to a set of arrows whose initial point is the origin.

example 6.6: Which parametric equations does the vector function $\mathbf{F}(u, v) = (u^2 + v^2)\,\mathbf{i} + u\mathbf{j} + v\mathbf{k}$ determine? What is $\mathbf{F}(1, 1)$?

solution: The parametric equations are $x(u, v) = u^2 + v^2$, $y(u, v) = u$,

$z(u, v) = v$. $\mathbf{F}(1, 1) = 2\mathbf{i} + \mathbf{j} + \mathbf{k}$. (See Figure 6.8.)

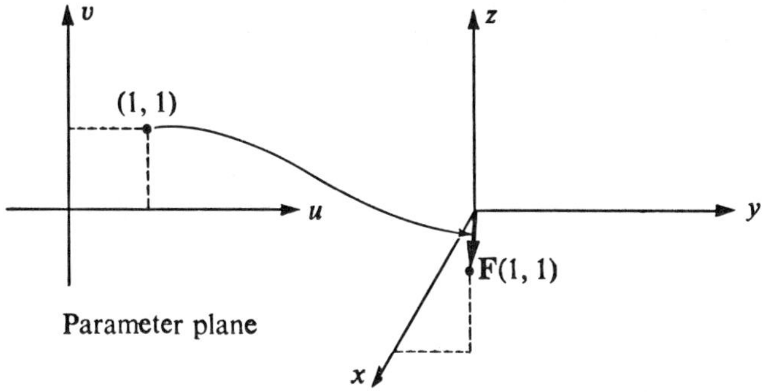

figure 6.8

6.4 vector calculus

The concepts of limit, continuity, and differentiability are introduced into vector functions analogously to real functions.

definition 6.3

A vector function $\mathbf{F}$ of a real variable t is said to have a limit $\mathbf{L}$ at t_0 if the difference

$$|\mathbf{L} - \mathbf{F}(t)|$$

is arbitrarily small whenever t is sufficiently close to t_0 but not equal to t_0. We write $\lim_{t \to t_0} \mathbf{F}(t) = \mathbf{L}$. (See Figure 6.9.)

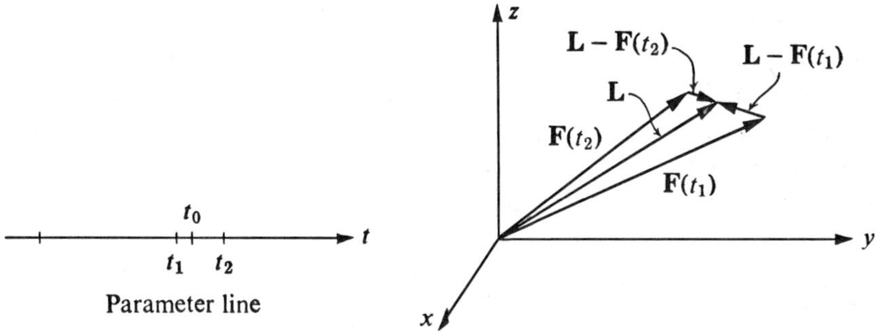

figure 6.9

definition 6.4

A vector function $\mathbf{F}$ of a real variable t is said to be *continuous* at t_0 if

$$\lim_{t \to t_0} \mathbf{F}(t) = \mathbf{F}(t_0)$$

where $\mathbf{F}(t_0)$ is presumed to be defined.

definition 6.5

A vector function $\mathbf{F}$ of a real variable t is said to have a *derivative* at t_0 (or to be *differentiable* at t_0) if

$$\lim_{t \to t_0} \frac{\mathbf{F}(t) - \mathbf{F}(t_0)}{t - t_0}$$

exists. If this limit does exist it is denoted by $\mathbf{F}'(t_0)$.

definition 6.6

A vector function $\mathbf{F}$ of two real variables (u, v) is said to be *continuous* at (u_0, v_0) if

$$\lim_{(u, v) \to (u_0, v_0)} \mathbf{F}(u, v) = \mathbf{F}(u_0, v_0),$$

where $\mathbf{F}(u_0, v_0)$ is presumed to exist.

definition 6.7

A vector function $\mathbf{F}$ of two real variables (u, v) is said to have a *partial derivative* at (u_0, v_0) if the

$$\lim_{u \to u_0} \frac{\mathbf{F}(u, v_0) - \mathbf{F}(u_0, v_0)}{u - u_0}$$

exists. If the limit does exist, it is written

$$\left.\frac{\partial \mathbf{F}}{\partial u}\right|_{u=u_0,\, v=v_0}$$

or $\mathbf{F}_u(u_0, v_0)$. $\mathbf{F}_v(u_0, v_0)$ is defined similarly.

theorem 6.2

Let $\mathbf{F}(t) = x(t)\,\mathbf{i} + y(t)\,\mathbf{j} + z(t)\,\mathbf{k}$. Then $\mathbf{F}$ is continuous (differentiable) at t_0 if and only if each of its component functions are continuous (differentiable) at t_0. Further, if $\mathbf{F}'(t_0)$ exists, then

$$\mathbf{F}'(t_0) = x'(t_0)\,\mathbf{i} + y'(t_0)\,\mathbf{j} + z'(t_0)\,\mathbf{k}.$$

A similar statement holds for vector functions of two or more real variables.

proof: See the exercises.

This theorem allows computation of the derivative of a vector valued function in terms of derivatives of the real valued component functions.

example 6.7: Let $\mathbf{F}(t) = t^2\mathbf{i} + t\mathbf{j} + t^3\mathbf{k}$. Find $\mathbf{F}'(t)$.

solution: Differentiating each of the components by well known rules from elementary calculus we obtain

$$\mathbf{F}'(t) = 2t\mathbf{i} + \mathbf{j} + 3t^2\mathbf{k}.$$

example 6.8: Find $\mathbf{r}_u \times \mathbf{r}_v$ if $\mathbf{r}(u, v) = (2 + \cos v)\cos u\,\mathbf{i} + (2 + \cos v)\,\mathbf{j}$.

solution:

$$\mathbf{r}_u = -(2 + \cos v)\sin u\,\mathbf{i}$$

$$\mathbf{r}_v = (-\sin v\cos u)\,\mathbf{i} - (\sin v)\,\mathbf{j}$$

$$\mathbf{r}_u \times \mathbf{r}_v = \begin{vmatrix} \mathbf{i} & \mathbf{j} & \mathbf{k} \\ -(2 + \cos v)\sin u & 0 & 0 \\ -\sin v\cos u & -\sin v & 0 \end{vmatrix}$$

$$= (\sin u \sin v)(2 + \cos v)\,\mathbf{k}.$$

The usual composite function rule holds and, in addition it is easy to show that if **F** and **G** are differentiable vector functions of t,

$$(\mathbf{F}\cdot\mathbf{G})' = \mathbf{F}'\cdot\mathbf{G} + \mathbf{F}\cdot\mathbf{G}' \qquad \text{and} \qquad (\mathbf{F}\times\mathbf{G})' = \mathbf{F}\times\mathbf{G}' + \mathbf{F}'\times\mathbf{G}.$$

example 6.9: Let $\mathbf{F}(t)$ be a vector function of constant magnitude. Show that $\mathbf{F}(t)$ and $\mathbf{F}'(t)$ are orthogonal for all values of t.

solution: Since $|\mathbf{F}(t)| = c$, it follows that $\mathbf{F}(t)\cdot\mathbf{F}(t) = c^2$. From the general rule for differentiating a dot product, $(\mathbf{F}\cdot\mathbf{F})' = 2\mathbf{F}\mathbf{F}'$. Hence, differentiating both sides of the equality $\mathbf{F}(t)\cdot\mathbf{F}(t) = c^2$, we obtain $2\mathbf{F}(t)\cdot\mathbf{F}'(t) = 0$. This shows that $\mathbf{F}(t)$ and $\mathbf{F}'(t)$ are orthogonal.

exercises for sections 6.3 and 6.4

1. Prove Theorem 6.2.
2. What vector function is equivalent to the set of parametric equations $x = \sin t$, $y = 2\cos t$, $z = 3$? What is the image of $t = 0$?
3. What vector function of two real variables is equivalent to the parametric equations $x = \cos u \sin v$, $y = \sin u \sin v$, $z = \cos v$? Is the function periodic? In what way?
4. Find $\mathbf{F}'(t)$ if $\mathbf{F}(t) = (\sin t)\,\mathbf{i} + (\cos^2 t)\,\mathbf{j} + t^2\mathbf{k}$.
5. Find an $\mathbf{F}(t)$ such that $\mathbf{F}'(t) = t\mathbf{i} + t^2\mathbf{j} + t^3\mathbf{k}$.

6. Show that if **F** and **G** are differentiable vector functions of t, then $(\mathbf{F}\cdot\mathbf{G})' = \mathbf{F}\cdot\mathbf{G}' + \mathbf{F}'\cdot\mathbf{G}$.
7. Show that if **F** and **G** are differentiable vector functions of t, then $(\mathbf{F}\times\mathbf{G})' = \mathbf{F}'\times\mathbf{G} + \mathbf{F}\times\mathbf{G}'$.
8. Find and prove a formula for the derivative of the scalar triple product of three differentiable vector functions of t.
9. Compute $|\mathbf{r}_u \times \mathbf{r}_v|$ when $\mathbf{r}(u, v) = (2\cos u)\,\mathbf{i} + (2\sin u)\,\mathbf{j} + v\mathbf{k}$.
10. Compute $|\mathbf{r}_u \times \mathbf{r}_v|$ when $\mathbf{r}(u, v) = x(u, v)\,\mathbf{i} + y(u, v)\,\mathbf{j} + z(u, v)\,\mathbf{k}$. (*Hint:* Use the result of Example 6.4.)

6.5 surfaces

In Section 6.3 we studied vector functions and implied that one of the primary applications for us will be to the representation of surfaces and curves. In this section we will study exactly how this application is made. Certainly, you should recall and review everything you can about surfaces and curves from elementary calculus. You are expected to have a grasp of at least some of the fundamentals of sketching surfaces. As a matter of course we will be reviewing some of the more familiar methods, and the examples should indicate about how much depth of knowledge is expected of you. Then there are a number of exercises, which could be counted as elementary calculus problems, on which to sharpen your skill.

definition 6.8

A *surface* is a set of points obtained by a mapping from a plane to three-dimensional space.

Surfaces tend to cause difficulty because of their different methods of representation. A surface can be represented (a) explicitly, (b) implicitly, and (c) vectorially or, equivalently, parametrically. To most of you, only (c) will be new.

To describe a surface explicitly, we use a function of two variables $f(x, y)$ and let $z = f(x, y)$. Then the mapping from the xy-plane into xyz-space describes a surface. Every function of two variables need not be interpreted as a surface, and indeed it is often useless to do so. The context will usually tell us if a function is being used to represent a surface.

example 6.10: Make a sketch of the first octant portion of the surface represented by $z = f(x, y) = 1 - x - y$.

solution: Since we (should) recognize $z = 1 - x - y$ as a plane, the traces in the coordinate planes and the intercepts on the coordinate axes will give an idea of the nature of the graph. The trace in the xy-plane is $x + y = 1$, in the xz-plane is $x + z = 1$, and in the yz-plane is $y + z = 1$. The intersections of these traces give the intercepts. The surface is sketched in Figure 6.10.

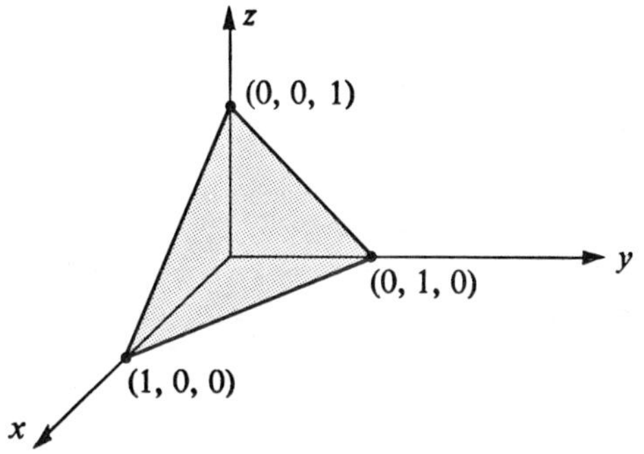

figure 6.10 $f(x, y) = 1 - x - y$

To describe a surface implicitly we use a function of three variables and equate it to a constant, $f(x, y, z) = c$. (In the next chapter we will call this a "constant level" surface.) The points with coordinates (x, y, z) which satisfy this equation represent a surface. In practice, it often turns out that the implicit relation given by $f(x, y, z) = c$ determines more than one value of z for each fixed pair (x, y). We usually by-pass this difficulty as we do for any "multiple valued" function by allowing the total surface to be the union of the surfaces defined by the several maps. If, for instance, the equation $f(x, y, z) = 0$ has two solutions z for each pair (x, y), then we can denote by $g(x, y)$ the smaller of these two solutions and by $h(x, y)$ the larger of the two. Then the implicit equation $f(x, y, z) = 0$ is equivalent to the two equations $z = g(x, y)$ and $z = h(x, y)$. The locus of $f(x, y, z) = 0$ is the union of the locus of $z = g(x, y)$ and $z = h(x, y)$.

example 6.11: Sketch the surface $f(x, y, z) = 4$ if $f(x, y, z) = x^2 + y^2 + z^2$.

solution: Solving for z we obtain two values; $z = -(4 - x^2 - y^2)^{1/2}$ is the smaller and $z = (4 - x^2 - y^2)^{1/2}$ is the larger of the two. The locus of the first one is a half sphere beneath the $z = 0$ plane and the second is a half sphere above. Hence the total locus of the given equation is a sphere of radius two centered at the origin. The surface is sketched in Figure 6.11.

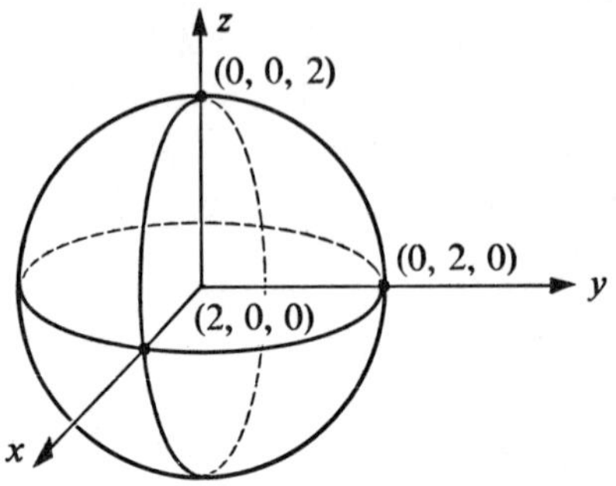

figure 6.11 $x^2 + y^2 + z^2 = 4$

When we equate a function of three variables to a constant a surface will not always result. For example, if $f(x, y, z) = x^2 + y^2 + z^2 + 1$ then $f(x, y, z) = 0$ does not define a surface since there are no ordered triples which satisfy the equation.

The use of vector functions of two variables to represent surfaces is quite natural. The trick is to think of the range of the vector function as consisting of position vectors (initial point at the origin). Then the loci of the tips of these vectors describe a surface. The vector form of a surface is not always the best for making a sketch. Sometimes we "eliminate the parameters" to obtain the relationship between x, y, and z.

example 6.12: Identify the surface defined vectorially by

$$\mathbf{r}(u, v) = u\mathbf{i} + v\mathbf{j} + (2 - u + v)\,\mathbf{k}.$$

solution: The vector equation is equivalent to the three equations $x = u$, $y = v$, and $z = 2 - u + v$, or $z = 2 - x + y$ which is easily identifiable as a plane. A normal vector to the plane at every point on the plane is $\mathbf{i} - \mathbf{j} + \mathbf{k}$.

example 6.13: Identify and sketch the surface defined by the vector function of two variables,

$$\mathbf{r}(u, v) = (\cos u)\,\mathbf{i} + (2 \sin u)\,\mathbf{j} + v\mathbf{k}.$$

solution: The vector function is equivalent to $x = \cos u$, $y = 2 \sin u$, $z = v$. Eliminating the parameter u from the first two equations is accomplished by dividing the second one by 2 and then squaring both and adding to obtain

$$x^2 + \frac{y^2}{4} = 1$$

which is an elliptic cylinder. The surface is sketched in Figure 6.12.

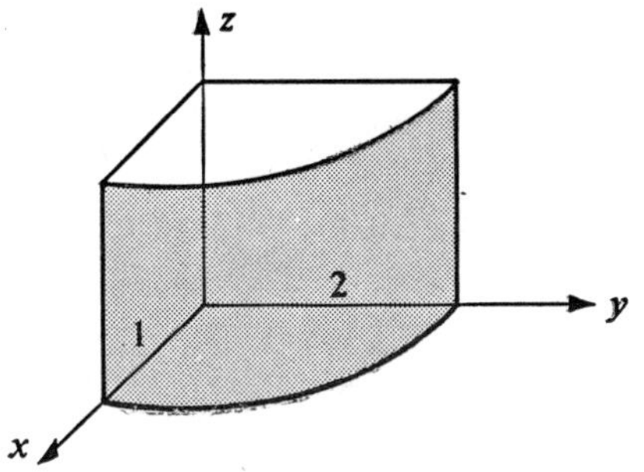

figure 6.12 $\mathbf{r}(u, v) = (\cos u)\,\mathbf{i} + (2 \sin u)\,\mathbf{j} + v\mathbf{k}$

In the process of eliminating the parameters, limitations on the values of the variables must be considered. For example, the vector functions $\mathbf{r}_1(u, v) = u\mathbf{i} + v\mathbf{j}$ and $\mathbf{r}_2(u, v) = u^2\mathbf{i} + v\mathbf{j}$ might both appear to represent the xy-plane (since $z(u, v) = 0$ in both cases). However, $\mathbf{r}_2(u, v)$ limits the values of x to those which are greater than or equal to zero. Therefore it represents only the right half of the plane.

6.6 curves

definition 6.9

A *curve* is a set of points obtained by a mapping from a line to three-dimensional space.

As with surfaces, curves can be described in several ways. We restrict ourselves to two fundamental methods: (a) as the intersection of two surfaces and (b) by a vector function of one real variable.

The representation of a curve as the intersection of two surfaces is particularly handy if you are called upon to sketch the curve, for then sometimes it is easier to get an idea of precisely "where the curve is" in space.

example 6.14: Sketch the curve given by the intersection of $4x + y + 2z = 4$ and $x^2 + y^2 = 1/4$.

solution: $4x + y + 2z = 4$ is a plane and $x^2 + y^2 = 1/4$ is a right circular cylinder whose axis is the z-axis. The first octant sketch of each is shown in Figure 6.13 along with the curve of intersection.

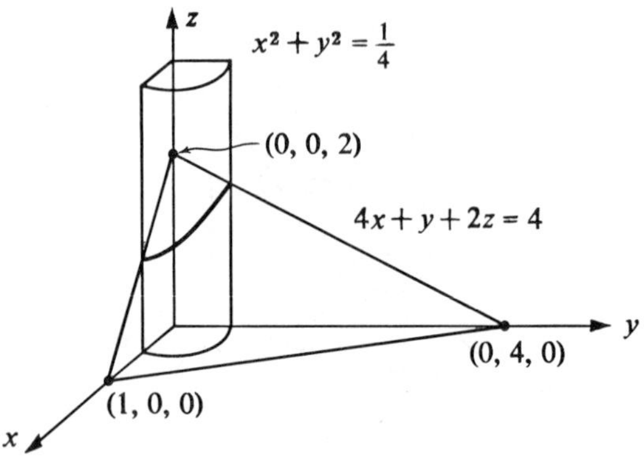

figure 6.13

When a curve is given as the intersection of two surfaces, it is often advisable to replace the given surfaces by simpler ones. Most often these will be surfaces in which one of the variables is missing (called "projecting cylinders" of the curve).

example 6.15: Sketch the line given by the intersection of the two planes $5x+4y+z=16$ and $x+4y-z=8$.

solution: By first subtracting the two equations and then adding them we find that the intersection of the given planes is the same as the pair of planes $2x+z=4$ and $3x+4y=12$. The sketch of both planes and the line of intersection is shown in Figure 6.14.

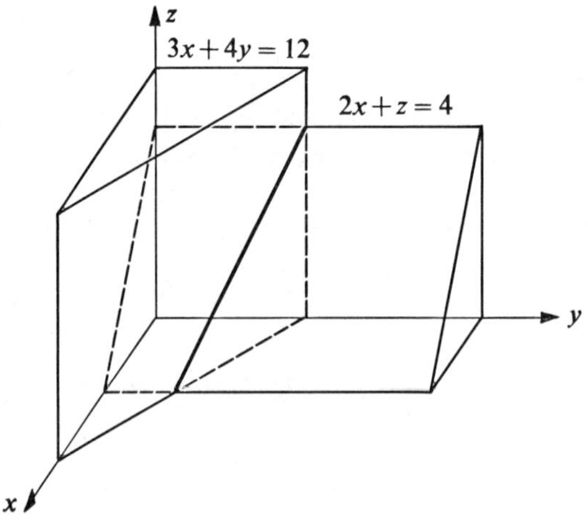

figure 6.14

example 6.16: Sketch the curve defined as the intersection of $x^2 + y^2 = 2x$ and $x^2 + y^2 + z^2 = 4$.

solution: The first of these is already a projecting cylinder. When we subtract the first from the second we obtain $z^2 = 4 - 2x$. The sketch of both surfaces and their indicated curve of intersection is shown in Figure 6.15.

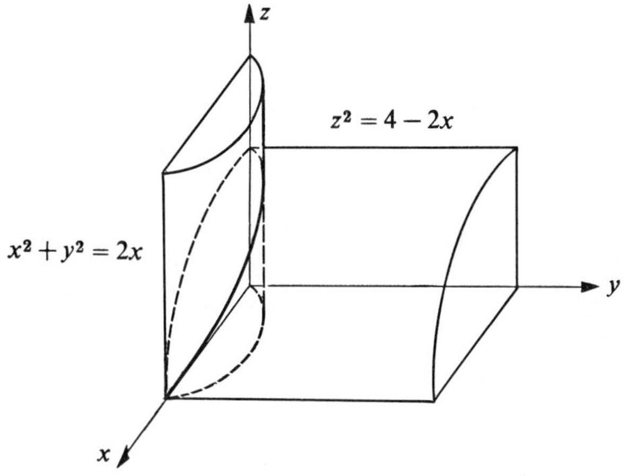

figure 6.15

If you understand how a vector function of two variables can be used to represent a surface, then it is an easy transition to use a vector function of one variable to describe a curve analytically. In succeeding sections we will see the usefulness of thinking of a vector function of one variable as being a curve. For now, we will give a few examples and discuss some related terminology.

example 6.17: Sketch the first octant portion of the curve described vectorially by $\mathbf{r}(t) = \cos t\,\mathbf{i} + \cos 2t\,\mathbf{j} + \sin t\,\mathbf{k}$.

solution: The vector function is equivalent to the three functions of t: $x(t) = \cos t$, $y(t) = \cos 2t$, $z(t) = \sin t$. The parameter t may be eliminated between the first and third of these equations by squaring and adding to obtain $x^2 + z^2 = 1$. Since $\cos 2t = 1 - 2\sin^2 t$, the elimination of the parameter from the second and third equations gives $y = 1 - 2z^2$. Thus, the given curve is the intersection of the two cylinders $x^2 + z^2 = 1$ and $y = 1 - 2z^2$. The curve is sketched in Figure 6.16.

example 6.18: Write a vector equation of a line passing through the point (x_1, y_1, z_1) and in the direction of the vector $\mathbf{M} = m_1\mathbf{i} + m_2\mathbf{j} + m_3\mathbf{k}$.

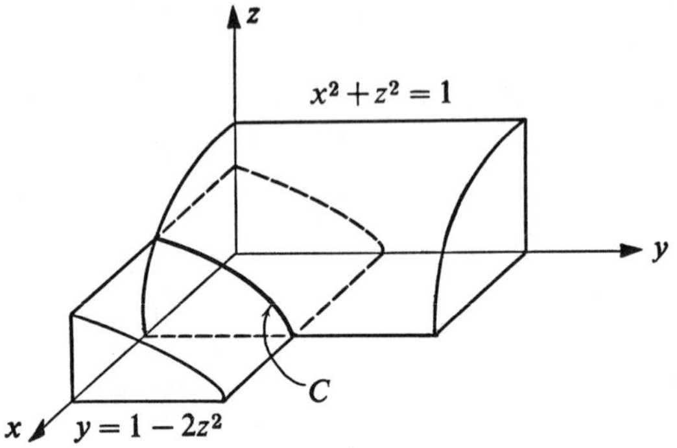

figure 6.16

solution: We seek a vector function of one variable (say $\mathbf{R}(t)$) such that for all values of t, the tip of $\mathbf{R}(t)$ is on the line. Let $\mathbf{B}$ be the vector from the origin to the point (x_1, y_1, z_1); that is, $\mathbf{B} = x_1\mathbf{i} + y_1\mathbf{j} + z_1\mathbf{k}$. Then the position vector to any point on the line is given by $\mathbf{R}(t) = \mathbf{B} + \mathbf{M}t$, which is equivalent to the three parametric equations $x = x_1 + m_1 t$, $y = y_1 + m_2 t$, $z = z_1 + m_3 t$. (See Figure 6.17.) Note that each of the component functions is linear. *This is characteristic of vector functions which represent lines.*

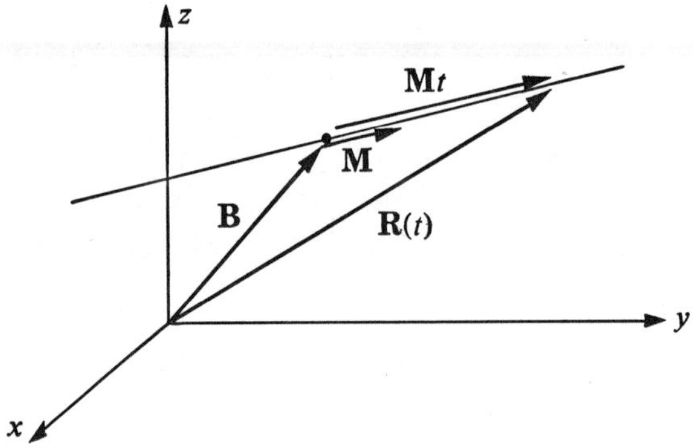

figure 6.17

example 6.19: Write a vector equation of a line passing through the points $(-1, 0, 2)$ and $(1, 4, 3)$.

solution: The vector in the direction of the line may be written $2\mathbf{i} + 4\mathbf{j} + \mathbf{k}$. Either of the two points may be chosen to give the vector $\mathbf{B}$. If $\mathbf{B} = -\mathbf{i} + 2\mathbf{k}$, the vector equation of the line is

$$\begin{aligned}\mathbf{R}(t) &= -\mathbf{i} + 2\mathbf{k} + (2\mathbf{i} + 4\mathbf{j} + \mathbf{k})\,t \\ &= (2t - 1)\,\mathbf{i} + 4t\mathbf{j} + (t + 2)\,\mathbf{k}.\end{aligned}$$

If **B** is chosen to be $\mathbf{i} + 4\mathbf{j} + 3\mathbf{k}$, then another equation of the same line (with a different parameter) is

$$\begin{aligned} \mathbf{F}(s) &= \mathbf{i} + 4\mathbf{j} + 3\mathbf{k} + (2\mathbf{i} + 4\mathbf{j} + \mathbf{k})\, s \\ &= (2s + 1)\,\mathbf{i} + 4\,(s + 1)\,\mathbf{j} + (s + 3)\,\mathbf{k}. \end{aligned}$$

A curve, $\mathbf{r}(t)$, defined on some interval of the parameter line is said to be *smooth* if $\mathbf{r}'(t)$ is continuous and nonzero on that interval. Graphically, smooth curves look just as you think they should, with no sharp corners or cusps. You will also be concerned with *piecewise smooth* curves, which are smooth except at a finite number of points. The reason for demanding a nonzero derivative on C will become clear in the next section.

The vector representation of a curve is usually not unique and a particular representation will *orient* the curve; that is, a direction for increasing t will be determined.

example 6.20: Show that the vector functions $\mathbf{r}_1(t) = (\sin t)\,\mathbf{i} + (\cos t)\,\mathbf{j}$ and $\mathbf{r}_2(t) = (\cos t)\,\mathbf{i} + (\sin t)\,\mathbf{j}$ both represent the same curve but orient it in opposite directions.

solution: By squaring and adding the component functions in each case, we see that both functions represent the circle $\{x^2 + y^2 = 1,\ z = 0\}$. For the first function, as t increases, the tip of the position vector rotates in a clockwise sense, while for the second function increasing t implies counterclockwise rotation of the tip.

example 6.21: Write an equation of the circle in the plane $z = 2$, centered at (0, 0, 2), of radius 3, oriented clockwise, as viewed from the origin.

solution: The parametric equations may be written

$$x = 3 \sin t, \qquad y = 3 \cos t, \qquad z = 2$$

and hence a vector function which represents the curve is

$$\mathbf{R}(t) = (3 \sin t)\,\mathbf{i} + (3 \cos t)\,\mathbf{j} + 2\mathbf{k}.$$

We will speak of a curve C being a *plane* curve if C lies in some plane in space; otherwise it is *twisted* or *skew*. For example, every line is a plane curve; every curve in the xy-plane is a plane curve. The following example will show you one of the most important twisted curves that you will meet.

example 6.22: Write the equation of the curve which lies on the right circular cylinder of radius a, and completes one revolution every b units.

Such a curve is called a *helix*. The particular helix we have in mind is shown in Figure 6.18.

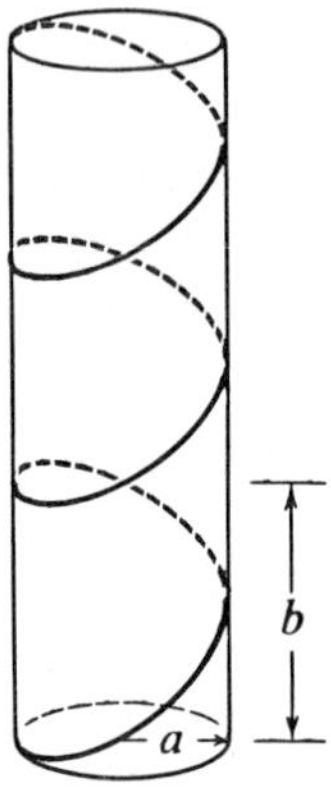

figure 6.18

solution: The projection of the helix onto the xy-plane is a circle of radius a, so that

$$x = a \cos t \qquad \text{and} \qquad y = a \sin t.$$

Since the value of z increases b units for every 2π increment of t, and in a linear fashion,

$$z = (1/2\pi)\, bt.$$

Hence, the curve may be written

$$\mathbf{r}(t) = a \cos t\mathbf{i} + a \sin t\mathbf{j} + (1/2\pi)\, bt\mathbf{k}.$$

A completely arbitrary elimination of the parameter will sometimes be misleading, for, as with surfaces, the parametric representation may limit the values of x or y or both.

example 6.23: Sketch the curve $\mathbf{r}(t) = \sin t\,\mathbf{i} + \sin t\,\mathbf{j}$.

solution: The parametric equations are $x(t) = \sin t$ and $y(t) = \sin t$, from which $x = y$. A rather careless person would conclude that the curve is the entire line $x = y$. However, the parametrization limits x and y between -1 and 1. Thus only that portion of the line $x = y$ is represented by $\mathbf{r}(t)$.

The portion of a curve between any two points on a nonintersecting part of a curve is called an *arc* of that curve. If a curve intersects itself, the points of intersection are called *multiple points.* A curve represented by a continuous one-to-one function is said to be *simple*. Note that a simple curve does not intersect itself. If a curve is simple except that its beginning and end points are the same, it is called a simple *closed* curve.

A curve is called *rectifiable* for $a \leqq t \leqq b$ if the integral

$$\int_a^b \left(\frac{d\mathbf{r}}{dt} \cdot \frac{d\mathbf{r}}{dt}\right)^{1/2} dt$$

exists, in which case the value of the integral is called the *length of arc* for that interval. The concept of rectifiability will be of little importance to us but we will have need of the *arc length function* defined by

$$s(t) = \int_a^t \left(\frac{d\mathbf{r}}{du} \cdot \frac{d\mathbf{r}}{du}\right)^{1/2} du.$$

For each value of t, the arc length function gives the length of the arc of the curve from some fixed point determined by $\mathbf{r}(a)$ to the tip of $\mathbf{r}(t)$. Occasionally, we will use the arc length, s, as the parameter in the representation of a curve.

example 6.24: Parametrize the curve $\mathbf{r}(t) = 5 \cos t\,\mathbf{i} + 5 \sin t\,\mathbf{j}$ in terms of arc length.

solution: This is a circle of radius 5 centered at the origin, and for circles of radius a the relation of angle to arc length is known to be $s = at$, or, in this case $s = 5t$. Therefore,

$$\mathbf{R}(s) = \mathbf{r}(t(s)) = 5 \cos\frac{s}{5}\,\mathbf{i} + 5 \sin\frac{s}{5}\,\mathbf{j}.$$

exercises for sections 6.5 and 6.6

1. Sketch the surface defined by:
   (a) $\mathbf{r}(u, v) = u\mathbf{i} + v^2\mathbf{j}$.
   (b) $\mathbf{r}(u, v) = 2u\mathbf{i} + v\mathbf{j} + (1 - 2u - 3v)\,\mathbf{k}$.
   (c) $\mathbf{r}(u, v) = v\mathbf{i} + (u + v - 1)\,\mathbf{j} + (u - v)\,\mathbf{k}$.
   (d) $\mathbf{r}(u, v) = (\cos u \cos v)\,\mathbf{i} + (\sin u \cos v)\,\mathbf{j} + (\sin v)\,\mathbf{k}$.
   (e) $\mathbf{r}(u, v) = (u \cos v)\,\mathbf{i} + (u \sin v)\,\mathbf{j}$.
   (f) $\mathbf{r}(u, v) = u^2\mathbf{i} + u^2\mathbf{j} + v\mathbf{k}$.
   (g) $\mathbf{r}(u, v) = (uv)\,\mathbf{i} + v\mathbf{j} + 2\mathbf{k}$.
   (h) $\mathbf{r}(u, v) = \cos u \cos v\,\mathbf{i} + \sin u \sin v\,\mathbf{j}$.
2. Sketch the surfaces:
   (a) $z = x^2 + 3y^2$. (b) $y = \sin x$. (c) $y = z^2$.

(d) $y^2+z^2=x^2$. (e) $y=2$. (f) $x+y=2$.
(g) $x^2+4y^2=4$. (h) $x=y+z$. (i) $x^2-y^2=1$.
(j) $z^2=4$. (k) $z=x^2+y$.

3. Sketch the curves defined by:

(a) $x+y+z=1$, $y+z=1$. (b) $x^2+y=^2 16$, $z=2$. (c) $x^2+y^2+z^2=4$, $x=y$.
(d) $x^2+y^2+z^2=4$, $x=1$. (e) $x^2+y^2+z^2=4$, $x=3$. (f) $x^2+y^2=4$, $z=3$.
(g) $x^2+y^2=4$, $x=2$. (h) $x^2=y$, $y=z$.

4. Sketch the curves defined by:

(a) $\mathbf{r}(t)=(\sin t)\,\mathbf{i}+(\cos t)\,\mathbf{j}+2\mathbf{k}$.
(b) $\mathbf{r}(t)=t^2\mathbf{i}+t\mathbf{j}+t\mathbf{k}$.
(c) $\mathbf{r}(t)=(t)^{1/2}\mathbf{i}+(t)^{1/2}\mathbf{j}$.
(d) $\mathbf{r}(t)=(2t-1)\,\mathbf{i}+t\mathbf{j}+(3t+1)\,\mathbf{k}$.
(e) $\mathbf{r}(t)=2\mathbf{i}+3t\mathbf{j}+t\mathbf{k}$.
(f) $\mathbf{r}(t)=t\mathbf{i}+2t\mathbf{j}$.

5. Express the surfaces of Exercise 2 in parametric form.

6. Express the curves of Exercise 3 in parametric form.

7. Express the curves of Exercise 4 as the intersection of two surfaces.

8. Write the vector equation of the line passing through the points (1, 0, 2) and (5, 1, 1).

9. Write the vector equation of the line passing through the points (1, 0, 0) and (0, 2, 0).

10. Write the vector equation of the elliptical helix which lies on a right elliptical cylinder, major semiaxis a, minor semiaxis b, and pitch c units.

11. If two vector functions represent the same line, how do they differ? (See Example 6.19.)

12. Write the vector equation of the parabola $y=x^2$, $z=0$ oriented from (2, 4, 0) to (0, 0, 0).

13. Write the vector equation of the circle centered at (0, 2, 2), of radius 2, in the plane $y=z$.

14. Write the vector equation of the plane $x-4y+2z=1$.

15. Write the vector equation of the sphere $x^2+(y-2)^2+(z-3)^2=4$.

16. Write the vector equation of the ellipsoid

$$\frac{x^2}{4}+\frac{y^2}{1}+\frac{z^2}{16}=1.$$

17. Write the vector equation of the surface $x^2 + 4y^2 - 4z^2 = 1$ in terms of trigonometric and hyperbolic functions.

6.7 tangents to curves

You have seen how to use vector functions of one real variable to represent curves. As a result, the derivative of such a vector function has a meaningful geometric interpretation. If $\mathbf{r}(t)$ represents C, then the tips of the vectors $\mathbf{r}(t_0 + \Delta t)$ and $\mathbf{r}(t_0)$ lie on the curve at points P and Q as shown in Figure 6.19. The difference between the two vectors is thus a secant vector

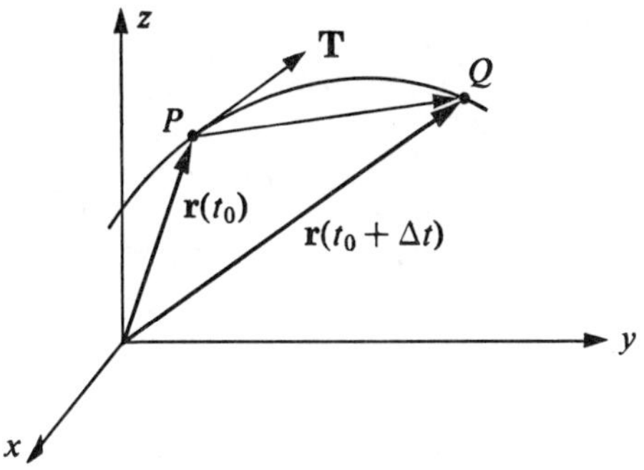

figure 6.19 *tangent and secant vectors*

pointing from P to Q. Since division by a positive real number does not affect the direction of a vector, the difference quotient

$$\frac{\mathbf{r}(t_0 + \Delta t) - \mathbf{r}(t_0)}{\Delta t}, \qquad \Delta t > 0$$

is a vector which also has direction from P to Q. As Δt approaches zero, this difference quotient vector approaches a direction tangent to C at P. Hence, if $\mathbf{r}'(t_0) \neq 0$, the direction of $\mathbf{r}'(t_0)$ will be tangential to C at P. The vector

$$\frac{\mathbf{r}'(t_0)}{|\mathbf{r}'(t_0)|}$$

is therefore a unit tangent vector to C which is denoted by $\mathbf{T}(t_0)$.

The vector $\mathbf{T}(t_0)$ tends to orient C at each point in the direction of increasing t. Notice that the definition of a smooth curve given in the preceding section essentially requires the existence of a tangent vector at each point which is a continuous function of the points on the curve. This property assures that smooth curves do not have the undesirable characteristics of corners or cusps.

example 6.25: Write the vector equation of the line tangent to the curve $\mathbf{r}(t) = \cos t\,\mathbf{i} + \sin t\,\mathbf{j}$ when $t = \pi/4$. (Write the equation of the line in the form $\mathbf{q}(w)$.)

solution: A vector tangent to C for any value of t is

$$\mathbf{r}'(t) = -\sin t\,\mathbf{i} + \cos t\,\mathbf{j},$$

and therefore a tangent vector at $\pi/4$ is $-(\sqrt{2}/2)\,\mathbf{i} + (\sqrt{2}/2)\,\mathbf{j}$. The point on the curve corresponding to $t = \pi/4$ is $(\sqrt{2}/2, \sqrt{2}/2)$. Using the result of Example 6.16 of the previous section, you can see that the equation of the line is

$$\begin{aligned}\mathbf{q}(w) &= \mathbf{r}\left(\frac{\pi}{4}\right) + \mathbf{r}'\left(\frac{\pi}{4}\right) w \\ &= \left(\frac{\sqrt{2}}{2}\mathbf{i} + \frac{\sqrt{2}}{2}\mathbf{j}\right) + \left(-\frac{\sqrt{2}}{2}\mathbf{i} + \frac{\sqrt{2}}{2}\mathbf{j}\right) w.\end{aligned}$$

(See Figure 6.20.)

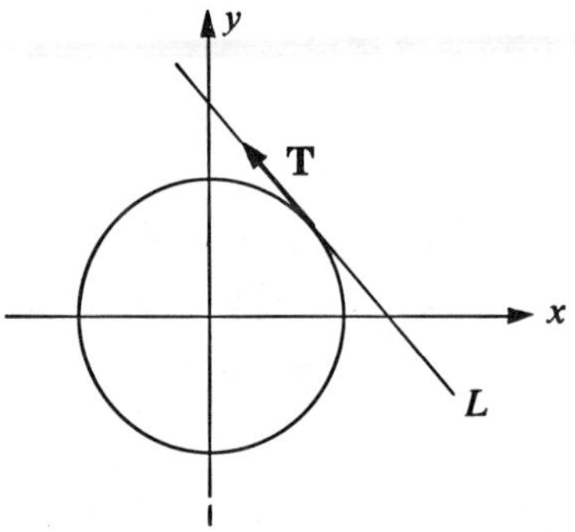

figure 6.20

If C is parametrized by arc length, s, say by the vector function $\mathbf{R}(s)$, then the difference quotient

$$\frac{\mathbf{R}(s + \Delta s) - \mathbf{R}(s)}{\Delta s}$$

has magnitude given by

$$\frac{\text{chord length}}{\text{arc length}}$$

which approaches 1 as Δs approaches 0. Hence, the unit tangent vector is given by

$$\mathbf{T} = \frac{d\mathbf{r}}{ds}.$$

Now notice that if C is parametrized by any variable, t, say by the vector function $\mathbf{r}(t)$, then the arc length is a function of t. (Precisely what this function is will not usually concern us.) Then, by the composite function rule,

$$\frac{d\mathbf{r}}{dt} = \frac{d\mathbf{r}}{ds}\frac{ds}{dt} = \mathbf{T}\frac{ds}{dt}$$

and thus

$$\left|\frac{d\mathbf{r}}{dt}\right| = \frac{ds}{dt},$$

that is, the magnitude of the first derivative is the first derivative of the arc length function with respect to the parameter, t.

Since $\mathbf{T}(t)$ is of constant magnitude, we have from Example 6.9 that $\mathbf{T}'(t)$ is perpendicular to $\mathbf{T}(t)$ for all t. The unit vector in the direction of $\mathbf{T}'(t)$ we denote by $\mathbf{N}(t)$, that is,

$$\mathbf{N}(t) = \frac{\mathbf{T}'(t)}{|\mathbf{T}'(t)|},$$

which is called the *principal normal* to the curve, C, at each point.

Sometimes, yet another unit vector, $\mathbf{B}$, is defined for every point on the curve by the equation

$$\mathbf{B}(t) = \mathbf{T}(t) \times \mathbf{N}(t).$$

The vector $\mathbf{B}$ is also normal to C at each point and is called the *unit binormal*. (See Figure 6.21.)

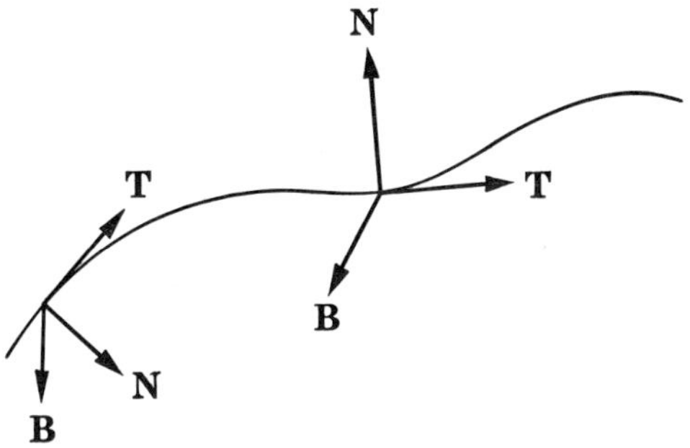

figure 6.21 *tangent, normal, and binormal*

The three unit vectors, **T**, **N**, and **B**, given by

$$\mathbf{T}(t) = \frac{\mathbf{r}'(t)}{|\mathbf{r}'(t)|}, \qquad \mathbf{N}(t) = \frac{\mathbf{T}'(t)}{|\mathbf{T}'(t)|}, \qquad \mathbf{B}(t) = \mathbf{T}(t) \times \mathbf{N}(t),$$

form an orthonormal set sometimes used as a basis for the set of arrows in three dimensions instead of $\{\mathbf{i}, \mathbf{j}, \mathbf{k}\}$. If C is nonlinear, the use of the basis $\{\mathbf{T}, \mathbf{N}, \mathbf{B}\}$ has the added complication of changing from one point to another.

example 6.26: Find **T**, **N**, and **B** at the point on the curve C corresponding to $t = 2$, for the curve described by

$$\mathbf{r}(t) = t^2\mathbf{i} + 3t\mathbf{j} - t^3\mathbf{k}.$$

solution: Since

$$\mathbf{r}(t) = t^2\mathbf{i} + 3t\mathbf{j} - t^3\mathbf{k},$$

we have that

$$\mathbf{r}'(t) = 2t\mathbf{i} + 3\mathbf{j} - 3t^2\mathbf{k},$$

so that

$$\mathbf{T}(t) = \frac{\mathbf{r}'(t)}{|\mathbf{r}'(t)|} = \frac{2t\mathbf{i} + 3\mathbf{j} - 3t^2\mathbf{k}}{(4t^2 + 9 + 9t^4)^{1/2}}.$$

To compute $\mathbf{N}(t)$ we first find $\mathbf{T}'(t)$,

$$\mathbf{T}'(t) = (2\mathbf{i} - 6t\mathbf{k})(4t^2 + 9 + 9t^4)^{-1/2} - \tfrac{1}{2}(4t^2 + 9 + 9t^4)^{-3/2}(8t + 36t^3)(2t\mathbf{i} + 3\mathbf{j} - 3t^2\mathbf{k}).$$

Evaluating at $t = 2$,

$$\mathbf{T}(2) = \frac{4\mathbf{i} + 3\mathbf{j} - 12\mathbf{k}}{13}$$

and

$$\begin{aligned}\mathbf{T}'(2) &= \frac{1}{13}\left[2\mathbf{i} - 12\mathbf{k} - \frac{152}{169}(4\mathbf{i} + 3\mathbf{j} - 12\mathbf{k})\right] \\ &= \frac{1}{13^3}(-270\mathbf{i} - 456\mathbf{j} - 204\mathbf{k}) \\ &= -\frac{6}{13^3}(45\mathbf{i} + 76\mathbf{j} + 34\mathbf{k}),\end{aligned}$$

from which

$$\begin{aligned}\mathbf{N}(2) &= \frac{\mathbf{T}'(2)}{|\mathbf{T}'(2)|} \\ &= \frac{1}{\sqrt{8957}}(-45\mathbf{i} - 76\mathbf{j} - 34\mathbf{k})\end{aligned}$$

and

$$\mathbf{B}(2) = \mathbf{T}(2) \times \mathbf{N}(2)$$
$$= \frac{-1}{13\sqrt{8957}}(1014\mathbf{i} - 676\mathbf{j} + 169\mathbf{k}).$$

If you carried out all the details of the previous example you will notice that the computation of $\mathbf{N}$ can be quite tedious. We can avoid the differentiation of the vector $\mathbf{T}$ if we are asked to find $\mathbf{N}$ at a *particular* point. We will have to find the (usually easier to compute) vector $\mathbf{r}''(t)$. To see this, recall that

$$\mathbf{r}'(t) = \mathbf{T}\frac{ds}{dt}.$$

Using the product rule,

$$\mathbf{r}''(t) = \frac{d}{dt}\left(\mathbf{T}\frac{ds}{dt}\right) = \mathbf{T}\frac{d^2s}{dt^2} + \frac{d\mathbf{T}}{dt}\frac{ds}{dt}.$$

Since $\frac{d\mathbf{T}}{dt}$ points in the direction of $\mathbf{N}$, the second derivative is always a linear combination of the normal and tangential vectors. The following example shows how this is put to use in finding a normal vector to C (from which, if desired, a unit normal could be found).

example 6.27: Find a normal vector to the curve of the previous example at the point corresponding to $t = 2$.

solution: From the given function, we obtain,

$$\mathbf{r}'(t) = \frac{d\mathbf{r}}{dt} = 2t\mathbf{i} + 3\mathbf{j} - 3t^2\mathbf{k}, \quad \mathbf{r}'(2) = 4\mathbf{i} + 3\mathbf{j} - 12\mathbf{k};$$

$$\mathbf{r}''(t) = \frac{d^2\mathbf{r}}{dt^2} = 2\mathbf{i} - 6t\mathbf{k}, \quad \mathbf{r}''(2) = 2\mathbf{i} - 12\mathbf{k}.$$

Since

$$\mathbf{T} = \frac{\mathbf{r}'(t)}{|\mathbf{r}'(t)|}, \quad \mathbf{T}(2) = \frac{4\mathbf{i} + 3\mathbf{j} - 12\mathbf{k}}{13}.$$

Also, since $\mathbf{r}''(t)$ is a linear combination of $\mathbf{T}$ and $\mathbf{N}$, $\mathbf{r}''(2) = c_1\mathbf{T}(2) + c_2\mathbf{N}(2)$ where c_1 is the scalar projection of $\mathbf{r}''(2)$ onto $\mathbf{T}$ and hence,

$$c_1 = \mathbf{r}''(2)\cdot\mathbf{T}(2) = \frac{8 + 144}{13} = \frac{152}{13}.$$

Therefore, a normal vector, $c_2\mathbf{N}$, is given by

$$\begin{aligned} c_2\mathbf{N} &= 2\mathbf{i} - 12\mathbf{k} - \tfrac{152}{13}\mathbf{T} \\ &= 2\mathbf{i} - 12\mathbf{k} - \tfrac{152}{169}(4\mathbf{i} + 3\mathbf{j} - 12\mathbf{k}) \\ &= -\tfrac{270}{169}\mathbf{i} - \tfrac{456}{169}\mathbf{j} - \tfrac{204}{169}\mathbf{k}, \end{aligned}$$

which is in the same direction as the normal obtained in the previous example.

6.8 curves on surfaces

Sometimes, you will need to know the special representation and terminology of curves which are constrained to lie on a surface defined by the vector function $\mathbf{r}(u, v)$. In particular, the kind of surface curves of most importance to us are the so-called *coordinate curves* obtained by holding either u or v constant, say $u = u_0$ or $v = v_0$. The coordinate curves are written vectorially either as $\mathbf{r}(u_0, v)$ or $\mathbf{r}(u, v_0)$.

example 6.28: Write the coordinate curves for the surface

$$\mathbf{r}(u, v) = (a \cos v \sin u)\,\mathbf{i} + (a \cos v \cos u)\,\mathbf{j} + (a \sin v)\,\mathbf{k}, \quad 0 \leqq u \leqq 2\pi, \; -\frac{\pi}{2} \leqq v \leqq \frac{\pi}{2},$$

and sketch one of each.

solution: The surface is a sphere centered at O with radius a. If u is constant, the coordinate curve is the intersection of a plane through the z-axis with the sphere; and if v is constant you will get the intersection of the sphere

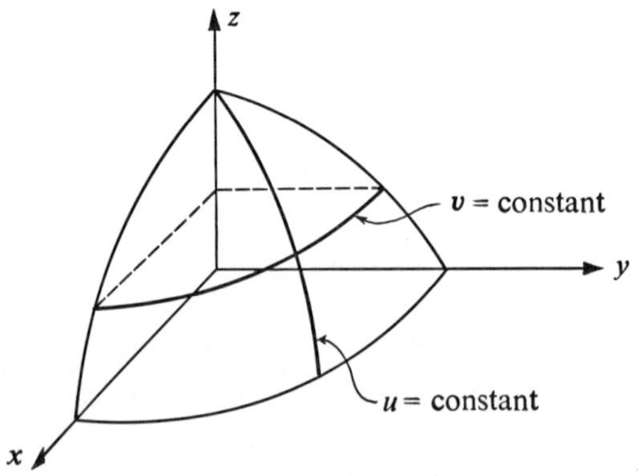

figure 6.22

with a plane parallel to the xy-plane. Written in vector form the coordinate curves are

$$\mathbf{r}(u_0, v) = a \cos v \sin u_0 \,\mathbf{i} + a \cos v \cos u_0 \,\mathbf{j} + a \sin v \,\mathbf{k},$$
$$\mathbf{r}(u, v_0) = a \cos v_0 \sin u \,\mathbf{i} + a \cos v_0 \cos u \,\mathbf{j} + a \sin v_0 \,\mathbf{k}.$$

Representative curves are sketched in Figure 6.22.

On a sphere, the coordinate curves are called meridians (if u is a constant) and parallels (if v is a constant). For example, the meridian with longitude $u_1 = (\pi/6)$ East has the equation

$$\mathbf{r}\left(\frac{\pi}{6}, v\right) = \left(\frac{a}{2}\cos v\right)\mathbf{i} + \left(\frac{a\sqrt{3}}{2}\cos v\right)\mathbf{j} + (a \sin v)\,\mathbf{k}$$

and the parallel $v_0 = (\pi/4)$ North may be represented by

$$\mathbf{r}\left(u, \frac{\pi}{4}\right) = \left(\frac{a}{\sqrt{2}}\sin u\right)\mathbf{i} + \left(\frac{a}{\sqrt{2}}\cos u\right)\mathbf{j} + \frac{a}{\sqrt{2}}\mathbf{k}.$$

example 6.29: Sketch the coordinate curves on the surface

$$\mathbf{r}(u, v) = u\mathbf{i} + v\mathbf{j} + (2 - u - v)\,\mathbf{k}.$$

solution: The surface is a plane and the coordinate curves are lines corresponding to the intersection of the plane with planes $x = u = \text{constant}$ and $y = v = \text{constant}$. Figure 6.23 shows what the situation looks like.

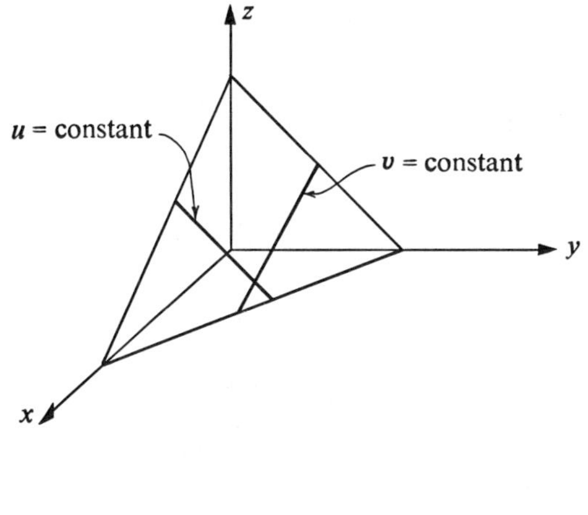

figure 6.23

example 6.30: Find and sketch the coordinate curves for the surface

$$\mathbf{r}(u, v) = (u^2 - v^2)\,\mathbf{i} + 2uv\mathbf{j}.$$

solution: The surface is obviously the plane $z = 0$ (the xy-plane). The vector equation is equivalent to the two real equations $x = u^2 - v^2$ and $y = 2uv$. If $u = u_0$, then $v = y/2u_0$ and $x = u_0^2 - (y^2/4u_0^2)$. If $v = v_0$, $x = (y^2/4v_0^2) - v_0^2$. In both cases we obtain parabolas with the vertex on the x-axis, but in one case the parabola opens to the left and in the other case it opens to the right, as shown in Figure 6.24.

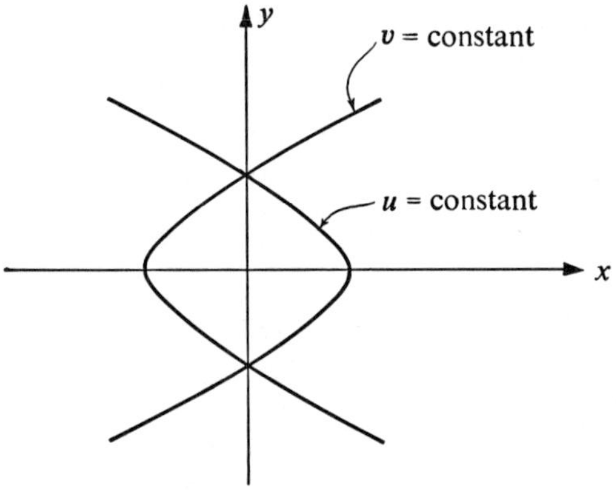

figure 6.24

example 6.31: Let a surface be defined by

$$\mathbf{r}(u, v) = (u^2 - v)\mathbf{i} + 2uv\mathbf{j} + u^3v^2\mathbf{k}.$$

Find a normal to the surface at the tip of $\mathbf{r}(u_0, v_0)$.

solution: $\mathbf{r}_u(u_0, v_0)$ is a vector tangential to the curve $\mathbf{r}(u, v_0)$ at the tip of $\mathbf{r}(u_0, v_0)$. Similarly, $\mathbf{r}_v(u_0, v_0)$ is tangential to the coordinate curve $\mathbf{r}(u_0, v)$ at that point. Therefore a vector normal to the surface is

$$\mathbf{r}_u(u_0, v_0) \times \mathbf{r}_v(u_0, v_0).$$

In this case

$$\mathbf{r}_u(u_0, v_0) = 2u_0\mathbf{i} + 2v_0\mathbf{j} + 3u_0^2v_0^2\mathbf{k}$$

and

$$\mathbf{r}_v(u_0, v_0) = -\mathbf{i} + 2u_0\mathbf{j} + 2u_0^3v_0\mathbf{k}.$$

Hence a normal vector is

$$-2u_0^3v_0^2\mathbf{i} - (3u_0^2v_0^2 + 4u_0^4v_0)\mathbf{j} + (4u_0^2 + 2v_0)\mathbf{k}.$$

The normal vector to a surface, S, is often used to define elementary properties of S. For example, if a unit normal is constant for the entire surface, this an essential property of a plane. If a normal vector determines a unique positive direction as it moves continuously over the surface, the surface is said to be *orientable*. We say that S is oriented by choosing one of the two possible normal directions to be positive. In this sense the normal vector to a surface plays the same role as the tangent vector to a curve.

A well known example of a nonorientable surface is a Möbius strip, shown in Figure 6.25. A model of a Möbius strip can be made by taking a

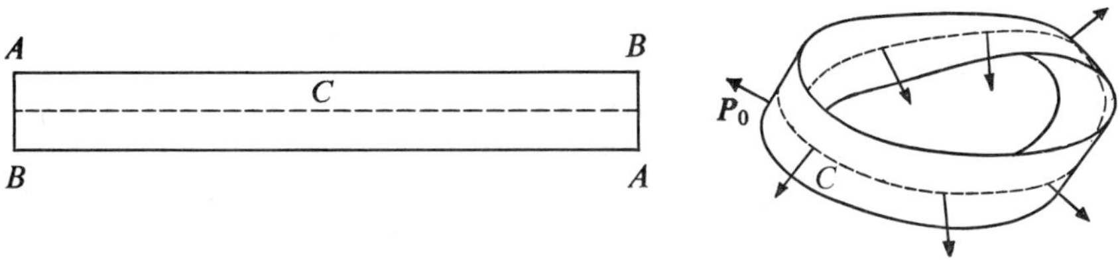

figure 6.25 *möbius strip*

rectangular piece of paper, twisting it, and sticking the shorter sides together so that A corresponds to A and B corresponds with B as in the figure. When a normal vector at P is displaced continuously along C until returning to P, it will be pointing in the opposite direction of its initial orientation.

If a surface is described by the vector function $\mathbf{r}(u, v)$ a normal is found by computing $\mathbf{r}_u \times \mathbf{r}_v$. Consequently, we usually demand that this quantity be nonzero. Further, we will restrict our discussions to those orientable surfaces for which the quantity $\mathbf{r}_u \times \mathbf{r}_v$ is a continuous function of u and v (or is at least piecewise continuous). Surfaces for which $\mathbf{r}_u \times \mathbf{r}_v$ is nonzero and continuous are said to be *smooth*. For example, the surface of a sphere is smooth while the surface of a cube is piecewise smooth.

6.9 dynamics of a particle

If a particle moves in space you may describe its motion with a vector function of one real variable, which in most cases is time. The path of the particle is a curve C and thus we can apply the results of the previous sections, especially that of Section 6.7, to analyze the movement.

The *velocity* and *acceleration* are defined to be the first and second derivatives with respect to time of the vector function,

$$\mathbf{v} = \frac{d\mathbf{r}}{dt} = \frac{dx}{dt}\mathbf{i} + \frac{dy}{dt}\mathbf{j} + \frac{dz}{dt}\mathbf{k}, \qquad \mathbf{a} = \frac{d^2\mathbf{r}}{dt^2} = \frac{d^2x}{dt^2}\mathbf{i} + \frac{d^2y}{dt^2}\mathbf{j} + \frac{d^2z}{dt^2}\mathbf{k}.$$

Since, from Section 6.7, $(d\mathbf{r}/dt) = \mathbf{T}(ds/dt)$, the direction of the velocity vector is always tangential to the path of the particle. The factor ds/dt, which is the magnitude of the velocity, is called the *speed* and is obviously equal to the square root of the sum of the squares of the component velocities,

$$\frac{ds}{dt} = \left[\left(\frac{dx}{dt}\right)^2 + \left(\frac{dy}{dt}\right)^2 + \left(\frac{dz}{dt}\right)^2\right]^{1/2}.$$

Also, we know that the second derivative may be written as the sum of a tangential and a normal component,

$$\begin{aligned}\mathbf{a} &= \frac{d^2s}{dt^2}\mathbf{T} + \left(\frac{ds}{dt}\right)^2 \kappa\mathbf{N}, \left(\kappa = \left|\frac{d\mathbf{T}}{ds}\right|\right) \\ &= \mathbf{a}_{\mathrm{T}} + \mathbf{a}_{\mathrm{N}},\end{aligned}$$

where $\mathbf{a}_{\mathrm{T}} = (\mathbf{a}\cdot\mathbf{T})\,\mathbf{T}$ and $\mathbf{a}_{\mathrm{N}} = (\mathbf{a}\cdot\mathbf{N})\,\mathbf{N}$. (See Figure 6.26.)

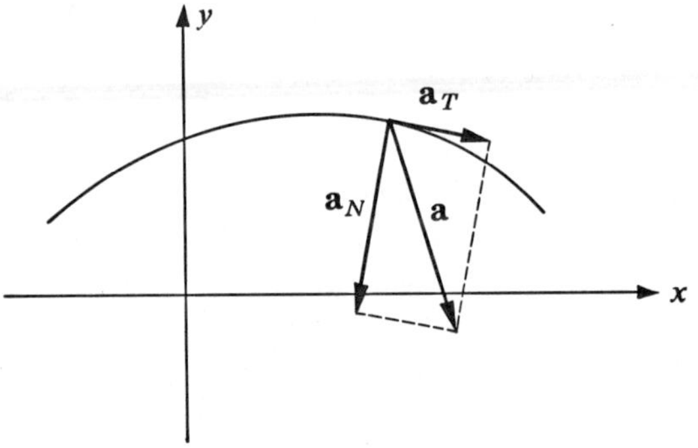

figure 6.26 *components of acceleration*

The scalar component normal to the path of the particle is often called the *centripetal acceleration*, and if this vanishes the value of the acceleration reduces to that obtained for rectilinear motion, d^2s/dt^2. You should be able to see how it is possible for a particle to move with constant speed and yet have nonzero acceleration.

example 6.32: The path of a particle is described by

$$\mathbf{r}(t) = \cos \pi t\,\mathbf{i} + \sin \pi t\,\mathbf{j} + t\mathbf{k}.$$

Find the velocity and acceleration vectors at $t = 1$. Also compute the speed, the unit tangent, normal, and binormal vectors corresponding to that value of t.

solution: Since $\mathbf{r}(t) = \cos \pi t\, \mathbf{i} + \sin \pi t\, \mathbf{j} + t\mathbf{k}$,

$$\mathbf{v}(t) = -\pi \sin \pi t\, \mathbf{i} + \pi \cos \pi t\, \mathbf{j} + \mathbf{k}$$
$$\mathbf{a}(t) = -\pi^2 \cos \pi t\, \mathbf{i} - \pi^2 \sin \pi t\, \mathbf{j}.$$

Evaluating at $t = 1$,

$$\mathbf{v}(1) = -\pi\mathbf{j} + \mathbf{k}, \qquad \text{and} \qquad \mathbf{a}(1) = \pi^2\, \mathbf{i},$$

from which the speed is π^2. The unit tangent vector

$$\mathbf{T} = \frac{\mathbf{v}(1)}{|\mathbf{v}(1)|} = \frac{-\pi\mathbf{j} + \mathbf{k}}{(\pi^2 + 1)^{1/2}}.$$

There are two, more or less obvious, approaches to obtain $\mathbf{N}$. Since $\mathbf{a}_\mathrm{T} = (\mathbf{a} \cdot \mathbf{T})\, \mathbf{T}$,

$$\mathbf{a}_\mathrm{N} = \mathbf{a} - (\mathbf{a} \cdot \mathbf{T})\, \mathbf{T} = \pi^2\mathbf{i} - 0 = \pi^2\mathbf{i}$$

and hence

$$\mathbf{N} = \frac{\mathbf{a}_\mathrm{N}}{|\mathbf{a}_\mathrm{N}|} = \mathbf{i}.$$

To obtain $\mathbf{N}$, we could have also used the fact that since $\mathbf{T}$ is a constant, $d\mathbf{T}/dt$ points in the direction of $\mathbf{N}$ (see Example 6.9), that is, we must perform the computation

$$\frac{d}{dt}\left(-\frac{\pi \sin \pi t\, \mathbf{i} + \pi \cos \pi t\, \mathbf{j} + \mathbf{k}}{(\pi^2 + 1)^{1/2}}\right)$$

and evaluate it at $t = 1$.

Since $\mathbf{T}$ and $\mathbf{N}$ are both known,

$$\mathbf{B} = \mathbf{T} \times \mathbf{N} = \frac{\mathbf{j} + \pi\mathbf{k}}{(\pi^2 + 1)^{1/2}}.$$

The path (or trajectory) may be described by a parametric representation using a parameter other than time. The most relevant case is when the trajectory is given in terms of arc length and the dependency of s on time is given by $s = h(t)$. In this way, the characteristics of the path, which are implicitly known from the path description $\mathbf{r}(s)$, are separated from the speed characteristics of the particle traversing the path. For example, the function $\mathbf{r}(s)$ might describe the properties of a road, while the function $h(t)$ would give the time behavior of the engine of the machine on the road just as a speedometer does.

Sometimes you are given the velocity vector, not necessarily in terms of time, and are asked to find the law governing the path of the particle.

example 6.33: Find an expression for the path of a particle whose velocity vector is $x\mathbf{i} + y\mathbf{j} + \mathbf{k}$.

solution: Let the position vector function which gives the path of the particle be denoted by

$$\mathbf{r}(t) = x(t)\,\mathbf{i} + y(t)\,\mathbf{j} + z(t)\,\mathbf{k}.$$

Then we have the following three differential equations to be satisfied simultaneously:

$$\frac{dx}{dt} = x, \qquad \frac{dy}{dt} = y, \qquad \frac{dz}{dt} = 1,$$

from which we obtain

$$x = c_1 e^t, \qquad y = c_2 e^t, \qquad z = t + c_3.$$

Therefore,

$$\mathbf{r}(t) = c_1 e^t \mathbf{i} + c_2 e^t \mathbf{j} + (t + c_3)\,\mathbf{k}.$$

The problem of finding the path if the velocity is known always reduces to the solution of three simultaneous ordinary differential equations. Generally, this may not be particularly easy but in this book you will not meet problems which cannot be done by the elementary methods of Section 3.11. In practice, we sometimes are satisfied to know that a solution exists even though we do not know the specific form.

exercises for sections 6.7–6.9

1. Find the vector equation of the line tangent to the curve $\mathbf{r}(t) = t^2\mathbf{i} + t\mathbf{j} + t^3\mathbf{k}$ at the point (4, 2, 8).
2. Find the unit tangent, normal, and binormal to the curve $\mathbf{r}(t) = t\mathbf{i} + t^2\mathbf{j} + t^3\mathbf{k}$ at (1, 1, 1).
3. Find the unit tangent, normal, and binormal to the curve $\mathbf{r}(t) = (\sin t)\,\mathbf{i} + (\cos t)\,\mathbf{j} + t\mathbf{k}$ at $t = 2\pi$.
4. Find the vector equation of the line tangent to the curve of Exercise 3, at $t = 2\pi$.

5. Find the coordinate curves on the surfaces of Exercise 1 of the previous exercise set.

6. Find a normal to the surface $\mathbf{r}(u, v) = 2u\mathbf{i} + v\mathbf{j} + (1 - 2u - 3v)\,\mathbf{k}$ at the point corresponding to (1, 2).

7. Find a normal to the surface $\mathbf{r}(u, v) = u^2\mathbf{i} + (uv - v^2)\,\mathbf{j} + (v^3 - 3u)\,\mathbf{k}$ at the point corresponding to $(-1, 1)$.

8. Find the velocity and acceleration at $t = 2$ of a particle which moves according to the law $\mathbf{r}(t) = (\sin^2 \pi t)\,\mathbf{i} + (\cos \pi t)\,\mathbf{j} + t^2\mathbf{k}$. What is the speed at that time? What are the tangential and normal vector components of acceleration at that time? Repeat for $t = 1/4$.

9. The velocity vector of a particle is given by $\mathbf{v} = xy\mathbf{i} + y\mathbf{j} + z\mathbf{k}$. Find the path of the particle as a function of time.

10. Same as Exercise 9 except use $\mathbf{v} = t^2\mathbf{i} + t\mathbf{j} - t^3\mathbf{k}$.

11. Same as Exercise 10 except use $\mathbf{v} = -2y\mathbf{i} + 2x\mathbf{j}$.

differentiation of fields

7

7.1 introduction

In this chapter you will learn some of the analysis used in discussing scalar and vector fields. This analysis has come to be known as "field theory" and is not restricted to any particular branch of engineering. Technically, Chapter 8 will also be on field theory; the difference is that in this chapter we will consider only differentiation of our fields, while the material of the next chapter will expose you to the integration of these fields.

7.2 scalar and vector fields

A *scalar field* is a function, f, whose domain is some subset of the points in three-dimensional space and whose range is a set of real numbers. Physically, this means that to each point, P, in space is assigned a real number. Temperature of some solid and density of some material are two ready examples of scalar fields. If f is a scalar field, then the equation which describes $f(P)$ is usually dependent upon some choice of coordinate system, but the actual value at P must be the same regardless of which coordinate system is used. We say that f must be *invariant* under changes of the coordinate system.

Perhaps the most important kind of scalar field is a *potential* field. Potential is a measure of work per some kind of unit. There are two basic kinds of potential: (a) *gravitational potential* which measures the work required to raise a unit mass from sea level to a certain altitude and (b) *electric potential* which measures the work per unit charge required to transport a test charge from one point to another. The potential at each point is a number (scalar) and not a vector.

A *vector field*, represented by $\mathbf{F}(P)$, is a function whose domain is some subset of the points in three-dimensional space and whose range is some subset of the set of arrows in three dimensions. Thus, at each point P, three numbers (the components) are required to describe the arrow. In other words, three functions (called the *component functions*) are required to specify the vector field. We write $\mathbf{F}(P) = f(x, y, z)\,\mathbf{i} + g(x, y, z)\,\mathbf{j} + h(x, y, z)\,\mathbf{k}$ to describe the field if the basis $\{\mathbf{i}, \mathbf{j}, \mathbf{k}\}$ is being used. More generally, the three component functions of a vector field must transform according to a definite law guaranteeing that the components with respect to a different basis will determine the same vector. We will not usually stop to prove that a given set of three functions qualifies as a vector field, for such is beyond the scope of this book. The best way to assure yourself that a given set of components is a vector field is to give a coordinate free definition.

In passing, we note that scalar fields and vector fields hardly exhaust the class of quantities of interest in applied mathematics and physics. In fact, there are quantities of a more complicated structure called tensors whose specification at a given point requires more than three numbers. These are discussed (albeit in an elementary way) in Chapter 9.

In describing a vector field, a vector arrow is given for each point in space. The arrow and the point go together; we do not move the arrow to the origin.

example 7.1: Describe the velocity field of a rotating body.

solution: Let $\boldsymbol{\omega}$ be a vector pointing in the direction of the axis of rotation with magnitude equal to the angular speed of the body. The vector $\boldsymbol{\omega}$ is

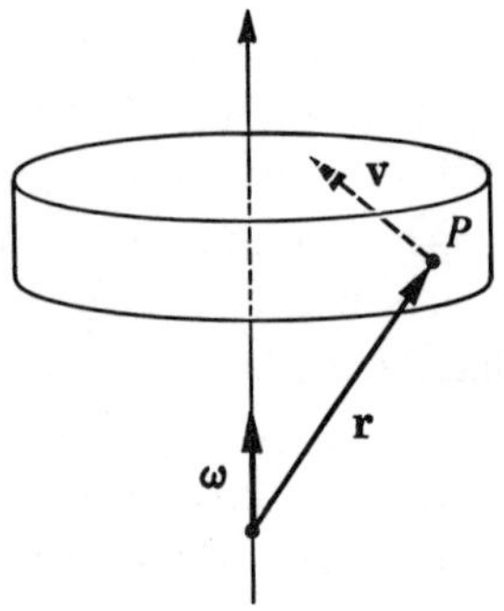

figure 7.1 *velocity field of rotating body*

called the *angular rotation vector*. If $\mathbf{r}$ is a vector from some point on the axis of rotation to a point P on the body, then since the magnitude of the velocity is $|\boldsymbol{\omega}|\,|\mathbf{r}|\sin\theta$ and the direction of motion is tangential, the velocity at that point may be represented by

$$\mathbf{v}(P) = \boldsymbol{\omega} \times \mathbf{r}.$$

(See Figure 7.1.) For example, if the angular rotation vector is $\boldsymbol{\omega} = 3\mathbf{k}$, then

$$\mathbf{v}(P) = -3y\mathbf{i} + 3x\mathbf{j}.$$

example 7.2: Describe the gravitational field due to a particle of mass M at a fixed point P_0. See Figure 7.2.

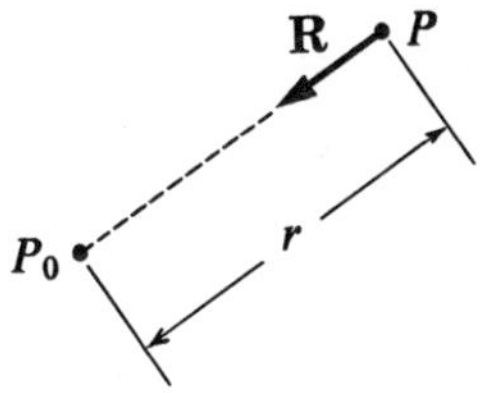

figure 7.2 *gravitational field*

solution: By the law of gravitation, the force at a point P is directed from P to P_0 and has magnitude inversely proportional to the square of the distance from P to P_0. If $\mathbf{R}$ is a unit vector directed from P to P_0, and r is the distance from P to P_0, then the vector $\mathbf{F}$,

$$\mathbf{F} = c\,\frac{\mathbf{R}}{r^2} \quad (c \text{ is a constant}),$$

describes the force field at every point P. In terms of x, y, z coordinates this would be

$$\mathbf{F}(x, y, z) = c\,\frac{-(x-x_0)\,\mathbf{i} - (y-y_0)\,\mathbf{j} - (z-z_0)\,\mathbf{k}}{[(x-x_0)^2 + (y-y_0)^2 + (z-z_0)^2]^{3/2}}$$

where c is a constant dependent upon the units chosen.

This is an example of a *central force field* since the field is directed toward one central point. Inverse square law fields and central fields are both of prime importance in physics and engineering applications.

example 7.3: Describe the electric field due to a positive charge of magnitude Q_1 located at the origin.

solution: The electric field **E** is defined to be the force per unit charge (newtons per coulomb) at a point. If a charge Q_2 is brought into the neighborhood of the origin it is acted upon by a force directed away from the origin (if Q_2 is positive). From electromagnetic theory we know that this force is given by

$$\mathbf{F} = \frac{Q_1 Q_2}{4\pi\varepsilon_0} \frac{\mathbf{R}}{r^2}, \qquad (\varepsilon_0 = \text{permittivity of vacuum})$$

where if mks units are used the value of ε_0 is $10^{-9}/36\pi$. Since **E** is the force per unit charge, it is given by

$$\mathbf{E} = \frac{Q_1}{4\pi\varepsilon_0} \frac{\mathbf{R}}{r^2}.$$

example 7.4: A negative charge of 10^{-8} coulomb is situated in the air at the origin. Find the electric field vector, **E**, at a point on the x-axis 3 meters from the origin.

solution: We use the result of the previous example with $\mathbf{R} = \mathbf{i}$, $r = 3$, $Q_1 = -10^{-8}$, and $\varepsilon_0 = 10^{-9}/36\pi$. Therefore

$$\mathbf{E}(3, 0) = -\mathbf{i} \frac{10^{-8}}{4\pi \dfrac{10^{-9}}{36\pi}} \frac{1}{9}$$

$$= -10\mathbf{i} \text{ newtons per coulomb}.$$

example 7.5: Find $\mathbf{E}(1, 0)$ if a charge of -2×10^{-9} coulombs is located one meter above the origin and another of magnitude 10^{-9} is located at the origin. (See Figure 7.3.)

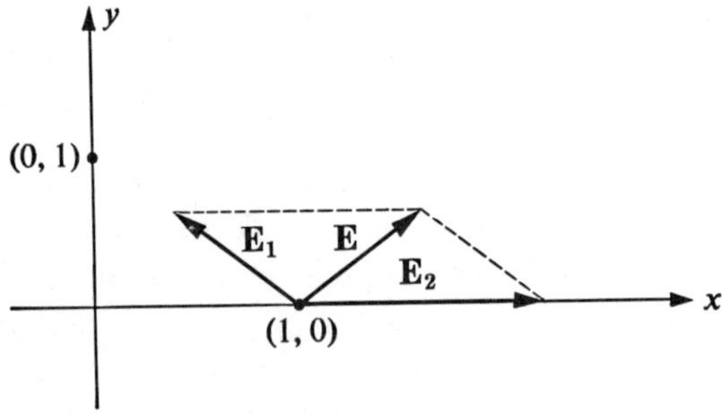

figure 7.3 *electric field*

solution: Let $\mathbf{E}_1$ be the electric field at (1, 0) due to the charge at (0, 1) and let $\mathbf{E}_2$ be the field due to the charge at the origin. The magnitude of $\mathbf{E}_1$ is given by

$$|\mathbf{E}_1| = \frac{2 \times 10^{-9}}{4\pi \dfrac{10^{-9}}{36\pi}} \frac{1}{2}$$

$$= 9 \text{ newtons per coulomb.}$$

Since the charge is negative, the field is directed from (1, 0) to (0, 1). Hence the unit vector, $\mathbf{R}$, is given by

$$\mathbf{R} = -\frac{\sqrt{2}}{2}\mathbf{i} + \frac{\sqrt{2}}{2}\mathbf{j}$$

and therefore, $\mathbf{E}_1(1, 0)$ is given by

$$\mathbf{E}_1 = \frac{9}{\sqrt{2}}(-\mathbf{i} + \mathbf{j}).$$

It is left for you to show that $\mathbf{E}_2 = 9\mathbf{i}$. The total electric field is obtained by adding the fields caused by each charge,

$$\begin{aligned} \mathbf{E} &= \mathbf{E}_1 + \mathbf{E}_2 \\ &= -6.36\mathbf{i} + 6.36\mathbf{j} + 9\mathbf{i} \\ &= 2.64\mathbf{i} + 6.36\mathbf{j} \text{ newtons per coulomb.} \end{aligned}$$

7.3 sketching scalar and vector fields

To give an idea of the variation of a scalar field we sometimes use a technique called *level surfaces*. (In two dimensions, *level curves*). A level surface of a scalar field f is the set of points with coordinates (x, y, z) such that $f(x, y, z) =$ a constant. Depending upon the application, you will hear level surfaces given other names. For example, if the scalar function is temperature, the level surfaces are called *isothermal surfaces*. If the scalar function is electric potential, then the constant level surfaces are called *equipotential surfaces*. If you carefully examine the definitions, you can see that the work required to move a charge along an equipotential surface is zero. We will return to this idea in the next section, but in a more general setting.

example 7.6: Sketch the level curves of the scalar function $f(x, y) = x - 2y + 5$ and the level surfaces of the scalar function $g(x, y, z) = x + 2y + z$.

solution: (a) The level curves of the first function are straight lines $x - 2y + 5 = c$. See Figure 7.4.

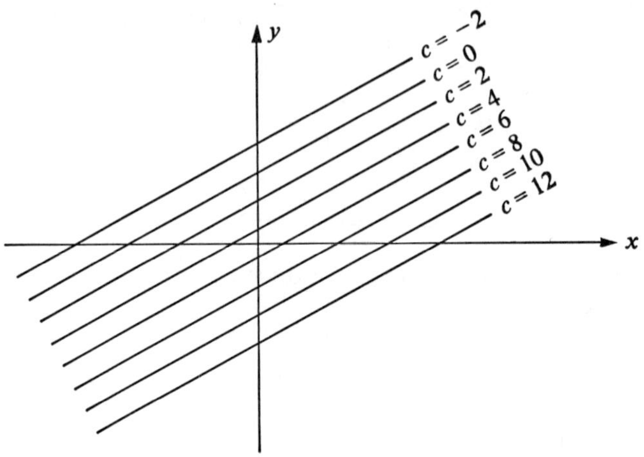

figure 7.4 *level curves of* $f(x, y) = x - 2y + 5$

(b) The level surfaces of the function g are the planes $x + 2y + z = c$. See Figure 7.5.

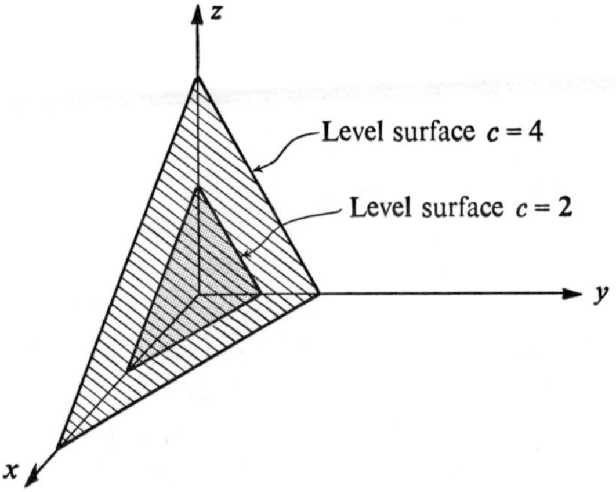

figure 7.5 *level surfaces for the field* $g(x, y, z) = x + 2y + z$

To give a picture of the variation of a vector field you must somehow tell how the magnitude *and* the direction of the field are changing. The procedure, although admittedly naive, is to sketch a "certain amount" of arrows in the space under consideration. To assist in the sketch, we sometimes calculate the *constant magnitude curves* for fields in two dimensions and *constant magnitude surfaces* for fields in three dimensions. These are the curves (or surfaces) on which the vector field has the same magnitude.

example 7.7: Sketch the vector field defined by $\mathbf{F}(x, y) = -y\mathbf{i} + x\mathbf{j}$.

solution: First, look back at Example 7.1, and notice that this is a special case of a velocity field of a rotating body. The constant level curves are in the xy-plane with $|\mathbf{F}|^2 = y^2 + x^2 = c^2$. This is a family of circles centered at the origin. In Figure 7.6 you will see some of the level curves and how they help in the sketching of the vector field.

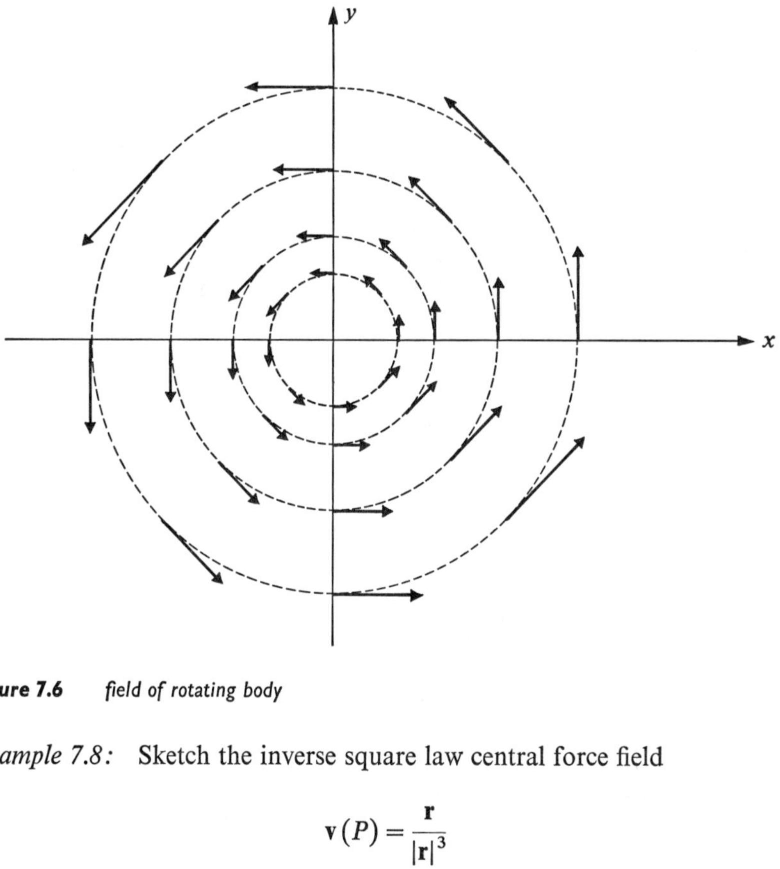

figure 7.6 *field of rotating body*

example 7.8: Sketch the inverse square law central force field

$$\mathbf{v}(P) = \frac{\mathbf{r}}{|\mathbf{r}|^3}$$

where $\mathbf{r}$ is a vector from the origin to a point, P.

solution: Note that this *is* an inverse *square* law field, even though it might not look like it since the radius in the denominator is taken to the third power. The constant level curves are

$$|\mathbf{v}| = \frac{1}{|\mathbf{r}|^2} = c.$$

Therefore the curves are circles centered at the origin. A sketch of the field is shown in Figure 7.7.

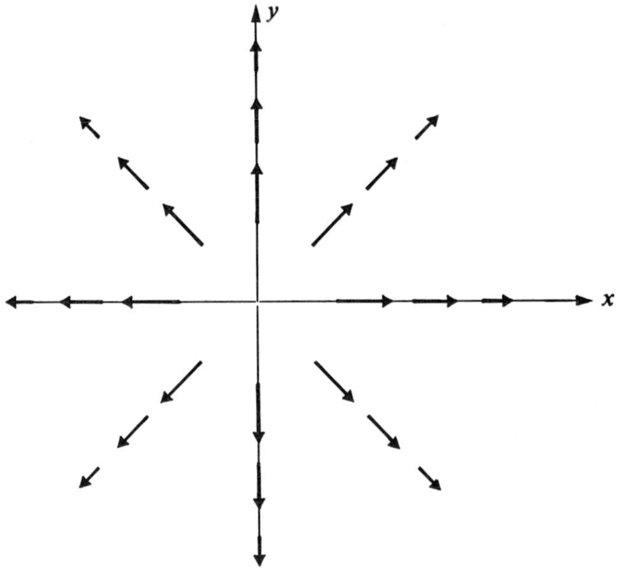

figure 7.7 *inverse square law field*

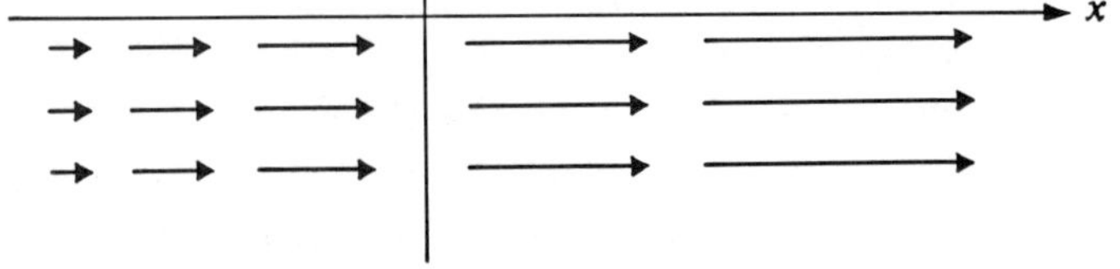

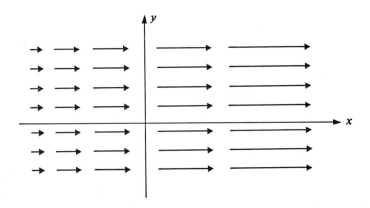

figure 7.8

example 7.9: Let **F** be the vector field sketched in Figure 7.8. Describe **F** functionally.

solution: This is an example of a field which is *unidirectional*; that is, a field in which all the vectors point in the same direction. In this case the field is always in the **i** direction. Since the field does not vary with y, it is a function of x alone. Therefore, functionally, the field may be described by the relation

$$\mathbf{F} = f(x)\,\mathbf{i}.$$

exercises for sections 7.1–7.3

1. Sketch constant level curves for the vector fields defined by:
   (a) $x^{1/2}\mathbf{i} + y\mathbf{j}$. (b) $x\mathbf{i} + y\mathbf{j}$.
   (c) $xy\mathbf{i}$. (d) $\sin x\,\mathbf{i} + \cos x\,\mathbf{j}$.
   (e) $x\mathbf{i}$. (f) $y\mathbf{i}$.
   (g) $x\mathbf{j}$. (h) $-x\mathbf{i} + y\mathbf{j}$.
   (i) $y\mathbf{i} + x\mathbf{j}$.
   Compare (b), (h), and (i).
2. Sketch the vector fields of Exercise 1 in a manner similar to the examples of Section 7.3. Compare the vector field of Example 7.7 to those of 2(b), 2(h), and 2(i).
3. Sketch constant level surfaces for the vector fields defined by:
   (a) $x\mathbf{i} + y\mathbf{j} + z\mathbf{k}$. (b) $y\mathbf{i} - z\mathbf{j} + x\mathbf{k}$. (c) $x\mathbf{i} + z^{1/2}\mathbf{j} + y\mathbf{k}$.
4. Sketch equipotential curves for the following potentials defined on the plane:
   (a) $f(x, y) = x^2 + y^2$. (b) $f(x, y) = x^2 + 4y^2$.
   (c) $f(x, y) = x^2 - y^2$. (d) $f(x, y) = xy$.
   (e) $f(x, y) = x^2 - y$.
5. Sketch equipotential surfaces for the following potentials:
   (a) $w(x, y, z) = x^2 + y^2 - z$. (b) $w(x, y, z) = x^2 - z$.

7.4 variation of a scalar field

Let $f(P)$ represent a scalar field and suppose you wish to know the rate of change of this field at a point P. For example you may want to examine the temperature variation in a building near an exit. The measure of the variation will obviously be dependent upon whether we make our computation toward the doorway or away from it, and in fact many other directions could be used to find the rate of change of the temperature at the point. Usually, therefore, the direction of the desired rate of change is specified, in which case we call the value so obtained the *directional derivative* of f at P in the direction given. The following example is in two dimensions, but it will give you some idea of the point we are trying to make.

example 7.10: Let $f(x, y) = x^2y$. What is the variation of $f(x, y)$ at (2, 1) in the direction 30° from the horizontal? (See Figure 7.9.)

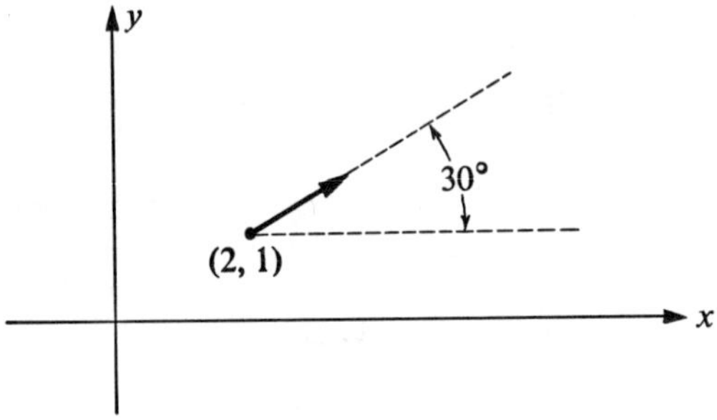

figure 7.9

solution: By definition, to find a rate of change we form the difference quotient of the change in the value of the function to the measure of the distance from (2, 1),

$$\frac{\Delta f}{\Delta s} = \frac{f(x, y) - f(2, 1)}{\sqrt{(\Delta x)^2 + (\Delta y)^2}}$$
$$= \frac{(2 + \Delta x)^2 (1 + \Delta y) - 4}{\sqrt{(\Delta x)^2 + (\Delta y)^2}}.$$

Since the variation is in a direction 30° from the horizontal, $\tan\theta = \sqrt{3}/3$ and hence $\Delta y = (\sqrt{3}/3)\,\Delta x$ and this last expression becomes

$$\frac{\Delta f}{\Delta s} = \frac{(4 + 4\,\Delta x + (\Delta x)^2)\left(1 + \frac{\sqrt{3}}{3}\,\Delta x\right) - 4}{\sqrt{\frac{4}{3}(\Delta x)^2}}$$

$$= \frac{4\,\Delta x + (\Delta x)^2 + 4\frac{\sqrt{3}}{3}\,\Delta x + 4\frac{\sqrt{3}}{3}(\Delta x)^2 + \frac{\sqrt{3}}{3}(\Delta x)^3}{\frac{2\sqrt{3}}{3}\,\Delta x}$$

$$= 2\sqrt{3} + \frac{\sqrt{3}}{2}\,\Delta x + 2 + 2\,\Delta x + \frac{(\Delta x)^2}{2}.$$

Letting $\Delta x \to 0$, we obtain the pointwise variation

$$\lim_{\Delta s \to 0} \frac{\Delta f}{\Delta s} = 2\sqrt{3} + 2.$$

For the sake of completeness we give a more precise definition of the directional derivative.

definition 7.1

The *directional derivative, df/ds, of a scalar field f in the direction from P to Q* at a point P is the value, if it exists, of

$$\lim_{\overline{QP}\to 0} \frac{f(Q)-f(P)}{\overline{QP}}.$$

Often, the direction from P to Q is specified by giving a unit vector, $\mathbf{U}$, in the direction from P to Q instead of giving another point Q. Or the direction may be given as tangential to some curve, or normal to some surface.

We see that there are infinitely many directional derivatives at a given point, each corresponding to a different direction. You will notice that the three partial derivatives of the function f are special cases of df/ds. What is not at all obvious is that the knowledge of the three partial derivatives is sufficient to specify completely *all* the directional derivatives at a given point.

To see this, let $\mathbf{U}$ represent the specified direction and let C be any curve through P with unit tangent vector $\mathbf{U}$ at P. (See Figure 7.10.) We may

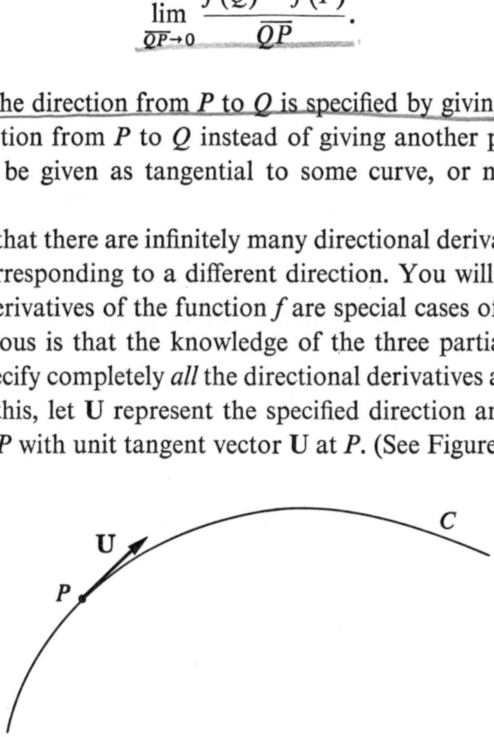

figure 7.10 *curve through P with tangent vector* **U** *at P*

suppose that C is represented by the vector function $\mathbf{r}(s)$ where s is the arc length. By use of the formula for the total derivative (See Section 2.5),

$$\begin{aligned}\frac{df}{ds} &= f_x\frac{dx}{ds}+f_y\frac{dy}{ds}+f_z\frac{dz}{ds}\\ &= \left(\frac{dx}{ds}\mathbf{i}+\frac{dy}{ds}\mathbf{j}+\frac{dz}{ds}\mathbf{k}\right)\cdot(f_x\mathbf{i}+f_y\mathbf{j}+f_z\mathbf{k}).\end{aligned}$$

The first quantity in parentheses may be recognized as the unit tangent vector to the curve C (See Section 6.7). The second quantity we denote by

∇f and hence,

$$\frac{df}{ds} = \mathbf{U}\cdot\nabla f\,.$$

Since the components of ∇f are the partial derivatives of f, we have shown that knowledge of the partial derivatives is sufficient to specify *all* directional derivatives. The result is independent of the curve C through P.

The quantity ∇f is called the *gradient vector field* derived from the scalar field, f. To show that the vector value, $\nabla f(P)$, is independent of the coordinate system (and thereby entitled to be called a vector) note that ∇f is an element of the equation

$$\frac{df}{ds} = \mathbf{U}\cdot\nabla f\,.$$

Since both df/ds and $\mathbf{U}$ are independent of the coordinate representation, ∇f must be also.

The formula

$$\frac{df}{ds} = \mathbf{U}\cdot\nabla f = |\nabla f|\cos(\nabla f, \mathbf{U})^{\dagger}$$

associates directions (given by $\mathbf{U}$) at a point with values of the directional derivative of f at the point. In effect, the vector ∇f replaces the infinity of scalars df/ds. Each of the scalars df/ds is a scalar component of the vector ∇f at P in a direction specified by $\mathbf{U}$. (See Figure 7.11.)

Since df/ds is a maximum when $\cos(\nabla f, \mathbf{U}) = 1$, *the direction of the maximum directional derivative at P is in the direction of the gradient vector at P. The magnitude of the maximum directional derivative is precisely the magnitude of the gradient vector.*

example 7.11: Let a scalar field be represented by

$$f(x, y, z) = 2xy^2 + z^2xy + x^2\,.$$

Find the value of the directional derivative of f at (1, 1, 2) in the direction of the vector $3\mathbf{i} - 2\mathbf{j} + 5\mathbf{k}$. What is the maximum value of the directional derivative at that point?

solution: In this case

$$\mathbf{U} = \frac{3\mathbf{i} - 2\mathbf{j} + 5\mathbf{k}}{\sqrt{38}}$$

and

$$\nabla f = (2y^2 + z^2y + 2x)\,\mathbf{i} + (4xy + xz^2)\,\mathbf{j} + 2xyz\mathbf{k}$$
$$\nabla f(1, 1, 2) = 8\mathbf{i} + 8\mathbf{j} + 4\mathbf{k}\,.$$

† By $\cos(\mathbf{A}, \mathbf{B})$ is meant the cosine of the angle between the vectors **A** and **B**.

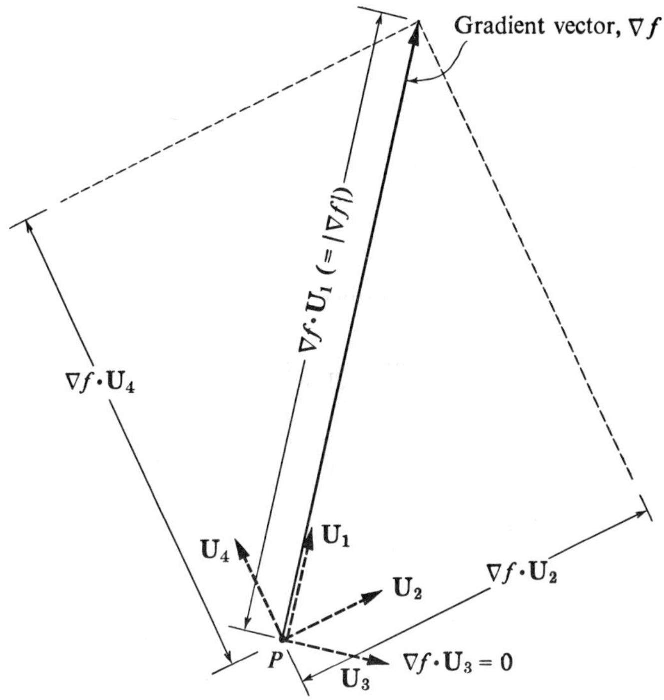

figure 7.11 *directional derivatives obtained from the gradient vector*

Therefore

$$\frac{df}{ds} = \frac{3\mathbf{i} - 2\mathbf{j} + 5\mathbf{k}}{\sqrt{38}} \cdot (8\mathbf{i} + 8\mathbf{j} + 4\mathbf{k})$$
$$= \frac{28}{\sqrt{38}}.$$

Also

$$\left.\frac{df}{ds}\right|_{\max} = |\nabla f| = \sqrt{8^2 + 8^2 + 4^2} = \sqrt{144} = 12.$$

While the application of the gradient vector to obtain the directional derivative of the scalar field is important, of more immediate concern is still another geometrical characterization of ∇f. Remember from Section 7.2 that the surfaces $f(x, y, z) = c$ are called constant level surfaces, which by their very definition are surfaces along which there is no change in the values of the scalar function. Hence, as long as **T** is tangential to such a surface,

$$\frac{df}{ds} = \nabla f \cdot \mathbf{T} = 0.$$

Therefore the vectors ∇f and $\mathbf{T}$ are perpendicular and since this must be true for all vectors tangent to the surface at P, ∇f is perpendicular to the constant level surface. This result leads to one of the more important uses of the gradient vector, that of computing normals to surfaces. (See Figure 7.12.)

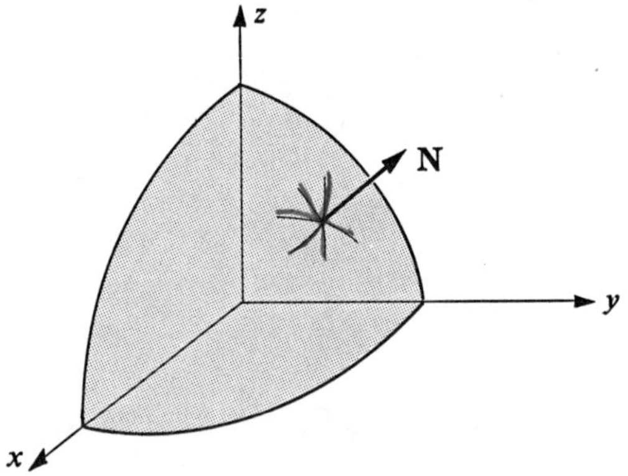

figure 7.12 *normal to surface*

example 7.12: Find a unit vector normal to the surface defined implicitly by $z^2 = x^2 + 2y^2$ at (1, 2, 3).

solution: Consider the scalar function $f(x, y, z) = x^2 + 2y^2 - z^2$. The constant level surface $f(x, y, z) = 0$ is the same surface as the one given in the statement of the problem. A unit vector normal to the surface is given by $\nabla f / |\nabla f|$ evaluated at (1, 2, 3). Since

$$\nabla f = 2x\mathbf{i} + 4y\mathbf{j} - 2z\mathbf{k}, \quad \mathbf{N} = \frac{2\mathbf{i} + 8\mathbf{j} - 6\mathbf{k}}{\sqrt{104}}.$$

Another (correct) answer would be the negative of this vector.

Vector fields which arise as the gradient of scalar fields are particularly important. We give them a special name.

definition 7.2

Let $\mathbf{F}$ be a vector field defined in some domain D. If there exists a scalar field, f, such that $\nabla f = \mathbf{F}$ throughout D, then $\mathbf{F}$ is said to be *conservative* in D. The scalar field f is called the scalar potential of $\mathbf{F}$.

In Chapter 8 you will see the appropriateness of the label "conservative" and some alternate definitions of the concept.

example 7.13: A given vector field is represented by $\mathbf{F} = (-\mathbf{r})/|\mathbf{r}|^3$, where $\mathbf{r} = x\mathbf{i} + y\mathbf{j} + z\mathbf{k}$. (See Example 7.8.) Show that this field is conservative in any region not including the origin by showing that its potential field is $f(x, y, z) = 1/|\mathbf{r}| = (x^2 + y^2 + z^2)^{-1/2}$.

solution: If the potential of the vector field is f, then the three partial derivatives of f must be the components of $(-\mathbf{r})/|\mathbf{r}|^3$. But it is just a matter of elementary differentiation to show

$$f_x = \frac{\partial}{\partial x}(x^2 + y^2 + z^2)^{-1/2} = \frac{-x}{|\mathbf{r}|^3},$$

$$f_y = \frac{-y}{|\mathbf{r}|^3},$$

and

$$f_z = \frac{-z}{|\mathbf{r}|^3}.$$

Therefore $\nabla f = \mathbf{F}$ and hence $\mathbf{F}$ is conservative.

example 7.14: A given electric field is represented by

$$\mathbf{E} = (z^2 - 2xy)\,\mathbf{i} - x^2\mathbf{j} + 2xz\mathbf{k}.$$

Show that this field is conservative everywhere by finding the electric potential.

solution: Let v be the electric potential. Then since $\nabla v = \mathbf{E}$, we have the following three conditions:

$$v_x = z^2 - 2xy; \qquad v_y = -x^2; \qquad v_z = 2xz.$$

Partially antidifferentiating the first of these we obtain

$$v(x, y, z) = z^2x - x^2y + T(y, z).$$

From this expression for v we may compute v_y,

$$v_y = -x^2 + T_y,$$

which by the second of our conditions is equal to $-x^2$. Therefore $T_y = 0$ and hence $T(y, z)$ is a function of z only, say $F(z)$. Thus

$$v(x, y, z) = z^2x - x^2y + F(z).$$

We compute v_z from this expression to obtain

$$v_z = 2xz + F'(z)$$

which by the third condition is equal to $2xz$. Therefore $F'(z) = 0$ and hence the scalar potential is

$$v(x, y, z) = z^2x - x^2y + (\text{a constant}).$$

The process of finding a scalar potential of a conservative vector field is exactly the same as that shown in Section 2.5 for finding a function u (of more than one real variable) if the total differential, du, is known. Hence, the scalar potential is to vector functions what the antiderivative is to real functions.

exercises for section 7.4

1. Find the gradient of the following scalar fields:
   (a) $f(x, y, z) = xyz$.
   (b) $f(x, y, z) = x^2 + y^2 + z^2$.
   (c) $f(x, y, z) = x\cos y + y\sin z + z\cos x$.
2. By using your knowledge of the gradient vector and its relation to constant level curves, find a unit vector normal to the plane curve $y = 3x^2 - 4$, $z = 0$ at $(1, -1, 0)$.
3. Let $f(x, y, z) = xy + yz + xz$. Find:
   (a) df/ds at $(1, 1, 3)$ in the direction toward $(1, 1, 1)$.
   (b) The maximum value of df/ds and give the direction in which df/ds is a maximum.
   (c) A vector normal to the surface $xy + yz + xz = 7$ at any point of the surface.
4. Assuming $\mathbf{F}$ is conservative, find a scalar potential:
   (a) $\mathbf{F} = \mathbf{i} + \mathbf{j} + \mathbf{k}$.
   (b) $\mathbf{F} = 2x\mathbf{i} + 2y\mathbf{j} + 2z\mathbf{k}$.
   (c) $\mathbf{F} = \dfrac{x\mathbf{i} + y\mathbf{j}}{(x^2 + y^2)^{1/2}}$.
   (d) $\mathbf{F} = (2xyz + x\sin 2x^2)\,\mathbf{i} + (x^2z - z\sin yz)\,\mathbf{j} + (x^2y - y\sin yz)\,\mathbf{k}$.
   (e) $\mathbf{F} = (2xy + z^2)\,\mathbf{i} + (2yz + x^2)\,\mathbf{j} + (2xz + y^2)\,\mathbf{k}$.
5. Is it possible to take the gradient of a vector field? Explain.
6. By using your knowledge of gradient vectors, find normals to the following surfaces at any point on the surface:
   (a) $x^2 + y^2 + z^2 = 2$.
   (b) $xyz = 1$.
   (c) $\cos x + \sin y + x\sin z = 1$.
7. The magnetic potential is given by $u(x, y, z) = x^2y + xyz$. Find the magnetic induction field vector $\mathbf{B} = \nabla u$.
8. Write the equation of the plane tangent to, and the line normal to, the following surfaces at the points indicated:
   (a) $x^2 + y^2 = z$ at the point $(1, 1, 2)$.
   (b) $y = \sin x$ at the point $(\pi/2, 1, 0)$.
   (c) $xyz = 8$ at the point $(2, 2, 2)$.

7.5 variation of a vector field: divergence

The study of the calculus of vector fields, corresponding to the differentiation of real functions, centers on two distinct types of variation. One of these, called the divergence of a vector field, is a scalar quantity. It is discussed in this section. The other, called the curl, is a vector variation and is the subject of Section 7.6.

Let $\mathbf{F}$ be a vector field with component functions $\{f, g, h\}$ with respect to the $\{\mathbf{i}, \mathbf{j}, \mathbf{k}\}$ basis. Then we can make the following definition.

definition 7.3

The divergence of the vector field $\mathbf{F}$ is the scalar field whose value at each point is given by the sum $f_x + g_y + h_z$. The divergence of $\mathbf{F}$ is denoted by $\nabla \cdot \mathbf{F}$ (read 'del dot $\mathbf{F}$'') or div$\mathbf{F}$.

example 7.15: Find the divergence of the vector field defined by

$$\mathbf{F} = 3xy\mathbf{i} + x^2y\mathbf{j} + y^2z\mathbf{k}.$$

solution: Since $f(x, y, z) = 3xy$, $g(x, y, z) = x^2y$, and $h(x, y, z) = y^2z$,

$$\nabla \cdot \mathbf{F} = 3y + x^2 + y^2.$$

The formula for the divergence might seem naturally associated with the notation $\nabla \cdot \mathbf{F}$ since you can think of the "operator" $\mathbf{i}\dfrac{\partial}{\partial x} + \mathbf{j}\dfrac{\partial}{\partial z} + \mathbf{k}\dfrac{\partial}{\partial z}$ as "dotting" with the vector field to give the scalar quantity indicated. Do not carry this thinking too far, though, since it would certainly be incorrect to think of the del operator as in any sense being an "arrow." (What is the direction of ∇?) The definition of the divergence is tied together with the coordinate system and thus at least apparently dependent upon it. Such dependence could, of course, mean that it really is not a scalar field. In the next paragraph we will attach some physical meaning to the sum of these partial derivatives.

Consider the motion of some substance (such as fluid, gas, electric charges, alpha particles) spreading through space. Then the velocities of the particles at each instant form a vector field, $\mathbf{V} = v_1\mathbf{i} + v_2\mathbf{j} + v_3\mathbf{k}$ where v_1, v_2, and v_3 are the component velocities. At a point P in space we impose a rectangular coordinate system so that without loss of generality P may be considered to have coordinates $(0, 0, 0)$.

We examine the flow of this fluid through a small rectangular parallelepiped of dimensions Δx, Δy, Δz with edges parallel to the coordinate axes.

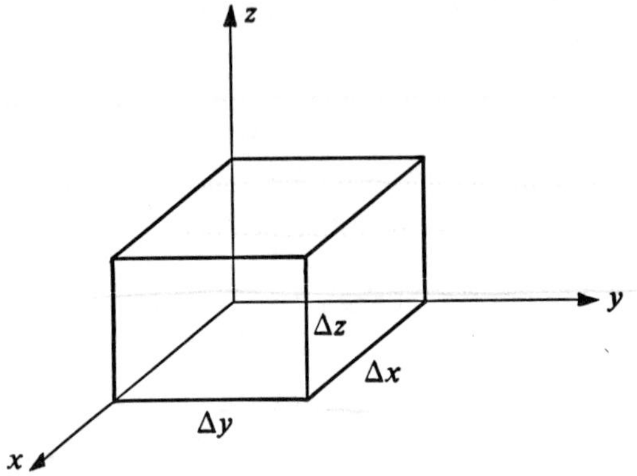

figure 7.13 *divergence*

(See Figure 7.13.) If ρ is the density of the quantity, then the vector quantity $\mathbf{F} = \rho\mathbf{V}$ has component functions ρv_1, ρv_2, and ρv_3.

We compute the change in mass per unit time within the parallelepiped by computing the net outward flow. The normal inward component of $\mathbf{F}$ at the back face is ρv_1 (evaluated at $(0, y, z)$) and the approximate normal outward component at the front face is given by

$$\rho v_1 + \frac{\partial \rho v_1}{\partial x} \Delta x.$$

This approximation is good if, for example, the value of Δx is small. Hence, the mass entering the back face per unit time, Δt, is given by

$$\rho v_1 (0, y, z)\,(\Delta y\,\Delta z)\,\Delta t.$$

Likewise, the mass leaving the front face is approximately

$$\rho v_1 (0, y, z)\,(\Delta y\,\Delta z)\,\Delta t + \left(\frac{\partial \rho v_1}{\partial x}\Delta x\right)(\Delta y\,\Delta z)\,\Delta t.$$

There are similar expressions for the mass entering and leaving the other four faces. Adding the outward flow of mass we obtain

$$\begin{aligned}\left(\begin{array}{l}\text{Approximate net change of outward}\\ \text{flow of mass in time interval } \Delta t\end{array}\right) &= \left(\frac{\partial \rho v_1}{\partial x} + \frac{\partial \rho v_2}{\partial y} + \frac{\partial \rho v_3}{\partial z}\right)\Delta x\,\Delta y\,\Delta z \\ &= \nabla\cdot(\rho\mathbf{V})\,\Delta x\,\Delta y\,\Delta z\end{aligned}$$

which means that the net outward flow of mass per unit volume in time interval Δt is $\nabla\cdot\rho\mathbf{V}$, that is, the divergence of the vector field $\rho\mathbf{V}$ at P.

This net change of mass per unit volume per unit time at a given point may be nonzero, either because of velocities changing through space, or because of a change in density due to a local creation (source) or destruction (sink) of substance. Thus in studying the flow of oxygen through the human body we may have a sink due to conversion into carbon dioxide by metabolism.

Let ϕ represent the algebraic difference between the source density and the sink density (that is, the difference between the rates for which the substance is being created or destroyed per unit time). Then the equation governing the flow of mass is

$$\nabla \cdot \rho \mathbf{V} = \phi - \frac{\partial \rho}{\partial t}.$$

In case there are no sources or sinks, we have the *continuity equation for compressible fluid flow* (also called the *conservation of mass equation*),

$$\nabla \cdot \rho \mathbf{V} = -\frac{\partial \rho}{\partial t}.$$

If the density is time invariant, that is, a steady flow, then the continuity equation becomes

$$\nabla \cdot \rho \mathbf{V} = 0.$$

Finally, if the fluid is incompressible ($\rho =$ a constant),

$$\nabla \cdot \mathbf{V} = 0$$

which is called the *condition for incompressibility* (in a field where $\phi = 0$).

In the case of electric fields, the sources and sinks are electric charges and the equation for the condition for incompressibility applies to the electric displacement vector **D**,

$$\nabla \cdot \mathbf{D} = \phi,$$

where ϕ is the charge density. Physically, this means that the electric displacement vector has zero divergence in charge free regions.

For the magnetic field vector, **B**, you might expect the sources to be magnetic poles. However, there are no free magnetic poles and hence $\nabla \cdot \mathbf{B} = 0$ always. The two equations which specify the pointwise behavior of the vector fields **B** and **D** are two from a set of four equations known as Maxwell's equations. What we have shown is how to derive the two divergence equations from the condition for the conservation of mass. In electromagnetic theory you usually begin with Maxwell's equations as the basis of the development.

example 7.16: Find the charge density at the point (1, 1, 2) if the electric displacement vector field is

$$\mathbf{D} = x^2 y\mathbf{i} + z^2\mathbf{j} + z^3\mathbf{k}.$$

solution: The scalar field $\nabla \cdot \mathbf{D}$ is equal to $(\partial/\partial x)\,(x^2y)+(\partial/\partial y)\,(z^2)+(\partial/\partial z)\,(z^3)$, which is equal to $2xy+3z^2$. Hence the value of the divergence of $\mathbf{D}$ at $(1, 1, 2)$ is 14, which by Maxwell's equation is the charge density in coulombs per cubic meter.

definition 7.4

A vector field which has zero divergence throughout a region is said to be *solenoidal* or *incompressible* in that region.

The name solenoidal derives from the fact that a magnetic field, which has zero divergence everywhere, is easily generated by a solenoid. The word incompressible is used since vector fields which have zero divergence satisfy the condition for incompressibility, $\nabla \cdot \mathbf{F} = 0$.

example 7.17: Show that a conservative vector field is solenoidal if the sum of the second partial derivatives of the scalar potential is zero.

solution: If $\mathbf{F}$ is a conservative vector field, then by definition there exists a scalar function f such that $\nabla f = \mathbf{F}$. Hence

$$\begin{aligned}\nabla \cdot \mathbf{F} &= \nabla \cdot \nabla f\\ &= \nabla \cdot (f_x\mathbf{i} + f_y\mathbf{j} + f_z\mathbf{k})\\ &= f_{xx} + f_{yy} + f_{zz}\end{aligned}$$

and therefore the divergence of $\mathbf{F}$ is zero (that is, $\mathbf{F}$ is solenoidal) if the sum of the second partial derivatives is zero.

Instead of writing $\nabla \cdot \nabla f$ as you did in the previous example, you will often use the notation $\nabla^2 f$ and call the resulting sum, the *Laplacian of f*, that is,

$$\nabla^2 f = f_{xx} + f_{yy} + f_{zz}.$$

Functions f such that $\nabla^2 f = 0$ (Laplace's differential equation) are called *harmonic*. What we have shown in the previous example is that the potential, if it exists, of any solenoidal field is harmonic. For example, the function $f(x, y, z) = (x^2+y^2+z^2)^{-1/2}$ is harmonic since it is the potential of the solenoidal field $\mathbf{F} = (-\mathbf{r})/(|\mathbf{r}|^3)$ (see Example 7.13 and the following).

example 7.18: Show that the vector field $(-\mathbf{r})/(|\mathbf{r}|^3)$ is solenoidal.

solution: To show that this vector field is solenoidal we must show that the divergence is zero,

$$\nabla\cdot\frac{-\mathbf{r}}{|\mathbf{r}|^3} = -\frac{\partial}{\partial x}\left[\frac{x}{(x^2+y^2+z^2)^{3/2}}\right] - \frac{\partial}{\partial y}\left[\frac{y}{(x^2+y^2+z^2)^{3/2}}\right] - \frac{\partial}{\partial z}\left[\frac{z}{(x^2+y^2+z^2)^{3/2}}\right]$$

$$= -\left[\frac{y^2+z^2-2x^2}{(x^2+y^2+z^2)^{5/2}}\right] - \left[\frac{x^2+z^2-2y^2}{(x^2+y^2+z^2)^{5/2}}\right] - \left[\frac{x^2+y^2-2z^2}{(x^2+y^2+z^2)^{5/2}}\right]$$

$$= 0.$$

exercises for section 7.5

1. Find the divergence of the field **F** at each point if **F** is defined by:
   (a) $xyz(\mathbf{i}+\mathbf{j}+\mathbf{k})$. (b) $(3x^2+6xz+3z^2)\,\mathbf{i}$.
2. Find a formula for the divergence of a vector field **F** if **F** is the product of a scalar field with another vector field.
3. Show directly that the function $f(x, y, z) = (x^2+y^2+z^2)^{-1/2}$ is harmonic.
4. Show that the function $e^x \cos y$ is harmonic. What solenoidal field does this potential generate?
5. Find the directional derivative of div**F** at (2, 1, 3) in the direction of the outer normal to the sphere $x^2+y^2+z^2=14$ where $\mathbf{F} = xz^2\mathbf{i} + zy^2\mathbf{j} + x^2z\mathbf{k}$.
6. Show that the steady fluid flow whose velocity vector is given everywhere by $x\mathbf{i}$ is compressible. Sketch this vector field in the xy-plane.
7. Show that the vector field $xz\mathbf{i} + yz\mathbf{j} - z^2\mathbf{k}$ is solenoidal everywhere.
8. An electric displacement vector field is described by $x^2\mathbf{i} + y^2\mathbf{j} + z^2\mathbf{k}$. With this field, are there any regions of space which are charge free?
9. Show that the vector field $y^2z\mathbf{i} + xz^3\mathbf{j} + y^2x^2\mathbf{k}$ is solenoidal everywhere and therefore could describe a magnetic vector field.
10. Review and discuss the concepts of conservation of mass, incompressible fields, solenoidal fields, conservative fields, harmonic functions, and Laplacian of f.

7.6 variation of a vector field: curl

In the previous section you learned how to compute one type of important variation of a vector field. Perhaps you will recall that the formula for divergence could very easily be remembered if you thought of ∇ as "dotting" the field. The other possibility is to have the del operator operate in a "crossing" manner.

definition 7.5

Let $\mathbf{F}$ be a vector field with component functions f, g, and h. Then curl $\mathbf{F}$ $=\nabla \times \mathbf{F}$ is given by

$$\nabla \times \mathbf{F} = (h_y - g_z)\,\mathbf{i} + (f_z - h_x)\,\mathbf{j} + (g_x - f_y)\,\mathbf{k}.$$

Symbolically,

$$\nabla \times \mathbf{F} = \begin{vmatrix} \mathbf{i} & \mathbf{j} & \mathbf{k} \\ \dfrac{\partial}{\partial x} & \dfrac{\partial}{\partial y} & \dfrac{\partial}{\partial z} \\ f & g & h \end{vmatrix}.$$

(This last expression is not a determinant but is a device which is often used to enable easy recall of the formula for curl $\mathbf{F}$. It is "evaluated" just like a third-order determinant.)

You may feel right at home calling curl $\mathbf{F}$ a vector field, but perhaps it would be better if you did have a bit of uneasiness. For the expression for the curl depends upon the coordinate system. When we discuss line integration in Chapter 8 we will give an alternate, and perhaps more popular, definition of curl which will be independent of the coordinate system. For the time being we will assume that the component functions with respect to different bases transform in just the right way to give us the independence of coordinates that we must have.

Traditionally, the concept of the curl of a vector field has been quite elusive to the beginner, not so much in his evaluating it (for that just involves a few partial derivatives) but in his understanding what it is. The following paragraphs and examples are intended to give you some feeling for curl rather than to increase your computational skill in working with it.

Let $\mathbf{F}$ be a vector force field in the xy-plane with component functions with respect to the $\{\mathbf{i}, \mathbf{j}\}$ basis denoted by f and g. We wish to determine the tendency of this force field to rotate the rectangle $ABCD$ shown in Figure 7.14. At point A the horizontal component of the field is given by $f(x_A, y_A)$. At point D the horizontal component is approximately $f(x_A, y_A) + f_y(x_A, y_A)\,\Delta y$. The difference between these two values gives the tendency of the horizontal

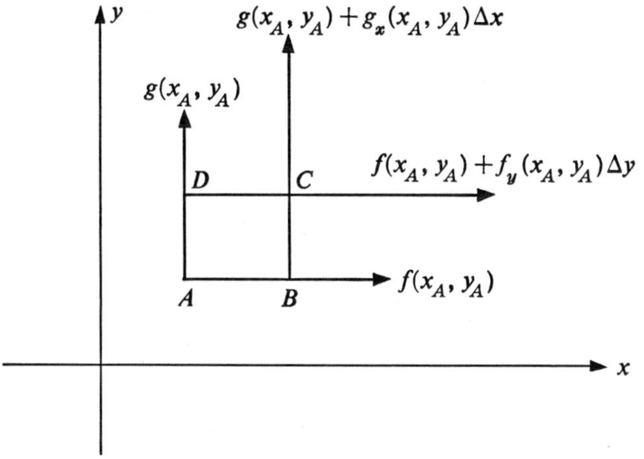

figure 7.14 curl

component of the field to rotate the element, in a clockwise direction if $f_y(x_A, y_A)$ is positive and counterclockwise if negative. Similarly, the change of the vertical component in the horizontal direction, which is g_x, gives a counterclockwise rotation if it is positive and clockwise if negative. Note that it is the variation of the component functions in a direction perpendicular to the direction of the field which yields this rotation. Thus, the difference $g_x - f_y$ will be a measure of the vector field's rotational force on the rectangle. In the usual manner, we allow the rectangle to shrink to zero to give a pointwise measure of the rotational force. The direction of the motion follows the convention of the "right hand rule" and hence we denote the pointwise rotational tendency as $\mathbf{k}(g_x - f_y)$.

More generally, in three dimensions you can probably convince yourself that the curl is a measure of the tendency of the vector field to cause a rotation at a point. Its direction is along the axis of rotation and in a direction corresponding to the familiar right hand rule. This should show you why the words curl and rot are familiar identifiers of $\nabla \times \mathbf{F}$. It should also show the reasonableness of the following definition.

definition 7.6

A vector field, **F**, is said to be *irrotational* at those points for which $\nabla \times \mathbf{F} = \mathbf{0}$.

example 7.19: Show that a conservative vector field is irrotational at each point where it is defined.

solution: If **F** represents the vector field, then by the definition of conservative (Definition 7.2), there is a scalar field f such that $\nabla f = \mathbf{F}$. Therefore

$$\nabla \times \mathbf{F} = \nabla \times \nabla f = \begin{vmatrix} \mathbf{i} & \mathbf{j} & \mathbf{k} \\ \dfrac{\partial}{\partial x} & \dfrac{\partial}{\partial y} & \dfrac{\partial}{\partial z} \\ f_x & f_y & f_z \end{vmatrix}$$

$$= (f_{zy} - f_{yz})\,\mathbf{i} + (f_{xz} - f_{zx})\,\mathbf{j} + (f_{yx} - f_{xy})\,\mathbf{k}.$$

If we assume that the second mixed partial derivatives of f are equal (a reasonable assumption for the applications to physics and engineering), then each of the components of the curl is zero and hence $\nabla \times \mathbf{F} = 0$. Thus, $\mathbf{F}$ is irrotational.

The curl of a vector field can be conveniently measured with a device called a "curl meter." This looks something like a paddle wheel with its fins causing the shaft of the meter to turn. (See Figure 7.15.) For example,

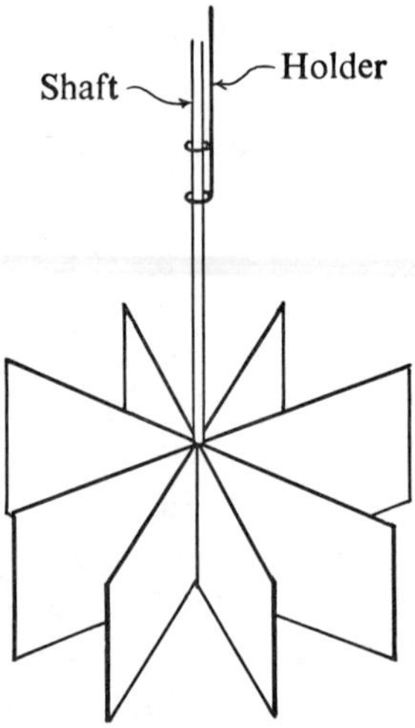

figure 7.15 *curlmeter*

to measure the curl of the vector field $\mathbf{F} = 2\mathbf{i}$, we insert the curl meter with its shaft as if it were coming out of the paper. See Figure 7.16. At the asterisk the curl meter will not rotate (and in fact nowhere will it rotate) and hence $\nabla \times \mathbf{F} = \mathbf{0}$ everywhere.

example 7.20: Use a curl meter to analyze the field $\mathbf{F} = \mathbf{i}\,(\sin y)$.

solution: Figure 7.17 is a sketch of the field. If a curl meter is inserted some-

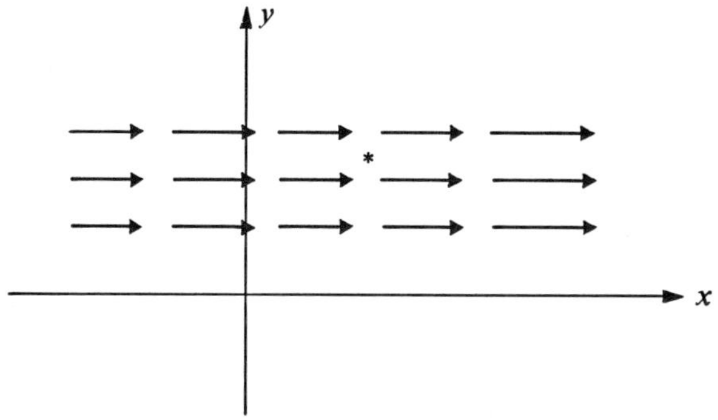

figure 7.16 vector field with zero curl

where in the field for $0 < y < \pi/2$, a clockwise spin is obtained which approaches 0 as y approaches $\pi/2$. For $y = \pi/2$, there is no spin and for $y > \pi/2$, the rotation is counterclockwise. Direct computation of curl **F** gives

$$\nabla \times \mathbf{F} = -(\cos y)\,\mathbf{k}$$

which is in agreement with these observations.

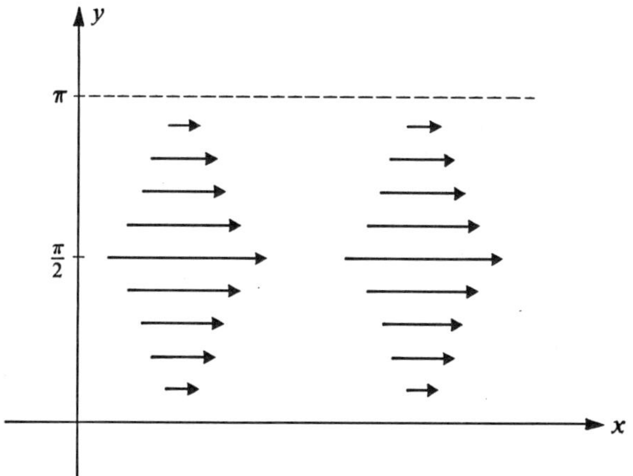

figure 7.17

The two fields illustrated in Figure 7.18 are examples of unidirectional fields with zero curl and zero divergence. Can you tell which is which?

example 7.21: Show that a vector field **G** which is the curl of another field is solenoidal.

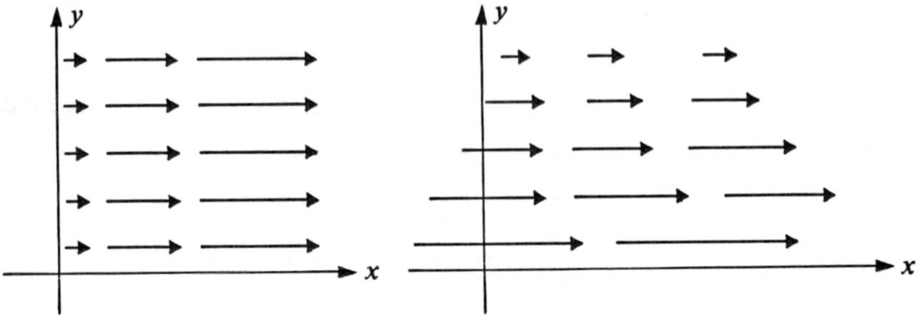

figure 7.18 *irrotational and solenoidal fields*

solution: Let $\mathbf{G} = \nabla \times \mathbf{F}$. Then $\nabla \cdot \mathbf{G} = \nabla \cdot \nabla \times \mathbf{F}$. If $\mathbf{F}$ has component functions f, g, and h, then

$$\begin{aligned} \nabla \cdot \mathbf{G} &= \frac{\partial}{\partial x}(h_y - g_z) + \frac{\partial}{\partial y}(f_z - h_x) + \frac{\partial}{\partial z}(g_x - f_y) \\ &= h_{yx} - g_{zx} + f_{zy} - h_{xy} + g_{xz} - f_{yz}. \end{aligned}$$

If we assume the second mixed partial derivatives of the component functions are equal, then $\nabla \cdot \mathbf{G} = \mathbf{0}$ and hence $\mathbf{G}$ is solenoidal.

The vector field $\mathbf{F}$ in the previous example is often called the *vector potential* of the field $\mathbf{G}$. In the applications, if we have a solenoidal field, we often postulate the existence of such a potential and find it very useful. The technique of finding the vector potential is not important to us and will be omitted.

exercises for section 7.6

1. Describe functionally the vector fields of Figure 7.18.
2. Find curl $\mathbf{F}$ if $\mathbf{F}$ is equal to:
   (a) $x\mathbf{i} + 3xy\mathbf{k}$. (b) $z\mathbf{i} + x\mathbf{j} + y\mathbf{k}$.
   (c) $x\mathbf{i} + y\mathbf{j} + z\mathbf{k}$. (d) $\dfrac{x\mathbf{i} + y\mathbf{j} + z\mathbf{k}}{(x^2 + y^2 + z^2)^{3/2}}$.
3. Are the concepts of irrotational and solenoidal mutually exclusive? Explain.
4. Is the curl of the divergence of a vector field defined? Is the divergence of the curl of a vector field defined? Explain completely.
5. Prove the identity $\nabla \cdot (\mathbf{V} \times \mathbf{W}) = \mathbf{W} \cdot (\nabla \times \mathbf{V}) - \mathbf{V} \cdot (\nabla \times \mathbf{W})$ using components of $\mathbf{V}$ and $\mathbf{W}$ with respect to the $\{\mathbf{i}, \mathbf{j}, \mathbf{k}\}$ basis.

6. Show that the cross product of two conservative fields is solenoidal. (*Hint:* Use the identity of the previous exercise.)

7. Show that a central force field is irrotational everywhere it is defined and differentiable. (*Hint:* a central force field may be described by $f(r)\mathbf{r}$.)

8. Sketch and discuss the following fields with respect to their solenoidal and irrotational properties (these fields were all sketched, if you did the exercises for Sections 7.1-7.3):

   (a) $-y\mathbf{i}$. (b) $x\mathbf{j}$. (c) $-y\mathbf{i}+x\mathbf{j}$.
   (d) $y\mathbf{i}+x\mathbf{j}$. (e) $-x\mathbf{i}+y\mathbf{j}$. (f) $x\mathbf{i}+y\mathbf{j}$.

9. Establish a "law" for unidirectional fields to be solenoidal or irrotational, after examining Figure 7.18. (Also see Exercise 11, below).

10. Determine a *necessary* condition for a vector field to have a vector potential.

11. Show that the curl of any unidirectional vector field is perpendicular to that direction.

integration of vector fields

8

8.1 *introduction*

In this chapter you will learn some of the so-called integral theorems of the more classical vector field theory. The relationships between the various types of boundary problems, del operations, and integrals are of far-reaching consequence.

In the beginning of the chapter we define line and surface integration under the assumption that you are familiar with the definitions of single and multiple Riemann definite integrals. You should review these definitions if you think it necessary. You will also be required to recall the technique of evaluating double and triple integrals as iterated integrals. The definitions of line and surface integrals of scalar fields will then be extended to integration over curves and surfaces in the presence of vector fields.

8.2 *line integration*

The type of integration you learned in elementary calculus was along the x-axis. This concept may be generalized to include movement along rather general curves. Our discussion will be limited to oriented smooth curves or at least to a finite collection of such curves. Besides the curve, C, we consider also a scalar valued function, f, assumed to be continuous on the curve. You will see almost immediately that the terminology "curve integration" would be more fitting, since normally the integration is along some nonlinear part of the curve. We shall maintain tradition at the expense of some ambiguity.

To define what we mean by "the line integral of a scalar function f along a curve C," we proceed in a manner analogous to that of a function of one real variable. As we will see later, there are several types of line integrals and the description given here is the one which defines $\int_C f(P)\,dx(P)$ or, as it is more commonly written, $\int_C f(P)\,dx$.

We assume that C is piecewise smooth and orientable and connects points R and Q in such a manner that the curve does not intersect itself, except that R and Q may coincide. We then subdivide C by a finite set of ordered points

$$B = \{P_0 = R, P_0', P_1, P_1', P_2, \cdots, P_i, P_i', P_{i+1}, \cdots, P_{n-1}, P_{n-1}', P_n = Q\}.$$

Figure 8.1 indicates the manner of this subdivision.

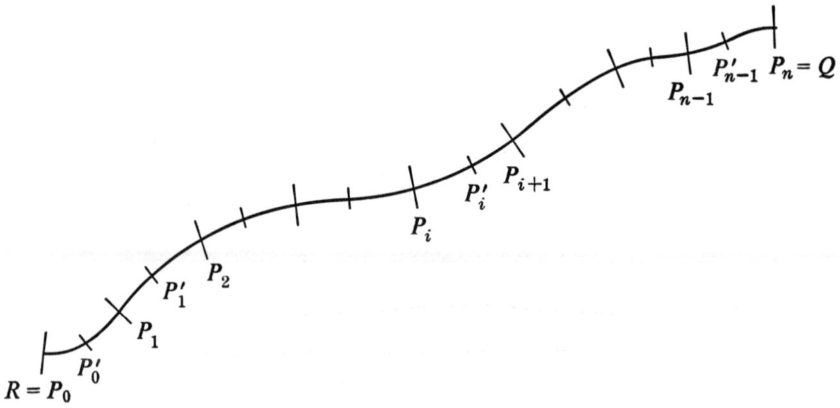

figure 8.1 partitioning a curve

The finite set of ordered points, which we will call a *partition* of C, may be chosen rather arbitrarily. Each partition also essentially defines a set consisting of the values of x at the points, P_i,

$$B_x = \{x_0, x_0', x_1, x_1', x_2, \cdots, x_i, x_i', x_{i+1}, \cdots, x_{n-1}, x_{n-1}', x_n\}.$$

For a given partition we evaluate the quantities $f(P_i')$ at each P_i', $\Delta x_i = x_{i+1} - x_i$, and the sum

$$J(B) = \sum_{i=0}^{n-1} f(P_i')\,\Delta x_i.$$

Also associated with each partition is a number called the *x-norm*, which is the largest of the Δx_i, and is denoted by $\Delta_x(B)$ or $\|\Delta x_i\|$.

To give you some idea of how a partition affects the sum $J(B)$, we consider two examples.

example 8.1: Find $J(B)$ in the case where the curve C is in the xy-plane and defined by $y = x^2$, the function is $f(x, y) = x + y + 2$, and the partition of the curve is given by $\{(0, 0), (1/4, 1/16), (1/4, 1/16), (1/2, 1/4), (1, 1)\}$. (See Figure 8.2.)

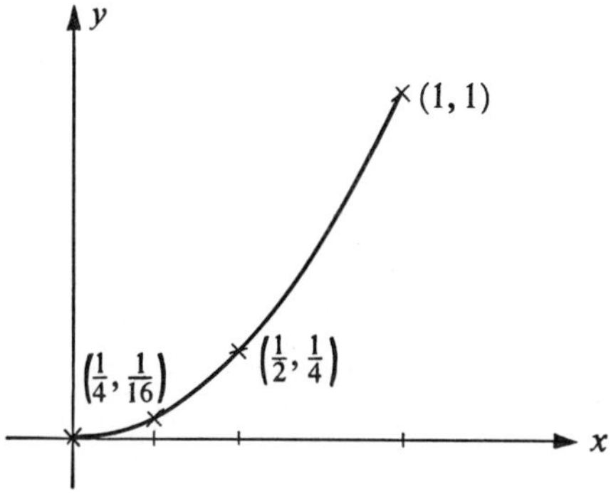

figure 8.2

solution: In this case, the set B_x is given by $\{0, 1/4, 1/4, 1/2, 1\}$ and hence $\Delta x_0 = 1/4$, $\Delta x_1 = 3/4$, $x_0' = 1/4$, and $x_1' = 1/2$. Therefore

$$\begin{aligned} J(B) &= f(P_0')\,\Delta x_0 + f(P_1')\,\Delta x_1 \\ &= (\tfrac{1}{4} + \tfrac{1}{16} + 2)\,\tfrac{1}{4} + (\tfrac{1}{2} + \tfrac{1}{4} + 2)\,\tfrac{3}{4} \\ &= \tfrac{169}{64}. \end{aligned}$$

The x-norm of B is 3/4.

example 8.2: Find $J(B)$ if the curve and function are as in the previous example but the partition is defined by the set

$$\{(0, 0), (\tfrac{1}{4}, \tfrac{1}{16}), (\tfrac{1}{4}, \tfrac{1}{16}), (\tfrac{1}{2}, \tfrac{1}{4}), (\tfrac{1}{2}, \tfrac{1}{4}), (\tfrac{3}{4}, \tfrac{9}{16}), (1, 1)\}.$$

(See Figure 8.3.)

solution: In this case, the set B_x is given by $\{0, 1/4, 1/4, 1/2, 1/2, 3/4, 1\}$ and hence $\Delta x_0 = 1/4$, $\Delta x_1 = 1/4$, $\Delta x_2 = 1/2$, $x_0' = 1/4$, $x_1' = 1/2$, and $x_2' = 3/4$. Therefore,

$$\begin{aligned} J(B) &= f(P_0')\,\Delta x_0 + f(P_1')\,\Delta x_1 + f(P_2')\,\Delta x_2 \\ &= (\tfrac{1}{4} + \tfrac{1}{16} + 2)\,\tfrac{1}{4} + (\tfrac{1}{2} + \tfrac{1}{4} + 2)\,\tfrac{1}{4} + (\tfrac{3}{4} + \tfrac{9}{16} + 2)\,\tfrac{1}{2} \\ &= \tfrac{187}{64}. \end{aligned}$$

The x-norm of the partition is 1/2.

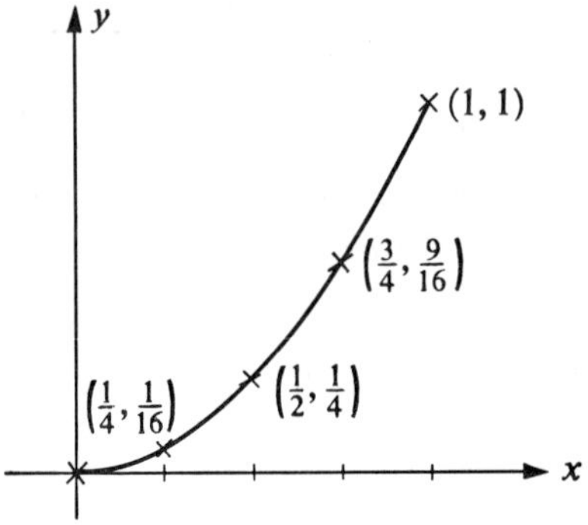

figure 8.3

The two examples should show that the value of $J(B)$ is dependent not only on the number of points chosen for the partition but upon how the selection is made. Moreover, the importance of the points P_i' and how they affect the sum is apparent. You would expect, correctly, that as different choices of B are made, the values of $J(B)$ can be made close together if the x-norm of the partition is small. To anticipate this we make the following definition.

definition 8.1

The limit of finite sums (if it exists) as the x-norm approaches 0 is called the *line integral of f along C with respect to x* and is denoted by

$$\int_C f(P)\,dx.$$

If C is a closed curve, we write

$$\oint_C f(P)\,dx.$$

More specifically, the line integral is a numerical quantity for which for any $\varepsilon > 0$ there exists a δ such that if $\Delta_x(B) < \delta$, then

$$\left|\int_C f(P)\,dx - J(B)\right| < \varepsilon.$$

Under fairly reasonable assumptions on the curve C and the function f, the line integral exists. We have the following theorem, which we will not prove.

theorem 8.1

If f is continuous on C and C is piecewise smooth, then

$$\lim_{\Delta_x(B)\to 0} J(B)$$

exists and is unique.

It should not be a difficult matter for you to define line integrals of the type $\int_C f(P)\,dy$ or $\int_C f(P)\,dz$, since in each case the method of definition is exactly analogous to $\int_C f(P)\,dx$. The difference is that the approximating finite sums are

$$\sum_{i=0}^{n-1} f(P_i')\,\Delta y_i$$

and

$$\sum_{i=0}^{n-1} f(P_i')\,\Delta z_i$$

respectively. Also, if the arc length function is known as a function of points on the curve, say $s(P)$, then the differences in arc length $\Delta s_i = s(P_i) - s(P_{i-1})$ may be computed. The limit of the corresponding finite sum

$$\sum_{i=0}^{n-1} f(P_i')\,\Delta s_i$$

is denoted by $\int_C f(P)\,ds$.

In Section 8.4 you will learn how to evaluate line integrals without resorting to the definition. Even so, a knowledge of the definition is essential for approximations to be made with the use of finite sums. Also, the definition of a line integral as a limit of approximating sums will make the following properties seem obvious:

1. $\int_C kf(P)\,dx = k\int_C f(P)\,dx.$
2. $\int_C [f(P)+g(P)]\,dx = \int_C f(P)\,dx + \int_C g(P)\,dx.$
3. $\int_C f(P)\,dx = -\int_{-C} f(P)\,dx$, where by $-C$ is meant the path which is the same set of points as C but with opposite orientation.
4. If C is the union of C_1 and C_2, then

$$\int_C f(P)\,dx = \int_{C_1} f(P)\,dx + \int_{C_2} f(P)\,dx.$$

(See Figure 8.4.)

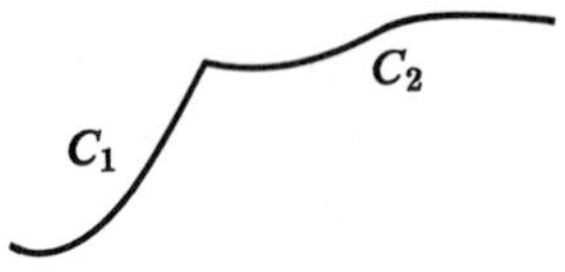

figure 8.4

8.3 line integral of a vector field

Let f, g, and h be component functions of a vector field, **F**. We often find it convenient to write the sum of the three line integrals

$$\int_C f(P)\,dx,\ \int_C g(P)\,dy,\ \int_C h(P)\,dz$$

as one line integral of a vector field. Let $d\mathbf{r}$ represent the vector expression $dx\,\mathbf{i} + dy\,\mathbf{j} + dz\,\mathbf{k}$. If x, y, and z are considered as functions of arc length, the expression for $d\mathbf{r}$ may be written

$$\begin{aligned} d\mathbf{r} &= \frac{dx}{ds}\,ds\,\mathbf{i} + \frac{dy}{ds}\,ds\,\mathbf{j} + \frac{dz}{ds}\,ds\,\mathbf{k} \\ &= \mathbf{T}\,ds \end{aligned}$$

where **T** is the unit vector tangent to C in the direction of increasing arc length. Hence the sum of the three integrals may be written

$$\int_C f(P)\,dx + \int_C g(P)\,dy + \int_C h(P)\,dz = \int_C \mathbf{F}\cdot\mathbf{T}\,ds = \int_C \mathbf{F}\cdot d\mathbf{r}.$$

For obvious reasons, this integral is called the *line integral of the tangential component of the vector field* **F**. Other than for a notational device, no new definitions are needed to define this integral since the three line integrals comprising the vector line integral have been defined previously.

Line integrals of vector fields are used to define physical quantities. For example, if a particle moves in a force field, **F**, then the *work*, W, done on the particle is given by

$$W = \int_C \mathbf{F}\cdot d\mathbf{r}.$$

This may be interpreted to be the integral of the tangential component of the force field acting on the particle. See Figure 8.5.

With the use of line integrals, an alternate (and perhaps more popular) definition of the elusive concept of the curl of a vector field is possible. Let

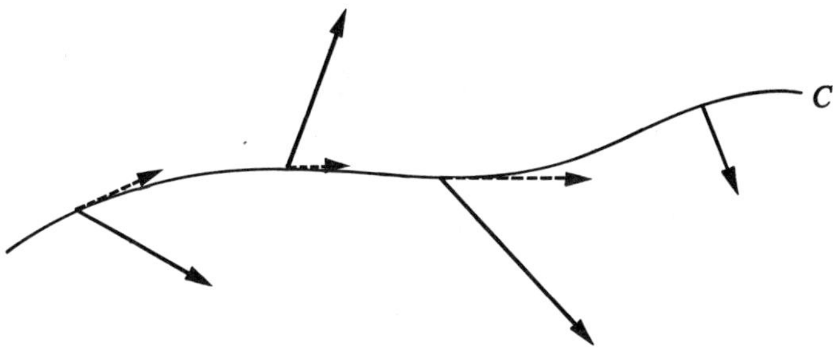

figure 8.5 *force field vectors (solid arrows) and components of the field tangential to C (dashed arrows)*

F be a vector field and $C(a, \Pi, P)$ a circle of radius a centered at P in the plane Π. Then the *circulation* of **F** around C is the line integral $\oint_C \mathbf{F}\cdot d\mathbf{r}$. Note that if **F** is a force field, then the circulation has the physical dimensions of work.

As you may find out by examples, for a fixed center, a fixed plane,

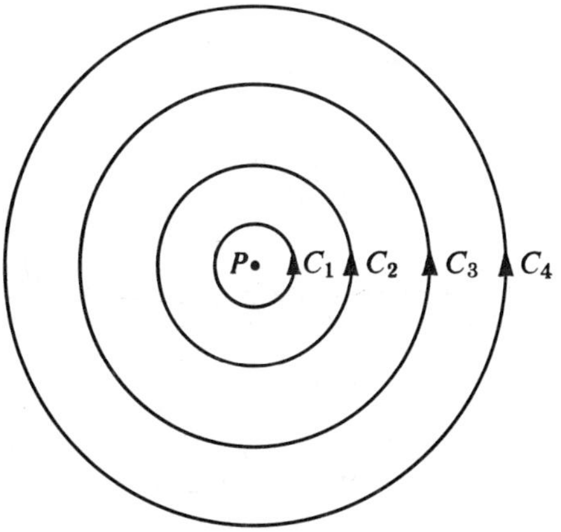

figure 8.6 *circulation of a vector field*

and variable radii, the circulation is roughly proportional to the square of the radius. (See Figure 8.6.) Therefore, we consider

$$\lim_{a\to 0} \frac{\text{circulation}}{\pi a^2}.$$

This limit will usually exist for each plane Π, but its value will change when the plane changes. To show simultaneously the direction of the normal to this plane and the value of the limit, we introduce a vector which will be called the *circulation density vector.*

1. This vector will have direction normal to the plane of C (in a direction, **N**, given by the right hand rule with respect to the direction of integration on C).
2. This vector will have magnitude equal to

$$\lim_{a \to 0} \frac{1}{\pi a^2} \oint_C \mathbf{F} \cdot d\mathbf{r}.$$

There are infinitely many of these circulation density vectors, one for each direction **N**. Out of this set of vectors only one is of practical importance.

definition 8.2

The *curl* of a vector field **F** at P is a vector with

(a) direction the same as that in which the magnitude of the circulation density vectors is maximal.

(b) magnitude equal to the maximum of the magnitudes of the circulation density vectors.

Hence, the curl is a vector whose direction gives the greatest positive value to

$$\lim_{a \to 0} \frac{1}{\pi a^2} \oint_C \mathbf{F} \cdot d\mathbf{r}$$

and whose magnitude is the value of this limit (as the direction changes). The fact that Definition 8.2 defines a unique vector will not be proven here.

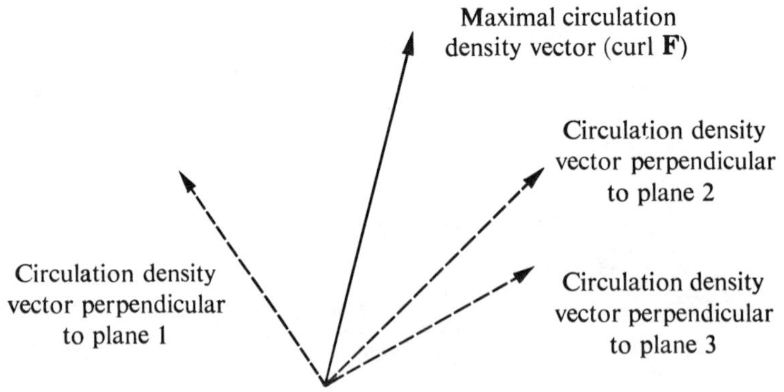

figure 8.7 *relation of curl to circulation density vectors*

The equivalence of this definition with Definition 7.5 in Section 7.6 is not at all obvious now but is more or less a consequence of Stokes' Theorem, to be given later in Section 8.10. You will find it useful to have both the integral and differential definitions available.

You can probably recognize that each of the circulation density vectors at a point P is the vector component of the curl vector in the direction of $\mathbf{N}$. Thus the circulation density vectors are unimportant since the curl determines each one. See Figure 8.7.

exercises for sections 8.1–8.3

1. Let C be any path connecting P_1 to P_2. Show by resorting to the definition of the line integral that

$$\int_C dx(P) = x(P_2) - x(P_1).$$

2. Generalize the result of Exercise 1; that is, let u be any real valued function defined on C. Prove that

$$\int_C du(P) = u(P_2) - u(P_1).$$

3. Use the definition of the line integral to approximate the work done in moving a particle along the path $x = y^2$ from the point $(1, 1)$ to $(9, 3)$ in the presence of the force field $\mathbf{F} = x^2\mathbf{i} + y^2\mathbf{j}$. Use the partition

$$B = \{(1, 1), (\tfrac{9}{4}, \tfrac{3}{2}), (4, 2), (\tfrac{25}{4}, \tfrac{5}{2}), (9, 3)\}.$$

4. Approximate $\int_C x^2y\,dx$ along a straight line path from $(1, 1)$ to $(5, 5)$. Use the partition

$$B = \{(1, 1), (\tfrac{3}{2}, \tfrac{3}{2}), (2, 2), (\tfrac{5}{2}, \tfrac{5}{2}), (3, 3), (\tfrac{7}{2}, \tfrac{7}{2}), (4, 4), (\tfrac{9}{2}, \tfrac{9}{2}), (5, 5)\}.$$

5. Approximate $\int_C x^2y\,dy$ with the same path and partition as in the previous exercise.

6. Approximate $\int_C x^2y\,ds$ with the same path and partition as in Exercise 4.

7. Approximate $\int_C y^2\,ds$ where C is the unit circle. Use the partition obtained by the 8 points, equally spaced on C, beginning at $(0, 1)$.

8. Repeat Exercise 7 for $\int_C y^2\,dy$.

9. Find the circulation density vector of the field $\mathbf{F} = y^2z\mathbf{i} + x^2\mathbf{j} + 2xy\mathbf{k}$ at the point $(-1, 3, 2)$ if the circulation is computed in the plane $x - 3y + z + 8 = 0$.

10. What is the maximum magnitude of the circulation density vector of the field $\mathbf{F} = 3yz\mathbf{i} - x^2z\mathbf{j} + xy\mathbf{k}$ at $(2, -1, 1)$? Write the equation of the plane in which the maximum circulation is computed.

8.4 evaluation of line integrals

The limit defining a line integral of a scalar function f along C is obviously quite complicated and would be impossible to compute directly except by numerical methods. Often, in practical situations, the curve may be parametrized and thus represented by a vector function $\mathbf{r}(t)$. Under those circumstances the evaluation of the line integral can be done by appeal to the following theorem.

theorem 8.2

Let C be represented by $\mathbf{r}(t)$ such that $\mathbf{r}'(t)$ exists and is nonzero over the interval $a \leqq t \leqq b$ for which the curve is defined. If f is continuous on C,

$$\int_C f(P)\,dx = \int_a^b f[x(t), y(t), z(t)]\frac{dx}{dt}\,dt.$$

The value obtained is independent of the method of parametrization.

Analogous theorems hold for $\int_C f(P)\,dy$, $\int_C f(P)\,dz$, and $\int_C f(P)\,ds$.

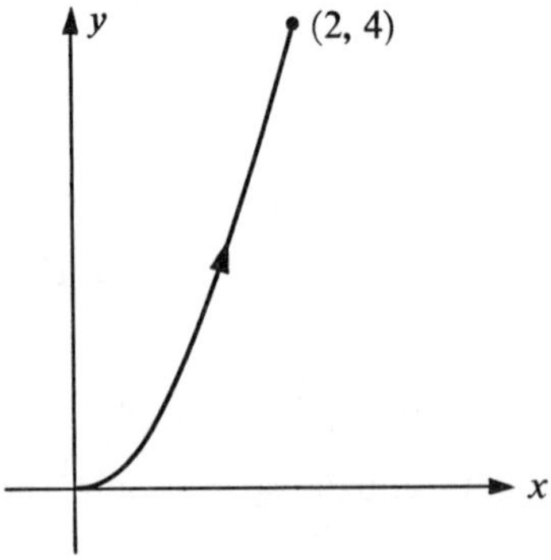

figure 8.8

example 8.3: Evaluate the line integral of the tangential component of the vector field $\mathbf{F} = x^2y\mathbf{i} + (x+y)\mathbf{j}$ along the curve $y = x^2$ from (0, 0) to (2, 4). (See Figure 8.8.)

solution: The curve C may be represented vectorially as $\mathbf{r}(t) = t\mathbf{i} + t^2\mathbf{j}$. Then

$$\begin{aligned}\int_C \mathbf{F}\cdot\mathbf{T}\,ds &= \int_C \mathbf{F}\cdot d\mathbf{r}\\ &= \int_C x^2y\,dx + \int_C (x+y)\,dy\\ &= \int_0^2 (t^2)(t^2)(1)\,dt + \int_0^2 (t+t^2)(2t)\,dt\\ &= \tfrac{32}{5} + (\tfrac{16}{3} + 8)\\ &= \tfrac{296}{15}.\end{aligned}$$

example 8.4: Let C be represented by $\mathbf{r}(t) = (1-t)\mathbf{i} + (1-t)\mathbf{j}$, $0 \leqq t \leqq 1$, and let $\mathbf{F}$ be as in the previous example. Evaluate

$$\int_C \mathbf{F}\cdot d\mathbf{r}.$$

(See Figure 8.9.)

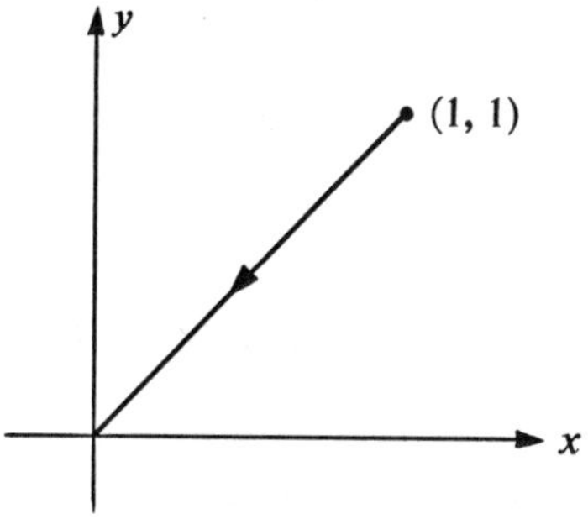

figure 8.9

solution:

$$\begin{aligned}\int_C \mathbf{F}\cdot d\mathbf{r} &= \int_C x^2y\,dx + \int_C (x+y)\,dy\\ &= \int_0^1 (1-t)^2(1-t)(-1)\,dt + \int_0^1 (1-t+1-t)(-1)\,dt\\ &= -\tfrac{5}{4}.\end{aligned}$$

example 8.5: Evaluate $\int_C \mathbf{F}\cdot d\mathbf{r}$ over the closed path consisting of $y = x^2$ and $y = x$, taken counterclockwise. Use the same field as in the two previous examples. (See Figure 8.10.)

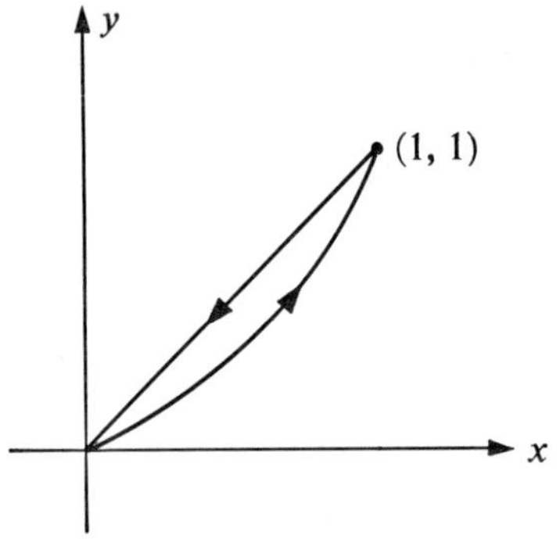

figure 8.10

solution: The given path is the union of the paths given by $\mathbf{r} = t\mathbf{i} + t^2\mathbf{j}$, $0 \leqq t \leqq 1$ and the path of the previous example. The value of $\int_C \mathbf{F}\cdot d\mathbf{r}$ over the first of these paths can be shown to be 41/30. By the additivity principle, the value of the integral is the sum of the values of the two separate integrals,

$$\oint_C \mathbf{F}\cdot d\mathbf{r} = \tfrac{41}{30} - \tfrac{5}{4}$$
$$= \tfrac{7}{60}.$$

Perhaps you thought initially that line integrals around closed paths should be zero. Example 8.5 should be evidence sufficient to keep you from ever believing that. The next example shows that a line integral around a closed path might indeed be zero. Then, in the following section, some necessary qualifications for this to be true are examined.

example 8.6: Let C be the circle of radius 1 in the plane $z = 1$ centered on the z-axis. Let $\mathbf{F} = (y + z)\,\mathbf{i} + (x + z)\,\mathbf{j} + (x + y)\,\mathbf{k}$. Compute $\oint_C \mathbf{F}\cdot d\mathbf{r}$ where the direction around the closed path is counterclockwise as viewed from the origin. (See Figure 8.11.)

solution: One convenient parametrization of C is $x = \cos t$, $y = \sin t$, and

$z = 1$. Using this, we get

$$\oint_C \mathbf{F}\cdot d\mathbf{r} = \oint_C (y+z)\,dx + \oint_C (x+z)\,dy + \oint_C (x+y)\,dz$$
$$= \int_0^{2\pi} (\sin t + 1)(-\sin t)\,dt + \int_0^{2\pi} (1+\cos t)(\cos t)\,dt.$$

(The third integral isn't shown since $dz/dt = 0$.) From this, it is an elementary problem to show that

$$\oint_C \mathbf{F}\cdot d\mathbf{r} = 0.$$

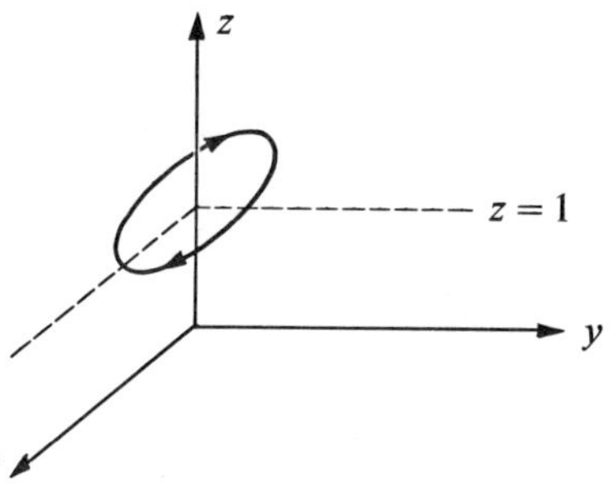

figure 8.11

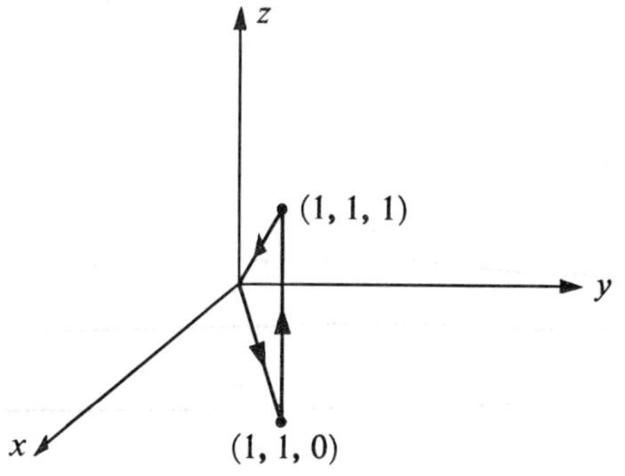

figure 8.12

example 8.7: A particle moves in a force field $\mathbf{F} = x\mathbf{i} - z\mathbf{j} + 2y\mathbf{k}$ along the path shown in Figure 8.12. Find the work done in moving the particle completely around the path.

solution: The path consists of three pieces which may be parametrized as follows:

$$\begin{aligned} \mathbf{r}(t) &= t\mathbf{i} + t\mathbf{j} & 0 \leqq t \leqq 1 \\ &= \mathbf{i} + \mathbf{j} + (t-1)\mathbf{k} & 1 \leqq t \leqq 2 \\ &= (3-t)\mathbf{i} + (3-t)\mathbf{j} + (3-t)\mathbf{k} & 2 \leqq t \leqq 3. \end{aligned}$$

(This is certainly not the only way to represent C and indeed may not be the simplest.) From the definition of work,

$$\begin{aligned} W &= \oint_C \mathbf{F}\cdot d\mathbf{r} \\ &= \oint_C x\,dx - \oint_C z\,dy + \oint_C 2y\,dz \\ &= \int_0^1 t\,dt + \int_1^2 2\,dt - \int_2^3 (3-t)\,dt + \int_2^3 (3-t)\,dt - \int_2^3 2(3-t)\,dt \\ &= \tfrac{1}{2} + 2 - \tfrac{1}{2} + \tfrac{1}{2} - 1 \\ &= \tfrac{3}{2}. \end{aligned}$$

8.5 independence of path

As you may have supposed, we have more than passing interest in determining conditions for which the integral of the tangential component of a vector field around a closed path is zero. Actually our discussion will concern those fields for which such an integration around *any* closed path is zero. In such fields, the value of a line integral over a path which is not closed is uniquely determined by the end points.

definition 8.3

The line integral $\int_C \mathbf{F}\cdot d\mathbf{r}$ is *independent of the path* connecting two points P and Q if the value of the integral is the same regardless of the path connecting the two points. If $\int_C \mathbf{F}\cdot d\mathbf{r}$ is independent of path, the usual notation is sometimes replaced by $\int_P^Q \mathbf{F}\cdot d\mathbf{r}$. (See Figure 8.13.)

The idea of independence of path is that a path connecting two points may be "continuously deformed" without affecting the value of the integral. As we will see the only stipulations are that the end points must remain fixed and the curve must stay entirely in a domain in which $\mathbf{F}$ is continuous. The concepts of independence of path and zero valued line integral around a closed path can be identified by using the following theorem.

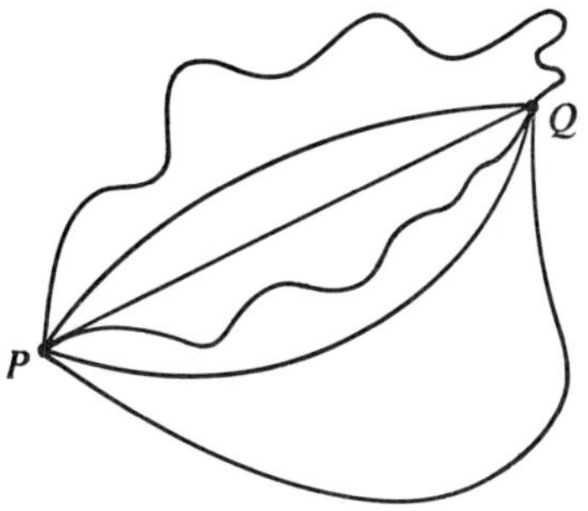

figure 8.13 deformation of path

theorem 8.3

A line integral $\int_C \mathbf{F}\cdot d\mathbf{r}$ is independent of the path connecting the end points of C in a domain† D if and only if $\oint_{C'} \mathbf{F}\cdot d\mathbf{r} = 0$ for all closed paths C' in D.

proof: Suppose $\int_C \mathbf{F}\cdot d\mathbf{r}$ is independent of path. Select two points on a closed curve C' and use them to cut C' into two arcs C_1 and C_2, where C_1 is from P to Q and C_2 is taken from Q to P. (See Figure 8.14.) Then

$$\oint_{C'} \mathbf{F}\cdot d\mathbf{r} = \int_{C_1} \mathbf{F}\cdot d\mathbf{r} + \int_{C_2} \mathbf{F}\cdot d\mathbf{r}.$$

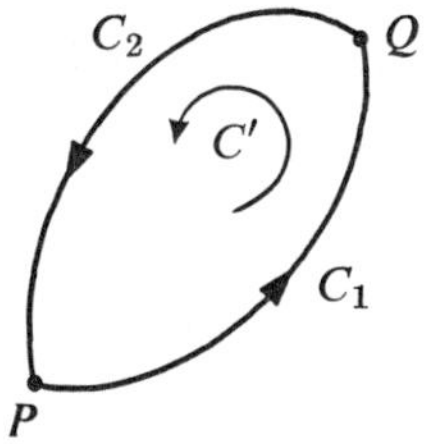

figure 8.14

Because of the independence of path,

$$\int_{C_1} \mathbf{F}\cdot d\mathbf{r} = \int_{-C_2} \mathbf{F}\cdot d\mathbf{r}$$

† For our purposes a domain is an open connected set. Loosely speaking, a connected set is one which is all in one piece, or one in which any two points may be connected by a path which remains in the set.

where the curve $-C_2$ is obtained from C_2 by a reversal of sense and thus is oriented from P to Q. Thus

$$\int_{C_1} \mathbf{F}\cdot d\mathbf{r} = -\int_{C_2} \mathbf{F}\cdot d\mathbf{r}$$

and hence

$$\oint_{C'} \mathbf{F}\cdot d\mathbf{r} = 0\,.$$

The converse of the theorem is proved similarly and is left for the exercises.

When a conservative vector field was defined in Section 7.4, you may have wondered about the aptness of the word conservative. The next theorem should give you a motive for describing vector fields with scalar potentials as "conservative." At the same time, the theorem also gives a well known method for evaluating a line integral of the tangential component of a conservative vector field. The method looks a bit like the fundamental theorem of calculus, and with good reason.

theorem 8.4

Let $\mathbf{F}$ be a vector field with continuous component functions f, g, and h. The line integral $\int_C \mathbf{F}\cdot d\mathbf{r}$ is independent of path in a domain D if and only if $\mathbf{F}$ is conservative. Further, if U is the scalar potential of $\mathbf{F}$ (that is, $\nabla U = \mathbf{F}$), and P and Q are endpoints of C, then

$$\int_C \mathbf{F}\cdot d\mathbf{r} = U(Q) - U(P)\,.$$

proof: First assume $\int_C \mathbf{F}\cdot d\mathbf{r}$ is independent of path. To show $\mathbf{F}$ to be conservative requires what is called a "bootstraps" operation. For the scalar potential U must actually be constructed. Let P_1 be a fixed point in the domain and define the function U to be

$$U(P) = \int_C \mathbf{F}\cdot d\mathbf{r}$$

where C is any path connecting P_1 to P. The fact that a function has truly been defined follows from the hypothesis.

To show that $\mathbf{F}$ is conservative we must show $U_x = f$, $U_y = g$, and $U_z = h$ where f, g, and h are the component functions of $\mathbf{F}$. We will show only the first of these three equalities; the others are proved similarly. To find U_x, first form the difference quotient obtained by evaluating U at the points

(x, y, z) and $(x + \Delta x, y, z)$,

$$\frac{U(x + \Delta x, y, z) - U(x, y, z)}{\Delta x}$$

where $U(x + \Delta x, y, z)$ is evaluated by using the same path from P_1 to (x, y, z) as is used for $U(x, y, z)$ together with a straight line path from (x, y, z) to $(x + \Delta x, y, z)$. (See Figure 8.15.) The numerator of the difference quotient

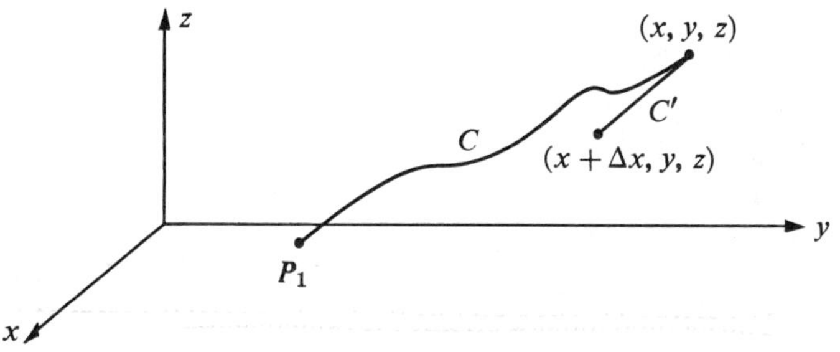

figure 8.15

equals $\int_{C'} \mathbf{F} \cdot d\mathbf{r}$ where C' is the straight line path connecting (x, y, z) to $(x + \Delta x, y, z)$. Since y and z are fixed on C', $\int_{C'} \mathbf{F} \cdot d\mathbf{r} = \int_{C'} f\,dx$. By the mean value theorem for integrals

$$\int_{C'} f\,dx = f(x_i, y, z)\,\Delta x$$

where x_i is between x and $x + \Delta x$. Therefore the difference quotient becomes

$$\frac{f(x_i, y, z)\,\Delta x}{\Delta x} = f(x_i, y, z).$$

Since the component functions are assumed to be continuous, the limit of the difference as Δx approaches 0 is $f(x, y, z)$, that is, $U_x = f$.

To prove the converse, assume that the scalar potential of $\mathbf{F}$ is U. Then

$$\begin{aligned}\int_C \mathbf{F} \cdot d\mathbf{r} &= \int_C f\,dx + g\,dy + h\,dz \\ &= \int_C U_x\,dx + U_y\,dy + U_z\,dz \\ &= \int_C dU \\ &= U(P_2) - U(P_1).\end{aligned}$$

(See Exercise 2, Exercises for Sections 8.1-8.3.)

The theorem not only tells us that a line integral of the tangential component of a conservative vector field is independent of path but also gives us a technique of evaluation based upon finding the scalar potential and computing the "potential difference."

example 8.8: Show that $\mathbf{F} = yz\mathbf{i} + xz\mathbf{j} + xy\mathbf{k}$ is conservative and then compute $\int_C \mathbf{F}\cdot d\mathbf{r}$ along any path connecting $(1, -1, 3)$ to $(5, 2, -1)$.

solution: It is easy to show that the scalar potential is the function $U(x, y, z) = xyz$ (see Section 7.4) and hence the value of the line integral is $U(5, 2, -1) - U(1, -1, 3) = -10 - (-3) = -7$.

theorem 8.5

If a line integral $\int_C \mathbf{F}\cdot d\mathbf{r}$ is independent of path in D, then $\nabla \times \mathbf{F} = \mathbf{0}$ throughout D.

proof: This theorem follows immediately from Theorem 8.4 and the knowledge that the curl of a field with a scalar potential is zero.

exercises for sections 8.4 and 8.5

1. Complete the proof of Theorem 8.3.
2. Evaluate $\int_C \mathbf{F}\cdot d\mathbf{r}$ if $\mathbf{F} = (2x + y)\,\mathbf{i} + x\mathbf{j}$ and C is:
   (a) the straight line from $(0, 0)$ to $(3, 2)$.
   (b) the y-axis from $(0, 0)$ to $(0, 2)$ and then the straight line from $(0, 2)$ to $(3, 2)$.
   (c) $\mathbf{r}(t) = 3t^2\mathbf{i} + 2t\mathbf{j}$ from $t = 0$ to $t = 1$.
3. Evaluate $\int_C \mathbf{F}\cdot d\mathbf{r}$ if $\mathbf{F} = x^2y\mathbf{i} + y^2z\mathbf{j} + z^2x\mathbf{k}$ and C is:
   (a) the straight line from $(0, 0, 0)$ to $(1, 3, 3)$.
   (b) the polygonal figure connecting $(0, 0, 0)$ to $(1, 0, 0)$ to $(1, 3, 0)$ to $(1, 3, 3)$.
   (c) $\mathbf{r}(t) = t\mathbf{i} + 3t^2\mathbf{j} + 3t^3\mathbf{k}$ from $t = 0$ to $t = 1$.
4. Find the work done in moving a particle in a force field $\mathbf{F} = x^2y\mathbf{i} - yz\mathbf{j} + z\mathbf{k}$ along the path $\mathbf{r}(t) = t\mathbf{i} + t^2\mathbf{j} + t^3\mathbf{k}$ from the origin to the point $(2, 4, 8)$. Repeat for the straight line connecting the two points.
5. Evaluate $\oint_C \mathbf{F}\cdot d\mathbf{r}$ around the complete unit circle in the plane $z = 0$, taken counterclockwise, if $\mathbf{F} = x^2\mathbf{i} + y\mathbf{j}$.

6. Let C be the unit semicircle $\mathbf{r}(t) = \cos t\mathbf{i} + \sin t\mathbf{j}$, $0 \leqq t \leqq \pi$. Evaluate $\int_C \mathbf{F} \cdot d\mathbf{r}$ if $\mathbf{F} = x^2y\mathbf{i} + (x+y)\mathbf{j}$. Repeat for the straight line connecting the two endpoints of C.

7. Evaluate $\oint_C \mathbf{F} \cdot \mathbf{T}\, ds$ if:
   (a) $\mathbf{F} = -3y\mathbf{i} + 3x\mathbf{j} + z\mathbf{k}$ and C is the unit circle centered at (0, 0, 1), in the plane $z = 1$, oriented counterclockwise as viewed from the origin.
   (b) $\mathbf{F} = y\mathbf{i} + xz^3\mathbf{j} - zy^3\mathbf{k}$ and C is the unit circle centered at (0, 0, -1) oriented counterclockwise as viewed from the origin.
   (c) $\mathbf{F} = \dfrac{-y\mathbf{i} + x\mathbf{j}}{x^2 + y^2}$, and C is the unit circle centered at (0, 0, 0) oriented counterclockwise as viewed from above.
   (d) $\mathbf{F} = (\sin z)\mathbf{i} - (\cos x)\mathbf{j} + (\sin y)\mathbf{k}$, and C is the boundary of the rectangle $0 \leqq x \leqq 1$, $0 \leqq y \leqq 2$, $z = 3$ oriented clockwise as viewed from the origin.

8. Evaluate $\int_C \mathbf{F} \cdot d\mathbf{r}$ if:
   (a) $\mathbf{F} = (y-7)\mathbf{i} + x\mathbf{j}$ and C is the path $\mathbf{r}(t) = \sin t^2\,\mathbf{i} + t\mathbf{j}$ connecting (0, 0) to $(1, \sqrt{\pi/2})$.
   (b) $\mathbf{F} = (2y^2 - 3x)\mathbf{i} - 4xy\mathbf{j}$ and C is the straight line connecting (0, 0) to (2, 2).
   (c) $\mathbf{F} = yz\mathbf{i} + xz\mathbf{j} + xy\mathbf{k}$ and C is the closed path $x = y^2$, $y = x^2$ taken counterclockwise in the plane $z = 2$.
   (d) $\mathbf{F} = (2xy + x^2)\mathbf{i} + x^2\mathbf{j}$ and C is the parabola connecting $(-1, 1)$ to (1, 1).
   (e) $\mathbf{F} = yz\cos(xz)\mathbf{i} + \sin(xz)\mathbf{j} + xy\cos(xz)\mathbf{k}$ and C is any path connecting (1, 1, 1) to (2, 3, 4).

9. Let $\mathbf{F} = (y^2 + 2xz)\mathbf{i} + (2xy - z)\mathbf{j} + (x^2 - y)\mathbf{k}$. Evaluate $\int_C \mathbf{F} \cdot d\mathbf{r}$ where C is:
   (a) the ellipse $4x^2 + y^2 = 1$ beginning at (0, 1) oriented clockwise, in the plane $z = 0$.
   (b) the path $\mathbf{r}(t) = \sin^3 \pi t/2\,\mathbf{i} + t\cos \pi t\,\mathbf{j} + t^2 \cos^2 \pi t\,\mathbf{k}$ from (0, 0, 0) to $(1, -1, 1)$.
   (c) the straight line path from (0, 0, 0) to (2, 4, 7).

10. Let $\mathbf{F} = 3z^2\mathbf{i} + 2y\mathbf{j} + 6xz\mathbf{k}$ and repeat the previous exercise.

8.6 surface area

In this section you will learn a "reasonable" definition of surface area in terms of elements on the parameter plane. The eventual understanding of the more general concept of a surface integral is almost completely dependent upon agreements reached with respect to an elemental piece of surface. There are several approaches to the problem of how to define surface area, some of them based upon practical considerations. First we will show you a method applicable to the most general kind of surface and then two other approaches that are used in specific cases.

Let S be a surface defined by $\mathbf{r}(u, v)$ where the (u, v) are restricted to a domain S'. Then S is the image of S'. The image of a rectangular element $\Delta S'(=\Delta u\,\Delta v)$ is a curvilinear parallelogram on S. (See Figure 8.16.) The

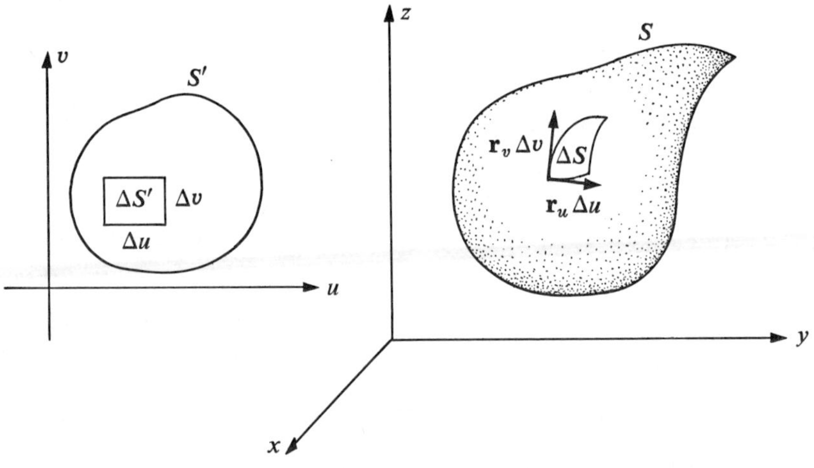

figure 8.16 parameter plane and surface

sides of the curvilinear parallelogram are parts of coordinate curves, $\mathbf{r}(u_0, v)$, $\mathbf{r}(u_0 + \Delta u, v)$, $\mathbf{r}(u, v_0)$, and $\mathbf{r}(u, v_0 + \Delta v)$. Thus the lengths of the sides are approximately $|\mathbf{r}_u\,\Delta u|$ and $|\mathbf{r}_v\,\Delta v|$ and therefore, also approximately,

$$\begin{aligned}\Delta S &= |(\mathbf{r}_u\,\Delta u) \times (\mathbf{r}_v\,\Delta v)| \\ &= |\mathbf{r}_u \times \mathbf{r}_v|\,\Delta u\,\Delta v.\end{aligned}$$

You should see, then, why it is a reasonable thing to *define* the total surface area to be

$$S = \iint_{S'} |\mathbf{r}_u \times \mathbf{r}_v|\,dA_{uv}$$

where dA_{uv} represents an element of area in the uv-plane. For reasons more apparent in a general setting, the notation for surface area is $\iint_S dS$. This notation will mean more to you after you read Section 8.7.

example 8.9: Find the surface area of the portion of the parabolic cylinder $y = x^2$ in the first octant bounded by the plane $z = 2$ and $y = 1/4$.

solution: The surface in question may be parametrized by letting $x = u$ and $z = v$ (the xz-plane is the parameter plane). Thus $\mathbf{r}(u, v) = u\mathbf{i} + u^2\mathbf{j} + v\mathbf{k}$ and, by the formula of Example 6.4

$$|\mathbf{r}_u \times \mathbf{r}_v|^2 = (\mathbf{r}_u \cdot \mathbf{r}_u)(\mathbf{r}_v \cdot \mathbf{r}_v) - (\mathbf{r}_u \cdot \mathbf{r}_v)^2$$
$$= 1 + 4u^2.$$

Therefore,

$$S = \iint_{S'} (1 + 4u^2)^{1/2}\, dA_{uv}, \qquad S' = \left\{(u, v) \,\middle|\, \begin{matrix} 0 \leqq u \leqq \frac{1}{2} \\ 0 \leqq v \leqq 2 \end{matrix}\right\}$$
$$= \int_0^2 \int_0^{1/2} (1 + 4u^2)^{1/2}\, du\, dv$$
$$= \frac{\sqrt{2}}{2} + \tfrac{1}{4} \ln(1 + \sqrt{2}).$$

You will see the formula for surface area written differently from the one we gave and thus you should see how the formula looks in some special cases. If the surface is represented in the form $z = h(x, y)$, then we may think of the surface S as covering some region S' in the xy-plane; that is, S' is the projection of S on the xy-plane. Let γ be the angle between the upward directed normal and the z-axis, and consider an element of surface ΔS. (See Figure 8.17.) Then ΔS and $\Delta S'$ are approximately related by

$$\Delta S' = \Delta S \cos\gamma,$$

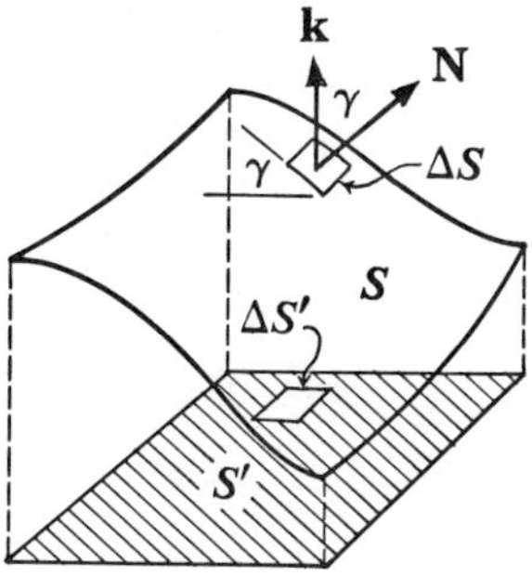

figure 8.17 *projection of surface on xy-plane*

or

$$\Delta S = \Delta S' \sec\gamma .$$

(See Figure 8.17.) We thus *define* the total surface area to be

$$S = \iint_{S'} \sec\gamma \, dA_{xy}$$

where S' is the projection of S on the xy-plane.

The fact that the two definitions are equivalent is shown by noting that the explicit equation $z = h(x, y)$ gives the vector representation $\mathbf{r}(u, v) = u\mathbf{i} + v\mathbf{j} + h(u, v)\,\mathbf{k}$. Hence

$$|\mathbf{r}_u \times \mathbf{r}_v| = \sqrt{h_u^2 + h_v^2 + 1}$$

from which

$$S = \iint_{S'} \sqrt{h_u^2 + h_v^2 + 1} \, du \, dv$$
$$= \iint_{S'} \sqrt{h_x^2 + h_y^2 + 1} \, dy \, dx .$$

For the surface defined by $z - h(x, y) = 0$, it is easy to show that a unit normal to the surface is

$$\mathbf{N} = \frac{-h_x\mathbf{i} - h_y\mathbf{j} + \mathbf{k}}{\sqrt{h_x^2 + h_y^2 + 1}}$$

from which

$$\cos\gamma = \frac{1}{\sqrt{h_x^2 + h_y^2 + 1}},$$

and hence the integrand of the integral over S' is just $\sec\gamma$ which proves the equivalence of the two definitions.

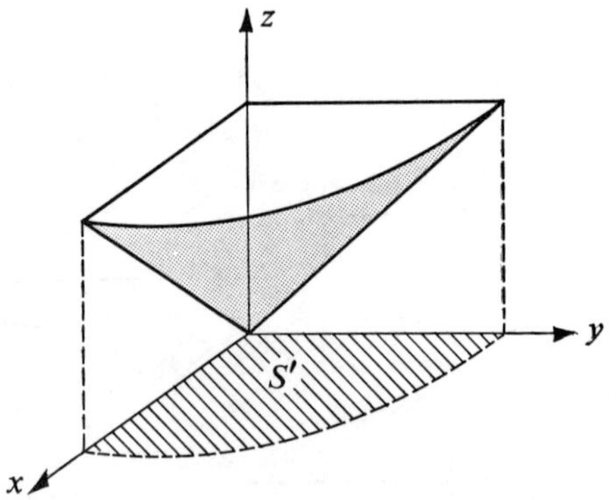

figure 8.18

example 8.10: Find the area of the surface $z^2 = x^2 + y^2, -1 \leqq z \leqq 1$. (See Figure 8.18.)

solution: Consider the part of the surface on which z is positive. To find a normal to the surface, we consider a constant level surface of the function $\phi(x, y, z) = x^2 + y^2 - z^2$. Then

$$\mathbf{N} = \frac{\nabla\phi}{|\nabla\phi|} = \frac{2x\mathbf{i} + 2y\mathbf{j} - 2z\mathbf{k}}{(4x^2 + 4y^2 + 4z^2)^{1/2}} = \frac{x\mathbf{i} + y\mathbf{j} - z\mathbf{k}}{z\sqrt{2}}.$$

Therefore

$$\sec\gamma = \frac{1}{|\mathbf{N}\cdot\mathbf{k}|} = \sqrt{2}.$$

The expression for the surface area corresponding to positive z is

$$S = \iint_{S'} \sqrt{2}\, dA_{xy} = \sqrt{2}\int_{-1}^{1}\int_{-(1-x^2)^{1/2}}^{(1-x^2)^{1/2}} dy\, dx = \pi\sqrt{2}.$$

The surface area for negative z obviously has the same value, and hence the total surface area is $2\pi\sqrt{2}$.

If S is represented in the form $y = g(x, z)$ or $x = f(y, z)$, the definition for surface area leads to the integrals $\iint_{S'} \sec\beta\, dA_{xz}$ and $\iint_{S'} \sec\alpha\, dA_{yz}$ where α, β, and the region of integration are analogous to the case described above. You will be asked to supply the details in the exercises. In all the cases, you will note that the problem of finding the value of the surface area is reduced to a problem of double integration in the parameter plane or (which is the same thing) in the plane onto which the surface projects.

There is another method often used for the computation of surface area which is quite straightforward but which requires some prior knowledge of the construction of an element of surface. This technique is especially valuable in connection with surfaces such as right circular cylinders or spheres. For example, if θ and z are the parameters for a cylinder of fixed radius a, then one could reason directly that the element of surface area is $a\, d\theta\, dz$. Similarly, one often reasons informally that the element of surface

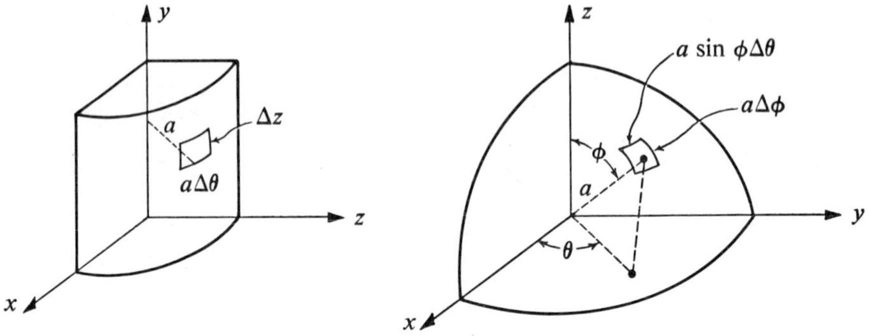

figure 8.19 *elements of surface area on a cylinder and sphere*

area on a sphere of fixed radius is $a^2 \sin\phi \, d\theta \, d\phi$. (See Figure 8.19.) The informal reasoning is done by appealing to the graphs of a right circular cylinder and a sphere but the validity of the reasoning is verified by considering the vector representations

$$\mathbf{r}(\theta, z) = (a\cos\theta)\,\mathbf{i} + (a\sin\theta)\,\mathbf{j} + z\mathbf{k}$$

for a cylinder and

$$\mathbf{r}(\theta, \phi) = (a\sin\phi\cos\theta)\,\mathbf{i} + (a\sin\phi\sin\theta)\,\mathbf{j} + (a\cos\phi)\,\mathbf{k}$$

for a sphere. The details are left for the exercises.

8.7 surface integrals

Integrals of functions on surfaces ("surface integrals") are defined just like the other integration processes except that the element of a partition is an elemental surface area. Let f be a scalar function defined and continuous on a surface S. After subdividing S into n subregions each of area ΔS_i, we pick a point within each subregion, P_i', and evaluate $f(P_i')$. If the limit of the finite sum

$$\sum_{i=1}^{n} f(P_i')\,\Delta S_i$$

exists as the largest of the ΔS_i approaches zero, the limit is called the *surface integral of the scalar function f over S* and is denoted by

$$\iint_S f(P)\,dS.$$

(Sometimes only one integral sign is used; the reason for the use of two is largely historical.)

In keeping with the tone of this book, we will not concern ourselves with examining the conditions to be imposed on the function and the surface, necessary for the surface integral to exist. Suffice it to say that if f is continuous and S is piecewise smooth then the integral of f on S exists.

The notation for a surface integral uses two integral signs just as the double integral in a plane. Do not confuse the two – conceptually they are quite different, though notationally similar.

In practice you will have to compute few, if any, surface integrals, not because of a lack of usefulness but only because lengthy calculations would take too much of your time and because the integrands often contain a square root which makes integration in terms of elementary functions impossible. However, you should have at least a passing acquaintance with some of the techniques. There are three complications of the procedure.

1. Finding the element of surface area, dS.
2. Finding the limits of integration for iterated integrals.
3. Finding antiderivatives for the integrands.

If you are reasonably familiar with the material of the previous section, then the only added complication is that you must now handle integrand functions other than a constant. If $f(x, y, z)$ is a scalar valued function and the surface S is represented by

$$\mathbf{r}(u, v) = x(u, v)\,\mathbf{i} + y(u, v)\,\mathbf{j} + z(u, v)\,\mathbf{k},$$

then the first step in the calculations is usually to find and simplify the function $g(u, v)$ defined by

$$g(u, v) = f\,[x(u, v), y(u, v), z(u, v)].$$

Thus we replace the function f defined over the whole space by the function g defined only for points on the parameter plane.

To summarize some of the foregoing, we now state a theorem which is the analogue of Theorem 8.2.

theorem 8.6

Let f be continuous on a smooth orientable[†] surface S. Then $\iint_S f(P)\,dS$ exists and equals

$$\iint_{S'_{uv}} f[x(u, v), y(u, v), z(u, v)]\,|\mathbf{r}_u \times \mathbf{r}_v|\,du\,dv.$$

† See Section 6.8.

As in the previous section, the factor $|\mathbf{r}_u \times \mathbf{r}_v|$ may be replaced by $\sec\gamma$ if the xy-plane serves as the parameter plane; or by some other appropriate factor dependent upon the method used to represent the surface. Sometimes, several different approaches are possible.

example 8.11: Evaluate the surface integral of the function

$$f(x, y, z) = \frac{2y^2 + z}{(4x^2 + 4y^2 + 1)^{1/2}}$$

over the surface of the paraboloid

$$x^2 + y^2 + z = 4$$

for which z is positive. (See Figure 8.20.)

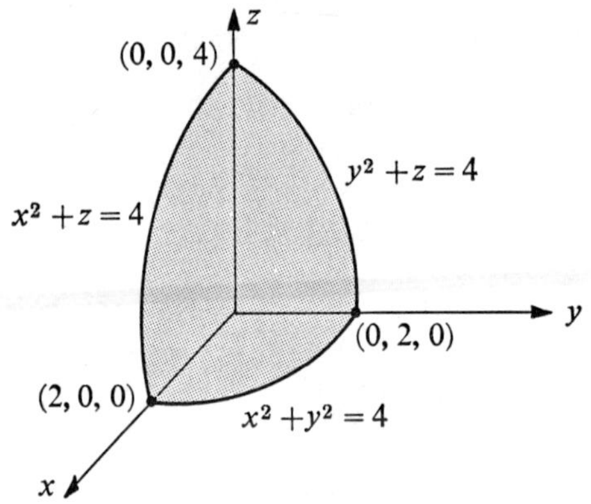

figure 8.20

solution: $$\iint_S f(x, y, z)\, dS = \iint_{S'_{xy}} \frac{2y^2 + z(x, y)}{(4x^2 + 4y^2 + 1)^{1/2}} \sec\gamma\, dy\, dx.$$

A normal to S is found by considering a constant level surface of the function $\phi(x, y, z) = x^2 + y^2 + z$.

$$\mathbf{N} = \frac{\nabla\phi}{|\nabla\phi|} = \frac{2x\mathbf{i} + 2y\mathbf{j} + \mathbf{k}}{(4x^2 + 4y^2 + 1)^{1/2}}.$$

$$\sec\gamma = \frac{1}{|\mathbf{N}\cdot\mathbf{k}|} = (4x^2 + 4y^2 + 1)^{1/2}.$$

Using this value in the double integral over the region S'_{xy} in the xy-plane, we obtain

$$\iint_S f(x, y, z)\, dS = \int_{-2}^{2} \int_{-(4-x^2)^{1/2}}^{(4-x^2)^{1/2}} (4 - x^2 + y^2)\, dy\, dx.$$

Using polar coordinates, this becomes,

$$\begin{aligned}\int_0^{2\pi} \int_1^2 (4 + r^2 - 2r^2 \cos^2\theta)\, r\, dr\, d\theta &= \int_0^{2\pi} \left(2r^2 + \frac{r^4}{4} - \frac{r^4}{2}\cos^2\theta\right)\Bigg|_0^2 d\theta \\ &= \int_0^{2\pi} (12 - 8\cos^2\theta)\, d\theta \\ &= 24\pi - 8\pi \\ &= 16\pi.\end{aligned}$$

In the next example, the surface is the same as for the one we just used when projecting onto the xy-plane. However, in this case we choose to project onto a different coordinate plane because of the nature of the function to be integrated. Projecting onto any other plane would make a very complicated problem!

example 8.12: Let $\mathbf{N}$ be the outward directed normal to the surface of the previous example, and let β be the angle between $\mathbf{N}$ and the positive y-axis. Let $f(x, y, z) = y \cos\beta$. Evaluate $\iint_S f(P)\, dS$.

solution:

$$\begin{aligned}\iint_S f(P)\, dS &= \iint_{S'_{xz}} y \cos\beta\, |\sec\beta|\, dx\, dz \\ &= \iint_{S'_{xz}} \pm\, y\, dx\, dz\end{aligned}$$

where the plus sign is used if the angle is acute and the minus sign otherwise. For the part of the surface on which y is positive, β is acute and when y is negative, β is obtuse. Hence

$$\iint_S f(P)\, dS = \iint_{S'_+} y(x, z)\, dA_{xz} - \iint_{S'_-} y(x, z)\, dA_{xz}$$

where S'_+ is the projection onto the xz-plane of the portion of S for which y is positive and S'_- is the projection corresponding to negative y.

Since for positive y, $y = (4 - x^2 - z)^{1/2}$
and for negative y, $y = -(4 - x^2 - z)^{1/2}$,

$$\begin{aligned}
\iint_S f(P)\,dS &= 2\iint_{S'} (4 - x^2 - z)^{1/2}\,dA_{xz} \\
&= 2\int_{-2}^{2}\int_0^{4-x^2} (4 - x^2 - z)^{1/2}\,dz\,dx \\
&= 2\left(\frac{-2}{3}\right)\int_{-2}^{2} (4 - x^2 - z)^{3/2}\Big|_0^{4-x^2}\,dx \\
&= \frac{4}{3}\int_{-2}^{2} (4 - x^2)^{3/2}\,dx \\
&= \frac{8}{3}\int_0^{2} (4 - x^2)^{3/2}\,dx.
\end{aligned}$$

Using the substitution $x = 2\sin\theta$, we obtain

$$\begin{aligned}
\iint_S f(P)\,dS &= \frac{128}{3}\int_0^{\pi/2} \cos^4\theta\,d\theta \\
&= \frac{128}{3}\left(\frac{3}{8}\theta + \frac{1}{4}\sin 2\theta + \frac{1}{32}\sin 4\theta\right)\Bigg|_0^{\pi/2} \\
&= 8\pi.
\end{aligned}$$

For surface integrals we have a property analogous to that of independence of path for line integrals.

definition 8.4

The surface integral

$$\iint_S f(P)\,dS$$

is said to be *independent of the surface* with boundary curve C if the numerical value is the same for each surface with the same boundary curve.

Some people have suggested that the concept of independence of surface can best be described by a butterfly net in which the surface (the net) may be deformed while holding the boundary curve (the rim) constant. See Figure 8.21.

In Section 8.10 we will determine a condition for a certain class of surface integrals to be independent of surface. For now we will be content with a theorem which resembles Theorem 8.3

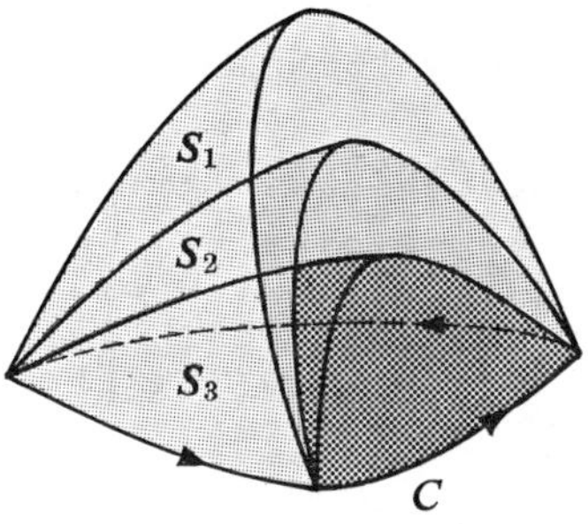

figure 8.21 *independence of surface*

theorem 8.7

A surface integral

$$\iint_S f(P)\, dS$$

is *independent of the surface*, *S*, with boundary curve *C* in a domain *D* if and only if

$$\iint_{S'} f(P)\, dS = 0$$

for all closed† surfaces, S', in *D*.

proof: You should prove the Theorem as an exercise for this section. (Compare the Proof of Theorem 8.3.)

8.8 surface integrals of vector fields

You may recall that the discussion of line integrals in a vector field was merely an agreement on the notation used to represent the integral of the *tangential* component of the field. In that case, the vector line integral could be written as the sum of three line integrals of the component functions. You will see now that we will make a similar agreement for the surface integral of the component of the field *normal* to the surface. As you could easily guess, this kind of surface integral is of foremost importance in the applications. It is denoted by

$$\iint_S \mathbf{F}\cdot\mathbf{N}\, dS$$

† A domain in three-dimensional space is *bounded* if it is contained within a sphere of finite radius. A surface is said to be *closed* if it is the boundary of some bounded domain.

and can be evaluated by finding the dot product of **F** and **N**, and integrating the resulting scalar function by the methods of the previous section.

example 8.13: Let $\mathbf{F} = y\mathbf{j} + z\mathbf{k}$ and let S be that portion of the surface of the paraboloid $x^2 + y^2 = 4 - z$ which lies above the xy-plane. If **N** is the upward directed normal to the surface, evaluate $\iint_S \mathbf{F}\cdot\mathbf{N}\,dS$. (See Figure 8.20.)

solution: By Example 8.11,

$$\mathbf{N} = \frac{2x\mathbf{i} + 2y\mathbf{j} + \mathbf{k}}{(4x^2 + 4y^2 + 1)^{1/2}}$$

and hence

$$\mathbf{F}\cdot\mathbf{N} = \frac{2y^2 + z}{(4x^2 + 4y^2 + 1)^{1/2}},$$

from which

$$\iint_S \mathbf{F}\cdot\mathbf{N}\,dS = 16\pi.$$

See Example 8.11 for the technique of evaluating this surface integral.

There is an alternate notation for

$$\iint_S \mathbf{F}\cdot\mathbf{N}\,dS$$

which can be a bit ambiguous but nonetheless is somewhat popular. If S is orientable, then at each point of S we can designate a positive unit normal **N** by $\mathbf{N} = \cos\alpha\,\mathbf{i} + \cos\beta\,\mathbf{j} + \cos\gamma\,\mathbf{k}$. If **F** has component functions f, g, and h, the surface integral can be written

$$\iint_S \mathbf{F}\cdot\mathbf{N}\,dS = \iint_S f\cos\alpha\,dS + \iint_S g\cos\beta\,dS + \iint_S h\cos\gamma\,dS.$$

The integrands of each of the three surface integrals on the right are similar to those used in Example 8.12. If we assume that S can be projected onto all three coordinate planes, then by reasoning similar to that used in Example 8.12,

$$\iint_S \mathbf{F}\cdot\mathbf{N}\,dS = \iint_{S'_{yz}} f\cos\alpha\,|\sec\alpha|\,dA_{yz} + \iint_{S'_{xz}} g\cos\beta\,|\sec\beta|\,dA_{xz}$$

$$+ \iint_{S'_{xy}} h\cos\gamma\,|\sec\gamma|\,dA_{xy}$$

$$= \pm\iint_{S'_{yz}} f\,dA_{yz} \pm \iint_{S'_{xz}} g\,dA_{xz} \pm \iint_{S'_{xy}} h\,dA_{xy}$$

where in each case the positive sign is used if the corresponding direction angle of the normal is acute and negative if it is obtuse. (Be sure that you can tell the difference between the surface integrals and the double integrals in the foregoing equation.)

Generally, the method described here is of little importance as an evaluation technique since the method of Example 8.13 will usually suffice. You should have a thorough understanding of the notation since it may confront you in the applications. Particularly, be sure you see how the surface integral of the normal component of the field can be expressed as the sum of three double integrals over regions which are projections of the surface on the coordinate planes.

example 8.14: Do Example 8.13 another way.

solution: If we let **N** be $\cos\alpha\,\mathbf{i} + \cos\beta\,\mathbf{j} + \cos\gamma\,\mathbf{k}$, then

$$\iint_S \mathbf{F}\cdot\mathbf{N}\,dS = \iint_S y\cos\beta\,dS + \iint_S z\cos\gamma\,dS.$$

The first of the surface integrals on the right has been evaluated in Example 8.12 and is equal to 8π. Since **N** is pointing upward, $\cos\gamma$ is positive and hence

$$\iint_S z\cos\gamma\,dS = \iint_{S'_{xy}} z(x, y)\,dA_{xy}$$
$$= \int_{-2}^{2}\int_{-(4-x^2)^{1/2}}^{(4-x^2)^{1/2}} (4 - x^2 - y^2)\,dy\,dx.$$

Using polar coordinates, this becomes,

$$= \int_0^{2\pi}\int_0^2 (4 - r^2)\,r\,dr\,d\theta$$
$$= \int_0^{2\pi}\left(2r^2 - \frac{r^4}{4}\right)\Big|_0^2\,d\theta$$
$$= \int_0^{2\pi} 4\,d\theta$$
$$= 8\pi$$

and hence $\iint_S \mathbf{F}\cdot\mathbf{N}\,dS = 16\pi$ as before.

exercises for sections 8.6–8.8

1. Show that the element of surface area for a sphere or a cylinder can be obtained formally from the definition and is the same as that which we derived heuristically.

2. Reason to the formulas for surface area $\iint_{S'_{xz}} \sec\beta \, dA_{xz}$ and $\iint_{S'_{yz}} \sec\alpha \, dA_{yz}$ without the benefit of the definition of surface area.

3. Using the general formula for surface area, find the area of the surface $x^2 + y^2 = 4$ from $z = 0$ to $z = 3$. Check by using the known result for the area of the curved surface of a right circular cylinder.

4. Find the surface area of the paraboloid of revolution $z = x^2 + y^2$ from $z = 0$ to $z = 2$ in two different ways.

5. Evaluate $\iint_S (x\mathbf{i} + y\mathbf{j} + z\mathbf{k}) \cdot \mathbf{N} \, dS$ over the plane bounded by the triangle whose vertices are the points $(2, 0, 0)$, $(0, 2, 0)$, and $(0, 0, 2)$ where the normal vector points away from the origin.

6. Evaluate $\iint_S \mathbf{F} \cdot \mathbf{N} \, dS$ over the hemisphere $z = (4 - x^2 - y^2)^{1/2}$ where $\mathbf{F} = xz\mathbf{i} + zx\mathbf{j} + xy\mathbf{k}$ and $\mathbf{N}$ is the upward unit normal.

7. Evaluate $\iint_S \mathbf{F} \cdot \mathbf{N} \, dS$ if:
   (a) $\mathbf{F} = y\mathbf{i} + 2x\mathbf{j} - z\mathbf{k}$ and S is the surface of the plane $2x + y = 6$ in the first octant cut off by the plane $z = 4$.
   (b) $\mathbf{F} = (x^2 + y^2)\,\mathbf{i} - 2x\mathbf{j} + 2yz\mathbf{k}$ and S is the surface of the plane $2x + y + 2z = 6$ in the first octant.
   (c) $\mathbf{F} = 2y\mathbf{i} - z\mathbf{j} + x^2\mathbf{k}$ and S is the surface of the parabolic cylinder $y^2 = 8x$ in the first octant bounded by the planes $y = 4$ and $z = 6$.
   (d) $\mathbf{F} = 6z\mathbf{i} + (2x + y)\,\mathbf{j} - x\mathbf{k}$ and S is the *entire* surface of the region bounded by the cylinder $x^2 + y^2 = 9$, $x = 0$, $y = 0$, $z = 0$, $z = 8$.

8. Let $\mathbf{F} = x\mathbf{i} + y\mathbf{j} + 2\mathbf{k}$. Evaluate $\iint_S \mathbf{F} \cdot \mathbf{N} \, dS$ over:
   (a) the surface of the unit cube $0 \leqq x \leqq 1$, $0 \leqq y \leqq 1$, $0 \leqq z \leqq 1$.
   (b) the portion of the plane $x + 2y + 3z = 6$ which lies in the first octant.
   (c) the entire surface of the unit sphere centered at the origin.
   (d) the portion of the cone $x^2 + y^2 - (1 - z)^2 = 0$ between $z = 0$ and $z = .1$

9. Find the surface area of the torus

$$\mathbf{r}(u, v) = (a + b\cos v)\cos u\,\mathbf{i} + (a + b\cos v)\sin u\,\mathbf{j} + b\sin v\,\mathbf{k},$$
$$0 < b < a$$
$$0 \leqq u \leqq 2\pi,\ 0 \leqq v \leqq 2\pi.$$

10. Evaluate $\iint_S z\,dS$ where S is the surface defined by $\mathbf{r}(u, v) = u\mathbf{i} + v\mathbf{j} + u^3\mathbf{k}$ for $0 < u < 1,\ 0 < v < 1$.
11. Find the surface area of the surface defined by $\mathbf{r}(u, v) = v \cos u\,\mathbf{i} + v \sin u\,\mathbf{j} + v^2\mathbf{k},\ 0 < u < 1,\ 0 < v < 1$.
12. Find the surface area of the surface defined by $\mathbf{r}(u, v) = u\mathbf{i} + v\mathbf{j} + u\mathbf{k},\ 0 < u < 1,\ 0 < v < 1$.
13. Evaluate $\iint_S (6x + z - y^2)\,dS$ where S is the surface defined by $\mathbf{r}(u, v) = u\mathbf{i} + v\mathbf{j} + u\mathbf{k},\ 0 \leqq u \leqq 1,\ 0 \leqq v \leqq 1$.
14. Evaluate $\iint_S (6x + y - x^2)\,dS$ where S is the surface defined by $\mathbf{r}(u, v) = u\mathbf{i} + u^2\mathbf{j} + v\mathbf{k},\ 0 \leqq u \leqq 1,\ 0 \leqq v \leqq 1$.

8.9 green's theorem in the plane

For the remainder of this chapter we will study what are called "transformation theorems" or, more classically, "vector integral theorems." Generally speaking, these theorems relate integrals of functions over closed sets to integrals of some differential operator of the function integrated over the interior of the set. A special case of this type of transformation is the relationship of a function evaluated at the endpoints of an interval to the derivative of the function integrated over the interval

$$f(b) - f(a) = \int_a^b \frac{df}{dx}\,dx.$$

The points a and b are the boundary of the interval (a, b) and in this case the differential operator is just the derivative.

We choose to limit our discussion to three main theorems, although others are a part of the exercises and examples. The most fundamental one is called Green's Theorem in the plane. It is particularly relevant since it immediately generalizes into Stokes' Theorem (Section 8.10) and the Divergence Theorem (Section 8.11).

The proof of Stokes' theorem usually requires Green's theorem as a lemma, and hence, Green's theorem is often called "Green's lemma."

theorem 8.8

Green's Theorem in the Plane. Let R be a domain in the xy-plane whose boundary is a piecewise smooth curve C, and let $\mathbf{F} = f\,\mathbf{i} + g\mathbf{j}$ be a vector field continuously differentiable on D. Then

$$\oint_C \mathbf{F}\cdot d\mathbf{r} = \iint_D (\nabla \times \mathbf{F})\cdot\mathbf{k}\, dA_{xy}$$

where the integration along C is such that D is on the left as one advances in the direction of integration.

proof: What we must prove is that

$$\oint_C f(P)\, dx + g(P)\, dy = \iint_D [g_x(x, y) - f_y(x, y)]\, dy\, dx,$$

a form in which Green's theorem usually appears.

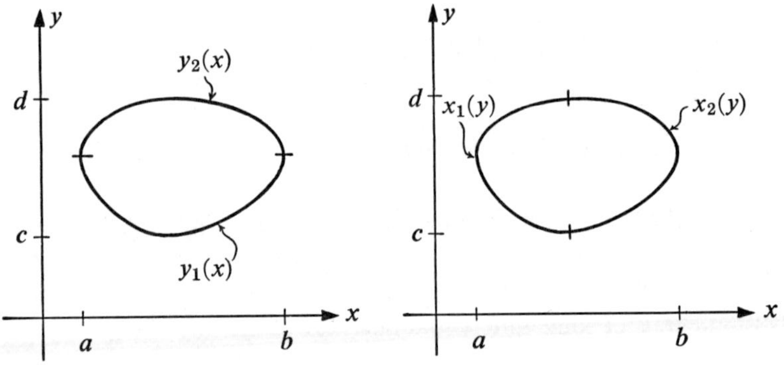

figure 8.22 *two views of a standard domain*

We first prove Green's theorem for the special case in which D is a "standard" domain.† See Figure 8.22. By the ordinary methods of evaluating a double integral as an iterated integral, we have

$$\iint_D f_y(x, y)\, dy\, dx = \int_a^b \left[\int_{y_1(x)}^{y_2(x)} f_y(x, y)\, dy\right] dx.$$

Carrying out the integration on the inside integral we obtain

$$\iint_D f_y(x, y)\, dy\, dx = \int_a^b [f(x, y_2(x)) - f(x, y_1(x))]\, dx$$

which, upon reversing the order of integration in the integral corresponding to the first term, becomes

$$= -\int_b^a f(x, y_2(x))\, dx - \int_a^b f(x, y_1(x))\, dx$$

† A standard domain is a domain D which can be represented in *both* the forms $a \leqq x \leqq b$, $y_1(x) \leqq y \leqq y_2(x)$; $c \leqq y \leqq d$, $x_1(y) \leqq x \leqq x_2(y)$.

which is just another way of writing the line integral

$$-\oint_C f(P)\,dx.$$

In a similar manner,

$$\begin{aligned}\iint_D g_x(x, y)\,dx\,dy &= \int_c^d \left[\int_{x_1(y)}^{x_2(y)} g_x(x, y)\,dx\right] dy\\ &= \int_c^d [g(x_2(y), y) - g(x_1(y), y)]\,dy\\ &= \oint_C g(P)\,dy,\end{aligned}$$

and adding the two results proves the theorem for the case in which D is standard.

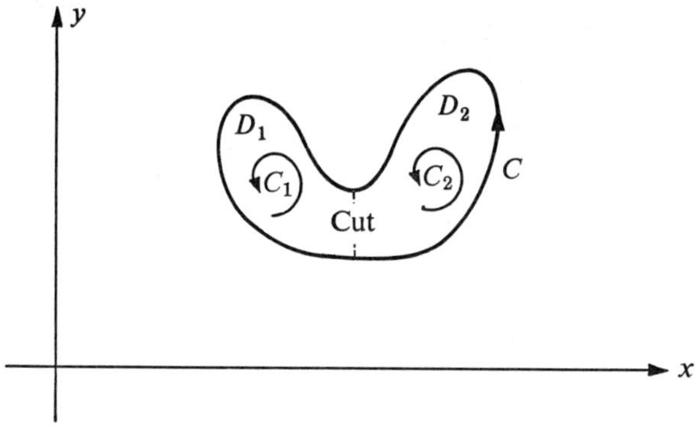

figure 8.23 decomposition of a nonstandard domain

Suppose D is not standard but can be decomposed into two pieces by means of a "cut." Figure 8.23 will give you an idea of how the cut should be introduced. Since the integral over the domain is additive,

$$\begin{aligned}&\iint_D [g_x(x, y) - f_y(x, y)]\,dy\,dx\\ &\quad= \iint_{D_1} [g_x(x, y) - f_y(x, y)]\,dy\,dx + \iint_{D_2} [g_x(x, y) - f_y(x, y)]\,dy\,dx\,.\end{aligned}$$

where D_1 and D_2 are both standard. By applying Green's theorem to each of these separately, we obtain

$$\begin{aligned}&\iint_D [g_x(x, y) - f_y(x, y)]\,dy\,dx\\ &\qquad= \oint_{C_1} f(P)\,dx + g(P)\,dy + \oint_{C_2} f(P)\,dx + g(P)\,dy.\end{aligned}$$

The cut is a part of both C_1 *and* C_2 but the integration is in opposite directions and hence

$$\iint_D [g_x(x, y) - f_y(x, y)]\, dy\, dx = \oint_C f(P)\, dx + g(P)\, dy .$$

This technique can be employed to prove Green's theorem for domains which can be decomposed by $n-1$ cuts into n standard domains.

example 8.15: Let $\mathbf{F} = y\mathbf{i} - x\mathbf{j}$ and let C be the circle of radius 1 centered at the origin. Compute

$$\oint_{C(\text{ccw})} \mathbf{F} \cdot d\mathbf{r}$$

by using Green's theorem.

solution: C can be represented by $\mathbf{r}(t) = \cos t\,\mathbf{i} + \sin t\,\mathbf{j}$. Then D is the domain interior to C and by Green's theorem

$$\oint_{C(\text{ccw})} \mathbf{F} \cdot d\mathbf{r} = \iint_D (-1-1)\, dA_{xy} = -2\pi .$$

example 8.16: Evaluate

$$\int_{C(\text{ccw})} \mathbf{F} \cdot d\mathbf{r}$$

where

$$\mathbf{F} = \frac{-y\mathbf{i} + x\mathbf{j}}{(x^2 + y^2)^{1/2}},$$

and C is the same as in the previous example.

solution: Green's theorem cannot be applied to this closed line integral, since C does not bound a region in which $\mathbf{F}$ is continuously differentiable. ($\mathbf{F}$ is not even defined at (0, 0).) The evaluation can be performed by the methods learned in Section 8.4,

$$\int_{C(\text{ccw})} \mathbf{F} \cdot d\mathbf{r} = \int_{C(\text{ccw})} \frac{-y\, dx + x\, dy}{(x^2 + y^2)^{1/2}} = \int_0^{2\pi} (\sin^2 t + \cos^2 t)\, dt = 2\pi .$$

As proved, Green's theorem in the plane applies to what are called *simply connected* domains in the plane. These are domains, D, which have the property of being the interior of simple closed curves C. A property of simply connected domains, often used as the defining property, is that every simple closed curve in D can be shrunk continuously to a point, the point also being in D. Loosely speaking we sometimes say that a simply connected domain is one "without holes in it." The distinction between a general domain and one which is simply connected can be of extreme importance. Later in this chapter we will use it to distinguish sharply between an irrotational and a conservative field. A domain which is not simply connected is called *multiply connected.*

To apply Green's theorem to multiply connected domains, proper interpretation must be made of the domain D, the boundary curve C, and the direction of the line integration on C. The case of a doubly connected domain is shown in Figure 8.24. The boundary of D is both the inside and

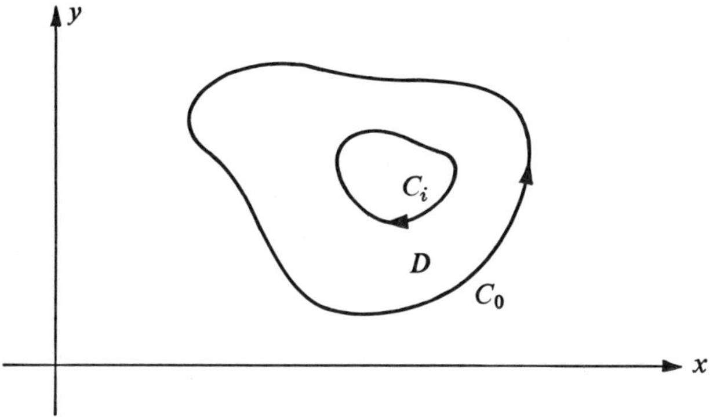

figure 8.24 *Green's theorem for doubly connected region*

outside curve and the direction of line integration is counterclockwise along the outside curve and clockwise along the inside curve. Hence, the statement of Green's theorem for the domain shown is

$$\iint_D [g_x(x, y) - f_y(x, y)]\, dy\, dx$$
$$= \oint_{C_0(\text{ccw})} f(P)\, dx + g(P)\, dy + \oint_{C_i(\text{cw})} f(P)\, dx + g(P)\, dy$$
$$= \int_{C_0(\text{ccw})} f(P)\, dx + g(P)\, dy - \int_{C_i(\text{ccw})} f(P)\, dx + g(P)\, dy.$$

The proof of Green's theorem for a domain such as the one shown in Figure 8.24 is analogous to that of extending Green's theorem to non-standard regions. The given domain is "cut" into simply connected pieces and the result of Green's theorem applied to each of the individual pieces. The details are left for Exercise 10 in the next set of exercises.

The application of Green's theorem may be other than as a transformation theorem. For example, the statement of Green's theorem for doubly connected domains can be considered to relate three numerical quantities so that if any two are known, then so is the remaining term. A case in point is in the evaluation of line integrals over "uncommon" closed curves. If we establish a domain D bounded by a "common" curve (such as a circle) and the given one, and if the double integral over the domain D is known, then the value of the line integral over the "uncommon" curve is known. Naturally f must satisfy the requirements of Green's theorem on D.

example 8.17: Assume that the area between the curve C in Figure 8.25 and the circle of radius 1 centered at the origin is 5π square units. Compute the value of $\int_{C'(\text{ccw})} \mathbf{F}\cdot d\mathbf{r}$ where $\mathbf{F}$ is as in Example 8.15.

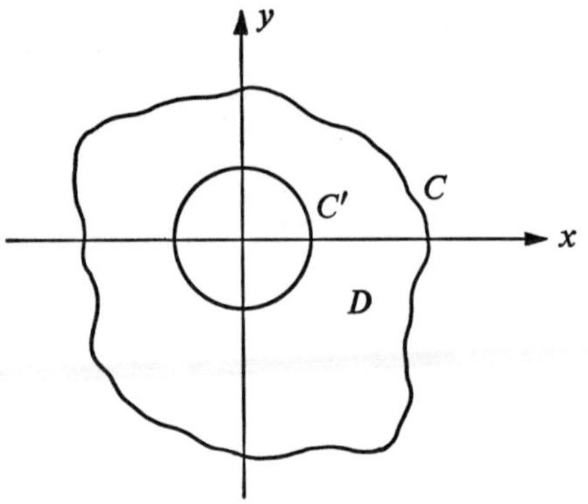

figure 8.25

solution: By Green's theorem for doubly connected domains,

$$\iint_D (\nabla \times \mathbf{F})\cdot\mathbf{k}\, dy\, dx = \oint_{C(\text{ccw})} \mathbf{F}\cdot d\mathbf{r} - \oint_{C'(\text{ccw})} \mathbf{F}\cdot d\mathbf{r}$$

where D is the domain between the two curves and C' is the circle of radius 1 centered at the origin. From Example 8.15, the value of the double integral is $-2(\text{Area}) = -2(5\pi) = -10\pi$. Also from that example, the value of the line integral over C' is -2π. Hence, the value of the given line integral is $-10\pi - 2\pi = -12\pi$.

If the value of the double integral is zero regardless of which domain D is established (such would be the case, for example, if the curl of the vector field is identically zero), the foregoing technique leads to another form of the *principle of deformation of path* alluded to in Section 8.5. For then, the uncommon curves may be deformed into common curves without affecting the value of the line integral so long as in the deforming process, no points of discontinuity of the function are included.

example 8.18: Evaluate

$$\oint_{C(\text{ccw})} \mathbf{F}\cdot d\mathbf{r}$$

where $\mathbf{F}$ is as in Example 8.16 but C is the curve defined by Figure 8.25.

solution: Consider the domain D between the given curve and the circle C' of radius 1 centered at the origin. In this (doubly connected) domain, the function $\mathbf{F}$ satisfies the requirements of Green's theorem and hence,

$$\iint_D (\nabla \times \mathbf{F})\cdot\mathbf{k}\, dy\, dx = \oint_{C(\text{ccw})} \mathbf{F}\cdot d\mathbf{r} - \oint_{C'(\text{ccw})} \mathbf{F}\cdot d\mathbf{r}$$

where C' is the notation used for the inner circle. Since $\nabla \times \mathbf{F} = \mathbf{0}$ throughout D (show this!), the double integral is zero and hence the value of the two line integrals is the same. From Example 8.16 we see that their common value is 2π.

Whereas Green's theorem can be considered the basic transformation of a double integral over a domain into a line integral around the boundary, other interesting transformations are immediate consequences. The following example shows how to transform the double integral of the Laplacian of a function into a line integral of the "normal derivative" of the function.

example 8.19: Let w be a continuous scalar function with first and second derivatives in some multiply connected domain D bounded by a curve C. Show

$$\iint_D \nabla^2 w\, dA_{xy} = \int_C \frac{\partial w}{\partial n}\, ds$$

where $\nabla^2 w = w_{xx} + w_{yy}$ and $\partial w/\partial n$ (the "normal derivative") is the directional derivative of w in the direction of the normal pointing to the exterior of the domain.

solution: If we let $f = -w_y$ and $g = w_x$, then $g_x - f_y = \nabla^2 w$ and therefore by Green's theorem

$$\iint_D \nabla^2 w\, dy\, dx = \oint_C f(P)\, dx + g(P)\, dy.$$

But

$$\begin{aligned}\int_C f(P)\,dx + g(P)\,dy &= \int_C \left(f\frac{dx}{ds} + g\frac{dy}{ds}\right) ds \\ &= \int_C \left(-w_y\frac{dx}{ds} + w_x\frac{dy}{ds}\right) ds \\ &= \int_C \nabla w\cdot\left(\frac{dy}{ds}\mathbf{i} - \frac{dx}{ds}\mathbf{j}\right) ds \\ &= \int_C \nabla w\cdot\mathbf{N}\,ds \\ &= \int_C \frac{\partial w}{\partial n}\,ds.\end{aligned}$$

(The fact that $(dy/ds)\,\mathbf{i} - (dx/ds)\,\mathbf{j}$ is the correct expression for $\mathbf{N}$ is left as an exercise.)

Our statement of Green's theorem related the double integral of the curl of the field to the line integral of the *tangential* component. The next example will show you that the double integral of the divergence is related to the line integral of the *normal* component of the field.

example 8.20: Let $\mathbf{V}$ represent a continuously differentiable vector field in the plane and let D and C satisfy the requirements of Green's theorem. Show

$$\oint_C \mathbf{V}\cdot\mathbf{N}\,ds = \iint_D \nabla\cdot\mathbf{V}\,dA_{xy}.$$

solution: Suppose $\mathbf{V}$ has component functions v_1 and v_2 and let $\mathbf{F} = -v_2\,\mathbf{i} + v_1\mathbf{j}$. First, observe that

$$\begin{aligned}\iint_D (\nabla\times\mathbf{F})\cdot\mathbf{k}\,dA_{xy} &= \iint_D \left(\frac{\partial v_1}{\partial x} + \frac{\partial v_2}{\partial y}\right) dA_{xy} \\ &= \iint_D \nabla\cdot\mathbf{V}\,dA_{xy}.\end{aligned}$$

Applying Green's theorem to the field $\mathbf{F}$,

$$\begin{aligned}\iint_D (\nabla\times\mathbf{F})\cdot\mathbf{k}\,dA_{xy} &= \int_C -v_2\,dx + v_1\,dy \\ &= \int_C \mathbf{V}\cdot\left(\frac{dy}{ds}\mathbf{i} - \frac{dx}{ds}\mathbf{j}\right) ds \\ &= \int_C \mathbf{V}\cdot\mathbf{N}\,ds\end{aligned}$$

and the result is proved.

exercises for section 8.9

1. Let $\mathbf{r} = x(s)\,\mathbf{i} + y(s)\,\mathbf{j}$ represent a closed curve C. Show that the vector $(dy/ds)\,\mathbf{i} - (dx/ds)\,\mathbf{j}$ is a vector normal to C which points to the exterior of the domain of which C is a boundary.

2. Show how to compute the area of a domain in terms of a line integral around its boundary. (*Hint:* Apply Green's theorem to the vector field $-y\mathbf{i}$).

3. Using direct calculations, evaluate $\oint_C \mathbf{F}\cdot d\mathbf{r}$ and then verify using Green's theorem:
   (a) $\mathbf{F} = y^2\mathbf{i} + x^2\mathbf{j}$; C is the boundary of the square $0 \leqq x \leqq 1, 0 \leqq y \leqq 1$.
   (b) $\mathbf{F} = (3x^2 + y)\,\mathbf{i} + 4y^2\mathbf{j}$; C is the boundary of the triangle with vertices at (0, 0), (1, 0), and (0, 2) taken clockwise.
   (c) $\mathbf{F} = (x^2 + y)\,\mathbf{i} - xy^2\mathbf{j}$; C is the boundary of the square whose vertices are (0, 0), (1, 0), (1, 1), and (0, 1) taken counterclockwise.

4. Apply the result of Example 8.19 to the function $\phi(x, y) = x$ on the unit square.

5. Show that if ϕ is harmonic in a domain D of the xy-plane, then

$$\iint_D (\phi_x^2 + \phi_y^2)\, dA_{xy} = \int_C \phi \frac{\partial \phi}{\partial n}\, ds$$

where C is the boundary of D. Apply this result to $\phi(x, y) = e^x \cos y$ on the unit square.

6. Evaluate

$$\int_C \frac{\partial \phi}{\partial n}\, ds$$

where $\phi(x, y)$ is equal to the following, and in each case C is the boundary of the unit square:
   (a) $x^2 + 2y^2$. (b) $x^3y - xy + y^2$. (c) $e^{2x} + e^y$.

7. Evaluate $\int_C \mathbf{F}\cdot d\mathbf{r}$ if:
   (a) $\mathbf{F} = -y^3\mathbf{i} + x^3\mathbf{j}$; C is the unit circle centered at the origin.
   (b) $\mathbf{F} = \cos x \sin y\,\mathbf{i} + \sin x \cos y\,\mathbf{j}$; C as in part (a).

8. Evaluate

$$\int_C x \frac{\partial x}{\partial n}\, ds$$

where C is the unit square taken counterclockwise.

9. In Example 8.20, how are the fields **F** and **V** related? Obtain a "div-curl" relationship between the two.
10. Prove Green's theorem for a domain like that of Figure 8.24. Assume that the theorem has been proved for simply connected domains.
11. Assume that the area between the closed curve C in Figure 8.25 and the unit circle is 25 square units. Compute

$$\oint_C \mathbf{F}\cdot d\mathbf{r}$$

if $\mathbf{F} = x^2\mathbf{i} + y^2\mathbf{j}$.
12. Repeat Exercise 11 for the field $\mathbf{F} = y\mathbf{i} - x\mathbf{j}$.

8.10 stokes' theorem

In Section 8.5 we showed how the value of $\int_C \mathbf{F}\cdot d\mathbf{r}$ is not affected by a deformation of path if **F** has a scalar potential. In this section we determine an analogous result for surface integrals. You will see that the value $\iint_S \mathbf{G}\cdot\mathbf{N}\,dS$ will be the same for different surfaces with the same boundary curve if **G** has a vector potential; that is, that there exists another vector field **F** such that $\nabla \times \mathbf{F} = \mathbf{G}$. The theorem which leads directly to this result is known as Stokes' Theorem, perhaps the most powerful theorem of classical analysis. Its modern day importance stems from the fact that it can be thought to be the generalization of the fundamental theorem of the calculus.

theorem 8.9

Stokes' Theorem. Let S be a smooth oriented surface bounded by a smooth simple closed curve C. Then, if **F** is a continuously differentiable vector field,

$$\oint_C \mathbf{F}\cdot d\mathbf{r} = \iint_S (\nabla \times \mathbf{F})\cdot\mathbf{N}\,dS$$

where **N** is a unit vector normal to S and the direction of integration on C is in accordance with the right hand rule relative to **N**.

proof: See Appendix B.

The extension of Stokes' Theorem to surfaces bounded by more than one simple closed curve is done in a manner similar to the extension of Green's Theorem to multiply connected domains in the plane. The statement of the theorem is essentially the same but the description and interpretation of the relationship between the direction of integration on the curve and the normal to S is more like the statement of Green's Theorem than that of Theorem 8.9. In specific practical circumstances, there is usually little difficulty in determining this relation and hence we will omit the rather awkward statement of this relation. You should at least try to determine a statement for a planar domain such as Figure 8.24 and then compare your result to that stated for Green's theorem in the plane.

Stokes' Theorem tells us that if the integrand of the surface integral is of the form of the normal component of the curl of a vector field, then it may be replaced by the line integral of the tangential component of that field along the path which bounds S. Hence the next theorem is an immediate consequence.

theorem 8.10

$\iint_S \mathbf{G}\cdot\mathbf{N}\, dS$ is independent of the surface bounded by a closed curve C if $\mathbf{G}$ has a vector potential where S and $\mathbf{N}$ satisfy the requirements of Stokes' Theorem.

proof: See Exercise 1 of the next set of exercises.

As a result of Theorem 8.7 and the preceding theorem, the following theorem is obvious.

theorem 8.11

If S and $\mathbf{N}$ satisfy the requirements of Stokes' Theorem, if $\mathbf{G}$ has a vector potential, and if S is a closed surface, then

$$\iint_S \mathbf{G}\cdot\mathbf{N}\, dS = 0.$$

Compare this to Theorem 8.4.

example 8.21: Evaluate the integral of the normal component of curl $\mathbf{F}$ over the hemisphere $z = (4 - x^2 - y^2)^{1/2}$ by Stokes' Theorem and then directly by computation of the surface integral. Let $\mathbf{F} = -y\mathbf{i} + x\mathbf{j} + \mathbf{k}$.

solution: The surface is bounded by a circle C of radius 2 centered at the origin which may be represented by

$$\mathbf{r}(t) = 2\,(\cos t\,\mathbf{i} + \sin t\,\mathbf{j}).$$

We wish to evaluate $\iint_S (\nabla \times \mathbf{F})\cdot\mathbf{N}\,dS$. First, by Stokes' Theorem,

$$\begin{aligned}\iint_S (\nabla \times \mathbf{F})\cdot\mathbf{N}\,dS &= \oint_C \mathbf{F}\cdot d\mathbf{r}\\ &= \oint_C -y\,dx + x\,dy + dz\\ &= \int_0^{2\pi} 4(\sin^2 t + \cos^2 t)\,dt\\ &= 8\pi.\end{aligned}$$

Evaluating directly,

$$\begin{aligned}\iint_S (\nabla \times \mathbf{F})\cdot\mathbf{N}\,dS &= \iint_S (2\mathbf{k})\cdot\mathbf{N}\,dS\\ &= \iint_S 2\cos\gamma\,|\sec\gamma|\,dA_{xy}.\end{aligned}$$

Since γ is acute everywhere on the hemisphere,

$$\begin{aligned}\iint_S (\nabla \times \mathbf{F})\cdot\mathbf{N}\,dS &= 2\,(\text{Area of projection of hemisphere on the } xy \text{ plane})\\ &= 8\pi.\end{aligned}$$

Thus, Stokes' Theorem gives another method of computing surface integrals of normal components of vector fields, *if* the vector field in the integrand has a vector potential. Unfortunately, it is not particularly easy to determine when a vector field has a vector potential (see Exercise 5 for a convenient necessary condition). Nor is it simple to find the vector potential if we know that one exists. It does happen quite often, as in the previous example, that the form of the vector field in the integrand is already expressed as the curl of another field.

Sometimes it is convenient to use the principle of independence of surface (Theorem 8.10) so as to integrate over an "easier" surface than the one given. Remember, this theorem cannot be applied unless the integrand is of a very specific form and the vector field has a vector potential.

example 8.22: Let $\mathbf{F} = 4y\mathbf{i} - x\mathbf{j} + 2z^3\mathbf{k}$, S the hemisphere of the previous problem and $\mathbf{N}$ the outward directed normal to S. Evaluate

$$\iint_S (\nabla \times \mathbf{F})\cdot\mathbf{N}\,dS$$

in *three* ways.

solution: (a) Integrate the expression as it stands. You can easily show that $\mathbf{N} = (x\mathbf{i} + y\mathbf{j} + z\mathbf{k})/2$ and that $\nabla \times \mathbf{F} = -6\mathbf{k}$, from which $(\nabla \times \mathbf{F})\cdot\mathbf{N} = -6z/2$. Therefore

$$\begin{aligned}\iint_S (\nabla \times \mathbf{F})\cdot\mathbf{N}\, dS &= \iint_S (-3z)\, dS \\ &= \iint_{S'_{xy}} (-3z)\, |\sec\gamma|\, dA_{xy} \\ &= \iint_{S'_{xy}} (-3z)\frac{2}{z}\, dA_{xy} \\ &= (-6)\,(\text{Area of projection of hemisphere on the } xy \text{ plane}) \\ &= -24\pi .\end{aligned}$$

(b) Use Stokes' Theorem to equate the given surface integral to $\oint_C \mathbf{F}\cdot d\mathbf{r}$, where C is the circle $x^2 + y^2 = 4$ in the xy-plane. Using the representation of C as given in the previous example,

$$\begin{aligned}\iint_S (\nabla \times \mathbf{F})\cdot\mathbf{N}\, dS &= \int_C \mathbf{F}\cdot d\mathbf{r} \\ &= \oint_C 4y\, dx - x\, dy + 2z^3\, dz \\ &= \int_0^{2\pi} (-16 \sin^2 t - 8 \cos^2 t)\, dt \\ &= -24\pi .\end{aligned}$$

(c) Use the fact that the vector field whose normal component we are integrating has a vector potential—in this case the vector potential of the integrated field is $\mathbf{F}$. Hence the integral is independent of the surface, so instead of integrating over the given surface, we use the domain in the xy-plane which is interior to the circle $x^2 + y^2 = 4$. Therefore,

$$\begin{aligned}\iint_S (\nabla \times \mathbf{F})\cdot\mathbf{N}\, dS &= \iint_{S'_{xy}} (-6)\, dA_{xy} \\ &= -24\pi .\end{aligned}$$

An argument for the proof of Stokes' Theorem, often used in books on electromagnetic theory, stems from the integral definition of the curl of a vector field. We will present just the highlights of this argument. Consider a square in the xy-plane and suppose a vector field $\mathbf{F}$, with component functions f and g, is defined at least on the square and its interior. If $\mathbf{F}$ is a force field, the work required to move a unit particle around the perimeter of the square is given by $\oint_C \mathbf{F} \cdot d\mathbf{r}$ where C is the perimeter. Now, divide the square into n^2 smaller squares in a manner similar to that shown in Figure 8.26.

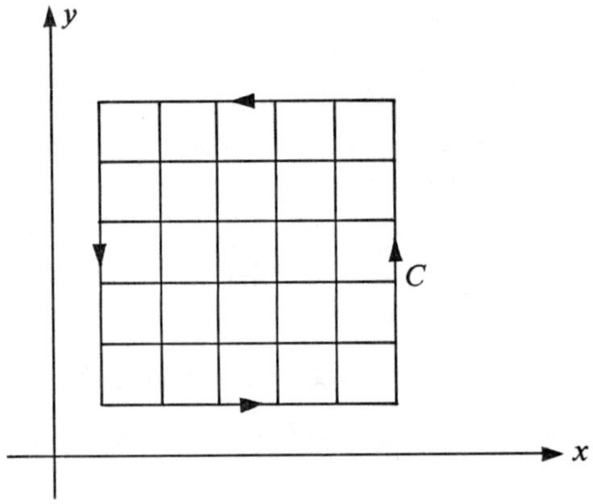

figure 8.26

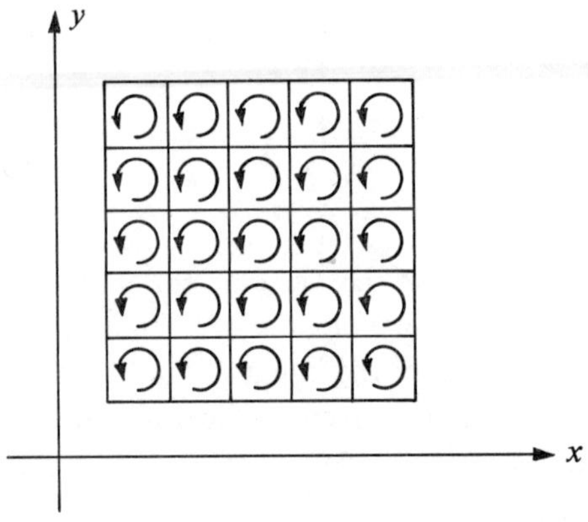

figure 8.27

If we compute the line integral $\oint_{C_i} \mathbf{F} \cdot d\mathbf{r}$ around the perimeter C_i of each of of the smaller squares (see Figure 8.27) it is obvious that the sum of these is equal to $\oint_C \mathbf{F} \cdot d\mathbf{r}$; that is,

$$\sum_{i=1}^{n^2} \oint_{C_i} \mathbf{F} \cdot d\mathbf{r} = \oint_C \mathbf{F} \cdot d\mathbf{r}.$$

Each of the line integrals $\oint_{C_i} \mathbf{F}\cdot d\mathbf{r}$ is approximately equal to $\mathbf{k}\cdot(\nabla\times\mathbf{F})|_{P_i}\Delta s_i$, where P_i is a point interior to the square. Therefore

$$\sum_{i=1}^{n^2} \mathbf{k}\cdot(\nabla\times\mathbf{F})|_{P_i}\Delta s_i \cong \oint_C \mathbf{F}\cdot d\mathbf{r}.$$

Now, it should seem at least plausible that the limit of the lefthand side as Δs_i approaches zero is the surface integral $\iint_S (\nabla\times\mathbf{F})\cdot\mathbf{k}\,dS$, where S is the interior of C. This rough argument could be extended to more general surfaces.

By Theorem 8.5, if $\mathbf{F}$ is conservative (or, which is the same thing, if $\oint_C \mathbf{F}\cdot d\mathbf{r} = 0$ for all closed paths C in D), then curl $\mathbf{F} = \mathbf{0}$ in D. With the assistance of Stokes' Theorem we can show the converse to be true if D is simply connected.†

theorem 8.12

Let D be simply connected and curl $\mathbf{F} = \mathbf{0}$ in D. Then

$$\oint_C \mathbf{F}\cdot d\mathbf{r} = 0$$

for all simple closed paths, C, in D.

proof: Let C be any simple closed path in D. Since D is simply connected we can find a surface, S, bounded by C. Hence, Stokes' Theorem may be applied to obtain

$$\oint_C \mathbf{F}\cdot d\mathbf{r} = \iint_S (\nabla\times\mathbf{F})\cdot\mathbf{N}\,dS$$
$$= \iint_S \mathbf{0}\cdot\mathbf{N}\,dS$$
$$= 0.$$

Theorem 8.12 is not true for other than simply connected domains. See Example 8.16 for a case where curl $\mathbf{F} = \mathbf{0}$ in a doubly connected domain in the plane but the line integral around a closed path is nonzero.

† A domain D in space is simply connected if for every simple closed curve C in D there exists a surface S with C as its boundary, that lies entirely in D. The interior of a sphere is simply connected; the domain between concentric spheres is simply connected; the interior of a torus is *not* simply connected.

To summarize, consider these properties of a vector field **F** which:

1. Has a scalar potential throughout D;
2. Is such that the integral $\int_C \mathbf{F}\cdot d\mathbf{r}$ is independent of the path connecting two fixed points P and Q in D;
3. Is such that the integral $\oint_C \mathbf{F}\cdot d\mathbf{r} = 0$ for all closed paths C in D;
4. Has curl $\mathbf{F} = \mathbf{0}$ throughout D.

These properties are all equivalent in a *simply connected domain.* If the domain is not simply connected, only the first three are equivalent. Any of the first three properties may be used to define the concept of a conservative field. In this book we used the first property for that purpose mainly because it arose in a natural way *before* the others.

When you study the use of vector analysis in other subject areas, you will find that the main use of Stokes' Theorem is not in computing surface integrals or line integrals, but rather in deriving other results. The next example will show you how Stokes' Theorem is put to good use in the theory of electricity and magnetism.

example 8.23: Beginning with Ampere's circuital law, derive Maxwell's equation relating the magnetic intensity vector to the current density.

solution: Ampere's circuital law says that $\int_C \mathbf{H}\cdot d\mathbf{r} = I$ where **H** is the magnetic intensity, C is a closed curve, and I is the current crossing *any* surface S bounded by C. (We say that I is the current "linking" C.) If **J** is the current density (the current crossing the unit area perpendicular to **J**) then the current across S is given by the surface integral $\iint_S \mathbf{J}\cdot\mathbf{N}\,dS$. Therefore by Ampere's law

$$\int_C \mathbf{H}\cdot d\mathbf{r} = \iint_S \mathbf{J}\cdot\mathbf{N}\,dS.$$

By Stokes' Theorem,

$$\int_C \mathbf{H}\cdot d\mathbf{r} = \iint_S (\nabla\times\mathbf{H})\cdot\mathbf{N}\,dS$$

and hence, we have Maxwell's equation

$$\nabla\times\mathbf{H} = \mathbf{J}.$$

exercises for section 8.10

1. Prove Theorem 8.10.

2. Prove Theorem 8.11.
3. Verify Stokes' Theorem for:
   (a) $\mathbf{F} = z\mathbf{i} + x\mathbf{j} + y\mathbf{k}$ taken over the hemisphere $x^2 + y^2 + z^2 = a^2, z \geqq 0$.
   (b) $\mathbf{F} = (2y + z)\,\mathbf{i} + (x - z)\,\mathbf{j} + (y - x)\,\mathbf{k}$ taken over the triangle cut from the plane $x + y + z = 1$ by the coordinate planes.
4. Evaluate $\int_C \mathbf{F} \cdot d\mathbf{r}$ if:
   (a) $\mathbf{F} = y\mathbf{i} + z\mathbf{j} + y\mathbf{k}$ and C is the intersection of $x^2 + y^2 + z^2 = 6z$ and $z = x + 3$ oriented clockwise as viewed from the origin.
   (b) $\mathbf{F} = -y\mathbf{i} + x\mathbf{j} + 2z\mathbf{k}$ and C is the unit circle in the plane $z = 2$, oriented clockwise as viewed from the origin.
   (c) $\mathbf{F} = y^2\mathbf{i} - x^2\mathbf{j} - (y + z)\,\mathbf{k}$ and C is the triangle in the xy-plane with vertices at (0, 0, 0), (1, 0, 0), and (1, 1, 0).
   (d) $\mathbf{F} = x^2\mathbf{i} + y^2\mathbf{j} + z^2\mathbf{k}$ and C is the intersection of the sphere of radius 2 centered at the origin and the cylinder $z = y^2$.
5. As was pointed out in this section, it is sometimes to your advantage to know if a vector field has a vector potential. Can you give a "test" which might be used to give a quick answer, even though the test might fail in some instances? (See Exercise 10 in the exercises for Section 7.6.)
6. Show that the integral of the normal component of the curl of any conservative field over any smooth surface is equal to zero.

8.11 *divergence theorem*

As you learned in Section 8.9, Green's Theorem has two important and fundamental interpretations in the plane:

1. The statement of Theorem 8.8 relates the curl of a vector field to the tangential component of the field around a closed path.
2. Example 8.20 shows how the divergence of the field may be related to the normal component of the field around a closed path.

The proper extension of the first of these relationships leads to Stokes' Theorem, which is virtually a restatement of Green's Theorem, with the difference that the surface and line integrals of Stokes' Theorem are not restricted to a plane. In this section we extend the relationship of the normal component and the divergence to three dimensions. The resulting theorem, called the *divergence* theorem, transforms a closed surface integral into a volume integral.

The divergence theorem is usually attributed to Gauss, although it is also called Green's Theorem in space. The proof is quite similar to Green's Theorem.

theorem 8.13

Divergence Theorem. Let D be a standard† domain in space bounded by a smooth orientable surface, S. Let $\mathbf{F}$ be a continuously differentiable vector function with component functions f, g, and h. Then

$$\iint_S \mathbf{F}\cdot\mathbf{N}\,dS = \iiint_D (\nabla\cdot\mathbf{F})\,dV$$

where $\mathbf{N}$ is the unit vector normal to S and pointing away from D.

proof: In rectangular coordinates the divergence theorem may be stated

$$\iint_S (f\cos\alpha + g\cos\beta + h\cos\gamma)\,dS = \iiint_D (f_x + g_y + h_z)\,dV.$$

As in the case of Green's Theorem we prove only one piece of the theorem. The rest of the proof would be repetitious. First we carry out one iteration on the triple integral $\iiint_D h_z\,dV$. Since D is z-standard,

$$\iiint_D h_z\,dV = \iint_{D_{xy}} \left(\int_{z_1(x,y)}^{z_2(x,y)} h_z\,dz\right) dA_{xy}$$

where D_{xy} is the projection of D onto the xy-plane. Therefore,

$$\iiint_D h_z\,dV = \iint_{D_{xy}} [h(x, y, z_2) - h(x, y, z_1)]\,dA_{xy}.$$

The surface integral $\iint_S h\cos\gamma\,dS$ can be written as the sum of three surface integrals,

$$\iint_S h\cos\gamma\,dS = \iint_{S_1} h\cos\gamma\,dS + \iint_{S_2} h\cos\gamma\,dS + \iint_{S_3} h\cos\gamma\,dS.$$

The direction angle, γ, is acute on S_2, obtuse on S_1, and equal to $\pi/2$ on S_3.

$$\iint_S h\cos\gamma\,dS = -\iint_{D_{xy}} h(x, y, z_1)\,dA_{xy} + \iint_{D_{xy}} h(x, y, z_2)\,dA_{xy}$$

† D is said to be "z-standard" if the boundary surface can be decomposed into a lower surface S_1 which has the equation $z = z_1(x, y)$, an upper surface $z = z_2(x, y)$, and a lateral surface S_3 with segments parallel to the z-axis. The surface S_3 may degenerate into a curve (as for example, a sphere). Then D is said to be "standard" if it is x-, y-, and z-standard.

which shows that

$$\iiint_D h_z\, dV = \iint_S h \cos\gamma\, dS.$$

In like manner

$$\iiint_D g_y\, dV = \iint_S g \cos\beta\, dS$$

and

$$\iiint_D f_x\, dV = \iint_S f \cos\alpha\, dS.$$

Adding the last three equations proves the theorem.

The divergence theorem is also applicable to domains which can be subdivided into finitely many standard domains. The procedure is similar to that followed in the proof of Green's Theorem for nonstandard domains.

If the domain D is bounded by two separate surfaces, as in the case of concentric spheres, notice that the normal to the surface must be taken as pointing "away from" D. Further, you should note carefully that the divergence theorem relates a *closed* surface integral to a volume integral. (See the footnote to Theorem 8.7 in Section 8.7.)

example 8.24: Evaluate $\iint_S \mathbf{F}\cdot\mathbf{N}\, dS$ if $\mathbf{F} = xz\mathbf{i} + yz\mathbf{j} + z^2\mathbf{k}$, S is the surface of the sphere $x^2 + y^2 + z^2 = 9$, and $\mathbf{N}$ is the outward directed normal to S.

solution: By the divergence theorem,

$$\begin{aligned}\iint_S \mathbf{F}\cdot\mathbf{N}\, dS &= \iiint_D (\nabla\cdot\mathbf{F})\, dV\\ &= \iiint_D 4z\, dx\, dy\, dz\\ &= 4V\bar{z}\end{aligned}$$

where $\bar{z}$ represents the third coordinate of the centroid of the sphere and thus $\bar{z} = 0$. Therefore the given surface integral has the value 0.

example 8.25: Evaluate $\iint_S \mathbf{F}\cdot\mathbf{N}\, dS$ if $\mathbf{F} = x\mathbf{i} + y\mathbf{j} + z\mathbf{k}$, S is the complete surface of the finite cylinder $x^2 + y^2 = 4$, $0 \leqq z \leqq 8$, and $\mathbf{N}$ is the outward directed normal. Work the problem in two ways.

solution: By the divergence theorem,

$$\iint_S \mathbf{F}\cdot\mathbf{N}\, dS = \iiint_D (\nabla\cdot\mathbf{F})\, dV = 3\iiint_D dV = 96\pi.$$

Directly,

$$\iint_S \mathbf{F}\cdot\mathbf{N}\,dS = \iint_{\substack{\text{Top}\\(\mathbf{N}=\mathbf{k})}} \mathbf{F}\cdot\mathbf{N}\,dS + \iint_{\substack{\text{Bottom}\\(\mathbf{N}=-\mathbf{k})}} \mathbf{F}\cdot\mathbf{N}\,dS + \iint_{\substack{\text{curved}\\\text{surface}\\(\mathbf{N}=(x\mathbf{i}+y\mathbf{j})/2)}} \mathbf{F}\cdot\mathbf{N}\,dS$$

$$= 32\pi + 0 + 64\pi = 96\pi\,.$$

If the surface is not closed, it may be possible to apply the divergence theorem if the value of the surface integral over an "easier" surface is known.

example 8.26: Evaluate $\iint_S \mathbf{F}\cdot\mathbf{N}\,dS$ if $\mathbf{F} = (\sin z)\,\mathbf{i} + y\mathbf{j} + (x^2 + y^2)\,\mathbf{k}$ and S is the part of the sphere $x^2 + y^2 + z^2 = 4$ for which $z \geqq 0$. Let $\mathbf{N}$ be the outward directed normal.

solution: We consider the domain D in space bounded by the given surface and the interior of the circle $x^2 + y^2 = 4$ for $z = 0$. Letting the interior of the circle be S' and applying the divergence theorem, we obtain

$$\iint_S \mathbf{F}\cdot\mathbf{N}\,dS + \iint_{S'} \mathbf{F}\cdot\mathbf{k}\,dS = \iiint_D \nabla\cdot\mathbf{F}\,dV\,.$$

Solving for the integral over S and performing the indicated operations,

$$\iint_S \mathbf{F}\cdot\mathbf{N}\,dS = \iiint_D dV - \iint_{S'} (x^2 + y^2)\,dy\,dx$$

$$= \tfrac{2}{3}\pi\,(2)^3 - 8\pi = -\tfrac{8}{3}\pi\,.$$

Probably the most important use for the divergence theorem is in deriving some very important equations and formulae. The following two examples will give you some idea.

example 8.27: Let $T(x, y, z, t)$ represent the temperature of a body at time t and at the point (x, y, z). The velocity of heat flow in the body is of the form $\mathbf{V} = -K\,\text{grad}\,T$ where K is the thermal conductivity of the body. Using the divergence theorem, derive the *heat equation*

$$T_t = \left(\frac{K}{\sigma\rho}\right)\nabla^2 T$$

where σ is the specific heat of the material of the body and ρ is its density.

solution: Let D represent the domain of the body and S its boundary surface. The amount of heat leaving D per unit time is

$$\iint_S \mathbf{V}\cdot\mathbf{N}\,dS\,.$$

Applying the divergence theorem to this surface integral,

$$\iint_S \mathbf{V}\cdot\mathbf{N}\,dS = -K\iiint (\nabla\cdot\nabla T)\,dV = -K\iiint \nabla^2 T\,dV\,.$$

The total amount of heat in R is given by $\sigma\rho\iiint_D T\,dV$. The time rate of change of this integral is equal to $\iint_S \mathbf{V}\cdot\mathbf{N}\,dS$. Assuming that the process of differentiation with respect to time and the spatial integration can be interchanged, we obtain

$$-\sigma\rho\iiint_D T_t\,dV = -K\iiint_D \nabla^2 T\,dV\,.$$

Since this equation holds anywhere in the body, and for any part of the body, it follows that

$$T_t = \left(\frac{K}{\sigma\rho}\right)\nabla^2 T\,.$$

example 8.28: Let f and g be scalar functions such that $\mathbf{F} = f\nabla g$ is continuously differentiable in some domain D. Prove Green's first two formulas:

1. $$\iiint_D [f\nabla^2 g + (\nabla f)\cdot(\nabla g)]\,dV = \iint_S f\frac{\partial g}{\partial n}\,dS\,.$$

2. $$\iiint_D (f\nabla^2 g - g\nabla^2 f)\,dV = \iint_S \left(f\frac{\partial g}{\partial n} - g\frac{\partial f}{\partial n}\right)dS\,.$$

solution: We apply the divergence theorem to the vector field $\mathbf{F}$. First, note that $\mathbf{F}\cdot\mathbf{N} = f(\partial g/\partial n)$ from the formula for the directional derivative. Then

$$\nabla\cdot\mathbf{F} = \nabla\cdot(f\nabla g) = f\nabla^2 g + (\nabla f)\cdot(\nabla g)\,.$$

Green's first formula now follows directly from the formula in the divergence theorem.

If we interchange the roles of f and g we obtain a similar formula and subtracting this from Green's first formula, we have

$$\iiint_D (f\nabla^2 g - g\nabla^2 f)\,dV = \iint_S \left(f\frac{\partial g}{\partial n} - g\frac{\partial f}{\partial n}\right)dS$$

which is Green's second formula.

exercises for section 8.11

Exercises 1–3 are review problems on triple integration.

1. Find the z-coordinate of the centroid of the volume bounded by the planes $x + y + z = 4$; $x = 0$, $y = 0$, and $z = 0$.
2. Find the volume bounded by the cylinder $z = 4/(y^2 + 1)$ and the plane $y = x$, $y = 3$, $x = 0$, and $z = 0$.
3. Find the moment of inertia with respect to the z-axis of the mass in the first octant bounded by the cylinder $x^2 + z^2 = 4$ and the planes $y = x$, $y = 0$, and $z = 0$, if the density is given by $2z$.
4. Evaluate $\iint_S \mathbf{F}\cdot\mathbf{N}\, dS$ if S is the surface of the sphere $x^2 + y^2 + z^2 = 16$, $\mathbf{N}$ is the unit normal pointing away from the origin, and:
   (a) $\mathbf{F} = x\mathbf{i} + y\mathbf{j} + z\mathbf{k}$.
   (b) $\mathbf{F} = xy^2\mathbf{i} + yz^2\mathbf{j} + zx^2\mathbf{k}$.
   (c) $\mathbf{F} = xy\mathbf{i} + yz\mathbf{j} + zx\mathbf{k}$.
   (d) $\mathbf{F} = yz\mathbf{i} + zx\mathbf{j} + xy\mathbf{k}$.
5. Show that if a vector field has a vector potential, the integral of the normal component of the field over a closed surface is zero.
6. Let D be a domain in three dimensions with a smooth orientable boundary surface S, $\mathbf{N}$ be the unit outward normal to S, V be the volume of the domain, and $(\bar{x}, \bar{y}, \bar{z})$ represent the centroid.
   (a) Let $\mathbf{F} = x\mathbf{i} + y\mathbf{j} + z\mathbf{k}$ and show $\iint_S \mathbf{F}\cdot\mathbf{N}\, dS = 3V$.
   (b) Let $\mathbf{F} = y^2\mathbf{i} + 2xy\mathbf{j} - xz\mathbf{k}$ and show $\iint_S \mathbf{F}\cdot\mathbf{N}\, dS = \bar{x}V$.
   (c) Let $\mathbf{F} = 2xz\mathbf{i} + 3yz\mathbf{j} + 4z^2\mathbf{k}$, and show $\iint_S \mathbf{F}\cdot\mathbf{N}\, dS = 13\,\bar{z}V$.
7. Show that if g is harmonic in D, then the integral of the normal derivative of g over the boundary of D is zero.
8. Same as Exercise 7(d), Exercises for Sections 8.6–8.8.
9. Evaluate $\iint_S \mathbf{F}\cdot\mathbf{N}\, dS$ if $\mathbf{F} = 4xz\mathbf{i} - y^2\mathbf{j} + yx\mathbf{k}$, where S is the complete surface of the cube bounded by the planes $x = 0$, $y = 0$, $z = 0$, $x = 1$, $y = 1$, $z = 1$, and $\mathbf{N}$ is the unit outward directed normal to S.

8.12 summary of techniques of evaluation

In this section we give a short summary of the techniques used for evaluating line and surface integrals of vector fields. Specifically, we restrict ourselves to line integrals of the tangential component of a vector field and to surface integrals of the normal component of a vector field.

To evaluate $\int_C \mathbf{F}\cdot d\mathbf{r}$, we consider first whether the field is conservative and if the curve is closed.

1. If $\mathbf{F}$ is conservative, it has a scalar potential ϕ and the line integral $\int_C \mathbf{F}\cdot d\mathbf{r}$ is independent of path.
   (a) If C is closed, then $\oint_C \mathbf{F}\cdot d\mathbf{r} = 0$.
   (b) If C is open with endpoints P_1 and P_2, then

$$\int_C \mathbf{F}\cdot d\mathbf{r} = \phi(P_2) - \phi(P_1).$$

2. If $\mathbf{F}$ is not conservative, the line integral is not independent of path.
   (a) If C is closed, then we apply Stokes' Theorem (or, in a particular case it will be Green's Theorem) to equate the line integral to the surface integral of the normal component of the curl.
   (b) If C is open,
      (1) The line integral may be evaluated by the method of Theorem 8.2.
      (2) A supplementary arc C' may be found which together with C forms a closed path. Then by applying Stokes' Theorem,

$$\int_C \mathbf{F}\cdot d\mathbf{r} = \iint_S (\nabla \times \mathbf{F})\cdot\mathbf{N}\, dS - \int_{C'} \mathbf{F}\cdot d\mathbf{r}$$

      where S is a surface of which $C \cup C'$ is the boundary.

To evaluate the surface integral of the normal component of a vector function, $\iint_S \mathbf{G}\cdot\mathbf{N}\, dS$, we decide initially (if possible) if the field has a vector potential and if the surface is closed.

1. If $\mathbf{G}$ has a vector potential, there is a field $\mathbf{F}$ such that $\mathbf{G} = \nabla \times \mathbf{F}$ and hence the integral is independent of surface.
   (a) If S is closed, then $\iint_S \mathbf{G}\cdot\mathbf{N}\, dS = 0$.
   (b) If S is open then the surface integral is equal to $\oint_C \mathbf{F}\cdot d\mathbf{r}$, where C is the boundary of S. (Then we may evaluate the line integral directly by using Theorem 8.2 or re-apply Stokes' Theorem to integrate $\mathbf{G}\cdot\mathbf{N}$ over another surface of which C is the boundary.)

2. If **G** does not have a vector potential, then the surface integral is not independent of surface.
   (a) If S is closed, we apply the divergence theorem to obtain

$$\iint_S \mathbf{G}\cdot\mathbf{N}\,dS = \iiint_D \nabla\cdot\mathbf{G}\,dV,$$

   where D is the domain of which S is the boundary.
   (b) If S is open,
      (1) The surface integral may be evaluated by the method of Theorem 8.6 or one of the ways related to it.
      (2) A supplementary surface, S', may be found which together with the given one comprises a closed surface. Then by applying the divergence theorem

$$\iint_S \mathbf{G}\cdot\mathbf{N}\,dS = \iiint_D \nabla\cdot\mathbf{G}\,dV - \iint_{S'} \mathbf{G}\cdot\mathbf{N}\,dS$$

      where D is the domain interior to $S \cup S'$.

exercises for section 8.12

In Exercises 1–17, compute $\int_C \mathbf{F}\cdot d\mathbf{r}$ where **F** and C are as given.

1. $\mathbf{F} = xy\mathbf{i} + x^3y^2\mathbf{j}$; C is the straight line from (1, 1, 1) to (3, 1, 4).
2. $\mathbf{F} = x\mathbf{i} + xy\mathbf{j} + x^2yz\mathbf{k}$; C is the circle in the xy-plane of radius 1 centered at (0, 1, 0) taken ccw, as viewed from above.
3. $\mathbf{F} = x\mathbf{i} + xy\mathbf{j} + x^2yz\mathbf{k}$; C is the ellipse in the xy-plane, $x^2 + 4(y-1)^2 = 1$, taken ccw as viewed from above.
4. $\mathbf{F} = (2xy + \cos x)\,\mathbf{i} + (2y + x^2)\,\mathbf{j}$; C is the path $x = \cos y$ from (1, 0) to $(1, 2\pi)$.
5. $\mathbf{F} = (y^2 - x^2)\mathbf{i} + 2xy\mathbf{j}$; C is the curve $\mathbf{r}(t) = \sec^2 t\,\mathbf{i} + \tan^2 t\,\mathbf{j}$, $0 \leqq t \leqq \pi/4$.
6. $\mathbf{F} = x^2y\mathbf{i} + z\mathbf{j} + y\mathbf{k}$; C is the straight line path from the origin to (2, 3, 4).
7. $\mathbf{F} = z\mathbf{i}$; C is the curve of intersection of $x^2 + y^2 + z^2 = 4$ and $x = y$.
8. $\mathbf{F} = yz\mathbf{i} + (xz + 1)\,\mathbf{j} + xy\mathbf{k}$; C is the curve defined by $\mathbf{r}(t) = (1 + \sin^3 t)\,\mathbf{i} + \tan^2(t/2)\,\mathbf{j} + (8t/\pi)\,\mathbf{k}$ from $t = 0$ to $t = \pi/2$.
9. $\mathbf{F} = 2\mathbf{i} + z\mathbf{j} + 3y\mathbf{k}$; C is the intersection of $x^2 + y^2 + z^2 = 1$ and $x = z$.
10. $\mathbf{F} = \cos x\,\mathbf{i} + \sin z\,\mathbf{k}$; C is any curve joining $(0, 1, \pi/2)$ and $(\pi, 0, -\pi/2)$.
11. $\mathbf{F} = xy\mathbf{i} + y^2z\mathbf{j} + xyz^2\mathbf{k}$; C is the straight line from (0, 0, 0) to (2, 5, 9).
12. $\mathbf{F} = 2xy\mathbf{i} + (2yz + x^2)\,\mathbf{j} + (y^2 + 2z)\,\mathbf{k}$; C is the curve defined by $\mathbf{r}(t) = t\mathbf{i} + \sin t\,\mathbf{j} + \sin t\,\mathbf{k}$ from (0, 0, 0) to $(\pi/2, 1, 1)$.

13. $\mathbf{F} = xy\mathbf{i} + y^2z\mathbf{j} + xyz^2\mathbf{k}$; C is the circle of radius 2 in the xy-plane centered at (3, 0, 0) taken ccw as viewed from above.

14. $\mathbf{F} = y^2\mathbf{i} + x^2\mathbf{j} - (x + z)\,\mathbf{k}$; C is the boundary of the triangle with vertices (0, 0, 0), (1, 0, 0), and (1, 1, 0) taken in the ccw sense as viewed from above.

15. $\mathbf{F} = yz\mathbf{i} + xz\mathbf{j} + xy\mathbf{k}$; C is the curve $\mathbf{r}(t) = \sin(\pi t/4)\,\mathbf{i} + \tan(\pi t/4)\,\mathbf{j} + \sec(\pi t/4)\,\mathbf{k}$ from (0, 0, 1) to $(\sqrt{2}/2, 1, \sqrt{2})$.

16. $\mathbf{F} = -y\mathbf{i} + 3x\mathbf{j} + z\mathbf{k}$; C is the curve $\mathbf{r}(t) = \cos t\,\mathbf{i} + \sin t\,\mathbf{j} + 3\mathbf{k}$.

17. $\mathbf{F} = (2xyz \sin xyz + x^2y^2z^2 \cos xyz)\,\mathbf{i} + (x^2z \sin xyz + x^3yz^2 \cos xyz)\,\mathbf{j} + (x^2y \sin xyz + x^3y^2z \cos xyz)\,\mathbf{k}$; C is the straight line from (0, 0, 0) to $(1/2, 1/2, \pi)$.

In Exercises 18–31 compute $\iint_S \mathbf{F}\cdot\mathbf{N}\,dS$ where $\mathbf{F}$, $\mathbf{N}$, and S are as given.

18. $\mathbf{F} = x^2\mathbf{i} + 2y\mathbf{j} + x^2y\mathbf{k}$, S is the surface of the unit cube lying between $x = 0$, $x = 1$, $y = 0$, $y = 1$, $z = 0$, and $z = 1$, and $\mathbf{N}$ is the unit outward directed normal.

19. $\mathbf{F} = 3\mathbf{i} + (y^2 - x^2)\,\mathbf{j} + 2z(2 - y)\,\mathbf{k}$, S is the complete surface of the sphere $x^2 + y^2 + z^2 = 1$, and $\mathbf{N}$ is the unit outward directed normal.

20. $\mathbf{F} = x\mathbf{i} + y\mathbf{j} + z\mathbf{k}$, S is the surface of $x^2 + y^2 + z^2 = 4$, and $\mathbf{N}$ is the unit outward directed normal.

21. $\mathbf{F} = (x^2 + y^2 + z^2)\,\mathbf{i} + xy\mathbf{j} + xy\mathbf{k}$, S is the surface of the five faces of the unit cube for which $z \neq 0$, and $\mathbf{N}$ is the unit outward directed normal.

22. $\mathbf{F} = y\mathbf{i} + x\mathbf{j} + y^2\mathbf{k}$, S is the surface of the ellipsoid $x^2 + y^2/4 + z^2/9 = 1$, and $\mathbf{N}$ is the unit normal directed outward from S.

23. $\mathbf{F} = (x + z)\,\mathbf{i} + (y + z)\,\mathbf{j} + (x + y)\,\mathbf{k}$, S is the surface of the sphere $x^2 + y^2 + z^2 = 4$, and $\mathbf{N}$ is the unit normal directed outward.

24. $\mathbf{F} = y\mathbf{j} + z\mathbf{k}$, S is the surface $x^2 + y^2 = 4 - z$, $z > 0$, and $\mathbf{N}$ is the unit upward directed normal to S.

25. $\mathbf{F} = x^2\mathbf{i} + y^2\mathbf{j} + 2z(xy - x - y)\,\mathbf{k}$, S is the surface of the unit cube, and $\mathbf{N}$ is the unit outward directed normal.

26. $\mathbf{F} = x\mathbf{i} + y\mathbf{j} + z\mathbf{k}$, S is the surface $x^2 + y^2 - z = 0$ for $0 \leqq z \leqq 4$, $\mathbf{N}$ is the upward directed unit normal.

27. $\mathbf{F} = \operatorname{curl}[-yx^2\mathbf{i} + (xy^2 - z)\,\mathbf{j} + x^2\mathbf{k}]$, S and $\mathbf{N}$ are as in Exercise 26.

28. $\mathbf{F} = (x^2/2 + \sin^2 z)\,\mathbf{i} + (yz + \cos x^2)\,\mathbf{j} + (x^2 + z)\,\mathbf{k}$, S is the surface of the six faces of the unit cube and $\mathbf{N}$ is the unit outward directed normal.

29. $\mathbf{F} = \operatorname{curl}[z^2\mathbf{i} + x^2\mathbf{j} + y^2\mathbf{k}]$, S is the surface of the sphere $x^2 + y^2 + z^2 = a^2$, and $\mathbf{N}$ is the unit outward directed normal.

30. $\mathbf{F} = x^2\mathbf{i}$, S is the surface of the plane $x + y + z = 1$ in the first octant, and $\mathbf{N}$ is the unit upward directed normal.

31. $\mathbf{F} = \text{curl}[-y\mathbf{i} + x\mathbf{j} + \mathbf{k}]$, S is the surface of the hemisphere $z = (4 - x^2 - y^2)^{1/2}$, and $\mathbf{N}$ is the unit upward directed normal.

tensor analysis

9

9.1 introduction

This is an *introductory* chapter to the topic of tensor analysis and is not a study in depth of the subject. It is intended to acquaint you with the conventions of notation and with some new terminology by placing vectors in a tensor setting. The jump from vectors to tensors is often quite difficult and we hope that in placing stress on some of the fundamentals it will be easier for you to learn more readily some of the advanced ideas when they are used in the applications. As usual, even though our discussion will be in two or three dimensions, most of the concepts may be easily extended to higher dimensions.

9.2 notation

The understanding of tensors requires the handling of many different coordinate systems all more or less simultaneously. There are several conventions of notation peculiar to the subject of tensor analysis which are designed to assist in this task. Unfortunately, these conventions often hinder rather than help the beginner. Often, one's first exposure to tensor analysis is so overshadowed by these conventions that the more fundamental discussions are obscured. If the notation confuses you, write out the expression in a more familiar form until you are better acquainted with the subject.

To designate the three-space variables, a superscript on one letter is used instead of using several letters. Thus, instead of x, y, z or u, v, w, the three variables are designated as x^1, x^2, x^3 or $\bar{x}^1, \bar{x}^2, \bar{x}^3$. It may be annoying to adjust to the new symbols. For example, the equation of the plane $z = 1 - x - y$ takes the form $x^3 = 1 - x^1 - x^2$ and the sphere $x^2 + y^2 + z^2 = 1$ becomes $(x^1)^2 + (x^2)^2 + (x^3)^2 = 1$. When the superscript notation is used, exponents will be a rarity, but when they do occur, they should always be written outside the parentheses. In this chapter we will not always use the superscript notation, but it will be more or less obvious when we do not use it.

In addition to the use of superscripted letters to designate the space variables, we will also use an indexing and summation scheme encountered in contemporary physical and mathematical sciences. This has come to be known as the Einstein convention or "umbral" notation.

1. Every letter index appearing once can take the value 1, 2, or 3. Thus the notation A_i stands for the three quantities A_1, A_2, and A_3. A_{ij} stands for nine quantities $A_{11}, A_{12}, A_{13}, A_{21}, A_{22}, A_{23}, A_{31}, A_{32}$, and A_{33}. With the use of this convention we are able to specify all three-space variables by x^i, since this stands for the three quantities x^1, x^2, x^3.
2. Every letter index appearing twice is to be regarded as being summed from one to three.† For example $A_{ii} = A_{11} + A_{22} + A_{33}$ and $A_iB^i = A_1B^1 + A_2B^2 + A_3B^3$. If the repeated letter is separated by a plus sign, it is not to be thought of as a dummy variable of summation. Hence

$$A_iB^j - A_jB^i \neq \sum_{i=1}^{3} \sum_{j=1}^{3} A_iB^j - A_jB^i.$$

example 9.1: Let $A_1 = 1$, $A_2 = -1$, $A_3 = 5$, $B^1 = 0$, $B^2 = 2$, and $B^3 = -2$. Find A_iB^i and A_iB^j.

solution: A_iB^i is the sum $A_1B^1 + A_2B^2 + A_3B^3 = 0 - 2 - 10 = -12$. A_iB^j is the set of nine quantities $A_1B^1 = 0$, $A_2B^1 = 0$, $A_3B^1 = 0$, $A_1B^2 = 2$, $A_2B^2 = -2$, $A_3B^2 = 10$, $A_1B^3 = -2$, $A_2B^3 = 2$, and $A_3B^3 = -10$. Often, the nine quantities A_iB^j are placed in matrix form,

$$(A_iB^j) = \begin{pmatrix} 0 & 2 & -2 \\ 0 & -2 & 2 \\ 0 & 10 & -10 \end{pmatrix}.$$

example 9.2: Show how the notation conventions may be used to designate the dot and cross product of two "arrows."

solution: Let A_i designate the components of the vector **A** and B^i the com-

† Some users of the convention would say the summation should not be performed unless the repeated index appears once as a superscript and once as a subscript.

ponents of the vector $\mathbf{B}$ both with respect to the $\{\mathbf{i}, \mathbf{j}, \mathbf{k}\}$ basis. Then $\mathbf{A} \cdot \mathbf{B} = A_i B^i$.

If we let $C_i^j = A_i B^j - A_j B^i$, then $\mathbf{A} \times \mathbf{B} = C_2^3\, \mathbf{i} + C_3^1\, \mathbf{j} + C_1^2\, \mathbf{k}$.

As is the case with the summation sign, the letter used for the index of summation has no effect on the actual sum,

$$A_i B^i = A_j B^j = A_k B^k.$$

If we do not wish the summation convention to apply, we will have to say so explicitly. Hence $A_i B^i$ (i not summed) will represent the three individual products $A_1 B^1$, $A_2 B^2$, and $A_3 B^3$.

9.3 basis vectors and coordinate systems

In this section we will examine how a given coordinate system can affect a choice of basis vectors for the set of geometric vectors and conversely, how the basis influences the coordinate system.

To establish a pattern where you can easily observe the interrelationship between a basis set and a coordinate system, consider the set of geometric vectors in the plane. A basis consists of any two linearly independent vectors. Each of the vectors can determine a coordinate axis if it is drawn from a common origin and we then let Ox^1 denote the axis in the direction $\mathbf{e}_1$ and Ox^2 in the direction $\mathbf{e}_2$. If the basis set is orthogonal then the coordinate system determined from it is called *rectangular.* Otherwise the coordinate system is *oblique.* (See Figure 9.1.)

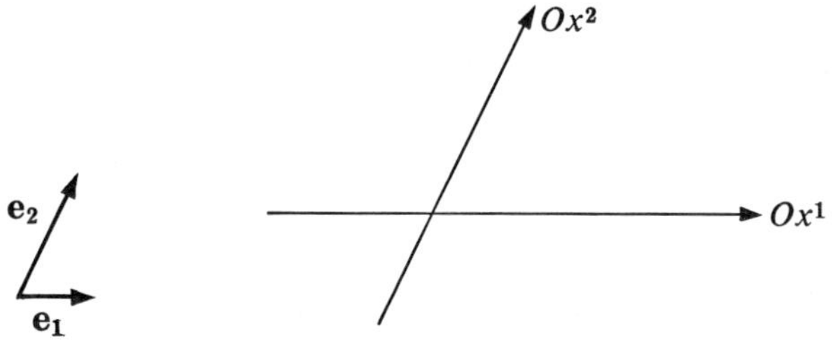

figure 9.1 *oblique coordinate system from nonorthogonal basis*

Conversely, a coordinate system may be used to determine a set of basis vectors. This choice may be made in many different ways out of which we choose two of the more obvious methods.

1. Choose the basis vectors along the coordinate axes. The vector along Ox^1 will be designated by $\mathbf{e}_1$ and along Ox^2 by $\mathbf{e}_2$. For now, the lengths are unspecified.
2. Choose basis vectors perpendicular to the coordinate axes. The vector perpendicular to Ox^1 will be designated by $\mathbf{e}^2$ and the one perpendicular to Ox^2 by $\mathbf{e}^1$. The lengths are determined from the rules

$$\mathbf{e}_1 \cdot \mathbf{e}^1 = 1 \qquad \text{and} \qquad \mathbf{e}_2 \cdot \mathbf{e}^2 = 1 .$$

(See Figure 9.2.)

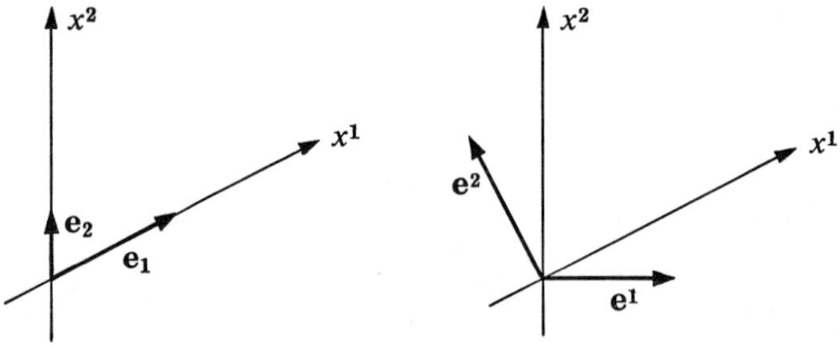

figure 9.2 *two methods of choosing basis vectors*

Because of the method of choosing the directions of $\mathbf{e}^1$ and $\mathbf{e}^2$, it should be obvious that $\mathbf{e}_1 \cdot \mathbf{e}^2 = \mathbf{e}_2 \cdot \mathbf{e}^1 = 0$. Combined with the choice of length for $\mathbf{e}^1$ and $\mathbf{e}^2$, we may state this with the one rule

$$\begin{aligned} \mathbf{e}_i \cdot \mathbf{e}^j &= 1, \quad \text{if} \quad i = j \\ &= 0, \quad \text{if} \quad i \neq j . \end{aligned}$$

In Section 9.4 we shall see that this is a property of what we will call *reciprocal* basis sets.

We now want to generalize the preceding to show the association between basis sets and general curvilinear coordinate systems. Our procedure will be parallel to the one for basis vectors and coordinate systems for the plane. Before doing this we shall review a few ideas from Chapter 6 on curves and surfaces.

A coordinate system for three-dimensional space will be defined in terms of curvilinear coordinates $\bar{x}^i$ by the three equations

$$x^i = x^i(\bar{x}^1, \bar{x}^2, \bar{x}^3)$$

where for now the x^i may be considered as the ordinary rectangular coordinates which you usually desginate as x, y, z. The variables $\bar{x}^1, \bar{x}^2, \bar{x}^3$ are called the curvilinear variables. We assume that the three equations have sufficient properties of continuity and differentiability to allow for an inverse relationship to hold between the variables too; that is, given an $(\bar{x}^1, \bar{x}^2, \bar{x}^3)$ the three equations give a point with coordinates (x^1, x^2, x^3) and conversely a point in space determines an ordered triple $(\bar{x}^1, \bar{x}^2, \bar{x}^3)$.

Recall from Section 6.8 that a coordinate curve is obtained by allowing only one of the curvilinear variables to change, and coordinate surfaces are obtained by allowing two of them to vary.

example 9.3: Find the coordinate curves and surfaces for the coordinate system defined by

$$\begin{aligned} x^1 &= 2\bar{x}^1 + 3\bar{x}^2 - 7\bar{x}^3 \\ x^2 &= -\bar{x}^1 - 2\bar{x}^2 + 5\bar{x}^3 \\ x^3 &= -\bar{x}^1 - \bar{x}^2 + 3\bar{x}^3 \end{aligned}$$

where the x^i are ordinary rectangular cartesian coordinates.

solution: Since the defining equations are linear, holding two of the variables $\bar{x}^i$ constant will yield straight lines in space. The coordinate surfaces, obtained by holding, in turn, $\bar{x}^1$, $\bar{x}^2$, and $\bar{x}^3$ constant are planes. They are found by solving the above three equations for the $\bar{x}^i$ variables

$$\begin{aligned} \bar{x}^1 &= x^1 + 2x^2 - x^3 \\ \bar{x}^2 &= 2x^1 + x^2 + 3x^3 \\ \bar{x}^3 &= x^1 + x^2 + x^3 . \end{aligned}$$

From this, you may readily see that planes parallel to the plane $x^1 + 2x^2 - x^3 = 0$ are the coordinate surfaces corresponding to $\bar{x}^1$ held constant. Planes parallel to $2x^1 + x^2 + 3x^3 = 0$ are the coordinate surfaces corresponding to $\bar{x}^2$ held constant and planes parallel to $x^1 + x^2 + x^3 = 0$ are the coordinate surfaces corresponding to $\bar{x}^3$ held constant. The coordinate surfaces do not intersect in right angles and thus the coordinate system is oblique.

example 9.4: Describe the coordinate system whose defining equations are

$$\begin{aligned} x^1 &= \bar{x}^1 \cos \bar{x}^2 \\ x^2 &= \bar{x}^1 \sin \bar{x}^2 \\ x^3 &= \bar{x}^3 \end{aligned}$$

where the x^i are the ordinary rectangular cartesian coordinates.

solution: This is obviously the cylindrical coordinate system written in the notation of this chapter. Its coordinate curves are lines perpendicular to and passing through the x^3-axis, circles centered on the x^3-axis parallel to the plane $x^3 = 0$, and lines parallel to the x^3-axis. The coordinate surfaces are planes through the x^3-axis and perpendicular to the plane $x^3 = 0$, right circular cylinders perpendicular to the plane $x^3 = 0$, and planes parallel to $x^3 = 0$.

If a point P has curvilinear coordinates $(\bar{c}^1, \bar{c}^2, \bar{c}^3)$, a vector from the origin to the point P is given by

$$\mathbf{r}(\bar{c}^1, \bar{c}^2, \bar{c}^3) = x^i(\bar{c}^1, \bar{c}^2, \bar{c}^3)\,\mathbf{e}_i$$

where the vectors $\mathbf{e}_i$ are an orthonormal basis. The $\bar{x}^1$-coordinate curve through the point P may then be represented vectorially by

$$\mathbf{r}(\bar{x}^1, \bar{c}^2, \bar{c}^3) = x^i(\bar{x}^1, \bar{c}^2, \bar{c}^3)\,\mathbf{e}_i.$$

By the technique of Section 6.7, we may find a vector tangent to this coordinate curve at P by differentiating $\mathbf{r}$ with respect to $\bar{x}^1$,

$$\bar{\mathbf{e}}_1 = \frac{\partial \mathbf{r}}{\partial \bar{x}_1} = \frac{\partial x^i}{\partial \bar{x}^1}\,\mathbf{e}_i.$$

In exactly the same way, vectors $\bar{\mathbf{e}}_2$ and $\bar{\mathbf{e}}_3$ are defined so that the general formula is given by

$$\bar{\mathbf{e}}_j = \frac{\partial \mathbf{r}}{\partial \bar{x}^j} = \frac{\partial x^i}{\partial \bar{x}^j}\,\mathbf{e}_i.$$

In addition to giving an explicit formula for the computation of the three vectors $\bar{\mathbf{e}}_j$ in terms of the $\mathbf{e}_i$, the three equations may be thought of as the relationship between the basis vectors $\mathbf{e}_i$ and a set of natural basis vectors $\bar{\mathbf{e}}_i$ determined at the point P by the curvilinear coordinate system. Except in rare instances which you will not encounter in practical applications, the three vectors $\bar{\mathbf{e}}_j$ are linearly independent and hence form a basis for the set of three-dimensional arrows. The three vectors $\bar{\mathbf{e}}_j$ may be different at each distinct point and hence the set is often called a *local basis* for the curvilinear coordinate system. This basis set generalizes the choice made in the plane along the coordinate axes.

example 9.5: Find $\bar{\mathbf{e}}_1$, $\bar{\mathbf{e}}_2$, and $\bar{\mathbf{e}}_3$ at each point for the coordinate systems of Examples 9.3 and 9.4.

solution: For the coordinate system of Example 9.3,

$$\bar{\mathbf{e}}_1 = \frac{\partial x^i}{\partial \bar{x}^1}\,\mathbf{e}_i = 2\mathbf{e}_1 - \mathbf{e}_2 - \mathbf{e}_3$$

$$\bar{\mathbf{e}}_2 = \frac{\partial x^i}{\partial \bar{x}^2}\,\mathbf{e}_i = 3\mathbf{e}_1 - 2\mathbf{e}_2 - \mathbf{e}_3$$

$$\bar{\mathbf{e}}_3 = \frac{\partial x^i}{\partial \bar{x}^3}\,\mathbf{e}_i = -\,7\mathbf{e}_1 + 5\mathbf{e}_2 + 3\mathbf{e}_3.$$

Note that the local basis vectors for this system do not vary from point to point which is an indication that the coordinate system is oblique, as opposed to curvilinear.

For the coordinate system of Example 9.4,

$$\begin{aligned}\bar{\mathbf{e}}_1 &= \cos \bar{x}^2\, \mathbf{e}_1 + \sin \bar{x}^2\, \mathbf{e}_2 \\ \bar{\mathbf{e}}_2 &= -\, \bar{x}^1 \sin \bar{x}^2\, \mathbf{e}_1 + \bar{x}^1 \cos \bar{x}^2\, \mathbf{e}_2 \\ \bar{\mathbf{e}}_3 &= \mathbf{e}_3 .\end{aligned}$$

The other method of choosing a basis at P, corresponding to what we did in the plane, is to find vectors normal to the coordinate surfaces, $\bar{x}^i =$ a constant at the point P. Perhaps the easiest way to compute a normal vector to a surface $\bar{x}^i =$ a constant is to compute $\nabla \bar{x}^i$. We denote this vector by $\bar{\mathbf{e}}^i$. Since $\bar{\mathbf{e}}_j$ is tangential to the surface $\bar{x}^i =$ a constant, $i \neq j$, the vectors $\bar{\mathbf{e}}_j$ and $\bar{\mathbf{e}}^i$ are perpendicular for $i \neq j$. In terms of the dot product,

$$\bar{\mathbf{e}}_j \cdot \bar{\mathbf{e}}^i = 0 \quad \text{for} \quad i \neq j .$$

example 9.6: Calculate the three vectors $\bar{\mathbf{e}}^i$ for the coordinate systems of Examples 9.3 and 9.4.

solution: In Example 9.3 we solved for the $\bar{x}^i$ in terms of the x^1, x^2, and x^3. It is a simple matter to calculate the gradient of each of these functions, since the $\mathbf{e}_i$ is assumed to be an orthonormal system.

$$\begin{aligned}\bar{\mathbf{e}}^1 &= \nabla \bar{x}^1 = \mathbf{e}^1 + 2\mathbf{e}^2 - \mathbf{e}^3 \\ \bar{\mathbf{e}}^2 &= \nabla \bar{x}^2 = 2\mathbf{e}^1 + \mathbf{e}^2 + 3\mathbf{e}^3 \\ \bar{\mathbf{e}}^3 &= \nabla \bar{x}^3 = \mathbf{e}^1 + \mathbf{e}^2 + \mathbf{e}^3\end{aligned}$$

For the coordinate system of Example 9.4, we first solve for $\bar{x}^i$ in terms of the x^i,

$$\begin{aligned}\bar{x}^1 &= [(x^1)^2 + (x^2)^2]^{1/2} \\ \bar{x}^2 &= \mathrm{Tan}^{-1} \frac{x^2}{x^1} \\ \bar{x}^3 &= x^3 .\end{aligned}$$

Hence,

$$\begin{aligned}\bar{\mathbf{e}}^1 &= \frac{x^1}{[(x^1)^2 + (x^2)^2]^{1/2}}\, \mathbf{e}^1 + \frac{x^2}{[(x^1)^2 + (x^2)^2]^{1/2}}\, \mathbf{e}^2 \\ \bar{\mathbf{e}}^2 &= \frac{-\, x^2}{(x^1)^2 + (x^2)^2}\, \mathbf{e}^1 + \frac{x^1}{(x^1)^2 + (x^2)^2}\, \mathbf{e}^2 \\ \bar{\mathbf{e}}^3 &= \mathbf{e}^3 .\end{aligned}$$

If a curvilinear coordinate system determines a locally orthogonal basis at each point it is called an *orthogonal curvilinear coordinate system.* The results of Examples 9.5 and 9.6 can be used to show that the cylindrical coordinate system is orthogonal—and, as you might have guessed, most of the important ones are.

To recapitulate, at each point P, a curvilinear coordinate system (or, in a particular case an oblique coordinate system) determines many types of basis sets of which the following two are the most important:

1. The three vectors tangential to the coordinate curves at P given by
$$\bar{\mathbf{e}}_j = \frac{\partial \mathbf{r}}{\partial \bar{x}^j}.$$
2. The three vectors which are normal to the coordinate surfaces at P given by
$$\bar{\mathbf{e}}^j = \nabla \bar{x}^j.$$

exercises for sections 9.1–9.3

1. Write out the expression for $A^{ij}B_j$ in three dimensions.
2. Write in Einstein summation notation $A^{12}B_1 + A^{22}B_2 + A^{32}B_3$.
3. The quantities A^{ij} and B_{ij} in matrix form with i as the row index and j the column index are represented
$$(A^{ij}) = \begin{pmatrix} 1 & 3 & 5 \\ 2 & -2 & 0 \\ 1 & 6 & -1 \end{pmatrix}, \qquad (B_{ij}) = \begin{pmatrix} 2 & 6 & 12 \\ 0 & 3 & -1 \\ 4 & 0 & 1 \end{pmatrix}.$$
Find $A^{ij}B_{jk}$ and $A^{ij}B_{ik}$.
4. Show that $\bar{\mathbf{e}}^j \cdot \bar{\mathbf{e}}_i = 1$ for $i = j$.
5. Let x^1, x^2, x^3 and $\bar{x}^1, \bar{x}^2, \bar{x}^3$ be respectively rectangular and oblique coordinates related by the equations
$$\begin{aligned} \bar{x}^1 &= 2x^1 \qquad\quad + x^3 \\ \bar{x}^2 &= x^1 + 2x^2 + 3x^3 \\ \bar{x}^3 &= x^1 + x^2 + x^3. \end{aligned}$$
Find the $\bar{\mathbf{e}}^i$ and the $\bar{\mathbf{e}}_i$.
6. Find the coordinate curves and surfaces for the coordinate system defined by the following equations where the x^i are rectangular coordinates.
$$\begin{aligned} x^1 &= \tfrac{1}{2}(\bar{x}^1)^2 - \tfrac{1}{2}(\bar{x}^2) \\ x^2 &= \bar{x}^1 \\ x^3 &= \bar{x}^3. \end{aligned}$$
Compute the $\bar{\mathbf{e}}^i$ and the $\bar{\mathbf{e}}_i$.

7. Same as Exercise 6, except use the curvilinear coordinate system defined by the equations

$$x^1 = \bar{x}^1 \cos \bar{x}^2 \cos \bar{x}^3$$
$$x^2 = \bar{x}^1 \sin \bar{x}^2 \cos \bar{x}^3$$
$$x^3 = \bar{x}^1 \sin \bar{x}^3 .$$

8. Same as Exercise 6, except use the curvilinear coordinate system defined by the equations

$$x^1 = \cosh \bar{x}^1 \cos \bar{x}^2$$
$$x^2 = \sinh \bar{x}^1 \sin \bar{x}^2$$
$$x^3 = \bar{x}^3 .$$

9. For the $\bar{\mathbf{e}}^i$ and $\bar{\mathbf{e}}_i$ of the cylindrical coordinate system, verify

$$\bar{\mathbf{e}}^i \cdot \bar{\mathbf{e}}_j = 1, \quad \text{for} \quad i = j$$
$$= 0, \quad \text{for} \quad i \neq j .$$

10. Same as Exercise 9, except use the coordinate system of Exercise 7.

11. Same as Exercise 9, except use the coordinate systems of Exercises 6 and 8.

9.4 reciprocal bases

Let $\mathbf{e}_i$ represent a basis for the space of arrows. In its most general setting the problem of finding the components of a vector $\mathbf{A}$ with respect to this basis reduces to projecting $\mathbf{A}$ onto the axes of the coordinate system determined by $\mathbf{e}_i$. Usually you will have to solve a system of three equations in three unknowns to find the components. We touched upon this method in Chapter 1 but an example is given here to refresh your memory.

example 9.7: Find the components of the vector $\mathbf{A} = 2\mathbf{i} - 3\mathbf{j} + \mathbf{k}$ with respect to the basis $\mathbf{e}_1 = \mathbf{i} + \mathbf{j}$, $\mathbf{e}_2 = \mathbf{i} + 2\mathbf{j} - \mathbf{k}$, and $\mathbf{e}_3 = -\mathbf{i} + \mathbf{j} + \mathbf{k}$.

solution: The components of $\mathbf{A}$ with respect to the $\mathbf{e}_i$ are constants A^i such that $\mathbf{A} = A^i\mathbf{e}_i$. Using the values given for the $\mathbf{e}_i$ leads to the vector equation

$$2\mathbf{i} - 3\mathbf{j} + \mathbf{k} = A^1(\mathbf{i} + \mathbf{j}) + A^2(\mathbf{i} + 2\mathbf{j} - \mathbf{k}) + A^3(-\mathbf{i} + \mathbf{j} + \mathbf{k})$$

which gives the three scalar equations

$$2 = A^1 + A^2 - A^3$$
$$-3 = A^1 + 2A^2 + A^3$$
$$1 = \quad - A^2 + A^3$$

from which we obtain $A^1 = 3$, $A^2 = -7/3$, and $A^3 = -4/3$.

As you probably already know, this problem can be eased considerably if the basis is an orthogonal one. Then, if

$$\mathbf{A} = A^i\mathbf{e}_i$$

we take the dot product of both sides with respect to $\mathbf{e}_j$ to obtain

$$\mathbf{e}_j\cdot\mathbf{A} = A^i(\mathbf{e}_i\cdot\mathbf{e}_j)$$

so that, since $\mathbf{e}_i\cdot\mathbf{e}_j = 0$ if $i \neq j$,

$$A^i = \frac{\mathbf{A}\cdot\mathbf{e}_i}{\mathbf{e}_i\cdot\mathbf{e}_i}.$$

This technique of finding the components of a vector with respect to a given basis by computing dot products is not applicable if the basis is nonorthogonal. However, this direct approach may be used in an analogous way if we have readily available a basis which is *reciprocal* to $\mathbf{e}_i$.

definition 9.1

A basis $\mathbf{e}^i$ is said to be reciprocal to a basis $\mathbf{e}_i$ if

$$\begin{aligned}\mathbf{e}_i\cdot\mathbf{e}^j &= 1, \quad \text{for} \quad i = j\\ &= 0, \quad \text{for} \quad i \neq j.\end{aligned}$$

Under the assumption that we have a reciprocal basis, the process of finding the components of $\mathbf{A}$ with respect to the $\mathbf{e}_i$ is immediate. We let

$$\mathbf{A} = A^i\mathbf{e}_i$$

and take the dot product of both sides with a member of the reciprocal basis

$$\mathbf{e}^j\cdot\mathbf{A} = A^i\mathbf{e}_i\cdot\mathbf{e}^j$$

which by the conditions imposed on the reciprocal basis reduces to

$$A^j = \mathbf{e}^j\cdot\mathbf{A}$$

and hence

$$\mathbf{A} = (\mathbf{e}^i\cdot\mathbf{A})\,\mathbf{e}_i\,.$$

To construct a basis explicitly reciprocal to a given basis $\mathbf{e}_i$, we proceed in the following way. The vector $\mathbf{e}^1$ must be perpendicular to $\mathbf{e}_2$ and $\mathbf{e}_3$ and therefore $\mathbf{e}^1 = m(\mathbf{e}_2 \times \mathbf{e}_3)$ where m is some constant. The scalar m is determined from the condition $\mathbf{e}_1\cdot\mathbf{e}^1 = 1$ and therefore

$$\mathbf{e}_1\cdot\mathbf{e}^1 = 1 = m(\mathbf{e}_1\cdot\mathbf{e}_2 \times \mathbf{e}_3).$$

Thus

$$m = \frac{1}{\mathbf{e}_1\cdot\mathbf{e}_2 \times \mathbf{e}_3}.$$

The formulas for $\mathbf{e}^2$ and $\mathbf{e}^3$ are derived in the same way.

You may recognize $\mathbf{e}_1 \cdot \mathbf{e}_2 \times \mathbf{e}_3$ as the volume of the parallelepiped spanned by the vectors $\mathbf{e}_1$, $\mathbf{e}_2$, $\mathbf{e}_3$ (see Exercise 17 in the exercises for Sections 6.1 and 6.2). In this case the formula for the reciprocal basis may be written

$$\mathbf{e}^1 = \frac{\mathbf{e}_2 \times \mathbf{e}_3}{V}$$

where V is the volume of the box shown in Figure 9.3. Similarly, $\mathbf{e}^i = \dfrac{\mathbf{e}_j \times \mathbf{e}_k}{V}$ where the cyclic ordering of the subscripts is important.

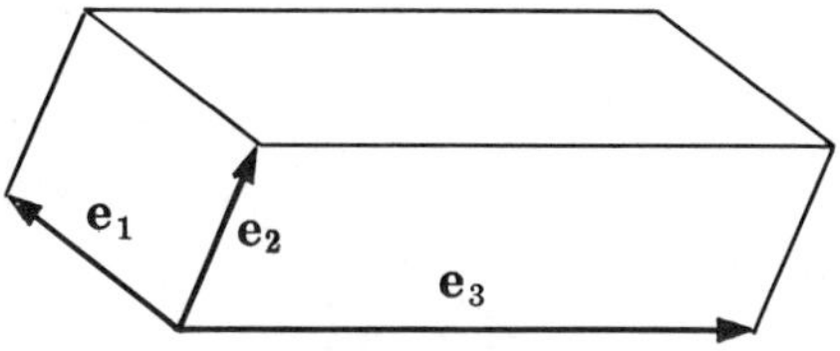

figure 9.3 *volume given by* $\mathbf{e}_1 \cdot \mathbf{e}_2 \times \mathbf{e}_3$

example 9.8: Find a basis reciprocal to the basis of Example 9.7 and find the components of $\mathbf{A}$ with respect to this basis.

solution: Applying the formulas just derived,

$$\mathbf{e}^1 = \frac{\mathbf{e}_2 \times \mathbf{e}_3}{V} = \frac{3\mathbf{i} + 3\mathbf{k}}{3} = \mathbf{i} + \mathbf{k}$$

$$\mathbf{e}^2 = \frac{\mathbf{e}_3 \times \mathbf{e}_1}{V} = \frac{-\mathbf{i} + \mathbf{j} - 2\mathbf{k}}{3}$$

$$\mathbf{e}^3 = \frac{\mathbf{e}_1 \times \mathbf{e}_2}{V} = \frac{-\mathbf{i} + \mathbf{j} + \mathbf{k}}{3}.$$

Let the components of $\mathbf{A}$ with respect to the basis $\mathbf{e}^i$ be denoted by A_i. We find A_i by taking the dot product of both sides of the equality $\mathbf{A} = A_i \mathbf{e}^i$ with $\mathbf{e}_j$,

$$\mathbf{e}_j \cdot \mathbf{A} = A_i (\mathbf{e}^i \cdot \mathbf{e}_j) = A_j .$$

Hence

$$A_1 = \mathbf{A} \cdot \mathbf{e}_1 = -1$$
$$A_2 = \mathbf{A} \cdot \mathbf{e}_2 = -5$$
$$A_3 = \mathbf{A} \cdot \mathbf{e}_3 = -4.$$

For us, the two most important bases are the two "natural" choices at each point of a curvilinear coordinate system $\bar{\mathbf{e}}_i$ and $\bar{\mathbf{e}}^i$. We have the following important theorem, which is easy to prove.

theorem 9.1

Let $\mathbf{r}(\bar{x}^1, \bar{x}^2, \bar{x}^3)$ define a curvilinear coordinate system with coordinate surfaces defined by $\bar{x}^j =$ a constant. At each point P where the two basis sets

$$\bar{\mathbf{e}}_j = \frac{\partial \mathbf{r}}{\partial \bar{x}^j} = \frac{\partial x^k}{\partial \bar{x}^i}\mathbf{e}_k \quad \text{and} \quad \bar{\mathbf{e}}^j = \nabla \bar{x}^j = \frac{\partial \bar{x}^i}{\partial x^k}\mathbf{e}^k$$

are defined, they are reciprocal.

proof: See Exercise 7 for this section, or Example 9.20.

exercises for section 9.4

1. Work Example 9.7 an easier way by using the results of Example 9.8.
2. Let V' be the volume spanned by the reciprocal basis and V the volume spanned by the local basis. Show $VV' = 1$.
3. Find the basis reciprocal to $\mathbf{e}_1 = \mathbf{i} + \mathbf{j} - \mathbf{k}$, $\mathbf{e}_2 = \mathbf{i} + \mathbf{j}$, and $\mathbf{e}_3 = \mathbf{i} - \mathbf{j} + 2\mathbf{k}$.
4. Find the components of the vector $\mathbf{A} = 7\mathbf{i} - 2\mathbf{j} + \mathbf{k}$ with respect to the basis $\mathbf{e}_i$ of the previous exercises and then with respect to the reciprocal basis $\mathbf{e}^i$.
5. Show that a basis and its reciprocal are the same if and only if it is orthonormal.
6. How are a basis and its reciprocal related if the basis is orthogonal but not orthonormal?
7. Prove Theorem 9.1.

9.5 covariant and contravariant components

In a generalized curvilinear coordinate system we have two natural choices of bases at each point which by Theorem 9.1 are reciprocal,

$$\bar{\mathbf{e}}_i = \frac{\partial x^k}{\partial \bar{x}^i}\mathbf{e}_k$$

$$\bar{\mathbf{e}}^i = \frac{\partial \bar{x}^i}{\partial x^k}\mathbf{e}^k.$$

In the previous sections $\mathbf{e}_k$ and $\mathbf{e}^k$ ("unbarred") were each assumed to be orthonormal sets and therefore equal to each other. In actuality, this restriction may be lifted and the two equations correctly relate the local basis and its reciprocal in any two rather arbitrary coordinate systems.

The two choices of bases at each point result in two natural sets of components for a vector **A**. The components with respect to the basis $\mathbf{e}^i$ are denoted by A_i and with respect to the basis $\mathbf{e}_i$ by A^i.

definition 9.2

The components A_i are called the *covariant* components of the vector **A**. The components A^i are called the *contravariant* components.

Note that the distinction between covariant and contravariant vanishes if the basis set $\mathbf{e}_i$ (and consequently the set $\mathbf{e}^i$) is orthonormal. If we write **A** in terms of its covariant components $\mathbf{A} = A_i\mathbf{e}^i$, we say that we have written **A** covariantly. Often, instead of calling it the covariant representation of **A**, we say (a bit erroneously) that A_i is a covariant vector. Similarly, the contravariant vector A^i is really the vector **A** represented by $A^i\mathbf{e}_i$. A_i and A^i represent the same vector with respect to two different bases, one being the reciprocal of the other. Thus a vector is uniquely determined by the three numbers A^i which it associates with the basis $\mathbf{e}_i$ or by the association A_i with $\mathbf{e}^i$.

example 9.9: At a given point in space, a local basis is given by $\mathbf{e}_1 = \mathbf{i} - \mathbf{j} + \mathbf{k}$, $\mathbf{e}_2 = \mathbf{i} + 2\mathbf{j} - \mathbf{k}$, and $\mathbf{e}_3 = -\mathbf{i} + \mathbf{j} + 2\mathbf{k}$. Write the vector $\mathbf{A} = 7\mathbf{i} - 3\mathbf{j} + 2\mathbf{k}$ both contravariantly and covariantly with respect to this basis.

solution: The basis reciprocal to the given basis is

$$\mathbf{e}^1 = \frac{5\mathbf{i} - \mathbf{j} + 3\mathbf{k}}{9}; \quad \mathbf{e}^2 = \frac{\mathbf{i} + \mathbf{j}}{3}; \quad \mathbf{e}^3 = \frac{-\mathbf{i} + 2\mathbf{j} + 3\mathbf{k}}{9}.$$

The covariant components may be found from the formula $A_i = \mathbf{A}\cdot\mathbf{e}_i$ to be $A_1 = 12$, $A_2 = -1$, $A_3 = -6$. The contravariant components are found from the formula $A^i = \mathbf{A}\cdot\mathbf{e}^i$ to be $A^1 = 44/9$, $A^2 = 4/3$, and $A^3 = -7/9$. Hence, the vector **A** written covariantly is

$$\mathbf{A} = 12\mathbf{e}^1 - \mathbf{e}^2 - 6\mathbf{e}^3$$

and contravariantly,

$$\mathbf{A} = \tfrac{44}{9}\mathbf{e}_1 + \tfrac{4}{3}\mathbf{e}_2 - \tfrac{7}{9}\mathbf{e}_3$$

The case of a vector **A** in two dimensions is shown in Figures 9.4 and 9.5. In Figure 9.4, **A** is decomposed into components along the coordinate axes, since those are contravariant components. In Figure 9.5, **A** is decomposed into covariant components which are perpendicular to the coordinate axes.

In passing we note that in addition to covariant and contravariant

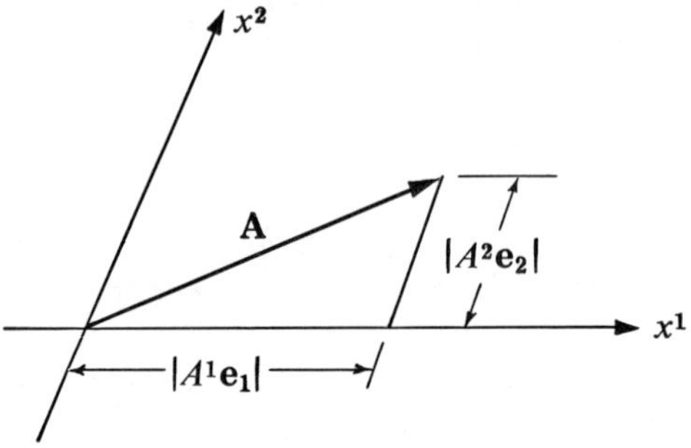

figure 9.4 contravariant components

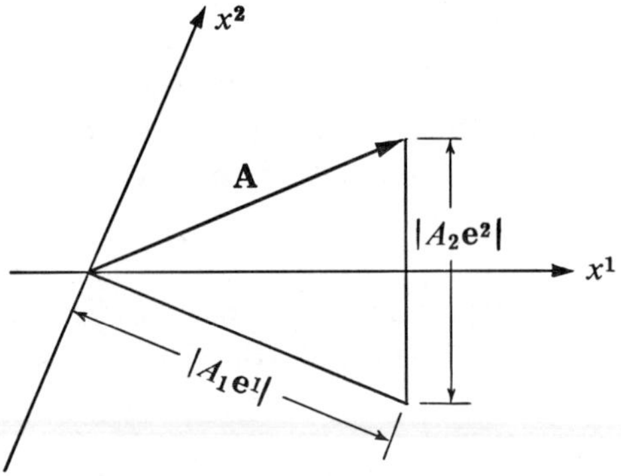

figure 9.5 covariant components

components, we are sometimes interested in the parallel and perpendicular projections of a vector onto the $\mathbf{e}_i$. We define a new unit basis as

$$\mathbf{e}_i^* = \frac{\mathbf{e}_i}{|\mathbf{e}_i|}$$

and its reciprocal

$$(\mathbf{e}^i)^* = \mathbf{e}^i\,|\mathbf{e}_i|$$

from which so-called *physical components* A_i^* and A^{i*} are defined by

$$\mathbf{A} = A_i^*\mathbf{e}^{i*} = A^{i*}\mathbf{e}_i^*.$$

It is mostly an exercise in unraveling the notation to show that

$$A^{i*} = A^i\,|\mathbf{e}_i| \quad \text{and} \quad A_i^* = \frac{A_i}{|\mathbf{e}_i|}.$$

(i not summed in either case.)

example 9.10: Find the physical components of **A** in Example 9.9.

solution:

$$A^{1*} = A^1\,|\mathbf{e}_1| = \tfrac{44}{9}(3)^{1/2} = \frac{44}{3(3)^{1/2}}$$

$$A^{2*} = A^2\,|\mathbf{e}_2| = \tfrac{4}{3}(6)^{1/2}$$

$$A^{3*} = A^3\,|\mathbf{e}_3| = -\tfrac{7}{9}(6)^{1/2}$$

$$A_1^* = \frac{12}{(3)^{1/2}}$$

$$A_2^* = -\frac{1}{(6)^{1/2}}$$

$$A_3^* = -(6)^{1/2}$$

Physical components are relatively unimportant in comparison to the use of contravariant and covariant components.

exercises for section 9.5

1. Let $\mathbf{e}_1 = \mathbf{i} - \mathbf{j} + \mathbf{k}$, $\mathbf{e}_2 = 2\mathbf{i} + 3\mathbf{j} - \mathbf{k}$, and $\mathbf{e}_3 = 5\mathbf{i} - 3\mathbf{j} + 2\mathbf{k}$ be a local basis at a point in space. Find the reciprocal basis. Write the vector $\mathbf{A} = 6\mathbf{i} + 5\mathbf{j} - 7\mathbf{k}$ both covariantly and contravariantly.
2. Find the physical components of $\mathbf{A} = 6\mathbf{i} + 5\mathbf{j} - 7\mathbf{k}$.
3. Find the covariant components at the point $(2, \pi/6, 1)$ of the vector $\mathbf{A} = 6\mathbf{i} + 5\mathbf{j} - 7\mathbf{k}$ in the cylindrical coordinate system. Find the contravariant components at that point, in that coordinate system.
4. Prove the formulas for the physical components of a vector in terms of the contravariant and covariant components.
5. Find the covariant and contravariant components of the vector $\mathbf{A} = 3\mathbf{i} - \mathbf{j} + 2\mathbf{k}$ at the point $(3, \pi/6, \pi/3)$ in the spherical coordinate system with respect to the local basis at that point. (See Exercise 7, Exercises for Sections 9.1–9.3 for the definition of a spherical coordinate system.)

9.6 *transformation laws*

As was pointed out earlier, you will often find it necessary to work with several coordinate systems simultaneously. For this reason (as well as others), it is imperative that you know the relationships between various quantities in the different coordinate systems. These relationships are called *transformation laws.*

Suppose we have two curvilinear coordinate systems with curvilinear coordinates x^i and $\bar{x}^i$. We already know how the basis vectors are related,

$$\bar{\mathbf{e}}_i = \frac{\partial x^k}{\partial \bar{x}^i} \mathbf{e}_k \quad \text{and} \quad \bar{\mathbf{e}}^i = \frac{\partial \bar{x}^i}{\partial x^k} \mathbf{e}^k.$$

In this section we show that these two transformation laws are sufficient to specify completely the law relating the covariant components in the two different coordinate systems and the law relating the contravariant components.

Consider a vector $\mathbf{A}$. If we equate its two contravariant representations in the two coordinate systems we have

$$A^i \mathbf{e}_i = \bar{A}^i \bar{\mathbf{e}}_i.$$

Taking the dot product of both sides with $\bar{\mathbf{e}}^k$ we obtain

$$A^i (\mathbf{e}_i \cdot \bar{\mathbf{e}}^k) = \bar{A}^i (\bar{\mathbf{e}}_i \cdot \bar{\mathbf{e}}^k).$$

Because of the relationship of a local basis to its reciprocal basis, the right side reduces to $\bar{A}^k$. On the left, we replace $\bar{\mathbf{e}}^k$ by

$$\frac{\partial \bar{x}^k}{\partial x^j} \mathbf{e}^j,$$

and since

$$\mathbf{e}_i \cdot \left(\frac{\partial \bar{x}^k}{\partial x^j} \mathbf{e}^j \right) = \frac{\partial \bar{x}^k}{\partial x^i},$$

we have

$$\bar{A}^k = \frac{\partial \bar{x}^k}{\partial x^i} A^i$$

which is the basic *contravariant transformation law.*

If we equate the two covariant representations $A_i \mathbf{e}^i = \bar{A}_i \bar{\mathbf{e}}^i$ and dot each side with $\bar{\mathbf{e}}_k$ we get

$$A_i \mathbf{e}^i \cdot \bar{\mathbf{e}}_k = \bar{A}_i \bar{\mathbf{e}}^i \cdot \bar{\mathbf{e}}_k = \bar{A}_k.$$

It is an easy matter to show that $\mathbf{e}^i \cdot \bar{\mathbf{e}}_k = \partial x^i / \partial \bar{x}^k$ and therefore

$$\bar{A}_k = \frac{\partial x^i}{\partial \bar{x}^k} A_i$$

which is the basic *covariant transformation law.*

9.7 relationship between covariant and contravariant components

In a generalized coordinate system, a vector $\mathbf{A}$ is uniquely determined by either its three covariant components A_i or its three contravariant components A^i. In the previous section, we saw that under a change of basis the two sets of components obey quite different transformation laws.

Now we will learn that the covariant and contravariant components in a particular coordinate system are related too. If we equate the contravariant and covariant representations of **A**, we obtain

$$A^i \mathbf{e}_i = A_i \mathbf{e}^i .$$

Now take the dot product of both sides with the vector $\mathbf{e}_j$,

$$A^i (\mathbf{e}_i \cdot \mathbf{e}_j) = A_i (\mathbf{e}^i \cdot \mathbf{e}_j) = A_j .$$

If, instead, you take the dot product of both sides with vector $\mathbf{e}^j$,

$$A^i (\mathbf{e}_i \cdot \mathbf{e}^j) = A^j = A_i (\mathbf{e}^i \cdot \mathbf{e}^j) .$$

It is customary to denote the nine quantities $\mathbf{e}_i \cdot \mathbf{e}_j$ by g_{ij} and the nine quantities $\mathbf{e}^i \cdot \mathbf{e}^j$ by g^{ij}.

Thus, we can write the relationships between the covariant and contravariant components in any particular coordinate system as

$$A_i = g_{ij} A^j \qquad \text{and} \qquad A^i = g^{ij} A_j .$$

Figure 9.6 is intended to be a diagrammatic picture of the relationships between different components of a vector, both within a coordinate system and between coordinate systems. Between coordinate systems, a transformation law relates the covariant components in one system to the covariant components in the other system, and is called the covariant transformation law. The same can be said for contravariant components. Within a particular

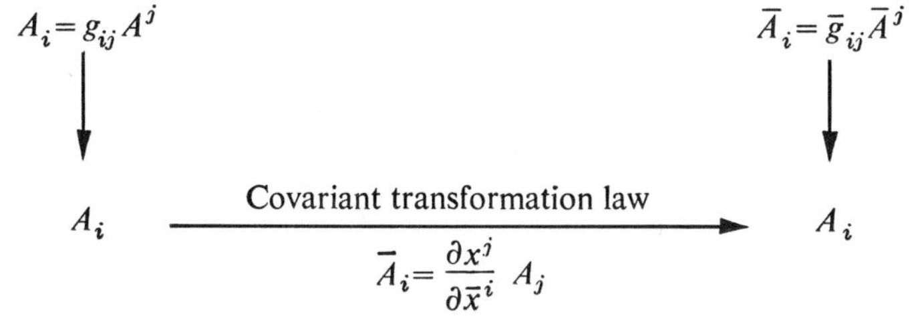

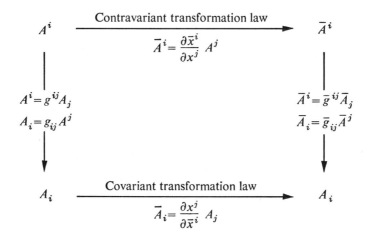

figure 9.6 *relationship between components*

coordinate system, the contravariant and covariant components are related by the g_{ij}'s.

The g_{ij}'s are of fundamental importance in the study of tensor analysis.

example 9.11: Find the g_{ij} if the basis is as given in Example 9.7.

solution:

$$g_{11} = \mathbf{e}_1 \cdot \mathbf{e}_1 = 2; \quad g_{12} = g_{21} = \mathbf{e}_1 \cdot \mathbf{e}_2 = 3; \quad g_{13} = g_{31} = \mathbf{e}_1 \cdot \mathbf{e}_3 = 0;$$
$$g_{22} = \mathbf{e}_2 \cdot \mathbf{e}_2 = 6; \quad g_{23} = g_{32} = \mathbf{e}_2 \cdot \mathbf{e}_3 = 0; \quad g_{33} = \mathbf{e}_3 \cdot \mathbf{e}_3 = 3.$$

We often write the set of g_{ij}'s in matrix form,

$$\mathbf{G} = \begin{pmatrix} 2 & 3 & 0 \\ 3 & 6 & 0 \\ 0 & 0 & 3 \end{pmatrix}.$$

If a coordinate system is specified, and the "natural" choice of basis is made at each point, a corresponding set of g_{ij}'s is determined at each point.

example 9.12: Find the g_{ij}'s at each point for a rectangular coordinate system.

solution: In this case the basis is orthonormal (and unchanging) at each point in space. Hence

$$\begin{aligned} g_{ij} &= 1, \quad i = j \\ &= 0, \quad i \neq j. \end{aligned}$$

In matrix form,

$$\mathbf{G} = \begin{pmatrix} 1 & 0 & 0 \\ 0 & 1 & 0 \\ 0 & 0 & 1 \end{pmatrix}.$$

example 9.13: Find the $\bar{g}_{ij}$'s at each point in space for a cylindrical coordinate system.

solution: The set of vectors $\bar{\mathbf{e}}_i$ was calculated in Example 9.5. Using those results, $\bar{g}_{11} = 1$, $\bar{g}_{22} = (\bar{x}^1)^2$, $\bar{g}_{33} = 1$, $\bar{g}_{ij} = 0$, $i \neq j$. In matrix form,

$$\overline{\mathbf{G}} = \begin{pmatrix} 1 & 0 & 0 \\ 0 & (\bar{x}^1)^2 & 0 \\ 0 & 0 & 1 \end{pmatrix}.$$

example 9.14: Derive the general transformation law for the g_{ij}'s.

solution:

$$\bar{g}_{ij} = \bar{\mathbf{e}}_i \cdot \bar{\mathbf{e}}_j = \left(\frac{\partial x^k}{\partial \bar{x}^i} \mathbf{e}_k\right) \cdot \left(\frac{\partial x^m}{\partial \bar{x}^j} \mathbf{e}_m\right)$$
$$= \frac{\partial x^k}{\partial \bar{x}^i} \frac{\partial x^m}{\partial \bar{x}^j} g_{km}.$$

If the "unbarred" system is rectangular cartesian,

$$\bar{g}_{ij} = \frac{\partial x^k}{\partial \bar{x}^i} \frac{\partial x^k}{\partial \bar{x}^j}.$$

(Note carefully that this *is* a sum.)

In passing we note that the nine quantities g_{ij} describe the fundamental metric properties that the basis $\mathbf{e}_i$ imposes on the space. To see this, let ds be the arc length between two "close" points x^i and $x^i + dx^i$ and let $d\mathbf{r}$ be the vector joining the two points. Assume that the contravariant components of $d\mathbf{r}$ are dx^i. Then,

$$(ds)^2 = d\mathbf{r} \cdot d\mathbf{r} = (\mathbf{e}_i\, dx^i) \cdot (\mathbf{e}_j\, dx^j)$$
$$= g_{ij}\, dx^i\, dx^j.$$

Thus if we find the element of arc length and "identify" coefficients, we have an alternate way of finding the g_{ij}'s.

example 9.15: Do Example 9.13 another way.

solution: Since $x^1 = \bar{x}^1 \cos \bar{x}^2$, $x^2 = \bar{x}^1 \sin \bar{x}^2$, and $x^3 = \bar{x}^3$,

$$dx^1 = \cos \bar{x}^2\, d\bar{x}^1 - \bar{x}^1 \sin \bar{x}^2\, d\bar{x}^2;$$
$$dx^2 = \sin \bar{x}^2\, d\bar{x}^1 + \bar{x}^1 \cos \bar{x}^2\, d\bar{x}^2;$$
$$dx^3 = d\bar{x}^3;$$
$$(dx^1)^2 = (\cos \bar{x}^2)^2 (d\bar{x}^1)^2 - 2\bar{x}^1 \cos \bar{x}^2 \sin \bar{x}^2\, d\bar{x}^1\, d\bar{x}^2 + (\bar{x}^1)^2 (\sin \bar{x}^2)^2 (d\bar{x}^2)^2;$$
$$(dx^2)^2 = (\sin \bar{x}^2)^2 (d\bar{x}^1)^2 + 2\bar{x}^1 \sin \bar{x}^2 \cos \bar{x}^2\, d\bar{x}^1\, d\bar{x}^2 + (\bar{x}^1)^2 (\cos \bar{x}^2)^2 (d\bar{x}^2)^2;$$
$$(dx^3)^2 = (d\bar{x}^3)^2.$$

Therefore, since $(ds)^2 = dx^i\, dx^i$, we add the last three equations to obtain

$$(ds)^2 = (d\bar{x}^1)^2 + (\bar{x}^1)^2 (d\bar{x}^2)^2 + (d\bar{x}^3)^2.$$

This last expression is equal to $\bar{g}_{ij}\, d\bar{x}^i\, d\bar{x}^j$ and thus we may identify the coefficients to obtain the values of $\bar{g}_{ij}$. Hence

$$\bar{g}_{11} = 1, \quad \bar{g}_{22} = (\bar{x}^1)^2, \quad \bar{g}_{33} = 1, \quad \text{and} \quad \bar{g}_{ij} = 0, \quad \text{for} \quad i \neq j.$$

example 9.16: A coordinate system for the plane is defined by $x^1 = \bar{x}^1 + \bar{x}^2$, $x^2 = 2\bar{x}^1 - \bar{x}^2$. Find the four quantities $\bar{g}_{ij}$, under the assumption that the unbarred system is rectangular cartesian.

solution:

$$\begin{aligned} dx^1 &= d\bar{x}^1 + d\bar{x}^2 \\ dx^2 &= 2d\bar{x}^1 - d\bar{x}^2 \\ (ds)^2 &= (dx^1)^2 + (dx^2)^2 \\ &= 5\,(d\bar{x}^1)^2 - 2\,d\bar{x}^1\,d\bar{x}^2 + 2\,(d\bar{x}^2)^2 \end{aligned}$$

and hence $\bar{g}_{11} = 5$, $\bar{g}_{12} = \bar{g}_{21} = -1$, $\bar{g}_{22} = 2$.

If the basis is orthogonal, then $g_{ij} = 0$ for $i \neq j$, in which case we often write

$$(ds)^2 = (h_i\,dx^i)^2$$

and then the $h_i (= \sqrt{g_{ii}})$ are called the *scale factors.*

example 9.17: Find the scale factors for the case of a rectangular coordinate system and a cylindrical coordinate system.

solution: In the first case

$$(ds)^2 = (dx^1)^2 + (dx^2)^2 + (dx^3)^2$$

from which $h_1 = h_2 = h_3 = 1$.

In the second case

$$(ds)^2 = (d\bar{x}^1)^2 + (\bar{x}^1)^2\,(d\bar{x}^2)^2 + (d\bar{x}^3)^2$$

from which $h_1 = h_3 = 1$ and $h_2 = \bar{x}^1$.

exercises for sections 9.6 and 9.7

1. What are the g_{ij} in the case of an orthogonal basis?
2. Find the g_{ij} in the case of spherical coordinates in two different ways.
3. The nine quantities $\mathbf{e}^i \cdot \mathbf{e}_j$ are denoted by g^i_j. Show that $g^i_j = 1$ if $i = j$, and 0 if $i \neq j$, for *any* coordinate system.
4. How are the g^{ii} (not summed) related to the g_{ii} (not summed) in the case of an orthogonal basis?
5. The covariant components at $(3, \pi/6, 2)$ of a vector $\mathbf{A}$ in the cylindrical coordinate system are $A_1 = 2$, $A_2 = -1$, $A_3 = 7$. Find the contravariant components of $\mathbf{A}$.

6. Find the covariant components of the vector **A** at $(2, \pi/6, 2)$ in the cylindrical coordinate system if its contravariant components are $A^1 = 1$, $A^2 = 7$, $A^3 = -2$.
7. A coordinate system is defined by

$$
\begin{aligned}
x^1 &= \bar{x}^1 + \bar{x}^2 + 3\bar{x}^3 \\
x^2 &= \bar{x}^1 - \bar{x}^2 - 2\bar{x}^3 \\
x^3 &= \bar{x}^1 + \bar{x}^2 .
\end{aligned}
$$

Find the $\bar{g}_{ij}$ for this coordinate system in two different ways.
8. The contravariant components of a vector **A** with respect to a local basis for the coordinate system defined in Exercise 7 are $A^1 = -2$, $A^2 = 1$, $A^3 = 4$. Find the covariant components.

9.8 tensors

In this section we will give a definition for a tensor. The material of the previous sections will probably motivate you to believe in the appropriateness of the definitions.

Tensors are used to describe physical phenomena, but a description in one coordinate system must be related by a specific law to the description of the same tensor in another coordinate system. It is this "transformation law" which is at the heart of the definition of a tensor.

Loosely speaking, tensors are ordered sets of numbers. Scalars and vectors are merely specific examples of tensors, and our first definition is one which allows a scalar (field) to be considered as a tensor (field). Recall that a scalar (field) is a quantity whose specification in any particular coordinate system requires only one number (at each point).

definition 9.3

By a tensor of rank† zero is meant a quantity uniquely determined (at a given point) in any coordinate system by a single real number. This number is called the "component" and is invariant under changes of the coordinate system.

Thus if ϕ and $\bar{\phi}$ are scalar functions describing the same tensor of rank zero in a different coordinate system, then for any point P, $\phi(x^i) = \bar{\phi}(\bar{x}^i)$ where x^i and $\bar{x}^i$ denote the coordinates of P in the two different systems.

On the other hand a vector requires three "components" (at each point). With respect to a fixed local basis we can, and do, specify two different

† Sometimes the word "order" is used instead of "rank."

kinds of components, one called covariant and the other called contravariant. The transformation laws specified in the definitions are precisely those derived in Section 9.6.

definition 9.4

A *covariant tensor (field) of rank one* is a set of three quantities (specified at each point) whose description in two different coordinate systems is related by the three equations

$$\bar{A}_i = \frac{\partial x^k}{\partial \bar{x}^i} A_k .$$

A *contravariant tensor (field) of rank one* is a set of three quantities (specified at each point) whose description in two different coordinate systems is related by the three equations

$$\bar{A}^i = \frac{\partial \bar{x}^i}{\partial x^k} A^k .$$

Tensors of rank one are called vectors and, as you learned in the previous section, if a rectangular cartesian coordinate system is used, the distinction between covariant and contravariant components vanishes. This is why tensors are not necessary to the discussion in more elementary analysis.

The indexing scheme here is important. Subscripts must be used for the covariant description of a vector and superscripts are used for a contravariant description. Since we are now making the indexing such a crucial part of the notation for tensors, we are forbidden to use either subscripts or superscripts for a set of three quantities until we have proven their tensor character.

example 9.18: Show that the superscripts used for the differentials, dx^i, are "correct."

solution: We must show that the differentials transform contravariantly. By a familiar formula for the differential,

$$dx^i = \frac{\partial x^i}{\partial \bar{x}^j} d\bar{x}^j$$

or, by interchange of notation, we obtain

$$d\bar{x}^i = \frac{\partial \bar{x}^i}{\partial x^j} dx^j$$

which is the contravariant transformation law.

example 9.19: Let ϕ be a scalar field. Show that the three partial derivatives form a tensor field of rank one and determine what kind it is.

solution: Let $A(i) = \partial\phi/\partial x^i$. We use the chain rule to find $\partial\bar{\phi}/\partial\bar{x}^i$,

$$\frac{\partial\bar{\phi}}{\partial\bar{x}^i} = \frac{\partial\bar{\phi}}{\partial x^j}\frac{\partial x^j}{\partial\bar{x}^i}$$

and since $\phi = \bar{\phi}$,

$$= \frac{\partial\phi}{\partial x^j}\frac{\partial x^j}{\partial\bar{x}^i}$$

or,

$$\bar{A}(i) = \frac{\partial x^j}{\partial\bar{x}^i} A(j),$$

which shows that the three quantities $A(i)$ are the components of a covariant tensor of rank one. You probably recognize this field as the gradient vector field. We say that the gradient of a scalar field is a covariant vector field, which means it is most conveniently represented covariantly.

In just the same way as for a tensor of rank zero or one, tensors of rank two are defined in terms of transformation laws relating components in one coordinate system to the components in some other coordinate system.

definition 9.5

A *contravariant tensor (field) of rank two* is a set of nine quantities (specified at each point) whose description in two different coordinate systems is related by the nine equations

$$\bar{A}^{ij} = \frac{\partial\bar{x}^i}{\partial x^m}\frac{\partial\bar{x}^j}{\partial x^n} A^{mn}.$$

Likewise we have *covariant tensors of rank two* whose components are related by the nine equations

$$\bar{A}_{ij} = \frac{\partial x^m}{\partial\bar{x}^i}\frac{\partial x^n}{\partial\bar{x}^j} A_{mn}.$$

There is also a *mixed tensor of rank two* whose components are related by the nine equations

$$\bar{A}^i_j = \frac{\partial\bar{x}^i}{\partial x^m}\frac{\partial x^n}{\partial\bar{x}^j} A^m_n.$$

The transformation law for the nine quantities g_{ij} which we found in Example 9.14 shows that they form a covariant tensor of rank two. It is called the "metric tensor."

example 9.20: Let $g^i_j = 1$ for $i = j$ and 0 for $i \neq j$. Show that this set of nine quantities is a mixed tensor of rank two.

solution: Note that $\partial\bar{x}^i/\partial\bar{x}^j = 1$ if $i = j$, and 0 if $i \neq j$, by the very definition of the partial derivative. Therefore,

$$\frac{\partial\bar{x}^i}{\partial\bar{x}^j} = \bar{g}^i_j.$$

Further, from the chain rule

$$\frac{\partial\bar{x}^i}{\partial\bar{x}^j} = \frac{\partial\bar{x}^i}{\partial x^m}\frac{\partial x^m}{\partial\bar{x}^j},$$

which by the definition of g^n_m may be written

$$\frac{\partial\bar{x}^i}{\partial\bar{x}^j} = \frac{\partial\bar{x}^i}{\partial x^n}\frac{\partial x^m}{\partial\bar{x}^j}\, g^n_m = \bar{g}^i_j$$

which proves the desired tensor character.

One of the exercises will ask you to show that the nine quantities g^{ij} is a contravariant tensor of rank two. What may be even more surprising to you is that the g_{ij}, g^i_j, and the g^{ij} are the covariant, mixed, and contravariant components of one and the same tensor. (See Borisenko and Tarapov, 1968.)

You should now be in a position to generalize things a little. A tensor of rank three would have 3^3 components, a tensor of rank four, 3^4 components, and a tensor of rank r, 3^r components. If the dimension of the space is changed, then so does the number of components. A tensor of rank r in a space of dimension n has n^r components.

Tensors of rank two are often written as (square) matrices, 2×2 in two dimensions, 3×3 in three dimensions, and $n \times n$ in n dimensions. This is more than just a notational convenience since some of the algebraic operations for tensors of rank two are easily translated into and performed by matrix operations.

exercises for section 9.8

1. A vector has covariant components $A_1 = -5$, $A_2 = 3$, $A_3 = 7$ at the point $(1, -1, 2)$ in rectangular cartesian coordinates. Find the covariant components of the same vector if cylindrical coordinates are used.
2. Write out completely (without the use of the summation convention) the definition of a covariant tensor of rank two in the plane.

3. A covariant tensor of rank two is represented in cartesian coordinates by

$$(A_{ij}) = \begin{pmatrix} 1 & -1 \\ 5 & 0 \end{pmatrix}$$

at the point (1, 1). Find the representation in polar coordinates of the tensor at that point.

4. Verify that the metric tensor obeys the rank two covariant transformation law by considering a transformation from rectangular coordinates to circular cylindrical.

5. Same as Exercise 4 but for the transformation from rectangular coordinates to spherical.

6. Find the components of the metric tensor in spherical coordinates by assuming them to be known for cylindrical coordinates and using the appropriate transformation from cylindrical to spherical.

7. Carefully define all possible rank three and four tensors in three-dimensional space.

8. Let A^i and B^j be contravariant tensors of rank one. Determine the tensor character (if any) of the nine quantitities

$$C(i, j) = A^i B^j - A^j B^i .$$

9.9 algebra of tensors

In this section we introduce some of the elementary algebraic operations on the set of tensors. Some of the operations are valid only between tensors of the same rank and dimension.

definition 9.6

A tensor **T** is *equal* to a tensor **S** if corresponding indexed quantities are equal in any one coordinate system (and hence in all coordinate systems).

definition 9.7

By the *sum* of two tensors **T** and **S** is meant the set of quantities obtained by adding like indexed components from **T** and **S**. The sum is denoted by $\mathbf{T} + \mathbf{S}$.

The fact that $\mathbf{T} + \mathbf{S}$ is a tensor of the same rank as each of the addends is easily proved, at least for special cases. (See the exercises.)

example 9.21: Let **T** be a tensor of rank two represented by the matrix

$\begin{pmatrix} xy & x^2 \\ -y^2 & x-y \end{pmatrix}$ and **S** be represented by $\begin{pmatrix} -xy & y^2 \\ -x^2 & x+y \end{pmatrix}$. Find **S** + **T** at the point (1, 1).

solution: Addition of tensors of rank two is just like matrix addition,

$$\mathbf{S}+\mathbf{T}=\begin{pmatrix} -xy & y^2 \\ -x^2 & x+y \end{pmatrix}+\begin{pmatrix} xy & x^2 \\ -y^2 & x-y \end{pmatrix}=\begin{pmatrix} 0 & x^2+y^2 \\ -x^2-y^2 & 2x \end{pmatrix}$$

which when evaluated at (1, 1) gives $\begin{pmatrix} 0 & 2 \\ -2 & 2 \end{pmatrix}$.

Only tensors of exactly the same rank and type may be added. Thus, two tensors of rank two may be added only if both are covariant, both mixed, or both contravariant. A covariant vector may not be added to a contravariant vector.

definition 9.8

The *outer product* of two tensors **S** and **T** denoted by **ST** is the set of quantities obtained by multiplying each component of **S** by each component of **T**.

If **S** has 3^r components and **T** has 3^s components, the outer product consists of 3^{r+s} components. The fact that the set of products so obtained is a tensor will not be proved in general but the following special case will give you the idea.

theorem 9.2

Let **T** = a contravariant tensor of rank two with components T^{ij}. Let **S** be a mixed tensor of rank two with components S^k_m. Then **TS** is a tensor of contravariant rank three and covariant rank one.

proof: Let $C(i, j, k, m) = T^{ij}S^k_m$ represent the eighty-one quantities in the unbarred coordinate system and $\bar{C}(i, j, k, m)$ represent them in the barred coordinate system. The fact that these eighty-one quantities make up the appropriate tensor is proven by showing that just the right transformation law is obeyed. Since we are given that **T** and **S** are tensors,

$$\bar{T}^{ij} = \frac{\partial \bar{x}^i}{\partial x^n}\frac{\partial \bar{x}^j}{\partial x^o} T^{no}$$

$$\bar{S}^k_m = \frac{\partial \bar{x}^k}{\partial x^p}\frac{\partial x^q}{\partial \bar{x}^m} S^p_q .$$

Hence,

$$\bar{C}(i,j,k,m) = \bar{T}^{ij}\bar{S}^k_m$$
$$= \frac{\partial \bar{x}^i}{\partial x^n}\frac{\partial \bar{x}^j}{\partial x^o}\frac{\partial \bar{x}^k}{\partial x^p}\frac{\partial x^q}{\partial \bar{x}^m} T^{no}S^p_q$$
$$= \frac{\partial \bar{x}^i}{\partial x^n}\frac{\partial \bar{x}^j}{\partial x^o}\frac{\partial \bar{x}^k}{\partial x^p}\frac{\partial x^q}{\partial \bar{x}^m} C(n,o,p,q)$$

which proves the theorem.

example 9.22: Let $\mathbf{U} = \mathbf{i} + 3\mathbf{k}$ and $\mathbf{V} = \mathbf{i} - \mathbf{j}$. Find the outer product of these two vectors.

solution: The outer product consists of the nine quantities U_iV_j. $U_1V_1 = 1$, $U_2V_1 = 0$, $U_3V_1 = 3$, $U_1V_2 = -1$, $U_2V_2 = 0$, $U_2V_3 = -3$, $U_3V_1 = 0$, $U_3V_2 = 0$, $U_3V_3 = 0$. We may write the outer product of two vectors in matrix form since such a product is a tensor of rank two,

$$\mathbf{UV} = (U_iV_j) = \begin{pmatrix} 1 & -1 & 0 \\ 0 & 0 & 0 \\ 3 & -3 & 0 \end{pmatrix}.$$

If we write $\mathbf{U}$ in "column" form and $\mathbf{V}$ in "row" form, the outer product takes the form

$$\begin{pmatrix} 1 \\ 0 \\ 3 \end{pmatrix} \begin{pmatrix} 1 & -1 & 0 \end{pmatrix} = \begin{pmatrix} 1 & -1 & 0 \\ 0 & 0 & 0 \\ 3 & -3 & 0 \end{pmatrix}$$

which can be computed as a matrix product.

There is yet another notation used for the outer product of two vectors which has traditionally been called *dyadic* notation. (Sometimes a tensor of rank two is called a *dyad*.) The nine components of a dyad are designated by **ii, ij, ik, ji, jj, jk, ki, kj,** and **kk**. If we use this notation for Example 9.22,

$$\mathbf{UV} = \mathbf{ii} - \mathbf{ij} + 3\mathbf{ki} - 3\mathbf{kj}.$$

For tensors of rank two and larger with mixed components, there is an operation which is performed on the tensor itself called *contraction*. For example, suppose a tensor has components designated by A^{ij}_{kmn}. A contraction of this tensor is obtained by equating a contravariant index to a covariant index. In this case there are six possible contractions: A^{ij}_{imn}, A^{ij}_{kin}, A^{ij}_{kmi}, A^{ij}_{jmn}, A^{ij}_{kjn}, and A^{ij}_{kmj}. Each of the contractions involves a sum over the equated index.

theorem 9.3

The contraction of a tensor is a tensor of rank two less than the given tensor.

proof: See the exercises for a proof in a specific case.

example 9.23: Let A_k^{ij} represent a tensor of rank three in the plane with components $A_1^{11}=2$, $A_1^{12}=0$, $A_1^{21}=-1$, $A_1^{22}=7$, $A_2^{11}=3$, $A_2^{12}=5$, $A_2^{21}=-2$, $A_2^{22}=4$. Find the contractions of A_k^{ij}.

solution: The contractions of A_k^{ij} are A_i^{ij} and A_j^{ij}.
A_i^{ij} has components

$$A_*^{*1} = A_1^{11} + A_2^{21} = 2 + (-2) = 0,$$
$$A_*^{*2} = A_1^{12} + A_2^{22} = 0 + 4 = 4.$$

A_j^{ij} has components

$$A_*^{1*} = A_1^{11} + A_2^{12} = 2 + 5 = 7,$$
$$A_*^{2*} = A_1^{21} + A_2^{22} = (-1) + 4 = 3.$$

Both contractions are of rank one.

definition 9.9

The *inner product* of two tensors **S** and **T** is found by forming the outer product, **ST**, and contracting the result by equating a contravariant index from one of the tensors to a covariant index from the other.

The inner product of two tensors is obviously a tensor (of two less than the rank of the outer product) since both the outer product and contraction produce tensors.

example 9.24: Let A^{ij} be a tensor of rank two in the plane represented by $\begin{pmatrix} 1 & -1 \\ 3 & -2 \end{pmatrix}$, and let B_k be a covariant tensor of rank one with components $B_1 = 1$ and $B_2 = 5$. Find all possible inner products of the two tensors.

solution: There are two possible inner products, $A^{ij}B_j$ or $A^{ij}B_i$. The components of $A^{ij}B_i$ are $A^{i1}B_i = 1 + 15 = 16$ and $A^{i2}B_i = -1 - 10 = -11$.

The components of $A^{ij}B_j$ are $A^{1j}B_j = -4$ and $A^{2j}B_j = -7$.

The most important inner products in tensor analysis are those taken with the fundamental metric tensors g_{ij} or g^{ij}. A tensor obtained by the process of inner multiplication of any tensor, say B^{ij}, with g_{lm} is called a tensor *associated* with the tensor B^{ij}. It is possible to interpret all tensors associated with a given tensor as really representing the same quantity but with respect to different bases. The idea is much the same as covariant and contravariant components representing the same vector. In fact, the associated vector to A_i is the contravariant vector A^i.

exercises for section 9.9

1. Show that if the components of a tensor are all zero in one coordinate system, they are all zero in every coordinate system.
2. Show that the sum of two tensors of covariant rank two is a tensor of covariant rank two.
3. Show that the contraction of a tensor of covariant rank one and contravariant rank two is a contravariant vector.
4. Let

$$\mathbf{S} = (S^{ij}) = \begin{pmatrix} x^2 & xy & yz \\ xy & y^2 & xz \\ yz & xz & z^2 \end{pmatrix}$$

and

$$\mathbf{T} = (T^{ij}) = \begin{pmatrix} x & y & z \\ y & y & x \\ x & x & z \end{pmatrix}.$$

Compute $\mathbf{S} + \mathbf{T}$ at $(1, -1, 3)$.

5. Find all possible inner products of the tensor $\mathbf{S}$ as given above and $\mathbf{V}$ with components $V_1 = z^2$, $V_2 = xy^2$, $V_3 = x^2$ at the point $(1, -1, 3)$.
6. How many inner products are possible between:
   (a) a covariant tensor of rank one and a contravariant tensor of rank one?
   (b) two covariant tensors of rank one?
   (c) a contravariant tensor of rank two and a covariant tensor of rank three?
7. Prove that the tensors A^i and A_i are *associated.*

9.10 covariant differentiation

In this section we will study how to describe the variation of a tensor field. For a scalar field, we already know that the three partial derivatives of the field will be adequate to describe the variation of the field at any point, since all directional derivatives are expressible in terms of the gradient vector.

A vector point function, **F**, can be represented by

$$\mathbf{F}(x^1, x^2, x^3) = F^i(x^1, x^2, x^3)\, \mathbf{e}_i$$

(contravariant representation)

or by

$$\mathbf{F}(x^1, x^2, x^3) = F_i(x^1, x^2, x^3)\, \mathbf{e}^i.$$

(covariant representation)

In effect, the three component functions F^i or F_i completely describe the field.

To describe the variation of **F** would require nine partial derivatives at each point, three for each component function. If we use the contravariant representation and use a *fixed* basis (such as with an oblique coordinate system) these nine partial derivatives $\partial F^i/\partial x^j$ can be shown to form the components of a mixed tensor of rank two and are denoted by $F^i{}_{,j}$.

definition 9.10

Let **F** be expressed contravariantly with respect to a *fixed* basis, $\mathbf{F} = F^i \mathbf{e}_i$. Then the nine quantities $\partial F^i/\partial x^j$ form a mixed tensor of rank two called the *covariant derivative* of the contravariant vector field F^i.

example 9.25: Find the covariant derivative of the contravariant vector field with components in the rectangular cartesian coordinate system given by

$$F^1 = x^2 y, \qquad F^2 = xyz, \qquad F^3 = y^2 z.$$

(The superscripts on the right sides of these equations are exponents.)

solution: Since the $\mathbf{e}_i$ are just the fixed orthonormal basis $\{\mathbf{i}, \mathbf{j}, \mathbf{k}\}$ the covariant derivative is found by computing the nine partial derivatives $\partial F^i/\partial x^j$. Since $F^i{}_{,j}$ is a tensor of rank two, we may represent it in matrix form,

$$(F^i{}_{,j}) = \begin{pmatrix} 2xy & x^2 & 0 \\ yz & xz & xy \\ 0 & 2yz & y^2 \end{pmatrix}.$$

To generalize the concept of the covariant derivative, consider the vector field $\mathbf{F} = F^i \mathbf{e}_i$ but now allow the $\mathbf{e}_i$ to vary from point to point. The variation of the field will be described by computing the variation in the direction of each of the three coordinate curves, $\partial \mathbf{F}/\partial x^k$. Thus,

$$\frac{\partial \mathbf{F}}{\partial x^k} = \frac{\partial F^i}{\partial x^k}\mathbf{e}_i + F^i \frac{\partial \mathbf{e}_i}{\partial x^k} = \frac{\partial F_i}{\partial x^k}\mathbf{e}^i + F_i \frac{\partial \mathbf{e}^i}{\partial x^k}.$$

definition 9.11

Let $\mathbf{F}$ be expressed contravariantly with respect to a variable basis, $\mathbf{F} = F^i\mathbf{e}_i$. Then the nine quantities $\mathbf{e}^j \cdot (\partial\mathbf{F}/\partial x^k)$ form a mixed tensor of rank two called the *covariant derivative of the contravariant vector field* F^i, denoted by $F^j{}_{,k}$.

Thus, the covariant derivative of the contravariant vector field is made up of the projections of the partial derivatives in the direction of the basis vectors $\mathbf{e}^j$.

In like manner, the covariant derivative of the covariant vector F_i is defined to be the covariant tensor of rank two whose components are

$$F_{j,k} = \mathbf{e}_j \cdot \frac{\partial\mathbf{F}}{\partial x^k}.$$

It follows from the definitions of the two types of covariant derivatives and from the relationship between reciprocal bases that

$$F^j{}_{,k} = \frac{\partial F^j}{\partial x^k} + F^i\left(\mathbf{e}^j \cdot \frac{\partial\mathbf{e}_i}{\partial x^k}\right),$$

$$F_{j,k} = \frac{\partial F_j}{\partial x^k} + F_i\left(\mathbf{e}_j \cdot \frac{\partial\mathbf{e}^i}{\partial x^k}\right).$$

Both formulas reduce to the partial derivative of the component functions of the vector field if the basis vectors do not vary. If they do vary, so-called "corrective" terms are added,

$$F^i\left(\mathbf{e}^j \cdot \frac{\partial\mathbf{e}_i}{\partial x^k}\right)$$

in the contravariant case and

$$F_i\left(\mathbf{e}_j \cdot \frac{\partial\mathbf{e}^i}{\partial x^k}\right)$$

in the covariant case. These terms may be thought to "make up for the curvilinearity" of the coordinate system.

Traditionally, the twenty-seven terms $\mathbf{e}^j \cdot (\partial\mathbf{e}_i/\partial x^k)$ have been denoted by $\{ik,j\}$ or Γ^j_{ik} and are called the Christoffel three-index symbols of the second kind.† While the latter notation is perhaps a bit more popular, it suggests a tensor character for the twenty-seven terms which is not true. It is easy to show that the twenty-seven terms are *not* components of a tensor since they are all zero in some coordinate systems, while in other coordinate systems they do not all vanish.

† Christoffel symbols of the first kind are defined by $[i,jk] = \mathbf{e}_i \cdot (\partial\mathbf{e}_j/\partial x^k)$.

example 9.26: Find the eight Christoffel symbols at each point of the plane associated with the polar coordinate system.

solution: Recall that for the polar coordinate system

$$\mathbf{e}_1 = \cos\theta\,\mathbf{i} + \sin\theta\,\mathbf{j}\,; \quad \mathbf{e}_2 = r\,(-\sin\theta\,\mathbf{i} + \cos\theta\,\mathbf{j}),$$

$$\mathbf{e}^1 = \cos\theta\,\mathbf{i} + \sin\theta\,\mathbf{j}\,; \quad \mathbf{e}^2 = \frac{1}{r}(-\sin\theta\,\mathbf{i} + \cos\theta\,\mathbf{j}).$$

From these, we calculate the eight symbols from the formula $\{ik, j\} = \mathbf{e}^j \cdot (\partial \mathbf{e}_i/\partial x^k)$,

$$\{11, 1\} = 0\,; \quad \{11, 2\} = 0\,; \quad \{12, 1\} = 0\,; \quad \{12, 2\} = \frac{1}{r}\,;$$

$$\{21, 1\} = 0\,; \quad \{21, 2\} = \frac{1}{r}\,; \quad \{22, 1\} = -\,r\,; \quad \{22, 2\} = 0.$$

The formulas for the two covariant derivatives are usually written in terms of the Christoffel symbols of the second kind. In one of the exercises, we will ask you to show that

$$\mathbf{e}_j \cdot \frac{\partial \mathbf{e}^i}{\partial x^k} = -\,\{jk, i\}$$

and hence the two covariant derivatives are given by

$$F_{j,k} = \frac{\partial F_j}{\partial x^k} - \{jk, i\}\,F_i$$

and

$$F^j{}_{,k} = \frac{\partial F^j}{\partial x^k} + \{ik, j\}\,F^i.$$

example 9.27: Find the covariant derivative of the vector function in the plane whose covariant representation in polar coordinates is $F_1 = r^2 \cos\theta$, $F_2 = r\sin^2\theta$.

solution: In the previous example, the necessary Christoffel symbols were computed. Therefore

$$\begin{aligned}
F_{1,1} &= 2r\cos\theta &&- \{11, i\}\,F_i = 2r\cos\theta \\
F_{1,2} &= -\,r^2\sin\theta &&- \{12, i\}\,F_i = -\,r^2\sin\theta - \sin^2\theta \\
F_{2,1} &= \sin^2\theta &&- \{21, i\}\,F_i = \sin^2\theta - \sin^2\theta = 0 \\
F_{2,2} &= r\sin 2\theta &&- \{22, i\}\,F_i = r\sin 2\theta + r^3\cos\theta.
\end{aligned}$$

In matrix form,

$$(F_{j,k}) = \begin{pmatrix} 2r\cos\theta & -r^2\sin\theta - \sin^2\theta \\ 0 & r\sin 2\theta + r^3\cos\theta \end{pmatrix}.$$

The covariant derivative of a vector field, which is a tensor field, may be used to express laws and formulas independent of the coordinate system. For example, the divergence of a vector field **A** is defined to be $\operatorname{div}\mathbf{A} = A^i{}_{,i}$. You can convince yourself that this is the "correct" statement for the generalized formula for the divergence by showing what it reduces to if rectangular cartesian coordinates are used.

The concept of covariant differentiation may be applied to tensor fields of rank higher than two. The generalization is straightforward and any treatise on tensor analysis will give the appropriate formulas.

exercises for section 9.10

1. Prove $\{jk, i\} = \{kj, i\}$. (*Hint:* $\dfrac{\partial \mathbf{e}_j}{\partial x^k} = \dfrac{\partial}{\partial x^k}\left(\dfrac{\partial \mathbf{r}}{\partial x^j}\right)$.)
2. Show that $\mathbf{e}_j \cdot \partial\mathbf{e}^i/\partial x^k = -\{jk, i\}$. (*Hint:* Start with the easily provable fact, $\dfrac{\partial}{\partial x^k}(\mathbf{e}_i \cdot \mathbf{e}^j) = 0$.)
3. Find the twenty-seven Christoffel symbols for cylindrical coordinates.
4. Find the twenty-seven Christoffel symbols for spherical coordinates.
5. Find the covariant derivative of the vector function with contravariant components in cylindrical coordinates,
$$F^1 = r^2 + z^2, \qquad F^2 = r\sin^2\theta, \qquad F^3 = rz^2\cos\theta.$$
6. Find the covariant derivative of the covariant vector with components in the spherical coordinate system,
$$F_1 = \rho^2\theta; \qquad F_2 = \theta^2\sin^2\phi; \qquad F_3 = \rho\theta\phi.$$
7. Show that $A^i{}_{,i}$ reduces to the well-known formula for the divergence in rectangular coordinates.
8. Find the expression for the divergence of a vector field in cylindrical coordinates; in spherical coordinates.
9. Show that the $\{jk, i\}$ are not components of a tensor.

fourier expansions

10

10.1 introduction

We often wish to express functions in a kind of standard method, which will not necessarily be the easiest for computational or analytical purposes. In elementary calculus you learned how to represent functions in a standard way as the limit of a sequence of polynomial functions. The limit of a sequence of this type is called a power series or, in a particular case, a Taylor series with center at $x = a$. In this chapter we will consider the basic concepts and techniques of representing rather arbitrary functions as the limit of a sequence of linear combinations (also called partial sums) of orthogonal sets.

Recall that one of the main limitations for a Taylor series representation is that the function must be infinitely differentiable at the point about which its Taylor series is written. Many of the functions useful in applied mathematics exhibit finite discontinuities and it is particularly to these functions that the techniques of this chapter are directed.

Representing a function in terms of a standard set of functions is often called an *expansion* in terms of the given set. "Fourier expansions" are always in terms of a set of orthogonal functions and thus are also called *orthogonal expansions.*

In elementary work, some set of trigonometric functions is usually used to write Fourier expansions. Hence, we will usually assume the terminology "Fourier expansion" to mean a Fourier trigonometric expansion. An example of a different kind of orthogonal set is the set of Legendre polynomials. A Fourier expansion in terms of this set is called a Fourier-Legendre expansion.

10.2 the space $PC[-L, L]$

From the general discussion of vectors in Chapter 1 you learned that the set of piecewise continuous functions on the interval $[a, b]$ can be considered a normed vector space with an inner product defined by

$$(f, g) = \int_a^b f(x)\, g(x)\, dx$$

and the corresponding norm

$$\|f\| = \left(\int_a^b [f(x)]^2\, dx\right)^{1/2}.$$

In this chapter we will be considering functions with the *periodic* property

$$f(x) = f(x + 2L),$$

where the quantity $2L$ is called a *period* of the function. Since we may restrict the discussion to any interval of length $2L$, we will discuss the space $PC[-L, L]$ without loss of generality.

Since we will be considering orthogonal expansions, you will find it important to know certain sets of trigonometric functions which are orthogonal on $[-L, L]$. The proof of the orthogonality is similar to Example 1.6. The sets are:

1. The set of cosine functions,

$$\left\{\cos \frac{n\pi x}{L}\right\}_{n=0}^{\infty}.$$

2. The set of sine functions,

$$\left\{\sin \frac{n\pi x}{L}\right\}_{n=1}^{\infty}.$$

3. The complete trigonometric set which is the union of the two sets described in 1. and 2.
4. Any subset of the three sets of 1., 2. and 3.

In Section 10.13 you will learn how other orthogonal sets arise quite naturally in applied mathematics. Most of the results of this chapter apply to expansions in terms of any orthogonal set although the examples will be in terms of one of the sets listed in the previous paragraph.

For theoretical purposes, using an orthonormal set instead of an orthogonal one is sometimes easier. This is not a significant restriction to the discussion since an orthonormal set is obtained from one which is orthogonal by dividing each function by its norm.

example 10.1: Find the orthonormal set corresponding to the complete trigonometric set on $[-L, L]$.

solution: We compute the norm of each of the functions of the orthogonal set,

$$\left\|\cos\frac{0\pi x}{L}\right\| = \|1\| = \left(\int_{-L}^{L} dx\right)^{1/2} = \sqrt{2L}$$

$$\left\|\cos\frac{n\pi x}{L}\right\| = \left(\int_{-L}^{L} \cos^2\left(\frac{n\pi x}{L}\right) dx\right)^{1/2} = \sqrt{L}$$

$$\left\|\sin\frac{n\pi x}{L}\right\| = \left(\int_{-L}^{L} \sin^2\left(\frac{n\pi x}{L}\right) dx\right)^{1/2} = \sqrt{L}.$$

Therefore, the orthonormal set is

$$\left\{\frac{1}{\sqrt{2L}}\right\} \cup \left\{\frac{\cos(n\pi x/L)}{\sqrt{L}}\right\}_{n=1}^{\infty} \cup \left\{\frac{\sin(n\pi x/L)}{\sqrt{L}}\right\}_{n=1}^{\infty}.$$

Sometimes a slightly different inner product is defined for function spaces. This results in different formulas for the norm of a function and for orthogonality.

definition 10.1

We say that (f, g) is a *weighted integral inner product* on the space $PC[a, b]$ with respect to the weight function $p(x)$, $(p(x) > 0)$, if

$$(f, g) = \int_a^b f(x)\, g(x)\, p(x)\, dx.$$

With respect to this inner product, the norm of f takes the form

$$\|f\| = \left(\int_a^b p(x)\,[f(x)]^2\, dx\right)^{1/2}.$$

A set of functions $\{g_n\}_{n=1}^{\infty}$ is orthogonal on $[a, b]$ with respect to the weight function $p(x)$ if

$$\int_a^b p(x)\, g_m(x)\, g_n(x)\, dx = 0 \qquad \text{for } n \neq m.$$

If the weight function is 1, the definitions reduce to those previously given.

10.3 two kinds of equality

Because of the introduction of the integral norm we are led to consider a new type of equality between functions.

definition 10.2

We say that the functions f and g are *equal in the norm*, or *equal in the mean*, if $\| f-g \| = 0$ in the space under consideration.

Thus two functions may be equal in the mean even if they are unequal at a finite number of points, since altering the integrand at finitely many points does not affect the value of the integral.

To avoid confusion of this new equality with the equality considered up to now, we give a name and definition to this old equality.

definition 10.3

Two functions, f and g, are *pointwise equal* if they have the same domains and if $f(x) = g(x)$ for every x in the domain.

example 10.2: Let $g(x) = x$ on $(-1, 1)$ and let $f(x) = g(x)$ except at $x = 1/4$, $1/2$ and $3/4$ where $f(x) = 0$. Determine if, and in what sense, the two functions are equal. (See Figure 10.1.)

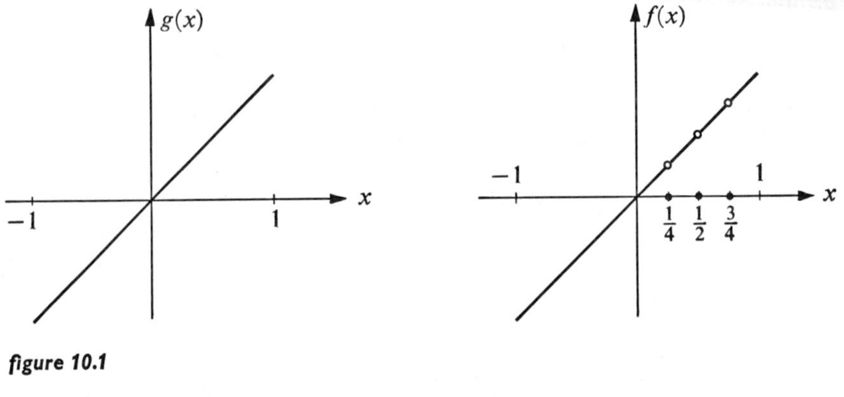

figure 10.1

solution: The functions are obviously not equal in the pointwise sense since g and f do not agree at three points. The norm, $\| f-g \|$, is zero. Therefore, g and f are equal in the mean.

Henceforth, the equals sign may have two meanings, although there is usually little difficulty in determining the sense in which a given equals sign is used. In cases of possible ambiguity, the notation $\doteq$ is often used to represent equality in the mean.

The two different kinds of equality lead to two types of convergence of a sequence of functions.

definition 10.4

We say the sequence of functions $\{g_n\}_{n=1}^{\infty}$ *converges in the mean* to the function f if

$$\lim_{n\to\infty} \|f - g_n\| = 0.$$

The same sequence is said to *converge pointwise* to f_L if

$$\lim_{n\to\infty} |f_L(x) - g_n(x)| = 0$$

for every x in the interval.

Pointwise convergence requires the convergence of many sequences of numbers, one for each value of x. It is discussed in elementary calculus and reviewed in Appendix A of this book. For convergence in the mean we require the one sequence of numbers

$$a_n = \left(\int_{-L}^{L} [f(x) - g_n(x)]^2\, dx\right)^{1/2}$$

to converge to zero. Figuratively, $a_n \to 0$ for convergence in the mean and $g_n(x) \to f(x)$ (for each x) for pointwise convergence.

example 10.3: Consider the sequence of functions $f_n(x) = x^n$ on [0, 1]. Analyze for mean convergence. (This same sequence of functions is analyzed for pointwise convergence in Example A.3, Appendix A.)

solution: The sequence converges in the mean to the "zero function" because

$$\begin{aligned}\|0 - x^n\| &= \left(\int_0^1 x^{2n}\, dx\right)^{1/2} \\ &= \left(\left.\frac{x^{2n+1}}{2n+1}\right|_0^1\right)^{1/2} \\ &= \left(\frac{1}{2n+1}\right)^{1/2}\end{aligned}$$

which approaches 0 as n becomes arbitrarily large.

The next example is introduced only to exhibit the striking difference between the two different types of convergence. From that standpoint, at least, it can be considered a "practical" example.

example 10.4: Let $\{f_n\}_{n=1}^{\infty}$ be a sequence of functions defined on the unit interval as follows:

$$\begin{aligned}
&f_1(x) = 1 \text{ over the entire interval.}\\
&f_2(x) = 1 \text{ on } [0, \tfrac{1}{2}] \text{ and 0 otherwise.}\\
&f_3(x) = 1 \text{ on } [\tfrac{1}{2}, 1] \text{ and 0 otherwise.}\\
&f_4(x) = 1 \text{ on } [0, \tfrac{1}{4}] \text{ and 0 otherwise.}\\
&\quad\vdots \qquad\qquad\qquad \vdots\\
&f_{2^n}(x) = 1 \text{ on } \left[0, \frac{1}{2^n}\right] \text{ and 0 otherwise.}\\
&f_{2^n+1}(x) = 1 \text{ on } \left[\frac{1}{2^n}, \frac{2}{2^n}\right] \text{ and 0 otherwise, etc.}
\end{aligned}$$

Analyze for the two types of convergence. (See Figure 10.2.)

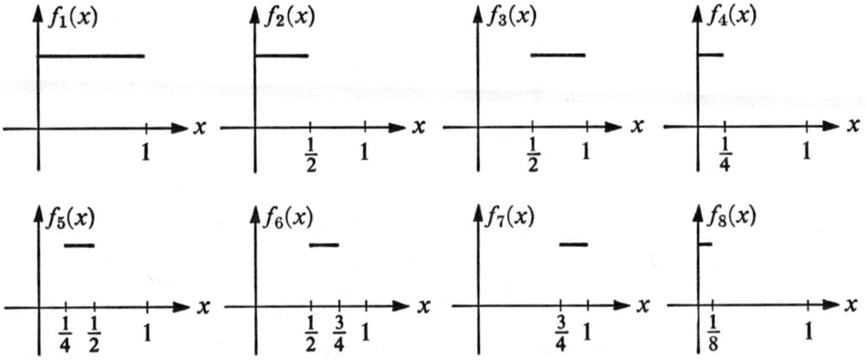

figure 10.2

solution: The sequence converges pointwise nowhere on the interval since for any value of N and for any fixed x on the interval, there exist $m > N$, and $n > N$, such that $f_m(x) = 0$ and $f_n(x) = 1$.

The sequence converges in the mean to the zero function because

$$\begin{aligned}
\|f_n - 0\|^2 &= \int_0^1 [f_n(x)]^2\, dx\\
&= \text{length of the interval on which } f_n(x) \text{ is nonzero.}
\end{aligned}$$

As $n \to \infty$, this length $\to 0$.

Remember, by definition, a series converges if its sequence of partial sums converges. Similarly, we now define convergence of a series in the mean by considering the sequence of partial sums.

definition 10.5

We say the infinite series of functions $\sum_{n=1}^{\infty} g_n$ *converges in the mean* to the function f if

$$\lim_{n\to\infty} \left\| f - \sum_{k=1}^{n} g_k \right\| = 0.$$

The same series is said to *converge pointwise* to the function f_L if

$$\lim_{n\to\infty} \left| f_L(x) - \sum_{k=1}^{n} g_k(x) \right| = 0$$

for every x in the domain of f_L.

We are now in a position to define a basis for the function space $PC[a, b]$.

definition 10.6

An orthogonal set $\{g_n\}_{n=1}^{\infty}$ on $PC[a, b]$ is a *basis set* if and only if each function in the space can be written *uniquely* in the form

$$f \doteq \sum_{n=1}^{\infty} c_n g_n.$$

The series $\sum_{n=1}^{\infty} c_n g_n$ is called the *Fourier series* of f with respect to the basis set $\{g_n\}_{n=1}^{\infty}$.

Note that a basis set is defined in terms of mean convergence. In this book we will assume without proof that certain basis sets exist. In particular, the *complete* trigonometric set is a basis for the function space $PC[-L, L]$. This assumption is of course crucial, and, having thus disposed of the greater difficulty, what is left for us to discuss becomes relatively trivial; that is, by the very nature of a basis, every piecewise continuous function has a Fourier expansion which converges in the mean to itself. More lengthy treatments of Fourier series *prove* that a set such as the complete trigonometric set is a basis for $PC[-L, L]$ in the sense of Definition 10.6. The main thing for you to know is that our assumption is far from being trivial.

It is important that you realize precisely which are the elements of the trigonometric basis set for a given interval. For example, the function $\sin x$

is a member of the set

$$\{\cos nx\}_{n=0}^{\infty} \cup \{\sin mx\}_{m=1}^{\infty}$$

which is a basis for $PC[-\pi, \pi]$. It is also a member of the set

$$\left\{\cos \frac{nx}{2}\right\}_{n=0}^{\infty} \cup \left\{\sin \frac{mx}{2}\right\}_{m=1}^{\infty}$$

which is a basis for $PC[-2\pi, 2\pi]$. However, it is not a member of the set

$$\{\cos 2nx\}_{n=0}^{\infty} \cup \{\sin 2mx\}_{m=1}^{\infty}$$

which is a basis for $PC[-\pi/2, \pi/2]$.

exercises for sections 10.1–10.3

1. Show that x^2 and x^4 are orthogonal on $[-1, 1]$ with respect to the weight function x.
2. Show that $\left\{\sin \frac{n\pi x}{L}\right\}_{n=1}^{\infty}$ is an orthogonal set on $[0, L]$. Find the corresponding orthonormal set.
3. Show that $L_0 = 1$, $L_1 = 1 - x$, $L_2 = 1 - 2x + (x^2/2)$ form an orthogonal set on $[0, \infty)$ with respect to the weight function $p(x) = e^{-x}$. L_0, L_1, and L_2 are the first three members of a set of polynomials called the Laguerre polynomials.
4. Show that $T_0 = 1$, $T_1 = x$, $T_2 = 2x^2 - 1$ are orthogonal on $[-1, 1]$ with respect to the weight function $p(x) = (1 - x^2)^{-1/2}$. T_0, T_1, T_2 are the first three Tchebichef polynomials.
5. Is it a "suitable" definition for equality in the norm to demand that $\|f\| = \|g\|$? Explain.
6. Let $f_n(x) = 0$, if $0 \leq x \leq 1/n$ or if $2/n \leq x \leq 1$, and let $f_n(x) = \sqrt{n}$ if $1/n < x < 2/n$. Show that the sequence $\{f_n\}_{n=1}^{\infty}$ converges pointwise to the zero function on $[0, 1]$ but does not converge in the mean to the zero function.
7. For the sequence of Example 10.4, sketch f_{63}, f_{100}, f_{200}. Derive a formula for the length of the interval on which $f_n(x) \neq 0$.
8. Is $\cos(n\pi x/4)$ a member of the trigonometric basis set on:
   (a) $PC[-1, 1]$; (b) $PC[-\pi, \pi]$; (c) $PC[-\frac{1}{4}, \frac{1}{4}]$;
   (d) $PC[-4, 4]$; (e) $PC[-8, 8]$.
   In general what characteristic must L have if $\cos(n\pi x/4)$ is to be a member of the trigonometric basis for $PC[-L, L]$?

9. Let B be a basis for $PC[a, b]$. Prove: The zero function is the only function orthogonal to every member of B.

10.4 fourier coefficients

The form of the representation of a function f in terms of an orthogonal set follows closely the techniques shown in Chapter 1 for finite dimensional vector spaces. The application to the infinite dimensional vector space $PC[a, b]$ may be considered a purely formal extension of finite dimensional methods.

Assume that the set $\{g_n\}_{n=1}^{\infty}$ is an orthogonal basis for the function space. Then if f is a member of the space, we write

$$f \doteq \sum_{k=1}^{\infty} c_k g_k .$$

The restrictions to be imposed on the function f to give meaningful pointwise convergence will be discussed later. For now, assume that the series converges to $f(x)$ for at least some values of x.

Assuming that we may take the inner product termwise (in the case of the integral inner product, this would be assuming term-by-term integration to be permissible), we take the inner product of both sides with respect to g_n to obtain

$$(g_n, f) = \left(g_n, \sum_{k=1}^{\infty} c_k g_k\right) = \sum_{k=1}^{\infty} (g_n, c_k g_k) = \sum_{k=1}^{\infty} c_k (g_n, g_k).$$

Since the set $\{g_n\}_{n=1}^{\infty}$ is orthogonal, $(g_k, g_n) = 0$ for $n \neq k$ and hence only the term $n = k$ remains in the series, $(g_n, f) = c_n (g_n, g_n)$, from which

$$c_n = \frac{(g_n, f)}{\|g_n\|^2} .$$

In terms of the integral inner product,

$$c_n = \frac{\displaystyle\int_a^b f(x)\, g_n(x)\, dx}{\displaystyle\int_a^b [g_n(x)]^2\, dx} .$$

The c_n's chosen in this manner are called the *Fourier coefficients of the function f with respect to the orthogonal set* $\{g_n\}_{n=1}^{\infty}$. They are analogous to the components of a vector with respect to a finite basis.

The name "Fourier coefficient" will also be given to a number c_n defined by

$$c_n = \frac{(g_n, f)}{\|g_n\|^2}$$

regardless of whether the orthogonal set $\{g_n\}_{n=1}^{\infty}$ is a basis or not, and thus regardless of whether f and $\sum_{n=1}^{\infty} c_n g_n$ are equal in the mean or not. We write

$$f \sim \sum_{n=1}^{\infty} c_n g_n$$

which is read, "the Fourier series of f corresponding to the orthogonal set $\{g_n\}_{n=1}^{\infty}$," only to indicate how the c_n's were obtained. It is possible for all the Fourier coefficients of a nonzero function with respect to an orthogonal set to be zero.

example 10.5: Let $f(x) = x$ be considered as a member of $PC[-\pi, \pi]$. Write the Fourier series of f corresponding to the orthogonal set $\{\cos nx\}_{n=1}^{\infty}$.

solution: The coefficients are

$$c_n = \frac{1}{\pi} \int_{-\pi}^{\pi} x \cos nx \, dx$$

which is easily shown to be zero for all n, either by integrating by parts or using the result that the integral of an odd function over an interval symmetric about the origin is zero. (This will be proved in the next section.) Hence,

$$x \sim \sum_{n=1}^{\infty} 0 \cdot \cos nx = 0.$$

Obviously the Fourier sereis does not represent the function in any meaningful way. The problem here is that the given orthogonal set is *not* a basis for $PC[-\pi, \pi]$.

Since the complete trigonometric set

$$\left\{\cos \frac{n\pi x}{L}\right\}_{n=0}^{\infty} \cup \left\{\sin \frac{n\pi x}{L}\right\}_{n=1}^{\infty}$$

is a basis for the space $PC[-L, L]$, any $f \in PC[-L, L]$ and its Fourier series are equal in the mean. Such series take on added importance and we shall therefore calculate the Fourier coefficients,

$$f \sim \sum_{n=0}^{\infty} a_n \cos\frac{n\pi x}{L} + \sum_{n=1}^{\infty} b_n \sin\frac{n\pi x}{L}.$$

$$a_0 = \frac{(1, f)}{\|1\|^2} = \frac{1}{2L}\int_{-L}^{L} f(x)\, dx,$$

$$a_n = \frac{\left(\cos\dfrac{n\pi x}{L}, f\right)}{\left\|\cos\dfrac{n\pi x}{L}\right\|^2} = \frac{1}{L}\int_{-L}^{L} f(x)\cos\frac{n\pi x}{L}\, dx,$$

$$b_n = \frac{\left(\sin\dfrac{n\pi x}{L}, f\right)}{\left\|\sin\dfrac{n\pi x}{L}\right\|^2} = \frac{1}{L}\int_{-L}^{L} f(x)\sin\frac{n\pi x}{L}\, dx.$$

The formulas for the space $PC[-\pi, \pi]$ are particularly simple,

$$f \sim \sum_{n=0}^{\infty} a_n \cos nx + \sum_{n=1}^{\infty} b_n \sin nx.$$

$$a_0 = \frac{1}{2\pi}\int_{-\pi}^{\pi} f(x)\, dx,$$

$$a_n = \frac{1}{\pi}\int_{-\pi}^{\pi} f(x)\cos nx\, dx,$$

$$b_n = \frac{1}{\pi}\int_{-\pi}^{\pi} f(x)\sin nx\, dx.$$

In passing we note that some engineering handbooks write the Fourier series of f in the form

$$f \sim \frac{a_0}{2} + \sum_{n=1}^{\infty} a_n \cos nx + \sum_{n=1}^{\infty} b_n \sin nx.$$

This makes the formula for a_0 exactly that given for a_n with $n=0$.

example 10.6: Find the Fourier trigonometric series of the function

$$\begin{aligned} f(x) &= 2, & -\pi \leqq x < 0 \\ &= 1, & 0 \leqq x \leqq \pi \end{aligned}$$

with respect to the trigonometric set.

solution:

$$a_0 = \frac{1}{2\pi}\int_{-\pi}^{\pi} f(x)\,dx = \frac{1}{2\pi}\int_{-\pi}^{0} 2\,dx + \frac{1}{2\pi}\int_{0}^{\pi} 1\,dx = \frac{3}{2};$$

$$a_n = \frac{1}{\pi}\int_{-\pi}^{0} 2\cos nx\,dx + \frac{1}{\pi}\int_{0}^{\pi} \cos nx\,dx = 0, \quad n \geqq 1;$$

$$\begin{aligned} b_n &= \frac{1}{\pi}\int_{-\pi}^{0} 2\sin nx\,dx + \frac{1}{\pi}\int_{0}^{\pi} \sin nx\,dx \\ &= \frac{1}{\pi}\left[\frac{-2\cos nx}{n}\bigg|_{-\pi}^{0} - \frac{\cos nx}{n}\bigg|_{0}^{\pi}\right] \\ &= \frac{1}{n\pi}[-2 + 2\cos n\pi - \cos n\pi + 1] \\ &= \frac{\cos n\pi - 1}{n\pi} = \frac{(-1)^n - 1}{n\pi}. \end{aligned}$$

Therefore

$$f \sim \frac{3}{2} - \frac{2}{\pi}\left(\sin x + \frac{\sin 3x}{3} + \frac{\sin 5x}{5} + \cdots\right).$$

example 10.7: Find the Fourier series of the function

$$\begin{aligned} f(x) &= 0, \qquad -2 \leqq x < 0 \\ &= x, \qquad 0 \leqq x \leqq 2. \end{aligned}$$

solution: In this case $L = 2$, and

$$a_0 = \frac{1}{4}\int_{0}^{2} x\,dx = \frac{x^2}{8}\bigg|_{0}^{2} = \frac{1}{2};$$

$$\begin{aligned} a_n &= \frac{1}{2}\int_{0}^{2} x\cos\frac{n\pi x}{2}\,dx \\ &= \frac{1}{2}\left[\frac{4}{n^2\pi^2}\cos\frac{n\pi x}{2} + \frac{2x}{n\pi}\sin\frac{n\pi x}{2}\bigg|_{0}^{2}\right] \\ &= \frac{2(\cos n\pi - 1)}{n^2\pi^2} = \begin{cases} 0, & n \text{ even}, n \neq 0, \\ \dfrac{-4}{n^2\pi^2}, & n \text{ odd}; \end{cases} \end{aligned}$$

$$\begin{aligned} b_n &= \frac{1}{2}\int_{0}^{2} x\sin\frac{n\pi x}{2}\,dx \\ &= \frac{1}{2}\left[\frac{4}{n^2\pi^2}\sin\frac{n\pi x}{2} - \frac{2x}{n\pi}\cos\frac{n\pi x}{2}\bigg|_{0}^{2}\right] \\ &= \frac{-2\cos n\pi}{n\pi} = \frac{(-1)^{n+1}2}{n\pi}. \end{aligned}$$

Therefore

$$f \sim \frac{1}{2} - \frac{4}{\pi^2}\left(\cos\frac{\pi x}{2} + \frac{1}{9}\cos\frac{3\pi x}{2} + \frac{1}{25}\cos\frac{5\pi x}{2} + \cdots\right)$$
$$+ \frac{2}{\pi}\left(\sin\frac{\pi x}{2} - \frac{1}{2}\sin \pi x + \frac{1}{3}\sin\frac{3\pi x}{2} - \cdots\right).$$

example 10.8: Find the Fourier series of the half-wave rectified sine wave shown below and described analytically by

$$f(x) = 0, \qquad -3 < x < 0$$
$$= \sin\frac{\pi x}{3}, \qquad 0 \leqq x \leqq 3.$$

(See Figure 10.3.)

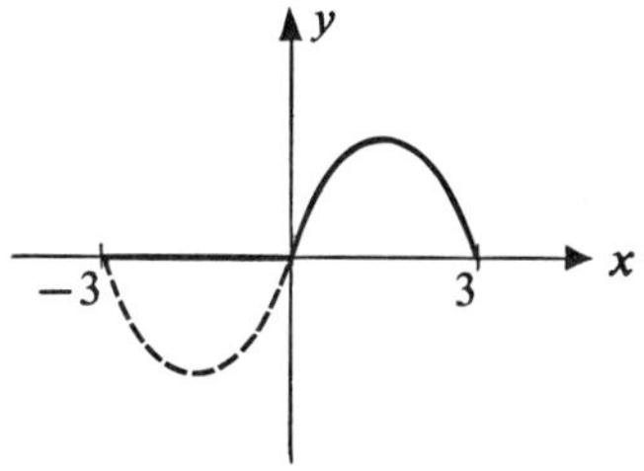

figure 10.3 *half-wave rectified sine wave*

solution:

$$a_0 = \frac{1}{6}\int_0^3 \sin\frac{\pi x}{3}\,dx = \frac{1}{\pi};$$

$$a_1 = \frac{1}{3}\int_0^3 \sin\frac{\pi x}{3}\cos\frac{\pi x}{3}\,dx = 0.$$

For $n \geqq 2$,

$$a_n = \frac{1}{3}\int_0^3 \sin\frac{\pi x}{3}\cos\frac{n\pi x}{3}\,dx$$
$$= \frac{1}{6}\int_0^3 \sin\frac{(n+1)\pi x}{3}\,dx - \frac{1}{6}\int_0^3 \sin\frac{(n-1)\pi x}{3}\,dx$$

$$= -\frac{1}{2\pi}\,\frac{\cos\dfrac{(n+1)\pi x}{3}}{n+1}\Bigg|_0^3 + \frac{1}{2\pi}\,\frac{\cos\dfrac{(n-1)\pi x}{3}}{n-1}\Bigg|_0^3$$

$$= \frac{1}{2\pi}\left(\frac{-\cos(n+1)\pi + 1}{n+1} + \frac{\cos(n-1)\pi - 1}{n-1}\right)$$

$$a_n = 0, \quad \text{if } n \text{ is odd, } n \neq 1;$$

$$= \frac{1}{\pi}\left(\frac{n-1-n-1}{n^2-1}\right) = \frac{-2}{\pi(n^2-1)}, \qquad \text{if } n \text{ is even;}$$

$$b_1 = \frac{1}{3}\int_0^3 \sin^2\left(\frac{\pi x}{3}\right) dx = \frac{1}{2}.$$

For $n \geqq 2$,

$$b_n = \frac{1}{3}\int_0^3 \sin\frac{\pi x}{3}\sin\frac{n\pi x}{3}\,dx$$

$$= \frac{1}{6}\int_0^3 \cos\frac{(n-1)\pi x}{3}\,dx - \frac{1}{6}\int_0^3 \cos\frac{(n+1)\pi x}{3}$$

$$= \frac{1}{2}\,\frac{\sin\dfrac{(n-1)\pi x}{3}}{(n-1)\pi}\Bigg|_0^3 - \frac{1}{2}\,\frac{\sin\dfrac{(n+1)\pi x}{3}}{(n+1)\pi}\Bigg|_0^3 = 0.$$

Therefore

$$f \sim \frac{1}{\pi} + \frac{1}{2}\sin\frac{\pi x}{3} - \frac{2}{\pi}\left(\frac{\cos\dfrac{2\pi x}{3}}{3} + \frac{\cos\dfrac{4\pi x}{3}}{15} + \frac{\cos\dfrac{6\pi x}{3}}{35} + \cdots\right).$$

exercises for section 10.4

1. Find the Fourier series of the function $f(x) = x^2$ with respect to the orthogonal set $\{\sin nx\}_{n=1}^{\infty}$ on $[-\pi, \pi]$. Then find the Fourier series with respect to $\{\cos nx\}_{n=0}^{\infty}$ on $[-\pi, \pi]$. What is the Fourier series of x^2 on $[-\pi, \pi]$ with respect to the complete trigonometric set?
2. Show how the Fourier coefficients of $f(x) = x^m$, m a positive integer, with respect to the complete trigonometric set on $[-L, L]$ are related to those with respect to the trigonometric set on $[-\pi, \pi]$.
3. Find the Fourier series of $f(x) = \cos^2 x$ on $[-\pi, \pi]$ with respect to the complete trigonometric set.

In Exercises 4–10, find the Fourier series of the given function with respect to the complete trigonometric set on the indicated interval.

4. $f(x)=1$, on $-\pi<x<0$;
$=2$, on $0<x<\pi$.

5. $|\cos x|$ on $[-\pi, \pi]$.

6. $|\sin x|$ on $[-\pi, \pi]$.

7. x^2 on $[-1, 1]$.

8. $f(x)=1$, on $[-\pi, 0)$;
$=-1$, on $[0, \pi]$.

9. $f(x)=0$, on $[-2, 0)$;
$=1$, on $[0, 2]$.

10. $\sin x$ on $[-\pi/2, \pi/2]$.

10.5 subspaces of PC[−L, L]

The function space $PC[-L, L]$ has two important subspaces, namely the set of all even functions $PC_e[-L, L]$ and the set of all odd functions $PC_o[-L, L]$. As you will see, the work to compute the Fourier coefficients for functions in these two subspaces is considerably reduced.

theorem 10.1

Let f be an odd function. Then

$$\int_{-L}^{L} f(x)\, dx = 0.$$

proof: From elementary algebra the fact that f is odd is expressed by the equation $f(x)=-f(-x)$. Further,

$$\int_{-L}^{L} f(x)\, dx = \int_{-L}^{0} f(x)\, dx + \int_{0}^{L} f(x)\, dx.$$

Using the change of variables $u=-x$ in the integral from $-L$ to 0,

$$\begin{aligned}\int_{-L}^{L} f(x)\, dx &= \int_{0}^{L} f(-u)\, du + \int_{0}^{L} f(x)\, dx \\ &= -\int_{0}^{L} f(u)\, du + \int_{0}^{L} f(x)\, dx \\ &= 0,\end{aligned}$$

as was to be shown.

theorem 10.2

If f is an even function, then the Fourier coefficients, b_n, are zero for all n. (The fact that f is even is described by the equation $f(x)=f(-x)$.)

proof:

$$b_n = \frac{1}{L}\int_{-L}^{L} f(x) \sin\frac{n\pi x}{L}\, dx.$$

Since the integrand is the product of an even function with an odd function, it is odd and hence the integral is zero by the preceding theorem.

theorem 10.3

If f is an odd function, then the Fourier coefficients, a_n, are zero for all n.

proof: The same reasoning is used as in Theorem 10.2.

example 10.9: Find the Fourier series for the function defined on the interval $[-\pi, \pi]$ by $f(x) = x$.

solution: Since the given function is odd, each a_n equals zero and

$$\begin{aligned} b_n &= \frac{1}{\pi}\int_{-\pi}^{\pi} x \sin nx \, dx \\ &= \frac{2}{\pi}\int_{0}^{\pi} x \sin nx \, dx \\ &= \frac{2}{\pi}\left(\frac{-x}{n}\cos nx + \frac{1}{n^2}\sin nx\right)\Bigg|_0^{\pi} \\ &= (-1)^{n+1}\frac{2}{n}. \end{aligned}$$

Therefore,

$$x \sim 2\sum_{n=1}^{\infty} \frac{(-1)^{n+1}}{n} \sin nx.$$

example 10.10: Find the Fourier series for the function $f(x) = |x|$ on $[-5, 5]$.

solution: Since the given function is even, all the b_n's $= 0$. Thus,

$$a_0 = \frac{1}{10}\int_{-5}^{5} |x| \, dx = \frac{1}{5}\int_{0}^{5} x \, dx = \frac{x^2}{10}\Bigg|_0^5 = \frac{5}{2};$$

$$\begin{aligned} a_n &= \frac{2}{5}\int_{0}^{5} x \cos \frac{n\pi x}{5} \, dx \\ &= \frac{2}{n\pi} x \sin \frac{n\pi x}{5}\Bigg|_0^5 + \frac{10}{n^2\pi^2} \cos \frac{n\pi x}{5}\Bigg|_0^5 \\ &= \frac{-20}{n^2\pi^2} \text{ for } n \text{ odd}, \\ &= 0 \text{ for } n \text{ even}. \end{aligned}$$

Therefore,

$$|x| \sim \frac{5}{2} - \frac{20}{\pi^2}\left(\cos\frac{\pi x}{5} + \frac{1}{9}\cos\frac{3\pi x}{5} + \frac{1}{25}\cos \pi x + \cdots\right).$$

There are other, not so obvious, examples of subspaces of $PC[-L, L]$ for which even more of the Fourier coefficients can be predetermined equal to zero. The following example is typical.

example 10.11: Let f be a function in $PC_o[-\pi, \pi]$ and symmetric with respect to the line $x = \pi/2$. Show that the only possible nonzero Fourier coefficients are b_n for n odd.

solution: Since f is odd, $a_n = 0$ for all n, and

$$\begin{aligned} b_n &= \frac{2}{\pi}\int_0^{\pi} f(x)\sin nx\, dx \\ &= \frac{2}{\pi}\int_0^{\pi/2} f(x)\sin nx\, dx + \frac{2}{\pi}\int_{\pi/2}^{\pi} f(x)\sin nx\, dx. \end{aligned}$$

Since f is symmetric with respect to the line $x = \pi/2$,

$$f(x) = f(\pi - x).$$

Hence the integral from $\pi/2$ to π becomes

$$\int_{\pi/2}^{\pi} f(\pi - x)\sin nx\, dx.$$

Let $u = \pi - x$ in this integral to obtain

$$\begin{aligned} b_n &= \frac{2}{\pi}\int_0^{\pi/2} f(x)\sin nx\, dx + \frac{2}{\pi}\int_0^{\pi/2} f(u)\sin n(\pi - u)\, du \\ &= \frac{2}{\pi}[1 + (-1)^{n+1}]\int_0^{\pi/2} f(x)\sin nx\, dx. \end{aligned}$$

Therefore, if n is even, $b_n = 0$, and if n is odd

$$b_n = \frac{4}{\pi}\int_0^{\pi/2} f(x)\sin nx\, dx.$$

example 10.12: Find the Fourier series of the function whose graph is shown in Figure 10.4.

solution: A basis for the space is

$$\left\{\cos\frac{n\pi x}{2}\right\}_{n=0}^{\infty} \cup \left\{\sin\frac{n\pi x}{2}\right\}_{n=1}^{\infty}.$$

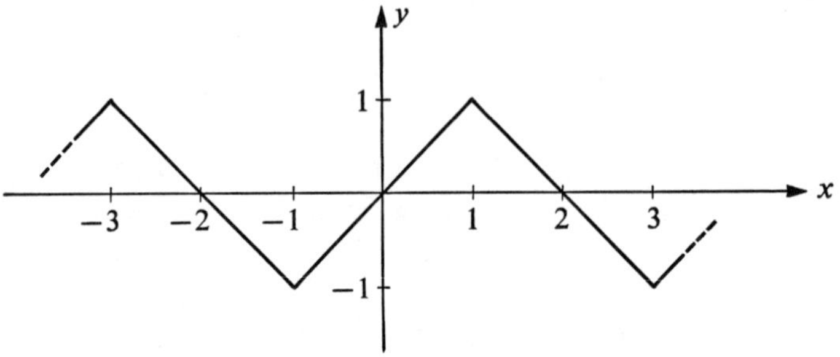

figure 10.4

The function is odd and symmetric about the line $x = 1$, and in a manner similar to the previous example it is easy to show that the only possible non-zero coefficients are given by

$$b_{2n+1} = \frac{4}{2}\int_0^1 f(x)\sin(2n+1)\frac{\pi x}{2}\,dx$$
$$= 2\int_0^1 x\sin(2n+1)\frac{\pi x}{2}\,dx.$$

Integrating by parts with $u = x$ and $dv = \sin(2n+1)\dfrac{\pi x}{2}\,dx$ gives

$$b_{2n+1} = \frac{-4x\cos(2n+1)\dfrac{\pi x}{2}}{(2n+1)\pi}\Bigg|_0^1 + \frac{4}{(2n+1)\pi}\int_0^1 \cos(2n+1)\frac{\pi x}{2}\,dx$$
$$= 0 + \frac{8}{\pi^2(2n+1)^2}\sin(2n+1)\frac{\pi x}{2}\Bigg|_0^1$$
$$= \frac{8(-1)^n}{\pi^2(2n+1)^2}.$$

Hence, the Fourier series is

$$f(x) \sim \frac{8}{\pi^2}\sum_{n=0}^{\infty}\frac{(-1)^n}{(2n+1)^2}\sin(2n+1)\frac{\pi x}{2}.$$

10.6 properties of fourier coefficients

The Fourier coefficients have some elementary properties, which while being almost trivially obvious serve to reduce the amount of computational work. Our first theorem is a case in point.

theorem 10.4

1. Let $f(x) = g(x) + h(x)$. Then the Fourier coefficients of the function f are the sum of the coefficients of g and of h.
2. The Fourier coefficients of $cf(x)$ are c times the Fourier coefficients of f.

proof: Follows directly from the formulas for the Fourier coefficients.

theorem 10.5

Let $g(x) - h(x)$ be a constant. Then the Fourier coefficients of g and h differ only in the value of a_0, the difference being precisely $g(x) - h(x)$.

proof: Consider the specific case of Fourier expansions on $PC[-\pi, \pi]$, and let $f(x) = g(x) + c$. Let a_n and b_n be the Fourier coefficients of g, and a_n' and b_n' be the Fourier coefficients of f, both with respect to the complete set of trigonometric functions on $[-\pi, \pi]$. Then,

$$\begin{aligned} a_n &= \frac{1}{\pi}\int_{-\pi}^{\pi} f(x)\cos nx\,dx \\ &= \frac{1}{\pi}\int_{-\pi}^{\pi} g(x)\cos nx\,dx + \frac{c}{\pi}\int_{-\pi}^{\pi} \cos nx\,dx \\ &= a_n' + 0 = a_n'. \end{aligned}$$

Similarly,

$$\begin{aligned} b_n &= b_n'; \\ a_0 &= \frac{1}{2\pi}\int_{-\pi}^{\pi} f(x)\,dx \\ &= \frac{1}{2\pi}\int_{-\pi}^{\pi} g(x)\,dx + \frac{c}{2\pi}\int_{-\pi}^{\pi} dx \\ &= a_0' + c. \end{aligned}$$

Thus, the theorem is proved.

theorem 10.6

Let $\{\phi_n\}_{n=1}^{\infty}$ be an orthogonal basis for $PC[a, b]$, and suppose $f(x) = \phi_j(x)$ for some j. Then the Fourier coefficients of f with respect to that basis set are all zero except c_j which is equal to 1.

proof: See Exercise 7 in the next set of exercises.

example 10.13: Find the Fourier series of $3x + \sin x$ considered as a member of $PC[-\pi, \pi]$.

solution: The Fourier coefficients of the function $f(x) = x$ are found in Example 10.9 to be

$$b_n = (-1)^{n+1}\left(\frac{2}{n}\right), \qquad a_n = 0 \quad \text{for all } n.$$

Hence, the Fourier coefficients of $f(x) = 3x$ are

$$b_n = (-1)^{n+1}\left(\frac{6}{n}\right), \qquad a_n = 0 \quad \text{for all n}.$$

Since $\sin x$ is a member of the trigonometric basis set on $PC[-\pi, \pi]$, the Fourier coefficients of $\sin x$ are

$$b_1 = 1 \qquad \text{and all the rest } 0.$$

Thus, the Fourier series for $3x + \sin x$ is

$$3x + \sin x \sim 7 \sin x + \sum_{n=2}^{\infty} \frac{6}{n}(-1)^{n+1} \sin nx.$$

example 10.14: Find the Fourier series of the function shown in Figure 10.5.

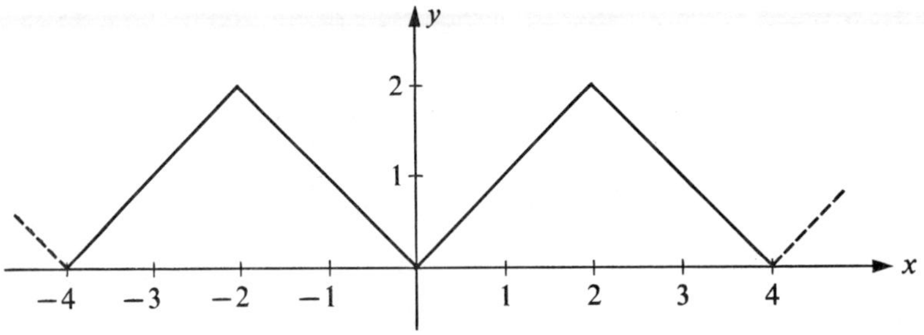

figure 10.5

solution: The given function is the same as that of Example 10.12 except that it is raised by 1 unit and shifted to the right by 1 unit. Hence the Fourier series of this function is

$$f(x) \sim 1 + \frac{8}{\pi^2} \sum_{n=0}^{\infty} \frac{(-1)^n}{(2n+1)^2} \sin (2n+1) \frac{\pi(x-1)}{2}$$

$$= 1 - \frac{8}{\pi^2} \sum_{n=0}^{\infty} \frac{1}{(2n+1)^2} \cos (2n+1) \frac{\pi x}{2}.$$

example 10.15: Find the Fourier series for the function $\sin(x/2)$:
(a) on $PC[-2\pi, 2\pi]$; (b) on $PC[-\pi, \pi]$.

solution: (a) The trigonometric basis set for $PC[-2\pi, 2\pi]$ is

$$\left\{\cos\frac{nx}{2}\right\}_{n=0}^{\infty} \cup \left\{\sin\frac{mx}{2}\right\}_{m=1}^{\infty}.$$

Note that $\sin(x/2)$ is a member of this basis set (for $m = 1$), and hence its Fourier series is just $\sin(x/2)$.

(b) The trigonometric basis set for $PC[-\pi, \pi]$ is

$$\{\cos nx\}_{n=0}^{\infty} \cup \{\sin mx\}_{m=1}^{\infty}.$$

The function is *not* a member of this basis set. It is an odd function and hence $a_n = 0$ for all n. Further,

$$\begin{aligned}
b_n &= \frac{2}{\pi}\int_0^{\pi} \sin\frac{x}{2}\sin nx\,dx \\
&= \frac{1}{\pi}\int_0^{\pi} \cos\left(n - \tfrac{1}{2}\right)x\,dx - \frac{1}{\pi}\int_0^{\pi}\cos\left(n + \tfrac{1}{2}\right)x\,dx \\
&= \frac{1}{\pi}\left[\frac{2}{2n-1}\sin\left(\frac{2n-1}{2}\right)x\Big|_0^{\pi} - \frac{2}{2n+1}\sin\left(\frac{2n+1}{2}\right)x\Big|_0^{\pi}\right] \\
&= \frac{2}{\pi}\left[\frac{(-1)^{n+1}}{2n-1} - \frac{(-1)^n}{2n+1}\right] \\
&= (-1)^{n+1}\left(\frac{2}{\pi}\right)\left[\frac{1}{2n-1} + \frac{1}{2n+1}\right] \\
&= (-1)^{n+1}\left(\frac{8}{\pi}\right)\frac{n}{4n^2-1}.
\end{aligned}$$

The Fourier series for the given function is

$$f(x) \sim \frac{8}{\pi}\sum_{n=0}^{\infty}(-1)^{n+1}\frac{n}{4n^2-1}\sin nx.$$

exercises for sections 10.5 and 10.6

1. Determine whether the following functions are even, odd, or neither:
(a) x. (b) x^2. (c) $x^2 + x$. (d) $|\sin x|$.
(e) $|\cos x|$. (f) $\cos x$. (g) $\sin x$. (h) $\sin 2x$.
(i) $\sin x + \cos x$. (j) e^x. (k) $\cosh x$. (l) $\sinh 2x$.
(m) x^{2n}. (n) x^{2n+1}.

2. Prove:
   (a) The product of two odd functions is even.
   (b) The product of an odd function and an even function is odd.
   (c) The product of two even functions is even.
3. Evaluate

$$\int_{-1}^{1} x \tan^8 x \cos^{2/3} x e^{x^2} \sin^2 x^4 \, dx.$$

4. Find the Fourier series of the function which is -1 on $[-\pi, 0)$ and 1 on $[0, \pi]$. Then find the Fourier series of the function which is $-1/2$ on $[-\pi, 0)$ and is $1/2$ on $[0, \pi]$.
5. Find the Fourier series of x and x^2 on $[-\pi, \pi]$ and then $2x + 3x^2$ on $[-\pi, \pi]$.
6. Find the Fourier series of the function $f(x) = 5 \cos 2x + \sin(x/2) - 3$ on $[-\pi, \pi]$.
7. Prove Theorem 10.6.
8. Find the Fourier series of $\cos x$:
   (a) on $[-\pi, \pi]$. (b) on $[-2\pi, 2\pi]$. (c) on $[-\pi/2, \pi/2]$.
9. Find a Fourier series which contains only sine terms and which is the Fourier series of $x - \pi$ on $[0, \pi]$.
10. Find a Fourier series which contains only cosine terms and which is the Fourier series of $x - \pi$ on $[\pi, 2\pi]$.

10.7 convergence of a fourier series

As we stated earlier, the very definition of a basis assures convergence in the mean of the Fourier series of a function with respect to an orthogonal set which is a basis. The series will converge in the mean to the function itself. Thus for all the examples of the previous sections in which a basis set is used for the expansion, $\doteq$ may be substituted for $\sim$.

The situation for pointwise convergence is not simple. In fact, one of the unsolved problems of mathematics is the determination of conditions which are both necessary and sufficient for the Fourier series to converge pointwise to the function. For example, there exist continuous functions whose Fourier series do not represent them. However, the class of functions which can be represented by their Fourier series is surprisingly large and general. The following theorem delineates one set of sufficient conditions, called the *Dirichlet conditions*.

theorem 10.7

Let f be bounded, piecewise continuous, and have a finite number of maxima and minima on $[-L, L]$. Then the corresponding Fourier series converges on $[-L, L]$. Its sum is $f(x)$ except at points of discontinuity where it converges to the average of the right- and left-hand limits.

proof: See Churchill (1941), listed in Appendix E.

The Dirichlet conditions make it clear that f need not be continuous in order to be represented by its Fourier series expansion. This means that a function may consist of a number of disjointed arcs of different curves, each defined by a different equation and still be represented by a Fourier series. Note that outside the interval $[-L, L]$ the Fourier series converges to the periodic extension of f. Hence if f is already periodic, then its Fourier series represents it everywhere except at its points of discontinuity. Further, the Fourier series of the function will converge to the average value at the discontinuities regardless of the definition (or lack of definition) of the function at these points.

example 10.16: Sketch the graph of the function to which the Fourier series of f converges if

$$f(x) = 2, \qquad -\pi < x < 0,$$
$$= 1, \qquad 0 \leqq x \leqq \pi.$$

(See Example 10.6.)

solution: The function is extended periodically and at the points of discontinuity, the series averages the right- and left-hand limits. (See Figure 10.6.)

figure 10.6

example 10.17: Show that

$$1 - \frac{1}{3} + \frac{1}{5} - \frac{1}{7} + \cdots + \frac{(-1)^{n+1}}{2n+1} + \cdots = \frac{\pi}{4}.$$

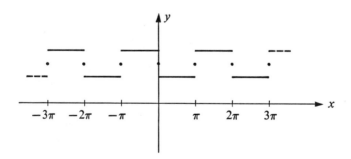

solution: We use the function of the previous example (and of Example 10.6) and Theorem 10.7 to write

$$f(x)=\frac{3}{2}-\frac{2}{\pi}\left(\sin x+\frac{\sin 3x}{3}+\frac{\sin 5x}{5}+\cdots\right)$$

where the equality is valid for all x except at the points of discontinuity where the series converges to 3/2.

Using the equality at the point $x=\pi/2$,

$$f\left(\frac{\pi}{2}\right)=1=\frac{3}{2}-\frac{2}{\pi}\left(\sin\frac{\pi}{2}+\frac{\sin(3\pi/2)}{3}+\frac{\sin(5\pi/2)}{5}+\cdots\right).$$

Transposing, and simplifying, we obtain

$$\frac{1}{2}=\frac{2}{\pi}\left(1-\frac{1}{3}+\frac{1}{5}-\frac{1}{7}+\cdots\right).$$

Therefore,

$$\frac{\pi}{4}=1-\frac{1}{3}+\frac{1}{5}-\frac{1}{7}+\cdots.$$

example 10.18: Sketch the graph of the function to which the Fourier series of Example 10.9 converges. Use the series to obtain the same result as in the previous example.

solution: The sketch of the function is shown in Figure 10.7. If we evaluate the series at $x=\pi/2$ and simplify, the same result as the one derived in the previous example is obvious.

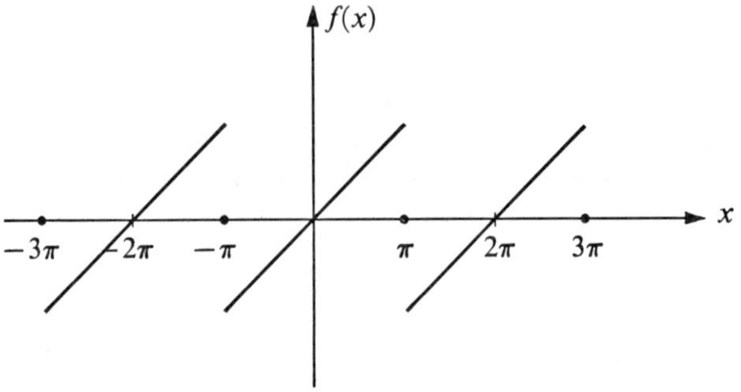

figure 10.7

example 10.19: Using the Fourier series of Example 10.8 show

$$\sum_{n=1}^{\infty}\frac{1}{4n^2-1}=\frac{1}{2}.$$

solution: The Fourier series of the function, derived in Example 10.8, converges pointwise to the function over the entire interval. In particular at $x = 3, f(3) = 0$, and thus

$$0 = \frac{1}{\pi} - \frac{2}{\pi}\left(\frac{1}{3} + \frac{1}{15} + \frac{1}{35} + \cdots + \frac{1}{4n^2 - 1} + \cdots\right)$$

and simplifying this expression yields the desired result.

In elementary calculus, you were usually asked to determine only if a series converges or diverges. Examples like the previous one show how sums for some infinite series can be determined. The trick is to find just the right function, find its Fourier series, and then evaluate it at just the right point.

Closely associated with the result of Theorem 10.7 is the following important theorem, which we state without proof.

theorem 10.8

Let f be integrable on $[-L, L]$. Then, for every x on $[-L, L]$

$$\int_{-L}^{x} f(t)\, dt = \int_{-L}^{x} [\text{the Fourier series of } f(t)]\, dt.$$

proof: See Churchill (1941).

example 10.20: Use the previous theorem to compute the Fourier series of the function $f(x) = |x| - \pi, \ -\pi < x < \pi$.

solution: Consider the function

$$\begin{aligned} g(x) &= -1, \qquad -\pi < x < 0, \\ &= 1, \qquad 0 < x < \pi. \end{aligned}$$

First note that

$$\int_{-\pi}^{x} g(t)\, dt = f(x).$$

Since g is odd, we need calculate only the b_n's,

$$\begin{aligned} b_n &= \frac{2}{\pi}\int_0^{\pi} \sin nx\, dx \\ &= \frac{2}{\pi}\frac{\cos nx}{n}\bigg|_{\pi}^{0} \\ &= 0 \text{ if } n \text{ is even}, \\ &= \frac{4}{n\pi} \text{ if } n \text{ is odd}. \end{aligned}$$

Hence

$$g(x) = \frac{4}{\pi} \sum_{n=0}^{\infty} \frac{\sin(2n+1)x}{2n+1}$$

where by Theorem 10.7, equality will hold throughout the interval if we define $g(0) = 0$.

Using Theorem 10.8,

$$f(x) = -\frac{4}{\pi} \sum_{n=0}^{\infty} \frac{\cos(2n+1)x}{(2n+1)^2} - \frac{\pi}{2}.$$

Roughly speaking, Theorem 10.8 says that the indefinite integral of a function is equal to the term-by-term indefinite integral of its Fourier series. Even roughly speaking, the same kind of theorem is not true for the derivative of a function. It is one of those times in mathematics when things happen a bit differently than perhaps you would suppose, or would wish.

To see this, consider the Fourier series for $f(x) = x$. Its Fourier series is computed in Example 10.9, so that on the open interval $(-\pi, \pi)$,

$$x = 2 \sum_{n=1}^{\infty} (-1)^{n+1} \frac{\sin nx}{n}.$$

The derivative of the left-hand side with respect to x is 1 and of the right-hand side is

$$2 \sum_{n=1}^{\infty} (-1)^{n+1} \cos nx.$$

This latter series is certainly not a representative of the function $f(x) = 1$, and hence the Fourier series of the derivative of the given function is not equal to the term-by-term derivative of the Fourier series. (Note that the Fourier series of the function $f(x) = 1$ *does* exist.)

The function $f(x) = x$ is not a "weird" function in any sense, although its periodic extension does have discontinuities at $x = \pm(2n+1)\pi$. You should be extremely careful in differentiating both sides of an equality involving Fourier series. The equality may not be maintained.

Conditions under which a Fourier series *may* be differentiated term by term to yield the derivative of the original function are given in the following theorem.

theorem 10.9

Let f and its first $n+1$ derivatives satisfy the Dirichlet conditions on $[-L, L]$. Let $f^{(n)}$ be continuous and $f^{(n+1)}(x)$ be bounded. Then the Fourier series for f can be differentiated term by term n times and the resulting series will converge to $f^{(n)}(x)$.

In applying this theorem you must consider the periodic extension of the function, not just the function as defined on $(-L, L)$. This will explain why an apparently continuous function such as $f(x) = x$ must be considered as discontinuous at $-L$ and L, and hence the differentiated series is not equal to the differentiated function.

exercises for section 10.7

1. Sketch the graph of the function to which the Fourier series of f converges on $[-3\pi, 3\pi]$ if
$$f(x) = 1, \text{ on } (-\pi, 0),$$
$$= 2, \text{ on } (0, \pi).$$
2. Repeat Exercise 1 for $f(x) = |\cos x|$ on $[-\pi, \pi]$.
3. Repeat Exercise 1 for $f(x) = |\sin x|$ on $[-\pi, \pi]$.
4. Repeat Exercise 1 for $f(x) = |x|$ on $[-\pi, \pi]$.
5. Repeat Exercise 1 for $f(x) = x$ on $[-\pi, \pi]$.
6. Sketch the graph of the function to which the Fourier series of f converges on $[-3, 3]$ if $f(x) = x^2$ on $(-1, 1)$.
7. Sketch the graph of the function to which the Fourier series of f converges on $[-6, 6]$ if
$$f(x) = 0, \text{ on } [-2, 0),$$
$$= 1, \text{ on } [0, 2].$$
8. Using the Fourier series of $f(x) = x^2$ on $PC[-\pi, \pi]$, show
$$\sum_{n=1}^{\infty} \frac{1}{n^2} = \frac{\pi^2}{6}; \quad \sum_{n=1}^{\infty} \frac{(-1)^{n+1}}{n^2} = \frac{\pi^2}{12}.$$

10.8 sine and cosine series

Let f be defined on $[0, L]$. The *even extension* of f to $[-L, L]$ is the function f_e such that $f_e(-x) = f(x)$ for x on $[0, L]$. The *odd extension* of f to $[-L, L]$ is a function f_o such that $f_o(-x) = -f(x)$ for x on $[0, L]$. The three functions f, f_e, and f_o all agree on $[0, L]$. Even if f is defined on $[-L, 0]$, it has no effect on f_e and f_o, for on that interval $f_e(x) = -f_o(x)$. (See Fig. 10.8.)

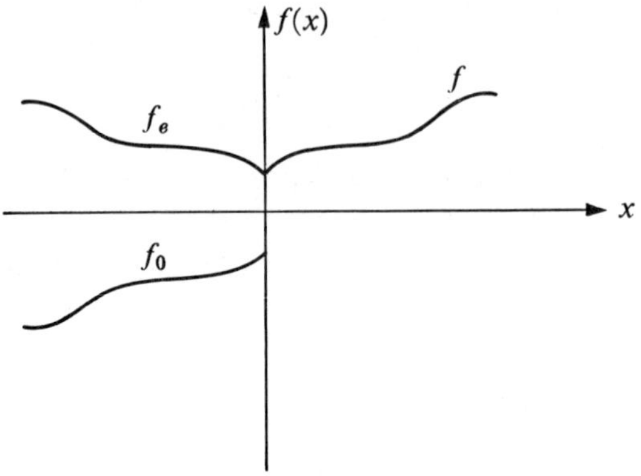

figure 10.8 even and odd extensions

example 10.21: Find f_e and f_o for the function defined on $[0, 1]$ by $f(x) = x$.

solution: $f_e(x) = x$ on $[0, 1]$, $\quad f_o(x) = x$ on $[0, 1]$,
$= -x$ on $[-1, 0]$. $\quad = x$ on $[-1, 0]$.
(See Figure 10.9.)

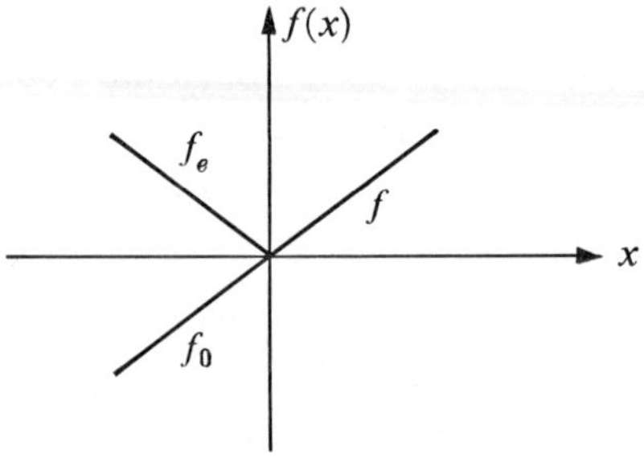

figure 10.9

example 10.22: Find f_e and f_o in the case $f(x) = x - x^2$ on $[0, 1]$. Sketch the odd and even periodic extensions.

solution: $f_e(x) = x - x^2$ on $[0, 1]$, $\quad f_o(x) = x - x^2$ on $[0, 1]$,
$= -x - x^2$ on $[-1, 0]$. $\quad = x + x^2$ on $[-1, 0]$.
(See Figure 10.10.)

The even extension of f, f_e, is a member of $PC_e[-L, L]$ and thus may be expressed in terms of the set $\{\cos(n\pi x/L)\}_{n=1}^{\infty}$; that is, the set of cosine functions is a basis for $PC_e[-L, L]$. Therefore, the following formulas

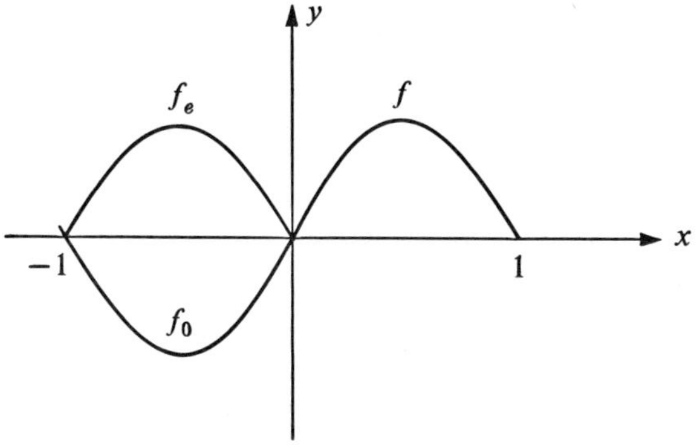

figure 10.10

are valid:

$$f_e \sim \sum_{n=0}^{\infty} a_n \cos \frac{n\pi x}{L};$$
$$a_0 = \frac{1}{L}\int_0^L f_e(x)\,dx = \frac{1}{L}\int_0^L f(x)\,dx;$$
$$a_n = \frac{2}{L}\int_0^L f(x)\cos\frac{n\pi x}{L}\,dx, \qquad n \neq 0.$$

The conditions for the series actually to represent f_e are the same as before but now we note that the series converges to the even periodic extension of f on $[0, L]$. This is true even if the original f is left undefined outside of the interval $[0, L]$. The Fourier series so obtained is called the *Fourier cosine series of f.*

example 10.23: Let f be defined on $[0, \pi]$ by the rule

$$\begin{aligned} f(x) &= x, \qquad \text{for } x \text{ on } \left[0, \frac{\pi}{2}\right] \\ &= 0, \qquad \text{for } x \text{ on } \left[\frac{\pi}{2}, \pi\right]. \end{aligned}$$

Write the Fourier cosine series of f.

solution:

$$a_0 = \frac{1}{\pi}\int_0^{\pi/2} x\,dx = \frac{\pi}{8};$$
$$a_n = \frac{2}{\pi}\int_0^{\pi/2} x\cos nx\,dx = \frac{1}{\pi n^2}\left(2\cos\frac{n\pi}{2} + n\pi\sin\frac{n\pi}{2} - 2\right);$$
$$f(x) = \frac{\pi}{8} + \frac{1}{\pi}\left[(\pi - 2)\cos x - \cos 2x - \left(\frac{\pi}{3} + \frac{2}{9}\right)\cos 3x + \cdots\right].$$

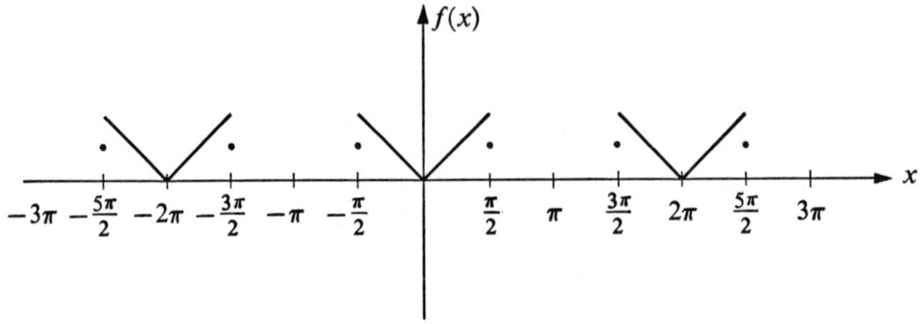

figure 10.11

Figure 10.11 shows the graph of the function to which the Fourier cosine series converges on $[-3\pi, 3\pi]$.

In the same way, the function f_o is a member of the subspace $PC_o[-L, L]$ and hence may be written in terms of the basis $\{\sin(n\pi x/L)\}_{n=1}^{\infty}$.

$$f_o \sim \sum_{n=1}^{\infty} b_n \sin\frac{n\pi x}{L}$$

where

$$b_n = \frac{2}{L}\int_0^L f_o(x)\sin\frac{n\pi x}{L}\,dx = \frac{2}{L}\int_0^L f(x)\sin\frac{n\pi x}{L}\,dx.$$

The Fourier series so obtained is called the *Fourier sine series of f*. If f satisfies the Dirichlet conditions, the series will actually represent f on $[0, L]$ and will converge to the odd periodic extension of f outside of $[0, L]$ (except at the points of discontinuity.)

example 10.24: Write the Fourier sine series of the function of the previous example.

solution:

$$\begin{aligned} b_n &= \frac{2}{\pi}\int_0^{\pi/2} x\sin nx\,dx \\ &= \frac{1}{\pi n^2}\left(2\sin\frac{n\pi}{2} - n\pi\cos\frac{n\pi}{2}\right). \end{aligned}$$

Therefore,

$$f(x) \sim \frac{1}{\pi}\left(2\sin x + \frac{\pi}{2}\sin 2x - \frac{2}{9}\sin 3x - \frac{\pi}{4}\sin 4x + \cdots\right).$$

Figure 10.12 shows the graph of the function to which the series converges on $[-3\pi, 3\pi]$.

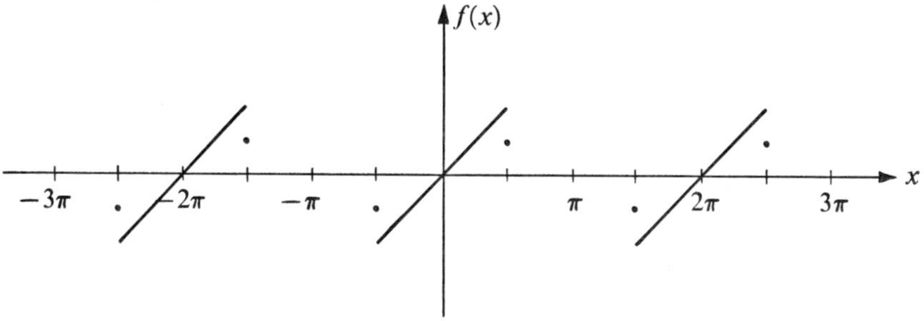

figure 10.12

10.9 which fourier expansion to use?

Suppose a function is defined on $[-L, L]$ but you desire a Fourier series representation only on $[0, L]$. Or suppose the function is defined only on $[0, L]$ but you may extend it to $[-L, 0]$ any way that you wish so long as you ultimately represent the portion on $[0, L]$ by some Fourier expansion.

1. If f *is* defined on $[-L, 0]$ you might write the Fourier series for f on the entire interval even though you would want to use only the portion on $[0, L]$.
2. If f is not defined on $[-L, 0]$ then you could use the odd extension of f and the resulting Fourier series is the Fourier sine series of the original function.
3. If f is not defined on $[-L, 0]$ you could use the even extension of f and the resulting Fourier series is the Fourier cosine series of the original function.
4. If f is not defined on $[-L, 0]$ you might define it to be zero there, since perhaps you would conclude that the computation of the Fourier coefficients would be easier.

Sketches of graphs for these choices are shown in Figure 10.13.

Aside from the motivation of writing the series with the least number of calculations, you would want to choose the expansion which results in the smoothest overall graph. For then it would seem reasonable that the series would converge "faster," that is, fewer terms of the series would be required to give a reasonable approximation to the limit function. Recall that "faster" convergence may be judged by the order† of the coefficients.

† We say that a_n is of the *order* of $1/n^p$ if $\lim_{n\to\infty} n^p a_n =$ a constant.

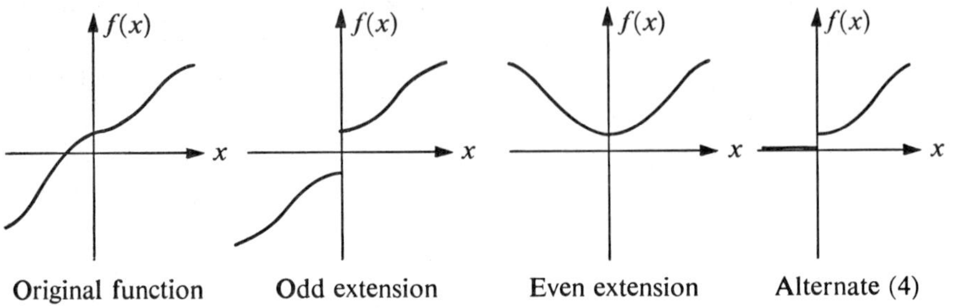

figure 10.13 possible extensions of f

theorem 10.10

Let f and its first k derivatives satisfy the Dirichlet conditions on $[-L, L]$. Then the Fourier coefficients of f are at most of the order of $(1/n)^{k+1}$.

More concisely, though less accurately, the theorem asserts that the smoother the function, the faster its Fourier expansion converges. Specifically, a function which is discontinuous will have coefficients of the order of $1/n$; a function which is continuous but has a discontinuous derivative will have coefficients of the order of $1/n^2$, and so forth. So, to decide which Fourier series to use in a practical sense, it is best to sketch the periodic extension, the odd periodic extension, and the even periodic extension of the function. The one with the smoothest graph will yield a series which converges most quickly.

example 10.25: If you are to represent the function $f(x) = x$ by a Fourier expansion on $[0, \pi]$, would a Fourier sine or cosine series be better?

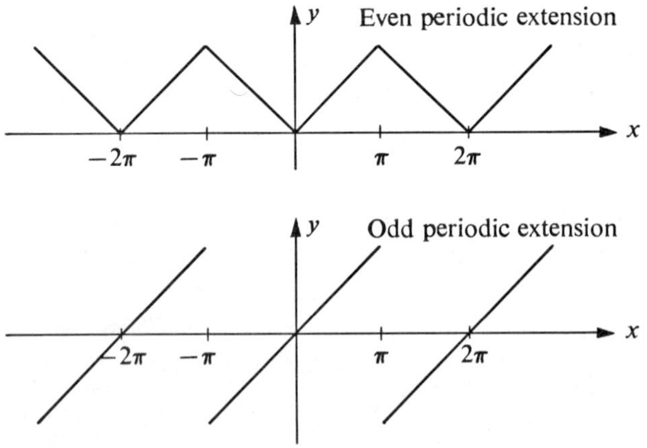

figure 10.14

solution: The graphs of the even and odd periodic extensions are shown in Figure 10.14. Since the graph of the function to which the cosine series converges is continuous and that to which the sine series converges is discontinuous, we should use the Fourier cosine series. The Fourier sine series is the same (in this case) as the Fourier series of the function $f(x) = x$ on $[-\pi, \pi]$. This was computed in Example 10.9 and found to be

$$f(x) = 2 \sum_{n=1}^{\infty} \frac{(-1)^{n+1}}{n} \sin nx.$$

The Fourier cosine series is easily computed and found to be

$$f(x) = \frac{\pi}{2} - \frac{4}{\pi} \sum_{n=0}^{\infty} \frac{\cos(2n+1)x}{(2n+1)^2}.$$

The nature of the coefficients verifies Theorem 10.10.

example 10.26: Let $f(x) = x - x^2$ on $[-1, 1]$. (See Figure 10.15.) Sketch the graphs of the functions: (a) to which the Fourier series of the function on $[-1, 1]$ converges on $[-3, 3]$; (b) to which the Fourier cosine series of the function on $[0, 1]$ converges on $[-3, 3]$; (c) to which the Fourier sine series of the function on $[0, 1]$ converges on $[-3, 3]$. Which representation would you rather use?

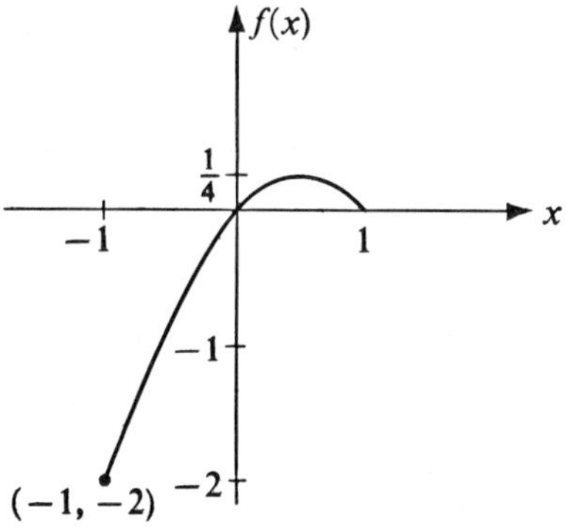

figure 10.15

solution: Figure 10.16 shows the three graphs. The first is the periodic extension of the given graph on $[-1, 1]$, the second is the even periodic extension of the given graph on $[0, 1]$, and the last is the odd periodic extension. Obviously, the Fourier sine series leads to the smoothest graph.

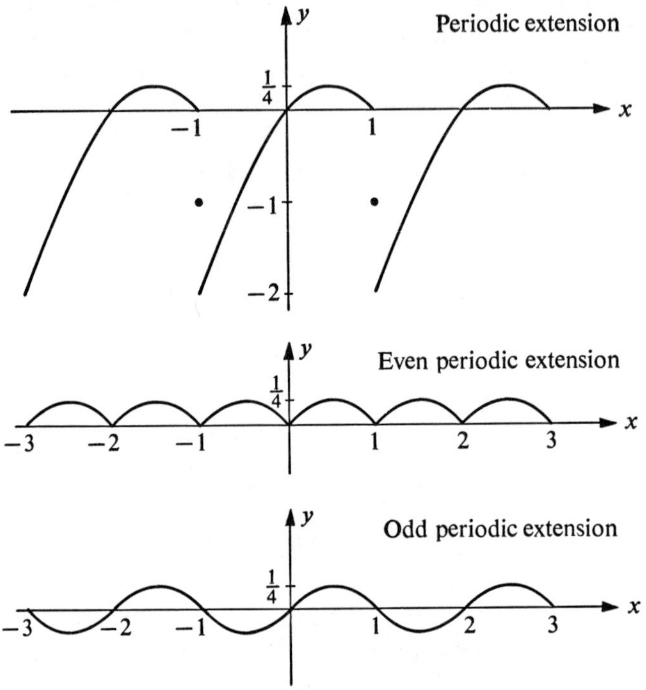

figure 10.16 three extensions for the function of example 10.26

exercises for sections 10.8 and 10.9

1. Find the functions f_o and f_e on $[-2, 2]$ if $f(x)$ is defined on $[0, 2]$ by:
   (a) $f(x) = x - 1, \quad 0 < x < 1;$ $\quad$ (b) $f(x) = x^2 + x.$
   $\quad\quad = 1 - x, \quad 1 < x < 2.$
   (c) $f(x) = \sin\dfrac{\pi x}{2}.$ $\quad$ (d) $f(x) = \cos\dfrac{\pi x}{2}.$
   (e) $f(x) = e^x.$
2. Find the Fourier sine and cosine series of the function $f(x) = \sin x$, $0 < x < \pi$. Sketch the graphs of the functions to which the two series converge on $[-3\pi, 3\pi]$.
3. Find the Fourier sine and cosine series of the function $f(x) = \cos x$, $0 < x < \pi$. Sketch the graphs of the functions to which the two series converge on $[-3\pi, 3\pi]$.
4. Find the Fourier sine and cosine series of the function $f(x) = x^2$ on $[0, 1]$.

5. Would you rather use a sine or a cosine series for the following functions whose definition on $[0, \pi]$ is:
(a) $f(x) = x$. (b) $f(x) = x^2$. (c) $f(x) = \sin\frac{x}{2}$.

10.10 approximation in the mean

The two types of convergence of a Fourier series to a function, f, lead to two meaningful approximations. In this section we will find the finite linear combination of an orthogonal set which best approximates f in the mean.

Let F_N be any linear combination of the first N functions of an orthogonal set $\{\phi_n\}$,

$$F_N(x) = \sum_{n=1}^{N} \alpha_n \phi_n(x).$$

We desire to choose the components α_n in such a way that F_N will approximate the function, f. One of the choices of the α_n's is the Fourier coefficients of the function, f, with respect to the given orthogonal set. In this case the linear combination is denoted by F_N^* and is called the *Fourier linear combination* of the ϕ_n.

$$F_N^*(x) = \sum_{n=1}^{N} c_n \phi_n(x)$$

where

$$c_n = \frac{(f, \phi_n)}{\|\phi_n\|^2}.$$

The concept of convergence in the mean leads to the following meaningful concept for approximating a function by a finite linear combination of a set which is orthogonal on (a, b).

definition 10.7

The *error in the mean*, e_N, is the difference in the mean between the function, f, and the linear combination, F_N,

$$e_N = \|f - F_N\| = \left(\int_a^b [f(x) - F_N(x)]^2\, dx\right)^{1/2}.$$

Practically, it is less cumbersome to work with the square of e_N, which we denote by E_N, called the *mean square error*.

example 10.27: Compute the mean square error if $\cos(\pi x/2)$ is used to approximate $1-x^2$ on $[-1, 1]$. See Figure 10.17. (Note that $\cos(\pi x/2)$ is not a member of the trigonometric basis for $PC[-1, 1]$.)

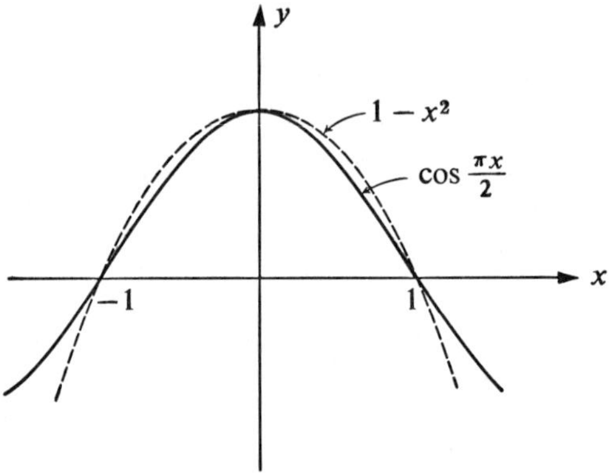

figure 10.17

solution: The mean square error is the difference in the mean (squared) of the function $\cos(\pi x/2)$ and $1-x^2$ on the interval $[-1, 1]$,

$$\begin{aligned}E_1 &= \left\| 1 - x^2 - \cos\frac{\pi x}{2} \right\|^2 \\ &= \int_{-1}^{1} \left(1 - x^2 - \cos\frac{\pi x}{2}\right)^2 dx \\ &= \int_{-1}^{1} \left(1 + x^4 + \cos^2\frac{\pi x}{2} - 2x^2 - 2\cos\frac{\pi x}{2} + 2x^2\cos\frac{\pi x}{2}\right) dx \\ &= 2 + \frac{2}{5} + 1 - \frac{4}{3} - \frac{8}{\pi} + \frac{8}{\pi} - \frac{64}{\pi^3} \\ &= \frac{31}{15} - \frac{64}{\pi^3} \\ &\approx 0.0026.\end{aligned}$$

example 10.28: A linear combination of the functions $\sin x$ and $\sin 2x$ is used to approximate the function defined on $[-\pi, \pi]$ as

$$\begin{aligned}f(x) &= -1, \text{ for } -\pi \leqq x \leqq 0; \\ &= 1, \quad \text{for} \quad 0 < x \leqq \pi.\end{aligned}$$

Determine which of the following two linear combinations is better in the sense of mean square error:

(a) $\sin x + \sin 2x$.
(b) The Fourier combination of $\sin x$ and $\sin 2x$ considered as elements of $PC[-\pi, \pi]$.

solution: We have

$$\begin{aligned} b_n &= \frac{2}{\pi}\int_0^\pi \sin nx\, dx \\ &= \frac{-2}{\pi}\frac{\cos nx}{n}\Big|_0^\pi \\ &= \frac{4}{n\pi}, \qquad n \text{ odd}, \\ &= 0, \qquad n \text{ even}. \end{aligned}$$

Therefore, $b_1 = 4/\pi$ and $b_2 = 0$, so that we must decide whether $\sin x + \sin 2x$ or $(4/\pi)\sin x$ best approximates f in the mean.

Let E_2 be the mean square error for the first approximation and let E_2^* be the mean square error when the Fourier combination is used. Then,

$$E_2 = 2\int_0^\pi (1 - \sin x - \sin 2x)^2\, dx = 2(2\pi - 4) \approx 4.56;$$

$$E_2^* = 2\int_0^\pi \left(1 - \frac{4}{\pi}\sin x\right)^2 dx = 2\left(\frac{\pi^2 - 8}{\pi}\right) \approx 1.19.$$

Thus, the Fourier combination is the better approximation in the sense of mean square error.

Naturally, we would usually wish to choose the linear combination of the orthogonal set which minimizes the mean square error. The particular linear combination which does this is called the *best approximation to f in the mean.* The previous example showed that the Fourier combination is better than another possibility in a particular case. The following theorem will show that this is not a random occurrence.

theorem 10.11

The mean square error of a linear approximation F_N of orthogonal functions to f on (a, b) is a minimum if and only if the coefficients of F_N are the Fourier coefficients of f relative to the orthogonal set $\{\phi_n\}_{n=1}^{\infty}$.

proof: By definition

$$\begin{aligned} E_N &= \int_a^b [f(x) - F_N(x)]^2\, dx \\ &= \int_a^b [f(x)]^2\, dx - 2\int_a^b f(x)\, F_N(x)\, dx + \int_a^b [F_N(x)]^2\, dx. \end{aligned}$$

The first of these integrals is $\|f\|^2$.

The second integral, when written in a form involving the finite combination of the orthogonal set becomes

$$-2\int_a^b f(x)\,F_N(x)\,dx = -2\sum_{n=1}^{N}\alpha_n\int_a^b f(x)\,\phi_n(x)\,dx$$
$$= -2\sum_{n=1}^{N}\alpha_n c_n\,\|\phi_n\|^2$$

where c_n is the Fourier coefficient of ϕ_n (which by Section 10.4 is equal to $(f, \phi_n)/\|\phi_n\|^2$) in the expansion of $f(x)$. Further,

$$\int_a^b [F_N(x)]^2\,dx = \int_a^b \left[\sum_{n=1}^{N}\alpha_n\phi_n(x)\right]^2 dx$$
$$= \int_a^b \sum_{n=1}^{N}\alpha_n^2[\phi_n(x)]^2\,dx + 2\sum_{n=1}^{N}\sum_{m=1}^{N}\alpha_n\alpha_m\int_a^b \phi_n(x)\,\phi_m(x)\,dx.$$

Since the set $\{\phi_n\}_{n=1}^{\infty}$ is orthogonal, each of the integrals in the double summation is zero. Therefore, the right-hand side reduces to

$$\sum_{n=1}^{N}(\alpha_n\,\|\phi_n\|)^2.$$

Therefore,

$$E_N = \|f\|^2 - 2\sum_{n=1}^{N}\alpha_n c_n\,\|\phi_n\|^2 + \sum_{n=1}^{N}(\alpha_n\,\|\phi_n\|)^2.$$

This formula for the mean square error is perfectly general and is valid for any choice of functions or orthogonal sets. If the Fourier combination is used, the coefficients α_n are equal to c_n. Denoting the mean square error in that case by E_N^*,

$$E_N^* = \|f\|^2 - \sum_{n=1}^{N}(c_n\,\|\phi_n\|)^2.$$

To show the mean square error is a minimum for the Fourier combination, we will show $E_N - E_N^* \geqq 0$,

$$E_N - E_N^* = \|f\|^2 - 2\sum_{n=1}^{N}\alpha_n c_n\,\|\phi_n\|^2 + \sum_{n=1}^{N}(\alpha_n\|\phi_n\|)^2 - \|f\|^2 + \sum_{n=1}^{N}(c_n\|\phi_n\|)^2$$
$$= \sum_{n=1}^{N}(\alpha_n^2 - 2\alpha_n c_n + c_n^2)\,\|\phi_n\|^2$$
$$= \sum_{n=1}^{N}(\alpha_n - c_n)^2\,\|\phi_n\|^2$$

which is $\geqq 0$ and hence $E_N \geqq E_N^*$ which shows that

$$(E_N)_{\min} = E_N^*.$$

The expression for the minimum mean square error may be easily determined for the case of the trigonometric orthogonal basis on $PC[-L, L]$. It is more convenient (as will become obvious) to use $2N+1$ terms of the basis and continue to denote the error by E_N^*. We have that

$$F_N^* = a_0 + \sum_{n=1}^{N} a_n \cos \frac{n\pi x}{L} + \sum_{n=1}^{N} b_n \sin \frac{n\pi x}{L}$$

where a_0, a_n, and b_n are the Fourier coefficients with respect to the trigonometric basis on $[-L, L]$ derived previously.

Further, since $\|1\|^2 = 2L$, $\left\|\cos \frac{n\pi x}{L}\right\|^2 = L$, and $\left\|\sin \frac{n\pi x}{L}\right\|^2 = L$, the value of the minimum mean square error is

$$E_N^* = \|f\|^2 - 2La_0^2 - \sum_{n=1}^{N} La_n^2 - \sum_{n=1}^{N} Lb_n^2 .$$

example 10.29: Compute the mean square error if three nonzero terms of its Fourier cosine series are used to approximate $f(x) = x$ on $[0, \pi]$. Do the same for the Fourier sine series.

solution: It is easy to show that for the cosine series,

$$a_0 = \frac{\pi}{2}; \quad a_{2n+1} = \frac{-4}{\pi(2n+1)^2}, \ n \geqq 0; \quad a_{2n} = 0, \ n \geqq 1 .$$

Hence we wish to approximate the function by the sum

$$F_3(x) \, (= F_4(x)) = \frac{\pi}{2} - \frac{4}{\pi} \cos x - \frac{4}{9\pi} \cos 3x .$$

The mean square error is

$$\begin{aligned} E_3^* \, (= E_4^*) &= 2 \int_0^{\pi} x^2 \, dx - (2\pi)\left(\frac{\pi^2}{4}\right) - \pi\left(\frac{16}{\pi^2} + \frac{16}{81\pi^2}\right) \\ &\approx 0.01 . \end{aligned}$$

For the sine series we have that $b_n = (-1)^{n+1}(2/n)$, so that we desire to approximate the given function by

$$2 \sin x - \sin 2x + (2/3) \sin 3x .$$

In this case the mean square error is

$$\begin{aligned} E_3^* &= 2 \int_0^{\pi} x^2 \, dx - \pi\left(4 + 1 + \tfrac{4}{9}\right) \\ &\approx 3.6 . \end{aligned}$$

A rather interesting theorem follows almost as a corollary to the preceding.

theorem 10.12

The Fourier coefficients of a function in $PC[a, b]$ with respect to the orthogonal set $\{\phi_n\}_{n=1}^{\infty}$ satisfy the inequality

$$\sum_{n=1}^{\infty} (c_n \|\phi_n\|)^2 \leqq \|f\|^2 \qquad \textit{(Bessel's Inequality)}$$

with equality (called *Parseval's Equality*) if and only if the orthogonal set is a basis. The c_n's are the Fourier coefficients of f with respect to $\{\phi_n\}_{n=1}^{\infty}$.

proof: From Theorem 10.11, $E_N \geqq 0$, and

$$E_N^* = \|f\|^2 - \sum_{n=1}^{N} (c_n \|\phi_n\|)^2,$$

it follows that

$$\|f\|^2 - \sum_{n=1}^{N} (c_n \|\phi_n\|)^2 \geqq 0.$$

Since this is true for any N, Bessel's Inequality is proven.

If the set $\{\phi_n\}_{n=1}^{\infty}$ is a basis, $\lim_{N \to \infty} E_N = 0$, so that

$$\lim_{N \to \infty} \left(\|f\|^2 - \sum_{n=1}^{N} (c_n \|\phi_n\|)^2 \right) = 0$$

and therefore,

$$\|f\|^2 = \sum_{n=1}^{\infty} (c_n \|\phi_n\|)^2,$$

which proves Parseval's Equality.

In terms of the trigonometric basis, Parseval's Equality takes the form

$$\|f\|^2 = 2La_0^2 + L \sum_{n=1}^{\infty} a_n^2 + L \sum_{n=1}^{\infty} b_n^2.$$

Sums of convergent series are obtainable from this equality.

example 10.30: Find the sum of

$$\sum_{n=1}^{\infty} \frac{1}{n^2}$$

by considering the Fourier sine series of $f(x) = x$ on $[0, \pi]$.

solution: The Fourier sine series for $f(x) = x$ is

$$\sum_{n=1}^{\infty} (-1)^{n+1} \left(\frac{2}{n}\right) \sin nx,$$

and hence, by Parseval's Equality

$$2 \int_0^{\pi} x^2 \, dx = \pi \sum_{n=1}^{\infty} \frac{4}{n^2}$$

or,

$$\frac{2}{3} \pi^3 = 4\pi \sum_{n=1}^{\infty} \frac{1}{n^2}$$

from which

$$\sum_{n=1}^{\infty} \frac{1}{n^2} = \frac{\pi^2}{6}.$$

Parseval's Equality in the case where the set $\{\phi_n\}_{n=1}^{\infty}$ is an orthonormal basis has a fascinating interpretation. In that case $\|\phi_n\| = 1$ and hence

$$\|f\|^2 = \sum_{n=1}^{\infty} c_n^2 .$$

Since the c_n's are computed with respect to a basis for $PC[a, b]$ they are often called *components* of f with respect to the basis. Thus, Parseval's Equality may be interpreted as the infinite-dimensional analogue of the situation in three dimensions

$$|\mathbf{A}|^2 = a_1^2 + a_2^2 + a_3^2$$

where a_1, a_2, and a_3 are the components of $\mathbf{A}$ with respect to some orthonormal basis for the set of three-dimensional geometric vectors. You can probably see why Parseval's Equality is sometimes called the "infinite-dimensional Pythagorean theorem."

10.11 pointwise approximation

The other kind of approximation, and one that may seem more natural is in the pointwise sense. In this sense we wish to have the graphs of the partial sums become like the given function.

example 10.31: Sketch the graph of the pointwise approximation given by the first three terms of the Fourier series for the function $f(x)=|x|$, $-\pi \leqq x \leqq \pi$.

solution: The Fourier coefficients are easily computed to be

$$a_0=\frac{\pi}{2}, \quad a_{2n+1}=\frac{-4}{(2n+1)^2\pi}, \quad a_{2n}=0, \text{ for } n \geqq 1;$$
$$b_n=0 \text{ for all } n.$$

The sum of the first three nonzero terms is

$$\frac{\pi}{2}-\frac{4}{\pi}\cos x-\frac{4}{9\pi}\cos 3x.$$

The graph of this sum is shown in Figure 10.18.

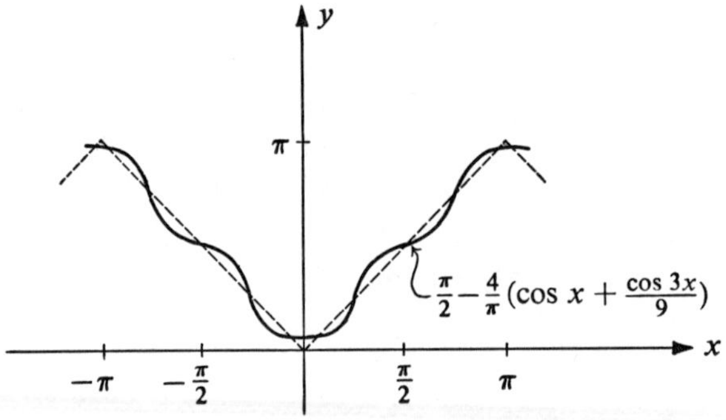

figure 10.18

The graphs of the partial sums of the Fourier series are called *approximating curves* to the function. The approximating curves exhibit a peculiarity around points of discontinuity known as the *Gibbs Phenomenon.*

example 10.32: Sketch the first three approximating curves to the function

$$f(x)=-1, \quad -\pi<x<0;$$
$$=1, \quad 0 \leqq x \leqq \pi.$$

solution: Since the function is odd, $a_n=0$ for all n. Further,

$$b_n=\frac{2}{\pi}\int_0^{\pi}\sin nx\,dx$$
$$=-\frac{2}{n\pi}\cos nx\Big|_0^{\pi}$$
$$=\frac{4}{n\pi}, \quad n \text{ odd},$$
$$=0, \quad n \text{ even}.$$

Therefore,

$$S_1 = \frac{4}{\pi}\sin x$$

$$S_2 = \frac{4}{\pi}\left(\sin x + \frac{\sin 3x}{3}\right)$$

$$S_3 = \frac{4}{\pi}\left(\sin x + \frac{\sin 3x}{3} + \frac{\sin 5x}{5}\right).$$

(See Figure 10.19.)

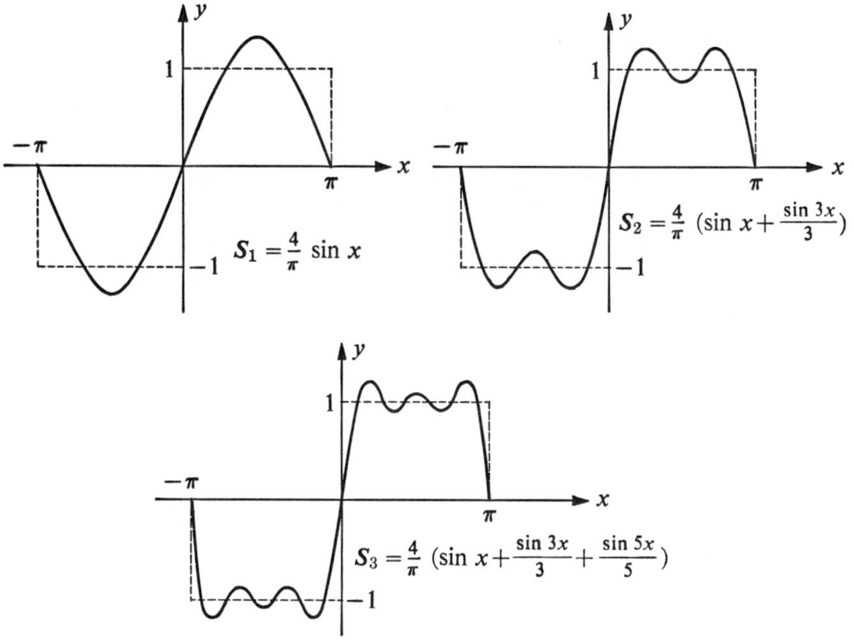

figure 10.19

Note that on either side of the discontinuity the approximating curve tends to "overshoot," a tendency which does not disappear as the peaks are shifted closer and closer to the discontinuity. Although the magnitude of the "overshoot" decreases, it never disappears completely. The Gibbs Phenomenon illustrates the impossibility of uniform convergence (see Section A.7) on any interval that contains a discontinuity.

exercises for sections 10.10 and 10.11

1. Find the mean square error if the function $\sin \pi x$ is used to approximate $x - x^2$ on $[0, 1]$.

2. Find which multiple of $\sin \pi x$ will give the best approximation in the mean to $x - x^2$ on $[0, 1]$. What is the mean square error in that case?
3. Which linear combination of the set $\{1, \cos x, \cos 2x\}$ will give the best approximation in the mean to the function $|x|$ on $[-\pi, \pi]$?
4. What is the minimum mean square error obtainable for the desired type of approximation in Exercise 3?
5. How many terms of the Fourier sine series should be used to approximate $f(x) = x$ on $[0, \pi]$ if a mean square error of $\leqq 1$ is desired?
6. Compute the form of Parseval's Equality in the case where the basis is the modified trigonometric set.
7. Use Parseval's Equality to show

$$\sum_{n=0}^{\infty} \frac{1}{(2n+1)^2} = \frac{\pi^2}{8}.$$

(*Hint:* consider the function $f(x) = -1, \quad -\pi < x < 0;$
$= 1, \quad 0 < x < \pi.$

8. Show

$$\sum_{n=1}^{\infty} \frac{1}{n^4} = \frac{\pi^4}{90}.$$

(*Hint:* consider $f(x) = x^2, \ -\pi < x < \pi.$)

9. Which multiple of $\cos(\pi x/2)$ will give the best approximation in the mean to $1 - x^2$ on $[-1, 1]$? (*Hint:* Use the *definition* of mean square error.)
10. Carefully sketch the first three approximating curves to the following functions:
    (a) $f(x) = |\sin x|, \ -\pi < x \leqq \pi.$
    (b) $f(x) = x^2, \ -1 < x < 1.$
    (c) $f(x) = x, \ -\pi < x < \pi.$
    (d) $f(x) = 0, \quad -2 \leqq x \leqq 0;$
    $= x, \quad 0 < x < 2.$

10.12 the fourier integral

The Fourier series representation of a function satisfying the Dirichlet conditions is adequate for periodic functions or for functions defined only on a finite interval. However, in numerous applied problems, the impressed force or voltage is nonperiodic and defined for all t. In this section we investigate the limit of a Fourier series as the period becomes unbounded, thereby obtaining a suitable representation of a nonperiodic function. We begin by showing the nature of a periodic function as the period becomes unbounded.

example 10.33: What function is defined by letting L become unbounded for the periodic function

$$\begin{aligned} f_L(x) &= 0, \quad -L < x < 0; \\ &= 1, \quad\quad 0 \leqq x < L? \\ f_L(x) &= f_L(x + 2L) \end{aligned}$$

solution:

$$\begin{aligned} \lim_{L\to\infty} f_L(x) &= 0, \quad x < 0; \\ &= 1, \quad x \geqq 0. \end{aligned}$$

(See Figure 10.20.)

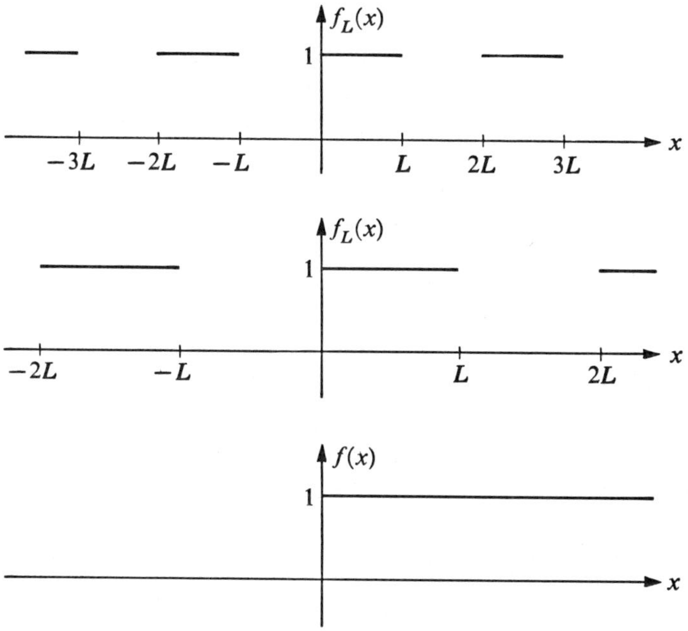

figure 10.20

example 10.34: Determine $f(x) = \lim_{L\to\infty} f_L(x)$, if

$$\begin{aligned} f_L(x) &= 0, \quad -L < x < -1; \\ &= 1, \quad -1 \leqq x \leqq 1; \\ &= 0, \quad\quad 1 < x < L, \end{aligned}$$

and $f_L(x)$ is periodic with period $2L$.

solution:

$$\lim_{L\to\infty} f_L(x) = 0, \quad |x| > 1;$$
$$= 1, \quad |x| \leqq 1.$$

(See Figure 10.21.)

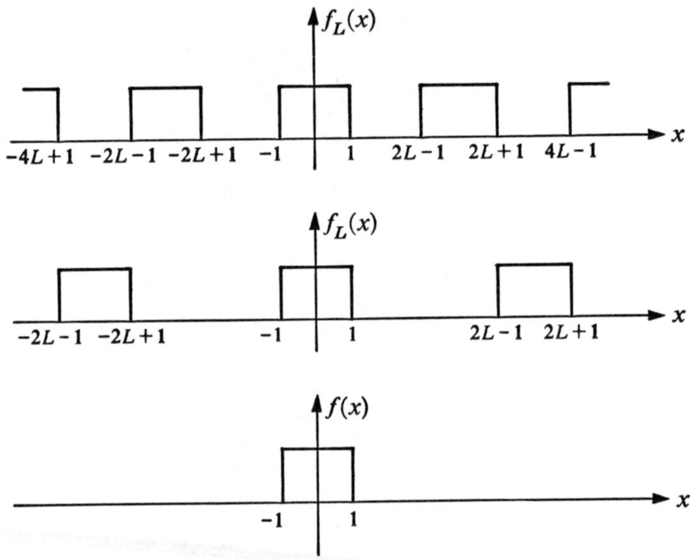

figure 10.21

Before taking the limit of a Fourier series as the period becomes infinite, it is convenient to have the series expressed in its exponential form. The Fourier trigonometric series for a function f_L with period $2L$ is given by

$$f_L(x) = a_0 + \sum_{n=1}^{\infty} a_n \cos\frac{n\pi}{L}x + b_n \sin\frac{n\pi}{L}x.$$

In Chapter 12 you will see that the real sine and cosine functions may be expressed in terms of the complex exponential,

$$\cos\frac{n\pi x}{L} = \frac{e^{in\pi x/L} + e^{-in\pi x/L}}{2}$$

$$\sin\frac{n\pi x}{L} = \frac{e^{in\pi x/L} - e^{-in\pi x/L}}{2i}$$

from which the Fourier series becomes

$$f_L(x) = a_0 + \sum_{n=1}^{\infty} \frac{a_n}{2}(e^{in\pi x/L} + e^{-in\pi x/L}) - i\frac{b_n}{2}(e^{in\pi x/L} - e^{-in\pi x/L})$$

$$= a_0 + \sum_{n=1}^{\infty} \left(\frac{a_n - ib_n}{2}\right)e^{in\pi x/L} + \left(\frac{a_n + ib_n}{2}\right)e^{-in\pi x/L}.$$

If we let

$$c_n = \frac{a_n - ib_n}{2}, \qquad c_{-n} = \frac{a_n + ib_n}{2}, \quad \text{and} \quad c_0 = a_0,$$

we obtain

$$f_L(x) = c_0 + \sum_{n=1}^{\infty} c_n e^{in\pi x/L} + \sum_{n=-1}^{-\infty} c_n e^{in\pi x/L}$$

$$= \sum_{n=-\infty}^{\infty} c_n e^{in\pi x/L}$$

where

$$c_n = \frac{1}{2L}\int_{-L}^{L} f_L(x)\, e^{-in\pi x/L}\, dx.$$

(Prove this last equation.)

Note the duality between the function and its Fourier coefficients with respect to the orthogonal set

$$\{e^{in\pi x/L}\}_{n=-\infty}^{\infty}$$

in the sense of similarity of formulas. This is made even more emphatic by writing the two equations in the equivalent form

$$f_L(x) = \frac{1}{\sqrt{2L}} \sum_{n=-\infty}^{\infty} c_n' e^{in\pi x/L}, \qquad c_n' = \frac{1}{\sqrt{2L}}\int_{-L}^{L} f_L(x)\, e^{-in\pi x/L}\, dx.$$

We will now show how this duality is extended as $L \to \infty$. We first write

$$f_L(x) = \sum_{n=-\infty}^{\infty} c_n e^{in\pi x/L}$$

$$= \sum_{n=-\infty}^{\infty} \left[\frac{1}{2L}\int_{-L}^{L} f_L(t)\, e^{-in\pi t/L}\, dt\right] e^{in\pi x/L}.$$

Let $\omega_n = n\pi/L$; then $\Delta\omega = \omega_{n+1} - \omega_n = \pi/L$ and

$$f_L(x) = \frac{1}{\sqrt{2\pi}} \sum_{n=-\infty}^{\infty} \left[\frac{1}{\sqrt{2\pi}} \int_{-L}^{L} f_L(t)\, e^{-i\omega_n t}\, dt \right] e^{i\omega_n x}\, \Delta\omega$$

$$= \frac{1}{\sqrt{2\pi}} \sum_{n=-\infty}^{\infty} F(\omega_n)\, e^{i\omega_n x}\, \Delta\omega .$$

As $\Delta\omega \to 0$ (that is, as $L \to \infty$), it seems plausible (and under rather general conditions it is true) that

$$f(x) = \lim_{L\to\infty} f_L(x) = \frac{1}{\sqrt{2\pi}} \int_{-\infty}^{\infty} F(\omega)\, e^{i\omega x}\, d\omega$$

where

$$F(\omega) = \frac{1}{\sqrt{2\pi}} \int_{-\infty}^{\infty} f(t)\, e^{-i\omega t}\, dt .$$

Writing the expression for $f(x)$ completely,

$$f(x) = \frac{1}{2\pi} \int_{-\infty}^{\infty} \int_{-\infty}^{\infty} f(t)\, e^{i\omega(x-t)}\, dt\, d\omega$$

which is called the Fourier *integral representation* of $f(x)$. This representation may be put into a form which is similar in appearance to that of a Fourier series. In terms of real trigonometric functions,

$$f(x) = \frac{1}{2\pi} \int_{-\infty}^{\infty} \int_{-\infty}^{\infty} f(t)\, [\cos\omega(x-t) + i \sin\omega(x-t)]\, dt\, d\omega$$

$$= \frac{1}{2\pi} \int_{-\infty}^{\infty} \int_{-\infty}^{\infty} f(t) \cos\omega(x-t)\, dt\, d\omega$$

$$+ i \int_{-\infty}^{\infty} \int_{-\infty}^{\infty} f(t) \sin\omega(x-t)\, dt\, d\omega .$$

The second of these two integrals is zero since the sine function is an odd function of ω. Therefore

$$f(x) = \frac{1}{2\pi} \int_{-\infty}^{\infty} \int_{-\infty}^{\infty} f(t) \cos\omega(x-t)\, dt\, d\omega .$$

Since the cosine function is an even function of ω,

$$\begin{aligned} f(x) &= \frac{1}{\pi}\int_0^{\infty}\int_{-\infty}^{\infty} f(t)\cos\omega(x-t)\,dt\,d\omega \\ &= \frac{1}{\pi}\int_0^{\infty}\int_{-\infty}^{\infty} f(t)\left[\cos\omega x\cos\omega t + \sin\omega x\sin\omega t\right]dt\,d\omega \\ &= \frac{1}{\sqrt{\pi}}\int_0^{\infty} A(\omega)\cos\omega x\,d\omega + \frac{1}{\sqrt{\pi}}\int_0^{\infty} B(\omega)\sin\omega x\,d\omega \end{aligned}$$

where

$$A(\omega) = \frac{1}{\sqrt{\pi}}\int_{-\infty}^{\infty} f(t)\cos\omega t\,dt,$$

$$B(\omega) = \frac{1}{\sqrt{\pi}}\int_{-\infty}^{\infty} f(t)\sin\omega t\,dt.$$

In this form, the Fourier integral representation is analogous to the Fourier (trigonometric) series.

Our approach has been largely heuristic but the following theorem, whose proof may be found in Churchill (1941), shows that the conditions for the validity of the representation are much the same as for infinite series.

theorem 10.13

If f satisfies the Dirichlet conditions on every finite interval and if $\int_{-\infty}^{\infty} |f(x)|\,dx$ exists, then at every point the Fourier integral of f converges to the average of the right- and left-hand limits of the function.

example 10.35: Find the Fourier integral representation of the function

$$\begin{aligned} f(x) &= 1, \quad -2 \leqq x \leqq 2; \\ &= 0, \quad \text{otherwise}. \end{aligned}$$

To what function does the Fourier integral converge?

solution:

$$\begin{aligned} A(\omega) &= \frac{1}{\sqrt{\pi}}\int_{-\infty}^{\infty} f(x)\cos\omega x\,dx \\ &= \frac{1}{\sqrt{\pi}}\int_{-2}^{2} \cos\omega x\,dx \\ &= \frac{1}{\sqrt{\pi}}\left.\frac{\sin\omega x}{\omega}\right|_{-2}^{2} \\ &= \frac{1}{\sqrt{\pi}}\frac{2\sin 2\omega}{\omega}; \end{aligned}$$

$$B(\omega) = 0;$$
$$f(x) = \frac{2}{\pi}\int_0^\infty \frac{\sin 2\omega \cos \omega x}{\omega}\, d\omega.$$

The integral converges to $f(x)$ everywhere except for $x = 2$ and $x = -2$. For these values the convergence is to 1/2.

With the use of Theorem 10.13 and a little ingenuity in selecting the function, some improper integrals can be evaluated which cannot be done any other way.

example 10.36: Evaluate the improper integral

$$\int_0^\infty \frac{\sin x}{x}\, dx.$$

solution: Consider the Fourier integral representation of the function $f(x) = 1$, $-1 \leqq x \leqq 1$ and 0 otherwise. In a manner much like the previous example we can show

$$f(x) = \frac{2}{\pi}\int_0^\infty \frac{\cos \omega x \sin \omega}{\omega}\, d\omega.$$

At $x = 0$, the improper integral is equal to 1. Therefore

$$\frac{2}{\pi}\int_0^\infty \frac{\sin \omega}{\omega}\, d\omega = 1,$$

or

$$\int_0^\infty \frac{\sin \omega}{\omega}\, d\omega = \frac{\pi}{2}.$$

This integral is evaluated in another way in Chapter 15, but it cannot be done by elementary methods since $\frac{\sin x}{x}$ does not have an elementary antiderivative.

Fourier series representations of periodic functions are easier to compute if the function is even or odd, and the same is true for Fourier integral representations of nonperiodic functions. If f is even, then $B(\omega) = 0$ and

$$f(x) = \frac{2}{\pi}\int_0^\infty \int_0^\infty f(t) \cos \omega x \cos \omega t\, dt\, d\omega.$$

If f is odd, then $A(\omega) = 0$ and

$$f(x) = \frac{2}{\pi}\int_0^\infty \int_0^\infty f(t) \sin \omega x \sin \omega t\, dt\, d\omega.$$

Sometimes the function is defined only on the positive x-axis and we consider either the odd extension or the even extension. In this case the integral representation of $f(x)$ is called the *Fourier cosine integral* or *Fourier sine integral*, respectively.

example 10.37: Find the Fourier cosine integral representation of $f(x)$ if

$$\begin{aligned} f(x) &= x, && 0 \leqq x \leqq 1; \\ &= 2 - x, && 1 \leqq x \leqq 2; \\ &= 0, && 2 \leqq x. \end{aligned}$$

solution:

$$\begin{aligned} f(x) &= \frac{2}{\pi}\int_0^\infty \left[\int_0^1 x \cos \omega x \, dx + \int_1^2 (2 - x) \cos \omega x \, dx\right] \\ &= \frac{2}{\pi}\int_0^\infty \frac{2\cos\omega - \cos 2\omega - 1}{\omega^2} \cos \omega x \, d\omega . \end{aligned}$$

As mentioned earlier, the function of x and the function of ω obtained when computing the Fourier integral representation have an unmistakable symmetry or duality. For the general Fourier integral,

$$f(x) = \frac{1}{\sqrt{2\pi}}\int_{-\infty}^{\infty} F(\omega)\, e^{i\omega x} \, d\omega ;$$

$$F(\omega) = \frac{1}{\sqrt{2\pi}}\int_{-\infty}^{\infty} f(x)\, e^{-i\omega x} \, dx .$$

The function F is called the *Fourier transform* of f. We obtain a similar duality from the sine integral by splitting the coefficient $2/\pi$ between the expression for $f(x)$ and $B(\omega)$. We obtain

$$f(x) = \frac{\sqrt{2}}{\sqrt{\pi}}\int_0^\infty B(\omega) \sin \omega x \, d\omega ;$$

$$B(\omega) = \frac{\sqrt{2}}{\sqrt{\pi}}\int_0^\infty f(x) \sin \omega x \, dx$$

and in this case the function B is called the *Fourier sine transform* of f. From the Fourier cosine integral,

$$f(x) = \frac{\sqrt{2}}{\sqrt{\pi}}\int_0^\infty A(\omega) \cos \omega x \, d\omega ;$$

$$A(\omega) = \frac{\sqrt{2}}{\sqrt{\pi}}\int_0^\infty f(x) \cos \omega x \, dx$$

where A is called the *Fourier cosine transform* of f.

The relationship between a function and its Fourier transform is called *reciprocal*; that is, if g is the transform of f, then f is the transform of g.

example 10.38: Find the Fourier sine transform of f if $f(x)=e^{-x}$, $x>0$, and show

$$\int_0^\infty \frac{x \sin ax}{x^2+1}\,dx = \frac{\pi}{2} e^{-a}, \qquad a>0.$$

solution:

$$B(\omega)=\frac{\sqrt{2}}{\sqrt{\pi}}\int_0^\infty e^{-x}\sin\omega x\,dx = \frac{\sqrt{2}}{\sqrt{\pi}}\frac{\omega}{\omega^2+1}.$$

The Fourier sine transform of B is

$$f(t)=\frac{2}{\pi}\int_0^\infty \frac{x}{x^2+1}\sin tx\,dx.$$

Since f satisfies the conditions of Theorem 10.13 we evaluate at $t=a$ to obtain

$$f(a)=e^{-a}=\frac{2}{\pi}\int_0^\infty \frac{x\sin ax}{x^2+1}\,dx, \qquad a>0,$$

which is what we needed to show.

Some very useful functions do not have Fourier transforms.

example 10.39: Show that the function of Example 10.33, the "unit step function," has no Fourier transform.

solution:

$$\begin{aligned}\sqrt{2\pi}\,F(\omega) &= \int_0^\infty e^{-i\omega x}\,dx \\ &= \lim_{t\to\infty} \left.\frac{e^{-i\omega x}}{-i\omega}\right|_0^t \\ &= \lim_{t\to\infty} \left.\frac{\cos\omega x - i\sin\omega x}{-i\omega}\right|_0^t \\ &= \lim_{t\to\infty} \frac{\cos\omega t - i\sin\omega t - 1}{-i\omega},\end{aligned}$$

and this limit does not exist because of the oscillatory nature of the sine and cosine functions.

The fact that such a simple function as the unit step function has no Fourier transform leads us to attempt to widen the class of functions to which the concept of an integral transformation could apply. A very popular method is to introduce the factor e^{-as} which leads to the concept of a Laplace transform as was discussed in Chapter 4.

exercises for section 10.12

1. Compute the Fourier transform for the following:

   (a) $f(x)=0, \quad x<0;$ $\quad = e^{-x}, x \geqq 0.$ (b) $f(x)=e^{-|x|}$.

   (c) $f(x)=1, 0<x<1;$ $\quad =0$, otherwise.

2. Compute the Fourier sine and cosine transforms of each of the functions of Exercise 1 from their definitions for $x \geqq 0$.

3. Find a Fourier integral representation for the function of Exercise 1(c). Show by direct calculation that the integral converges to 1/2 at $x=0$ and $x=1$. Use the result of Example 10.36.

4. Under the (correct) assumption that

$$\int_0^\infty \frac{\cos t}{\sqrt{t}}\, dt = \frac{\sqrt{\pi}}{\sqrt{2}}$$

   show that $f(x)=1/\sqrt{x}$ is equal to its Fourier cosine integral even though $\int_0^\infty |f(x)|\, dx$ does not exist. This shows the conditions of Theorem 10.13 to be sufficient but not necessary.

5. Show that $1/\sqrt{1-x}$ is equal to its Fourier cosine integral.

6. By using the function of Exercise 1(a), prove the formula

$$\int_0^\infty \frac{\cos \alpha x + \alpha \sin \alpha x}{1+\alpha^2}\, d\alpha = \begin{cases} 0 & \text{for } x<0, \\ \pi/2 & \text{for } x=0, \\ \pi e^{-x} & \text{for } x>0. \end{cases}$$

7. By applying the Fourier sine integral to the function $e^{-x}\cos x$ show

$$\int_0^\infty \frac{\omega^3 \sin \omega x}{\omega^4+4}\, d\omega = \frac{\pi}{2} e^{-x}\cos x, \quad \text{if } x>0.$$

8. By applying the Fourier sine integral to the function which is equal to $\sin x$ on $0 \leqq x \leqq \pi$ and zero otherwise, show that

$$\int_0^\infty \frac{\sin \omega x \sin \pi\omega}{1-\omega^2}\, d\omega = \frac{\pi}{2}\sin x, \quad 0 \leqq x \leqq \pi;$$
$$= 0, \quad x>\pi.$$

9. Show that the "solution" to the integral equation

$$\int_0^\infty f(x)\sin\omega x\,dx = 1, \quad 0<\omega<\pi$$
$$= 0, \quad \omega>\pi$$

is the function

$$f(x) = \frac{2}{\pi}\frac{(1-\cos\pi x)}{x}, \qquad x>0.$$

10. Show that the integral equation

$$\int_0^\infty f(x)\cos\omega x\,dx = e^{-\omega}$$

has the solution

$$f(x) = \frac{2}{\pi}\frac{1}{1+x^2}, \qquad x>0.$$

11. Show that the function $1/\sqrt{x}$ is its own Fourier cosine transform.

10.13 the sturm-liouville problem

It may surprise you to learn that orthogonal sets of functions arise in a very natural way in applied mathematics. In this section we will show how orthogonal sets evolve from the boundary value problem on the interval $[a, b]$ of the form

$$\frac{d}{dx}(r(x)\,y') + (q(x) + \lambda p(x))\,y = 0,$$
$$c_1 y(a) + c_2 y'(a) = 0; \quad c_1 \text{ and } c_2 \text{ not both zero}.$$
$$c_3 y(b) + c_4 y'(b) = 0; \quad c_3 \text{ and } c_4 \text{ not both zero}.$$

This differential equation with the boundary conditions at a and b as given is known as the Sturm-Liouville differential equation with boundary values. Notice that if $c_2 = c_4 = 0$, the boundary conditions become $y(a) = y(b) = 0$; that is, the solution function crosses the x-axis at $x = a$ and $x = b$.

example 10.40: Show that the boundary value problem $y'' + \lambda y = 0$ with the boundary values $y(0) = y(\pi) = 0$ is a Sturm-Liouville problem.

solution: If, in the general problem, we let $a = 0$, $b = \pi$, $r(x) = p(x) = 1$, and $q(x) = 0$, the given problem is seen to "fit" the general one.

The constant λ in the Sturm-Liouville problem may take on any real value. Thus, one Sturm-Liouville problem is really many boundary value problems, one for each value of λ. For some values of λ the resulting boundary value problem has no solution. Those values of λ for which there is a solution to the corresponding boundary value problem are called *eigenvalues*. The corresponding solutions are called *eigensolutions* or *eigenfunctions*.

example 10.41: Find the eigenvalues and eigenfunctions of the Sturm-Liouville problem $y'' + \lambda y = 0$; $y(0) = y(\pi) = 0$.

solution: If $\lambda = 0$, then the boundary value problem has the form $y'' = 0$, with the boundary values as given. Solving the differential equation we obtain $y = ax + b$, where a and b are constants to be determined from the boundary conditions. We find that $a = b = 0$, so the eigenfunction corresponding to $\lambda = 0$ is $y = 0$ on the interval $[0, \pi]$. This is called the trivial eigenfunction.

If $\lambda < 0$, say $\lambda = -\alpha^2$, the differential equation becomes $y'' - \alpha^2 y = 0$. The solution is $y = ae^{\alpha x} + be^{-\alpha x}$ where, once more, a and b are determined from the boundary conditions. Again, $a = b = 0$ for all values of α.

If $\lambda > 0$, say $\lambda = \beta^2$, the differential equation becomes $y'' + \beta^2 y = 0$, whose solution is $y = a \cos \beta x + b \sin \beta x$. Applying the boundary conditions,

$$0 = a \cos \beta 0 + b \sin \beta 0,$$
$$0 = a \cos \beta \pi + b \sin \beta \pi.$$

From the first of these equations $a = 0$, for all β. The second equation is then $b \sin \beta\pi = 0$. If β is nonintegral, $b = 0$; but if β is an integer, b may be any real number, and hence nontrivial solutions exist. The eigenvalues for the nontrivial eigenfunctions are thus $\lambda = 1, 4, 9, 16, 25, \cdots$. The eigenfunctions are $\sin x$, $\sin 2x$, $\sin 3x$, $\sin 4x$, $\sin 5x, \cdots$.

The set of eigenfunctions of the previous example forms an orthogonal set on $[0, \pi]$. The following theorem gives the result in the general case.

theorem 10.14

Let p, q, and r of the Sturm-Liouville problem be continuous on $[a, b]$. Let y_m and y_n be eigensolutions corresponding to distinct eigenvalues λ_m and λ_n. Then y_m and y_n are orthogonal with weight function p on $[a, b]$.

proof: By hypothesis y_m and y_n are solutions to the Sturm-Liouville bound ary value problem with $\lambda = \lambda_m$ and $\lambda = \lambda_n$ respectively. Hence

$$\frac{d}{dx}(ry'_m) + (q + \lambda_m p)\, y_m = 0$$
$$\frac{d}{dx}(ry'_n) + (q + \lambda_n p)\, y_n = 0.$$

Multiplying the first of these by y_n and the second by $-y_m$ and adding,

$$(\lambda_m - \lambda_n)\, p y_m y_n = \frac{d}{dx}\left(r y_m y_n' - r y_m' y_n\right).$$

Integrating over the interval $[a, b]$ we obtain

$$(\lambda_m - \lambda_n) \int_a^b p y_m y_n \, dx = r\left(y_m y_n' - y_m' y_n\right)\Big|_a^b$$
$$= 0.$$

(See the exercises.)

Therefore, since $\lambda_m \neq \lambda_n$,

$$\int_a^b p y_m y_n \, dx = 0$$

which is the desired orthogonality condition.

corollary: 1. If $r(a) = 0$, then no boundary condition is needed at $x = a$ in the Sturm-Liouville problem for the orthogonality condition to hold.
2. If $r(b) = 0$, no boundary condition is needed at $x = b$.
3. If $r(a) = r(b)$, then the "regular" boundary conditions can be replaced by $y(a) = y(b)$ and $y'(a) = y'(b)$. (These are known as "mixed" boundary conditions.)

proof: See Exercise 2 of the next set of exercises.

example 10.42: The differential equation

$$\frac{d}{dx}(1 - x^2)\, y' + \lambda y = 0$$

is called Legendre's differential equation for

$$\lambda = 0, 2, 6, 12, \cdots, n(n+1), \cdots.$$

For each n, the differential equation has a polynomial solution which we denote by P_n, called the Legendre Polynomial of order n.

Show that the set of all Legendre polynomials forms an orthogonal set on $[-1, 1]$.

solution: This is a Sturm-Liouville differential equation with $r(x) = 1 - x^2$, $q(x) = 0$, and $p(x) = 1$. Since $r(1) = r(-1) = 0$, no boundary conditions are necessary for the orthogonality property of the Legendre polynomials to hold, as a direct consequence of Theorem 10.14 and its corollary.

exercises for section 10.13

1. Show $r(y_m y_n' - y_m' y_n)|_a^b = 0$ in the proof of Theorem 10.14.
2. Prove the corollary to Theorem 10.14.
3. Solve the Sturm-Liouville problem $y'' + \lambda y = 0$, $y'(0) = 0$, $y'(L) = 0$.
4. Solve the Sturm-Liouville problem $y'' + \lambda y = 0$, $y(0) = 0$, $y'(L) = 0$.
5. Find a Sturm-Liouville problem whose eigenfunctions are the modified trigonometric set
$$\{1\} \cup \left\{\cos \frac{n\pi x}{L}\right\}_{n=1}^{\infty} \cup \left\{\sin \frac{n\pi x}{L}\right\}_{n=1}^{\infty}.$$
6. Sketch the first four Legendre polynomials on $[-1, 1]$ and find their norm on that interval,
$$P_0(x) = 1, \quad P_1(x) = x, \quad P_2(x) = \frac{3x^2 - 1}{2}, \quad P_3(x) = \frac{5x^3 - 3x}{2}.$$
7. Find the best approximation in the mean to $f(x) = x^2$ by the set $\{P_0, P_1, P_2, P_3\}$, on the interval $[-1, 1]$.
8. Find the best approximation in the mean to $f(x) = \sin x$ by the set of Exercise 7.

partial differential equations

11

11.1 introduction

In this chapter you will be exposed to some elementary techniques of solving equations that involve partial derivatives of some unknown function. A partial differential equation is one which involves several independent variables $x, y, z, \cdots$, a function u of these variables, and the partial derivatives $u_x, u_y, u_{xx}, u_{yy}, u_{xz}, \cdots$ of the function. The order of the highest derivative is called the *order of the differential equation.* Except for a few easy problems to the contrary, we will limit ourselves to equations in which the unknown function and its partial derivatives appear linearly. Such an equation is called a *linear partial differential equation.*

The most general second-order linear partial differential equation in two independent variables is

$$au_{xx} + bu_{xy} + cu_{yy} + du_x + eu_y + fu + g = 0$$

where $a, b, c, d, e, f,$ and g are known functions of x and y. If g is the zero function, then the equation is called *homogeneous*; otherwise it is said to be *nonhomogeneous.* Linear partial differential equations have a relatively simple general theory, and are the kind you meet the most in applications.

A function u is called a *solution* to a partial differential equation in some domain D of the space of the independent variables if the equation becomes an identity when u and its partial derivatives are substituted into the equation. This verification that a given function is, or is not, a solution is easy if we know how to differentiate the function.

example 11.1: Show that $u_1(x, y) = x^2 - y^2$ and $u_2(x, y) = e^x \cos y$ are solutions to the two-dimensional Laplace's equation

$$u_{xx} + u_{yy} = 0.$$

solution: Calculating the partial derivatives of u_1 and u_2,

$$(u_1)_x = 2x, \qquad (u_1)_{xx} = 2, \qquad (u_1)_y = -2y, \qquad (u_1)_{yy} = -2;$$
$$(u_2)_x = e^x \cos y, \qquad (u_2)_{xx} = e^x \cos y,$$
$$(u_2)_y = -e^x \sin y, \quad (u_2)_{yy} = -e^x \cos y.$$

In both cases it is easy to see that the sum of the second partial derivatives is zero and hence both u_1 and u_2 satisfy the differential equation. Note that any linear combination of u_1 and u_2 is also a solution.

The set of all solutions to a partial differential equation is called the *solution set*, a set which may be quite large. Usually, we are interested in any *complete* set of linearly independent solutions; that is, a set from which all other solutions can be derived.

Even this set may be large as is the case with solutions to Laplace's equation (see Example 11.1) whose members are called harmonic functions (if the second partial derivatives are continuous). The defining equation of the solution set is called the *general solution.* From it any *particular solution* can be obtained.

Some simple equations can be solved in a straightforward manner, as in the following example.

example 11.2: Solve the partial differential equation $u_x = y \sin x$.

solution: The general solution is obtained by antidifferentiating partially with respect to x,

$$u(x, y) = -y \cos x + f(y)$$

where f is an "arbitrary" function. Any two solutions differ by a function of y.

example 11.3: Solve the equation $u_{xx} = x$, considered in three-dimensional space.

solution: The general solution is found by two antidifferentiations with respect to x,

$$u(x, y, z) = \frac{x^3}{6} + xf(y, z) + g(y, z)$$

where f and g are arbitrary functions of two variables.

11.2 boundary value problems

From the two previous examples one might conjecture that the solution set of an nth-order linear partial differential equation is in some sense a family of functions depending on n arbitrary functions in the same way that an nth-order linear ordinary differential equation is dependent on two arbitrary constants. This statement is roughly correct, but it will not be a source of concern for us to make it more precise since we rarely have to find general solutions.

In practice, you are often asked to find not the general solution to a partial differential equation but a particular one arising from a physical situation. In some cases, this particularization is obtained by specifying the value of the solution at points on the boundary; in others, when the time t is one of the variables the values of the function and its derivative(s) at $t = 0$, called *initial conditions*, are prescribed. A differential equation with some specified auxiliary conditions for the solution to satisfy is called a *boundary value problem.*

In the study of ordinary differential equations, the procedure for the solution of an initial value problem or boundary value problem is to find the general solution set and then determine values of the arbitrary constants so that the initial values or boundary values are specified. In the solution of partial differential equations, this is usually not true. For even when the general solution is known, it still may be prohibitively difficult to find the "arbitrary" functions to match the auxiliary conditions.

Frequently, at least in elementary work, you will find that the procedure in solving boundary value problems is to assume a certain form for the particular solution immediately. We pick a form to which the auxiliary conditions are easily adapted, and, while the picking of the "best" form for

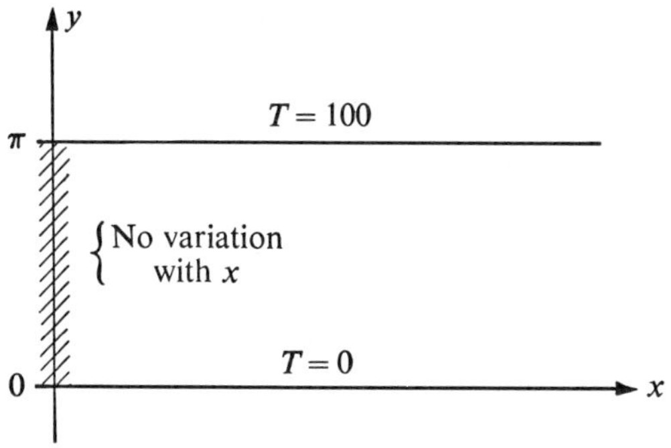

figure 11.1

these purposes might be quite difficult in general, it is relatively straightforward in some very important cases.

example 11.4: The temperature in a two-dimensional region satisfies Laplace's equation $T_{xx} + T_{yy} = 0$ with boundary conditions as shown in Figure 11.1. Formulate the mathematical model and solve the boundary value problem assuming a linear variation. (A boundary value problem whose differential equation is Laplace's equation is known as a "Dirichlet problem.")

solution: The mathematical model is the differential equation with the boundary conditions stated in terms of T,

$$T_{xx} + T_{yy} = 0; \qquad T(x, 0) = 0, \qquad T(x, \pi) = 100, x > 0;$$
$$T_x(0, y) = 0, \qquad 0 \leqq y \leqq \pi.$$

As a result of the assumed linearity,

$$T(x, y) = ax + by + c.$$

Since $T_x(0, y) = 0$, $a = 0$. Applying the boundary conditions at $y = 0$ and $y = \pi$,

$$b = \frac{100}{\pi} \quad \text{and} \quad c = 0.$$

Therefore, the solution is

$$T(x, y) = \frac{100}{\pi} y.$$

It is a theorem, proved in more advanced discussions of the subject of partial differential equations, that the solution to a boundary value problem like that of the previous example is unique. Hence we need concern ourselves with the technique of obtaining one solution to a boundary value problem, since if we do get one, it will be the only one.

example 11.5: Consider the boundary value problem

$$u_{xx} = u_{tt}$$
$$u(x, 0) = 2 \sin 3\pi x, \qquad u_t(x, 0) = 0 \text{ (initial conditions)};$$
$$u(0, t) = u(2, t) = 0 \text{ (boundary conditions)}.$$

Verify that one form of a solution is the product of a trigonometric function in x by a trigonometric function in t, and attempt to match the boundary conditions to this form.

solution: We assume the solution to be of the form

$$u(x, t) = (A \sin \alpha x + B \cos \alpha x)(C \sin \beta t + D \cos \beta t).$$

Partial differentiation of $u(x, t)$ twice with respect to x and twice with respect to t gives

$$u_{xx}(x, t) = -\alpha^2 u(x, t); \qquad u_{tt}(x, t) = -\beta^2 u(x, t)$$

and thus $\alpha = \beta$, since these two partial derivatives must be equal. Thus the trigonometric function in t must have the same angular frequency as the trigonometric function in x. The values of A, B, C, D, and α are determined from auxiliary conditions. The boundary condition $u(0, t) = 0$ gives $B = 0$. For the boundary condition $u(2, t) = 0$, we get

$$u(2, t) = A \sin 2\alpha (C \sin \alpha t + D \cos \alpha t) = 0$$

which means that

$$A \sin 2\alpha = 0$$

which is satisfied for

$$\alpha = \frac{n\pi}{2}, \qquad \text{for } n \text{ an integer}.$$

Applying the second initial condition gives $C = 0$ so that the solution is of the form

$$u(x, t) = AD \sin \frac{n\pi x}{2} \cos \frac{n\pi t}{2}.$$

The other initial condition gives

$$2 \sin 3\pi x = AD \sin \frac{n\pi x}{2}$$

from which $AD = 2$ and $n = 6$. Therefore, the solution is

$$u(x, t) = 2 \sin 3\pi x \cos 3\pi t.$$

In a later section, you will see that our choice for an assumed form need not be as specialized as in the preceding example. Rather than assume a product of trigonometric functions in x and t, we will assume a product of any function of x by any function of t.

11.3 superposition

Initially, in the study of homogeneous partial differential equations, it proves beneficial to reason analogously to the theory of ordinary differential equations. In ordinary differential equations, we know that a linear combination of any two solutions to a linear homogeneous equation is also a solution.

theorem 11.1

Let u_1 and u_2 be solutions to the differential equation

$$au_{xx} + bu_{xy} + cu_{yy} + du_x + eu_y + fu = 0,$$

where $a, b, c, d, e,$ and f are functions of x and y. Then $Au_1 + Bu_2$ is also a solution for any two constants A and B.

proof: See Exercise 8 in the next set of exercises.

We will be interested in the analog of Theorem 11.1 for nonfinite solution sets.

example 11.6: Find solutions to $u_x + u_y = 0$ and show that the infinite analog of Theorem 11.1 holds in at least one particular case.

solution: It is easy to show that each of the functions

$$u_n(x, y) = (x - y)^n$$

is a solution to the partial differential equation. Restricting n to the non-negative integers, we form the infinite series of functions

$$\sum_{n=0}^{\infty} c_n u_n(x, y) = \sum_{n=0}^{\infty} c_n (x - y)^n .$$

This series will converge to e^{x-y} if we choose c_n to be $1/n!$. Further, e^{x-y} is also a solution to the given partial differential equation. As a matter of fact, if we choose c_n to be the coefficients of *any* series expansion $f(t) = \sum_{n=0}^{\infty} c_n t^n$ of a function of one real variable, then the infinite series of functions $\sum_{n=0}^{\infty} c_n (x - y)^n$ so obtained converges to $f(x - y)$. The function $f(x - y)$ can easily be verified to be a solution to the given differential equation.

The principle of superposition will be needed for the solutions to certain types of boundary value problems also. If a set of solutions satisfies auxiliary conditions which require the function to vanish at a point, the condition is said to be *homogeneous*. If u_1 and u_2 are solutions to a homogeneous boundary value problem whose differential equation is linear, the linear combinations of u_1 and u_2 satisfy not only the differential equation but also all *the homogeneous auxiliary conditions.* You will be asked to prove this statement in the exercises.

exercises for sections 11.1–11.3

1. Verify that the following functions satisfy Laplace's differential equation:
   (a) $u(x, y) = x^2 - y^2$. (b) $u(x, y) = \mathrm{Tan}^{-1}\dfrac{y}{x}$.
   (c) $u(x, y) = ax + by + c$. (d) $u(x, y) = \cos x \cosh y$.
2. Find a partial differential equation satisfied by:
   (a) $u(x, y) = \sin(x^2 - y^2)$. (b) $u(x, y) = \cos(x^2 + y^2) + x^2$.
3. Show that $f(w)$ is a solution to $yu_x - xu_y = 0$ if $w = x^2 + y^2$.
4. Show that $u(x, y) = f(w) + g(v)$ is a solution to $u_{xx} = u_{yy}$ if $w = x + y$ and $v = x - y$.
5. Find the general solution to:
   (a) $u_y = 0$. (b) $u_{xx} + u = 0$.
   (c) $u_{xxx} + u_x = 0$. (d) $u_{xxy} = 1$.
6. Solve the two-dimensional Laplace's equation with the boundary conditions $u(0, y) = u(L, y) = 0$, $u_y(x, L) = (\pi/L) \sinh \pi \sin(\pi x/L)$, $u_y(x, 0) = 0$.
7. By assuming a product of trigonometric functions as a solution, solve $u_{xx} = u_{tt}$ if $u(x, 0) = 3 \sin 10\pi x$, $u_t(x, 0) = 0$, $u(0, t) = u(5, t) = 0$.
8. Prove Theorem 11.1.
9. Prove $u_n(x, y) = (x - y)^n$ satisfies the equation $u_x + u_y = 0$ for every n. Can you find an infinite series of these functions, different from the one in Example 11.6 whose sum is also a solution?
10. Find the solution of $u_{xx} - u = 0$ if $u(0, y) = y$, $u_x(0, y) = \sin y$.
11. Prove the last statement of Section 11.3 for the special case of Laplace's two-dimensional equations and the boundary conditions $u(x, 0) = u(x, L) = 0$.
12. Describe a "physical problem" with boundary values like the ones described in Figure 11.1.

11.4 three methods

When a homogeneous linear partial differential equation has constant coefficients, there are three convenient methods which you will find are commonly used in the applications.

Method 1. Recall that every homogeneous linear ordinary differential equation with constant coefficients has a solution of the form e^{mx} where m may be complex. To use the same idea for partial differential equations, we assume the solution to be of the form e^{rx+sy}. As with ordinary differential equations, we are then left with an algebraic equation to solve.

example 11.7: Solve the partial differential equation $u_y = 2u_x$, by assuming a solution of exponential form.

solution: Substituting e^{rx+sy} into the given equation, we obtain

$$s\, e^{rx+sy} = 2r\, e^{rx+sy}$$

or

$$s = 2r.$$

Thus the solutions of exponential type are

$$u_r(x, y) = c_r\, e^{r(x+2y)}$$

where r and c_r may be complex numbers, whose values are determined from auxiliary conditions. Other solutions may be obtained by superimposing these functions in various ways. Naturally, there may be solutions which are not of the assumed exponential form. For example, $u(x, y) = x + 2y$ is also a solution, but it cannot be obtained by summing those of exponential type. (Actually, any function of $x + 2y$ is also a solution.)

example 11.8: Solve the boundary value problem

$$u_y = 2u_x; \quad u(x, 0) = \cosh x, \; u(0, y) = \cosh 2y.$$

solution: The boundary conditions suggest an exponential form for the solution, so we use the result of Example 11.7. Applying the first condition, we see that the particular solution must be a sum of two exponentials,

$$u(x, y) = \frac{e^{(x+2y)} + e^{-(x+2y)}}{2} = \cosh(x + 2y).$$

This solution also satisfies the other boundary condition.

You may have thought that the previous example was a bit "trumped up," for if the second boundary condition had been just slightly different, we would have been forced to abandon this technique. This, of course, may occur, in which case another form should be tried. Just because an exponential solution does not fit a particular boundary value problem, you cannot assume the problem to be unsolvable.

Method 2. Closely related to the first method is the method of separation of variables or the *product method.* We seek possible solutions which are products of the functions of the separate variables

$$u(x, y) = X(x)\, Y(y).$$

example 11.9: Solve Laplace's equation in two dimensions, $\nabla^2 u = 0$, by assuming a product solution.

solution: We must solve the equation $u_{xx} + u_{yy} = 0$ by substituting $u(x, y) = X(x)\, Y(y)$,

$$X''(x)\, Y(y) + X(x)\, Y''(y) = 0.$$

Equivalently,

$$-\frac{X''(x)}{X(x)} = \frac{Y''(y)}{Y(y)}.$$

The left member of this equation is independent of y while the right member is independent of x. Thus the two members, known to be equal, are independent of *both* x and y. Therefore they are constant, which we denote by λ. Hence we have the two ordinary equations

$$\frac{X''}{X} = \lambda \quad \text{and} \quad \frac{Y''}{Y} = -\lambda.$$

The solutions will be trigonometric in y and exponential in x if $\lambda > 0$ and just the reverse if $\lambda < 0$. If $\lambda = 0$, both X and Y are linear functions. Therefore, the solutions to Laplace's equation (of the form of a product of a function of x and a function of y) are of the form $e^{ky} \cos kx$, $e^{ky} \sin kx$, $e^{-ky} \cos kx$, $e^{-ky} \sin kx$, $e^{kx} \cos ky$, $e^{kx} \sin ky$, etc.

example 11.10: Solve the boundary value problem $u_{xx} + u_{yy} = 0$ where $u(0, y) = 0$, $u(2, y) = 0$, $u(x, 0) = \sin \pi x - 3 \sin 5\pi x$, $u_y(x, 0) = 0$.

solution: First, we find those product solutions of the form of the previous example which satisfy the homogeneous boundary conditions. Then, by the remarks at the end of Section 11.3, any linear combination of these functions may be used in an attempt to satisfy the remaining, nonhomogeneous, condition. Consider, therefore, the function defined by

$$u(x, y) = (A \sin kx + B \cos kx)(C \cosh ky + D \sinh ky).$$

Since $u(0, y) = 0$, we have

$$u(0, y) = B(C \cosh ky + D \sinh ky) = 0,$$

which implies $B=0$. Since $u(2, y)=0$,

$$u(2, y)=A \sin 2k(C \cosh ky+D \sinh ky)=0,$$

which implies that $k=n\pi/2$, for n an integer. Thus,

$$u(x, y)=\sin \frac{n\pi x}{2}\left(C' \cosh \frac{n\pi y}{2}+D' \sinh \frac{n\pi y}{2}\right)$$

(where $C'=AC$ and $D'=AD$). From this expression, we find u_y,

$$u_y(x, y)=\frac{n\pi}{2} \sin \frac{n\pi x}{2}\left(C' \sinh \frac{n\pi y}{2}+D' \cosh \frac{n\pi y}{2}\right).$$

Since $u_y(x, 0)=0$, we have

$$u_y(x, 0)=\frac{n\pi}{2} \sin \frac{n\pi x}{2} D'=0$$

which implies $D'=0$. Hence, to satisfy the nonhomogeneous condition, we may use a linear combination of terms of the type

$$C_n' \sin \frac{n\pi x}{2} \cosh \frac{n\pi y}{2},$$

where n is an integer. This is easily done, by inspection, by choosing

$$C_2'=1, \quad C_{10}'=-3, \quad \text{and} \quad C_n'=0 \quad \text{for all other } n.$$

Therefore, the solution is

$$u(x, y)=\sin \pi x \cosh \pi y-3 \sin 5\pi x \cosh 5\pi y.$$

The method of separation of variables can be used to find solutions to some equations with nonconstant coefficients. If the substitution $u(x, y) = X(x)\,Y(y)$ leads to an equation in a form in which the functions X and Y may be separated, then the differential equation is said to be *separable.*

example 11.11: Solve $x^2u_{xx}+xu_x-u_y=0$ by assuming a product solution.

solution: Substituting $u(x, y)=X(x)\,Y(y)$ into the differential equation we obtain

$$x^2X''Y+xX'Y-XY'=0.$$

This leads to the equation in which the variables are separated,

$$\frac{x^2X''+xX'}{X}=\frac{Y'}{Y}.$$

By reasoning similar to that of Example 11.9, the two members of this equation have a common value which is constant and which we denote by λ, thereby obtaining the two ordinary differential equations,

$$x^2X'' + xX' - \lambda X = 0,$$
$$Y' - \lambda Y = 0.$$

The first of these equations is a "Cauchy-Euler" equation whose general solution may be found by a substitution $x = e^v$ (see the exercises for Section 3.7).

If $\lambda > 0$, say $\lambda = k^2$, then $X = c_1x^k + c_2x^{-k}$.

If $\lambda < 0$, say $\lambda = -k^2$, then $X = c_3 \cos(k \ln x) + c_4 \sin(k \ln x)$.

The second of the equations has a general solution $Y = C_1e^{k^2y}$ or $Y = C_2e^{-k^2y}$ if $\lambda = k^2$ and $-k^2$ respectively. Hence, terms of the form $e^{k^2y}x^{\pm k}$, $e^{-k^2y} \cos(k \ln x)$, and $e^{-k^2y} \sin(k \ln x)$ make up solutions to the given partial differential equation.

Note that despite the many solutions (one for each choice of k, c_1, c_2, c_3, c_4) which are factorable, there are many other solutions which are not factorable, be it only the obvious solution $e^yx + e^{4y}x^2$.

On the one hand, Method 1 is an unsatisfactory approach to the solution of the previous example since the solutions are of other than exponential form. On the other hand, there are equations with constant coefficients which are not separable, even though solutions exist which are factorable. You should try separating the variables when assuming a product solution to the innocent looking equation

$$u_{xx} + u_{xy} + u_{yy} = 0.$$

Method 3. A differential equation of the form

$$Au_{xx} + Bu_{xy} + Cu_{yy} = 0$$

can be put into the form $U_{vw} = 0$ with a linear change of the independent variables $v = x + my$; $w = x + ny$. The function $u(x, y)$ is a composite of the function $U(v, w)$ and the two functions $v(x, y)$ and $w(x, y)$ given by the relationship

$$u(x, y) = U(v(x, y), w(x, y)) = U(x + my, x + ny).$$

The idea is to select m and n so that in the transformed differential equation, the coefficients of U_{vv} and U_{ww} are zero. The method, at least in some particular applications, has come to be called *D'Alembert's method.*

In substituting into the differential equation for the partial derivatives, note that by using the chain rule,

$$u_x = U_v v_x + U_w w_x = U_v + U_w\,,$$
$$u_{xx} = U_{vv} v_x + U_{vw} w_x + U_{wv} v_x + U_{ww} w_x$$
$$= U_{vv} + 2U_{vw} + U_{ww}\,,$$

with similar formulas for u_{xy} and u_{yy}.

example 11.12: Solve the differential equation $u_{xx} - u_{xy} - 2u_{yy} = 0$.

solution: Using the change of variables, $v = x + my$, $w = x + ny$; and with repeated use of the chain rule,

$$u_{xx} = U_{vv} + 2U_{vw} + U_{ww}\,,$$
$$u_{xy} = mU_{vv} + (m + n)\,U_{vw} + nU_{ww}\,,$$
$$u_{yy} = m^2 U_{vv} + 2mnU_{vw} + n^2 U_{ww}\,.$$

The differential equation becomes

$$U_{vv}(1 - m - 2m^2) + U_{vw}(2 - m - n - 4mn) + U_{ww}(1 - n - 2n^2) = 0\,.$$

Therefore, we force m and n to be roots of the quadratic equation $1 - t - 2t^2 = 0$; that is, -1 and $1/2$. With this choice of m and n the differential equation can be written $U_{vw} = 0$ whose solution is

$$U(v, w) = g(v) + h(w)$$

where g and h are arbitrary twice differentiable functions. These functions are determined by auxiliary conditions and not by the partial differential equation. Hence, the solution has the form

$$u(x, y) = g(x - y) + h(x + \tfrac{1}{2}y)\,.$$

example 11.13: Use D'Alembert's method to solve the *wave equation* $u_{tt} = c^2 u_{xx}$ with the boundary conditions $u(x, 0) = f(x)$ on $[0, L]$, $u_t(x, 0) = 0$, $u(0, t) = 0 = u(L, t)$.

solution: First, we find $u(x, t)$ on $[0, L]$. Using the same technique as in the previous example you can show

$$u(x, t) = g(x + ct) + h(x - ct)$$

for arbitrary functions g and h.

From the conditions at $t = 0$,

$$u(x, 0) = g(x) + h(x) = f(x)\,,$$
$$u_t(x, 0) = cg'(x) - ch'(x) = 0\,.$$

Therefore, $g'(x) = h'(x)$ and hence $g(x) = h(x) + k$ where k is some constant. Using this in the first of the two equations,

$$h(x) + k + h(x) = f(x)$$

from which

$$h(x) = \frac{f(x) - k}{2}$$

and

$$g(x) = \frac{f(x) + k}{2}.$$

Therefore, the solution for $0 \leqq x \leqq L$ is

$$u(x, t) = \frac{f(x + ct) + f(x - ct)}{2}.$$

To find $u(x, t)$ for x not on the interval $[0, L]$, note that the condition $u(0, t) = 0$ gives

$$f(ct) + f(-ct) = 0;$$

that is,

$$f(ct) = -f(-ct),$$

which says that outside of $[0, L]$, f must be chosen to be the odd extension of the given function as defined on $[0, L]$. Thus the function is determined on $[-L, L]$. The condition $u(L, t) = 0$ gives

$$f(L + ct) = -f(L - ct).$$

Since f is odd, the second member of this equation may be written $f(-L + ct)$ and therefore

$$f(ct + L) = f(ct - L),$$

and this means that f should be chosen to be periodic with period $2L$. Hence, the formula for $u(x, t)$ outside of $[0, L]$ is the same as that on $[0, L]$ with the understanding that the function f is odd and periodic with period $2L$. In the Section 11.6 we will see that $f(x + ct)$ may be interpreted as a wave "traveling" to the left and $f(x - ct)$ as a wave "traveling" to the right.

exercises for section 11.4

1. Find solutions by assuming an exponential solution:
   (a) $u_x + 2u_y - u = 0$. (b) $u_{xx} - u_{yy} + u_y = 0$.
   (c) $u_x + u_y + u_z + u = 0$. (d) $u_{xy} + u_x + u_{yy} = 0$.
2. Do the problems of Exercise 1 assuming a product solution.

3. Solve by D'Alembert's method:
   (a) $u_{xx} - 5u_{xy} + 6u_{yy} = 0.$ (b) $u_{xx} + 4u_{yy} = 0.$
   (c) $u_{xy} - u_{yy} = 0.$ (d) $u_{xx} = 25u_{yy}.$
4. In D'Alembert's method, it may happen that $m = n$. Show that in this case $u = xf(x + my)$ is also a solution to $Au_{xx} + Bu_{xy} + Cu_{yy} = 0.$
5. Solve by assuming an exponential solution, $u_{xx} + u_{xy} + u_{yy} = 0$. Is this a separable differential equation?
6. Solve in any way you can:
   (a) $x^2u_{xy} + 3y^2u = 0.$ (b) $u_x + u_y = 0.$
   (c) $u_{xxxx} + yu_{xxy} = 0.$ (d) $u_{xy} - u = 0.$
   (e) $u_{xx} - (1 + y^2)\, u_{xy} = 0.$ (f) $u_{xyz} - xyzu = 0.$
   (g) $u_x + u_y = xu + yu.$

11.5 the wave equation: derivation

You are probably aware that completely different physical phenomena satisfy the same mathematical laws. In this section you will see how the wave equation is derived in two widely varied applications.

Consider the problem of a string stretched taut between two points and then given an initial deflection or "plucked." The transverse ("up and down") motion is small, in fact much smaller than the sketch in Figure 11.2

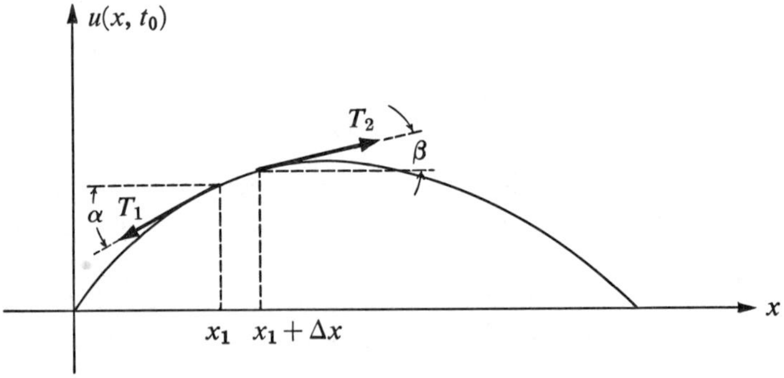

figure 11.2 *transverse motion of vibrating string*

would seem to imply, and the tension in the string is assumed to be large with respect to other forces.

To determine the equation for the transverse vibration we examine a small piece of the string between x_1 and $x_1 + \Delta x$. Since there is no horizontal movement,

$$T_1 \cos\alpha = T_2 \cos\beta = T = \text{constant}.$$

The difference of the vertical forces is equal to the mass of the element times the transverse acceleration,

$$T_2 \sin\beta - T_1 \sin\alpha = \rho\, \Delta s\, u_{tt}$$

where ρ is the mass per unit length and Δs is the length of the small piece under consideration. Since the vibration is small, $\Delta s \approx \Delta x$. Dividing each term of the second equation by the equal quantities of the first equation,

$$\tan\beta - \tan\alpha = \frac{\rho\, \Delta x}{T} u_{tt}.$$

Dividing by Δx and using the slope interpretation of the derivative,

$$\frac{u_x(x_1 + \Delta x, t) - u_x(x_1, t)}{\Delta x} = \frac{\rho}{T} u_{tt}.$$

Letting Δx approach 0, this becomes

$$u_{xx} = \frac{\rho}{T} u_{tt}$$

which is often written

$$u_{tt} = c^2 u_{xx}.$$

This partial differential equation is called the wave equation, and it, along with the auxiliary conditions

$$\begin{aligned} u(x, 0) &= f(x), && \text{initial deflection;} \\ u_t(x, 0) &= g(x), && \text{initial velocity;} \\ u(0, t) &= u(L, t) = 0, && \text{both ends fixed,} \end{aligned}$$

is the mathematical model for the vibrating string problem.

Next consider the flow of electricity in a long cable or transmission line assumed to be imperfectly insulated so that there is both capacitance

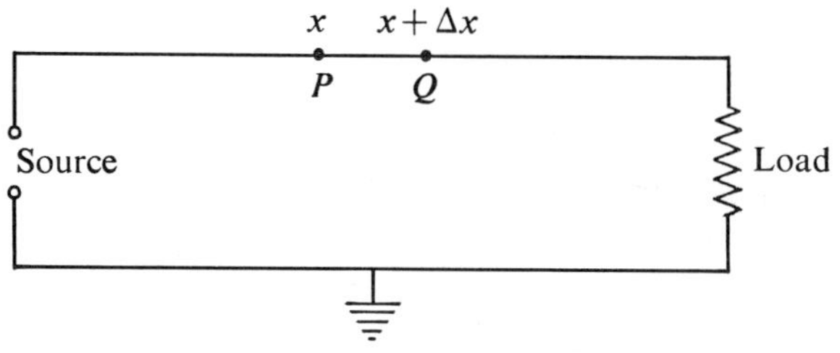

figure 11.3 *transmission line*

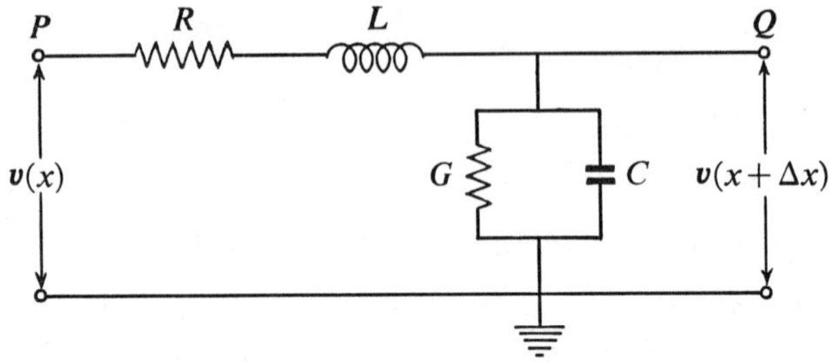

figure 11.4 schematic per unit length

and leakage of current to ground. Figure 11.3 gives a generalized schematic of the system and Figure 11.4 shows a representative portion of the transmission line for a small length Δx. Assuming R, L, G, and C to be the ordinary electric parameters (resistance, inductance, conductance, and capacitance) per unit length, the values of the lumped components are as shown.

The voltage drop from P to Q is given by

$$v(x+\Delta x) - v(x) = -(R\,\Delta x)\,i - (L\,\Delta x)\,i_t,$$

and the current loss is given by

$$i(x+\Delta x) - i(x) = -(G\,\Delta x)\,v - (C\,\Delta x)\,v_t.$$

In both equations, divide by Δx, and let Δx approach 0, to obtain the so-called *transmission line equations*

$$v_x = -Ri - Li_t,$$
$$i_x = -Gv - Cv_t.$$

These equations are often written in slightly different form. If you differentiate the first equation with respect to x and the second with respect to t and eliminate the variable i between the two, you get

$$v_{xx} = RGv + RCv_t + LGv_t + LCv_{tt}.$$

Reversing the procedure for the current

$$i_{xx} = RGi + RCi_t + LGi_t + LCi_{tt}.$$

Thus, the current and voltage satisfy exactly the same partial differential equation, known as the *wave equation of the transmission line.*

For high frequencies, the term involving the second derivative with respect to time dominates and the equations become

$$v_{xx} = LCv_{tt}$$
$$i_{xx} = LCi_{tt}$$

which are called the high frequency line equations (also the "lossless" line equations). The equations have precisely the same form as the one derived for the vibrating string problem. The actual mathematical model for the two situations will usually be quite different, though, since the vibrating string conditions of zero deflection at the endpoints of the line have no realistic analog on a transmission line.

11.6 solution by d'alembert's method

In Example 11.13 we solved a partial differential equation with boundary values which we now recognize as the wave equation with boundary values corresponding to the vibrating string problem. In that example, you saw that the deflection of the string is given by

$$u(x, t) = \frac{f(x + ct) + f(x - ct)}{2}$$

where f is understood to be the odd periodic extension of the initial deflection on $[0, L]$. The graph of $f(x + ct)$ is the same as that of $f(x)$ except that it is translated to the left by ct units. Similarly, the graph of $f(x - ct)$ is the translation of $f(x)$ to the right by ct units. Figure 11.5 shows an initial deflection, its odd periodic extension, and that extension shifted one unit to the right and left. (See p. 394.)

The functions $f(x + ct)$ and $f(x - ct)$ are said to "travel" since with increasing t, the first wave "moves" to the left and the second one to the right. The physical traveling of the two waves is a very illuminating result and is one of the major reasons for the importance of D'Alembert's method in the solution of the wave equation.

Hence, to find $u(x, t)$ on $[0, L]$ at a given instant of time, t_0, we may proceed:

1. Graphically, by plotting the two waves

$$\tfrac{1}{2}f(x + ct_0), \qquad \tfrac{1}{2}f(x - ct_0)$$

   and adding the ordinates on $[0, L]$.
2. Analytically, by computing the analytic expressions for $f(x + ct)$ and $f(x - ct)$ on $[0, L]$.

By displaying the deflection (or finding the analytic description of the deflection) at several different times you can get a good idea of the nature of the vibration.

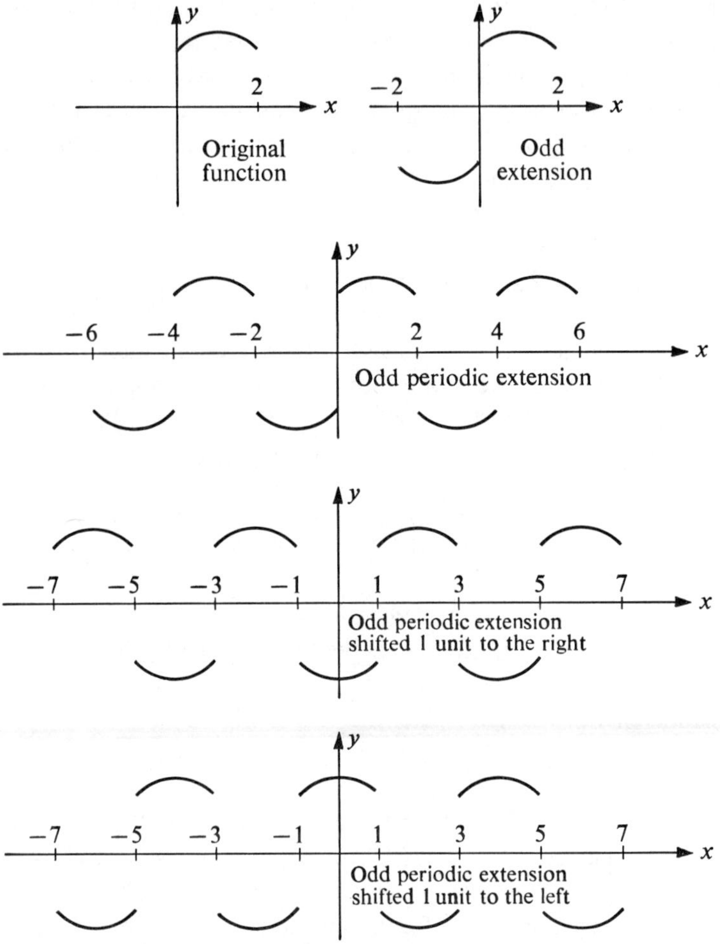

figure 11.5 *traveling waves*

example 11.14: Plot solutions to the vibrating string problem corresponding to a triangular initial deflection on [0, 2] given by the formula

$$\begin{aligned} f(x) &= x, & 0 \leqq x \leqq 1; \\ &= -x+2, & 1 \leqq x \leqq 2. \end{aligned}$$

Assume $c = 1$ and plot the deflection at $t = 0$, 1/3, 2/3, and 1. Verify the sketches by computing the deflection analytically.

solution: To sketch the solution curve at a particular value of t we sum the ordinates of the initial waveform translated to the right by t units with the ordinates of a left-translated waveform (and divide by two). The results are

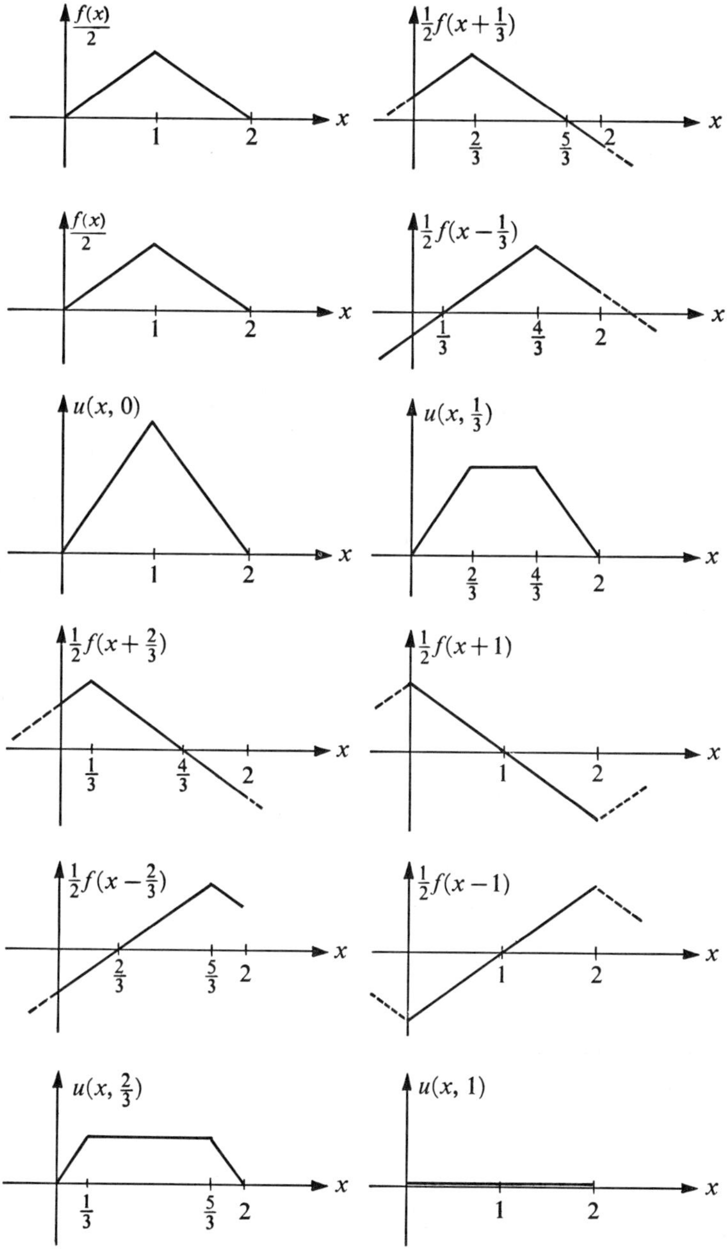

figure 11.6

shown in Figure 11.6. Since the initial deflection is defined so explicitly, an analytic solution is possible.

1. At $t=0$,

$$u(x,0)=\frac{f(x)+f(x)}{2}=f(x), \quad \text{on}\,[0,2].$$

2. At $t=1/3$,

$$u(x,\tfrac{1}{3})=\tfrac{1}{2}(f(x+\tfrac{1}{3})+f(x-\tfrac{1}{3})),$$

where

$$\begin{aligned} f(x+\tfrac{1}{3}) &= x+\tfrac{1}{3}, & 0 \leqq x \leqq \tfrac{2}{3}, \\ &= -x+\tfrac{5}{3}, & \tfrac{2}{3} \leqq x \leqq 2; \\ f(x-\tfrac{1}{3}) &= x-\tfrac{1}{3}, & 0 \leqq x \leqq \tfrac{4}{3}, \\ &= -x+\tfrac{7}{3}, & \tfrac{4}{3} \leqq x \leqq 2. \end{aligned}$$

Hence,

$$\begin{aligned} u(x,\tfrac{1}{3}) &= x, & 0 \leqq x \leqq \tfrac{2}{3}; \\ &= \tfrac{2}{3}, & \tfrac{2}{3} \leqq x \leqq \tfrac{4}{3}; \\ &= -x+2, & \tfrac{4}{3} \leqq x \leqq 2. \end{aligned}$$

3. At $t=2/3$,

$$\begin{aligned} f(x+\tfrac{2}{3}) &= x+\tfrac{2}{3}, & 0 \leqq x \leqq \tfrac{1}{3}, \\ &= -x+\tfrac{4}{3}, & \tfrac{1}{3} \leqq x \leqq 2; \\ f(x-\tfrac{2}{3}) &= x-\tfrac{2}{3}, & 0 \leqq x \leqq \tfrac{5}{3}, \\ &= -x+\tfrac{8}{3}, & \tfrac{5}{3} \leqq x \leqq 2. \end{aligned}$$

Hence,

$$\begin{aligned} u(x,\tfrac{2}{3}) &= x, & 0 \leqq x \leqq \tfrac{1}{3}; \\ &= \tfrac{1}{3}, & \tfrac{1}{3} \leqq x \leqq \tfrac{5}{3}; \\ &= -x+2, & \tfrac{5}{3} \leqq x \leqq 2. \end{aligned}$$

4. At $t=1$,

$$\begin{aligned} f(x+1) &= -x+1, & 0 \leqq x \leqq 2; \\ f(x-1) &= x-1, & 0 \leqq x \leqq 2. \end{aligned}$$

Hence

$$u(x,1)=0, \qquad 0 \leqq x \leqq 2.$$

exercises for sections 11.5 and 11.6

1. Using the transmission line equations, use the procedure outlined to derive the wave equation for the transmission line.

In Exercises 2–5, plot solutions in a manner similar to Example 11.14 for initial deflection functions as given.

2. For a string of length 3,

$$\begin{aligned} f(x) &= x, & 0<x<1; \\ &= 1, & 1<x<2; \\ &= -x+3, & 2<x<3. \end{aligned}$$

3. For a string of length 6, $f(x) = x(6 - x)$, $0 < x < 6$.
4. For a string of length 4,

$$\begin{aligned} f(x) &= 0, & 0 < x < 1; \\ &= x - 1, & 1 < x < 2; \\ &= -x + 3, & 2 < x < 3; \\ &= 0, & 3 < x < 4. \end{aligned}$$

5. For a string of length 1, $f(x) = \sin \pi x$, $0 < x < 1$.

6–8. Verify analytically the solutions to Exercises 2–4.

11.7 product solution

The wave equation with the homogeneous boundary conditions of the vibrating string problem may also be solved by assuming the solution to be the form of a product of a function of x by a function of t. We substitute $u(x, t) = X(x)\,T(t)$ into the equation to obtain

$$\frac{X''}{X} = \frac{\ddot{T}}{c^2 T}$$

which yields immediately the two ordinary differential equations

$$X'' - \lambda X = 0 \qquad \text{and} \qquad \ddot{T} - c^2 \lambda T = 0.$$

The homogeneous boundary conditions of u may be easily interpreted in terms of the so-called "space function," X,

$$\left.\begin{aligned} u(0, t) = X(0)\,T(t) = 0 \\ u(L, t) = X(L)\,T(t) = 0 \end{aligned}\right\} \text{for all } t.$$

This means $X(0) = X(L) = 0$ if a nontrivial function T is desired. Hence, we are led to the Sturm-Liouville problem

$$X'' - \lambda X = 0,$$
$$X(0) = X(L) = 0.$$

Up to multiplicative constants, the only nontrivial solutions to this problem are $X_n(x) = \sin(n\pi x/L)$, corresponding to eigenvalues $\lambda_n = -n^2\pi^2/L^2$. (See Section 10.13.) Moreover, when $\lambda_n = -n^2\pi^2/L^2$, the general solution of

$$\ddot{T} - c^2 \lambda_n T = 0$$

is

$$T_n(t) = C_n \cos \frac{cn\pi t}{L} + D_n \sin \frac{cn\pi t}{L}.$$

Forming the product of $X_n(x)$ and $T_n(t)$, we see that each of the functions

$$u_n(x, t) = \sin \frac{n\pi x}{L} \left(C_n \cos \frac{cn\pi t}{L} + D_n \sin \frac{cn\pi t}{L} \right)$$

satisfies the one-dimensional wave equation and the given set of homogeneous boundary conditions. Not only each $u_n(x, t)$ but also any sum of such functions and even any convergent infinite series of such functions satisfies the differential equation and the homogeneous boundary conditions.

It remains to match sums of the u_n's to the nonhomogeneous initial conditions. Such a matching may be possible by inspection.

example 11.15: Solve the vibrating string problem by the product method for a string of length 5 if the initial deflection is given by $3 \sin 2\pi x + \sin 4\pi x - 0.1 \sin 20\pi x$. Assume zero initial velocity.

solution: Choose u_{10}, u_{20}, and u_{100} with $C_{10} = 3$, $C_{20} = 1$ and $C_{100} = -0.1$. The sum of these three functions evaluated at $t = 0$ will match the initial deflection function. Since $u_t(x, 0) = 0$, it follows that $D_n = 0$, for all n. Hence, the solution is

$$u(x, t) = 3(\sin 2\pi x)(\cos 2\pi ct) + (\sin 4\pi x)(\cos 4\pi ct) - 0.1(\sin 20\pi x)(\cos 20\pi ct).$$

In general, no finite sum of the u_n will satisfy the initial conditions. In such cases we form an infinite series

$$\begin{aligned} u(x, t) &= \sum_{n=1}^{\infty} u_n(x, t) \\ &= \sum_{n=1}^{\infty} \sin \frac{n\pi x}{L} \left(C_n \cos \frac{n\pi ct}{L} + D_n \sin \frac{n\pi ct}{L} \right) \end{aligned}$$

and choose C_n and D_n so that the initial conditions are satisfied.

To satisfy the condition for initial deflection, $u(x, 0) = f(x)$,

$$u(x, 0) = \sum_{n=1}^{\infty} C_n \sin \frac{n\pi x}{L} = f(x).$$

This condition will be satisfied if the C_n are chosen in such a way that the series converges pointwise to the function f on $[0, L]$. This is the familiar problem of expanding a function by a Fourier sine series on $[0, L]$. Therefore, if we let

$$C_n = \frac{2}{L} \int_0^L f(x) \sin \frac{n\pi x}{L} \, dx,$$

the series will converge to f on the interval $[0, L]$ if f satisfies the Dirichlet conditions. (See Theorem 10.7.)

The technique for determining the D_n is much the same. Assuming that the series for $u(x, t)$ can be differentiated term by term with respect to t, we equate this series evaluated at $t = 0$ to the initial velocity function,

$$g(x) = u_t(x, 0) = \sum_{n=1}^{\infty} \frac{cn\pi}{L} D_n \sin \frac{n\pi x}{L}$$

and so, $(cn\pi/L)\, D_n$ must be the nth Fourier sine series coefficient of the function g. Hence,

$$D_n = \frac{2}{cn\pi} \int_0^L g(x) \sin \frac{n\pi x}{L}\, dx.$$

example 11.16: Do Example 11.14 by the product method.

solution: We need only calculate the Fourier sine coefficients of the given triangular function,

$$\begin{aligned} C_n &= \int_0^1 x \sin \frac{n\pi x}{L}\, dx + \int_1^2 (-x + 2) \sin \frac{n\pi x}{L}\, dx \\ &= \frac{8}{n^2\pi^2} \sin \frac{n\pi}{2} \\ &= \pm \frac{8}{n^2\pi^2}, \quad n \text{ odd}; \\ &= 0, \quad n \text{ even}. \end{aligned}$$

Therefore, since all the $D_n = 0$,

$$u(x, t) = \frac{8}{\pi^2} \sum_{n=1}^{\infty} \frac{(-1)^n}{(2n+1)^2} \sin \frac{(2n+1)\pi x}{2} \cos \frac{(2n+1)\pi t}{2}.$$

The frequency of the time factor in the first nonzero u_n is called the *fundamental frequency*. If $u_1 \neq 0$, the fundamental frequency is $c/2L$; all other frequencies are integral multiples of this frequency. The fundamental frequency determines the fundamental *tone* of the vibration; its integral multiples are called *overtones* or harmonics.

The solution of the vibrating string problem with zero initial velocity by the product method technique can be identified with that obtained by D'Alembert's method. The solution by the product method gives

$$u(x, t) = \sum_{n=1}^{\infty} C_n \sin \frac{n\pi x}{L} \cos \frac{cn\pi t}{L}$$

where the C_n's are the Fourier sine series coefficients of the initial deflection function on $[0, L]$.

We use the trigonometric identity

$$\sin \frac{n\pi x}{L} \cos \frac{cn\pi t}{L} = \frac{1}{2}\left[\sin \frac{n\pi}{L}(x - ct) + \sin \frac{n\pi}{L}(x + ct)\right]$$

to write

$$u(x, t) = \tfrac{1}{2} \sum_{n=1}^{\infty} C_n \sin \frac{n\pi}{L}(x - ct) + \tfrac{1}{2} \sum_{n=1}^{\infty} C_n \sin \frac{n\pi}{L}(x + ct).$$

Both these series are obtainable from the series for $f(x)$ by substituting $x - ct$ and $x + ct$ for x. Hence,

$$u(x, t) = \tfrac{1}{2}[f(x + ct) + f(x - ct)],$$

as desired.

exercises for section 11.7

Solve the vibrating string problem by the product method if the initial deflection function is as given in each exercise. Assume zero initial velocity.

1. For a string of length 1, $u(x, 0) = \sin \pi x$, $0 < x < 1$.
2. For a string of length 1, $u(x, 0) = x(1 - x)$, $0 < x < 1$.
3. For a string of length 2, $u(x, 0) = \sin 4\pi x - \sin 100\pi x$, $0 < x < 2$.
4. For a string of length 2, $u(x, 0) = x, \quad 0 < x < 1,$
   $= -x + 2, \quad 1 < x < 2.$
5. Same as Exercise 2 in the previous set of exercises.

11.8 the heat equation

The mathematical law which governs the flow of heat in a body may be derived by use of the divergence theorem. Let D be a domain in space which the body occupies, let S be its boundary surface, and T the temperature function of space and time. From physics we know that the velocity of heat flow, $\mathbf{v}$, is given by

$$\mathbf{v} = -K\nabla T$$

where K is the thermal conductivity of the body.

The amount of heat *leaving* the body is given by

$$\iint_S (\mathbf{v} \cdot \mathbf{N})\, dS$$

where **N** is a unit vector normal to the surface. The amount of heat *in* the body is given by

$$\sigma\rho \iiint_D T\,dV$$

where σ is the specific heat of the body and ρ is its density. The time rate of decrease of the heat in the body is equal to the amount of heat leaving the body,

$$-\frac{\partial}{\partial t}\left[\sigma\rho \iiint_D T\,dV\right] = -K \iint_S (\nabla T \cdot \mathbf{N}).$$

Applying the divergence theorem to the closed surface integral and assuming that the operations of space integration and differentiation with respect to time can be interchanged,

$$\sigma\rho \iiint_D T_t\,dV = K \iiint_D \nabla^2 T\,dV.$$

This equality holds not only for the domain D occupied by the whole body but for any subdomain; the equality of integrals for any D implies the equality of the integrands. Therefore,

$$T_t = \frac{K}{\sigma\rho}\nabla^2 T.$$

In one dimension, for example for the longitudinal heat flow in a thin bar, the heat equation is

$$T_t = \frac{K}{\sigma\rho} T_{xx}.$$

In Section 11.5 we showed how the high frequency transmission line equations have the same form as the wave equation. If, in the wave equation for the transmission line (see Section 11.5), we let G and L be negligibly small, which is the same effect as for low frequencies, we obtain

$$v_{xx} = RCv_t$$

and

$$i_{xx} = RCi_t.$$

These two equations are known as the *low frequency transmission line equations* or, perhaps more commonly, as the *telegraph equations.* Both have the same form as the partial differential equation governing heat flow in a body.

In the following example, we show how to solve the heat equation if it has homogeneous boundary conditions.

example 11.17: Find the temperature in a thin bar of length L whose ends are kept frozen and whose sides are well insulated if the bar has some initial heat distribution described by the temperature $T = f(x)$ at each point x.

solution: The mathematical model for this problem is given by the heat equation

$$T_t = c^2 T_{xx}$$

and the auxiliary conditions

$$T(0, t) = T(L, t) = 0,$$
$$T(x, 0) = f(x).$$

Assuming a product solution $u(x, t) = X(x)\,Y(t)$ and substituting into the differential equation, we obtain

$$X\dot{Y} = c^2 X'' Y$$

or

$$\frac{X''}{X} = \frac{\dot{Y}}{c^2 Y}.$$

As before, this leads to the two ordinary differential equations,

$$X'' + \lambda X = 0$$
$$\dot{Y} + c^2 \lambda Y = 0.$$

The homogeneous boundary conditions are translated into conditions on X such that $X(0) = X(L) = 0$. These, together with the first differential equation, give a now familiar Sturm-Liouville problem which has eigensolutions $X_n(x) = \sin(n\pi x/L)$ corresponding to the eigenvalues $\lambda_n = n^2\pi^2/L^2$, n an integer. With that choice for λ_n, the second differential equation becomes

$$\dot{Y} + \left(\frac{cn\pi}{L}\right)^2 Y = 0$$

whose general solution is

$$Y_n(t) = B_n\, e^{-(cn\pi/L)^2 t}$$

where B_n is an arbitrary constant. Hence, the functions

$$T_n(x, t) = B_n \sin\frac{n\pi x}{L}\, e^{-(cn\pi/L)^2 t}$$

satisfy the heat equation and the homogeneous boundary conditions. Generally speaking, to find a function which also satisfies the initial temperature distribution, we must form an infinite series of functions,

$$T(x, t) = \sum_{n=1}^{\infty} T_n(x, t).$$

For $t=0$, this becomes

$$T(x,0)=\sum_{n=1}^{\infty} T_n(x,0)$$
$$=\sum_{n=1}^{\infty} B_n \sin\frac{n\pi x}{L}.$$

Hence, the B_n should be chosen to be the Fourier sine series coefficients of the initial temperature distribution function.

In a body in which the temperature does not change with time, the so-called steady state condition is governed by $\nabla^2 T=0$, or in one dimension, $T_{xx}=0$. Sometimes, we need the steady state temperature since it may form the initial temperature distribution for the next period.

The following example is typical of how to handle a boundary value problem in which the boundary conditions are *not* homogeneous. The fundamental approach is to recast the problem in terms of a different variable in which the boundary conditions, or at least most of them, are homogeneous.

example 11.8: Find the temperature in the bar of the previous example if prior to $t=0$, one end of the bar is kept in boiling water and is then put into water whose temperature is 30 °C.

solution: Prior to time $=0$, under steady state conditions, the temperature distribution is governed by Laplace's equation $u_{xx}=0$ with boundary conditions $u(0,t)=0$ and $u(L,t)=100$. The fact that $u_{xx}=0$ implies the linearity of any solution and hence, prior to $t=0$, the solution is

$$u(x,t)=\frac{100}{L}x.$$

(See Example 11.4.)

For $t \geqq 0$, we must solve the boundary value problem

$$u_t=c^2u_{xx}; \quad u(0,t)=0,\ u(L,t)=30,\ u(x,0)=\frac{100}{L}x.$$

The boundary conditions of this problem are not homogeneous and hence a sum of solutions satisfying the differential equation will not also satisfy the boundary conditions. Therefore, we introduce another function which satisfies $u_{xx}=0$ as $t\to\infty$ (called the *steady state* solution for $t>0$), and denote it by $u(x,\infty)$. $u(x,\infty)$ is found just like $u(x,t)$ for $t<0$,

$$u(x,\infty)=\frac{30}{L}x.$$

Now, letting $u(x, t) = u(x, \infty) + v(x, t)$, we see that v must satisfy the heat equation with homogeneous boundary conditions,

$$v(0, t) = u(0, t) - u(0, \infty) = 0 - 0 = 0;$$
$$v(L, t) = u(L, t) - u(L, \infty) = 30 - 30 = 0.$$

The initial condition for the function v is

$$\begin{aligned} v(x, 0) &= u(x, 0) - u(x, \infty) \\ &= \frac{100}{L}x - \frac{30}{L}x \\ &= \frac{70}{L}x. \end{aligned}$$

The solution for v is the same as for T in the previous example,

$$v(x, t) = \sum_{n=1}^{\infty} B_n \sin \frac{n\pi x}{L} e^{-(cn\pi/L)^2 t}$$

where B_n are the Fourier sine coefficients of the initial temperature distribution function. In this particular case,

$$B_n = \frac{2}{L}\int_0^L \frac{70}{L} x \sin \frac{n\pi x}{L}\, dx = \frac{140}{n\pi}(-1)^{n+1}.$$

Therefore the temperature function $u(x, t)$ is determined in terms of the steady state formula $u(x, \infty)$ and the *transient* distribution $v(x, t)$.

exercises for section 11.8

1. A thin bar 2 meters long which is insulated on its longitudinal surface has a steady state distribution

$$\begin{aligned} f(x) &= x, & 0 < x < 1; \\ &= 2 - x, & 1 < x < 2. \end{aligned}$$

At time $t = 0$, the ends are immersed in ice water. Find the temperature distribution $T(x, t)$ for $t > 0$.

2. Repeat Exercise 1, if the distribution prior to $t = 0$ is given by $f(x) = x(2 - x)$, $0 < x < 2$.

3. The ends of a thin bar are kept at 10° and 90°. What is the steady state temperature distribution?

4. The ends of the bar in Exercise 3 are suddenly changed to 20° and 70°. Find the temperature distribution.

complex functions

12

12.1 the complex field

You will find the subject matter of this chapter to be almost totally different from the preceding ones. We will maintain our relatively unsophisticated approach and avoid meticulous argumentation wherever possible.

You probably already know the fundamental algebraic operations with complex numbers. The field of complex numbers is defined as a set of ordered pairs of real numbers, denoted by $\{z \mid z = (x, y)\}$ on which there are operations of addition and multiplication.

definition 12.1

$$(x_1, y_1) + (x_2, y_2) = (x_1 + x_2, y_1 + y_2)$$
$$(x_1, y_1) \cdot (x_2, y_2) = (x_1x_2 - y_1y_2, x_1y_2 + x_2y_1).$$

Both operations are associative and commutative, and multiplication is distributive over addition. Further, there is an additive identity $(0, 0)$ and a multiplicative identity $(1, 0)$. Each complex number (x, y) has an additive inverse $(-x, -y)$ and every number except $(0, 0)$ has a multiplicative inverse $(x/(x^2 + y^2), -y/(x^2 + y^2))$. Each of these properties is easy to prove. The following example shows how one of them is justified.

example 12.1: Verify that

$$\left(\frac{x}{x^2 + y^2}, \frac{-y}{x^2 + y^2}\right)$$

is the multiplicative inverse of (x, y).

solution: Using the definition of multiplication,

$$(x, y)\cdot\left(\frac{x}{x^2+y^2}, \frac{-y}{x^2+y^2}\right)$$
$$= \left(x\frac{x}{x^2+y^2} + y\frac{y}{x^2+y^2},\ x\frac{-y}{x^2+y^2} + y\frac{x}{x^2+y^2}\right)$$
$$= (1, 0)$$

which was to be proved.

The properties mentioned in the previous paragraph are sufficient to justify the use of the word "field" for the set of complex numbers. Most of these properties are direct consequences of the corresponding properties for the field of real numbers.

The real number $x = \text{Re}\, z$ in the complex number $z = (x, y)$ is called the *real part of z*. The real number $y = \text{Im}\, z$ is called the *imaginary part of z*. Instead of writing $z = (x, y)$, a much more common notation is $z = x + iy$. By convention, if $y = 1$, then y may be omitted. For example, (0, 1) is written just as i, (5, 1) as $5 + i$. Further, if $y = 0$, we delete iy and if $x = 0$, we delete $x +$.

The computation of $(0, 1)^2$ (or, if you wish, i^2) may be performed from Definition 12.1 by letting $x_1 = x_2 = 0$, and $y_1 = y_2 = 1$, to obtain $(0, 1)^2 = (-1, 0)$. Using the notation conventions mentioned earlier, we write

$$i^2 = -1.$$

This result allows computations with complex numbers to be made without continual reference to the definition of multiplication. The algebra of complex numbers becomes identical to that of real binomials if you use the fact that $i^2 = -1$. For example,

$$(3 + 4i)(7 + 6i) = (21 + 28i + 18i + 24i^2)$$
$$= 21 + 46i + 24i^2 = -3 + 46i.$$

12.2 the complex plane

Since complex numbers are ordered pairs of real numbers, some two-dimensional configuration is necessary to represent them graphically. The cartesian coordinate plane is often used for this purpose, in which case it is called the complex plane. Then, the x- and y-axes are called the real and imaginary axes respectively. The point with coordinates (x, y) corresponds to the complex number $z = x + iy$. The operation of addition is conveniently interpreted graphically to be the same as the familiar parallelogram law for addition of arrows. (See Figure 12.1.)

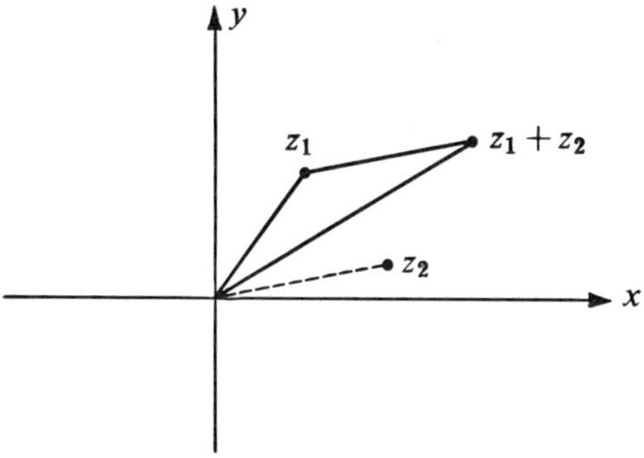

figure 12.1 *addition of complex numbers*

A complex number $z = x + iy$ may be represented in *polar form* by use of the transformation equations $x = r\cos\theta$, $y = r\sin\theta$. Then

$$\begin{aligned} z = x + iy &= r\cos\theta + ir\sin\theta \\ &= r(\cos\theta + i\sin\theta) \\ &= r\,\mathrm{cis}\,\theta \end{aligned}$$

where the last equality is used to define $\mathrm{cis}\,\theta$. The number r is called the magnitude or *modulus* of the complex number z and is denoted by $|z|$; it is always considered to be positive. The number θ is called the *argument* of z and is denoted by $\arg z$.

If the polar form of a complex number is used, the point in the complex plane corresponding to $z = r\,\mathrm{cis}\,\theta$ is denoted by (r, θ). In this manner the modulus of z is the measure of the distance from the origin to the point and the argument of z is the measure of the counterclockwise angle from the real axis to a line drawn from the origin to the point.

The following example shows how convenient it is to multiply complex numbers when they are expressed in polar form.

example 12.2: Develop a formula for the product of z_1 and z_2 in polar form.

solution: Let $z_1 = r_1\,\mathrm{cis}\,\theta_1$ and $z_2 = r_2\,\mathrm{cis}\,\theta_2$. Then

$$\begin{aligned} z_1 z_2 &= (r_1\cos\theta_1 + ir_1\sin\theta_1)(r_2\cos\theta_2 + ir_2\sin\theta_2) \\ &= r_1 r_2(\cos\theta_1 + i\sin\theta_1)(\cos\theta_2 + i\sin\theta_2) \\ &= r_1 r_2[(\cos\theta_1\cos\theta_2 - \sin\theta_1\sin\theta_2) + i(\sin\theta_1\cos\theta_2 + \sin\theta_2\cos\theta_1)] \\ &= r_1 r_2\,\mathrm{cis}(\theta_1 + \theta_2) \end{aligned}$$

which is to say that the modulus of the product is the product of the individual moduli and an argument of the product is the sum of arguments of the factors.

Graphically, multiplication of z_1 by z_2 results in a rotation of a line through z_1 by an angle equal to an argument of z_2. The modulus is expanded or contracted depending on whether $|z_2| > 1$ or < 1. (See Figure 12.2.)

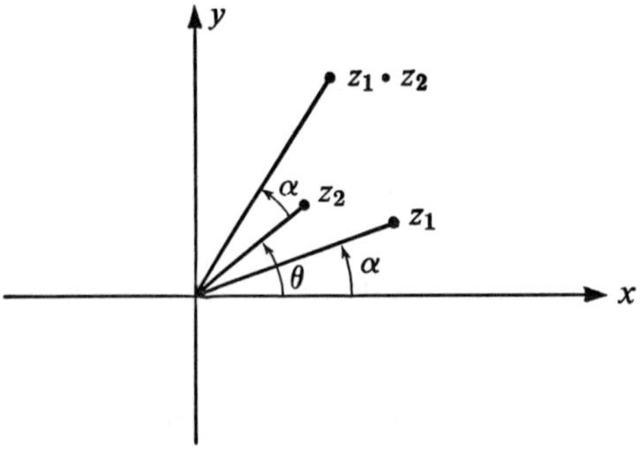

figure 12.2 *multiplication of complex numbers*

Note that a given complex number has infinitely many arguments, all differing by multiples of 2π. Sometimes, we limit $\arg z$ to a particular 2π span called the *principal values* of $\arg z$ and denoted by $\operatorname{Arg} z$. In this book, the values of $\operatorname{Arg} z$ will be $-\pi < \theta \leqq \pi$. We will have to say precisely when $\arg z$ is being limited to its principal values because there are many important occasions in which it is not. For instance, in taking the product of two complex numbers it is often inconvenient to have the argument of the product expressed in terms of a value between $-\pi$ and π.

The graphical interpretation of complex numbers allows us to accept the truth of an analytic statement on the basis of a geometric argument.

example 12.3: Show graphically that $|z_1 + z_2| \leqq |z_1| + |z_2|$. When does equality occur?

solution: Since $|z_1|$ and $|z_2|$ are sides of a triangle, Figure 12.1 suggests that the third side is always less than the sum of the other two. Because of the nature of the geometry connected with this inequality, it is called the *triangle inequality*. The degenerate case occurs when $\arg z_1 = \arg z_2$, in which case $|z_1| + |z_2| = |z_1 + z_2|$. (See Figure 12.3.)

For each complex number $z = x + iy$, the corresponding number

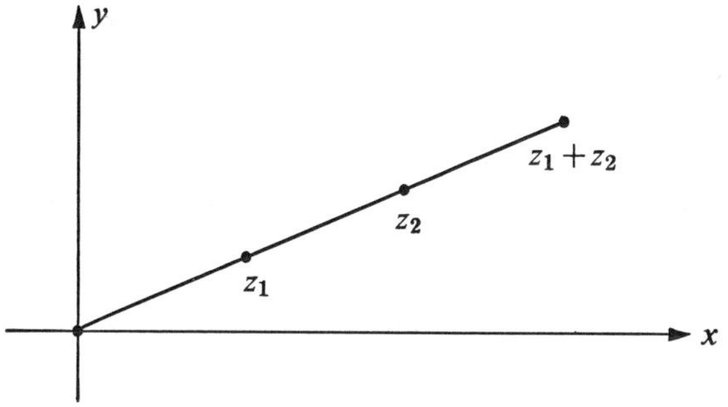

figure 12.3 *degenerate case of triangle inequality*

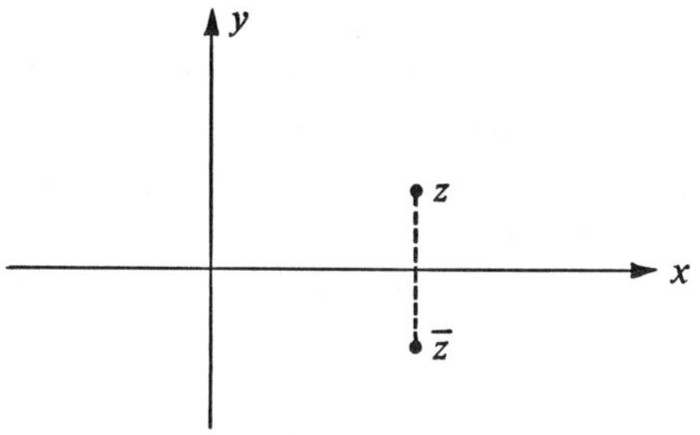

figure 12.4 *complex conjugate*

$x - iy$ is called the *conjugate* of z and is denoted by $\bar{z}$. Geometrically, $\bar{z}$ is the reflection of z in the real axis. Note that $z\bar{z} = (x + iy)(x - iy) = x^2 + y^2 = |z|^2$, an important and often used equality. (See Figure 12.4.)

Sets of points in the complex plane are sometimes represented by using magnitude and inequality notation. You should understand that $|z - a|$ is the distance to z from the point a.

example 12.4: Describe the sets of points $\{z \mid |z| = 1\}$, $\{z \mid |z-1| = |z-2|\}$, and $\{z \mid |z-i| < 2\}$.

solution: The equation $|z| = 1$ represents all points that are 1 unit away from the origin, which you recognize as a circle of radius 1 centered at the origin. The equation $|z - 1| = |z - 2|$ represents all points that are equidistant from the points (1, 0) and (2, 0). It is the perpendicular bisector of the line segment joining the two points. The inequality $|z - i| < 2$ represents those points which are less than 2 units away from the point (0, 1). It is the interior of the circle of radius 2 centered at (0, 1). (See Figure 12.5.)

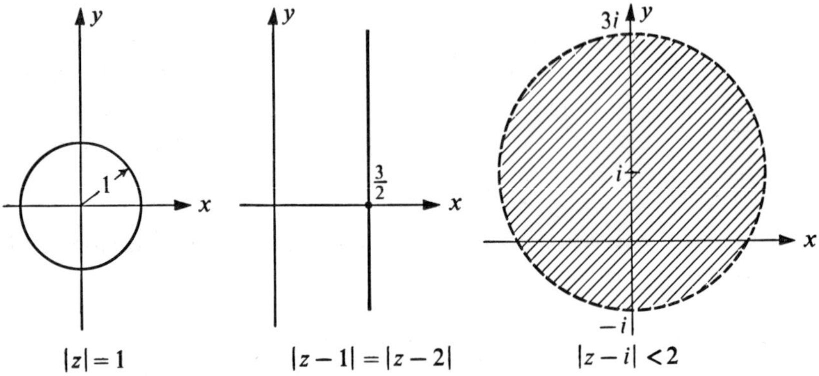

figure 12.5

A set S is *open* in the complex plane if every point of the set has some circular neighborhood which lies entirely in S. It is *closed* if its complement is open. A set is *bounded* if all its points lie within a circle of sufficiently large radius. An open set is *connected* if any two of its points can be joined by a broken line of finitely many linear segments all of whose points belong to the set. An open connected set is called a *domain.*

example 12.5: Describe the sets $\{z \mid |z-1| \geqq 1\}$, $\{z \mid 0 < |z| < 1\}$, and $\{z \mid |\text{Re}\, z| > 1\}$. (See Figure 12.6.)

solution: The first set is closed, since its complement $\{z \mid |z| < 1\}$ is open. It is unbounded and connected. The second set is open, is connected, and is bounded. The third set is open, unbounded, but not connected. Only the second set is a domain.

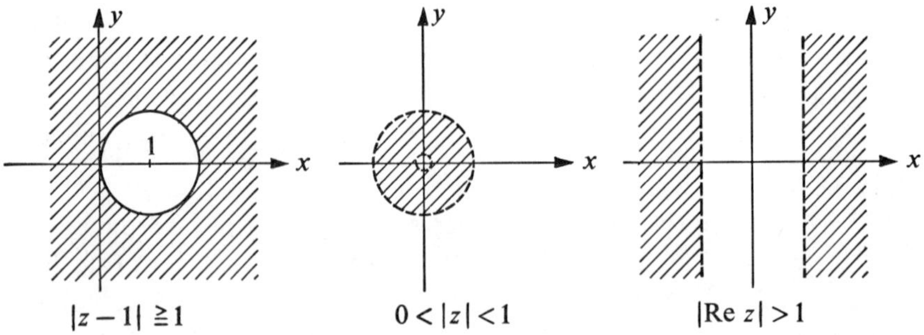

figure 12.6

12.3 functions and mappings

A complex function of a complex variable is a set of ordered pairs of complex numbers (z, w), where to each $z = x + iy$ in a set D of complex numbers (called the domain of f) there corresponds a unique $w = u + iv$. Functions are most often specified either by a rule in terms of z or a rule giving u and v in terms of x and y.

example 12.6: Let $f(z) = z^2$. Give the equivalent method of specifying this function. Repeat for $g(z) = 1/z$.

solution: $f(z) = z^2 = (x + iy)^2 = x^2 - y^2 + 2xyi$. Hence, the function may be written equivalently in terms of $u(x, y) = x^2 - y^2$ and $v = 2xy$. Also

$$g(z) = \frac{1}{z} = \frac{\bar{z}}{z\bar{z}} = \frac{x - iy}{x^2 + y^2}.$$

Hence

$$u = \frac{x}{x^2 + y^2}; \qquad v = \frac{-y}{x^2 + y^2}.$$

example 12.7: Let $w = u + iv = x^2 + y^2 + 2xyi$. Find $w = f(z)$ in terms of z and $\bar{z}$.

solution: We use the identities $x = (z + \bar{z})/2$ and $y = (z - \bar{z})/2i$ and substitute them into the given rule,

$$\begin{aligned} w &= \left(\frac{z + \bar{z}}{2}\right)^2 + \left(\frac{z - \bar{z}}{2i}\right)^2 + 2i\left(\frac{z + \bar{z}}{2}\right)\left(\frac{z - \bar{z}}{2i}\right) \\ &= z\bar{z} + \frac{z^2 - (\bar{z})^2}{2}. \end{aligned}$$

The geometric analog of a complex function of a complex variable is called a *mapping*. The domain of a complex mapping is a set of points in the z-plane and the range of values is a set of points in the w-plane.

We are sometimes asked to determine the image of point sets under complex mappings.

example 12.8: Let f be the mapping defined by $f(z) = z^2$. Find the image of the unit square $0 \leqq x \leqq 1$, $0 \leqq y \leqq 1$.

solution: Since $w = u + iv = x^2 - y^2 + i2xy$, it follows that $u(x, y) = x^2 - y^2$ and $v(x, y) = 2xy$.

The image of the line segment $y=0$, $0 \leqq x \leqq 1$, is given parametrically by

$$\left.\begin{cases} u(x,0) = x^2 \\ v(x,0) = 0 \end{cases}\right\} 0 \leqq x \leqq 1$$

which is the set of points on the u-axis between 0 and 1.
The image of the line segment $x=1$, $0 \leqq y \leqq 1$, is

$$\left.\begin{cases} u(1,y) = 1 - y^2 \\ v(1,y) = 2y \end{cases}\right\} 0 \leqq y \leqq 1$$

which is the portion of the parabola $u = 1 - (v^2/4)$, $0 \leqq v \leqq 2$.
The image of $y=1$, $0 \leqq x \leqq 1$, is

$$\left.\begin{cases} u(x,1) = x^2 - 1 \\ v(x,1) = 2x \end{cases}\right\} 0 \leqq x \leqq 1$$

which is the portion of the parabola $u = (v^2/4) - 1$, $0 \leqq v \leqq 2$.
The image of the segment $x=0$, $0 \leqq y \leqq 1$ is the set of points on the u-axis between -1 and 0. (See Figure 12.7.)

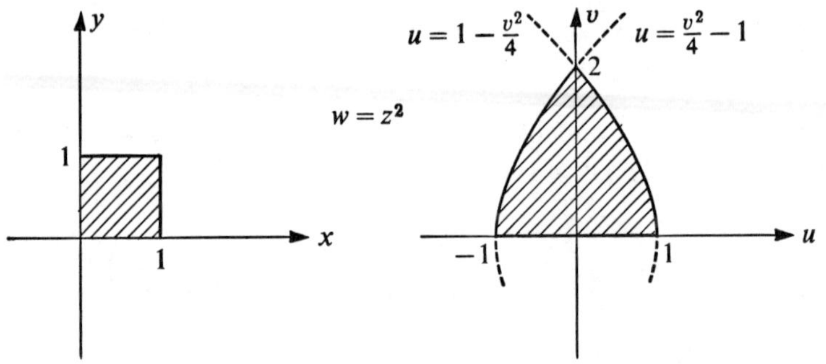

figure 12.7

example 12.9: Find the preimage of $|w| =$ a constant under the mapping $w = x \operatorname{cis} y$.

solution: Since $|w| = x$, the preimages of the circles centered at the origin in the w-plane are lines parallel to the imaginary axis. (See Figure 12.8.)

example 12.10: Find the image of the circle $|z| = 2$ under the mapping $w = 1/z$.

solution: Since $|w| = 1/|z|$, the circle $|w| = 1/2$ is the image of $|z| = 2$. (See Figure 12.9.) This problem may also be approached by finding the real and

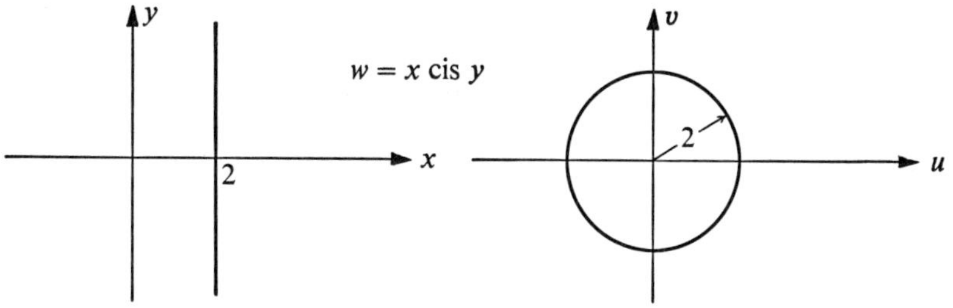

figure 12.8

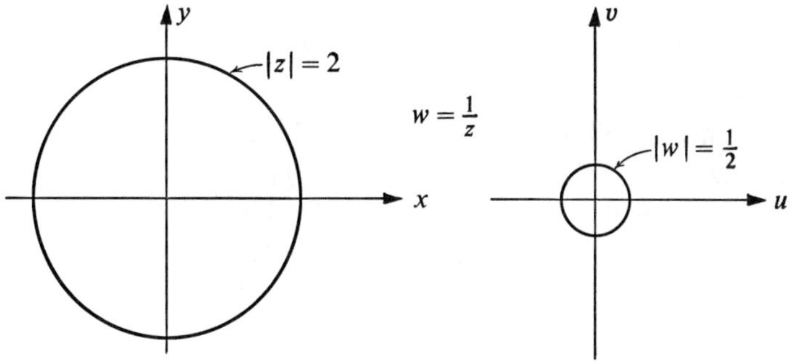

figure 12.9

imaginary parts of the mapping function. Thus

$$w = \frac{1}{z} = \frac{\bar{z}}{z\bar{z}} = \frac{\bar{z}}{|z|^2} = \frac{x - iy}{x^2 + y^2} = \frac{x - iy}{4}.$$

Therefore, $u = x/4$ and $v = -y/4$ from which

$$x^2 + y^2 = 16u^2 + 16v^2 = 4.$$

Hence, the image of the circle is

$$u^2 + v^2 = \tfrac{1}{4}$$

which is another way of writing $|w| = 1/2$.

exercises for sections 12.1–12.3

1. Find the value of each of the following expressions:
   (a) $(6 + 5i) \times (7 + 2i)$. (b) i^3.
   (c) $\left(5 \text{ cis } \frac{\pi}{6}\right)\left(2 \text{ cis } -\frac{\pi}{2}\right)$. (d) $\left(2 \text{ cis } \frac{\pi}{6}\right)^3$.

(e) $\left(2 \operatorname{cis} \frac{5\pi}{6}\right)^3$. (f) $(2+i)^2$.

(g) $(5-2i)^{-1}$. (h) $\operatorname{Re} z^2$.

(i) $|z|^2$. (j) $\frac{i}{(i-1)(i-2)(i-3)}$.

2. Write the expression for z in polar form if:
(a) $z = 1 - i\sqrt{3}$. (b) $z = 2i$. (c) $z = 5$.

3. Give a graphical description of what it means to multiply a complex number by i.

4. Show that if $z_1 z_2 = 0$, then either $z_1 = 0$ or $z_2 = 0$.

5. Show that:
(a) $z + \bar{z} = 2 \operatorname{Re} z$. (b) $z - \bar{z} = 2i \operatorname{Im} z$.

6. Describe the following sets by means of graphs:
(a) $\{z \mid |z-1| \leqq 1\}$. (b) $\{z \mid |z-1| = |z-2|\}$.
(c) $\{z \mid |z-3| \geqq 1\}$. (d) $\{z \mid \operatorname{Im} z = \operatorname{Re} z\}$.
(e) $\{z \mid |z-1| = 2\,|z-2|\}$. (f) $\{z \mid \pi/6 \leqq \arg z \leqq \pi/2\}$.
(g) $\{z \mid \arg z = 3\}$. (h) $\left\{z \,\middle|\, \left|\frac{1}{z}\right| \leqq 1\right\}$.
(i) $\{z \mid \operatorname{Re} z = 1/2\}$. (j) $\{z \mid 2 \leqq |z-1| \leqq 3\}$.

7. Find the images of the following regions under the given mapping:
(a) $\{z \mid y = x\}$, $w = f(z) = z + 1$.
(b) $\{z \mid y = x^2\}$, $w = f(z) = z^2$.
(c) $\{z \mid |z| < 2\}$, $w = f(z) = z^2$.
(d) $\left\{z \,\middle|\, |z| < 1,\ 0 < \arg z < \frac{\pi}{2}\right\}$; $f(z) = z^2$.
(e) $\{z \mid \operatorname{Re} z > 1\}$, $f(z) = 1/z$.

8. Find the images of the indicated region under the given mapping:
(a) $\{z \mid y = x\}$, $w = x^2 + iy^2$.
(b) $\{z \mid 0 \leqq x \leqq \pi/2,\ 0 \leqq y \leqq \pi/2\}$, $w = \sin x \cosh y + i \cos x \sinh y$.
(c) $\{z \mid 0 \leqq x \leqq \pi/2,\ 0 \leqq y \leqq \pi/2\}$, $w = e^x \cos y + i\, e^x \sin y$.
(d) $\{z \mid |z| < 2\}$, $w = x/(x^2+y^2) - i[y/(x^2+y^2)]$.
(e) $\{z \mid 0 < r < 3,\ 0 < \operatorname{Arg} z < \pi/3\}$, $w = \log_e |z| + i \operatorname{Arg} z$.

9. Show that the size of the angle of intersection of the curves $y = (x/\sqrt{3}) - 1$ and $y = x - \sqrt{3}$ is "preserved" by the mapping $w = z^2$, in the sense that the angle between the tangent lines at the point of intersection in the w-plane is the same as the angle between the lines in the z-plane.

10. Show that:
(a) $|\operatorname{Im} z| \leqq |z|$. (b) $|\operatorname{Re} z| \leqq |z|$.

11. Define what is meant by a complex function of a real variable. Give some examples.
12. Express as a function of z and $\bar{z}$:
    (a) $w = x^2 + y^2$. (b) $w = y + ix$.
    (c) $w = (x^3 - xy^2 - 2xy^2) + i(2x^2y + yx^2 - y^3)$.
13. Find $w = u + iv$ if the function in terms of z and $\bar{z}$ is:
    (a) $\bar{z}^2 + z$. (b) $z\bar{z} + 5z$. (c) $(z+1)/(\bar{z}^2+1)$.
14. Describe the set of points represented by $|az+b| = |cz+d|$. Specifically consider the cases $a = c$ and $a \neq c$.

12.4 derivative of a complex function

The definitions of limit, continuity, and derivative of a function of a complex variable have forms which correspond closely to those for functions of a real variable. You should be wary not to presume familiarity with the concepts since the interpretation of the definition is different in the complex plane.

example 12.11: Show that $\lim\limits_{z \to 2} z^2 = 4$.

solution: We must show there exists a δ such that $|z^2 - 4|$ is less than any arbitrary ε if $|z-2| < \delta$. If we first restrict z such that $|z-2| < 1$, then

$$|z^2 - 4| = |(z-2)(z+2)| = |z-2|\,|z+2| < 5|z-2| < 5\delta$$

which will be less than ε if $\delta < \varepsilon/5$. Therefore, choose $\delta < 1$ or $< \varepsilon/5$, whichever is smaller. (See Figure 12.10.)

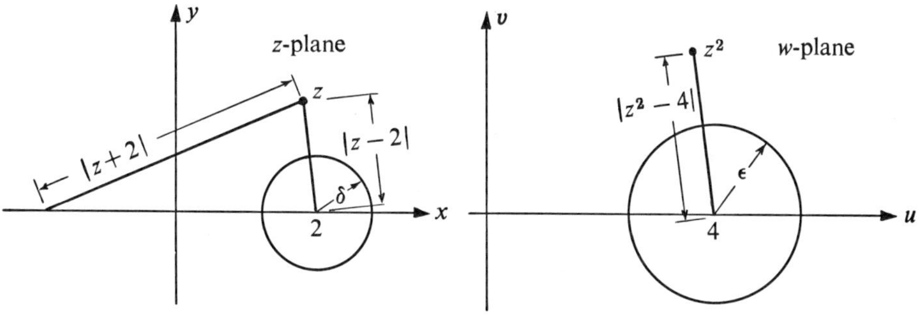

figure 12.10

To show that $\lim_{z \to z_0} f(z)$ does not exist, it is sufficient to show that the value of the limit is dependent upon the manner in which z_0 is approached.

example 12.12: Show that $\lim_{z \to 0} \operatorname{Re} z/z$ does not exist.

solution: If z approaches 0 along the positive y-axis, then $z = y$, and $\operatorname{Re} z = 0$. Hence,

$$\lim_{\substack{z \to 0 \\ x=0}} \frac{\operatorname{Re} z}{z} = \lim_{y \to 0} \frac{0}{y} = 0 .$$

If the approach is along the positive x-axis, then $z = x$, $\operatorname{Re} z = x$, and

$$\lim_{\substack{z \to 0 \\ y=0}} \frac{\operatorname{Re} z}{z} = \lim_{x \to 0} \frac{x}{x} = 1 .$$

Since the same definition is used for the derivative of a function of a complex variable as for a function of a real variable, the same *general* formulas and theorems are true. In fact, any theorem involving differentiation which uses only the abstract definition will be true in complex analysis also. In particular:

1. The derivative of a sum is the sum of the derivatives.
2. The derivative of a constant times a function is the constant times the derivative of the function.
3. The product and quotient rules for differentiation are valid.
4. The derivative of a composite function, $w(u(z))$ is given by the chain rule,

$$\frac{dw}{dz} = \frac{dw}{du}\frac{du}{dz}$$

where w and u are both functions which have derivatives.
5. The power rule is the same,

$$\frac{dz^n}{dz} = nz^{n-1} .$$

These five general rules are sufficient to enable you to differentiate any rational function of a complex variable. The techniques are the same as in elementary calculus.

The rules given above do not apply to powers of the conjugate of z

since, as we will now show, the function $f(z) = \bar{z}$ does not have a derivative anywhere.

example 12.13: Show that $f(z) = \bar{z}$ does not have a derivative anywhere.

solution: We must show that the limit of the difference quotient does not exist.

$$\begin{aligned}\lim_{\Delta z \to 0} \frac{f(z + \Delta z) - f(z)}{\Delta z} &= \lim_{\Delta z \to 0} \frac{(x + \Delta x) - i(y + \Delta y) - (x - iy)}{\Delta x + i\Delta y} \\ &= \lim_{\Delta z \to 0} \frac{\Delta x - i\Delta y}{\Delta x + i\Delta y} \\ &= \lim_{\Delta z \to 0} \operatorname{cis}(-2 \arg \Delta z),\end{aligned}$$

whose value is dependent upon the path of approach of Δz to 0. For example, the value for the approach through values of $\Delta z = \Delta x$ is 1 and if $\Delta z = i\,\Delta y$, and $\Delta y \to 0$, the limit is -1. Hence, the limit of the difference quotient does not exist for any value of z.

When a complex function is defined in terms of two real functions, we will usually want to differentiate it directly, rather than to re-express it in terms of the variable z. The following theorem not only gives the formula for this case, but also yields a most interesting and famous set of conditions for the derivative of a function of a complex variable to exist.

theorem 12.1

Let $f(z) = u(x, y) + iv(x, y)$ and suppose $f'(z)$ exists at $z_0 = x_0 + iy_0$. Then

$$\begin{aligned}f'(z_0) &= u_x(x_0, y_0) + iv_x(x_0, y_0) \\ &= v_y(x_0, y_0) - iu_y(x_0, y_0).\end{aligned}$$

proof: We approach z_0 in two different ways, one horizontal ($\Delta y = 0$) and one vertical ($\Delta x = 0$). (See Figure 12.11.) Since $f'(z_0)$ exists, the limit of the difference quotient is the same in both cases. For the first approach,

$$\begin{aligned}&\lim_{\substack{\Delta z \to 0 \\ \Delta y = 0}} \frac{f(z_0 + \Delta z) - f(z_0)}{\Delta z} \\ &\quad= \lim_{\Delta x \to 0} \frac{u(x_0 + \Delta x, y_0) - u(x_0, y_0) + iv(x_0 + \Delta x, y_0) - iv(x_0, y_0)}{\Delta x} \\ &\quad= \lim_{\Delta x \to 0} \frac{u(x_0 + \Delta x, y_0) - u(x_0, y_0)}{\Delta x} + i \lim_{\Delta x \to 0} \frac{v(x_0 + \Delta x, y_0) - v(x_0, y_0)}{\Delta x} \\ &\quad= u_x(x_0, y_0) + iv_x(x_0, y_0).\end{aligned}$$

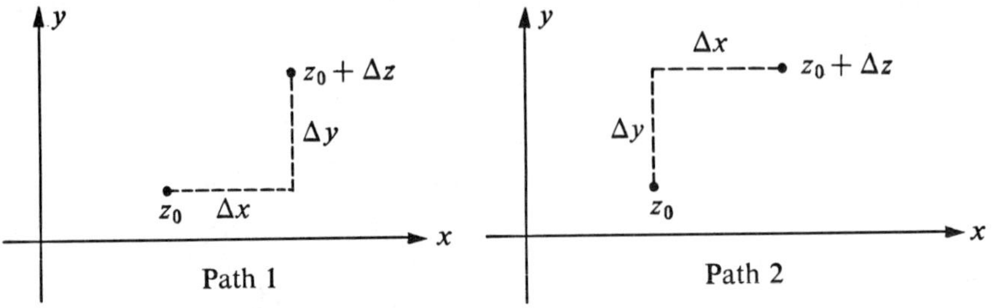

figure 12.11

For the other approach,

$$\begin{aligned} f'(z_0) &= \lim_{\Delta y \to 0} \frac{u(x_0, y_0 + \Delta y) - u(x_0, y_0) + iv(x_0, y_0 + \Delta y) - iv(x_0, y_0)}{i\,\Delta y} \\ &= \lim_{\Delta y \to 0} \frac{u(x_0, y_0 + \Delta y) - u(x_0, y_0)}{i\,\Delta y} + i \lim_{\Delta y \to 0} \frac{v(x_0, y_0 + \Delta y) - v(x_0, y_0)}{i\,\Delta y} \\ &= \frac{1}{i} u_y + \frac{i}{i} v_y \\ &= v_y(x_0, y_0) - iu_y(x_0, y_0). \end{aligned}$$

corollary: Let $f(z) = u + iv$ and suppose $f'(z_0)$ exists. Then at (x_0, y_0),

$$u_x = v_y; \qquad v_x = -u_y.$$

The two partial differential equations of the corollary are called the *Cauchy-Riemann* conditions. We have shown them to be conditions necessary for the existence of $f'(z_0)$, but it is also true that these conditions are sufficient, *if* the partial derivatives of u and v are assumed to be continuous. This is a remarkably easy test for the existence of f' at a point.

example 12.14: Let $f(z) = e^x \cos y + i\,e^x \sin y$. Where does $f'(z)$ exist?

solution:

$$u_x = e^x \cos y = v_y \qquad \text{and} \qquad v_x = e^x \sin y = -u_y.$$

These equations hold everywhere in the complex plane and therefore f' exists everywhere. Moreover, $f'(z) = f(z)$.

example 12.15: Let

$$g(z) = \frac{x^3}{3} + i\left(y - \frac{y^3}{3}\right).$$

Where does $g'(z)$ exist?

solution: $u_x = x^2$, $v_y = 1 - y^2$, $u_y = 0$, $v_x = 0$. The Cauchy-Riemann conditions are satisfied by those points for which $x^2 = 1 - y^2$, and no others. Therefore $g'(z)$ exists for all points on the unit circle.

example 12.16: Let $h(z) = x^2 + y^2$. Where does $h'(z)$ exist?

solution: $u_x = 2x$, $u_y = 2y$, $v_x = 0$, $v_y = 0$. The derivative exists where $2x = 0$ and $2y = 0$, which is at the origin only.

The Cauchy-Riemann equations are sufficient for a function to have a derivative at a point only if the four partial derivatives are continuous. The following example is intended to exhibit this fact.

example 12.17: Let

$$f(z) = \frac{x^3 - y^3}{x^2 + y^2} + i\,\frac{x^3 + y^3}{x^2 + y^2}, \qquad z \neq 0$$

$$f(0) = 0.$$

Show that the derivative of f does not exist at $z = 0$ even though the Cauchy-Riemann conditions hold at that point.

solution: It is easy to show that at $z = 0$, $u_x = 1 = v_y$, $u_y = -1 = -v_x$. (Use the definition of the partial derivative to verify this fact.) Thus, if the derivative at $z = 0$ exists, the limit of the difference quotient must be $1 + i$ regardless of the path used to approach 0. There are many paths for which this is not so. For example, consider the path $x = 2y$. On this path,

$$\frac{f(0 + \Delta z) - f(0)}{\Delta z} = \frac{\dfrac{(\Delta x)^3 - (\Delta y)^3}{(\Delta x)^2 + (\Delta y)^2} + i\,\dfrac{(\Delta x)^3 + (\Delta y)^3}{(\Delta x)^2 + (\Delta y)^2}}{\Delta x + i\,\Delta y}$$

$$= \frac{\dfrac{7(\Delta y)^3}{5(\Delta y)^2} + i\,\dfrac{9(\Delta y)^3}{5(\Delta y)^2}}{(\Delta y)(2 + i)}$$

$$= \frac{7 + 9i}{5(2 + i)}$$

$$= \frac{23 + 11i}{25}.$$

Hence, the limit of the difference quotient is dependent on the method of approach to zero and the derivative at $z = 0$ does not exist.

The existence of f' at isolated points is of little practical importance. We are usually interested in those points for which f' exists not only at a point but in some neighborhood about the point. We formalize this in the following definition.

definition 12.2

A function f is said to be *analytic* at the point z_0 if f' exists at z_0 and at every point in some neighborhood of z_0. If f is analytic at each point in a set D, then it is said to be *analytic on the set.*
If f is analytic for all points in a neighborhood of a but not at a, then the point a is called an isolated *singularity* of the function. There is no requirement that the function exist at a singular point.

The function $f(z)=1/(z-1)$ has a singularity at $z=1$, since it is obviously nonanalytic at the point, but $f'(z)$ does exist everywhere else.

To determine if a complex function is analytic at a point, it is insufficient to show the existence of the derivative there. For, in addition to f' existing at the point, it must also exist in some neighborhood about the point. To determine where a function is analytic, we first locate the points where f' exists, after which the points of analyticity are easily determined.

example 12.18: Determine where the functions of Examples 12.14–12.16 are analytic.

solution: Since $f'(z)$ exists everywhere, $f(z)$ is analytic everywhere. The function $g(z)$ is analytic nowhere since $g'(z)$ exists only on the unit circle. At no point where it exists does it have a neighborhood where $g'(z)$ exists. The function $h(z)$ is analytic nowhere since $h'(z)$ exists only at the origin.

The seemingly small requirement of demanding the existence of the derivative in a neighborhood rather than just at a point has much more strength than one might realize. Much of the study of complex variable theory is limited to the study of analytic functions.

One immediate property of analytic functions concerns the real and imaginary parts.

theorem 12.2

Both the real and imaginary parts of an analytic function are solutions to Laplace's equation.

proof: See Exercise 17 in the next set of exercises.

A solution to Laplace's equation which has continuous second-order partial derivatives is called a *harmonic function* (see Section 7.5). If $u+iv$ is analytic, then u and v are called *conjugate* harmonic functions. A conjugate of a given harmonic function may be found easily using the Cauchy-Riemann equations.

example 12.19: Determine an analytic function whose real part is $u(x, y) = x^2 - y^2$.

solution: Let the analytic function be $f(z) = x^2 - y^2 + iv(x, y)$. From the Cauchy-Riemann equations, $v_x = 2y$ and $v_y = 2x$, from which $v(x, y) = 2xy$ (+a constant). Therefore,

$$f(z) = x^2 - y^2 + i2xy.$$

exercises for section 12.4

1. Show that $\lim_{z \to 1+i} z^2 = 2i$.
2. Show that

$$f(z) = \frac{\operatorname{Im} z^2}{|z|^2}, \quad z \neq 0,$$
$$= 0, \quad z = 0$$

is discontinuous at the origin.

3. Prove the quotient rule for the derivative of a function of a complex variable which can be expressed as the ratio of two other differentiable functions.
4. Find $f'(z)$ if:

(a) $f(z) = z^3 + z$. (b) $f(z) = \dfrac{z^2 + 2z}{z^4 - 1}$. (c) $f(z) = \dfrac{1}{z}$.

(d) $f(z) = \dfrac{az + b}{cz + d}$. (e) $f(z) = (z^{15} + 3z^{10} + 17)^{18}$.

5. Find dw/dz if $w =$ the following:
   (a) $\sin x \cos y + i \cos x \sinh y$.
   (b) $\log_e |z| + i \operatorname{Arg} z$.
   (c) $\cos x \cosh y - i \sin x \sinh y$.
6. Where (if anywhere) are the following functions analytic?
   (a) $f(z) = |z|^2$. (b) $f(z) = x^2 + iy^2$. (c) $f(z) = (x - iy)^2$.
   (d) $f(z) = z\bar{z}$. (e) $f(z) = \operatorname{Arg} z$. (f) $f(z) = \dfrac{z+4}{z-i}$.
   (g) $f(z) = x^3 + 3y + i(y^3 - 3x)$.

7. Show that $f(z) = \log_e |z| + i \arg z$ is discontinuous along the negative real axis, if $\arg z$ is limited to its principal values.

8. Show that $f(z) = |z|^{1/2} \operatorname{cis}(\arg z/2)$ is discontinuous along the negative real axis if $\arg z$ is limited to its principal values.

9. Find an analytic function whose real part is:
   (a) $e^x \cos y$. (b) $x^3 - 3xy^2$. (c) y.
   (d) xy. (e) $\operatorname{Arg} z$. (f) $\log_e |z|$.

10. Explain clearly the difference between a function being analytic at a point and having a derivative there.

11. Show that each of the following functions is harmonic and find a complex analytic function $f(z) = u + iv$:
    (a) $u = x$. (b) $v = x$. (c) $u = e^x \sin y$.
    (d) $v = \log_e |z|$. (e) $u = \dfrac{x}{x^2 + y^2}$. (f) $u = \sin x \cosh y$.
    (g) $u = xy + x + y$.

12. Show that the conjugate harmonic of a function is unique up to an arbitrary constant.

13. Find the most general harmonic function of the form of a second-degree polynomial in x and y which has no linear or constant terms.

14. Show that if $f(z)$ is analytic with constant modulus, then $f(z)$ is a constant.

15. Prove the following statement. If we let $f(z)$ be analytic in a domain, D, then $f'(z)$ vanishes everywhere in D if and only if $f(z)$ is a constant in that domain.

16. Sketch the families of curves

$$\begin{Bmatrix} u = u(x, c) \\ v = v(x, c) \end{Bmatrix} \quad \text{and} \quad \begin{Bmatrix} u = u(k, y) \\ v = v(k, y) \end{Bmatrix}$$

    and in each case show that they are orthogonal if:
    (a) $f(z) = z$.
    (b) $f(z) = \sin x \cosh y + i \cos x \sinh y$.
    (c) $f(z) = e^x \cos y + i\, e^x \sin y$.

17. Prove Theorem 12.2.

12.5 integral and fractional powers of z

In this and the following sections we will introduce some of the more important complex functions. In most cases, if we can, we will call the complex function by the same name as that to which the function reduces when z is real. Do not be misled by this into believing that you are familiar with the particular function of a complex variable just because you know its properties when z is real. Some properties of these functions are significantly different from their real counterparts.

1. Integral Powers of z. This class of functions is denoted by $w = f(z) = z^n$, n a nonnegative integer. A function of this type is analytic in the entire z-plane. If we write $z = r \operatorname{cis} \theta$ and $w = R \operatorname{cis} \phi$, then the expression for this function becomes

$$R \operatorname{cis} \phi = r^n \operatorname{cis} n\theta$$

or

$$\left\{\begin{matrix} R = r^n \\ \phi = n\theta \end{matrix}\right\}.$$

Hence the mapping defined by $w = z^n$ maps the sector $0 \leqq \theta \leqq 2\pi/n$ onto the entire w-plane. (See Figure 12.12.)

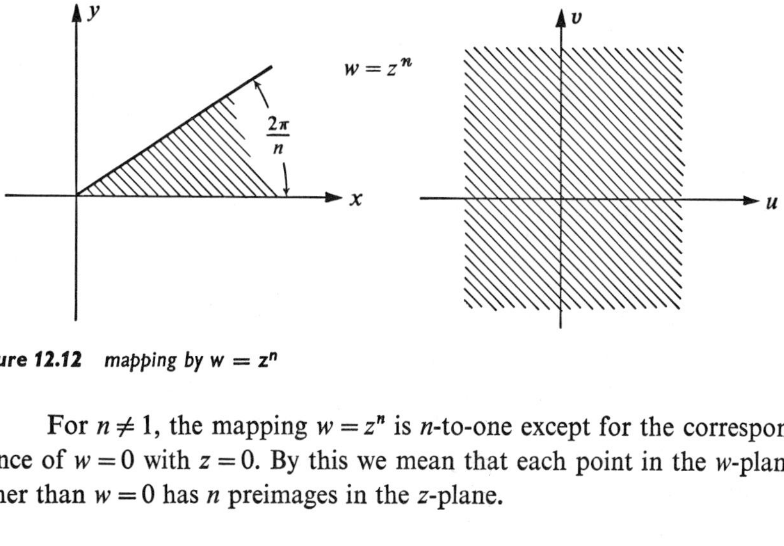

figure 12.12 *mapping by* $w = z^n$

For $n \neq 1$, the mapping $w = z^n$ is n-to-one except for the correspondence of $w = 0$ with $z = 0$. By this we mean that each point in the w-plane other than $w = 0$ has n preimages in the z-plane.

example 12.20: Find the preimages of the point

$$w = 32\left(\frac{1 + i}{\sqrt{2}}\right)$$

under the mapping $w = z^5$.

solution: We look for values of r and θ which satisfy the equation $r^5 \operatorname{cis} 5\theta = 32 \operatorname{cis} \pi/4$. Hence $r^5 = 32$ and $5\theta = \pi/4 + 2k\pi$, from which $r = 2$ and $\theta = \pi/20$ (+ integral multiples of $2\pi/5$). The distinct points which these equations give are

$$2\,\text{cis}\left(\frac{\pi}{20}+\frac{2\pi}{5}\right),\quad 2\,\text{cis}\left(\frac{\pi}{20}+\frac{4\pi}{5}\right),\quad 2\,\text{cis}\left(\frac{\pi}{20}+\frac{6\pi}{5}\right),$$
$$2\,\text{cis}\left(\frac{\pi}{20}+\frac{8\pi}{5}\right),\quad\text{and}\quad 2\,\text{cis}\left(\frac{\pi}{20}\right).$$

Figure 12.13 shows the points corresponding to these values. They are located on the circle of radius 2 at equally spaced intervals of $2\pi/5$.

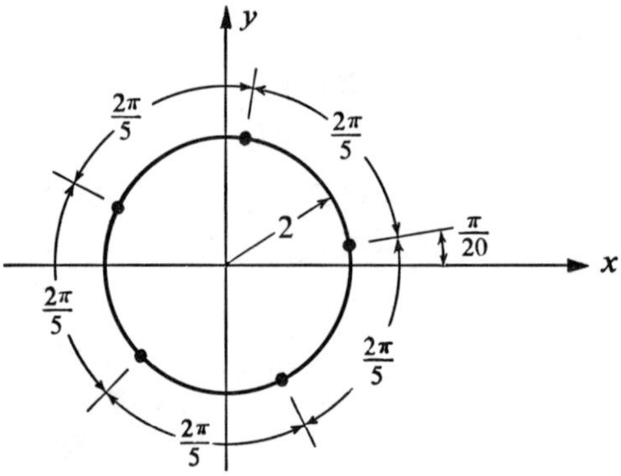

figure 12.13

2. Fractional Powers of z. This relation, denoted by $w=z^{1/n}$, is defined by the equation $w^n=z$. Classically, it has been called a "multi-valued" function, but the more contemporary approach is to agree that n *branch functions* are defined by the equation. Each of the branch functions is defined by limiting $\arg z$ to some interval of 2π. The rays which emanate from the origin at those intervals separate the domains of the branch functions and are called *branch cuts.* Different choices of branch cuts will result in different branch functions. The branch cuts all meet at the same point (the origin), called a *branch point* of the function.

example 12.21: Show two ways of defining branch functions from the relation $w=z^{1/3}$.

solution: By placing branch cuts at $\arg z=0$, 2π, 4π, and 6π we obtain the three distinct functions,

$$\begin{aligned} f_1(z)&=z^{1/3}, & 0&\leqq\theta<2\pi;\\ f_2(z)&=z^{1/3}, & 2\pi&\leqq\theta<4\pi;\\ f_3(z)&=z^{1/3}, & 4\pi&\leqq\theta<6\pi. \end{aligned}$$

By placing branch cuts at $-\pi, \pi, 3\pi$, and 5π we obtain the three functions

$$g_1(z) = z^{1/3}, \qquad -\pi \leqq \theta < \pi;$$
$$g_2(z) = z^{1/3}, \qquad \pi \leqq \theta < 3\pi;$$
$$g_3(z) = z^{1/3}, \qquad 3\pi \leqq \theta < 5\pi.$$

In each of the cases, the range values of the three branch functions taken together completely "cover" the w-plane, although in a somewhat different arrangement, as seen in Figure 12.14.

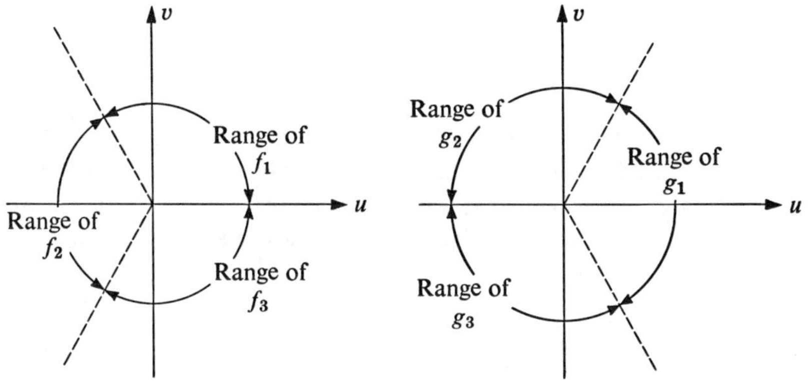

figure 12.14

An extension of $\arg z$ to higher values gives a repetition of the functional values from the initial three functions.

example 12.22: Evaluate $(-i)^{1/3}$ for each of branch functions of the previous example. Display graphically.

solution:

$$f_1(-i) = f_1\left(1 \text{ cis } \frac{3\pi}{2}\right) = 1 \text{ cis } \frac{\pi}{2} = g_2(-i).$$

$$f_2(-i) = f_2\left(1 \text{ cis } \frac{7\pi}{2}\right) = 1 \text{ cis } \frac{7\pi}{6} = -\frac{\sqrt{3}}{2} - i\frac{1}{2} = g_3(-i).$$

$$f_3(-i) = f_3\left(1 \text{ cis } \frac{11\pi}{2}\right) = 1 \text{ cis } \frac{11\pi}{6} = \frac{\sqrt{3}}{2} - i\frac{1}{2} = g_1(-i).$$

(See Figure 12.15.)

In all cases, the n-values of $z^{1/n}$ will form the same set regardless of the method of defining branch functions. From a practical standpoint, if you are asked to evaluate $z^{1/n}$ for a particular value of z_0, you are expected to

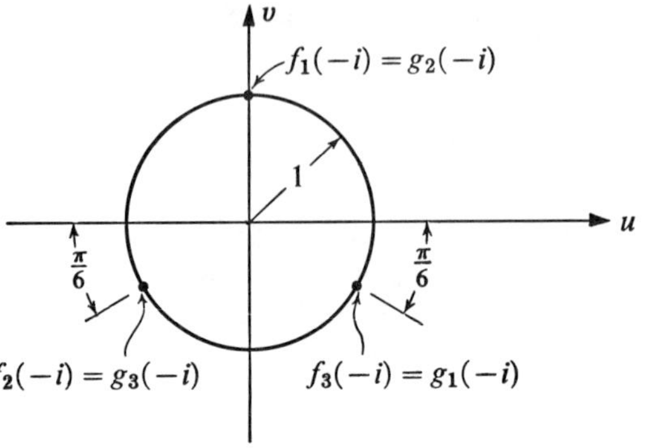

figure 12.15

give all n distinct values. Graphically, the n-values are equally spaced at intervals of $2\pi/n$ radians on a circle of radius $R = |z_0|^{1/n}$.

Each of the n branch functions maps the complete z-plane onto a sector of angular measurement $2\pi/n$ radians in the w-plane. (See Figure 12.16.)

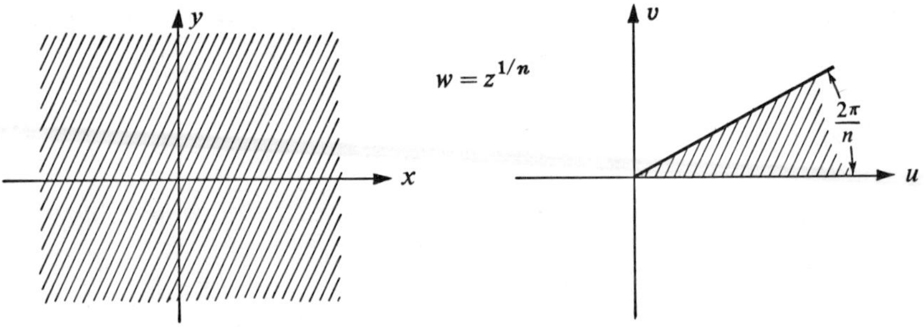

figure 12.16 *mapping by* $w = z^{1/n}$

In effect, we sometimes think of the z-plane as if overlaid by n layers, one layer for the domain of each branch function. Moreover, these layers should be connected in a helicoidal arrangement such that as we turn around the origin at each revolution we move up one layer. However, after n revolutions the connection should be such (and this is physically impossible) that the top layer continues into the lower layer. All of the planes together represent the complete domain of the function $w = z^{1/n}$. Such a graphical representation of a domain (or a range) is called an *n-sheeted Riemann surface*. (See Figure 12.17.) With the assistance of Riemann surfaces, the maps of functions which are not one-to-one may be interpreted as being a one-to-one pairing of points. There is an obvious correspondence between points on the w-plane and on the n-sheets.

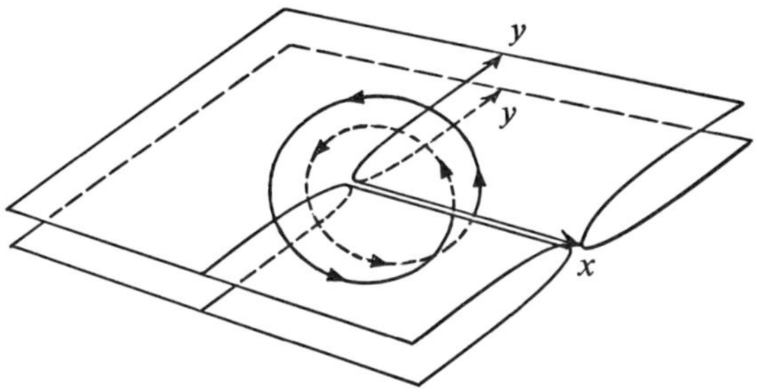

figure 12.17 two-sheeted Riemann surface

12.6 the exponential function

In defining the complex exponential function, we ask that it satisfy certain familiar properties, which you usually tend to associate with its real counterpart. We denote the complex exponential by $\exp z$ in order to make a clear distinction between the new complex function being defined and the real function e^x (the two notations are often used interchangeably), which is well known from elementary calculus. We ask that:

1. $\exp z$ be analytic for all z.
2. $\frac{d}{dz}(\exp z) = \exp z$.
3. $\exp z$ reduce to the real exponential function when $\operatorname{Im} z = 0$.

The idea of forcing a definition to satisfy some well-known properties is fairly standard practice and should remove the apparent randomness of the definition. It goes almost without saying that other approaches are possible.

Let $\exp z = u + iv$. By using the three properties above, we will obtain expressions for u and v. First, using property 1 and the Cauchy-Riemann conditions,

$$(u + iv) = (u + iv)' = u_x + iv_x = v_y - iu_y.$$

Differentiating again,

$$(u + iv) = (u + iv)'' = u_{xx} + iv_{xx} = - u_{yy} - iv_{yy}$$

from which

$$u = u_x \qquad \text{and} \qquad u = - u_{yy}.$$

The first of these equations has the solution

$$u = g(y)\, e^x.$$

Because of the second equation, the function g must satisfy

$$g'' + g = 0,$$

an ordinary differential equation whose solution is

$$g(y) = A\cos y + B\sin y.$$

Therefore,

$$u(x, y) = e^x (A\cos y + B\sin y).$$

Since $v = -u_y$,

$$v(x, y) = e^x (A\sin y - B\cos y).$$

When $z = x + i0$, $\exp z = e^x + i0$. Thus,

$$e^x = e^x A \quad \text{and} \quad 0 = e^x(-B)$$

from which $A = 1$ and $B = 0$. Therefore, the complex exponential function is defined to be

$$\begin{aligned}\exp z &= e^x \cos y + i\, e^x \sin y \\ &= e^x \operatorname{cis} y.\end{aligned}$$

In polar form,

$$\exp z = R \operatorname{cis} \phi = e^x \operatorname{cis} y$$

or,

$$\begin{aligned}R &= |\exp z| = e^x, \\ \phi &= \arg \exp z = y.\end{aligned}$$

Note that $\exp i\theta = \operatorname{cis}\theta$, a fact which often motivates the writing of a complex number z with polar coordinates (r, θ) in the form

$$z = r \exp i\theta.$$

From the relation $\exp i\theta = \operatorname{cis}\theta$, the so-called Euler formulas for the real trigonometric functions in terms of the complex exponential are derived,

$$\cos\theta = \frac{\exp i\theta + \exp(-i\theta)}{2},$$

$$\sin\theta = \frac{\exp i\theta - \exp(-i\theta)}{2i}.$$

The formula $\exp(z_1 + z_2) = (\exp z_1)(\exp z_2)$ is true, although it could really not be anticipated considering that this formula was not used as a hypothesis motivating our definition. The proof of the formula is a straightforward use of the definitions, a few trigonometric identities, and a few real exponential identities.

The complex exponential has some unfamiliar properties too, in the sense that the real exponential behaves differently. One example, which may surprise you, is that of the periodicity of $\exp z$. If the argument of z is increased by $2\pi i$, then

$$\begin{aligned}\exp(z+2\pi i) &= e^x \operatorname{cis}(y+2\pi)\\ &= e^x \operatorname{cis} y\\ &= \exp z\,.\end{aligned}$$

Therefore, $\exp z$ is periodic with *imaginary* period $2\pi i$.

The periodicity implies that the mapping defined by $w = \exp z$ is many-to-one. The function will assume all its values in any horizontal strip of width 2π. In particular, the strip defined by $-\pi < y \leqq \pi$ is called the *fundamental region* of $\exp z$. This choice is consistent with that made for the principal values of $\arg z$. A different choice of principal values of the argument will result in a different fundamental region for $\exp z$.

Since $|w| = e^x$, and $\arg w = y$, the vertical lines $x = c$ are mapped onto circles about the origin and horizontal lines are mapped onto rays through the origin. The region to the left of a vertical line, $x = c$, maps into the interior of the circle of radius e^c. (See Figure 12.18.) The origin in the w-plane has no preimage; that is, $\exp z$ is never 0.

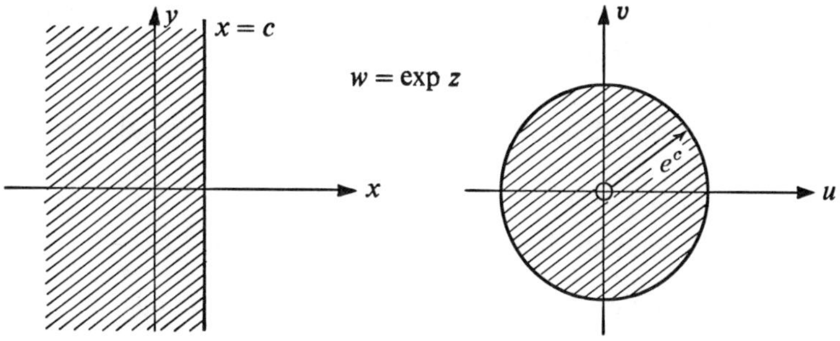

figure 12.18 *mapping by w = exp z*

example 12.23: Find the image of the region $x \geqq 0$, $0 \leqq y \leqq \pi$ under the mapping defined by $w = \exp z$.

solution: Since $w = R \operatorname{cis} \phi = e^x \operatorname{cis} y$, then $R = e^x$ and $\phi = y$. Therefore, if $x \geqq 0$, $R \geqq 1$ and if $0 \leqq y \leqq \pi$, then $0 \leqq \phi \leqq \pi$. The region in question is shown in Figure 12.19.

example 12.24: Find the image of the rectangular region $a \leqq x \leqq b$, $c \leqq y \leqq d$.

solution: The lines $x = a$ and $x = b$ are mapped onto the circles $R = e^a$ and

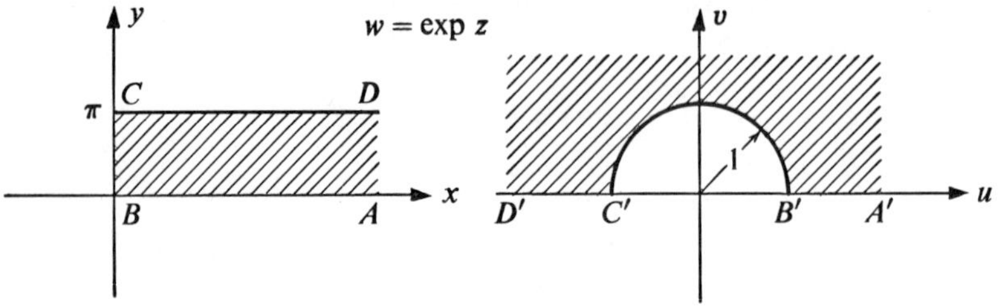

figure 12.19

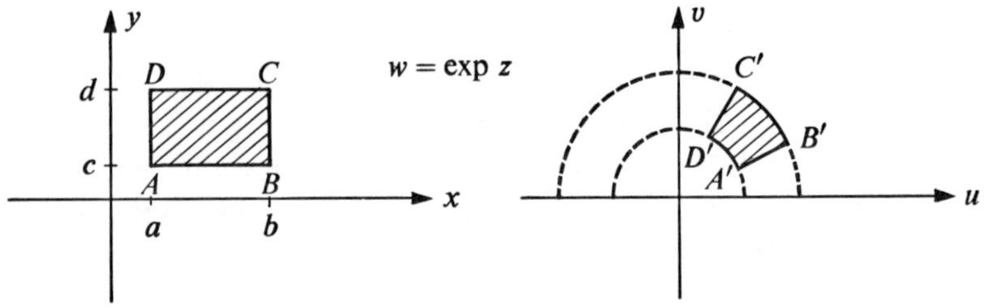

figure 12.20

$R = e^b$. The lines $y = c$ and $y = d$ are mapped onto the rays $\phi = c$ and $\phi = d$. The region is shown in Figure 12.20.

Earlier, we said that $\exp z$ is never 0. However, it does take on negative values.

example 12.25: Find all values of z such that $\exp z = -1$.

solution: The equation $\exp z = -1$ yields two real equations,

$$\left\{\begin{matrix} e^x \cos y = -1 \\ e^x \sin y = 0 \end{matrix}\right\}.$$

The second of these is satisfied only if $y = n\pi$, n an integer, for which the first equation becomes $e^x(-1)^n = -1$. This is true only if $x = 0$ and n is an odd integer. Therefore, $\exp z = -1$ for $z = 0 + (2n+1)\pi i$.

exercises for sections 12.5 and 12.6

1. Compute and display graphically:
   (a) $(-1)^{1/4}$. (b) $(8i)^{1/3}$. (c) $(-32)^{1/5}$. (d) $(-4i)^{1/2}$.
2. Solve the following equations and display the roots graphically:
   (a) $z^4 + 256 = 0$. (b) $z^2 - 6z + 12 = 0$. (c) $z^8 + 15z^4 = 16$.

3. Define two branch functions for $f(z) = z^{1/2}$. Where is the branch cut? Show that each of the branch functions is discontinuous at its branch cut.
4. Define branch functions for $w = z^{1/5}$ such that the branch cut is along the positive imaginary axis. How many sheets does the Riemann surface of this relation have?
5. Show that $\exp(z_1 + z_2) = (\exp z_1)(\exp z_2)$.
6. Find the image of the following under the exponential map $w = e^z$:
   (a) The line $x = 2$. (b) The line $y = 2$.
   (c) The line $x = y$. (d) The region $x \leqq 0$.
7. Solve the equations:
   (a) $\exp z = -2$. (b) $\exp z = 0$. (c) $\exp z = i$. (d) $\exp z = 1$.
8. Find the preimage of the interior of the unit circle under the mapping $w = e^z$.
9. Show that if $f(z) = \exp z$, then $f(\bar{z}) = \overline{f(z)}$.
10. Show that the exponential function is bounded along the imaginary axis.

12.7 complex trigonometric functions

The Euler formulas of the previous section motivate the following definitions for functions which we denote by $\sin z$ and $\cos z$:

$$\begin{cases} \sin z = \dfrac{\exp(iz) - \exp(-iz)}{2i} \\[2ex] \cos z = \dfrac{\exp(iz) + \exp(-iz)}{2}. \end{cases}$$

Since the exponential function is analytic everywhere, so are $\sin z$ and $\cos z$. It is an easy application of the definitions to show that

$$\frac{d}{dz}(\sin z) = \cos z \quad \text{and} \quad \frac{d}{dz}(\cos z) = -\sin z,$$

a result which should surprise you since, at least superficially, the functions $\sin z$ and $\cos z$ have little relation to the corresponding real functions. The same identities which hold for the real trigonometric functions are also true for their complex counterparts but *must* be proven independently. Once the three identities

$$\sin^2 z + \cos^2 z = 1,$$

$$\sin(z_1 \pm z_2) = \sin z_1 \cos z_2 \pm \sin z_2 \cos z_1$$

and

$$\cos(z_1 \pm z_2) = \cos z_1 \cos z_2 \mp \sin z_1 \sin z_2$$

are established, then most of the others will follow by techniques common to elementary trigonometry.

example 12.26: Find $\sin(iy)$ and $\cos(iy)$ where y is real.

solution:

$$\sin(iy) = \frac{\exp i(iy) - \exp - i(iy)}{2i} = - i\,\frac{e^{-y} - e^{y}}{2} = i \sinh y.$$

$$\cos(iy) = \frac{\exp i(iy) + \exp - i(iy)}{2} = \frac{e^{-y} + e^{y}}{2} = \cosh y.$$

This example shows that the complex trigonometric functions are unbounded.

Both the sine and cosine functions are periodic with real period 2π.

To interpret the mapping $w = \sin z$, it is convenient to have the real and imaginary parts of the function expressed in terms of real functions. To do this, we use the identity for the sine of the sum of two complex numbers,

$$\sin z = \sin(x + iy) = \sin x \cos iy + \cos x \sin iy.$$

Using the result of Example 12.26,

$$\sin z = \sin x \cosh y + i \cos x \sinh y.$$

Similarly,

$$\cos z = \cos x \cosh x - i \sin x \sinh y.$$

example 12.27: Determine the image of the semi-infinite strip

$$-\frac{\pi}{2} \leqq x \leqq \frac{\pi}{2}, \quad y \geqq 0$$

under the mapping $w = \sin z$.

solution: When $x = -\pi/2$, $u(-\pi/2, y) = -\cosh y$, $v(-\pi/2, y) = 0$. Therefore, the image of the left boundary is the half-line $v = 0$, $u \leqq -1$.

When $x = \pi/2$, $u(\pi/2, y) = \cosh y$, $v(\pi/2, y) = 0$. Hence, the image of the right boundary is the half-line $v = 0$, $u \geqq 1$.

When $y = 0$, $u(x, 0) = \sin x$, $v(x, 0) = 0$, and thus the image of the line segment $-\pi/2 \leqq x \leqq \pi/2$, $y = 0$ is the line segment $-1 \leqq u \leqq 1$, $v = 0$. The given infinite strip is mapped onto the half-plane $y \geqq 0$. (See Figure 12.21.) We sometimes say the mapping "turns down" the sides of the strip. We may like to visualize this mapping in two stages:

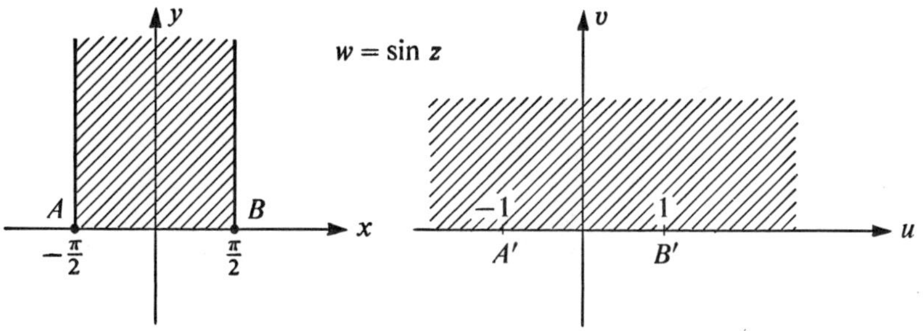

figure 12.21

1. First we stretch the strip horizontally by rotating the two vertical sides outward around the two points $\pm\pi/2+0i$ until they reach the horizontal positions along the real x-axis. Thus the strip has been mapped onto the upper half-plane.
2. Within this half-plane we do certain "shrinkages" which move the points $\pm\pi/2+0i$ into $\pm 1+0i$ and we do certain "stretchings" which move distant points much farther away (for large z, we have roughly $|w|=e^{|z|}$).

example 12.28: Determine the image of the square *ABCDEF* shown in Figure 12.22, under the mapping $w=\sin z$.

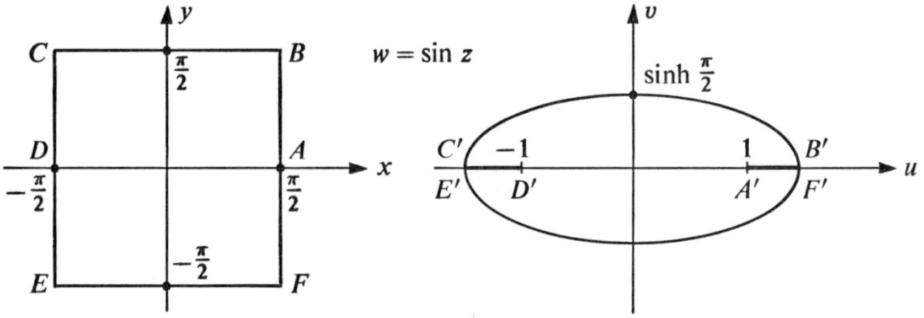

figure 12.22

solution: The parametric equations of the image of the line $x=\pi/2$ are

$$\begin{Bmatrix} u=\cosh y \\ v=0 \end{Bmatrix}.$$

Since y varies from $-\pi/2$ to $\pi/2$, the image of $\overline{FAB}$ is the line segment from $(\cosh\pi/2, 0)$ to $(1, 0)$ and then back to $(\cosh\pi/2, 0)$, that is, twice the line segment. (See Figure 12.22.) The parametric equations of the image of the line $y = \pi/2$ are

$$\begin{cases} u = \sin x \cosh \pi/2 \\ v = \cos x \ \sinh \pi/2 \end{cases}.$$

Eliminating the parameter we obtain

$$\frac{u^2}{\cosh^2 \pi/2} + \frac{y^2}{\sinh^2 \pi/2} = 1\,.$$

This is an ellipse with major semi-axis $\cosh \pi/2$ and minor semi-axis $\sinh \pi/2$. The distance from the origin to either focus is 1. The remainder of the analysis is left as an exercise.

12.8 inverse functions

By properly restricting the domain of the variable so that a function is one-to-one, we can define a function inverse to the one given. We will carry out some of the details of this process in the case of the complex logarithm function.

The complex logarithm, $w = \ln z$, is defined as the relation inverse to the complex exponential. Thus $w = \ln z$ means $z = \exp w$. Since the polar representation of z is $r \operatorname{cis} \theta$ we have

$$r \operatorname{cis} \theta = e^u \operatorname{cis} v$$

from which

$$r = e^u \quad \text{and} \quad \theta = v\,.$$

If we denote the real natural logarithm by $\log_e$, the complex logarithm can be written

$$w = \ln z = \log_e |z| + i\theta = \log_e |z| + i \arg z$$

where θ is any argument of z.

Note that $\ln z$ has many branch functions each determined by one period of the exponential. If $\arg z$ is limited to its principal values, or, which is the same thing, if $\exp z$ is limited to its fundamental period, the fundamental branch of $\ln z$ is defined and written as $\operatorname{Ln} z$. Thus,

$$\operatorname{Ln} z = \log_e |z| + i \operatorname{Arg} z\,.$$

All other values of $\ln z$ are related to this fundamental branch by the equation

$$\ln z = \operatorname{Ln} z + 2n\pi i\,.$$

The real logarithm of a negative number is undefined, but this is not the case for the complex logarithm.

example 12.29: Compute $\ln(-1)$ and $\operatorname{Ln}(-1)$.

solution: The *principal* polar coordinates of (-1) are $r=1$, $\theta=\pi$, and *all* polar coordinates of (-1) are $r=1$, $\theta=(2n+1)\pi$. Therefore,

$$\operatorname{Ln}(-1) = \log_e|-1| + i\pi = 0 + i\pi = i\pi,$$
$$\ln(-1) = i\pi + 2n\pi i = (2n+1)\pi i.$$

The branch functions of $\ln z$ all map the complete z-plane minus the origin onto a horizontal strip in the w-plane of width 2π. The domains of the various branch functions are separated by the branch cut along the negative real axis.

Since there are infinitely many branches to the complex logarithm, the mapping $w=\ln z$ may be interpreted as a map from an infinite-sheeted Riemann surface onto the w-plane.

12.9 the power function

In Section 12.5 you learned that z^n and $z^{1/n}$ were defined in essentially the same way as in real analysis. Now we will give meaning to the more general power of z, z^c, where c is any complex number. Once again, the method used is similar to the definition of x^b, where x and b are real,

$$z^c = \exp(c \ln z).$$

Perhaps you may recall that this is the method used to define irrational powers of real numbers. In the case of real numbers, both the real exponential and the real logarithm are single-valued so that x^b is single-valued. However, the complex logarithm is multi-valued and $\exp z$ is periodic so the precise behavior of z^c will depend on the nature of the exponent, c. If c is an integer, we write z^n and if c is rational, the function is written $z^{p/q}$. Both have been previously defined in Section 12.5 and we will now show that our present definition is consistent.

case 1: If c is a positive integer, then

$$\begin{aligned} z^n &= \exp(n \ln z) \\ &= \exp(n(\log_e|z| + 2m\pi i + i \operatorname{Arg} z)) \\ &= \exp(n \log_e|z| + 2mn\pi i + in \operatorname{Arg} z). \end{aligned}$$

Since n is an integer, so is $2mn$, and because of the periodicity of $\exp z$, the term $2mn\pi i$ drops out so that

$$\begin{aligned} z^n &= \exp(n \log_e |z| + in \operatorname{Arg} z) \\ &= |z|^n \operatorname{cis}(n \operatorname{Arg} z) \end{aligned}$$

which is the same value previously assigned to z^n.

case 2: Let $c = 1/n$. Then

$$\begin{aligned} z^{1/n} &= \exp\left(\frac{1}{n} \log_e |z| + \frac{2m\pi i}{n} + i \frac{\operatorname{Arg} z}{n}\right) \\ &= \exp\left(\frac{1}{n} \log_e |z| + i \frac{\operatorname{Arg} z}{n}\right) \exp\left(\frac{2m\pi i}{n}\right) \\ &= [\text{Principal value of } z^{1/n}] \operatorname{cis} \frac{2m\pi}{n} \end{aligned}$$

which yields distinct values for $m = 0, 1, 2, \cdots, (n-1)$. These are the same n-values determined by the definition used in Section 12.5.

Generally, z^c has infinitely many values. For example, if c is a real irrational number, then $\arg z^c = c \operatorname{Arg} z + 2mc\pi$. Since $2mc$ is never an integer, infinitely many values of $\arg z^c$ are obtained. This property sharply distinguishes the complex quantity z^c from the real function x^c, even when z is a real number.

example 12.30: Calculate $1^{\sqrt{2}}$.

solution: Considered as a problem in real analysis, $1^{\sqrt{2}} = 1$, since "1 to any power is 1."

In complex analysis,

$$\begin{aligned} |1^{\sqrt{2}}| &= 1 \\ \arg(1^{\sqrt{2}}) &= \sqrt{2} \cdot 0 + 2\sqrt{2}\, m\pi \\ &= 2\sqrt{2}\, m\pi. \end{aligned}$$

Therefore,

$$\begin{aligned} 1^{\sqrt{2}} &= \exp(2\sqrt{2}\, m\pi i) \\ &= \operatorname{cis} 2\sqrt{2}\, m\pi, \end{aligned}$$

where m is an integer. Thus, there are infinitely many values of $1^{\sqrt{2}}$. Graphically, they are all located on a circle of radius 1 centered at the origin.

In summary, z^c is single-valued if c is an integer; is multiple but finitely many-valued if c is a real rational number; and is infinitely many-valued in all other cases.

example 12.31: Calculate i^i.

solution:

$$\begin{aligned} i^i &= \exp(i \ln i) \\ &= \exp[i(\pi/2\, i + 2m\pi i)] \\ &= e^{-\pi/2 + 2m\pi}, \quad m \text{ an integer}. \end{aligned}$$

Note that the values of i^i are real, which may surprise you (although you might say at this stage that nothing does!).

To conclude this section we consider the formula for e^c where e is the real irrational number (approximately equal to 2.7) and c is any complex number. By definition

$$\begin{aligned} e^c &= \exp(c \ln e) \\ &= \exp[c(\log_e e + 0 + 2n\pi i)] \\ &= \exp[c(1 + 2n\pi i)] \end{aligned}$$

which means that e^c is (as is usually the case for the complex power) many-valued. For the case $n = 0$,

$$e^c = \exp c$$

or

$$[\text{Principal value of } e^c] = \exp c.$$

Hence very often (probably most often), the exponential function is written as e^z instead of $\exp z$. It is understood that when used to represent the exponential function, e^z is not multiple valued.

exercises for sections 12.7–12.9

1. Show that every vertical strip of width π is mapped by the sine function onto the complete w-plane.
2. Consider the ellipse in the w-plane which is the image of $y = c$, under the mapping $w = \sin z$. Show that the distance from the origin to the focus is independent of c. Show that all horizontal lines map onto a set of confocal ellipses.
3. Determine the image of the region $|x| \leqq c$, $|y| \leqq c$, where $0 < c < \pi/2$, under the mapping $w = \sin z$.
4. Solve the equations:
   (a) $\sin z = 2$. (b) $\cos z = 1/2$.
   (c) $\cos z = i$. (d) $\sin z = \cosh 2$.
5. Determine the preimage of the real axis under the mapping $w = \sin z$.

6. Show that $\sin z$ and $\cos z$ are periodic.
7. Prove the three identities mentioned at the beginning of Section 12.7.
8. Compute $|\sin z|$ and $|\cos z|$.
9. Show that $\cos \bar{z} = \overline{\cos z}$, and $\sin \bar{z} = \overline{\sin z}$.
10. Calculate and display graphically:
    (a) $\ln 1$. (b) $\ln e^i$. (c) $\ln i$.
11. Solve the equations:
    (a) $\ln z = -1$. (b) $\ln z = i$. (c) $\ln z = 0$.
12. (a) What is $e^{\ln z}$ if e names the exponential function?
    (b) What is $e^{\ln z}$ if e is the real number?
    (c) What is $\ln e^z$ if e names the exponential function?
    (d) What is $\ln e^z$ if e is the real number?
13. Compute:
    (a) $(1+i)^{-i}$. (b) i^{1-i}. (c) 1^i. (d) $1^{\sqrt{3}}$.
14. Show that $\operatorname{Ln} z$ is discontinuous at its branch cut.
15. Define the function inverse to the sine function. Show that it is many-valued. Write it in terms of the complex logarithm. (*Hint:* Use the definition of $\sin w$.)
16. Define $\sinh z$ and $\cosh z$. Find the real and imaginary parts.

mapping by analytic functions

13

13.1 the point at infinity

In this chapter we consider complex mappings by analytic functions. Several other related topics will also be introduced which you may find interesting in themselves.

The behavior of a function, f, for large values of z is called the *behavior of f near ∞_z*. To be precise, let $z' = 1/z$ and consider $f(1/z')$ for values of z' near 0. In particular, we make the following definition.

definition 13.1

The *value* of the function f at ∞_z, denoted by $f(\infty_z)$, is given by

$$f(\infty_z) = \lim_{z' \to 0} f\left(\frac{1}{z'}\right).$$

The set of complex numbers, C, along with the "special" symbol ∞_z is called the *extended complex number system*. In this sense ∞_z is treated as any other complex number.

example 13.1: Is the function $f(z) = z$ analytic at ∞_z? What is $f(\infty_z)$?

solution:

$$f\left(\frac{1}{z'}\right) = \frac{1}{z'},$$

and this function is certainly nonanalytic for $z' = 0$. Hence the given function is nonanalytic at ∞_z. Since

$$\lim_{z'\to 0} \frac{1}{z'} = \infty,$$

we may write $f(\infty_z) = \infty_w$.

example 13.2: Repeat Example 13.1 for the function

$$f(z) = \frac{z-2}{3z+1}.$$

solution:

$$f\left(\frac{1}{z'}\right) = \frac{1-2z'}{3+z'}.$$

This function is a rational function of z' and is defined and analytic for $z' = 0$. Since

$$\lim_{z'\to 0} \frac{1-2z'}{3+z'} = \frac{1}{3},$$

it follows that $f(\infty_z) = 1/3$.

13.2 curves in the complex plane

Recall that a curve may be represented parametrically by two equations $x = x(t)$ and $y = y(t)$. In vector analysis we showed (Chapter 6) how these two equations are equivalent to one vector function of one real variable, $\mathbf{F}(t)$, called the vector representation of the curve. In the same way, a complex function of one real variable,

$$z(t) = x(t) + iy(t),$$

is a convenient method of describing a curve.

The function, $z(t)$, has its domain on some parameter line and its range is a set of complex numbers which give points (on the curve) in the complex plane.

example 13.3: Use a complex function of one real variable to represent the curve $y = x^2$.

solution: Letting $x = t$, then $y = t^2$, so the function $z(t) = t + it^2$ may be used to represent the parabola.

example 13.4: Write the equation for the curve whose representation in terms of a complex function is $z(t) = t + (mt + b)i$.

solution: Since $x = t$ and $y = mt + b$, then $y = mx + b$ which is a straight line with slope m and y-intercept b.

Consider a mapping from the z-plane to the w-plane. We are often interested in the image not only of a given domain under such a mapping but also of some curve in that domain. The image of a curve, C, represented by $z(t)$ is the curve C^* in the w-plane given by the function $w = w(z(t))$.

example 13.5: Find the image of the line $z(t) = t + (1+t)i$ under the mapping $w = z^2$. Sketch both $z(t)$ and $w(z(t))$.

solution:

$$w(z(t)) = (t + (1+t)\,i)^2 = (-1-2t) + 2t(1+t)\,i\,.$$

Therefore, $u(t) = -1 - 2t$; $v(t) = 2t(1+t)$. Eliminating the parameter we obtain

$$v = 2\left(\frac{u+1}{-2}\right)\left(1 + \frac{u+1}{-2}\right) = \frac{u^2-1}{2}.$$

(See Figure 13.1.)

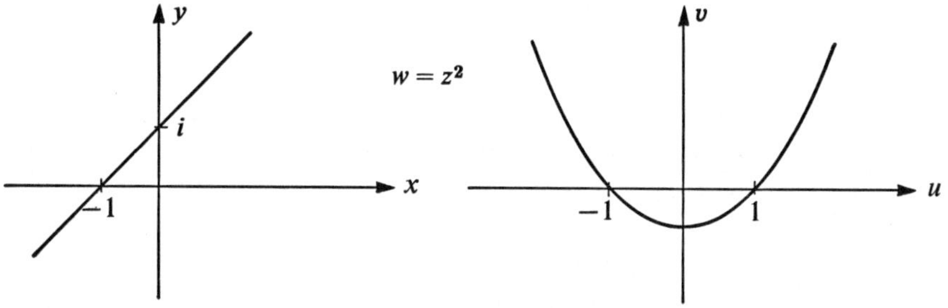

figure 13.1

As in vector analysis, the direction of a tangent vector to a curve is defined as the limiting direction of a vector from the point $z(t_0) = z_0$ to another point on the curve $z(t_0 + \Delta t)$. The vector corresponding to the difference $z(t_0 + \Delta t) - z(t_0)$, and the vector corresponding to the difference quotient

$$\frac{z(t_0 + \Delta t) - z(t_0)}{\Delta t}$$

have the same direction. Therefore, $\dot{z}(t_0)$, (that is, dz/dt evaluated at t_0) will represent a tangent to the curve at z_0. The angle between this vector and the positive real axis is $\operatorname{Arg}\dot{z}(t_0)$. (See Figure 13.2.)

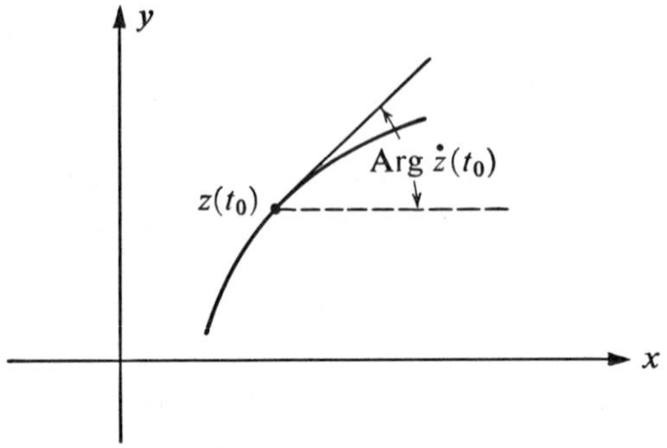

figure 13.2 angle between curve and tangent vector

13.3 conformal mapping

Generally, the image of a region under a mapping will be significantly distorted. A mapping by an analytic function has the important property of not distorting "locally." Specifically, angles between oriented curves are preserved.

definition 13.2

A mapping defined by $w = f(z)$ is said to be *conformal* at z_0, or *angle preserving*, if the angle between any two oriented curves in the z-plane at z_0 is the same as the angle between the images of these two curves at $w(z_0)$.

example 13.6: Let C_1 be defined by $z_1(t) = t + it^2$ and C_2 by $z_2(t) = t + it$. Find the images of these two curves under the mapping $w = z^2$. Show that the mapping $w = z^2$ preserves the angle between the two curves at the point of intersection corresponding to $t = 1$, but *not* at the point corresponding to $t = 0$.

solution: Let C_1^* and C_2^* be the images of C_1 and C_2 under the mapping.

$$C_1^*: \quad w_1(t) = (t^2 - t^4) + i2t^3,$$
$$C_2^*: \quad w_2(t) = 2t^2 i.$$

To show that the mapping preserves the angle of intersection at P, we must show $\psi_z = \psi_w$, or, which is the same thing for our purposes, $\tan\psi_z = \tan\psi_w$.

$$\tan\psi_z = \frac{\tan\theta_2 - \tan\theta_1}{1 + \tan\theta_2 \tan\theta_1}$$

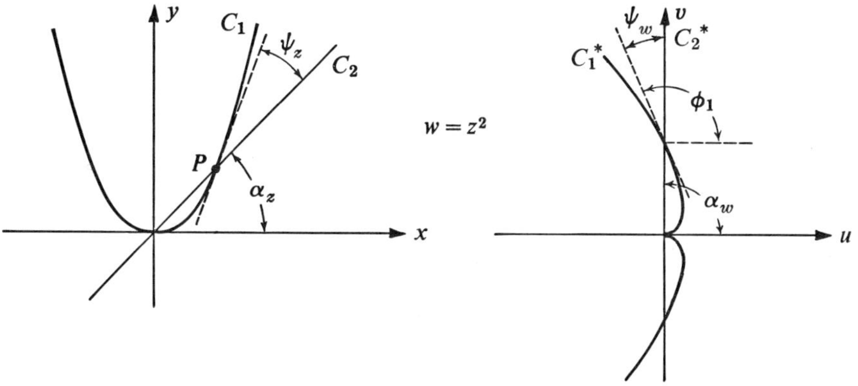

figure 13.3 *preservation of angles*

$$= \frac{(2) - (1)}{(1) + (2)\,(1)} = \frac{1}{3};$$

$$\tan \psi_w = -\cot \phi_1 = \tfrac{1}{3}.$$

It is obvious from Figure 13.3 that $\alpha_z = 45°$ and $\alpha_w = 90°$, which shows that the angle of intersection is not preserved at the point of intersection corresponding to $t = 0$.

In Example 13.6 we have shown how a mapping preserves an angle between two specific curves at a point, P. To show it to be conformal at P we would have to show this to be true for *every* pair of curves intersecting at P. The following theorem considers the general case.

theorem 13.1

Let $w = f(z)$ be analytic. Then the mapping defined by $w = f(z)$ is conformal at those points for which $f'(z) \neq 0$.

proof: To prove this theorem we must show that tangents to *all* curves which pass through z_0 are rotated by the same amount. Then the angle between any two curves will be preserved. Let C be any curve in the z-plane which passes through z_0. Then $\dot{z}(t_0)$ represents a vector tangent to C at z_0. Let C^* be the image of C, $w(z_0)$ the image of z_0, and $\dot{w}(z(t_0))$ the image of the tangent vector under the mapping $w = f(z)$. (See Figure 13.4.) To show that the tangent vectors are rotated by a fixed constant, independent of C, we find the relation between $\arg \dot{w}$ and $\arg \dot{z}$. By the chain rule,

$$\dot{w}(z(t_0)) = f'(z_0)\,\dot{z}(t_0).$$

Hence, if $f'(z_0) \neq 0$,

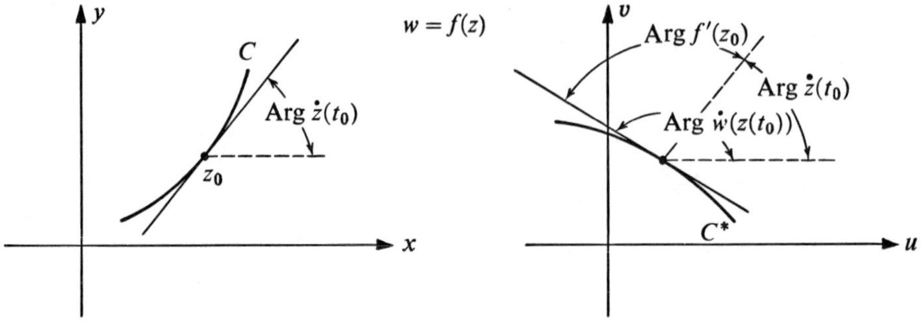

figure 13.4 *preservation of angles*

$$\text{Arg}\, \dot{w} = \text{Arg}\, f' + \text{Arg}\, \dot{z}$$

or

$$\text{Arg}\, \dot{w} - \text{Arg}\, \dot{z} = \text{Arg}\, f'.$$

Since $\text{Arg}\, f'$ is a constant dependent only on the mapping and the point, the tangents at z_0 are all rotated by a fixed amount.

example 13.7: Where is the mapping given by $w = z^2$ conformal?

solution: Since the mapping is analytic everywhere, and the derivative is nonzero everywhere except at the origin, the mapping is conformal everywhere except at the origin.

example 13.8: Where is the mapping defined by $w = (5z + 1)/(2z - 1)$ conformal?

solution: Since

$$\frac{dw}{dz} = \frac{-7}{(2z - 1)^2},$$

the function which defines the mapping is analytic everywhere except at the singularity, $z = 1/2$. Since dw/dz is never zero, the mapping is conformal everywhere except at $z = 1/2$.

exercises for sections 13.1–13.3

1. Find the images of the following sets of points under the mapping $w = 1/z$:
   (a) The curve $y = x^2 - 1$. (b) The circle $|z - 1| = 5$.
   (c) The set $\{z \mid x > 1, y > 0\}$. (d) The infinite strip $0 < y < (1/2)$.
   (e) The circle $|z| = 1$. (f) The circle $|z - 1| = 1$.
2. How much does $w = 1/z$ rotate tangents to curves at $z = i$? At $z = 1$? At $z = -i$?

3. Find the images of the following sets of points under the indicated mappings:
   (a) $x=\text{constant}$ and $y=\text{constant}$, under $w=iz$.
   (b) $x=y$, under $w=z^2$. (c) $x=y$, under $w=iz^2$.
   (d) $-\pi/6<\arg z<\pi/12$, $1<|z|<4$, under $w=z^2$.
   (e) $x=y$, under $w=x^2+iy^2$. (f) $0<x, 0<y<2$, under $w=iz-1$.
   (g) $x=y$ under $w=\sin z$. (h) $x=y$ under $w=e^z$.
4. Where are the following mappings conformal?
   (a) $w=z+\dfrac{1}{z}$. (b) $w=z^2+\dfrac{1}{z^2}$. (c) $w=z^3-1$.
   (d) $w=az^2+bz+c$. (e) $w=e^z$. (f) $w=\sin z$.
   (g) $w=\cos z$. (h) $w=1/z$.
5. How much does the mapping $w=z^2$ rotate the tangents to the curves at:
   (a) $1+i$. (b) i. (c) -3. (d) $-1/2$.
6. Show that the "magnification factor" (equal to $|f'(z)|$) is given by the square root of
$$\begin{vmatrix} u_x & u_y \\ v_x & v_y \end{vmatrix}.$$
7. What is/are the image(s) of ∞_z under the map
$$f(z)=\frac{z^2+z-1}{1-z^2}?$$
   What points map onto ∞_w? Is the function analytic at ∞_z?
8. Determine the image of the lines $y=x+1$ and $y=3(x-1)$ under the mapping $w=z^3$. Exhibit, by using these two curves, that the mapping is conformal at the point of intersection.
9. Determine if the following functions are analytic at ∞:
   (a) $w=e^z$. (b) $w=z^3-z^{-1}$. (c) $w=z^2e^{i/z}$. (d) $w=\dfrac{1}{z^2-1}$.
10. Represent each of the following curves by a complex function of one real variable:
    (a) The line $2x+3y-1=0$. (b) The ellipse $x^2+4y^2=4$.
    (c) The circle $(x-1)^2+(y+1)^2=4$.
    (d) The hyperbola $x^2-y^2=1$.
11. Find the image of each of the curves of the previous exercise under the mapping $w=z^2$.
12. Represent the tangent lines to the following curves at the indicated points in terms of a complex function of one real variable:

(a) The parabola $y = x^2$ at the point (2, 4).
(b) The curve $z(t) = t^3 + i e^t$ at $(1, e)$.
(c) The circle $x^2 + y^2 = 4$ at (0, 2).

13.4 bilinear transformations

You will find it convenient to use conformal mappings which are defined by functions which are one-to-one, since then an inverse function from the w-plane to the z-plane can be defined. Most often we are content with the ability to limit the domain of the function sufficiently so that the map is at least locally one-to-one. For example, the function $w = z^2$ is not one-to-one and the inverse relation is not single-valued. The function is locally one-to-one if we consider, for example, the region $\operatorname{Im} z > 0$.

A wide class of functions for which no such limitation of domain is necessary is called the set of *bilinear functions* or transformations. (The use of the term bilinear in this case is awkward since it is a bit inconsistent with the usual use of the term in other areas of mathematics.)

A bilinear transformation is defined by the equation

$$w = \frac{az + b}{cz + d}, \qquad ad - bc \neq 0,$$

where the constants a, b, c, and d may be real or complex. Solving for z, we obtain

$$z = \frac{-dw + b}{cw - a}$$

which shows that the mapping is one-to-one.

The point at infinity in the z-plane maps onto $w = a/c$ and the pre-image of ∞_w is $z = -d/c$. Hence, the point $z = -d/c$ is an isolated singularity of the bilinear function. Since

$$\frac{dw}{dz} = \frac{ad - bc}{(cz + d)^2}$$

and since $ad - bc \neq 0$, the mapping is conformal everywhere except at its singularity.

There are three linear mappings which "generate" the others.

1. The mapping $w = z + b$ may be regarded as a *translation* in the direction defined by the argument of b. If $b = 0$ the mapping $w = z$ "fixes" every point. (A point is said to be "fixed" if it is mapped onto itself.) The mapping $w = z$ is called the *identity* mapping.

2. The mapping $w = az$ *rotates* points through an angle equal to Arg a, followed by a dilation or expansion by the factor $|a|$.
3. The mapping $w = 1/z$ is called an *inversion* in the unit circle followed by a *reflection* in the real axis. In polar form the map $w = 1/z$ is $R \operatorname{cis} \phi = (1/r) \operatorname{cis}(-\theta)$. Hence, $R = 1/r$ (inversion) and $\phi = -\theta$ (reflection). (See Figure 13.5.)

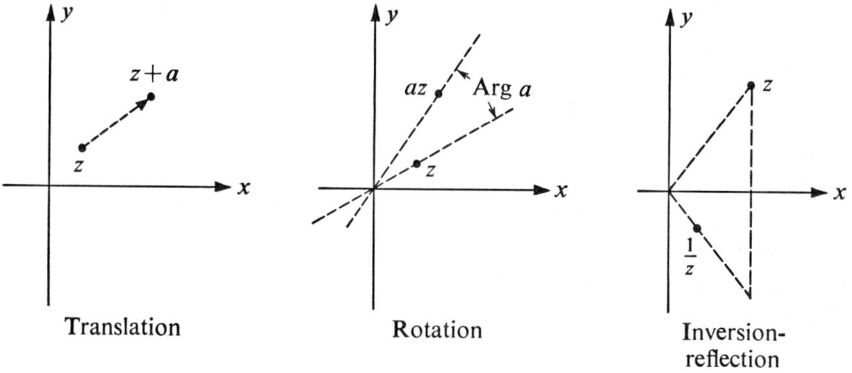

figure 13.5 *elementary mappings*

All bilinear mappings can be obtained by composition of translation, rotation, and inversion-reflection mappings. To see this, note that if $c \neq 0$, then

$$w = \frac{az+b}{cz+d} = \frac{a}{c} + \frac{bc-ad}{c(cz+d)}.$$

Then, the following composition of the three elementary bilinear mappings gives the most general bilinear map.

$$z \to cz \to cz + d \to \frac{1}{cz+d} \to \frac{\frac{bc-ad}{c}}{cz+d} \to \frac{a}{c} + \frac{bc-ad}{c(cz+d)}.$$

Rotation, Translation, Inversion-Reflection, Rotation, Translation

The case $c = 0$ is left for the exercises.

Practically, bilinear transformations are important because they "preserve" lines and circles in the sense of the following theorem.

theorem 13.2

A bilinear transformation maps the totality of circles and lines onto circles and lines.

proof: A bilinear map is obtained by composition of translation, rotation, and inversion-reflection. Obviously, a translation or a rotation takes a circle onto a circle and a line onto a line. So the theorem will be proved if the map $w = 1/z$ takes a circle onto a circle or a line and a line onto a circle or a line.

(a) A circle can be represented in the form $|z - a| = r$. The image curve is then

$$\left|\frac{1}{w} - a\right| = r \quad \text{or} \quad \left|\frac{1 - aw}{w}\right| = r,$$

from which

$$\left|w - \frac{1}{a}\right| = \left|\frac{r}{a}\, w\right|.$$

This is a circle if $r \neq a$ and a line if $r = a$. (See Exercise 14 in the previous set of exercises.)

(b) A line may be represented in the form $|z - a| = |z - b|$, so that the image curve is

$$\left|\frac{1}{w} - a\right| = \left|\frac{1}{w} - b\right|$$

so that $|aw - 1| = |bw - 1|$, and thus

$$|a|\left|w - \frac{1}{a}\right| = |b|\left|w - \frac{1}{b}\right|.$$

This is a line if $|a| = |b|$, and a circle otherwise, and hence the theorem is proved.

example 13.9: Consider the bilinear transformation defined by

$$w = \frac{z - 1}{z - 2}.$$

Find the image of (a) the circle $|z + 1| = 1$; (b) the circle $|z + 1| = 3$; (c) the line $|z - 1| = |z - 2|$.

solution: We first find the inverse transformation by solving for z,

$$z = \frac{2w - 1}{w - 1}.$$

(a) The image of $|z + 1| = 1$ is

$$\left|\frac{2w - 1}{w - 1} + 1\right| = 1.$$

Simplifying, this becomes $|3w - 2| = |w - 1|$ which is a circle centered at (5/8, 0) with radius 1/8. (The details of the calculations are left for you.)

(b) The image of $|z + 1| = 3$ is

$$\left|\frac{2w - 1}{w - 1} + 1\right| = 3,$$

which is $|3w - 2| = |3w - 3|$. This is the perpendicular bisector of the line joining 2/3 to 1, that is, $u = 5/6$.

(c) The image of $|z - 1| = |z - 2|$ is

$$\left|\frac{2w - 1}{w - 1} - 1\right| = \left|\frac{2w - 1}{w - 1} - 2\right|,$$

or $|w| = 1$ which is a circle of radius 1 centered at the origin.

To determine a specific bilinear transformation, the four constants a, b, c, d must be given. Actually, since

$$\frac{az + b}{cz + d} = \frac{a}{c}\frac{\left(z + \frac{b}{a}\right)}{\left(z + \frac{d}{c}\right)}$$

only three of the constants are independent so three conditions are sufficient to specify completely the bilinear transformation. This informal reasoning leads to the following theorem.

theorem 13.3

A linear transformation is uniquely determined by its effect on three points.

If two of the range points specified are 0_w and ∞_w, the linear transformation can be determined almost by inspection.

example 13.10: Find the linear transformation which maps 0_z, ∞_z, and a third point, say z_3, onto 0_w, ∞_w, and w_3.

solution: The fact that 0 and ∞ are so-called "fixed" points of the transformation means that the form of the defining equation is $w = cz$. Since the image of z_3 is w_3, $c = w_3/z_3$. Therefore,

$$w = \frac{w_3}{z_3} z.$$

example 13.11: Find the linear transformation which takes 1_z, 2_z, and 3_z onto 0_w, ∞_w, and 1_w.†

solution: The fact that 1_z and 2_z are the pre-images of 0_w and ∞_w determines that the transformation has the form

$$w = c\,\frac{z-1}{z-2}.$$

The constant c is determined from the other condition. Hence

$$1 = c\,\frac{3-1}{3-2} = 2c,$$

which means $c = 1/2$. Therefore,

$$w = \frac{1}{2}\left(\frac{z-1}{z-2}\right).$$

example 13.12: Same as the previous example, except that ∞_z is mapped onto 1_w.

solution: The form of the defining equation is the same. Since ∞_z is mapped onto 1_w,

$$1 = \lim_{z\to\infty} c\,\frac{z-1}{z-2},$$

and hence $c = 1$. Therefore,

$$w = \frac{z-1}{z-2}.$$

So, generally, we consider it an "easy" problem to map three points in the extended z-plane onto 0_w, ∞_w, and some third point.

To find the linear tranformation which takes any three points onto any three points, we first make two of these easy transformations from the z-plane to a Z-plane and from the w-plane onto a W-plane. One map takes z_1, z_2, and z_3 onto 0_Z, ∞_Z, and $Z_3(=(z_3 - z_1)/(z_3 - z_2))$ and the other takes w_1, w_2, w_3 onto 0_W, ∞_W, and $W_3(=(w_3 - w_1)/(w_3 - w_2))$. Then the map $W = (W_3/Z_3)\,Z$ maps 0_Z, ∞_Z and Z_3 onto 0_W, ∞_W, and W_3. (See Example 13.10.) The map from the z-plane to the w-plane is, in implicit form,

$$\frac{w-w_1}{w-w_2}\,\frac{w_3-w_2}{w_3-w_1} = \frac{z-z_1}{z-z_2}\,\frac{z_3-z_2}{z_3-z_1}.$$

† The subscripts z and w indicate the plane in which the given complex number is being considered.

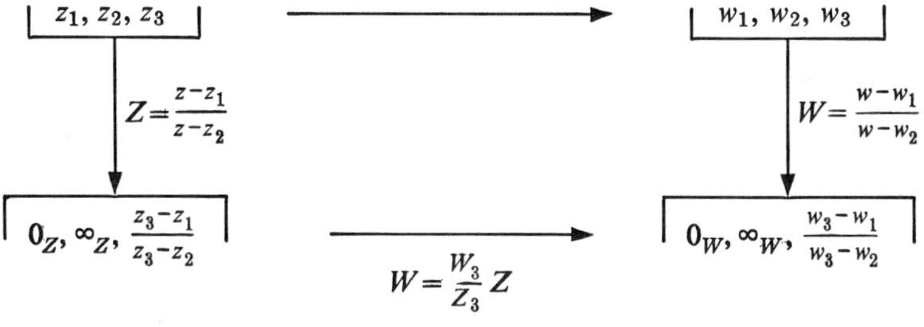

figure 13.6 *mapping of three points onto three points*

(See Figure 13.6.)

This formula is valid only if all the points are finite. You should not memorize the formula but rather learn the technique of first making the auxiliary maps. Then, if any of the points are not finite, it is a simple matter to modify the maps Z or W.

example 13.13: Find the linear transformation which takes the point 0 onto $-i$ and fixes -1 and 1.

solution: The map Z, which takes -1_z, 0_z, and 1_z onto 0_Z, ∞_Z, and $(2/1)_Z$ is given by $Z = (z+1)/z$. The map W which takes -1_w, $-i_w$, and 1_w onto 0_W, ∞_W, and $2/(1+i)$ is $W = (w+1)/(w+i)$. Since $W = (W_3/Z_3)\,Z$, the desired map is defined implicitly by

$$\frac{w+1}{w+i} = \frac{2/(1+i)}{2/1}\,\frac{z+1}{z}$$

or, explicitly,

$$w = \frac{z-i}{-iz+1}.$$

example 13.14: Find the linear transformation which takes 0_z, ∞_z, and 2_z onto 1_w, 3_w, and ∞_w.

solution: The map Z which takes 0_z, ∞_z, and 2_z onto 0_Z, ∞_Z, and 2_Z is the identity map. The map $W = (w-1)/(w-3)$ takes 1_w, 3_w, ∞_w onto 0_W, ∞_W, 1_W. Therefore, the map is defined implicitly by

$$\frac{w-1}{w-3} = \frac{1}{2}\,z$$

or

$$w = \frac{3z-2}{z-2}.$$

13.5 special linear mappings

In this section we particularize the preceding discussion to two specific types of linear mappings. These typify the application of linear maps.

A. *mapping of the upper half-plane onto the unit circle:* The region in the z-plane is described by the inequality $\operatorname{Im} z > 0$, and in the w-plane by $|w| < 1$. There are many ways of accomplishing a map of the upper half-plane onto the unit circle, but clearly, the boundary in the z-plane which is the real axis must correspond to the circle $|w| = 1$. In addition to making this correspondence of boundaries, you must map the upper half-plane onto the interior of the circle. A knowledge of the image of one point is sufficient, since the image of the entire upper half-plane will be either the interior or the exterior of the circle $|w| = 1$.

example 13.15: Show that the mapping given in Example 13.13 is a mapping from the upper half-plane onto the interior of the unit circle. What are the images of the lines $y = $ a constant? Write the equation for the image of the line $y = 1$.

solution: Since three points on the real line map onto three points on the unit circle, it is clear from Theorems 13.2 and 13.3 that the entire real axis maps onto the circle $|w| = 1$. It is easy to check that $z = i$ maps onto the origin and hence the upper half-plane is mapped onto the interior of the unit circle. Since ∞_z maps onto $w = i$, the images of all lines in the z-plane pass through $w = i$. Thus, in particular, the lines $y = c$ are mapped onto circles through $w = i$.

To find the image of $y = 1$, we could proceed in one of two ways:

1. Find two other points on the image circle which will give sufficient information to write its equation.
2. The equation of the line $y = 1$ can be written $|z| = |z - 2i|$. If we solve the transformation $w = (z - i)/(-iz + 1)$ for z we obtain $z = (w + i)/(iw + 1)$ and thus the equation of the image of the line is

$$\left|\frac{w+i}{iw+1}\right| = \left|\frac{w+i}{iw+1} - 2i\right|$$

or

$$|w + i| = |3w - i| .$$

In terms of u and v, the equation of the circle is

$$u^2 + (v - \tfrac{1}{2})^2 = \tfrac{1}{4}.$$

(See Figure 13.7.)

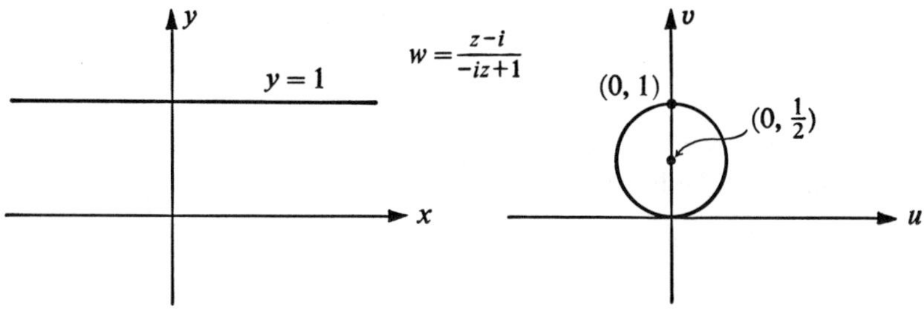

figure 13.7

More generally, to map the upper half-plane onto the unit circle with a linear transformation, the three points on the real axis in the z-plane should have image points on the circle $|w| = 1$. Moreover, to assure that the image of the upper half-plane is the interior (and not the exterior) of the circle, the ordering of points must be the same in both planes. Thus, if $z_1 = x_1$, $z_2 = x_2$, $z_3 = x_3$ are points on the real axis with $x_1 < x_2 < x_3$, then the image points w_1, w_2, and w_3 must be in cyclic counterclockwise order, otherwise the mapping will be onto the exterior of the circle. (See Figure 13.8.)

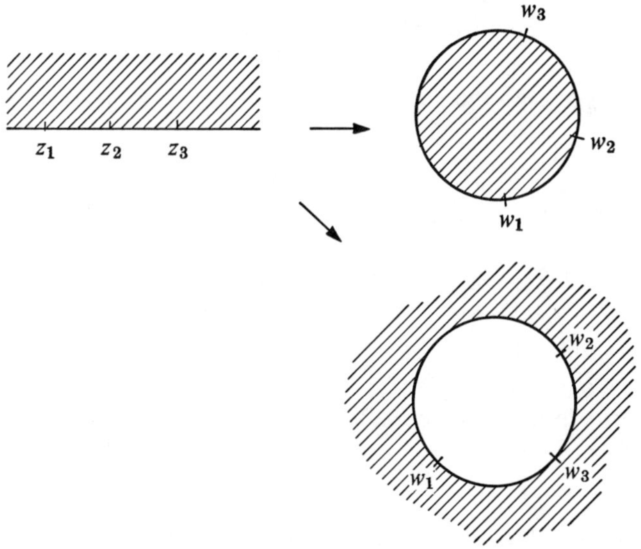

figure 13.8 mapping onto the interior and the exterior of the unit circle

B. *mapping of the upper half-plane onto the upper half-plane:* A mapping which takes the real axis onto the real axis will map the upper half-plane in the z-plane onto either the upper or lower half-plane in the w-plane. Hence, it is sufficient to check the mapping of a single point after the boundaries are matched to determine whether $\text{Im}\, z > 0$ is mapped onto $\text{Im}\, w > 0$ or $\text{Im}\, w < 0$.

example 13.16: Map the upper half-plane onto the upper half-plane such that the image of the unit interval is the positive real axis.

solution: This may be accomplished by fixing 0 and mapping 1_z onto ∞_w. Hence, $w = \pm z/(z-1)$. Since the image of i_z is

$$\pm \frac{i}{i-1} = \pm \frac{1-i}{2},$$

the negative sign is required to place the point in the upper half of the w-plane. Therefore,

$$w = \frac{z}{1-z}$$

is the required map.

Mapping three points on the real z-axis onto three points on the real w-axis may be done in several ways. To assure that the image of the upper half-plane is the upper half of the w-plane, the relative cyclic ordering arrangement of the three points in the two planes must be the same. That is, if x_1, x_2, x_3 are mapped onto u_1, u_2, u_3 and if $x_1 < x_2 < x_3$, then the upper half-plane is mapped onto the upper half-plane if $u_1 < u_2 < u_3$, or $u_2 < u_3 < u_1$, or $u_3 < u_1 < u_2$.

exercises for sections 13.4 and 13.5

1. Find the linear transformation which maps:
   (a) $2, i, -2$ onto $1, i, -1$. (b) $-i, 0, i$ onto $-1, i, 1$.
   (c) $i, 1, \infty$ onto $1, i, \infty$. (d) $1, 2, \infty$ onto $\infty, 5, 2$.
   (e) $0, 1, 2$ onto $0, 1, 2$.
2. Find the fixed points of the linear transformation:
   (a) $w = \dfrac{z+1}{z-1}$. (b) $w = \dfrac{3z-1}{z}$. (c) $w = z$.
3. What kind of a linear transformation has 3 fixed points?
4. Determine the general formula for a linear transformation specified by three points being mapped onto three points if:
   (a) one of the points z_i is ∞_z.
   (b) one of the points w_i is ∞_w.
   (c) a combination of (a) and (b).

5. Find a linear transformation whose fixed points are:
   (a) 1 and -1. (b) i and $-i$.
6. Find the image of the line $x = 5$ under the mapping $(z-1)/(z+1)$.
7. Represent

$$w = \frac{5z-1}{2z+1}$$

   as a composition of elementary mappings of translation, inversion, and rotation.
8. Show that

$$w = \frac{z-z_0}{\bar{z}_0 z - 1}, \qquad |z_0| < 1$$

   maps the unit disk onto the unit disk.
9. Find the linear mapping which takes $-1, 0, \infty$ onto $-1, \infty, -2$. What is the image of the upper half-plane?
10. Find the linear mapping which maps the real axis onto the unit circle and which takes $1, 2, \infty$ onto $i, -1, -i$. What is the image of the upper half-plane?
11. Find a linear transformation which fixes -1 and 1 and which maps the upper half-plane onto the interior of the unit circle.
12. Find an analytic function which maps $1 \leqq y \leqq x+2$, onto the interior of the unit circle.
13. Find an analytic function which maps the region $0 \leqq \arg z \leqq \pi/4$ onto $|w| \leqq 1$.
   (*Hint* for Exercises 12 and 13: Use a nonlinear mapping followed by a linear mapping.)
14. Show that the linear map $w = (az+b)/d$ can be written as a composition of elementary linear mappings.

13.6 table of mappings

Appendix C is a table of mappings of the z-plane onto the w-plane. More extensive treatments have larger tables to which you can refer. It is quite helpful to have these mappings available when making applications as in the next section. Corresponding parts of boundaries are indicated by letters. You should verify each of the mappings.

13.7 application to boundary value problems

Conformal mappings can be used to great advantage to solve some two-dimensional boundary value problems. We will need the following theorem whose proof is left to the exercises. The importance of the theorem will become obvious.

theorem 13.4

Let $fz = u(x, y) + iv(x, y)$ be a conformal map from the z-plane to the w-plane with inverse map $f^{-1}(w) = x(u, v) + iy(u, v)$. Let $\phi(x, y)$ be a scalar function defined in some region, R_z, of the z-plane with the image of the scalar function, denoted by $\Phi(u, v) = \phi(x(u, v), y(u, v))$, being defined in some region, R_w, in the w-plane.

(a) If ϕ is harmonic in the region R_z, then Φ is harmonic in R_w.

(b) If $\phi(x, y) = c$ along some curve C_z, then $\Phi(u, v) = c$ along C_w, where C_w is the image of C_z under f.

(c) If $\partial\phi/\partial n = 0$ along a curve C_z, then $\partial\Phi/\partial n = 0$ along C_w. The derivatives $\partial\phi/\partial n$ and $\partial\Phi/\partial n$ are called the normal derivatives of the scalar functions along the curves.

Roughly speaking the theorem says that Laplace's equation and boundary conditions of the type $\phi(x, y) = c$ and $\partial\phi/\partial n = 0$ are invariant under a conformal mapping. As you will see in the examples, a Dirichlet problem with boundary conditions of the kind mentioned is sometimes solved by inspection if we can find a conformal map to modify the region. Difficulties can and do arise in expressing the function in terms of the "old" variables; that is, they arise technically in actually finding the inverse of the mapping function.

example 13.17: Find the steady state temperature distribution in the plate illustrated in Figure 13.9.

solution: The mathematical model consists of the differential equation

$$T_{xx} + T_{yy} = 0,$$

with boundary conditions

$$T(0, y) = 100, \quad y > 1;$$

$$T(x, 0) = 0, \quad x > 1;$$

$$\frac{\partial T}{\partial n} = 0, \quad \text{on the arc}.$$

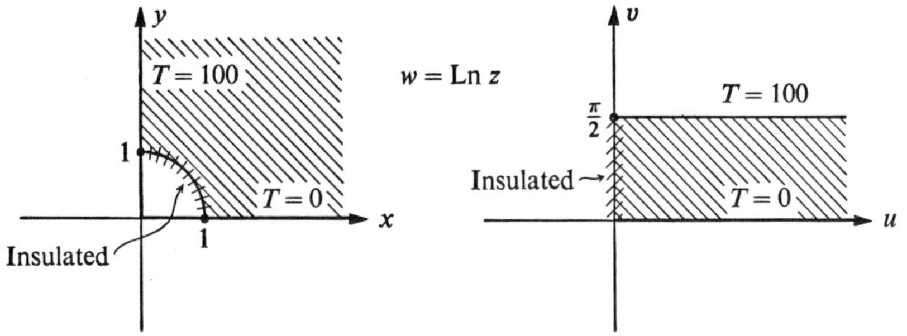

figure 13.9

By examining the table of mappings in Appendix 3, we see that the function defined by $w = \text{Ln}\, z$ will map the given region onto the rectangular region shown in the w-plane in Figure 13.9. In the uv-plane the problem has the form

$$T_{uu} + T_{vv} = 0$$

where

$$T(u, 0) = 0, \qquad u > 0;$$

$$T\left(u, \frac{\pi}{2}\right) = 100, \qquad u > 0;$$

$$\frac{\partial T}{\partial x} = 0, \qquad 0 < v < \frac{\pi}{2}.$$

The solution can now be determined by inspection. (See Example 11.4.)

$$T(u, v) = \frac{200}{\pi} v.$$

Since $x = e^u \cos v$ and $y = e^u \sin v$, $v = \text{Tan}^{-1} y/x$. Hence,

$$T(x, y) = \frac{200}{\pi} \text{Tan}^{-1} \frac{y}{x}.$$

Sometimes, a sequence of mappings is required. Example 13.18 is typical.

example 13.18: Find the steady state temperature in the region bounded by the lines $y = 0$ and $y = \pi$, for $x \geqq 0$, if the boundary conditions are $T = 0$ for $y = 0$ and $y = \pi$, and $T = 50$ for $x = 0$.

solution: We define mappings z_1, z_2, and w to convert the given region sequentially into the one shown in the w-plane as in Figure 13.10,

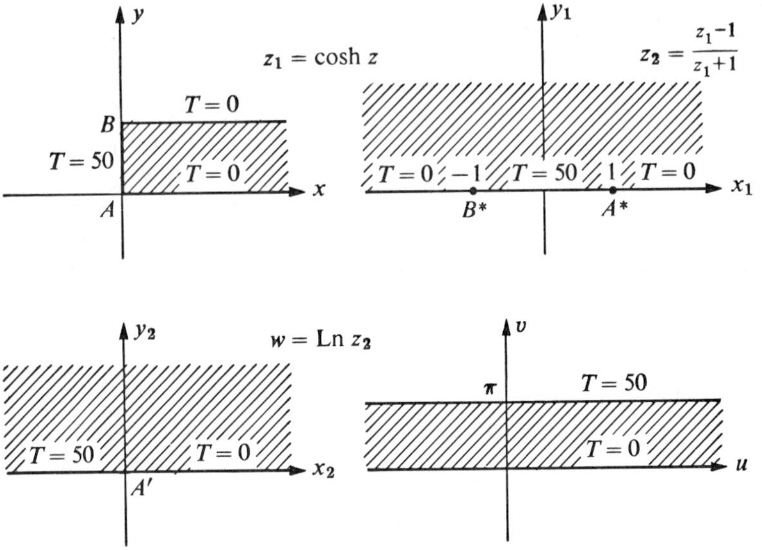

figure 13.10

$$z_1 = \cosh z = \cosh x \cos y + i \sinh x \sin y;$$

$$z_2 = \frac{z_1 - 1}{z_1 + 1}; \qquad w = \text{Ln}\, z_2 .$$

In the w-plane, the solution is

$$T(u, v) = \frac{50}{\pi} v .$$

In the z_1-plane,

$$\begin{aligned} T(x_1, y_1) &= \frac{50}{\pi} (\text{Arg}(z_1 - 1) - \text{Arg}(z_1 + 1)) \\ &= \frac{50}{\pi} \left(\text{Tan}^{-1} \frac{y_1}{x_1 - 1} - \text{Tan}^{-1} \frac{y_1}{x_1 + 1} \right) \\ &= \frac{50}{\pi} \text{Tan}^{-1} \frac{2y_1}{x_1^2 + y_1^2 - 1} . \end{aligned}$$

In the z-plane,

$$\begin{aligned} T(x, y) &= \frac{50}{\pi} \text{Tan}^{-1} \frac{2 \sinh x \sin y}{\sinh^2 x - \sin^2 y} \\ &= \frac{50}{\pi} \text{Arg} (\sinh x + i \sin y)^2 \\ &= \frac{100}{\pi} \text{Tan}^{-1} \frac{\sin y}{\sinh x} . \end{aligned}$$

example 13.19: Find the electrostatic potential in the region illustrated in Figure 13.11. What are the equipotential curves?

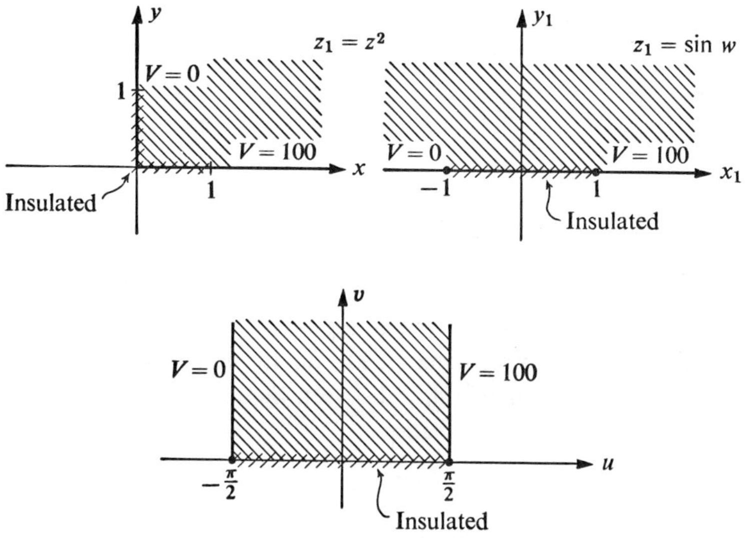

figure 13.11

solution: The mathematical model for the problem in the z-plane is

$$V_{xx} + V_{yy} = 0$$

where

$$\begin{aligned} V(x, 0) &= 100, \quad x > 1; \\ V(0, y) &= 0, \quad y > 1; \\ V_x(0, y) &= 0, \quad 0 < y < 1; \\ V_y(x, 0) &= 0, \quad 0 < x < 1. \end{aligned}$$

The mappings $z_1 = z^2$ and $w = \mathrm{Sin}^{-1} z_1$ transform the given region into the one shown below for the w-plane. The mathematical model in the z_1-plane is

$$V_{x_1x_1} + V_{y_1y_1} = 0$$

where

$$\begin{aligned} V_{y_1} &= 0, \quad -1 < x_1 < 1, \quad y_1 = 0; \\ V(x_1, 0) &= 100, \quad x_1 > 1; \\ V(x_1, 0) &= 0, \quad x_1 < -1. \end{aligned}$$

In the w-plane,

$$V_{uu} + V_{vv} = 0$$

where

$$V\left(\frac{\pi}{2},\ v\right)=100, \quad v>0;$$

$$V\left(\frac{-\pi}{2},\ v\right)=0, \quad v>0;$$

$$V_v(u,0)=0, \quad -\frac{\pi}{2}<u<\frac{\pi}{2}.$$

By inspection,

$$V(u,v)=\frac{100}{\pi}u+50.$$

The equipotential curves in the uv-plane are the lines $u=c$, $0<c<\pi/2$. The pre-images of these lines are determined from the equations $x_1=\sin c \cosh v$ and $y_1=\cos c \sinh v$. Eliminating v between these two equations gives the equipotential curves in the z_1-plane,

$$\frac{x_1^2}{\sin^2 c}-\frac{y_1^2}{\cos^2 c}=1.$$

This family of hyperbolas is confocal with common foci at (1, 0) and (− 1, 0). If we use the definition of the hyperbola ("the difference of the distances of any point on the hyperbola from the two foci is a constant") we obtain

$$\sqrt{(x_1+1)^2+y_1^2}-\sqrt{(x_1-1)^2+y_1^2}=2\sin c.$$

Substituting the appropriate values for x_1 and y_1 from the relation $z_1=z^2$,

$$\sqrt{(x^2-y^2+1)^2+4x^2y^2}-\sqrt{(x^2-y^2-1)^2+4x^2y^2}=2\sin c$$

which are the equipotential curves in the z-plane. Since the lines $u=c$ are the equipotential curves in the w-plane,

$$u=\operatorname{Sin}^{-1}\tfrac{1}{2}\left[\sqrt{(x_1+1)^2+y_1^2}-\sqrt{(x_1-1)^2+y_1^2}\right].$$

Therefore,

$$\begin{aligned}V(x,y)&=50+\frac{100}{\pi}u\\&=50+\frac{100}{\pi}\operatorname{Sin}^{-1}\tfrac{1}{2}\left[\sqrt{(x^2-y^2+1)^2+4x^2y^2}\right.\\&\qquad\qquad\left.-\sqrt{(x^2-y^2-1)^2+4x^2y^2}\right].\end{aligned}$$

exercises for section 13.7

1. Find the potential distribution in the upper half-plane, if the potential along the x-axis is given by

$$\begin{aligned} v(x,0) &= 0, & x &< -a; \\ &= 100, & -a &< x < a; \\ &= 0, & x &> a. \end{aligned}$$

2. Find the potential distribution in the first quadrant if the potential along the positive coordinate axes is given by

$$\begin{aligned} v(x,0) &= 0, \quad 0 < x < 2; & v(0,y) &= 0, \quad 0 < y < 1; \\ &= 100, \quad 2 < x. & &= 100, \quad 1 < y. \end{aligned}$$

(*Hint:* Let $z_1 = z^2$; $z_2 = (z_1 + 1)/(4 - z_1)$; $z_2 = \exp w$.)

3. Find the steady state temperature distribution in the first quadrant if $T = 100°$ along the positive x-axis and $T = 0°$ along the positive y-axis. What are the isotherms? (*Hint:* Let $w = \operatorname{Ln} z$.)

4. Find the temperature distribution in the right half-plane for $-a < y < a$, if

$$\begin{aligned} T(0,y) &= 100, & -a &< y < a; \\ T(x,a) &= T(x,-a) = 0, & x &> 0. \end{aligned}$$

(*Hint:* Let $z_1 = i\pi z/2a$; $z_2 = \sin z_1$; $z_3 = (z_2 - 1)/(z_2 + 1)$; $w = \operatorname{Ln} z_3$.)

5. Find the temperature distribution in the unit circle if the upper half of the boundary is maintained at 100° and the lower half at 0°. What are the isotherms? (*Hint:* Consider the map $z = (w - i)/(w + i)$.)

6. Find the steady state temperature in the first quadrant if

$$\begin{aligned} T(x,0) &= 0°, & x &> 0; \\ T(0,y) &= 100°, & 5 &< y; \\ T_x(0,y) &= 0, & 0 &< y < 5. \end{aligned}$$

(*Hint:* Let $z_1 = iz/5$; $z_2 = \sin w$.)

7. Show that $w = z + 1/z$ maps the exterior of the circle $|z| = 1$ in the upper half-plane onto the upper half-plane. Use this result to find the steady state temperature distribution in the part of the upper half-plane exterior to the unit circle, if $T = 100°$ along the linear portion and 0° along the circular portion of the boundary. (*Hint:* Let $z_1 = \frac{1}{2}[z + 1/z]$; $w = (1 + z_1)/(1 - z_1)$.)

8. Prove Theorem 13.4.

integration in the complex plane

14

14.1 introduction

While it should not be too unexpected to learn that the concepts of integration and differentiation in the complex plane are related, the strength of this relationship may surprise you. The theory of complex integrals plays a major role in establishing results that concern themselves almost completely with differentiation of complex functions. The relationship between integration and differentiation of an analytic function is both profound and fundamental.

The most important result is the famous Cauchy-Goursat theorem from which follow the Cauchy integral formula and the formulas for differentiation of an analytic function in terms of a closed line integral.

14.2 line integral

The line integral of a function, f, of a complex variable is defined analogously to the real line integral of a scalar function (see Section 8.2). Let C be a piecewise smooth, orientable curve which connects points $z = a$ and $z = b$. Subdivide C with a partition B,

$$B = \{a = z_0, z'_0, z_1, z'_1, z_2, z'_2 \cdots z_i, z'_i, z_{i+1} \cdots z_{n-1}, z'_{n-1}, z_n = b\}.$$

(See Figure 14.1.)

For a given partition we evaluate the quantities $f(z'_i)$ at each z'_i,

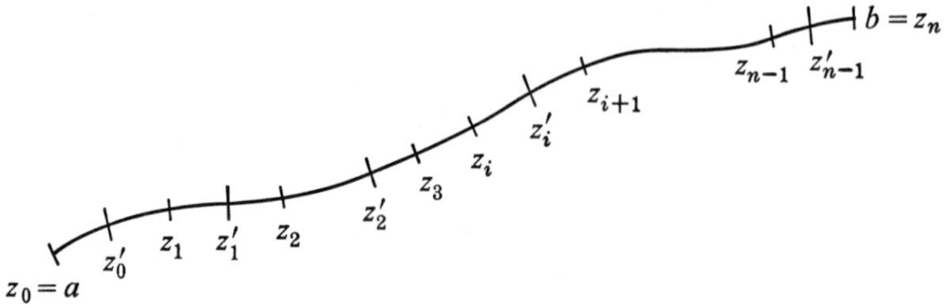

figure 14.1 partitioning of C

$\Delta z_i = z_{i+1} - z_i$, the finite sum

$$S(B) = \sum_{i=0}^{n-1} f(z'_i)\,\Delta z_i,$$

and the real numbers $|\Delta z_i|$. The largest of the real numbers $|\Delta z_i|$ is called "the norm of the partition B" and is denoted by $\Delta(B)$ or $\|\Delta z_i\|$. The value of $S(B)$ is dependent on the partitioning process, but as the norm of the partition is made smaller and smaller, we would expect the finite sums to approach a limit.

definition 14.1

The limit of finite sums (if it exists) as the norm of the partition approaches 0,

$$\lim_{\|z_i\|\to 0} S(B),$$

is called the *line integral of f along C* and is denoted by

$$\int_C f(z)\,dz.$$

If C is closed, we write

$$\oint_C f(z)\,dz.$$

example 14.1: Let C be any path connecting the point $z=a$ to the point $z=b$. Evaluate

$$\int_C dz.$$

(See Figure 14.2.)

solution: For any partition of the curve

$$\begin{aligned}&\{a = z_0, z'_0 \cdots z_i, z'_i, z_{i+1} \cdots z_n = b\},\\ S(B) &= (z_1 - a) + (z_2 - z_1) + \cdots + (z_{i+1} - z_i) + \cdots + (b - z_{n-1})\\ &= b - a\end{aligned}$$

figure 14.2 three paths from $z = a$ to $z = b$

and this result is independent of the partition used. Therefore,

$$\int_C dz = b - a\,.$$

Note that the value of the integral is dependent only on the endpoints of the curve and not on the curve itself.

In complex integration, the integrals

$$\int_C f(z)\,dx, \quad \int_C f(z)\,dy, \quad \text{and} \quad \int_C f(z)\,ds$$

are also defined but only the latter is of interest. It is usually denoted by

$$\int_C f(z)\,|dz|\,.$$

Hence, in the previous example, $\int_C |dz|$ is the length of the path connecting a to b.

example 14.2: Evaluate

$$\oint_C dz \quad \text{and} \quad \oint_C |dz|$$

where C is the closed circle of radius 1 connecting (1, 0) to itself.

solution: From the previous example

$$\oint_C dz = 0$$

since the path is closed and therefore $a = b$. Since

$$\oint_C |dz|$$

is the length of the path, its value is 2π.

A complex line integral can be expressed in terms of real line integrals of scalar functions. In fact, the result of the following theorem is sometimes used as the definition of the line integral of a function of a complex variable.

theorem 14.1

Let $f(z) = u(x, y) + iv(x, y)$. Then

$$\int_C f(z)\, dz = \int_C u(x, y)\, dx - v(x, y)\, dy + i \int_C u(x, y)\, dy + v(x, y)\, dx.$$

proof: For any partition, B, let $z_k' = x_k' + iy_k'$ and $\Delta z_k = \Delta x_k + i\,\Delta y_k$. Then

$$S(B) = \sum_{k=1}^{n} [u(x_k', y_k') + iv(x_k', y_k')][\Delta x_k + i\,\Delta y_k]$$
$$= \sum_{k=1}^{n} u(x_k', y_k')\,\Delta x_k - v(x_k', y_k')\,\Delta y_k + i[u(x_k', y_k')\,\Delta y_k + v(x_k', y_k')\,\Delta x_k]$$

which gives the desired result when we take the limit as $\|\Delta z_k\| \to 0$. (Note that $\|\Delta x_k\| \to 0$ and $\|\Delta y_k\| \to 0$ when $\|\Delta z_k\| \to 0$.)

Since the complex line integral may be expressed in terms of real line integrals, the theorems developed for the real case are also true for the complex case. In particular,

1. If f is continuous on C and C is piecewise smooth, then the complex line integral exists (Theorem 8.1).
2. If the curve C is parametrized by $x = x(t)$, $y = y(t)$, if dx/dt and dy/dt are not both zero, and if f is continuous,

$$\int_C f(z)\, dz = \int_{t_0}^{t_1} \left[u(x(t), y(t)) \frac{dx}{dt} - v(x(t), y(t)) \frac{dy}{dt} \right] dt$$
$$+ i \int_{t_0}^{t_1} \left[u(x(t), y(t)) \frac{dy}{dt} + v(x(t), y(t)) \frac{dx}{dt} \right] dt$$

   where C is a path connecting a and b, $a = x(t_0) + iy(t_0)$, and $b = x(t_1) + iy(t_1)$ (Theorem 8.2.). The substitution in terms of the parameter may sometimes be made directly into the complex function, in which case we have the following formula:

$$\int_C f(z)\, dz = \int_{t_0}^{t_1} f(z(t)) \frac{dz}{dt}\, dt$$

   where C is represented by $z(t) = x(t) + iy(t)$.
3. The value of the line integral is independent of the parametrization (Theorem 8.2). The value usually *is* dependent on the curve used to connect the two points.

example 14.3: Evaluate

$$\oint_C \frac{1}{z}\,dz$$

where C is the circle $|z| = 1$ taken once in the counterclockwise direction starting from $z = 1$.

solution: C may be parametrized by $z(t) = \cos t + i \sin t$ from which $dz/dt = -\sin t + i \cos t$. Then

$$\begin{aligned}\int_C \frac{1}{z}\,dz &= \int_0^{2\pi} \frac{-\sin t + i\cos t}{\cos t + i \sin t}\,dt \\ &= i\int_0^{2\pi} dt \\ &= 2\pi i.\end{aligned}$$

Another possible parametrization is $z(t) = e^{it}$. In this case,

$$\begin{aligned}\int_C \frac{1}{z}\,dz &= \int_0^{2\pi} \frac{ie^{it}}{e^{it}}\,dt \\ &= 2\pi i.\end{aligned}$$

The following properties of the complex line integral are restatements of the same properties for real line integrals as given in Section 8.2:

1. $\int_C kf(z)\,dz = k\int_C f(z)\,dz.$
2. $\int_C [f(z) + g(z)]\,dz = \int_C f(z)\,dz + \int_C g(z)\,dz.$
3. $\int_C f(z)\,dz = -\int_{-C} f(z)\,dz$
   where by $-C$ is meant the path which is the same set of points as C but with opposite orientation.
4. If C is the union of C_1 and C_2, then

$$\int_C f(z)\,dz = \int_{C_1} f(z)\,dz + \int_{C_2} f(z)\,dz.$$

example 14.4: Evaluate $\int \text{Re}\,z\,dz$ along:

(a) The straight line path from 0 to $1 + i$.
(b) The rectangular path from 0 to 1 and then to $1 + i$.

solution: (a) One parametrization of the first path is $z(t) = (1 + i)\,t$,

$0 \leqq t \leqq 1$. Thus,

$$\int_C x\,dz = \int_0^1 (1+i)\,t\,dt = \frac{1+i}{2}.$$

(b) The second path may be written,

$$\begin{aligned} z(t) &= t, & 0 \leqq t \leqq 1; \\ &= 1 + (t-1)\,i, & 1 \leqq t \leqq 2. \end{aligned}$$

Therefore,

$$\int_C x\,dz = \int_0^1 t\,dt + i \int_1^2 dt = \tfrac{1}{2} + i.$$

Sometimes a numerical estimate of the magnitude of the line integral of f is all that you will need. The following theorem is valuable in making such estimates.

theorem 14.2

Let L be the length of C and M an upper bound for $|f(z)|$ for z on C. Then

$$\left|\int_C f(z)\,dz\right| \leqq ML.$$

proof: For any partition,

$$\begin{aligned} |S(B)| &= \left|\sum_{k=0}^{n-1} f(z_k')\,\Delta z_k\right| \\ &\leqq \sum_{k=0}^{n-1} |f(z_k')\,\Delta z_k| \text{ (because of the triangle inequality)} \\ &= \sum_{k=0}^{n-1} |f(z_k')|\,|\Delta z_k| \leqq M \sum_{k=0}^{n-1} |\Delta z_k| \leqq ML. \end{aligned}$$

By reasoning similar to that used in the proof of Theorem 14.2, we also have the following related theorem.

theorem 14.3

$$\left|\int_C f(z)\,dz\right| \leqq \int_C |f(z)|\,|dz|.$$

example 14.5: Find an upper bound for

$$\left|\int_C \operatorname{Ln}(z+1)\,dz\right|$$

where C is the line segment from i to $2+i$. (See Figure 14.3.)

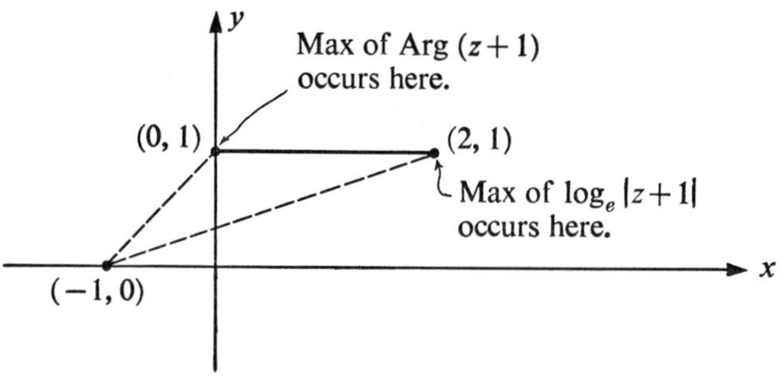

figure 14.3

solution:

$$\left|\int_C \operatorname{Ln}(z+1)\,dz\right| \leqq [\operatorname{Max}|\operatorname{Ln}(z+1)|]\,[\text{distance from } i \text{ to } 2+i].$$

A bound for $|\operatorname{Ln}(z+1)|$ on the line segment is found by finding bounds on the real and imaginary parts of $\operatorname{Ln}(z+1)$,

$$|\operatorname{Ln}(z+1)| \leqq \big|\log_e|z+1|\big| + |\operatorname{Arg}(z+1)|.$$

Since $|z+1|$ represents the distance from $z=-1$, the maximum value of $|z+1|$ is $\sqrt{10}$ (which occurs at $z=2+i$) and the maximum value of $\operatorname{Arg}(z+1)$ is $\pi/4$ (which occurs at $z=i$). Hence,

$$\int_C \operatorname{Ln}(z+1)\,dz \leqq \left[\log_e\sqrt{10}+\frac{\pi}{4}\right][2]$$
$$\approx (1.15+0.79)2 = (1.94)2$$
$$= 3.88.$$

Better estimates are possible.

exercises for sections 14.1 and 14.2

1. Consider the integral $\int_C z\,dz$ where C is any path connecting a to b.
   (a) Find S if the points of evaluation are the left-hand endpoints of the subinterval in a given partition.
   (b) Find the finite sum if the right-hand endpoints are used, and call it S_n^*.
   (c) Sum $S_n + S_n^*$. Under the assumption that the given integral exists, then
   $$\lim_{n\to\infty}(S_n+S_n^*) = 2\int_C z\,dz.$$
   Use this technique to evaluate the given integral.

2. Evaluate $\int_C dz$ and $\int_C |dz|$ where C is the perimeter of the square $-1 \leqq x \leqq 1$, $-1 \leqq y \leqq 1$.

3. Same as Exercise 2 except for C use the straight line path connecting $1 + i$ to $7 + 6i$.

4. Evaluate $\int_C |z|^2 \, dz$ if C is:
   (a) The path $y = x^2$ from (1, 1) to (2, 4).
   (b) The straight line path connecting the two points.

5. Find a bound on $\left|\int_C e^z \, dz\right|$ where C is the part of the circle $|z| = 1$ connecting (1, 0) to (0, 1).

6. Evaluate $\oint_C (z - z_0)^{-1} \, dz$ where C is a circle of any radius centered at z_0 taken once in the counterclockwise direction. Repeat, if C is the same path but taken twice in the clockwise direction.

7. Find a better estimate for the integral of Example 14.5 by subdividing the given path into two.

8. Evaluate $\int_C \operatorname{Re} z \, dz$ where C is:
   (a) The straight line path from 0 to i.
   (b) The circular path from 0 to i which is the right-hand side of the circle $|z - (1/2)\,i| = 1/2$.
   (c) The parabola $y = 2x^2 + 1$ from (1, 3) to (2, 9).
   (d) The circle $|z| = 7$, taken once in the counterclockwise direction.
   (e) The ellipse $4x^2 + y^2 = 4$ taken once in the clockwise direction.

9. By using Theorem 14.2 find bounds on:
   (a) $\left|\int_C \frac{1}{z^2} \, dz\right|$ where C is the line segment from i to $2 + i$.
   (b) $\left|\int_C e^z \, dz\right|$ where C is the upper semicircular path connecting 0 to 2π over $|z - \pi| = \pi$.
   (c) $\left|\int_C \frac{\sin z}{z} \, dz\right|$ where C is the straight line path from π to 2π.
   (d) $\left|\int_C \operatorname{Ln} z \, dz\right|$ where C is the straight line path from 1 to $1 + i$.
   (e) $\left|\int_C z^{-2} \, dz\right|$ where C is the straight line path from -1 to -2.

10. Evaluate $\int_C \sin \pi z \, dz$ over the straight line paths from:

(a) 1 to $1+3i$. (b) i to $3+i$.

11. Evaluate $\int_C \left(\frac{1}{z-2} + \frac{1}{(z-2)^2} \right) dz$ where C is the circle $|z-2| = 3$ taken once in the counterclockwise direction.

12. Evaluate $\int_C (z^2 - z^{-1} + z^{-2}) \, dz$ where C is the unit circle taken once in the clockwise direction.

14.3 fundamental theorem of complex integration

The theory of integration of real functions of a real variable is highlighted by the relationship between antidifferentiation and integration. The theory of line integration in the presence of vector fields is highlighted by a similar theorem (Theorem 8.4) stating conditions under which the integral is dependent only on the endpoints of the curve and the fact that the value of the integral is equal to the potential difference of the two endpoints. A similar theorem, called the fundamental theorem of complex line integration, is true for domains in the complex plane.

Before proving the fundamental theorem, we need to review the concept of independence of path (analogous to Definition 8.3 for vector fields) and how it relates to integrals around closed paths (analogous to Theorem 8.3.)

definition 14.2

The line integral $\int_C f(z) \, dz$ is *independent of the path* connecting two points in a domain D if the value of the integral is the same for all smooth paths connecting the points. We say that the value of the integral depends only the endpoints of C.

theorem 14.4

$\int_C f(z) \, dz$ is independent of path in a domain D if and only if $\oint_C f(z) \, dz = 0$ for all closed paths in D.

proof: Prove the theorem as an exercise. (*Hint:* See Theorem 8.3.)

The principle of independence of path may be interpreted in terms of a continuous deformation of a path. If the path is continuously deformed, if the endpoints are kept fixed, and if the curve stays entirely within the domain for which the independence of path property is valid, then the value of the line integral is unchanged.

theorem 14.5

Fundamental theorem of complex line integration. Let f be continuous on a domain D. Then $\int_C f(z)\,dz$ is independent of the path connecting two points in D if and only if there exists a function, F, such that $F'(z)=f(z)$ for all z in D.

proof: Suppose $\int_C f(z)\,dz$ is independent of path in D and let z_0' be in D. Consider any path C, lying in D, which connects the fixed point z_0' to any other point z. Define F by the rule

$$F(z_0)=\int_C f(z)\,dz.$$

Because of the independence of path, this is a well-defined value dependent only on z_0. We now show $F'(z)=f(z)$.

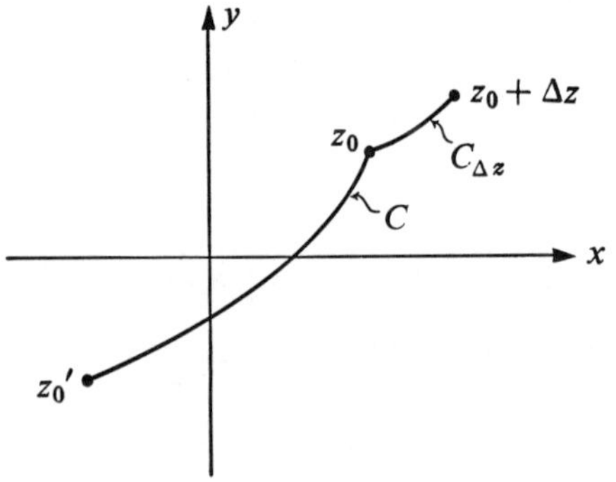

figure 14.4

Let $C_{\Delta z}$ be any path from z_0 to $z_0+\Delta z$. (See Figure 14.4.) Forming the difference quotient,

$$\frac{F(z_0+\Delta z)-F(z_0)}{\Delta z}=\frac{1}{\Delta z}\int_{C_{\Delta z}} f(z)\,dz.$$

To show that the limit of this difference quotient is $f(z_0)$, we subtract $f(z_0)$

from it and then go to the limit,

$$\frac{F(z_0+\Delta z)-F(z_0)}{\Delta z}-f(z_0)=\frac{1}{\Delta z}\int_{C_{\Delta z}} f(z)\,dz-\frac{f(z_0)}{\Delta z}\int_{C_{\Delta z}} dz$$
$$=\frac{1}{\Delta z}\int_{C_{\Delta z}} (f(z)-f(z_0))\,dz.$$

Since f is continuous on $C_{\Delta z}$, the difference $|f(z)-f(z_0)|$ can be made smaller than any arbitrary ε if $|z-z_0|$ is made small enough. Using Theorem 14.2,

$$\frac{F(z_0+\Delta z)-F(z_0)}{\Delta z}-f(z_0)<\frac{1}{\Delta z}\varepsilon L<\frac{1}{\Delta z}\varepsilon(\Delta x+\Delta y)<2\varepsilon.$$

Since ε is arbitrarily small, the limit of the left side is zero and therefore $F'(z_0)=f(z_0)$.

Conversely, suppose an antiderivative for f exists throughout D. We denote it by $F(z)=U(x,y)+iV(x,y)$. Therefore, $U_x=V_y=u$ and $V_x=-U_y=v$, and hence

$$\int_C f(z)\,dz=\int_C U_x\,dx+U_y\,dy+i\int_C V_x\,dx+V_y\,dy$$
$$=\int_{t_1}^{t_2}\frac{d}{dt}U(x,y)\,dt+i\int_{t_1}^{t_2}\frac{d}{dt}V(x,y)\,dt$$

where t_1 and t_2 are the parameter values corresponding to the endpoints (x_2,y_2) and (x_1,y_1) of the curve. Then

$$\int_C f(z)\,dz=U(x_2,y_2)-U(x_1,y_1)+i(V(x_2,y_2)-V(x_1,y_1))$$
$$=F(z_2)-F(z_1)$$

which tells us that the value of the integral is dependent only on the difference of the antiderivative values at the endpoints of C.

The antiderivative of f must be single-valued throughout D. Occasionally, this will mean using a single-valued branch of a multiple-valued relation. These branches are discontinuous at their branch cuts and hence, paths which cross the branch cut should be avoided or another branch with a different cut should be used.

example 14.6: Evaluate $\int_C z^2\,dz$ where C is the path

$$z(t)=\sin^2\pi t+i\cos 2\pi t$$

from $t=0$ to $t=1/2$.

solution: The endpoints of the curve are i and $1-i$ and an antiderivative of z^2 throughout the plane is $z^3/3$. Therefore,

$$\int_C z^2\,dz = \frac{z^3}{3}\bigg|_i^{1-i} = \frac{(1-i)^3}{3} - \frac{i^3}{3} = \frac{-2-i}{3}.$$

example 14.7: Evaluate $\int_C z^m\,dz$ where C is the unit circle and m is an integer.

solution: If $m \neq -1$, a single-valued antiderivative throughout the domain excluding the origin is $z^{m+1}/(m+1)$. Therefore $\int_C z^m\,dz = 0$ for $m \neq -1$. If $m = -1$, the value of the integral is $2\pi i$ (see Example 14.4.).

You may think that

$$\int_C \frac{1}{z}\,dz$$

should be independent of path, or, equivalently, the value of

$$\oint_C \frac{1}{z}\,dz$$

around any closed path not passing through the origin should be zero. For, you might reason, $\ln z$ is an antiderivative of $1/z$ and Theorem 14.4 may be applied. However, for any single-valued branch of $\ln z$, there is a ray from the origin which is a branch cut and thus a line of discontinuity. If the integration is over a path enclosing the origin, this line of discontinuity is crossed, and Theorem 14.4 does not apply. Any single-valued branch of $\ln z$ can be used to evaluate line integrals of the type

$$\int_C \frac{1}{z}\,dz$$

if C does not cross a branch cut.

example 14.8: Evaluate $\oint_C \frac{1}{z}\,dz$, where C is the unit circle taken in the counterclockwise direction, by the method of finding antiderivatives.

solution: We split the circle into upper and lower semicircles C_u and C_L, and then

$$\oint_C \frac{1}{z}\,dz = \int_{C_u} \frac{1}{z}\,dz + \int_{C_L} \frac{1}{z}\,dz.$$

Over C_u, we use the antiderivative whose branch cut is the negative imaginary axis,

$$F_1(z) = \log_e|z| + i \arg z\,; \; -\frac{\pi}{2} < \arg z \leqq \frac{3\pi}{2}.$$

Over C_L, we place the branch cut on the positive y-axis,

$$F_2(z) = \log_e|z| + i \arg z\,; \quad \frac{\pi}{2} < \arg z \leqq \frac{5\pi}{2}.$$

Using the result of Theorem 14.4 on each integral,

$$\begin{aligned}\oint_C \frac{1}{z}\,dz &= F_1(-1) - F_1(1) + F_2(1) - F_2(-1)\\ &= \pi i - 0 + 2\pi i - \pi i\\ &= 2\pi i.\end{aligned}$$

exercises for sections 14.3

1. Evaluate $\int_C \sin z\,dz$ where C is the path $y = x^2$ from $z = 0$ to $z = 1 + i$.
2. Evaluate $\int_C z^{-1}\,dz$ where C is the right semicircular path $|z| = 1$ taken counterclockwise from $-i$ to i.
3. Evaluate $\int_C z^{1/2}\,dz$ where C is the upper semicircle $|z| = 1$ from 1 to -1 and for $z^{1/2}$ we choose the branch obtained from the principal values of $\arg z$.
4. Repeat Exercise 3 using the branch of $z^{1/2}$ corresponding to $\pi < \arg z \leqq 3\pi$.
5. Evaluate $\int_C z^2\,dz$ where C is the path $z(t) = t^3 + \sin^2 \pi t$ from $t = 1$ to $t = 2$.
6. Evaluate $\int_C z^2\,dz$ where C is the straight line path connecting 0 to $1 + i$.
7. Evaluate Exercises 10, 11, and 12 of the previous exercise set by the methods of this section.
8. Evaluate the following integrals over straight line paths connecting the two points indicated:

   (a) $\int_C \sin 2z\,dz$, from π to 3π. (b) $\int_C z \sin z\,dz$, from 0 to 1.

(c) $\int_C \frac{1}{z(z+1)}\,dz$, from i to $3+i$.

(d) $\int_C \frac{1}{z}\,dz$, from 1 to $3+i$.

(e) $\int_C e^{3zi}\,dz$, from $1+2\pi i$ to $1-2\pi i$.

(f) $\int_C (z^3+z^2+z+1)\,dz$, from 0 to $3i$.

(g) $\int_C \frac{1}{z^2+4}\,dz$, from 0 to i.

14.4 the cauchy-goursat theorem

In the previous section we saw that if f has a continuous antiderivative throughout a domain, D, and if C is any closed path in D, then

$$\oint_C f(z)\,dz = 0.$$

In this section we will show that if the domain, D, is simply connected† and if f is analytic in D, then this same result is true without the *assumption* that f has an antiderivative. Later, using this theorem, we will be able to *prove* that antidifferentiability of f is a direct consequence of f being analytic in a simply connected domain D.

The Cauchy-Goursat Theorem, sometimes simply called Cauchy's theorem, is one of the cornerstones to mathematical progress. The result was first published by Cauchy in 1825 under the hypothesis not only that f be analytic (that is, f' exists) throughout a simply connected domain, D, but also that f' be continuous. The proof which we give in this section is due to Cauchy. It is a straightforward application of Green's theorem and the Cauchy-Riemann conditions.

A new idea for the proof was introduced by Goursat over fifty years later in 1883 which avoided the continuity assumption on f'. The resulting proof was quite delicate, and modified versions are found in most modern texts on functions of a complex variable. The theorem is a dramatic example of the importance of assuming as few hypotheses as necessary. For, by not making the *assumption* of continuity for f', it is possible to *prove* that f' is continuous.

† A domain, D, in the plane is simply connected if it is the interior of a simple closed curve. Otherwise, it is called multiply connected. See Section 8.9 for another characterization.

theorem 14.6

Let $f(z) = u(x, y) + iv(x, y)$ be analytic in a simply connected bounded domain D, and let C be a simple closed rectifiable curve in D. Then

$$\oint_C f(z)\,dz = 0.$$

proof: (Due to Cauchy)

$$\oint_C f(z)\,dz = \oint_C u(x, y)\,dx - v(x, y)\,dy + i \oint_C u(x, y)\,dy + v(x, y)\,dx.$$

If we assume that f' is continuous in D, then Green's theorem in the plane may be applied to each of the two real line integrals.

$$\oint_C f(z)\,dz = \iint_R [-v_x(x, y) - u_y(x, y)]\,dx\,dy + \iint_R [u_x(x, y) - v_y(x, y)]\,dx\,dy$$

where R is the domain interior to C. Because of the Cauchy-Riemann conditions, the integrands of each of the double integrals is zero and hence

$$\oint_C f(z)\,dz = 0.$$

corollary: If f is analytic in a simply connected domain D, then f has a continuous antiderivative throughout D.

proof: See the next set of exercises. (*Hint:* Consider this theorem and the two theorems of the previous section.)

example 14.9: Evaluate $\oint_C e^z\,dz$ for any closed path C in the finite z-plane.

solution: Since e^z is analytic everywhere, $\oint_C e^z\,dz = 0$.

example 14.10: Evaluate $\oint_C \frac{dz}{z}$ if C is the circle $|z - 2| = 1$.

solution: The value is zero, since the function is analytic within and on a circle of radius 1 centered at $z = 2$.

14.5 principle of deformation of path

In this section we show the circumstances in which the shape of a closed path does not affect the value of the line integral. As one immediate consequence, if the integrand is analytic, rather arbitrary closed curves may be replaced by circles (in fact *any* other closed curve is allowable).

We state and prove the theorem for the case in which f is analytic between two closed curves. The extension to other types of domains is obvious.

theorem 14.7

Let C_0 and C_i be two closed curves in a domain D (see Figure 14.5) and suppose f is analytic at all points between and on C_0 and C_i. Then

$$\oint_{\substack{C_0 \\ (\text{ccw})}} f(z)\,dz = \oint_{\substack{C_i \\ (\text{ccw})}} f(z)\,dz.$$

proof: A picture of the situation is shown in Figure 14.5 where we have assumed at least the possibility of f being nonanalytic at P. We make two

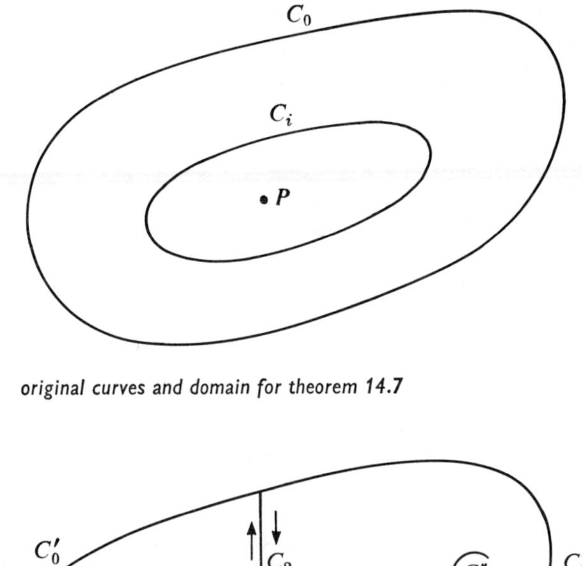

figure 14.5 *original curves and domain for theorem 14.7*

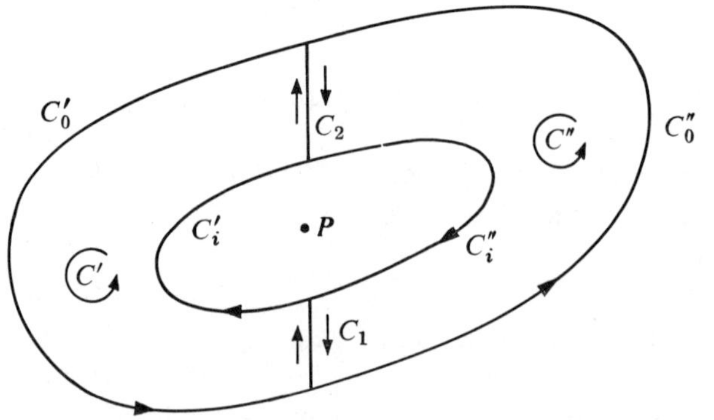

figure 14.6 *domain after insertion of cuts*

cuts C_1 and C_2 as shown in Figure 14.6, thereby establishing curves C' and C'' which bound a simply connected domain in which f is analytic. Therefore, the Cauchy-Goursat theorem may be applied to each one.

$$\oint_{\substack{C' \\ (\text{ccw})}} f(z)\,dz = \int_{\substack{C'_0 \\ (\text{ccw})}} f(z)\,dz + \int_{\substack{C_1 \\ (\text{up})}} f(z)\,dz + \int_{\substack{C'_i \\ (\text{cw})}} f(z)\,dz + \int_{\substack{C_2 \\ (\text{up})}} f(z)\,dz = 0$$

$$\oint_{\substack{C'' \\ (\text{ccw})}} f(z)\,dz = \int_{\substack{C''_0 \\ (\text{ccw})}} f(z)\,dz + \int_{\substack{C_1 \\ (\text{down})}} f(z)\,dz + \int_{\substack{C''_i \\ (\text{cw})}} f(z)\,dz + \int_{\substack{C_2 \\ (\text{down})}} f(z)\,dz = 0$$

where the sense of the integration over each curve is as indicated. Adding the members of the last two equalities, we obtain

$$\oint_{\substack{C_0 \\ (\text{ccw})}} f(z)\,dz + \oint_{\substack{C_i \\ (\text{cw})}} f(z)\,dz = 0$$

which means that

$$\oint_{\substack{C_0 \\ (\text{ccw})}} f(z)\,dz = \oint_{\substack{C_i \\ (\text{ccw})}} f(z)\,dz$$

and the theorem is proved.

The extension of this idea to a situation which would require more cuts is shown in Figure 14.7. It is easy to show that

$$\oint_{\substack{C_1 \\ (\text{ccw})}} f(z)\,dz = \oint_{\substack{C_2 \\ (\text{ccw})}} f(z)\,dz + \oint_{\substack{C_3 \\ (\text{ccw})}} f(z)\,dz.$$

The content of the previous theorem is a special case of the principle of deformation of path: If a path C_0 is obtained from C_i by a continuous deformation, the value of the line integral will not change, as long as any of the deformed paths do not pass through a point where $f(z)$ is not analytic. The principle, in slightly different form, was discussed in Section 14.3.

example 14.11: Let C be *any* closed path which encloses the origin once. Evaluate $\oint_C \frac{1}{z}\,dz$.

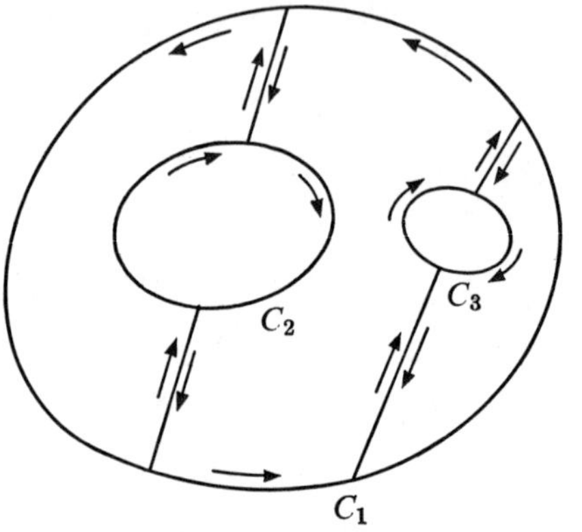

figure 14.7 *domain requiring four cuts*

solution: By the principle of deformation of path,

$$\oint_C \frac{1}{z}\,dz = \oint_{C_1} \frac{1}{z}\,dz$$

where C_1 is the unit circle. By Example 14.8 the value of the integral is $2\pi i$.

example 14.12: Evaluate

$$\oint_C \frac{2z-1}{z^2-z}\,dz$$

where C is a curve which contains the points 0 and 1 in its interior.

solution: By the usual partial fractions expansion,

$$\oint_C \frac{2z-1}{z^2-z}\,dz = \oint_C \frac{dz}{z} + \oint_C \frac{dz}{z-1}.$$

The first integral has value $2\pi i$ from the previous example. In the second integral, we replace the curve C by a circle C_2 of radius 1 about $z = 1$. This circle may be represented by $z(t) = 1 + e^{it}$. With this parametrization it is easy to show that the value of the second integral is also $2\pi i$ and hence

$$\oint_C \frac{2z-1}{z^2-z}\,dz = 4\pi i.$$

example 14.13: Evaluate

$$\oint_C \frac{2z-1}{z^2-z}\,dz$$

where C is the circle $|z| = 1/2$. (See Figure 14.8.)

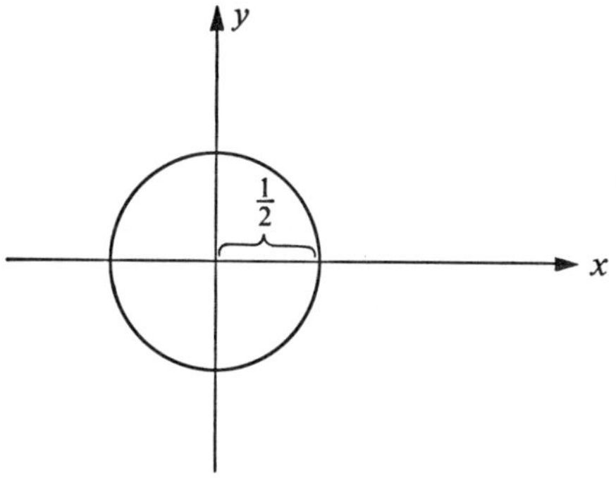

figure 14.8

solution: As in the previous example,

$$\oint_C \frac{2z-1}{z^2-z}\,dz = \oint_C \frac{dz}{z} + \oint_C \frac{dz}{z-1}.$$

The first integral has value $2\pi i$ as before, but the second integral is zero since the integrand is analytic within and on C. Therefore

$$\oint_C \frac{2z-1}{z^2-z}\,dz = 2\pi i.$$

14.6 cauchy's integral formula

This is the second major consequence of Cauchy's theorem. The formula yields a means of representing an analytic function in terms of a line integral.

theorem 14.8

Let f be analytic in a simply connected domain D and let z_0 be a point of D. If C is a simple closed path in D which encloses z_0 and is taken in the counterclockwise sense, then

$$\frac{1}{2\pi i}\oint_C \frac{f(z)}{z-z_0}\,dz = f(z_0).$$

(See Figure 14.9.)

proof:

$$\oint_C \frac{f(z)}{z-z_0}\,dz = \oint_C \frac{f(z)-f(z_0)}{z-z_0}\,dz + f(z_0)\oint_C \frac{dz}{z-z_0}.$$

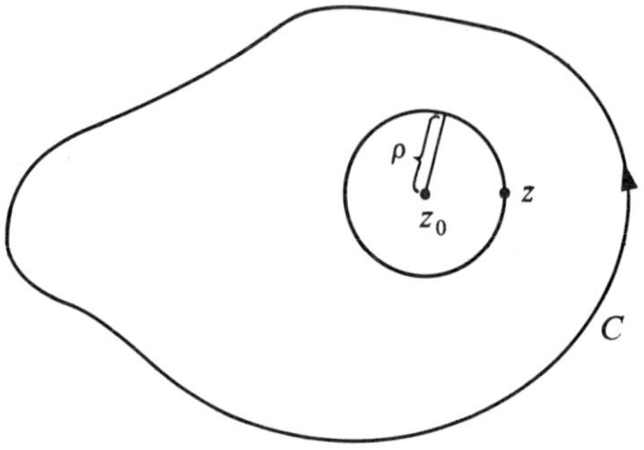

figure 14.9

The value of

$$\oint_C \frac{dz}{z - z_0} = 2\pi i$$

by the principle of deformation of path and the fact that the value is $2\pi i$ if the path is a circle. The proof will be completed if we show the value of the first integral to be zero. Replace C by a circle of radius ρ centered at z_0. By the continuity of f, there exists a δ such that $|f(z) - f(z_0)| < \varepsilon$ if $|z - z_0| < \delta$. Therefore we choose ρ to be less than δ and we have

$$\left| \int_{|z-z_0|=\rho} \frac{f(z) - f(z_0)}{z - z_0} \, dz \right| \leq \int_{|z-z_0|=\rho} \frac{|f(z) - f(z_0)|}{|z - z_0|} |dz|$$
$$\leq \frac{\varepsilon}{\rho} \int_{|z-z_0|=\rho} |dz| = 2\pi\varepsilon .$$

Since ε can be made arbitrarily small, the value of the integral is zero and the theorem is proved.

example 14.14: Find the value of

$$\oint \frac{z^2 + 1}{z^2 - 1} \, dz$$

where C is:

(a) $|z - 1| = 1$. (b) $|z - 1/2| = 1$. (c) $|z + 1| = 1$. (d) $|z - i| = 1$.

solution: (a) $$\oint_C \frac{z^2 + 1}{z^2 - 1} \, dz = \oint_C \frac{f(z)}{z - 1} \, dz$$

where $f(z) = (z^2 + 1)/(z + 1)$. Since f is analytic within and on the simply connected domain bounded by C and since $z = 1$ is in D,

$$\oint_C \frac{z^2+1}{z^2-1}\,dz = 2\pi i\, f(1) = 2\pi i .$$

(b) This answer is the same as part (a) by the principle of deformation of path.

(c) In this case we can write

$$\oint_C \frac{z^2+1}{z^2-1}\,dz = \oint_C \frac{g(z)}{z+1}\,dz$$

where $g(z) = (z^2+1)/(z-1)$. The function g is analytic within the circle $|z+1| = 1$, and thus

$$\oint_C \frac{z^2+1}{z^2-1}\,dz = 2\pi i\, g(-1) = -2\pi i .$$

(d) The value is zero by the Cauchy-Goursat theorem, since the integrand is analytic within and on the circle of radius 1 centered at $z = i$.

The Cauchy integral formula tells us that all values of an analytic function in a simply connected domain are determined once the functional values on the curve which bounds the domain are defined. The same statement is true for real harmonic functions.

Cauchy's integral formula may be extended to other than simply connected domains by use of cuts in a manner similar to the previous section. For example, if f is analytic between and on C_0 and C_i, and z_0 is any point in the domain bounded by the two curves, then

$$f(z_0) = \frac{1}{2\pi i}\int_{\substack{C_0\\(\text{ccw})}} \frac{f(z)}{z-z_0}\,dz - \frac{1}{2\pi i}\int_{\substack{C_i\\(\text{ccw})}} \frac{f(z)}{z-z_0}\,dz .$$

(See Figure 14.10.)

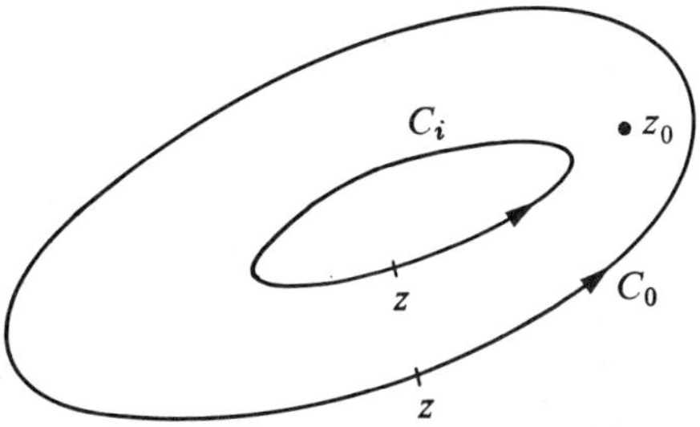

figure 14.10 *Cauchy's integral formula for non-simply connected domain*

14.7 derivatives of an analytic function

We now use the Cauchy integral formula to show that an analytic function has derivatives of all order. The actual formulas for the derivatives are not important computationally so the theorem's importance is in the statement of existence. Note that the only assumption necessary for f is that f' exist throughout D. No continuity assumptions are necessary for f'. This is particularly crucial since we will then be able to show that f' *is* continuous, a situation significantly different from that for functions of a real variable.

We will not prove the theorem since the technique of proof is similar to that used in the previous two theorems. Formally, the formulas may be obtained by "differentiating inside the integral sign."

theorem 14.9

If f is analytic in D, then it has derivatives of all order. The specific formula for the nth derivative at the point z_0 in D is given by the line integral around a closed path in D which encloses z_0,

$$f^{(n)}(z_0) = \frac{n!}{2\pi i} \oint_C \frac{f(z)}{(z - z_0)^{n+1}}\, dz.$$

example 14.15: Evaluate

$$\oint_C \frac{z^4 + 1}{(2z + 1)^3}\, dz$$

where C is the unit circle.

solution:

$$\oint_C \frac{z^4 + 1}{(2z + 1)^3}\, dz = \frac{1}{8} \oint_C \frac{z^4 + 1}{(z + \frac{1}{2})^3}\, dz = \frac{1}{8} \oint_C \frac{f(z)}{(z + \frac{1}{2})^3}\, dz,$$

where $f(z) = z^4 + 1$. Since

$$\oint_C \frac{f(z)}{(z + \frac{1}{2})^3}\, dz = \frac{2\pi i}{2!} f''(-\tfrac{1}{2}) = \pi i (12z^2)\big|_{z=-\frac{1}{2}} = 3\pi i,$$

the given integral is equal to $3\pi i/8$.

With the use of Theorem 14.9 several very important corollaries can be proven.

corollary 1 (*Morera*): Let f be continuous in a simply connected domain D. Then, $\oint_C f(z)\,dz = 0$ around all closed paths in D if and only if f is analytic in D.

proof: The "if" part of the theorem is just Cauchy's theorem. If $\oint_C f(z)\,dz = 0$ around all closed paths in D, then the line integral $\int_C f(z)\,dz$ is independent of path and hence f has an antiderivative, F, throughout D. Since F is analytic in D, it has derivatives of all order; in particular F'' exists and is equal to f'. Hence f is analytic in D.

corollary 2 (*Cauchy's Inequality*): Let C be a circle of radius r with center at z_0 and let $M = \max_{z \text{ on } C} |f(z)|$. Then

$$|f^{(n)}(z_0)| \leqq \frac{n!M}{r^n}.$$

proof:

$$\begin{aligned} |f^{(n)}(z_0)| &= \frac{n!}{2\pi}\left|\oint_C \frac{f(z)\,dz}{(z-z_0)^{n+1}}\right| \\ &\leqq \frac{n!}{2\pi} M \frac{1}{r^{n+1}} 2\pi r \\ &= \frac{n!M}{r^n}. \end{aligned}$$

Cauchy's inequality, like many things in mathematics, is important because of results to which it leads. For example, we have the following corollary which is often called Liouville's theorem.

corollary 3: If f is analytic and bounded in absolute value for all finite z, then $f(z)$ is a constant.

proof: By Cauchy's Inequality as applied to the first derivative, $|f'(z)| \leqq M/r$. If M is bounded, say by M_1, then $|f'(z)| \leqq M_1/r$ which goes to 0 as r becomes unbounded. Thus $f'(z) = 0$ which means $f(z) =$ a constant.

Liouville's theorem may be used to prove the fundamental theorem of algebra. To prove this theorem by algebraic methods is quite difficult but it is an easy consequence of the results of this section.

theorem 14.10

Fundamental Theorem of Algebra. If $P(z)$ is a polynomial in z, not a constant,

$$P(z) = \sum_{k=0}^{n} a_k z^k, \qquad n \geqq 1,\ a_n \neq 0,$$

then $P(z) = 0$ for at least one value of z.

proof: Suppose $P(z) \neq 0$ for all z. Then, the function

$$f(z) = \frac{1}{P(z)}$$

is everywhere analytic. Also, $|f(z)|$ approaches 0 as $|z| \to \infty$, so that $|f(z)|$ is bounded for all finite z. By Liouville's theorem, $f(z)$ is a constant. But this would mean $P(z)$ is also a constant which is an obvious contradiction to the hypothesis. Hence, $P(z) = 0$ for at least one value of z.

exercises for sections 14.4–14.7

1. Evaluate

$$\int_C \frac{z+2}{z^2+z}\,dz$$

where C is as follows and where the curves are taken in a counterclockwise sense:

(a) $|z| = 1/2$. (b) $|z-1| = 3$. (c) $|z-1| = 1/2$. (d) $|z-i| = 1/2$.

2. When does Cauchy's theorem apply to $\oint_C z^m\,dz$? Explain.

3. Evaluate $\int_C \frac{dz}{z}$ where $C = C_1 \cup C_2$, C_1 is the circle $|z| = 1$ taken counterclockwise and C_2 is the circle $|z| = 2$ taken clockwise.

4. Evaluate

$$\int_C \frac{dz}{(z-1)(z-2)}$$

where C is as follows and all the curves are taken in a counterclockwise sense:

(a) $|z| = 1$. (b) $|z-1| = 1/2$. (c) $|z| = 3/2$.
(d) $|z-1| = 1$. (e) $|z-1| = 2$. (f) $|z| = 4$.

5. Evaluate $\int_C \frac{e^z}{z}\,dz$ where C is the circle $|z| = 2$ taken counterclockwise.

6. Evaluate $\int_C \frac{dz}{z^2(z^2+9)}$ where C is:

(a) $|z-1|=1$. (b) $|z-1|=2$, taken in a counterclockwise sense.

7. Evaluate

$$\int_C \frac{z^2-1}{z^2+1}\,dz$$

where C is as follows and all the curves are taken in a counterclockwise sense:

(a) $|z-1|=1$. (b) $|z-i|=1$. (c) $|z+i|=1$.
(d) $|z-2i|=2$. (e) $|z|=1/2$.

8. Evaluate the following integrals over $|z|=1$, in a counterclockwise sense:

(a) $\oint_C \frac{e^z}{z}\,dz$. (b) $\oint_C \frac{e^z-1}{z}\,dz$. (c) $\oint_C \frac{\cos z}{z}\,dz$.

(d) $\oint_C \frac{z^2-1}{z^2+2}\,dz$. (e) $\oint_C \frac{z^3}{\left(z-\frac{i}{2}\right)}\,dz$. (f) $\oint_C \frac{\cos z}{z^2}\,dz$.

(g) $\oint_C \frac{\cos z}{z^4}\,dz$. (h) $\oint_C \frac{z^8+1}{(z-\frac{1}{2})^7}\,dz$. (i) $\oint_C \frac{e^z}{z^4}\,dz$.

(j) $\oint_C \frac{\sin z}{z^4}\,dz$.

9. Exhibit a function of a real variable for which $f'(x_0)$ exists but f' is not continuous at x_0. Repeat for complex functions.

complex series

15

15.1 introduction

You have seen that an analytic function of a complex variable has immediate application to conformal mapping and consequently to solving the Dirichlet problem in two dimensions. This chapter may seem a bit more theoretically oriented since we will show how convergent power series and analytic functions are identified. One often used approach to the theory of analytic functions is to define an analytic function as one having a power series representation.

Our approach in the early part of this chapter will be sketchy, since many of the results are precisely the same as for real functions and real series. In many cases, even the proofs have the same form.

15.2 elementary definitions

We begin by making several definitions which are similar to those in Appendix A.

definition 15.1

A complex sequence is a complex function whose domain is the set of positive integers. The range values are called "terms" of the sequence. We denote them by $z_1, z_2, z_3, \cdots, z_n, \cdots$.

As in the real case, the sequences of most interest are those whose terms have a limiting value in the sense of the following definition.

definition 15.2

A complex sequence $\{z_n\}$ is said to converge to c, if for all $\varepsilon > 0$, there exists N such that if $n > N$, then $|z_n - c| < \varepsilon$. We write

$$\lim_{n \to \infty} z_n = c.$$

A sequence which does not converge is said to diverge. (See Figure 15.1.)

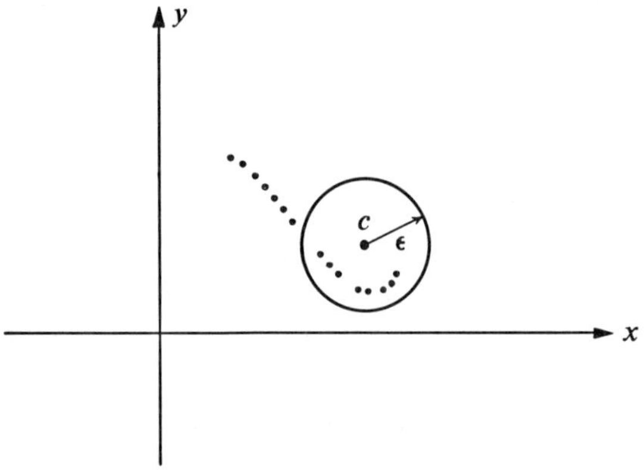

figure 15.1 *convergence of a complex sequence*

example 15.1: Determine if the sequence whose general term is $z_n = (2n + 1)/ni$ converges.

solution: Since

$$\lim_{n \to \infty} \frac{2n + 1}{ni} = -2i,$$

the sequence converges to $-2i$.

Terms of a sequence are sometimes written $z_n = x_n + iy_n$. The limit of each of the real sequences $\{x_n\}_{n=1}^{\infty}$ and $\{y_n\}_{n=1}^{\infty}$ may often be found easily. The knowledge of these two limits is sufficient to determine the limit of the complex sequence $\{z_n\}_{n=1}^{\infty}$.

theorem 15.1

A sequence of terms $z_n = x_n + iy_n$ converges to $c = a + ib$ if and only if $\lim_{n \to \infty} x_n = a$ and $\lim_{n \to \infty} y_n = b$.

proof: Prove the theorem as an exercise.

The definition of a complex infinite series is essentially the same as Definition A.13.

definition 15.3

An infinite series is an indicated limit of a sequence of partial sums

$$S_n = \sum_{k=1}^{n} z_k .$$

The limit is called the "sum" of the series and $\lim_{n\to\infty} S_n$ is denoted by

$$\sum_{k=1}^{\infty} z_k .$$

Theorem 15.1 has its analogue for series in the following theorem.

theorem 15.2

Let $z_k = x_k + iy_k$. Then $\sum_{k=1}^{\infty} z_k$ converges to $a + ib$ if and only if $\sum_{k=1}^{\infty} x_k = a$ and $\sum_{k=1}^{\infty} y_k = b$.

Theorems 15.1 and 15.2 allow the application of virtually all the theory of real sequences and real series to the complex case.

To determine if a series converges, we analyze it directly or else apply one of the "tests."

example 15.2: Consider

$$\sum_{k=0}^{\infty} z^k = 1 + z + z^2 + \cdots + z^k + \cdots .$$

Show that this series converges for $|z| < 1$ and diverges otherwise.

solution: $S_n = 1 + z + z^2 + \cdots + z^n$ and $zS_n = z + z^2 + z^3 + \cdots + z^{n+1}$. Therefore

$$S_n - zS_n = 1 - z^{n+1}$$

and

$$S_n = \frac{1 - z^{n+1}}{1 - z} .$$

Thus,

$$\begin{aligned} \lim_{n\to\infty} S_n &= \frac{1}{1 - z}, \qquad \text{if} \quad |z| < 1 ; \\ &= \infty , \qquad \text{if} \quad |z| > 1 . \end{aligned}$$

If $|z| = 1$, then $z = e^{i\theta}$ and

$$S_n = \frac{1 - e^{-in\theta}}{1 - e^{i\theta}}$$

which has no limit.

The series in the previous example is called the *geometric series.* It is very important, since many series which are not strictly geometric can be analyzed using the known results for geometric series.

example 15.3: Find the value of $\sum_{k=0}^{\infty} (-z)^k$.

solution: By letting the $(-z)$ of this example take the place of z in the previous one, the series is seen to converge to $1/(1+z)$ for $|z| < 1$.

definition 15.4

A series $\sum_{k=1}^{\infty} z_k$ is said to *converge absolutely* if the real series of positive terms $\sum_{k=1}^{\infty} |z_k|$ converges.

The importance of absolute convergence stems from the following two theorems whose proofs we omit.

theorem 15.3

If the series $\sum_{k=1}^{\infty} z_k$ is absolutely convergent, it is convergent.

theorem 15.4

Comparison Test. If $\sum_{k=1}^{\infty} b_k$ converges and if $|z_k| < b_k$ for all k beyond some N, then $\sum_{k=1}^{\infty} z_k$ is convergent.

The following two tests for convergence of a series are the same as for real series.

Ratio Test. If

$$\lim_{n \to \infty} \left| \frac{z_{n+1}}{z_n} \right| = L,$$

then the series converges if $L < 1$ and diverges if $L > 1$.

Root Test. If

$$\lim |z_n|^{1/n} = L,$$

then the series converges if $L < 1$ and diverges if $L > 1$.

Although other tests are available, these two will be sufficient for most elementary applications. They both have the nice feature of being easy to apply.

15.3 series of variable terms

If the terms of a series are variable, we apply the definitions, theorems, and tests of the previous section and attempt to determine the values of z for which the series converges. The set of values of z for which a series of variable terms converges is called the *domain of convergence* of the series. Usually, we are content to find the domain of absolute convergence.

example 15.4: Find the domain of absolute convergence of the series

$$\sum_{k=1}^{\infty} e^{kz}.$$

solution: Since

$$\left|\frac{e^{(k+1)z}}{e^{kz}}\right| = |e^z|,$$

this series converges for those values of z for which $|e^z| < 1$. Since $|e^z| = e^x$, the solution set for this inequality is $\operatorname{Re} z < 0$, that is in the left half-plane.

example 15.5: Find the domain of absolute convergence of the series

$$\sum_{k=1}^{\infty} \frac{z^k}{k^2(1+z)^k}.$$

solution: Applying the ratio test, we obtain the domain of convergence to be those z for which $|z/(1+z)| < 1$, or, $|z| < |1+z|$. This inequality is satisfied for $\operatorname{Re} z > -1/2$.

example 15.6: Determine the domain of absolute convergence for the series

$$\sum_{k=1}^{\infty} \left(\frac{3kz}{5k+1}\right)^k.$$

solution: Applying the root test, the domain of absolute convergence is for $|3z/5| < 1$, that is, for $|z| < 5/3$.

15.4 power series

For our purposes, the most important series of variable terms are series of nonnegative powers of $(z - a)$. These are called *power series* with center $z = a$.

The convergence behavior of power series is more easily characterized than a more general series of variable terms. Examples 15.2 and 15.3 show how power series converge in a circle of finite radius. The other possibilities are exhibited in the following two examples.

example 15.7: Determine the domain of convergence of the power series

$$\sum_{n=1}^{\infty} \frac{z^n}{n!}.$$

solution: By the ratio test

$$\left|\frac{z^{n+1}}{(n+1)!}\,\frac{n!}{z^n}\right| = \left|\frac{z}{n+1}\right| \to 0$$

as $n \to \infty$, and thus the series converges for all finite z.

example 15.8: Determine the domain of convergence of $\sum_{n=1}^{\infty} n!\, z^n$.

solution: The ratio of succeeding terms is

$$\frac{(n+1)!\,z^{n+1}}{n!\,z^n} = (n+1)\,z$$

whose absolute value $\to \infty$ as $n \to \infty$. Hence the series converges for $z = 0$ and nowhere else.

The following theorem characterizes the general situation.

theorem 15.5

Every power series

$$\sum_{n=1}^{\infty} c_n (z - a)^n$$

behaves in one of the following three ways·

1. It converges for $z = a$ only.
2. It converges absolutely for all values of z.
3. There exists a positive number R such that the series converges absolutely for $|z - a| < R$ and diverges for $|z - a| > R$.

The circle $|z - a| = R$ is called the *circle of convergence* and R is called the *radius of convergence*. The ratio test may be conveniently used to find the radius of convergence, but the behavior on the circle $|z - a| = R$ must be determined by some other method. The analysis of what precisely does happen on the circle of convergence can be very subtle, and will not be considered in this text. A series could diverge everywhere on the circle (as in the case $\sum_{n=0}^{\infty} z^n$) or converge everywhere (as in the case $\sum_{n=1}^{\infty} z^n/n^2$).

example 15.9: Find the circle of convergence of the power series

$$\sum_{n=1}^{\infty} \frac{(z+4)^{2n-1}}{2n-1}.$$

solution: The ratio of two succeeding terms is

$$\left|\frac{(z+4)^{2n+1}(2n-1)}{(2n+1)(z+4)^{2n-1}}\right|$$

which in the limit is $|z+4|^2$. Hence, the series converges for $|z+4| < 1$. (See Figure 15.2.)

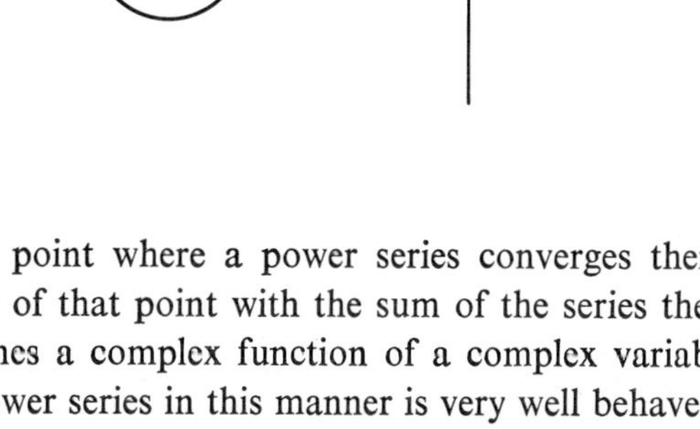

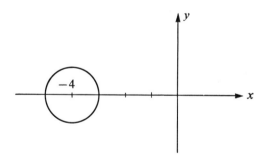

figure 15.2

At each point where a power series converges there is a natural correspondence of that point with the sum of the series there. This correspondence defines a complex function of a complex variable. A function defined by a power series in this manner is very well behaved.

theorem 15.6

Suppose $\sum_{n=1}^{\infty} a_n z^n$ converges for $|z| < R$. Then, in this domain the power series represents an analytic function. The derivative is found by term-by-term differentiation and the derived series converges in the same domain.

example 15.10: Let $f(z) = \sum_{n=1}^{\infty} nz^{n-1}$. What is the domain of f? Find the infinite series for f'.

solution: By the ratio test, this series converges for $|z| < 1$ and this is the domain of f. Moreover, within this circle,

$$f'(z) = \sum_{n=2}^{\infty} n(n-1)\, z^{n-2}.$$

These results are true only for a power series, not for more general series of variable terms. Other series do not necessarily converge in circles about a point; much more complicated patterns of convergence are possible.

In this section you have learned roughly that "every power series is an analytic function." In the next section, we will show that every analytic function can be written as a power series. The concepts of convergent power series and analytic functions will thus be identified. Sometimes, this characterization of analytic functions proves more useful than the definition we used initially.

exercises for sections 15.1–15.4

In Exercises 1–10 examine for convergence or divergence of the sequences.

1. $z_n = (-1)^n + \dfrac{1}{2^n}$.
2. $z_n = \dfrac{(-1)^n}{2^n}$.
3. $z_n = \dfrac{\cos n\pi}{n}$.
4. $z_n = \sqrt{\sqrt{n}}$.
5. $z_n = i^n$.
6. $z_n = \dfrac{i^n}{n}$.
7. $z_n = \dfrac{2n^2 + n}{3n^2 - 1}$.
8. $z_{2n} = 2, \quad z_{2n+1} = \dfrac{2n+1}{1+n^2}$.
9. $z_{2n} = \dfrac{2n+1}{3n}, \quad z_{2n+1} = \dfrac{2n}{3n-5}$.
10. $z_{2n} = \dfrac{1}{2n}, \quad z_{2n+1} = \dfrac{1}{3n}$.

In Exercises 11–20 examine for convergence or divergence of the series.

11. $\sum_{n=1}^{\infty} \frac{n+1}{n2^n}$.

12. $\sum_{n=0}^{\infty} \frac{1}{n!}$.

13. $\sum_{n=1}^{\infty} \frac{n}{\sqrt{n}}$.

14. $\sum_{n=1}^{\infty} \frac{1}{\sqrt{n}}$.

15. $\sum_{n=0}^{\infty} \frac{1}{2^n}$.

16. $\sum_{n=1}^{\infty} 2^n$.

17. $\sum_{n=1}^{\infty} \frac{2^n}{n!}$.

18. $\sum_{n=1}^{\infty} \frac{2n+1}{n^2}$.

19. $\sum_{n=1}^{\infty} \frac{2n+1}{n^3}$.

20. $\sum_{n=1}^{\infty} n^p$.

In Exercises 21–30 determine the domain of absolute convergence.

21. $\sum_{n=1}^{\infty} \frac{\sin nz}{n^2}$.

22. $\sum_{n=1}^{\infty} e^{n^2 z}$.

23. $\sum_{n=1}^{\infty} \left(\frac{z}{5+3z}\right)^n \frac{1}{n^2}$.

24. $\sum_{n=1}^{\infty} \left(\frac{4z}{z+2}\right)^n \frac{1}{n^3}$.

25. $\sum_{n=1}^{\infty} 3z^n$ and $\sum_{n=1}^{\infty} (3z)^n$.

26. $\sum_{n=1}^{\infty} \left(\frac{2z+5}{z}\right)^n \frac{1}{n^4}$.

27. $\sum_{n=1}^{\infty} n^2 (z-a)^n$.

28. $\sum_{n=1}^{\infty} (z-a)^{n^2}$.

29. $\sum_{n=1}^{\infty} \frac{(-1)^n z^n}{2^n}$.

30. $\sum_{n=1}^{\infty} \left(\frac{z-4}{3}\right)^n$.

15.5 power series representation of a function

In this section you will learn how the familiar Taylor expansion of a real function has its analog in complex analysis. There are some very important theoretical consequences of the results.

Assume f is analytic at $z = a$. To represent $f(z)$ as a series in powers of $(z - a)$, we consider a circle C in the neighborhood of the point a, and

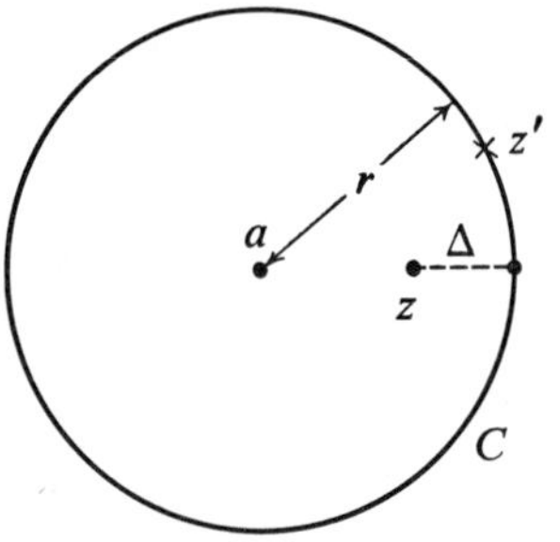

figure 15.3

let z be a point of this neighborhood. (See Figure 15.3.) By Cauchy's integral formula, the value of the function at z is given by

$$f(z) = \frac{1}{2\pi i} \oint_C \frac{f(z')}{z' - z}\, dz' .$$

The essence of Taylor's theorem is to convert Cauchy's integral representation of $f(z)$ to a power series representation in the same domain.

theorem 15.7

Let f be analytic in a neighborhood of $z = a$. Then $f(z)$ may be written

$$f(z) = \sum_{n=0}^{\infty} A_n (z - a)^n$$

where

$$A_n = \frac{1}{2\pi i} \oint_C \frac{f(z')}{(z' - a)^{n+1}}\, dz' .$$

The series is unique and converges to $f(z)$ in the largest circle around a as center such that $f(z)$ is analytic throughout the interior. (We sometimes say the series converges "out to the nearest singularity of f.")

proof: Note that

$$\frac{1}{z' - z} = \frac{1}{(z' - a) - (z - a)} = \frac{1}{(z' - a)\left(1 - \dfrac{z - a}{z' - a}\right)} .$$

By the familiar expression for a geometric progression with "common ratio" $(z-a)/(z'-a)$,

$$\frac{1}{z'-z}=\frac{1}{z'-a}\left[1+\frac{z-a}{z'-a}+\left(\frac{z-a}{z'-a}\right)^2+\cdots+\left(\frac{z-a}{z'-a}\right)^n+\frac{\left(\dfrac{z-a}{z'-a}\right)^{n+1}}{1-\dfrac{z-a}{z'-a}}\right]$$

$$=\frac{1}{z'-a}+\frac{z-a}{(z'-a)^2}+\frac{(z-a)^2}{(z'-a)^3}+\cdots+\frac{(z-a)^n}{(z'-a)^{n+1}}+\frac{\left(\dfrac{z-a}{z'-a}\right)^{n+1}}{z'-z}\ .$$

Therefore, since

$$f(z)=\frac{1}{2\pi i}\oint_C \frac{f(z')}{z'-z}\,dz',$$

$$\begin{aligned}f(z)=\frac{1}{2\pi i}\oint_C \frac{f(z')}{z'-a}\,dz'+\frac{z-a}{2\pi i}\oint_C \frac{f(z')}{(z'-a)^2}\,dz'\\ +\frac{(z-a)^2}{2\pi i}\oint_C \frac{f(z')}{(z'-a)^3}\,dz'+\cdots+\frac{(z-a)^n}{2\pi i}\oint_C \frac{f(z')}{(z'-a)^{n+1}}\,dz'\\ +\frac{(z-a)^{n+1}}{2\pi i}\oint_C \frac{f(z')}{(z'-a)^{n+1}(z'-z)}\,dz'.\end{aligned}$$

The last term is denoted by R_n and is called the *remainder*. Our proof will be completed if we show that $|R_n|$ can be made as small as desired. Since f is analytic on C, $|f(z)|$ is bounded, say by M. The point z is not on the circle, and hence there is a minimum value for $|z'-z|$, which we denote by Δ. The ratio $|(z-a)/(z'-a)|$ has a constant value, ρ, which is less than 1. Therefore,

$$\begin{aligned}|R_n|&=\frac{1}{2\pi}\left|\oint_C \frac{f(z')}{z'-z}\left(\frac{z-a}{z'-a}\right)^{n+1}dz'\right|\\ &\leqq\frac{1}{2\pi}\oint_C \left|\frac{f(z')}{z'-z}\right|\left|\frac{z-a}{z'-a}\right|^{n+1}|dz'|\\ &\leqq\frac{1}{2\pi}\frac{M}{\Delta}\rho^{n+1}2\pi r\end{aligned}$$

which $\to 0$ as $n\to\infty$, and thus the theorem is proved.

The formula for the coefficients, A_n, in the Taylor series takes on a more familiar look as a result of the following corollary.

corollary: With the hypothesis the same as in the previous theorem,

$$A_n=\frac{f^{(n)}(a)}{n!}.$$

proof: Since f is analytic at $z = a$, the formula for the derivative given by Theorem 14.9 may be used to give the desired result.

example 15.11: Find the Taylor series for the function $f(z) = 1/(1 - z)$ about $z = 0$. Where does the series converge? Where does the series represent the function?

solution: Since $f(z) = (1 - z)^{-1}$ and $f^{(n)}(z) = n!(1 - z)^{-n-1}$, $f^{(n)}(0) = n!$ and therefore

$$f(z) = \sum_{n=0}^{\infty} z^n.$$

The series converges and represents the function in a circular neighborhood about $z = 0$ extending to the circle $|z| = 1$ which passes through the singularity of f. The series diverges (and consequently could not represent $f(z)$) for $|z| > 1$.

example 15.12: Find the Taylor series for $\sin z$ about $z = 0$.

solution: Since $f(z) = \sin z$, $f(0) = 0$, $f'(0) = 1$, $f''(0) = 0$, $f'''(0) = -1$, etc. Therefore, the Taylor series for $\sin z$ is

$$\sum_{n=0}^{\infty} \frac{z^{2n+1}(-1)^n}{(2n+1)!}.$$

Since $\sin z$ is analytic everywhere, the series converges and represents $\sin z$ for all finite z.

Example 15.12 shows that the form of the Taylor series for the complex sine function has the same *form* as the real series for the real sine function. This is true in the general case since the form of the coefficients is the same in the real as in the complex case.

example 15.13: Find the Taylor series for $1/(1 - z)$ about $z = i$.

solution: It is easy to show that $f^{(n)}(z) = (1 - z)^{-n-1}$. Therefore $f^{(n)}(i) = (1 - i)^{-n-1}$. Therefore the Taylor series is

$$\sum_{n=0}^{\infty} (1 - i)^{-n-1}(z - i)^n.$$

The series represents the function within a circle passing through the nearest (and only) singularity of f. The domain of representation is $|z - i| < \sqrt{2}$. (See Figure 15.4.)

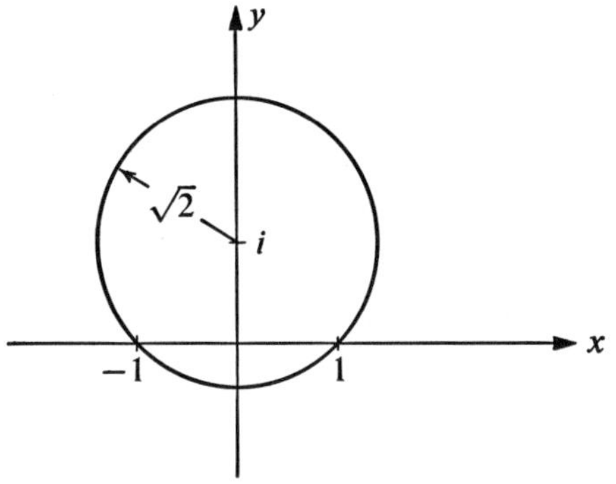

figure 15.4

Even though the formula for the coefficients is simplified when expressed in terms of the derivative, there are other practical methods for obtaining a Taylor series representation. The uniqueness of the representation allows us to find the series in any convenient way. A method often used is to write the given function in a form in which the results for geometric series may be applied.

example 15.14: Write the Taylor series for $1/(1+z^2)$ about $z=0$.

solution: Substitute $-z^2$ for z in the known expansion of $1/(1-z)$ to obtain

$$\frac{1}{1+z^2}=\sum_{n=0}^{\infty}(-z^2)^n=\sum_{n=0}^{\infty}(-1)^n z^{2n}.$$

The series represents the function for $|z^2|<1$, that is, inside a circle of radius 1 centered at the origin.

example 15.15: Find the series for $1/(1+z)$ about $z=1$.

solution:

$$\frac{1}{1+z}=\frac{1}{2+(z-1)}=\frac{1}{2\left(1+\dfrac{z-1}{2}\right)}.$$

Substitute $-(z-1)/2$ for the z of the geometric series to obtain

$$\frac{1}{1+z}=\frac{1}{2}\sum_{n=0}^{\infty}(-1)^n\left(\frac{z-1}{2}\right)^n.$$

The representation is valid for $|(z-1)/2|<1$, that is, $|z-1|<2$.

If the function is rational, you may wish to find the partial fractions decomposition before using other methods.

example 15.16: Find the Taylor series for $1/[(1+z)(2+z)]$ about $z=0$.

solution: By elementary methods,

$$\frac{1}{(1+z)(2+z)}=\frac{1}{1+z}-\frac{1}{2+z}.$$

The Taylor series for $1/(1+z)$ is $\sum_{n=0}^{\infty}(-1)^n z^n$ and the representation is valid for $|z|<1$. Also,

$$\frac{1}{2+z}=\frac{1}{2\left(1+\frac{z}{2}\right)}=\frac{1}{2}\sum_{n=0}^{\infty}(-1)^n\left(\frac{z}{2}\right)^n,$$

valid for $|z|<2$. Therefore, the Taylor series for the given function is

$$\frac{1}{(1+z)(2+z)}=\sum_{n=0}^{\infty}(-1)^n z^n-\frac{1}{2}(-1)^n\left(\frac{z}{2}\right)^n=\sum_{n=0}^{\infty}(-1)^n z^n\left(1-\frac{1}{2^{n+1}}\right)$$

which is a valid representation for $|z|<1$.

In addition to using the fact that the term-by-term derivative of the series converges to the derivative of the function it is also true that the antiderivative of the function is equal to the sum of the antiderivatives.

example 15.17: Write the Taylor series for $\text{Tan}^{-1}z$ about $z=0$.

solution: Antidifferentiating both sides of the equality

$$\frac{1}{1+z^2}=\sum_{n=0}^{\infty}(-1)^n z^{2n}$$

we obtain

$$\text{Tan}^{-1}z=\sum_{n=0}^{\infty}\frac{(-1)^n z^{2n+1}}{2n+1}.$$

The representation is valid for $|z|<1$.

example 15.18: Find the Taylor series for $1/(1+z)^2$ about $z=0$.

solution: The Taylor series for $1/(1+z)$ is $\sum_{n=0}^{\infty}(-1)^n z^n$.

By differentiating both sides,

$$\frac{-1}{(1+z)^2} = \sum_{n=1}^{\infty} (-1)^n nz^{n-1}$$

and therefore

$$\frac{1}{(1+z)^2} = \sum_{n=1}^{\infty} (-1)^{n-1} nz^{n-1}.$$

The representation is valid for $|z| < 1$.

The previous examples show why power series are best discussed in the complex plane. The reason that the series of the function $f(x) = 1/(1+x^2)$ has an interval of convergence restricted to $(-1, 1)$ is inexplainable when only real functions are considered. In the complex plane, we see that the power series for $1/(1+z^2)$ converges out to the nearest singularity of which there are two, located at i and $-i$. Thus the barrier of the circle $|z| = 1$ is invisible when only real functions are discussed.

exercises for section 15.5

In Exercises 1–21 find the Taylor series of the given function about $z = a$ and determine the radius of convergence R.

1. $\cos z, \quad a = 0.$
2. $\sin z, \quad a = 0.$
3. $e^{-z}, \quad a = 0.$
4. $\frac{1}{z}, \quad a = 1.$
5. $\frac{1}{1-z}, \quad a = i.$
6. $f(z) = \frac{\sin z}{z}, \quad z \neq 0;$ $\quad = 1 \text{ for } z = 0; \quad a = 0.$
7. $\frac{1}{z^2}, \quad a = 1.$
8. $\frac{1}{1-z^3}, \quad a = 0.$
9. $\sin z^2, \quad a = 0.$
10. $e^{z^2}, \quad a = 0.$
11. $z^2, \quad a = 1.$
12. $\frac{1}{3z+1}, \quad a = 2.$
13. $z - 1, \quad a = -1.$
14. $z^2 - 8z + 2, \quad a = 0.$
15. $z^2 - 8z + 2, \quad a = 1.$
16. $\frac{1}{2z+i}, \quad a = 2.$
17. $\frac{1}{z^2-4}, \quad a = 1.$
18. $\frac{1}{z^2+4}, \quad a = 1.$

19. $\dfrac{1}{(1+z)^4}$, $a=0.$

20. $\dfrac{1}{(z+3)(z+2)}$, $a=-1.$

21. $\dfrac{1}{z^4-1}$, $a=0.$

In Exercises 22–25, find the first few terms of the Taylor series of the given function about $a=0$.

22. $e^{1/1+z}$.

23. $\cos(1-z)$.

24. $e^{\sin z}$.

25. $(1+z)^{1/2}$.

15.6 laurent series

The Taylor series of a function is a suitable representation of a function around points where f is analytic. There are two major deficiencies:

1. No power series representation is possible about a point of singularity.
2. Even if the power series is written about a point where the function is analytic, the representation is valid only within a circle out to the nearest singularity.

A more general type of series expansion of a function, which includes power series as a special case, is of the form

$$\sum_{n=-\infty}^{\infty} A_n(z-a)^n,$$

and is called a *Laurent series*. In this section we will show how to represent a function in terms of such a series about a point where a function is non-analytic. The representation takes the form such that if f *is* analytic at a, the Laurent series reduces to the Taylor series, that is, $A_n=0$ for $n<0$.

theorem 15.8

(*Laurent.*) Let f be analytic on two concentric circles centered at a and in the annulus between them. (See Figure 15.5.) Then $f(z)$ can be represented by a Laurent series with coefficients

$$A_n=\frac{1}{2\pi i}\oint_{\substack{C\\(\text{ccw})}} \frac{f(z')}{(z'-a)^{n+1}}\,dz'.$$

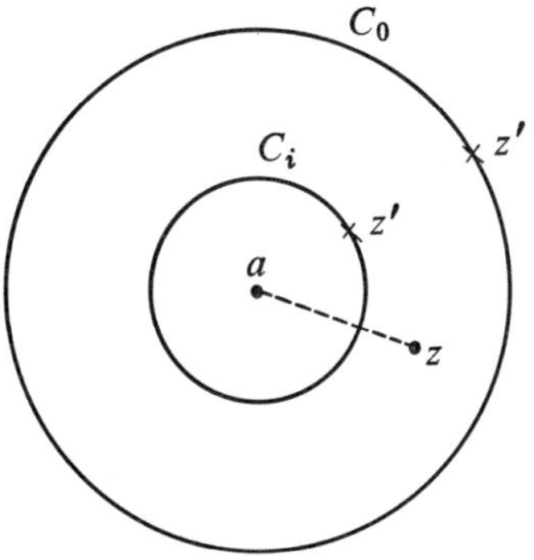

figure 15.5

C is any simple closed path which lies in the annulus and encloses the inner circle. The series converges and represents $f(z)$ in the annulus obtained by continuously increasing C_0 and decreasing C_i until each of the two circles reaches a point of singularity of f.

proof: The proof parallels that of Taylor's theorem so most of the details are left to the exercises. Let z be a point in the annulus. By Cauchy's integral formula applied to doubly connected domains (Section 14.6),

$$f(z) = \frac{1}{2\pi i}\oint_{\substack{C_0 \\ (\text{ccw})}} \frac{f(z')}{z'-z}\,dz' - \frac{1}{2\pi i}\oint_{\substack{C_i \\ (\text{ccw})}} \frac{f(z')}{z'-z}\,dz'$$
$$= g(z) - h(z),$$

an equality which has the effect of defining $g(z)$ and $h(z)$. Since z lies inside of C_0, the first integral is the same as the type considered in the proof of Taylor's theorem. Therefore,

$$g(z) = \frac{1}{2\pi i}\oint_{\substack{C_0 \\ (\text{ccw})}} \frac{f(z')}{z'-z}\,dz' = \sum_{n=0}^{\infty} A_n(z-a)^n.$$

By the principle of deformation of path, C_0 may be deformed to any path C in the annulus. Hence, the form of the coefficients is verified for $n \geqq 0$. The series which defines $g(z)$ converges within a circle obtained by enlarging C_0 out to the "next" singularity of f.

It remains to show that A_n has the correct form for $n<0$.

$$
\begin{aligned}
-h(z) &= \frac{-1}{2\pi i}\oint_{\substack{C_i\\(\text{ccw})}} \frac{f(z')}{z'-z}\,dz' \\
&= \frac{1}{2\pi i}\oint_{\substack{C_i\\(\text{ccw})}} \frac{f(z')}{(z-a)\left(1-\dfrac{z'-a}{z-a}\right)}\,dz' \\
&= \frac{1}{2\pi i}(z-a)^{-1}\oint_{\substack{C_i\\(\text{ccw})}} f(z')\,dz' + \frac{(z-a)^{-2}}{2\pi i}\oint_{\substack{C_i\\(\text{ccw})}} f(z')(z'-a)\,dz' \\
&\quad + \frac{(z-a)^{-3}}{2\pi i}\oint_{\substack{C_i\\(\text{ccw})}} f(z')(z'-a)^2\,dz' + \cdots \\
&\quad + \frac{(z-a)^{-m}}{2\pi i}\oint_{\substack{C_i\\(\text{ccw})}} f(z')(z'-a)^{m-1}\,dz' \\
&\quad + \frac{1}{2\pi i}\oint_{\substack{C_i\\(\text{ccw})}} \frac{f(z')}{z-z'}\left(\frac{z'-a}{z-a}\right)^m dz', \qquad \text{where } m>0.
\end{aligned}
$$

By a technique similar to that used in Theorem 15.7 you can show that the last term goes to zero as $m\to\infty$. Further, for $m>0$,

$$
A_m = \frac{1}{2\pi i}\oint_{\substack{C_i\\(\text{ccw})}} f(z')(z'-a)^{m-1}\,dz',
$$

so that if $n=-m$,

$$
A_n = \frac{1}{2\pi i}\oint_{\substack{C_i\\(\text{ccw})}} \frac{f(z')}{(z'-a)^{n+1}}\,dz', \qquad n<0.
$$

By the principle of deformation of path C_i may be deformed to any path C within the annulus. Hence A_n has the correct form for $n<0$.

The series for $h(z)$ will converge outside of the circle C_i since $|(z'-a)/(z-a)|<1$ only if $|z'-a|<|z-a|$. The circle C_i may be shrunk until a singularity of f is crossed. The domain common to g and h is the open annulus characterized in the statement of the theorem.

If the only singularity of f is at $z=a$, then the series converges to $f(z)$ for all $z\neq a$. The coefficients of the Laurent expansion cannot be expressed in terms of the derivative of f at a, since f is not analytic there.

Just as for Taylor series, the Laurent representation in a given annulus is unique. This allows the computation of the A_n by methods other than appeal to the formula.

example 15.19: Find all Laurent series for $f(z) = z^2 e^{1/z}$ with center at $z = 0$.

solution: The only singularity for f is at $z = 0$, the center of the series. The Laurent expansion for $e^{1/z}$ about 0 is

$$\sum_{n=0}^{\infty} \frac{1}{n!\,z^n},$$

valid for $|z| > 0$. Hence the expansion for $z^2 e^{1/z}$ is

$$\sum_{n=0}^{\infty} \frac{1}{n!\,z^{n-2}} = z^2 + z + \frac{1}{2} + \frac{1}{6z} + \frac{1}{24z^2} + \cdots.$$

The representation is valid for $|z| > 0$.

example 15.20: Find all Laurent expansions for $f(z) = 1/(z-2)$:
(a) with center at $z = 0$. (b) with center at $z = 2$.

solution: (a) With center at $z = 0$, there is a (Taylor) representation valid for $0 \leqq |z| < 2$,

$$\frac{1}{z-2} = \frac{-1}{2\left(1 - \dfrac{z}{2}\right)} = -\frac{1}{2}\sum_{n=0}^{\infty}\left(\frac{z}{2}\right)^n.$$

For $|z| > 2$,

$$\frac{1}{z-2} = \frac{1}{z\left(1 - \dfrac{2}{z}\right)} = \frac{1}{z}\sum_{n=0}^{\infty}\left(\frac{2}{z}\right)^n = \frac{1}{z}\sum_{n=0}^{-\infty}\left(\frac{z}{2}\right)^n.$$

(b) With center at $z = 2$, the Laurent representation for the function given is $1/(z-2)$, that is, the rule as given is the desired Laurent form with $z = 2$ as the center.

example 15.21: Find all Laurent expansions for $\dfrac{1}{(z-1)(z-2)}$:

(a) with center at $z = 0$. (b) with center at $z = 1$.

solution: (a) By a partial fractions expansion,

$$\frac{1}{(z-1)(z-2)} = -\frac{1}{z-1} + \frac{1}{z-2}.$$

Since the function has singularities at 1 and 2, there are Laurent expansions about 0 valid for $0 \leqq |z| < 1$, $1 < |z| < 2$, and $|2| < z$. (See Figure 15.6.) For $0 \leqq |z| < 1$,

$$\frac{-1}{z-1} = \sum_{n=0}^{\infty} z^n \quad \text{and} \quad \frac{1}{z-2} = -\frac{1}{2}\sum_{n=0}^{\infty}\left(\frac{z}{2}\right)^n.$$

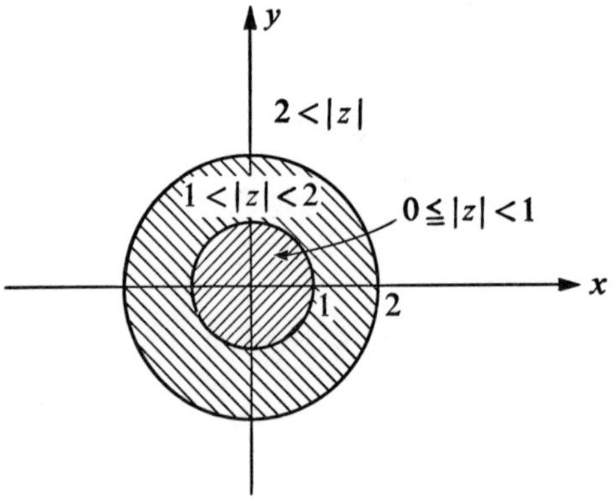

figure 15.6

The first of these is valid for $0 \leqq |z| < 1$ and the second for $0 \leqq |z| < 2$. Hence

$$fz = \sum_{n=0}^{\infty} z^n\left(1 - \frac{1}{2^{n+1}}\right), \quad \text{for } 0 \leqq |z| < 1.$$

For $1 < |z| < 2$,

$$\frac{-1}{z-1} = -\sum_{n=0}^{\infty} (z)^{-n-1}, \quad \text{for } 1 < |z|.$$

Hence

$$f(z) = -\sum_{n=0}^{\infty}\left(\frac{1}{z^{n+1}} + \frac{z^n}{2^{n+1}}\right), \quad \text{for } 1 < |z| < 2.$$

For $|z| > 2$,

$$\frac{1}{z-2} = \sum_{n=0}^{\infty} \frac{2^n}{z^{n+1}}$$

and hence,

$$f(z) = \sum_{n=0}^{\infty} \frac{(2^n - 1)}{z^{n+1}}, \quad \text{for } 2 < |z|.$$

(b) About $z=1$, there will be two representations, one valid for $0<|z-1|<1$ and the other for $1<|z-1|$ (see Figure 15.7),

$$\frac{1}{z-2}=\frac{1}{(z-1)-1}=\frac{-1}{1-(z-1)}=-\sum_{n=0}^{\infty}(z-1)^n.$$

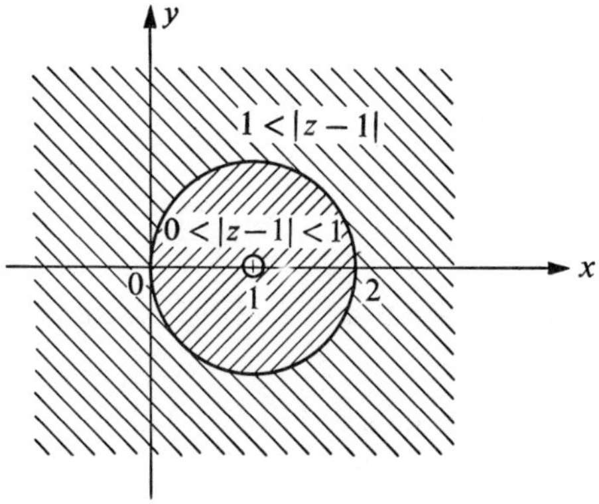

figure 15.7

Therefore

$$f(z)=-\sum_{n=0}^{\infty}(z-1)^n-\frac{1}{z-1},\qquad \text{for } 0<|z-1|<1.$$

For $|z-1|>1$, we have

$$\frac{1}{z-2}=\frac{1}{(z-1)\left(1-\dfrac{1}{z-1}\right)}=\frac{1}{z-1}\sum_{n=0}^{\infty}(z-1)^{-n}.$$

Hence

$$\begin{aligned} f(z)&=\sum_{n=0}^{\infty}(z-1)^{-n-1}-(z-1)^{-1}\\ &=\sum_{n=1}^{\infty}\frac{1}{(z-1)^{n+1}}. \end{aligned}$$

exercises for section 15.6

Find all Laurent series of the following functions about the specified centers. Give the domain of representation in each case.

1. $z^3 e^{1/z}, \quad a = 0.$

2. $\dfrac{1}{z^2+1}, \quad a = 1.$

3. $\dfrac{1}{z^4-1}, \quad a = 0.$

4. $\dfrac{1}{z^2}, \quad a = 1.$

5. $\dfrac{1}{z^3}, \quad a = 1.$

6. $\dfrac{1}{z^4}, \quad a = 1.$

7. $\dfrac{1}{z^2(z^2+1)}, \quad a = 0.$

8. $\dfrac{e^z}{(z-1)^2}, \quad a = 1.$

9. $\dfrac{3z^2+z}{z^3-z}, \quad a = -1.$

10. Find the first three nonzero terms of the Laurent series of $\cot z$ valid for $0 < |z| < \pi$.

11. Find all Laurent expansions of $1/(z^2-1)$ about:
    (a) 0. (b) 1. (c) -1.

12. Find all Laurent expansions of $1/z$ with center at:
    (a) 0. (b) 1. (c) 2.

13. Find all Laurent series for $1/[z(z-i)(z+1)]$ with center at:
    (a) 0. (b) i. (c) 1.

14. In the proof of Laurent's theorem, show that the last integral goes to zero as $m \to \infty$.

15.7 poles and zeros

Recall that the complex number $z = a$ is said to be a *zero* of a function if $f(a) = 0$. A zero of a function is said to be *isolated* if $f(a) = 0$ and there is a neighborhood of a such that $f(z) \neq 0$ in that neighborhood. It is possible to show that if a function is analytic, then its zeros are isolated.

With the Laurent expansion of a function we can further classify the zeros. An analytic function is said to have a *zero of the Nth order* at $z = a$ if $A_N \neq 0$ and $A_n = 0$ for $0 \leqq n < N$ in the Laurent expansion. Equivalently, f has a zero of the Nth order if $f^{(n)}(a) = 0$ for $n < N$ and $f^{(N)}(a) \neq 0$.

Singularities are also classified. If f has an isolated singularity at $z = a$, then in the neighborhood of $z = a$ the function has a Laurent representation

$$f(z) = g(z) + h(z) = \sum_{n=0}^{\infty} A_n (z-a)^n + \sum_{n=-1}^{-\infty} A_n (z-a)^n .$$

The series for $h(z)$ is called the *principal part of* $f(z)$ at $z=a$. The principal part is used to classify singularities.

1. If the principal part is zero, then the singularity is said to be *removable.* The singularity is removed by defining $f(a)$ to be A_0.
2. If the principal part has finitely many terms, the singularity is called a *pole.* In this case, the function may be written

$$f(z)=g(z)+A_{-1}(z-a)^{-1}+A_{-2}(z-a)^{-2}+\cdots+A_{-n}(z-a)^{-n}.$$

   The value of n is called the *order* of the pole.
3. If the principal part has infinitely many nonzero terms, the singularity is called *essential.*

example 15.22: Classify the poles and zeros of the functions:

(a) $f(z)=e^{1/z}$. (b) $f(z)=\dfrac{1}{z(z-2)^5}+\dfrac{3}{(z-2)^3}$.

(c) $f(z)=2z+6z^3$.

solution: (a) The Laurent expansion for the function $e^{1/z}$ with center $z=0$ is given by

$$\sum_{n=0}^{\infty}\frac{z^{-n}}{n!}$$

and so the singularity at $z=0$ is essential. There are no zeros since the exponential function is never zero.

(b) By inspection, there is a simple pole at $z=0$ and a pole of order 5 at $z=2$. The function is nonzero in the finite z-plane. To evaluate $f(\infty)$ we first find $f(1/z')$,

$$f\left(\frac{1}{z'}\right)=\frac{1}{\left(\dfrac{1}{z'}\right)\left(\dfrac{1}{z'}-2\right)^5}+\frac{3}{\left(\dfrac{1}{z'}-2\right)^3}$$

$$=\frac{(z')^3}{(1-2z')^3}\left[\frac{(z')^3}{(1-2z')^2}+3\right]$$

and from this we can see that there is a zero of order 3 at ∞.

(c) The given function obviously has no poles in the finite z-plane, but since

$$f\left(\frac{1}{z'}\right)=\frac{2}{z'}+\frac{6}{(z')^3}$$

the function has a pole of order 3 at ∞. There are simple zeros at 0, $i/\sqrt{3}$, and $-i/\sqrt{3}$.

exercises for section 15.7

Classify the poles and zeros.

1. $1 - z^4$.
2. z^4.
3. $\sin z$.
4. $\dfrac{1+z^2}{1-z}$.
5. $\tan^2 z$.
6. $z^3 \sin^2 1/z$.
7. $e^{2z} - e^z$.
8. $\dfrac{(z-6)^4}{z}$.
9. $\dfrac{z-1}{z^2+1}$.
10. $z + 5$.
11. $z^2 + 1$.
12. $\dfrac{1+z}{z^2}$.
13. $ze^{1/z}$.
14. $\cos z$.
15. $z + \dfrac{1}{z}$.
16. z.
17. $\dfrac{\sin z}{z}$.
18. $\dfrac{\sin z}{z^2}$.
19. $\dfrac{1}{z+1} + \left(\dfrac{1}{z-1}\right)^3 + \left(\dfrac{1}{z-2}\right)^2$.
20. $\dfrac{\sin z^2}{z}$.

15.8 residues

If f has an isolated singularity at $z = a$, then in the neighborhood of $z = a$, the function has the Laurent representation

$$f(z) = \sum_{-\infty}^{\infty} A_n (z-a)^n$$

where the coefficients A_n have been derived previously. In particular,

$$A_{-1} = \frac{1}{2\pi i} \oint_{\substack{C \\ (\text{ccw})}} f(z)\, dz$$

where C is a closed curve enclosing no singularities of f other than the one at $z = a$. From this we have

$$\oint_{\substack{C \\ (\text{ccw})}} f(z)\, dz = 2\pi i A_{-1}$$

which gives another technique for evaluating complex line integrals over closed curves. Once the value of A_{-1} is known, the value of the integral is determined.

definition 15.5

The number A_{-1}, which is the coefficient of $1/(z-a)$ in the Laurent expansion of the function with center at $z=a$, is called the *residue* of the function at $z=a$.

example 15.23: Evaluate $\oint_C z^{-4} \sin z\, dz$ where C is the unit circle taken in the counterclockwise sense. (See Figure 15.8.)

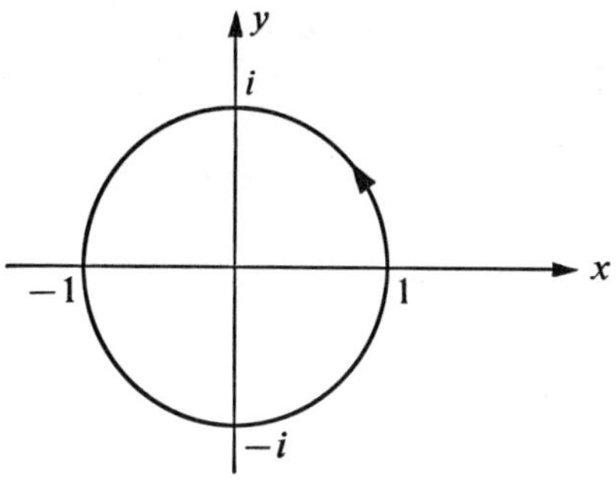

figure 15.8

solution: The only pole of the integrand is of order 3 at $z=0$. (Prove this!) The Laurent series for $z^{-4} \sin z$ about $z=0$ is

$$z^{-3} - \frac{z^{-1}}{3!} + \frac{z}{5!} - \frac{z^3}{7!} + \cdots.$$

Therefore $A_{-1} = -1/6$ and hence

$$\oint_C z^{-4} \sin z\, dz = -\frac{\pi i}{3}.$$

example 15.24: Evaluate $\oint_C \cot z\, dz$ where C is the circle $|z|=1$ taken clockwise.

solution: This is an example where the computation of the entire Laurent series would be tedious. Since only A_{-1} is required, a "long division" process may be used on the quotient of the two series.

$$\cot z = \frac{\cos z}{\sin z} = \frac{1 - \dfrac{z^2}{2} + \dfrac{z^4}{6} + \cdots}{z - \dfrac{z^3}{6} + \dfrac{z^5}{5!} + \cdots} = \frac{1}{z} + \cdots$$

which means that the residue of $\cot z$ at 0 is 1. Therefore,

$$\oint_C \cot z \, dz = 2\pi i$$

if C is taken in a counterclockwise sense. Therefore,

$$\oint_{\substack{C \\ (\text{cw})}} \cot z \, dz = -2\pi i .$$

Some special techniques are available for computing A_{-1} other than by finding the complete Laurent expansion. In particular, if the pole is simple, the residue may be found easily. When the function has a simple pole at $z = a$, the Laurent expansion has the form

$$f(z) = \frac{A_{-1}}{z-a} + A_0 + A_1(z-a) + A_2(z-a)^2 + \cdots .$$

If we multiply both sides of this equality by $z - a$ and take the limit as z approaches a we obtain

$$\lim_{z \to a} (z-a) f(z) = A_{-1} = \operatorname*{Res}_{z=a} f(z).$$

example 15.25: Let

$$f(z) = \frac{(z-1)(z-2)}{z(z-3)}.$$

Find the residue of $f(z)$ at $z = 0$ and $z = 3$.

solution: Since the poles are both simple, the formula just derived may be applied.

$$\operatorname*{Res}_{z=0} f(z) = \lim_{z \to 0} z f(z) = \frac{(-1)(-2)}{(-3)} = -\frac{2}{3}.$$

$$\operatorname*{Res}_{z=3} f(z) = \lim_{z \to 3} (z-3) f(z) = \frac{(2)(1)}{3} = \frac{2}{3}.$$

The expression for the residue at simple poles can be simplified even further in the case where $f(z)$ can be expressed as the quotient of two analytic functions. This method is especially appropriate if it is inconvenient or impractical to factor the denominator into linear factors. Suppose f has a simple pole at $z = a$ and that $f(z) = h(z)/q(z)$ where h and q have no common zeros. The fact that f has a simple pole at $z = a$ means $q(a) = 0$ and $q'(a) \neq 0$. Therefore,

$$\begin{aligned}\operatorname*{Res}_{z=a} f(z) &= \lim_{z\to a} (z-a)\frac{h(z)}{q(z)}\\ &= \lim_{z\to a} \frac{h(z)}{\dfrac{q(z)-q(a)}{z-a}}\\ &= \frac{h(a)}{q'(a)}.\end{aligned}$$

example 15.26: Find

$$\operatorname*{Res}_{z=i} \frac{1}{z^4-1}.$$

solution: The pole at $z = i$ is simple. Hence

$$\operatorname*{Res}_{z=i} \frac{1}{z^4-1} = \frac{1}{4z^3}\bigg|_{z=i} = -\frac{1}{4i} = \frac{i}{4}.$$

example 15.27: Find $\operatorname*{Res}_{z=0} \cot z$ by the foregoing technique.

solution:

$$\operatorname*{Res}_{z=0} \cot z = \operatorname*{Res}_{z=0} \frac{\cos z}{\sin z} = \frac{\cos 0}{\dfrac{d}{dz}(\sin z)|_{z=0}} = 1.$$

This is the residue at any pole for $\cot z$.

Variations of the above techniques may be used for poles of higher order. Rather than deriving formulas applicable to these more infrequent cases, we will be content with an example.

example 15.28: Find

$$\operatorname*{Res}_{z=1} \frac{z^2}{(z-1)^3(z+2)}.$$

solution: Since there is a third-order pole at $z = 1$, the Laurent series for $f(z)$ with center at 1 is of the form

$$\frac{z^2}{(z-1)^3(z+2)} = \frac{A_{-3}}{(z-1)^3} + \frac{A_{-2}}{(z-1)^2} + \frac{A_{-1}}{z-1} + A_0 + A_1(z-a) + \cdots.$$

If we multiply both sides by $(z-1)^3$ we have

$$\frac{z^2}{z+2} = A_{-3} + A_{-2}(z-1) + A_{-1}(z-1)^2 + A_0(z-1)^3 + \cdots.$$

Differentiating both sides twice and taking the limit as $z \to 1$, we obtain

$$\lim_{z\to 1} \frac{d^2}{dz^2}\left(\frac{z^2}{z+2}\right) = 2A_{-1}$$

and therefore

$$\operatorname*{Res}_{z=1} f(z) = \tfrac{4}{27}.$$

exercises for section 15.8

In Exercises 1–10 find the residues at each of the poles of the given functions.

1. $\dfrac{e^z}{(z-1)^2}$.
2. $\dfrac{\sin z}{z+2i}$.
3. $\tan z$.
4. $z^4 e^{1/z}$.
5. $\sec z$.
6. $e^{1/z}$.
7. $\dfrac{1}{z^4+1}$.
8. $\dfrac{1}{z^7+i}$.
9. $\dfrac{3+4z}{z^3+3z^2+2z}$.
10. $\dfrac{1}{z^2}$.

Evaluate the following integrals over the unit circle taken counterclockwise.

11. $\displaystyle\oint_C \frac{z+1}{z(z+2)}\,dz$.
12. $\displaystyle\oint_C \frac{z+1}{2z+1}\,dz$.
13. $\displaystyle\oint_C \frac{dz}{\sin z}\,dz$.
14. $\displaystyle\oint_C \frac{dz}{\cos z}$.
15. $\displaystyle\oint_C \frac{e^z}{z}\,dz$.
16. $\displaystyle\oint_C \frac{z^2+4}{(z-2)^2}\,dz$.
17. Evaluate

$$\oint_C \frac{z^3+1}{(z+1)(z^2+9)}\,dz$$

where C is the square with vertices at 2, $2i$, -2, and $-2i$ taken in the counterclockwise direction.

15.9 the residue theorem

Virtually all the previous techniques of integration around closed paths of functions with isolated singularities are special cases of the following theorem. It is the major elementary application of residues.

theorem 15.9

Let f be analytic inside a simple closed path C except for finitely many singular points $a_1, a_2, a_3, \cdots, a_m$ inside C. (See Figure 15.9.) Then

$$\oint_{\substack{C \\ \text{(ccw)}}} f(z)\,dz = 2\pi i \sum_{j=1}^{m} \operatorname*{Res}_{z=a_j} f(z).$$

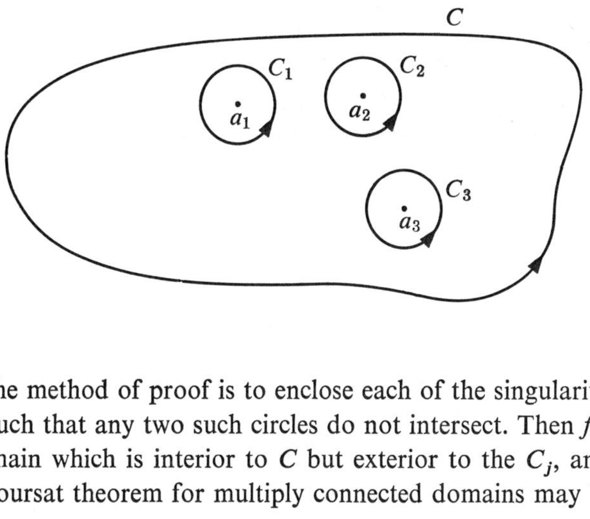

figure 15.9

proof: The method of proof is to enclose each of the singularities a_j by a circle C_j such that any two such circles do not intersect. Then f is analytic in the domain which is interior to C but exterior to the C_j, and thus the Cauchy-Goursat theorem for multiply connected domains may be applied.

$$\oint_{\substack{C \\ \text{(ccw)}}} f(z)\,dz + \sum_{j=1}^{m} \oint_{\substack{C_j \\ \text{(cw)}}} f(z)\,dz = 0$$

so that

$$\oint_{\substack{C \\ \text{(ccw)}}} f(z)\,dz = \sum_{j=1}^{m} \oint_{\substack{C_j \\ \text{(ccw)}}} f(z)\,dz.$$

Each of the integrals around the closed paths C_j is recognizable as $2\pi i \operatorname*{Res}_{z=a_j} f(z)$ and the theorem is proved.

example 15.29: Evaluate

$$\oint_C \frac{z^2 - 2}{z(z^2 - 1)}\,dz$$

where C is as follows, each taken in the counterclockwise sense (see Figure 15.10):

(a) the circle $|z| = 2$.   (b) the circle $|z-1| = 1/2$.

(c) the circle $|z| = 1$.   (d) the circle $|z-2| = 1/2$.

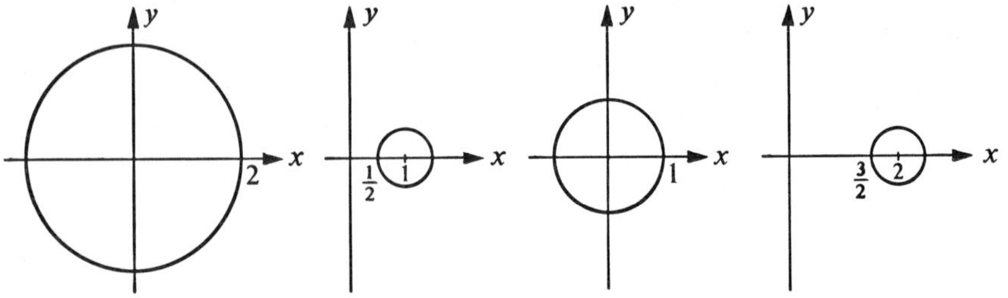

figure 15.10

solution: (a) The integrand has singularities at 0, 1, and -1. The path of integration encloses all the singularities. Therefore

$$\oint_{|z|=2} \frac{z^2-2}{z(z^2-1)}\,dz = 2\pi i \sum_{j=1}^{3} \operatorname*{Res}_{z=a_j} f(z)$$
$$= 2\pi i\left[\operatorname*{Res}_{z=0} f(z) + \operatorname*{Res}_{z=1} f(z) + \operatorname*{Res}_{z=-1} f(z)\right]$$
$$= 2\pi i(2 - \tfrac{1}{2} - \tfrac{1}{2}) = 2\pi i.$$

(b) The circle $|z| = 1/2$ encloses only the singularity at $z = 1$. Therefore

$$\oint_{|z|=\frac{1}{2}} \frac{z^2-2}{z(z^2-1)}\,dz = 2\pi i \operatorname*{Res}_{z=1} f(z)$$
$$= -\pi i.$$

(c) The path $|z| = 1$ passes through one of the singularities of the integrand and therefore the integral is undefined.

(d) The circle $|z-2| = 1/2$ encloses no singularities and hence, from the Cauchy-Goursat Theorem, the value of the integral is zero.

exercises for section 15.9

Evaluate the following integrals counterclockwise around the indicated circles.

1. $\displaystyle\oint_C \frac{dz}{\sin z}$, $|z| = 6$.   2. $\displaystyle\oint_C \frac{dz}{\cos z}$, $|z| = 10$.

3. $\displaystyle\oint_C \frac{e^z\,dz}{z(z+2)}, \quad |z| = 3.$ 4. $\displaystyle\oint_C \frac{dz}{1+3z^2}, \quad |z| = 1.$

5. $\displaystyle\oint_C \frac{z\,dz}{4+z^2}, \quad |z| = 3.$ 6. $\displaystyle\oint_C \frac{(z+2)^3}{z^4+4z^3+5z^2}\,dz, \quad |z| = 1.$

7. $\displaystyle\oint_C \frac{z^2+1}{(z+1)(z-3)}\,dz, \quad |z| = 5.$ 8. $\displaystyle\oint_C \tan z\,dz, \quad |z| = 1.$

9. $\displaystyle\oint_C \tan z\,dz, \quad |z| = 3.$ 10. $\displaystyle\oint_C \frac{1}{e^z}\,dz, \quad |z| = 1.$

11. Evaluate the integral

$$\oint_C \frac{dz}{(z+1)(z+2)(z+3)}$$

where C is as follows, taken counterclockwise:
(a) $|z| = 4$. (b) $|z| = 1/2$. (c) $|z| = 3/2$. (d) $|z| = 5/2$.

12. Evaluate the integral

$$\oint_C \frac{(z+2)^2}{2z^3+z}\,dz$$

where C is as follows, taken clockwise:
(a) $|z| = 1/2$. (b) $|z| = 2$. (c) $|z-1| = 3$. (d) $|z-1| = 1$.

13. Evaluate the integral

$$\oint_C \frac{6z^2+3z+1}{z^2(z^3+1)}\,dz$$

where C is as follows, taken counterclockwise:
(a) $|z| = 1/2$. (b) $|z-1| = 3$. (c) $|z-2| = 3$.
(d) $|z-3| = 1$. (e) $|z+1| = 1/2$.

14. Evaluate the integral

$$\oint_C \frac{(z^2+2)^3}{z^2-2z}\,dz$$

where C is as follows, taken counterclockwise:
(a) $|z| = 1$. (b) $|z| = 3$. (c) $|z| = 4$. (d) $|z-1| = 3$.
(e) $|z-2| = 1$.

15.10 evaluation of real integrals

Surprisingly, the most immediate application of the residue theorem is in the evaluation of real integrals. In making the application, the trick is to express the real integral in terms of a complex integral around a closed path. The general theory here is somewhat limited since it is more a matter of technique than anything else.

A. rational trigonometric functions: Real integrals whose integrands are rational functions of $\sin\theta$ and $\cos\theta$ and whose limits of integration are from 0 to 2π may be evaluated by using the residue theorem after an initial substitution $z = e^{i\theta}$. The path of integration in the complex plane is $|z| = 1$. After substitution, the integrand is a rational function of z.

example 15.30: Evaluate

$$\int_0^{2\pi} \frac{d\theta}{5 - 3\cos\theta}.$$

solution: We first write $\cos\theta$ in terms of $e^{i\theta}$,

$$\int_0^{2\pi} \frac{d\theta}{5 - 3\cos\theta} = \int_0^{2\pi} \frac{d\theta}{5 - 3\dfrac{e^{i\theta} + e^{-i\theta}}{2}}.$$

Letting $z = e^{i\theta}$, the real integral is transformed into an integral in the complex plane over the unit circle,

$$\begin{aligned}
\int_0^{2\pi} \frac{d\theta}{5 - 3\cos\theta} &= \oint_{|z|=1} \frac{\dfrac{dz}{iz}}{5 - \frac{3}{2}(z + z^{-1})} \\
&= \frac{1}{i}\oint_{|z|=1} \frac{dz}{5z - \frac{3}{2}z^2 - \frac{3}{2}} \\
&= -\frac{2}{i}\oint_{|z|=1} \frac{dz}{3z^2 - 10z + 3} \\
&= -\frac{2}{i}\oint_{|z|=1} \frac{dz}{(3z - 1)(z - 3)}.
\end{aligned}$$

The integrand has simple poles at $z = 3$ and $z = 1/3$. Hence,

$$\begin{aligned}
\int_0^{2\pi} \frac{d\theta}{5 - 3\cos\theta} &= 2\pi i\left(-\frac{2}{i}\right)\left(\operatorname*{Res}_{z=1/3} \frac{1}{(3z - 1)(z - 3)}\right) \\
&= -4\pi\left(-\frac{1}{8}\right) = \frac{\pi}{2}.
\end{aligned}$$

(see Figure 15.11).

If the interval of integration for the real integral is not from 0 to 2π, the integrand may sometimes be modified before applying the substitution.

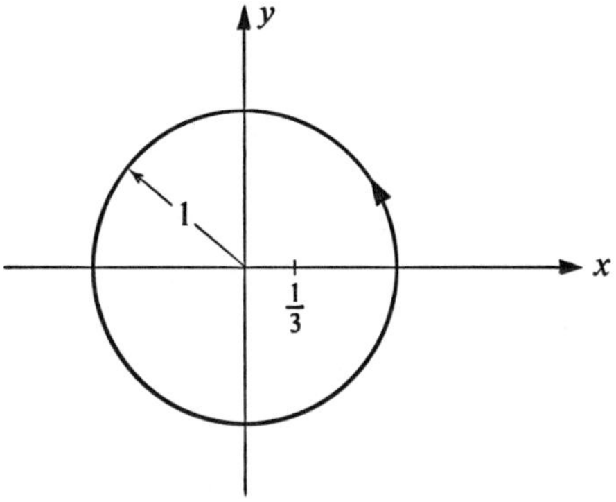

figure 15.11

example 15.31: Evaluate

$$\int_0^{\pi} \frac{d\theta}{5 - 3\cos\theta}.$$

solution: Because the values taken by the cosine function from 0 to π are the same as the values from π to 2π (although in a different order), we may write

$$\int_0^{\pi} \frac{d\theta}{5 - 3\cos\theta} = \frac{1}{2}\int_0^{2\pi} \frac{d\theta}{5 - 3\cos\theta} = \frac{\pi}{4}.$$

B. rational functions of x: In Appendix A, Section 4, you will find the following facts:

1. The improper integral $\int_{-\infty}^{\infty} f(x)\,dx$ converges in the ordinary sense to a number L (called its *value*) if

$$\lim_{\substack{A\to-\infty \\ B\to\infty}} \int_A^B f(x)\,dx = L.$$

2. The improper integral $\int_{-\infty}^{\infty} f(x)\,dx$ converges in the Cauchy sense to a number L_c (called the *Cauchy principal value*, C.P.V.) if

$$\lim_{A\to-\infty} \int_{-A}^{A} f(x)\,dx = L_c.$$

Note that for ordinary convergence, the limiting processes at the upper and lower limit are independent of one another whereas in finding the C.P.V. they are dependent.

3. If an improper integral converges in the ordinary sense, it converges in Cauchy sense and moreover, the value and the C.P.V. are the same.

In this section we show how the Cauchy principal value of improper integrals with rational integrands may be determined. In many cases, since the integrals will actually converge in the usual sense, the C.P.V. obtained will also be the value in the ordinary sense.

In the following general outline of the technique, we assume that $f(x)$ is of the order of $1/x^p$ for $p > 1$. For rational functions this means that the polynomial in the denominator has degree at least two greater than the degree of the polynomial in the numerator.

To find C.P.V. $\int_{-\infty}^{\infty} f(x)\,dx$ by *contour integration*, we specify

(a) the function of a complex variable which serves as the integrand, and
(b) the path or "contour" of integration.

In the case where the real integrand is a rational function of x we simply change x to z and use a path of integration which bounds the semicircular region in the upper half-plane. Then

$$\oint_C f(z)\,dz = \int_{-R}^{R} f(x)\,dx + \int_\Gamma f(z)\,dz$$

where Γ is the semicircular path shown in Figure 15.12. Hence

$$\int_{-R}^{R} f(x)\,dx = \int_C f(z)\,dz - \int_\Gamma f(z)\,dz$$

and therefore

$$\text{C.P.V.} \int_{-\infty}^{\infty} f(x)\,dx = 2\pi i \sum \operatorname{Res} f(z) - \lim_{R\to\infty} \int_\Gamma f(z)\,dz$$

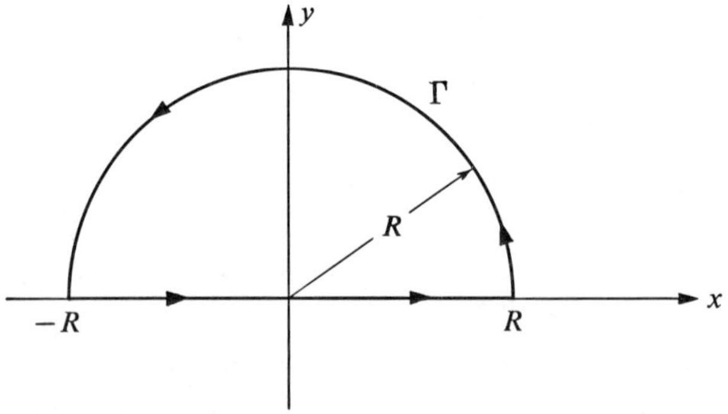

figure 15.12 *semicircular contour*

where the summation is over the points in the upper half-plane which are singularities of $f(z)$. In the last integral on the right represent Γ by $z = Re^{it}$. Then

$$\int_{\Gamma} f(z)\,dz = i\int_{0}^{\pi} f(Re^{it})\,Re^{it}\,dt.$$

Therefore,

$$\left|\int_{\Gamma} f(z)\,dz\right| \leqq \int_{0}^{\pi} |f(Re^{it})|\,R\,|e^{it}\,dt|.$$

If we assume $f(x)$ to be of the order of $1/x^p\,(p > 1)$ for large x,

$$\left|\int_{\Gamma} f(z)\,dz\right| \leqq \frac{M}{R^p}\,R\int_{0}^{\pi} |e^{it}\,dt| = \frac{M}{R^p}\,R\pi = \frac{M\pi}{R^{p-1}}$$

which approaches 0 as R becomes unbounded, and therefore,

$$\text{C.P.V.}\int_{-\infty}^{\infty} f(x)\,dx = 2\pi i \sum \operatorname{Res} f(z).$$

example 15.32: Evaluate

$$\int_{0}^{\infty} \frac{dx}{1+x^4}.$$

solution:

$$\int_{0}^{\infty} \frac{dx}{1+x^4} = \frac{1}{2}\,\text{C.P.V.}\int_{-\infty}^{\infty} \frac{dx}{1+x^4}$$

$$\text{C.P.V.}\int_{-\infty}^{\infty} \frac{dx}{1+x^4} = 2\pi i \sum \operatorname{Res}\left(\frac{1}{1+z^4}\right).$$

$1/(1+z^4)$ has simple poles in the upper half-plane at $z = e^{\pi i/4}$ and $z = e^{3\pi i/4}$. Since

$$\operatorname*{Res}_{z=z_j}\left(\frac{1}{1+z^4}\right) = \frac{1}{4z_j^3},$$

$$\text{C.P.V.}\int_{-\infty}^{\infty} \frac{dx}{1+x^4} = 2\pi i\left(\frac{1}{4e^{3\pi i/4}} + \frac{1}{4e^{\pi i/4}}\right)$$

$$= \frac{2\pi i}{4}\left(-e^{\pi i/4} + e^{-\pi i/4}\right)$$

$$= \pi \sin\frac{\pi}{4} = \pi/\sqrt{2}.$$

Since the given integral actually does converge,

$$\int_{0}^{\infty} \frac{dx}{1+x^4} = \frac{\pi}{2\sqrt{2}}.$$

C. fourier integrals: An important variation of the type of integral discussed in the previous section is that in which the rational integrand is multiplied by either a sine or a cosine function. Integrals of this type are called Fourier integrals, and with the same hypothesis on the rational function as before, only a slight modification in procedure is necessary.

The same kind of countour as in the section on "Rational functions of x" is suitable, but the function to integrate is not obtained by substituting z for x since both $\sin z$ and $\cos z$ are unbounded in the upper half-plane. Instead, we consider $\int_{-\infty}^{\infty} f(x)\, e^{isx}\, dx$ (where f is a rational function) and take the real and imaginary parts. In the latter integral, we let $x = z$ and then

$$\oint_C f(z)\, e^{isz}\, dz = \oint_C f(z) \cos sz\, dz + i \oint_C f(z) \sin sz\, dz.$$

Just as in the previous section,

$$\int_{-R}^{R} f(x)\, e^{isx}\, dx = \oint_C f(z)\, e^{isz}\, dz - \int_{\Gamma} f(z)\, e^{isz}\, dz.$$

Since $|e^{isz}| = e^{-sy} < 1$ if $y > 0$, the proof that

$$\lim_{R\to\infty} \left| \int_{\Gamma} f(z)\, e^{isz}\, dz \right| = 0$$

is essentially the same as in the previous case and therefore,

$$\text{C.P.V.} \int_{-\infty}^{\infty} f(x)\, e^{isx}\, dx = 2\pi i \sum \operatorname{Res} f(z)\, e^{isz}.$$

With the given hypothesis, the improper integral exists. Hence

$$\int_{-\infty}^{\infty} f(x) \cos sx\, dx = \operatorname{Re}\left(2\pi i \sum \operatorname{Res} f(z)\, e^{isz}\right)$$

$$\int_{-\infty}^{\infty} f(x) \sin sx\, dx = \operatorname{Im}\left(2\pi i \sum \operatorname{Res} f(z)\, e^{isz}\right).$$

example 15.33: Evaluate

$$\int_{-\infty}^{\infty} \frac{\cos 3x}{x^2+4}\, dx.$$

solution:

$$\begin{aligned}\int_{-\infty}^{\infty} \frac{\cos 3x}{x^2+4}\,dx &= \text{Re}\left(2\pi i \sum \text{Res}\,\frac{e^{3iz}}{z^2+4}\right)\\ &= \text{Re}\left(2\pi i\,\frac{e^{3i(2i)}}{2(2i)}\right)\\ &= \frac{\pi e^{-6}}{2}.\end{aligned}$$

D. integrands with poles on the real axis: If the integrand has a pole on the x-axis the contour is modified to circumvent these points by "indenting." Specifically, suppose you are to evaluate $\int_{-\infty}^{\infty} f(x)\,dx$ or $\int_{-\infty}^{\infty} f(x)\,e^{isx}\,dx$. As a bit of a variation, assume that f is a rational function with a denominator exactly *one* degree higher than the numerator, and with a simple pole at the origin. (See Figure 15.13.) Then,

$$\oint_C f(z)\,e^{isz}\,dz = \int_{-R}^{-r} f(x)\,e^{isx}\,dx + \int_{\lambda} f(z)\,e^{isz}\,dz + \int_{r}^{R} f(x)\,e^{isx}\,dx + \int_{\Gamma} f(z)\,e^{isz}\,dz.$$

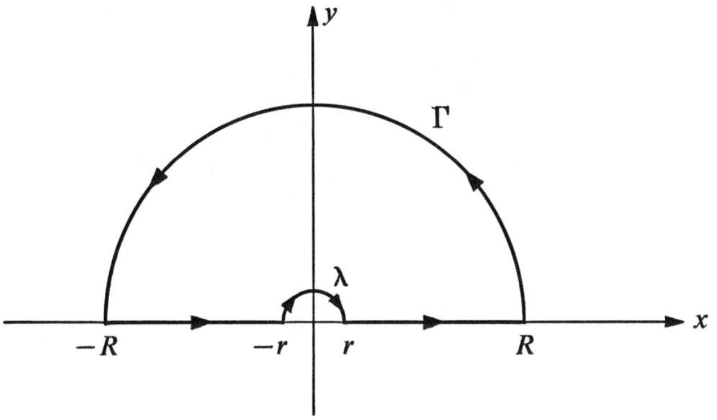

***figure 15.13** indented contour*

Since the pole at the origin is simple, the function has a Laurent series with center at 0 of the form

$$f(z) = \frac{\operatorname*{Res}_{z=0} f(z)\,e^{isz}}{z} + g(z)$$

where $g(z)$ is analytic at $z=0$. On λ, the path may be represented by $z(t)=re^{it}$,

$$\int_\lambda f(z)\,e^{isz}\,dz = \int_\lambda \frac{\operatorname*{Res}_{z=0}[f(z)\,e^{isz}]}{z}\,dz + \int_\lambda g(z)\,dz$$

$$= \operatorname*{Res}_{z=0}[f(z)\,e^{isz}] \int_\pi^0 i\,dt + \int_\pi^0 g(re^{it})\,(ire^{it})\,dt\,.$$

Since $g(z)$ is analytic at $z=0$, it is bounded near the origin and hence the last of these integrals can be made as small as desired by making r small enough. Hence

$$\lim_{r\to 0}\left|\int_\lambda f(z)\,e^{isz}\,dz\right| = -\pi i \operatorname*{Res}_{z=0} f(z)\,e^{isz}\,.$$

On Γ, the path can be written $z(t)=Re^{it}$,

$$\int_\Gamma e^{isz} f(z)\,dz = \int_0^\pi f(Re^{it})\,e^{isRe^{it}} iRe^{it}\,dt\,.$$

$$|e^{isRe^{it}}| = |e^{isR(\cos t + i\sin t)}| = e^{-sR\sin t}\,.$$

Therefore,

$$\left|\int_\Gamma e^{isz} f(z)\,dz\right| = \left|\int_0^\pi f(Re^{it})\,e^{isR(\cos t+i\sin t)}\,iRe^{it}\,dt\right|$$

$$\leqq \int_0^\pi e^{-sR\sin t}\varepsilon R\,dt$$

where ε can be made arbitrarily small since $f(z)$ is a rational function in which the numerator is dominated by the denominator. Using the fact that $\sin t$ assumes the same values on 0 to $\pi/2$ as on $\pi/2$ to π, this last integral is equal to $\int_0^{\pi/2} 2e^{-sR\sin t}\,\varepsilon R\,dt$. Using the fundamental trigonometric inequality $\sin t \geqq 2t/\pi$,

$$\left|\int_\Gamma e^{isz} f(z)\,dz\right| \leqq 2R\varepsilon \int_0^{\pi/2} e^{-2Rst/\pi}\,dt$$

$$= 2R\varepsilon\left(\frac{-\pi}{2Rs}\right) e^{-2Rst/\pi}\Big|_0^{\pi/2}$$

$$= \frac{\pi\varepsilon}{s}(1-e^{-Rs}) < \frac{\pi\varepsilon}{s}\,.$$

Therefore we have shown that the integral over $\Gamma\to 0$ as $R\to\infty$. Hence,

$$\text{C.P.V.}\int_{-\infty}^{\infty} f(x)\,e^{isx}\,dx = \pi i \operatorname*{Res}_{z=0} f(z)\,e^{isz} + 2\pi i \sum \operatorname{Res} f(z)\,e^{isz}\,.$$

example 15.34: Evaluate

$$\text{C.P.V.} \int_{-\infty}^{\infty} \frac{\sin x}{x}\, dx .$$

solution: We integrate e^{iz}/z over the semicircular contour indented at 0. The residue of the integrand at 0 is 1. The integrand has no other poles.

$$\text{C.P.V.} \int_{-\infty}^{\infty} \frac{e^{ix}}{x}\, dx = \pi i$$

which means that

$$\text{C.P.V.} \int_{-\infty}^{\infty} \frac{\sin x}{x}\, dx = \pi .$$

E. other types of contours: The choice of contours is much more varied than you may have been led to believe from the previous sections. In this section you will see two examples in which the contour of integration used is significantly different.

example 15.35: Evaluate

$$\int_{-\infty}^{\infty} \frac{e^{mx}}{1+e^{x}}\, dx, \qquad 0 < m < 1 .$$

solution: Choose $e^{mz}/(1+e^{z})$ as the function of a complex variable to integrate and let C be the contour shown in Figure 15.14.

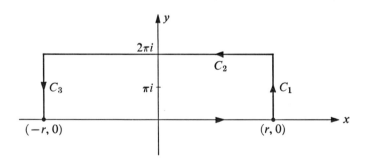

figure 15.14

The function $e^{mz}/(1+e^{z})$ has only the one pole at $z = \pi i$ within the contour. Since

$$\operatorname*{Res}_{z=\pi i} \frac{e^{mz}}{1+e^{z}} = \frac{e^{m\pi i}}{e^{\pi i}} = -\, e^{m\pi i},$$

we have by the residue theorem

$$\oint_C \frac{e^{mz}}{1+e^z}\,dz = -\,e^{m\pi i}.$$

Further,

$$\oint_C \frac{e^{mz}}{1+e^z}\,dz = \int_{C_1} \frac{e^{mz}}{1+e^z}\,dz + \int_{C_2} \frac{e^{mz}}{1+e^z}\,dz + \int_{C_3} \frac{e^{mz}}{1+e^z}\,dz + \int_{-r}^{r} \frac{e^{mx}}{1+e^x}\,dx.$$

On C_1, $z = r + iy$, so that

$$\left|\int_{C_1} \frac{e^{mz}}{1+e^z}\,dz\right| \leqq \int_{C_1} \frac{|e^{mz}|}{|1+e^z|}\,|dz| \leqq \frac{e^{mr}2\pi}{e^r - 1} = \frac{2\pi e^{(m-1)r}}{1-e^{-r}}.$$

Since $m < 1$, this becomes arbitarily small as r becomes unbounded.

A similar argument shows that

$$\lim_{r\to\infty}\left|\int_{C_3} \frac{e^{mz}}{1+e^z}\,dz\right| = 0.$$

On C_2, $z = x + 2\pi i$. Therefore

$$\int_{C_2} \frac{e^{mz}}{1+e^z}\,dz = \int_{r}^{-r} \frac{e^{mx}e^{2m\pi i}}{1+e^x e^{2\pi i}}\,dx = -\,e^{2m\pi i}\int_{-r}^{r} \frac{e^{mx}}{1+e^x}\,dx$$

and therefore

$$\oint_C \frac{e^{mz}}{1+e^z}\,dz = \lim_{r\to\infty}(1 - e^{2m\pi i})\int_{-r}^{r} \frac{e^{mx}}{1+e^x}\,dx = -\,2\pi i e^{m\pi i},$$

from which we obtain

$$\int_{-\infty}^{\infty} \frac{e^{mx}}{1+e^x}\,dx = \frac{2\pi i e^{m\pi i}}{e^{2m\pi i}-1} = \frac{\pi}{\sin m\pi}.$$

example 15.36: Evaluate

$$\int_0^{\infty} \frac{x^{p-1}}{1+x}\,dx, \qquad 0 < p < 1.$$

solution: It is possible to change variables in the integral of the previous example to obtain the one given here. We will give a different method of evaluation in which we are able to exhibit another kind of contour. (See Figure 15.15.) Since $z^{p-1}/(1+z)$ has a pole at $z = -1$, and nowhere else,

$$\oint_C \frac{z^{p-1}}{1+z}\,dz = 2\pi i \operatorname*{Res}_{z=-1} \frac{z^{p-1}}{1+z} = 2\pi i (e^{\pi i})^{p-1} = -\,2\pi i e^{\pi i p}.$$

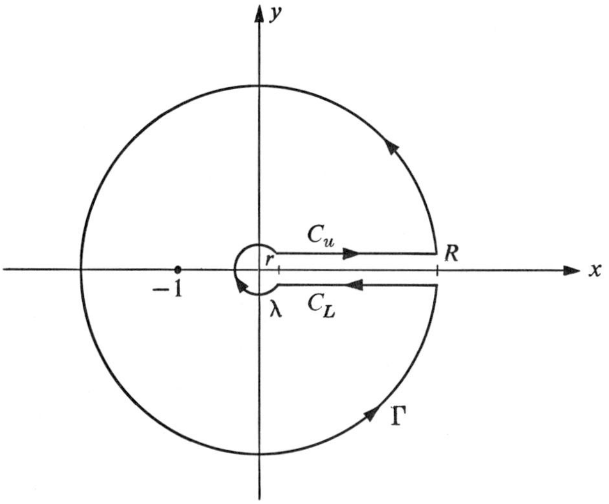

figure 15.15

Also

$$\int_C \frac{z^{p-1}}{1+z}\,dz = \left(\int_{C_u} + \int_{\Gamma} + \int_{\lambda} + \int_{C_L}\right) \frac{z^{p-1}}{1+z}\,dz.$$

For z^{p-1}, we choose the branch cut along the positive real axis with $0 \leqq \arg z < 2\pi$. On the upper ray from the origin, C_u, $z^{p-1} = x^{p-1}$. On C_L,

$$\begin{aligned} z^{p-1} &= e^{(p-1)\ln z} \\ &= e^{(p-1)\log_e x} e^{(p-1)2\pi i} \\ &= x^{p-1} e^{2\pi i p}. \end{aligned}$$

With the given limitations on the parameter p, it is relatively easy to show that the integrals over λ and Γ can be made arbitrarily small as $r \to 0$ and $R \to \infty$. Hence,

$$\begin{aligned} \oint_C \frac{z^{p-1}}{1+z}\,dz &= \lim_{\substack{r\to 0 \\ R\to\infty}} \int_{C_u} + \int_{C_L} \frac{z^{p-1}}{1+z}\,dz \\ &= \lim_{\substack{r\to 0 \\ R\to\infty}} \left[\int_r^R \frac{x^{p-1}}{1+x}\,dx + e^{2\pi i p} \int_R^r \frac{x^{p-1}}{1+x}\,dx\right] \end{aligned}$$

and therefore,

$$(1 - e^{2\pi i p}) \int_0^\infty \frac{x^{p-1}}{1+x}\,dx = -\,2\pi i e^{\pi i p}$$

from which

$$\int_0^\infty \frac{x^{p-1}}{1+x}\,dx = \frac{\pi}{\sin \pi p}.$$

exercises for section 15.10

Evaluate the following real integrals.

1. $\displaystyle\int_0^{2\pi} \frac{dx}{3+\cos x}.$

2. $\displaystyle\int_0^{2\pi} \frac{dx}{a+\cos x}, \quad a > 1.$

3. $\displaystyle\int_0^{2\pi} \frac{dx}{1+a\cos x}, \quad 0 < a < 1.$

4. $\displaystyle\int_0^{2\pi} \frac{\sin^2 x}{5-4\cos x}\,dx.$

5. $\displaystyle\int_0^{2\pi} \frac{\cos^2 3x}{5-4\cos 2x}\,dx.$

6. $\displaystyle\int_0^{\pi/2} \frac{dx}{1+\sin^2 x}.$

7. $\displaystyle\int_0^{\pi/2} \frac{dx}{a+\sin^2 x}, \quad a > 0.$

8. $\displaystyle\int_0^{2\pi} \frac{dx}{1-2m\cos x+m^2}, \quad 0 < m < 1.$

Find the Cauchy Principal Value of the following improper integrals.

9. $\displaystyle\int_{-\infty}^{\infty} \frac{dx}{1+x^4}.$

10. $\displaystyle\int_{-\infty}^{\infty} \frac{x^5}{1+x^{10}}\,dx.$

11. $\displaystyle\int_{-\infty}^{\infty} \frac{x}{(x^2-2x+2)^2}\,dx.$

12. $\displaystyle\int_{-\infty}^{\infty} \frac{x^2}{x^4+6x^2+13}\,dx.$

13. $\displaystyle\int_{-\infty}^{\infty} \frac{dx}{(x^2+1)^2(x^2+9)}.$

14. $\displaystyle\int_{-\infty}^{\infty} \frac{\cos 3x}{(x^2+9)^2}\,dx.$

15. $\displaystyle\int_{-\infty}^{\infty} \frac{\cos x}{x^2+1}\,dx.$

16. $\displaystyle\int_{-\infty}^{\infty} \frac{\sin 3x}{1+x^4}\,dx.$

17. $\displaystyle\int_{-\infty}^{\infty} \frac{dx}{(x+1)(x^2+2)}.$

18. $\displaystyle\int_{-\infty}^{\infty} \frac{\sin^2 sx}{x^2}\,dx, \quad s > 0.$

19. $\displaystyle\int_{-\infty}^{\infty} \frac{\cos mx}{e^x+e^{-x}}\,dx, \quad m > 0.$

20. $\displaystyle\int_{-\infty}^{\infty} \frac{x^2}{(x^2+1)^3}\,dx.$

21. Do Example 15.36 by changing variables in the integral of Example 15.35.

special functions

16

16.1 introduction

We often encounter functions which are nonelementary in their definition and conception. These functions can be very useful, but there is sometimes a lack of appreciation and understanding of them since they are defined differently from the more familiar functions. For lack of a better word these functions are called "special" and in this chapter we will study just a few of them. The ones we consider are really not any more "special" than many others but are representative of the types which do occur.

16.2 functions defined by improper integrals

In Appendix A, Section 5, the general idea of an improper integral with a parameter is summarized. An integral of this type defines a function on the domain of convergence. Although other improper integrals are important, our concern is with those functions defined in terms of an integral which is improper due to an infinite upper limit,

$$F(x) = \int_a^{\infty} f(x, t)\, dt.$$

We usually are concerned with the domain not only of convergence, but also of uniform convergence (see Definition A.11), since on the latter domain:

1. The function is continuous.
2. The integral of the function over the interval $[c, d]$ may be found by reversing that process with the improper integration; that is,

$$\int_c^d F(x)\,dx = \int_a^\infty \int_c^d f(x, t)\,dx\,dt.$$

3. If the improper integral of partial derivatives converges uniformly, then the derivative of the function is found by partially differentiating the integrand; that is,

$$F'(x) = \int_0^\infty \frac{\partial f}{\partial x}(x, t)\,dt.$$

16.3 the gamma function

The gamma function, denoted by $\Gamma(x)$, is defined by

$$\Gamma(x) = \int_0^\infty t^{x-1} e^{-t}\,dt.$$

The gamma function is sometimes called the *generalized factorial function.* To see why, consider the function defined by

$$F(u) = \int_0^\infty e^{-ut}\,dt = \frac{1}{u}.$$

The improper integral converges uniformly to $1/u$ on any finite interval $[c, d]$, $c > 0$ (see Example A.9). Since the improper integral of partial derivatives,

$$\int_0^\infty te^{-ut}\,dt,$$

converges uniformly, it follows from Theorem A.13 that

$$\int_0^\infty te^{-ut}\,dt = \frac{1}{u^2}.$$

With similar reasoning,

$$\int_0^\infty t^2 e^{-ut}\,dt = \frac{2}{u^3},$$

$$\int_0^\infty t^3 e^{-ut}\,dt = \frac{3!}{u^4},$$

$$\int_0^\infty t^4 e^{-ut}\,dt = \frac{4!}{u^5},$$

and, in general,

$$\int_0^\infty t^n e^{-ut}\,dt = \frac{n!}{u^{n+1}}.$$

In this last expression, we let $u = 1$, to obtain

$$\int_0^\infty t^n e^{-t}\,dt = n!.$$

We thus call the improper integral $\int_0^\infty t^n e^{-t}\,dt$ the *factorial function.* As defined, the value of n is restricted to positive integers. However, since the improper integral has meaning if n is any real number, $x > -1$, we write

$$x! = \int_0^\infty t^x e^{-t}\,dt.$$

It is thus obvious that the "extended" factorial function and the gamma function are one and the same except for a shift of variables; that is,

$$\Gamma(x) = (x-1)!.$$

In passing we note that when $x = 0$, the factorial function gives

$$0! = \int_0^\infty e^{-t}\,dt = 1,$$

which is why, in the more elementary definition of the factorial, $0!$ is defined to be 1.

In elementary courses, $n!$ is defined to be the product of the integers, $n(n-1)(n-2)\cdots(3)(2)(1)$ from which we obtain the so-called *factorial property,* $n! = n(n-1)!$. If x is an integer, $x = n$, we therefore have immediately,

$$\Gamma(n+1) = n! = n(n-1)! = n\Gamma(n).$$

The gamma function has this property for all $x > 0$ as we will now show.

$$\Gamma(x+1) = \int_0^\infty t^x e^{-t}\,dt$$

$$= \lim_{B\to\infty} \int_0^B t^x e^{-t}\,dt.$$

An integration by parts yields

$$\Gamma(x+1) = \lim_{B\to\infty} (-t^x e^{-t})\Big|_0^B + x\int_0^\infty t^{x-1} e^{-t}\,dt = x\Gamma(x),$$

which verifies the property.

Tables of values of the gamma function have been compiled. Because of the factorial property, values of the function need be tabulated only over any interval of unit length. The interval usually chosen for this tabulation is $1 \leqq x \leqq 2$. A short table is shown below.

table iii
Values of the Gamma Function for $1 \leqq x \leqq 2$

x	$\Gamma(x)$
1.0	1.000
1.1	0.951
1.2	0.918
1.3	0.897
1.4	0.887
1.5	0.886
1.6	0.894
1.7	0.909
1.8	0.931
1.9	0.962
2.0	1.000

In Section 16.4 (Example 16.11), we will show that $\Gamma(1/2) = \sqrt{\pi}$. Until then, we will assume the correctness of that result and use it when necessary.

example 16.1: Compute $\Gamma(7/2)$.

solution:

$$\Gamma(\tfrac{7}{2}) = \tfrac{5}{2}\Gamma(\tfrac{5}{2}) = \tfrac{5}{2}\,\tfrac{3}{2}\Gamma(\tfrac{3}{2}) = \tfrac{5}{2}\,\tfrac{3}{2}\,\tfrac{1}{2}\Gamma(\tfrac{1}{2}) = \tfrac{15}{8}\sqrt{\pi}.$$

example 16.2: Evaluate

$$\int_0^\infty e^{-h^2x^2}\,dx.$$

solution: Let $y = h^2x^2$. Then $dy = 2h^2x\,dx$, and the given integral becomes

$$\frac{1}{2h}\int_0^\infty y^{-1/2}e^{-y}\,dy = \frac{1}{2h}\Gamma\left(\frac{1}{2}\right) = \frac{\sqrt{\pi}}{2h}.$$

example 16.3: Evaluate

$$\int_0^\infty x^4 e^{-x^3}\, dx.$$

solution: Let $y = x^3$. Then the integral becomes

$$\tfrac{1}{3}\int_0^\infty y^{2/3} e^{-y}\, dy = \tfrac{1}{3}\Gamma(\tfrac{5}{3}) = \tfrac{1}{3}(0.9) = 0.3.$$

example 16.4: Evaluate

$$\int_0^1 x^2 \left(\log \frac{1}{x}\right)^3 dx.$$

solution: Let $y = -\log x$. Then the given integral becomes

$$-\int_\infty^0 e^{-3y} y^3\, dy = \int_0^\infty e^{-3y} y^3\, dy$$
$$= \tfrac{1}{27}\int_0^\infty e^{-3y}(3y)^3\, dy.$$

In this integral, let $t = 3y$ to obtain

$$\frac{1}{3^4}\int_0^\infty e^{-t} t^3\, dt = \frac{1}{3^4}\Gamma(4) = \frac{2}{27}.$$

It is possible to show that the defining integral for the gamma function converges for $x > 0$ and converges uniformly on any interval $[c, d]$ for which $c > 0$. Hence:

1. The domain of the gamma function is $\{x \mid x > 0\}$.
2. The gamma function is continuous for $\{x \mid x > 0\}$.
3. Since the improper integral of partial derivatives of the integrand function converges uniformly on $[c, d]$, the derivative of the gamma function is obtained by partially differentiating the integrand with respect to x. Similar statements may be made for higher order derivatives.

With the aid of Table III and with the factorial property, you can graph the gamma function. A knowledge of the sense of concavity of the graph is found from the second derivative. Thus

$$D_x\Gamma(x) = \int_0^\infty t^{x-1} \ln t\, e^{-t}\, dt,$$

$$D_x^2\Gamma(x) = \int_0^\infty t^{x-1} (\ln t)^2\, e^{-t}\, dt.$$

Since the integrand in the last integral is positive over the interval of integration, $D_x^2\Gamma(x) > 0$, and hence the graph of the gamma function is concave up everywhere.

For x near zero, the gamma function is unbounded. To see this, note that

$$\Gamma(x) > \int_0^1 t^{x-1} e^{-t}\, dt > e^{-1} \int_0^1 t^{x-1}\, dt = \frac{1}{ex}$$

which $\to \infty$ as $x \to 0$.

From the factorial property,

$$\lim_{x\to\infty} \Gamma(x) = \lim_{x\to\infty} (x-1)\,\Gamma(x-1) = \infty .$$

The graph is shown in Figure 16.1.

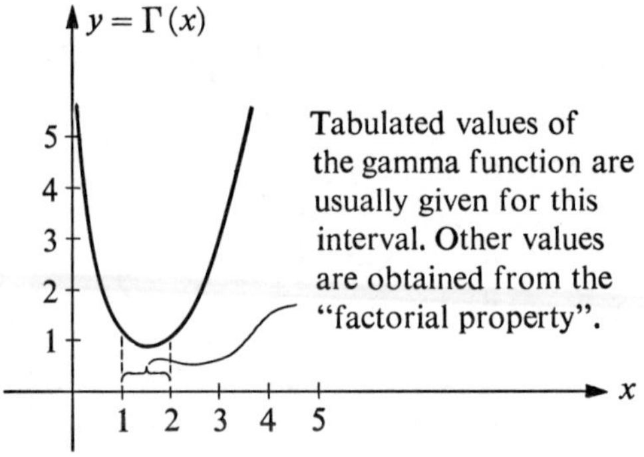

figure 16.1 *the gamma function*

We may extend the definition of the gamma function to values of x other than positive real values by postulating the preservation of the factorial property. In this way, all values of the gamma function for negative values of x are written in terms of the values for $0 < x < 1$, and of course, ultimately in terms of the values for $1 < x < 2$. For example,

if $-1 < x < 0$, then

$$\Gamma(x) = \frac{\Gamma(x+1)}{x} \qquad \text{(note that } 0 < x+1 < 1\text{);}$$

if $-2 < x < -1$, then

$$\Gamma(x) = \frac{\Gamma(x+2)}{x(x+1)} \qquad \text{(note that } 0 < x+2 < 1\text{);}$$

if $-3 < x < -2$, then

$$\Gamma(x) = \frac{\Gamma(x+3)}{x(x+1)(x+2)} \quad \text{(note that } 0 < x+3 < 1\text{);}$$

$$\vdots \qquad \vdots$$

if $-m < x < -m+1$, then

$$\Gamma(x) = \frac{\Gamma(x+m)}{x(x+1)\cdots(x+m-1)} \quad \text{(note that } 0 < x+m < 1\text{).}$$

The preservation of the factorial property does not allow for a definition of the gamma function at the nonpositive integers. Near every negative integer, the function becomes unbounded. The values are alternately positive and negative as determined by the denominator in the defining expression since the numerator is always positive. The graph of the extended function is shown in Figure 16.2.

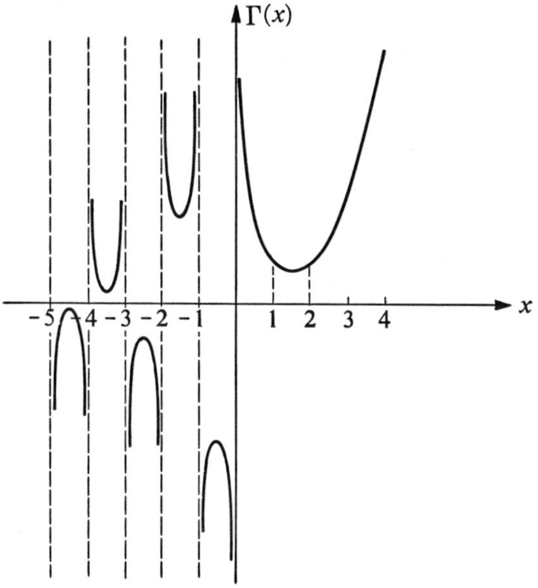

figure 16.2 *extended gamma function*

example 16.5: Compute $\Gamma(-3/2)$.

solution:

$$\Gamma(-\tfrac{3}{2}) = \frac{\Gamma(\tfrac{1}{2})}{(-\tfrac{3}{2})(-\tfrac{1}{2})} = \frac{4\sqrt{\pi}}{3}.$$

The gamma function is extended to the complex plane by considering the same (real) improper integral as before with z substituted for x,

$$\Gamma(z) = \int_0^\infty e^{-t} t^{z-1}\, dt.$$

Just as in the real case, this integral converges for $\operatorname{Re} z > 0$. Once again, the function is extended into the left half of the complex plane by the factorial property

$$\Gamma(z) = \frac{\Gamma(z+1)}{z}.$$

example 16.6: Express $\Gamma(-5/2+3i)$ in terms of a value of the gamma function with positive real argument.

solution:

$$\Gamma(-\tfrac{5}{2}+3i) = \frac{\Gamma(\tfrac{1}{2}+3i)}{(-\tfrac{1}{2}+3i)(-\tfrac{3}{2}+3i)(-\tfrac{5}{2}+3i)}.$$

By the nature of the defining equation, the gamma function has simple poles at the points corresponding to the negative integers.

example 16.7: Find the residue of the complex gamma function at $z=-1$.

solution: Since the singularity at $z=-1$ is a simple pole,

$$\begin{aligned}
\operatorname*{Res}_{z=-1} \Gamma(z) &= \lim_{z\to -1} (z+1)\Gamma(z) \\
&= \lim_{z\to -1} \frac{(z+1)\Gamma(z+1)}{z} \\
&= \lim_{z\to -1} \frac{(z+1)\Gamma(z+2)}{z(z+1)} \\
&= \frac{\Gamma(1)}{-1} = -1.
\end{aligned}$$

exercises for sections 16.1–16.3

1. Evaluate $\displaystyle\int_0^1 (\log x)^{1/3}\, dx$.
2. Evaluate $\displaystyle\int_0^\infty x^p e^{-x}\, dx$.
3. Evaluate $\displaystyle\int_0^\infty x^{3/4} e^{-x}\, dx$.
4. Evaluate $\displaystyle\int_0^1 x^4 \left(\log \frac{1}{x}\right)^2 dx$.
5. Evaluate $\operatorname{Res}\Gamma(z)$ at *any* negative integer.

6. Show that $\Gamma(n+\frac{1}{2}) = \dfrac{(2n)!\sqrt{\pi}}{2^{2n}n!}$.

7. Show that the integral defining $x!$ converges for $x > -1$.

8. Let

$$F(x) = \int_0^\infty \frac{te^{-xt}}{t^3+1}\,dt.$$

Show that the function is defined for $x > 0$ and that the improper integral converges uniformly on any interval $[c, d]$, $c \geqq 0$. Determine $F'(x)$.

9. Let

$$F(x) = \int_0^\infty \frac{\sin xt}{t^2+1}\,dt.$$

What is the domain of F and where does the improper integral converge uniformly?

10. Let

$$F(x) = \frac{2}{\pi}\int_0^\infty \frac{dt}{1+t^2x^2}.$$

Show that the improper integral converges uniformly to $1/x$ on $[c, d]$, $c > 0$. Show that

$$\int_0^\infty \frac{\tan^{-1} bx - \tan^{-1} ax}{x}\,dx = \frac{\pi}{2}\log\frac{b}{a},\ a > 0, b > 0.$$

(*Hint:* Consider the integral, $\int_a^b F(x)\,dx$.)

11. Make a sketch of the factorial function.

16.4 the beta function

In this section we consider a function of *two* variables defined by an improper integral. The so-called *beta function*, $B(x, y)$, is defined in several equivalent ways; in each case the function is initially defined only in the first quadrant, with extensions possible by methods similar to that used in Section 16.3.

1. $B(x, y) = \displaystyle\int_0^1 t^{x-1}(1-t)^{y-1}\,dt.$

2. $B(x, y) = 2\displaystyle\int_0^{\pi/2} (\sin\theta)^{2x-1}(\cos\theta)^{2y-1}\,d\theta.$

3. $B(x, y) = \int_0^\infty \frac{u^{x-1}}{(1+u)^{x+y}} \, du.$

4. $B(x, y) = \frac{\Gamma(x)\,\Gamma(y)}{\Gamma(x+y)}.$

The fact that these four "definitions" are equivalent is the content of the next theorem.

theorem 16.1

Definitions 1–4 above are equivalent.

proof: If in Definition 1 we let $t = \sin^2\theta$ we obtain Definition 2. If in Definition 1 we let $t = u/(1+u)$, we obtain Definition 3. The details are left as an exercise. To show the fact that the beta function is expressible in terms of gamma functions, we let $t = s^2$ in the defining integral for $\Gamma(x)$,

$$\Gamma(x) = \int_0^\infty t^{x-1} e^{-t} \, dt;$$

$$\Gamma(x) = 2\int_0^\infty s^{2x-1} e^{-s^2} \, ds;$$

$$\Gamma(y) = 2\int_0^\infty t^{2y-1} e^{-t^2} \, dt.$$

Therefore,

$$\Gamma(x)\,\Gamma(y) = 4\int_0^\infty \int_0^\infty s^{2x-1} t^{2y-1} e^{-(s^2+t^2)} \, dt \, ds.$$

Using polar coordinates,

$$\Gamma(x)\,\Gamma(y) = 4\int_0^{\pi/2} \int_0^\infty r^{2(x+y-1)} e^{-r^2} (\cos\theta)^{2x-1} (\sin\theta)^{2y-1} r \, dr \, d\theta$$

$$= \left[2\int_0^{\pi/2} (\cos\theta)^{2x-1} (\sin\theta)^{2y-1} \, d\theta\right] \left[2\int_0^\infty r^{2(x+y)-1} e^{-r^2} \, dr\right]$$

$$= B(x, y)\,\Gamma(x+y)$$

which proves the theorem.

example 16.8: Compute

$$\int_0^\infty \frac{t^3}{(1+t)^5} \, dt.$$

solution: This is just the beta function (Definition 3) with $x+y=5$, $x-1=3$. Therefore $x=4$, $y=1$ and hence the given integral is equal to

$$B(4, 1)=\frac{\Gamma(4)\,\Gamma(1)}{\Gamma(4+1)}=\frac{3!\,1!}{4!}=\frac{1}{4}.$$

example 16.9: Compute

$$\int_0^1 x^{-1/5}(1-x)^{-2/5}\,dx.$$

solution: The given integral is equal to

$$B(\tfrac{4}{5}, \tfrac{3}{5})=\frac{\Gamma(\frac{4}{5})\,\Gamma(\frac{3}{5})}{\Gamma(\frac{7}{5})}=1.95.$$

example 16.10: Compute

$$\int_0^2 x^2(2-x)^9\,dx.$$

solution: The given integral is equal to

$$2^9\int_0^2 x^2\left(1-\frac{x}{2}\right)^9 dx.$$

If we let $y=x/2$, this becomes

$$\begin{aligned}
&=2^{12}\int_0^1 y^2(1-y)^9\,dy\\
&=2^{12}B(3, 10)\\
&=2^{12}\frac{\Gamma(3)\,\Gamma(10)}{\Gamma(13)}\\
&=2^{12}\frac{2!\,9!}{12!}=\frac{2^{10}}{165}.
\end{aligned}$$

From the definition of the beta function in terms of the gamma function,

$$B(x, y)=\frac{\Gamma(x)\,\Gamma(y)}{\Gamma(x+y)},$$

we see that $B(x, y)=B(y, x)$. We say the beta function is *symmetric* in its variables. Consider the values of the beta function when $x+y=1$. By Definition 3,

$$B(x, 1-x)=\int_0^\infty \frac{u^{x-1}}{1+u}\,du.$$

In the previous chapter, the value of this integral was computed to be $\pi/\sin \pi x$, $0 < x < 1$. (See Example 15.36.) Hence,

$$B(x, 1-x) = \frac{\pi}{\sin \pi x}, \qquad 0 < x < 1.$$

example 16.11: Compute $\Gamma(1/2)$.

solution: Using the previous equation along with the definition of the beta function in terms of the gamma function,

$$B(\tfrac{1}{2}, \tfrac{1}{2}) = \Gamma(\tfrac{1}{2})\,\Gamma(\tfrac{1}{2}) = \frac{\pi}{\sin \frac{\pi}{2}} = \pi,$$

and hence

$$\Gamma(\tfrac{1}{2}) = \sqrt{\pi}$$

We used this result in some examples in Section 16.3.

With the use of the beta function, certain integrals of powers of sines and cosines may be evaluated. Let

$$I = \int_0^{\pi/2} \sin^\alpha u \cos^\beta u \, du, \qquad \alpha = 2x - 1 \quad \text{and} \quad \beta = 2y - 1.$$

Then,

$$I = \tfrac{1}{2} B\left(\frac{\alpha + 1}{2}, \frac{\beta + 1}{2}\right).$$

example 16.12: Evaluate

$$\int_0^{\pi/2} \sin u \cos^2 u \, du.$$

solution:

$$I = \frac{1}{2} B\left(1, \frac{3}{2}\right) = \frac{1}{2} \frac{\Gamma(1)\,\Gamma(3/2)}{\Gamma(5/2)} = \frac{1}{3}.$$

16.5 functions defined by indefinite integrals

There is another method of defining functions which is closely allied to the use of improper integrals. Let f be an integrable function over the interval $[c, d]$. Then the indefinite integral

$$\int_c^x f(t)\, dt$$

defines a function of x for every x within the interval of integration of the function f.

Sometimes this technique is used in elementary calculus to define the natural logarithm, by letting $f(t) = 1/t$. Then

$$\ln x = \int_1^x \frac{1}{t}\, dt.$$

(See Figure 16.3.) The usual properties of the logarithm (and its graph) are directly derivable from this definition. In the exercises you will be asked to carry out some of the details but you can consult some elementary calculus textbooks for assistance. As you know, the values of $\ln x$ are tabulated and appear in most any set of mathematical tables.

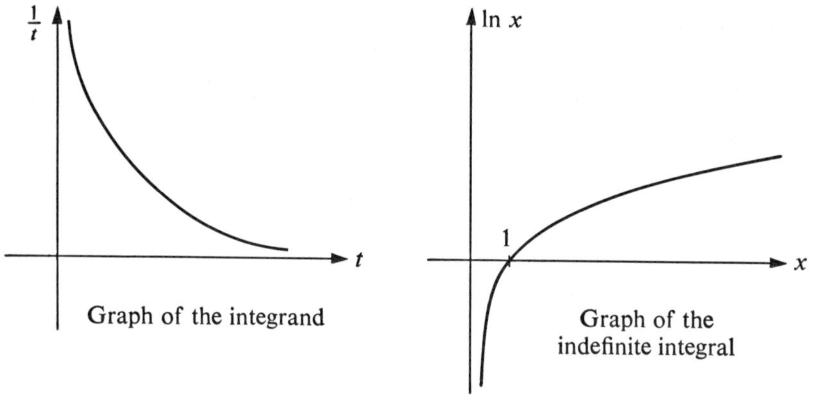

figure 16.3 *definition of* $\ln x$

In this section we wish to take at least superficial notice of a few other functions which are defined similarly. These functions are not as familiar as $\ln x$, but, like $\ln x$, their values are tabulated and available in some handbooks.

One of the more important functions defined by an indefinite integral is called the *error function*,

$$\operatorname{erf} x = \frac{2}{\sqrt{\pi}} \int_0^x e^{-t^2}\, dt.$$

Note that

$$\lim_{x\to\infty} \operatorname{erf} x = \int_0^\infty e^{-t^2}\, dt.$$

If we substitute $u = t^2$ into this improper integral, then

$$\lim_{x\to\infty} \operatorname{erf} x = \frac{2}{\sqrt{\pi}} \int_0^\infty \frac{e^{-u} u^{-1/2}}{2} \, du = \frac{1}{\sqrt{\pi}} \Gamma(\tfrac{1}{2}) = 1,$$

which shows why the "normalizing factor" $2/\sqrt{\pi}$ is used. Table IV shows the values of $\operatorname{erf} x$, for $0 \leqslant x \leqslant 2.7$. For $x > 2.7$, the value of $\operatorname{erf} x$ is within 3 decimal points of 1. In Figure 16.4 is a sketch of $\operatorname{erf} x$ and the integrand function e^{-x^2}.

table iv

Values of erf *x*

x	0	0.1	0.2	0.3	0.4	0.5	0.6
erf x	0.000	0.112	0.223	0.329	0.428	0.521	0.604
x	0.7	0.8	0.9	1.0	1.1	1.2	1.3
erf x	0.678	0.742	0.797	0.843	0.880	0.910	0.934
x	1.4	1.5	1.6	1.7	1.8	1.9	2.0
erf x	0.952	0.966	0.976	0.984	0.989	0.993	0.995
x	2.1	2.2	2.3	2.4	2.5	2.6	2.7
erf x	0.996	0.998	0.998	0.999	0.999	0.999	0.999

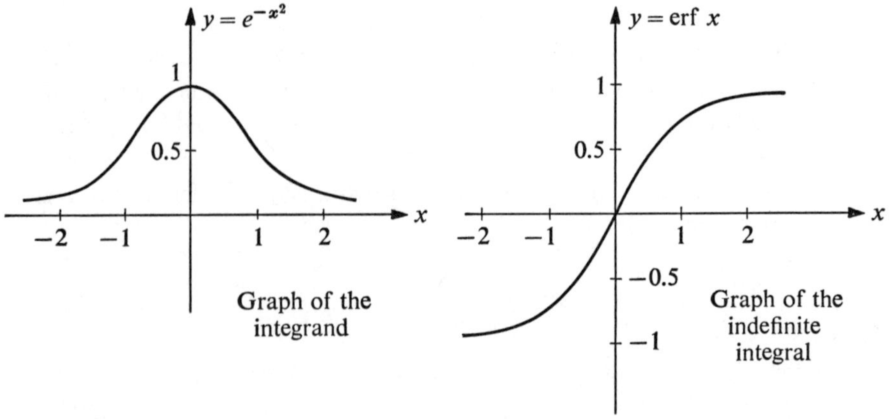

figure 16.4

In some applications it is not erf$x$ but a related function which is of interest and whose values are tabulated. In statistics the so-called cumulative distribution function of the standardized normal distribution is given by

$$\Phi(x) = \frac{1}{\sqrt{2\pi}} \int_{-\infty}^{x} e^{-t^2/2}\, dt.$$

It is easy to show that $\lim_{x\to\infty} \Phi(x) = 1$ and that $\Phi(0) = 1/2$. (See Exercise 17 in the next set of exercises.) Hence, if $x > 0$,

$$\Phi(x) = \frac{1}{\sqrt{2\pi}} \int_{0}^{x} e^{-t^2/2}\, dt + \tfrac{1}{2}.$$

Letting $u^2 = t^2/2$,

$$\Phi(x) = \frac{1}{\sqrt{\pi}} \int_{0}^{x/\sqrt{2}} e^{-u^2}\, du + \tfrac{1}{2}$$
$$= \tfrac{1}{2} \operatorname{erf}\left(\frac{x}{\sqrt{2}}\right) + \tfrac{1}{2}$$

which shows how to find values of the cumulative distribution function when a table for erf$x$ is given. Conversely, it is easy to show

$$\operatorname{erf} x = 2\Phi(x\sqrt{2}) - 1.$$

In passing we mention a few other functions defined by indefinite integrals.

$\operatorname{Si}(x) = \int_0^x \frac{\sin t}{t}\, dt$ is called the sine integral function.

$S(x) = \int_0^x \sin t^2\, dt$ is called a Fresnel integral.

$\operatorname{Ei}(x) = \int_0^x \frac{e^{-t}}{t}\, dt$ is called the exponential integral function.

exercises for sections 16.4 and 16.5

1. Using the result of Exercise 6 of the previous exercise set show that:

(a) $B(n, \tfrac{1}{2}) = \dfrac{2^{2n}(n!)^2}{n(2n)!}$. (b) $B(n + \tfrac{1}{2}, \tfrac{1}{2}) = \dfrac{(2n)!}{2^{2n-1}(n!)^2}\dfrac{\pi}{2}$.

2. Show that

$$\int_0^{\pi/2} \sin^n x\, dx = \tfrac{1}{2} B(n + \tfrac{1}{2}, \tfrac{1}{2}).$$

Evaluate in the case $n = 2$.

Express the following integrals in terms of beta functions and then evaluate by using gamma functions.

3. $\displaystyle\int_0^1 \frac{x^4}{(1-x^2)^{1/2}}\,dx.$

4. $\displaystyle\int_0^1 \frac{1}{(1-x^4)^{1/2}}\,dx.$

5. $\displaystyle\int_0^\infty \frac{dx}{1+x^4}.$

6. $\displaystyle\int_0^\infty \frac{x}{(1+x^3)^2}\,dx.$

7. $\displaystyle\int_0^1 \frac{dx}{\sqrt{x(1-x)}}.$

8. $\displaystyle\int_0^2 \frac{x^2}{\sqrt{2-x}}\,dx.$

9. $\displaystyle\int_0^{\pi/2} \sin^5\theta\cos^7\theta\,d\theta.$

10. $\displaystyle\int_0^{\pi/2} \frac{d\theta}{\sqrt{\sin\theta}}.$

11. $\displaystyle\int_0^{\pi/2} (\sin\theta)^{1/2}(\cos\theta)^{3/2}\,d\theta.$

12. Find the first quadrant area bounded by the curve $x^3 + y^3 = 8$ in terms of beta functions.

13. Compute the length of arc of the curve $r^2 = 16\cos 2\theta$ in terms of beta functions.

14. Show that erf (x) is an odd function of x.

15. Determine the points of inflection of the integrand function of erf (x).

16. Evaluate, using a table of error functions:

(a) $\displaystyle\int_0^2 e^{-x^2}\,dx.$ (b) $\operatorname{erf}(0.6).$

(c) $\displaystyle\frac{2}{\sqrt{\pi}}\int_{1.5}^\infty e^{-x^2}\,dx.$ (d) $\displaystyle\int_1^{1.5} e^{-x^2/2}\,dx.$

17. Prove that $\lim_{x\to\infty}\Phi(x) = 1$ and that $\Phi(0) = 1/2$.

18. Using the indefinite integral definition of $\ln x$, show that:

(a) $\ln 1 = 0.$ (b) $\ln(ab) = \ln a + \ln b.$ (c) $\ln x^p = p\ln x.$

19. Fill in the details of the proof of Theorem 16.1.

16.6 functions defined by differential equations

This section could have been titled "Functions Defined by Infinite Series." In Chapter 5 we saw that many differential equations may be solved by assuming that the solution function can be written as an infinite power series about some point. The differential equation essentially defines a function since the coefficients of the power series are determined by forcing the power series to be a solution.

While solutions in infinite series form are usually "nonelementary," this does not curtail their usefulness. Tables of values are constructed so that the functions become (in time and with use) as familiar as, for instance, the trigonometric functions. In fact the sine and cosine functions could very easily be considered as being essentially defined by the differential equation $y'' + k^2 y = 0$.

The name given to the solution function is the same as the name of the differential equation from which it arises. Thus the differential equation $y'' + k^2 y = 0$ is the trigonometric differential equation; the differential equation $y'' - k^2 y = 0$ might be called the hyperbolic differential equation.

The infinite series form of the solution function is usually the more desireable for the actual construction of a table of values. Moreover, the infinite series solution might not be expressible in terms of elementary functions.

16.7 bessel's functions

In this section we apply the technique of assuming an infinite series solution to a differential equation which arises very naturally in several important applications in which cylindrical symmetry is assumed. There are other, equivalent, ways of stating the differential equation, but we begin with

$$x^2 y'' + xy' + (x^2 - m^2)\, y = 0 .$$

This differential equation is called *Bessel's equation of order m*. Note that the word "order" does not refer to the differential equation itself which is obviously of second order for each value of m.

Since $x = 0$ is a regular singular point (see Section 5.5), it is necessary to assume that the solution is of the form of a Frobenius series rather than a power series about the origin.

Letting $y = x^r \sum_{n=0}^{\infty} c_n x^n$ and substituting into the differential equation, we obtain

$$\sum_{n=0}^{\infty} [(n+r)(n+r-1)\, c_n + (n+r)\, c_n - m^2 c_n]\, x^{n+r} + \sum_{n=0}^{\infty} c_n x^{n+r+2} = 0 ,$$

or,

$$c_0(r^2 - m^2) + c_1[(r+1)^2 - m^2]\,x + \sum_{n=2}^{\infty} [((n+r)^2 - m^2)\,c_n + c_{n-2}]\,x^{n+r} = 0.$$

To determine the coefficients, c_n, we equate each of the coefficients of powers of x in this equation to zero.

Coefficient of x^0: $c_0(r^2 - m^2) = 0$, which implies that $r = \pm m$ with c_0 "arbitrary."

Coefficient of x^1: $c_1[(r+1)^2 - m^2] = 0$, which implies that $c_1 = 0$ (since $r \neq m - 1$).

Coefficient of x^{n+r}: $[(n+r)^2 - m^2]\,c_n + c_{n-2} = 0$, for $n > 2$, which means

$$c_n = \frac{-1}{(n+r)^2 - m^2}\,c_{n-2}$$

a recursion relationship valid for $n \geqq 2$ and hence all the coefficients are determined. Although c_0 is arbitrary, you will find it customary† to require

$$c_0 = \frac{1}{2^m \Gamma(m+1)}$$

from which the general coefficient formula may be written,

$$c_{2n} = \frac{(-1)^n}{2^{2n+m} n!\,\Gamma(m+n+1)}$$

$$c_{2n+1} = 0.$$

Therefore, one solution to Bessel's equations, called *Bessel's function of the first kind*, is

$$J_m(x) = \sum_{n=0}^{\infty} \frac{(-1)^n}{2^{2n+m} n!\,\Gamma(m+n+1)}\,x^{2n+m}.$$

Actually, as long as m is an integer, any solution of this *type*, regardless of the choice of c_0, is called a Bessel function of the first kind.

Since Bessel's equation is a second-order differential equation we would naturally expect another solution function, not a constant multiple of the first. Further, it would seem only natural that the "other solution" be found by letting $r = -m$, and using the same recursion relationship as before. This is valid as long as m is not an integer, but (as is often the "practical"

† See Section 16.12.

case) if m is an integer, then, as we will now show, the solution J_{-m} is linearly dependent on J_m. By definition, $J_{-m}(x)$ is given by

$$J_{-m}(x) = \sum_{n=0}^{\infty} \frac{(-1)^n}{2^{2n-m} n! \Gamma(n-m+1)} x^{2n-m}.$$

If we define

$$\frac{1}{\Gamma(-1)} = \frac{1}{\Gamma(-2)} = \cdots = 0,$$

then,

$$J_{-m}(x) = \sum_{n=m}^{\infty} \frac{(-1)^n}{2^{2n-m} n! \Gamma(n-m+1)} x^{2n-m}.$$

In this power series, use a shift of index by letting $k = n - m$,

$$\begin{aligned} J_{-m}(x) &= \sum_{k=0}^{\infty} \frac{(-1)^{k+m}}{2^{2k+m}(k+m)! \Gamma(k+1)} x^{2k+m} \\ &= (-1)^m J_m(x). \end{aligned}$$

For emphasis, we repeat that this is *not* true if m is *not* an integer.

If m *is* an integer, a second solution to Bessel's equation, which is independent of the first, may be found by the method of variation of parameters and will include a term involving the logarithm of x. Consequently the "second" solution to Bessel's equation when m is an integer is unbounded near the origin and is useful in applications only for $x \neq 0$.

More generally, any solution to Bessel's equation which is continuous for all values of x is called a *Bessel function of the first kind.* A solution to Bessel's equation which is continuous for all x except $x = 0$ and which becomes unbounded for x near 0, is known as a *Bessel function of the second kind.* Any linear combination of a Bessel function of the first kind with one of the second kind is a solution to Bessel's equation.

Although $J_{-m}(x)$ is a satisfactory second solution for nonintegral m, it is not what is usually tabulated, which is

$$N_m(x) = Y_m(x) = \frac{(\cos \pi m)\, J_m(x) - J_{-m}(x)}{\sin \pi m}.$$

This function is known either as the Neumann function (in which case it is denoted by N_m) or as the Weber function (in which case the notation Y_m is used). It may seem strange to take such a peculiar combination to be the "second" solution when the J_{-m} is so easily available. However, the Neumann functions have the added feature that, while undefined for integral m, the limit as m approaches integral values *is* a correct "second" solution for that integer. From another viewpoint, it is just as if we said that the solutions to $y'' + y = 0$ were the functions $\sin x$ and $(\sin x + \cos x)/2$ instead of $\sin x$ and $\cos x$. Tabulating either of the two sets of values would be sufficient and only the practical applications would determine the "best" solutions to use.

As you will see later, some Bessel functions have at least a remote similarity to the sine and cosine functions. We can even notice this now by observing how much the first few terms of the functions J_0 and J_1 resemble the cosine and the sine.

$$J_0(x) = 1 - \frac{x^2}{2^2 1!} + \frac{x^4}{2^4 2!} - \frac{x^6}{2^6 3!} + \frac{x^8}{2^8 4!} - \cdots .$$

$$J_1(x) = \frac{x}{2} - \frac{x^3}{2^3 1!\,2!} + \frac{x^5}{2^5 3!\,2!} - \frac{x^7}{2^7 5!\,3!} + \frac{x^9}{2^9 7!\,4!} - \cdots .$$

Values of these functions are tabulated in Appendix D.

example 16.13: Write $J_{1/2}$ in terms of a function of x times a sine function.

solution:

$$J_{1/2}(x) = \sum_{n=0}^{\infty} \frac{(-1)^n x^{2n+1/2}}{2^{2n+1/2} n!\, \Gamma(n+\frac{3}{2})}$$

$$= \frac{2^{1/2} x^{-1/2}}{\pi^{1/2}} \sum_{n=0}^{\infty} \frac{(-1)^n x^{2n+1}}{(2n+1)!}$$

$$= \left(\frac{2}{\pi x}\right)^{1/2} \sin x .$$

16.8 modified bessel functions

In the applications, Bessel's equation appears in a form slightly different from the one presented in the previous section. The equation is called *Bessel's modified equation* and the solutions are, in turn, called *modified Bessel functions.* This distinction is not always carefully honored.

example 16.14: Find the general solution to the differential equation $x^2 y'' + xy' + (x^2 s^2 - m^2)\, y = 0$.

solution: This equation differs from Bessel's equation only in that sx takes the place of x. Letting $u = sx$ we obtain

$$\left(\frac{u}{s}\right)^2 \left(\frac{d^2y}{du^2} s^2\right) + \left(\frac{u}{s}\right)\left(\frac{dy}{du} s\right) + (u^2 - m^2)\, y = 0$$

which is exactly Bessel's equation. Therefore, the general solution is

$$y = c_1 J_m(sx) + c_2 Y_m(sx).$$

If, in Example 16.14, s is permitted to assume complex values, the solutions to Bessel's modified equation are in terms of Bessel functions with pure imaginary arguments. The procedure we follow is analogous to the case of the real exponential function with complex argument. In that case we express (or define) e^{ix} as a linear combination of trigonometric functions. Here, in order to "simplify" matters, we define the following two functions:

$$I_m(x) = i^{-m} J_m(ix),$$

$$K_m(x) = \frac{\pi}{2}\, i^{m+1} \left[J_m(ix) + i Y_m(ix)\right].$$

In this manner, solutions are expressed in terms of I_m and K_m.

exercises for sections 16.6–16.8

1. Show that $J_0(0) = 1$ and $J_1(0) = 0$.
2. Show that $J_{-1/2}(x) = (2/\pi x)^{1/2} \cos x$.
3. Show that $c_1 I_m + c_2 K_m$ is a solution to $x^2 y'' + xy' - (x^2 + m^2)\, y = 0$.

In Exercises 4–7, solve in terms of Bessel functions.

4. $x^2 y'' + xy' + (x^2 - \frac{1}{4})\, y = 0.$
5. $x^2 y'' + xy' - (x^2 + 9)\, y = 0.$
6. $x^2 y'' + xy' + \frac{1}{4}(x^2 - 1)\, y = 0.$
7. In each of the preceding exercises write the first few terms of each of the solution functions.
8. Using the tabulated values, make a rough sketch of $J_0(x)$ and $J_1(x)$ for $0 < x < 10$.
9. Show that for small x, $J_0(x) \approx 1 - (x^2/4)$ and use this to determine $J_0(x)$ for $x = 0.1, 0.2, 0.3$, and compare to the tabulated values.
10. Show that $J_n(x)$ is even when n is an even integer and an odd function when n is an odd integer.

16.9 some fundamental identities

Because the Bessel functions of the first kind are bounded near the origin, they are easily the most useful. In the following sections we will restrict our discussion to this type of Bessel function.

Bessel functions of the first kind obey some rather important identities. These are relations between Bessel functions of different orders and their derivatives.

identity 1

$$xJ_n'(x) = nJ_n(x) - xJ_{n+1}(x).$$

proof:

$$J_n(x) = \sum_{k=0}^{\infty} \frac{(-1)^k}{k!\,\Gamma(n+k+1)\,2^{n+2k}}\, x^{n+2k}.$$

Under the (correct) assumption that term-by-term differentiation is permissible,

$$J_n'(x) = \sum_{k=0}^{\infty} \frac{(n+2k)(-1)^k}{k!\,\Gamma(n+k+1)\,2^{n+2k}}\, x^{n+2k-1}.$$

Therefore,

$$xJ_n'(x) = nJ_n(x) + x\sum_{k=1}^{\infty} \frac{(-1)^k}{(k-1)!\,\Gamma(n+k+1)2^{n+2k-1}}\, x^{n+2k-1}.$$

Shifting index in the sum by letting $k = s+1$,

$$\begin{aligned} xJ_n'(x) &= nJ_n(x) - x\sum_{s=0}^{\infty} \frac{(-1)^s}{s!\,\Gamma(n+s+2)\,2^{n+2s+1}}\, x^{n+2s+1} \\ &= nJ_n(x) - xJ_{n+1}(x), \end{aligned}$$

as was to be shown.

identity 2

$$xJ_n'(x) = -\,nJ_n(x) + xJ_{n-1}(x).$$

proof: The details are left as an exercise (see Exercise 1 in the next set of exercises) but you should first let $n + 2k = 2(n+k) - n$ in the series expression for $xJ_n'(x)$ and use the fact that $\Gamma(n+k+1) = (n+k)\,\Gamma(n+k)$.

identity 3

Differentiation Formula.

$$J_n'(x) = \frac{J_{n-1}(x) - J_{n+1}(x)}{2}.$$

proof: Eliminate J_n between Identities 1 and 2.

identity 4

Recursion Formula.

$$J_n(x) = \frac{x}{n}\left(\frac{J_{n+1}(x) + J_{n-1}(x)}{2}\right)$$

or, which is the same thing,

$$J_{n+1}(x) = \frac{2n}{x} J_n(x) - J_{n-1}(x).$$

proof: Eliminate $J_n'(x)$ between Identities 1 and 2.

example 16.15: Derive a differentiation formula for $J_0(x)$.

solution: Using Identity 3,

$$J_0'(x) = \frac{J_{-1}(x) - J_1(x)}{2}$$

and then, applying the fact that $J_{-n}(x) = (-1)^n J_n(x)$, we obtain

$$J_0'(x) = J_{-1}(x) = -J_1(x).$$

example 16.16: Evaluate

$$\int_0^x tJ_0(t)\,dt.$$

solution: Note that

$$\frac{d}{dx}(xJ_1(x)) = xJ_1'(x) + J_1(x).$$

Using Identity 2 with $n = 1$, $xJ_1'(x) = -J_1(x) + xJ_0(x)$, which means that

$$\frac{d}{dx}(xJ_1(x)) = -J_1(x) + xJ_0(x) + J_1(x) = xJ_0(x).$$

Therefore,

$$\int_0^x tJ_0(t)\,dt = [tJ_1(t)]\Big|_0^x = xJ_1(x).$$

example 16.17: Express J_3 in terms of J_0 and J_1.

solution: Using Identity 4 with $n=2$,

$$J_3(x) = \frac{4}{x} J_2(x) - J_1(x)$$
$$= \frac{4}{x}\left[\frac{2}{x} J_1(x) - J_0(x)\right] - J_1(x)$$
$$= \left[\frac{8}{x^2} - 1\right] J_1(x) - \frac{4}{x} J_0(x).$$

The recursion formulae and the differentiation formula point up the importance of knowing well the properties of at least two of the Bessel functions, usually $J_0(x)$ and $J_1(x)$. For our purposes, we will only attempt to study $J_0(x)$ more carefully. Further, we will be satisfied to express an answer in terms of $J_0(x)$ and $J_1(x)$.

example 16.18: Find

$$\int x^{-1} J_1(x)\, dx.$$

solution: By letting $u = xJ_1(x)$ and $dv = x^{-2}\, dx$ in the formula for integration by parts, then $du = xJ_0(x)\, dx$, and $v = -x^{-1}$ from which,

$$\int x^{-1} J_1(x)\, dx = -x^{-1} x J_1(x) + \int x^{-1} x J_0(x)\, dx$$
$$= -J_1(x) + \int J_0(x)\, dx.$$

example 16.19: Evaluate

$$\int x^{-2} J_2(x)\, dx.$$

solution: Using the formula for integration by parts with $u = x^2 J_2(x)$ and $dv = x^{-4}\, dx$, then $du = x^2 J_1(x)\, dx$ (see Exercise 8, Exercises 16.9–16.12) and $v = -x^{-3}/3$, from which,

$$\int x^{-2} J_2(x)\, dx = -\frac{x^{-3}}{3}(x^2 J_2(x)) + \frac{1}{3}\int x^{-1} J_1(x)\, dx.$$

Now, if we use the result of the previous example, we may write this as

$$\int x^{-2} J_2(x)\, dx = -\frac{J_2(x)}{3x} - \frac{J_1(x)}{3} + \frac{1}{3}\int J_0(x)\, dx$$

$$= \frac{-2J_1(x)}{3x^2} - \frac{J_1(x)}{3} + \frac{J_0(x)}{3x} + \frac{1}{3}\int J_0(x)\, dx.$$

16.10 approximations to the bessel functions

In Section 16.7 we alluded to the fact that we might be able to make some very rough approximations to the Bessel functions of the first kind if some restrictions are made as to where the given approximation is to be used. In this section we will show how to approximate the Bessel functions when the argument is very small (that is, near zero) or very large.

By a rigorous argument, it is possible to show that, for large values of x,

$$J_m(x) = \left(\frac{2}{\pi x}\right)^{1/2} \cos\left(x - \frac{2m+1}{4}\pi\right).$$

By a less rigorous, but far easier, argument we can show the same fundamental result; that is, the Bessel functions behave something like a damped cosine function when the value of x is large. To see this, consider Bessel's differential equation

$$x^2 y'' + xy' + (x^2 - m^2)\, y = 0$$

in the form

$$y'' + \frac{1}{x} y' + \left(1 - \frac{m^2}{x^2}\right) y = 0.$$

We might anticipate the solution for large x by discarding the term m^2/x^2. The differential equation then becomes

$$y'' + \frac{1}{x} y' + y = 0.$$

Letting $u = yx^{1/2}$ (a change in the dependent variable), the differential equation becomes

$$u'' + \left(\frac{1}{4x^2} + 1\right) u = 0.$$

Once again, we ignore the term involving x^{-2} to obtain

$$u'' + u = 0$$

whose solution is

$$u = c_1 \cos x + c_2 \sin x$$

and thus the approximate solution to Bessel's equation for large values of x is

$$y = x^{-1/2}(c_1 \cos x + c_2 \sin x) = Ax^{-1/2} \cos(x + \beta),$$

which is what we wanted to show.

For small values of x, we examine the solution itself and discard all terms after the first,

$$J_m(x) \approx \frac{x^m}{2^m \Gamma(m+1)}.$$

This approximation is only good for small values of x. Notice that all Bessel functions of nonzero order are zero for $x=0$ and that $J_0(0)=1$.

16.11 generating function; bessel's integral representation

If the Laurent series of a function has coefficients which are, in turn, functions of another parameter, the Laurent series is said to *generate* the set of coefficient functions. The original function itself is then called a *generating function* for those coefficient functions. In this section we will see how to construct a generating function for the set of Bessel functions.

Consider the exponential functions $e^{xz/2}$ and $e^{-x/2z}$. The Laurent expansions for the two functions about $z=0$ are

$$e^{xz/2} = \sum_{k=0}^{\infty} \frac{(xz/2)^k}{k!} \quad \text{and} \quad e^{-x/2z} = \sum_{m=0}^{\infty} \frac{(-x/2z)^m}{m!}.$$

When we multiply the two series together, we obtain

$$e^{\frac{x}{2}(z-z^{-1})} = \sum_{k=0}^{\infty} \sum_{m=0}^{\infty} \frac{(-1)^m}{m!\,k!} \left(\frac{x}{2}\right)^{k+m} z^{k-m}.$$

The coefficient of the z^0 term is made up of those terms for which $k=m$ and is, therefore,

$$\sum_{k=0}^{\infty} \frac{(-1)^k x^{2k}}{2^{2k}(k!)^2}$$

which you should be able to recognize as $J_0(x)$.

More generally, the coefficient of the term z^n is found from those terms for which $k-m=n$ and is, therefore,

$$\sum_{m=0}^{\infty} \frac{(-1)^m x^{2m+n}}{(m+n)!\,m!\,2^{2m+n}}$$

which you can recognize as $J_n(x)$. In summary, the coefficients in the Laurent expansion of the generating function are precisely the Bessel functions of integral order, that is,

$$e^{\frac{x}{2}(z-z^{-1})} = \sum_{-\infty}^{\infty} J_n(x)\, z^n .$$

You will find that Bessel functions of *integral* order are often defined by this generating function. That is why, in Section 16.7 when discussing the infinite series definition of the Bessel functions, we indicated it was "customary" to define c_0 to be $1/(2^n\Gamma(n+1))$. This choice of c_0 makes the two definitions agree for integral orders.

example 16.20: Express the Bessel functions of integral order in terms of a definite integral with a parameter. This form is called *Bessel's Integral Form.*

solution: Let $z = e^{i\theta}$ in the generating function. Then

$$\begin{aligned} e^{\frac{x}{2}(e^{i\theta}-e^{-i\theta})} &= e^{ix\sin\theta} \\ &= \cos(x\sin\theta) + i\sin(x\sin\theta). \end{aligned}$$

Therefore,

$$\begin{aligned} \cos(x\sin\theta) + i\sin(x\sin\theta) &= \sum_{-\infty}^{\infty} J_n(x)\,(\cos\theta + i\sin\theta)^n \\ &= \sum_{-\infty}^{\infty} J_n(x)\cos n\theta + i\sum_{-\infty}^{\infty} J_n(x)\sin n\theta . \end{aligned}$$

Since $J_{-n}(x) = (-1)^n J_n(x)$ and using the fact that $\cos n\theta = \cos(-n\theta)$ and $\sin n\theta = -\sin(-n\theta)$, we have, upon equating the real and imaginary parts of this equation,

$$\cos(x\sin\theta) = J_0(x) + 2\sum_{n=1}^{\infty} J_{2n}(x)\cos 2n\theta$$

$$\sin(x\sin\theta) = 2\sum_{n=1}^{\infty} J_{2n-1}(x)\sin(2n-1)\theta.^{\dagger}$$

Multiply the first of these equations by $\cos k\theta$ and integrate from 0 to π to obtain

$$\begin{aligned} \frac{1}{\pi}\int_0^{\pi} \cos k\theta\cos(x\sin\theta)\,d\theta &= J_k(x), \qquad \text{for } k = 0, 2, 4, \cdots; \\ &= 0, \qquad \text{for } k = 1, 3, 5, \cdots. \end{aligned}$$

† Note that these are the Fourier cosine and sine series of $\cos(x\sin\theta)$ and $\sin(x\sin\theta)$ respectively.

Multiply the second of the equations by $\sin k\theta$ and integrate from 0 to π,

$$\frac{1}{\pi}\int_0^{\pi} \sin k\theta \sin(x \sin\theta)\, d\theta = J_k(x), \qquad \text{for } k = 1, 3, 5, \cdots;$$
$$= 0, \qquad \text{for } k = 0, 2, 4, \cdots.$$

Adding these last two equations we obtain the desired result,

$$J_k(x) = \frac{1}{\pi}\int_0^{\pi} \cos(k\theta - x\sin\theta)\, d\theta, \qquad \text{for } k \text{ a nonnegative integer.}$$

16.12 zeros of the bessel functions

You have seen in Section 16.10 that the Bessel functions behave something like damped harmonic functions. Consequently we would suspect that the graphs of the Bessel functions should cross the x-axis infinitely often and at intervals of length approximately equal to π. In this section, we will show how to prove the first part of this conjecture for $J_0(x)$. The method of proof is the same for other integral order Bessel functions; only the details are more cumbersome. By concentrating on $J_0(x)$ some of the notation is less complicated.

theorem 16.2

$J_0(x)$ has infinitely many zeros.

proof: From Bessel's integral form, with $n = 0$, $J_0(x) = \frac{1}{\pi}\int_0^{\pi} \cos(x \sin\theta) d\theta.$

In Exercise 15 you will show

$$J_0(x) = \frac{1}{\pi}\int_0^{\pi} \cos(x \cos\theta)\, d\theta.$$

(See Exercises for Sections 16.9–16.12.) Suppose x is confined to alternate intervals of length $\pi/2$; that is, $x = m\pi + (\pi t/2)$, $0 \leqq t \leqq 1$. The intervals of interest are shown as cross-hatched in Figure 16.5. The content of the proof is to show that the graph of $J_0(x)$ crosses the x-axis in the intervals which are not cross-hatched.

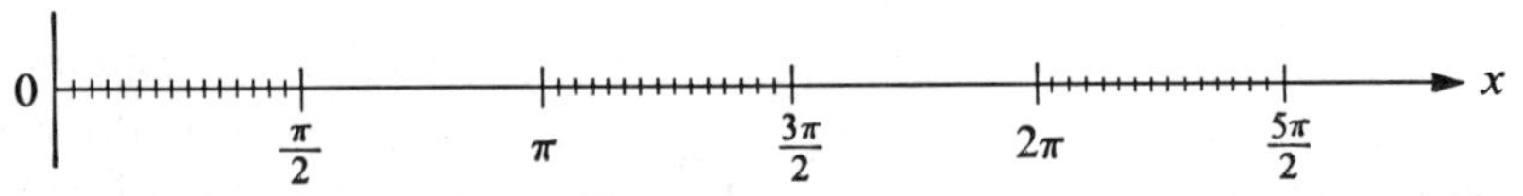

figure 16.5

In Bessel's integral form, make the change of variable defined implicitly by $\cos\theta = (\pi u/2x)$. Then

$$
\begin{aligned}
J_0(x) &= \frac{1}{\pi}\int_{-2x/\pi}^{2x/\pi} \cos(\pi u/2)\,\frac{\pi}{2x}\,\frac{1}{[1-(\pi^2/4x^2)\,u^2]^{1/2}}\,du \\
&= \frac{1}{2x}\int_{-2x/\pi}^{2x/\pi} \frac{\cos(\pi u/2)}{[1-(\pi u/2x)^2]^{1/2}}\,du \\
&= \frac{2}{\pi}\int_{0}^{2x/\pi} \frac{\cos(\pi u/2)}{[(2x/\pi)^2-u^2]^{1/2}}\,du\,.
\end{aligned}
$$

Using the fact that $2x/\pi = 2m + t$, this becomes

$$
J_0\left(\frac{\pi t}{2} + m\pi\right) = \frac{2}{\pi}\int_0^{2m+t} \frac{\cos(\pi u/2)}{[(2m+t)^2 - u^2]^{1/2}}\,du\,.
$$

The interval $0 \leqq x \leqq \pi/2$ is considered by letting $m = 0$. In that case the value of the Bessel function of order zero on that interval is given by

$$
J_0\left(\frac{\pi t}{2}\right) = \frac{2}{\pi}\int_0^{t} \frac{\cos(\pi u/2)}{[t^2 - u^2]^{1/2}}\,du\,, \qquad 0 \leqq t \leqq 1\,.
$$

Since the integrand is positive for all values of t, it is obvious that the function $J_0\left(\frac{\pi t}{2}\right)$ is positive in the first cross-hatched interval.

The value of the function in the second cross-hatched interval is determined by letting $m = 1$,

$$
\begin{aligned}
J_0\left(\frac{\pi t}{2} + \pi\right) &= \frac{2}{\pi}\int_0^{2+t} \frac{\cos(\pi u/2)}{[(2+t)^2 - u^2]^{1/2}}\,du \\
&= \frac{2}{\pi}\left[\int_0^1 + \int_1^2 + \int_2^{2+t}\right] \frac{\cos(\pi u/2)}{[(2+t)^2 - u^2]^{1/2}}\,du\,.
\end{aligned}
$$

The substitution of $u = 1 - s$ into the first of these integrals and of $u = s + 1$ into the second one leads to the following sum:

$$
\begin{aligned}
&\frac{2}{\pi}\int_0^1 \left[\frac{\cos(\pi s/2 - \pi/2)}{[(2+t)^2 - (1-s)^2]^{1/2}} + \frac{\cos(\pi s/2 + \pi/2)}{[(2+t)^2 - (1+s)^2]^{1/2}}\right] ds \\
&\qquad = \frac{2}{\pi}\int_0^1 \frac{\sin \pi s/2\,[\sqrt{(2+t)^2 - (1+s)^2} - \sqrt{(2+t)^2 - (1-s)^2}]}{\sqrt{(2+t)^2 - (1-s)^2}\,\sqrt{(2+t)^2 - (1+s)^2}}\,ds\,.
\end{aligned}
$$

The integrand is negative because the second radical in the numerator is larger than the first one. Hence the entire integral is negative. It is also easy to show that the integral from 2 to $2 + t$ is negative and hence $J_0(x)$ is negative over $\pi \leqq x \leqq 3\pi/2$.

Since $J_0(x)$ is continuous, its graph must cross the x-axis someplace on the interval $\pi/2 < x < \pi$. It is also obvious that the function changes from positive to negative at that point.

The proof that there is a zero in each of the non-crosshatched intervals proceeds in exactly the same way, alternately concluding that the function is positive and negative in the cross-hatched regions. The proof is not too complicated and may be found in Churchill (1941).

corollary: The Bessel functions of all integral orders have infinitely many zeros.

proof: Suppose $J_0(x_1) = J_0(x_2) = 0$. (See Figure 16.6.) By Rolle's theorem, $J_0'(x)$ vanishes at least once for some x between x_1 and x_2. But, since $J_0'(x) = -J_1(x) = J_{-1}(x)$ there is at least one zero of $J_1(x)$ (and one of $J_{-1}(x)$) between x_1 and x_2. Hence, there must be infinitely many zeros for $J_1(x)$. The proof for all $J_n(x)$ follows by induction. (See Churchill (1941).)

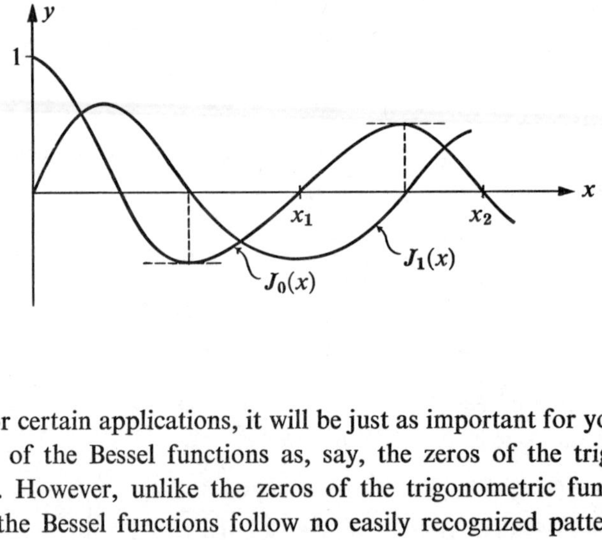

figure 16.6

For certain applications, it will be just as important for you to know the zeros of the Bessel functions as, say, the zeros of the trigonometric functions. However, unlike the zeros of the trigonometric functions, the zeros of the Bessel functions follow no easily recognized pattern. Hence, it is necessary to have them tabulated. Table V shows the first five zeros of $J_0(x)$ and $J_1(x)$.

Our original conjecture that the distance between the zeros approaches π as x becomes infinite is true but we will not give a proof.

With the information on the zeros of the functions, relatively good graphs of J_0 and J_1 may be sketched. (See Figure 16.7.)

table v

$J_0(x_m)=0$	
m	x_m
1	2.4
2	5.5
3	8.7
4	11.8
5	14.9

$J_1(x_m)=0$	
m	x_m
1	0
2	3.8
3	7.0
4	10.2
5	13.3

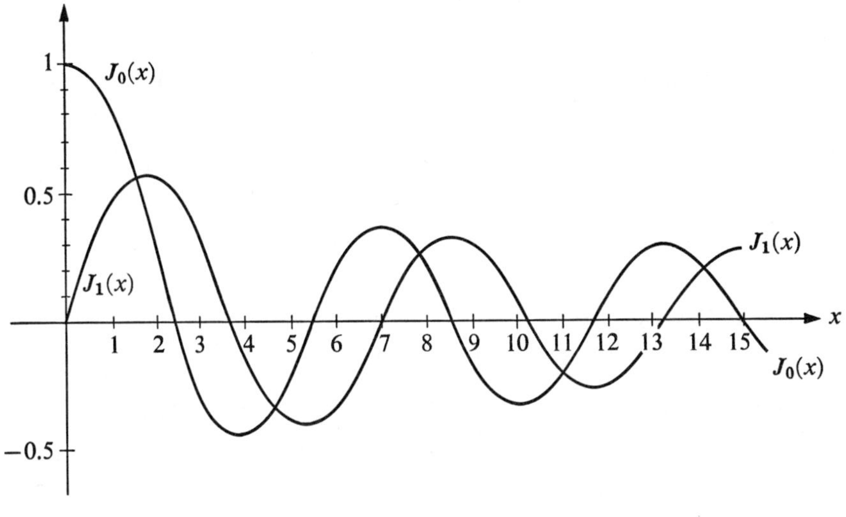

figure 16.7

exercises for sections 16.9–16.12

1. Complete the proofs of Identities 2,3, and 4, Section 16.9.
2. Show that if b is a zero of $J_n(x)$, then $-b$ is also a zero.
3. Show that $J_2(x) = -\frac{1}{x} J_0'(x) + J_0''(x)$.
4. Express J_4 in terms of J_0 and J_1.
5. Show that

$$J_2'(x) = \tfrac{1}{2}(J_1(x) - J_3(x)) = \left(1 - \frac{4}{x^2}\right) J_1(x) + \frac{2}{x} J_0(x).$$

6. Show that

$$J_3'(x) = \left(1 - \frac{12}{x^3}\right) J_2(x) + \frac{3}{x} J_1(x).$$

7. Show that

$$\int y J_n^2(y)\, dy = \tfrac{1}{2}(y^2 - n^2) J_n^2(y) + \tfrac{1}{2}[y J_n'(y)]^2.$$

(*Hint:* Use integration by parts along with a knowledge of the differential equation satisfied by $J_n(y)$.)

8. Show that $\int x^n J_{n-1}(x)\, dx = x^n J_n(x) + c.$

9. Show that $\int x^{-n} J_{n+1}(x)\, dx = -x^{-n} J_n(x) + c.$

Evaluate the following $\left(\int J_0(x)\, dx \text{ may be left unevaluated}\right)$.

10. $\int J_2(x)\, dx.$

11. $\int x J_1(x)\, dx.$

12. $\int J_3(x)\, dx.$

13. $\int x^2 J_0(x)\, dx.$

14. $\int x^3 J_0(x)\, dx.$

15. Show that $J_0(x) = \dfrac{1}{\pi}\displaystyle\int_0^\pi \cos(x \cos\theta)\, d\theta.$

16. Continue with the proof of Theorem 16.2 into the next "crosshatched" interval; that is, show that $J_0(x)$ is positive on the interval $2\pi \leqq x \leqq 5\pi/2$.

17. Prove the "boundedness" property for the Bessel functions of integral order; that is, show $J_n(x) \leqq 1$, $n = 0, 1, 2, 3, \cdots$. (*Hint:* Use Bessel's integral form for J_n.)

18. Show that $\lim_{n\to\infty} J_n(x) = 0$ for all x. (*Hint:* Consider the coefficients in the Fourier cosine and sine series of the generating function.)

16.13 fourier-bessel series

In Chapter 10 you learned that elements of $PC(a, b)$ could be "expanded" in terms of a basis on that space. At that time we considered almost exclusively the basis of the set of trigonometric functions. Certain sets of Bessel functions will also serve as a basis for some function spaces. We will not prove that any given set is a basis but we will show what to expect in the way of orthogonality.

If n is a nonnegative integer, the functions $J_n(sx)$ satisfy Bessel's modified equation

$$x^2 J_n''(sx) + xJ_n'(sx) + (s^2x^2 - n^2)\,J_n(sx) = 0,$$

or

$$[xJ_n'(sx)]' + \left(-\frac{n^2}{x} + s^2x\right) J_n(sx) = 0.$$

This will be a Sturm-Liouville problem with $\lambda = s^2, p(x) = x, r(x) = x$, and $q(x) = -n^2/x$, if the proper boundary values are chosen. Since $r(0) = 0$, we need only one homogeneous boundary condition for some finite value of x, say R. Hence, we set $J_n(sR) = 0$. This gives

$$s_{1n}R = \alpha_{1n}, s_{2n}R = \alpha_{2n}, \cdots, s_{mn}R = \alpha_{mn}, \cdots$$

where the numbers α_{mn} are the zeros of J_n. Hence the values of $s_{1n}, s_{2n}, \cdots,$ $s_{mn} \cdots$ are determined

$$s_{1n} = \frac{\alpha_{1n}}{R}, s_{2n} = \frac{\alpha_{2n}}{R}, \cdots, s_{mn} = \frac{\alpha_{mn}}{R}, \cdots.$$

The s_{mn} are the "eigenvalues" of the Sturm-Liouville problem and by the result of Section 10.12, the set

$$\{J_n(s_{mn}x)\}_{m=1}^{\infty}$$

is an orthogonal set on $PC(0, R)$ with respect to the weight function x; that is,

$$\int_0^R xJ_n(s_{mn}x)\,J_n(s_{qn}x)\,dx = 0, \qquad q \neq m.$$

Since n is fixed, each of the Bessel functions, along with its zeros, completely determines the following orthogonal sets:

$$\{J_0(s_{m0}x)\}_{m=1}^{\infty}, \{J_1(s_{m1}x)\}_{m=1}^{\infty}, \{J_2(s_{m2}x)\}_{m=1}^{\infty}, \cdots.$$

Each of the sets is a basis for the space $PC(0, R)$. Therefore, just like the Fourier expansion of a function in terms of the trigonometric set, we can expect to express rather general functions in terms of these orthogonal sets.

You should note carefully precisely which sets of Bessel functions are orthogonal. For example, the set of *all* Bessel functions of integral order is *not* one of the basis sets being considered. To apply the general theory of orthogonal expansions, we must find the norm of each of the functions $J_n(s_{mn}x)$ on $(0, R)$.

$$\|J_n(s_{mn}x)\|^2 = \int_0^R xJ_n^2(s_{mn}x)\,dx$$

which, by a change of variables $y = s_{mn}x$ becomes

$$\|J_n(s_{mn}x)\|^2 = \frac{1}{(s_{mn})^2}\int_0^{s_{mn}R} yJ_n^2(y)\,dy.$$

Using Exercise 7 of the previous exercise set to obtain an antiderivative for $yJ_n^2(y)$, this integral is equal to

$$\frac{1}{(s_{mn})^2}\frac{(y^2-n^2)}{2}J_n^2(y)\Bigg|_0^{s_{mn}R} + \frac{1}{2}\left[\frac{y}{s_{mn}}J_n'(y)\right]^2\Bigg|_0^{s_{mn}R}.$$

Because the boundary conditions at 0 and R are assumed to be homogeneous, the first term vanishes, so that

$$\|J_n(s_{mn}x)\|^2 = \tfrac{1}{2}R^2[J_n'(s_{mn}R)]^2.$$

By using Identity 1 in Section 16.9, with $x = R$, this becomes

$$\|J_n(s_{mn}x)\|^2 = \tfrac{1}{2}R^2J_{n+1}^2(s_{mn}R).$$

Thus,

$$\|J_n(s_{mn}x)\| = \frac{1}{\sqrt{2}}R\,|J_{n+1}(s_{mn}R)|.$$

For example,

$$\|J_0(s_{m0}x)\| = \frac{R}{\sqrt{2}}|J_1(s_{m0}R)|, \quad \text{where the } s_{m0}R \text{ are the zeros of } J_0;$$

$$\|J_1(s_{m1}x)\| = \frac{R}{\sqrt{2}}|J_2(s_{m1}R)|, \quad \text{where the } s_{m1}R \text{ are the zeros of } J_1;$$

$$\|J_2(s_{m2}x)\| = \frac{R}{\sqrt{2}}|J_3(s_{m2}R)|, \quad \text{where the } s_{m2}R \text{ are the zeros of } J_2.$$

From the formula for the norm, the orthonormal sets corresponding to the orthogonal sets of Bessel functions on $(0, R)$ may now be written

$$\left\{\frac{J_0(s_{m0}x)}{\frac{R}{\sqrt{2}}|J_1(s_{m0}R)|}\right\}_{m=1}^{\infty}, \left\{\frac{J_1(s_{m1}x)}{\frac{R}{\sqrt{2}}|J_2(s_{m1}R)|}\right\}_{m=1}^{\infty},$$

$$\left\{\frac{J_2(s_{m2}x)}{\frac{R}{\sqrt{2}}|J_3(s_{m2}R)|}\right\}_{m=1}^{\infty}, \ldots, \left\{\frac{J_n(s_{mn}x)}{\frac{R}{\sqrt{2}}|J_{n+1}(s_{mn}R)|}\right\}_{m=1}^{\infty}, \ldots.$$

example 16.21: Write the first three elements of the set of Bessel functions of order 0 which are orthonormal on (0, 3).

solution: From the table of zeros given in Section 16.12, $s_{10}R = 2.4$, $s_{20}R = 5.5$, $s_{30}R = 8.7$, and hence $s_{10} = 0.8$, $s_{20} = 1.8$, $s_{30} = 2.9$. From the table of values for $J_1(x)$ given in Appendix 4, $J_1(2.4) = 0.52$, $J_1(5.5) = -0.34$, and $J_1(8.7) = 0.27$. Thus, the first few terms of the orthonormal set are

$$\left\{ \frac{J_0(2.4x)}{\frac{3}{\sqrt{2}}(0.52)}, \frac{J_0(5.5x)}{\frac{3}{\sqrt{2}}(0.34)}, \frac{J_0(8.7x)}{\frac{3}{\sqrt{2}}(0.27)}, \ldots \right\}.$$

An expansion of a function of $PC(0, R)$ in terms of Bessel functions of order n is called a *Fourier-Bessel (order n) expansion.* The results of Chapter 10 apply to this case since we are assuming that each of the orthogonal sets is a basis for $PC(0, R)$. Hence, the Fourier-Bessel series converges in the mean to the function. To determine the coefficients in the expansion, the results of Chapter 10 tell us that if

$$f(x) = \sum_{j=1}^{\infty} c_j \phi_j,$$

then the Fourier coefficients are given by

$$c_j = \frac{(f, \phi_j)}{\|\phi_j\|^2}.$$

In the case of the sets of Bessel functions, the inner product is with respect to the weight function x, so that

$$c_j = \frac{1}{\|J_n(s_{jn}x)\|^2} \int_0^R xf(x)\, J_n(s_{jn}x)\, dx.$$

Using the values of $\|J_n(s_{jn}x)\|$ derived previously, the Fourier-Bessel expansion is written

$$f(x) = \sum_{j=1}^{\infty} A_j J_n(s_{jn}x)$$

where

$$A_j = \frac{2}{R^2 J_{n+1}^2(s_{jn}R)} \int_0^R xf(x)\, J_n(s_{jn}x)\, dx.$$

The coefficients A_j are often called the *Fourier-Bessel coefficients.*

example 16.22: Expand the function defined by $f(x) = 1$ on $PC(0, 2)$ in a series of Bessel functions of order zero.

solution: The Fourier-Bessel coefficients are

$$A_j = \frac{2}{4J_1^2(2s_{j0})}\int_0^2 xJ_0(s_{j0}x)\,dx$$
$$= \frac{1}{2J_1^2(2s_{j0})\,s_{j0}^2}\int_0^{2s_{j0}} uJ_0(u)\,du\,.$$

Using the result of Example 16.16,

$$A_j = \frac{1}{2J_1^2(2s_{j0})\,s_{j0}^2}\,[uJ_1(u)]_0^{2s_{j0}}$$
$$= \frac{1}{2J_1^2(2s_{j0})\,s_{j0}^2}\,2s_{j0}J_1(2s_{j0})$$
$$= \frac{1}{s_{j0}J_1(2s_{j0})}.$$

Therefore,

$$1 \sim \sum_{j=1}^{\infty} \frac{J_0(s_{j0}x)}{s_{j0}J_1(2s_{j0})}$$

where $2s_{j0}$ are the zeros of the Bessel function of order zero.

example 16.23: Expand the function of $PC(0, 1)$ defined by $f(x) = x$ in a series of Bessel functions of order 1.

solution: For the Bessel functions of order 1, the Fourier-Bessel coefficients with $R = 1$ are

$$A_j = \frac{2}{J_2^2(s_{j1})}\int_0^1 x^2J_1(s_{j1}x)\,dx$$
$$= \frac{2}{J_2^2(s_{j1})(s_{j1})^3}\int_0^{s_{j1}} u^2J_1(u)\,du$$
$$= \frac{2}{J_2^2(s_{ji})(s_{j1})^3}\,[u^2J_2(u)]_0^{s_{j1}}$$
$$= \frac{2}{s_{j1}J_2(s_{j1})}.$$

(See Exercise 8 of the previous set of Exercises.) Thus:

$$x \sim \sum_{j=1}^{\infty} \frac{2}{s_{j1}J_2(s_{j1})}\,J_1(s_{j1}x)$$

where the s_{j1} are the zeros of the order 1 Bessel function.

16.14 the vibrating membrane

It was Daniel Bernoulli who first used Bessel functions as long ago as 1732 in the study of the problem of small oscillations of the hanging chain. Since then, there have been a number of classical problems whose solutions demand the use of Bessel functions. They are also used in more modern analysis, such as in the description of the flux distribution in a nuclear reactor. In this and the following section, you will see two relatively elementary important applications.

Consider the vibrations of a taut circular membrane of radius R clamped along its circumference. It is a part of empirical knowledge that the differential equation describing small transverse vibrations is the two-dimensional wave equation

$$u_{tt} = c^2 \nabla^2 u$$

where c is a physical constant of the membrane. We introduce a system of cylindrical coordinates (r, θ) with the origin at the center of the membrane. (See Figure 16.8.) The u-axis is considered to be perpendicular to the equilibrium plane of the membrane. The wave equation then takes the form

$$u_{tt} = c^2 \left(u_{rr} + \frac{1}{r} u_r + \frac{1}{r^2} u_{\theta\theta} \right).$$

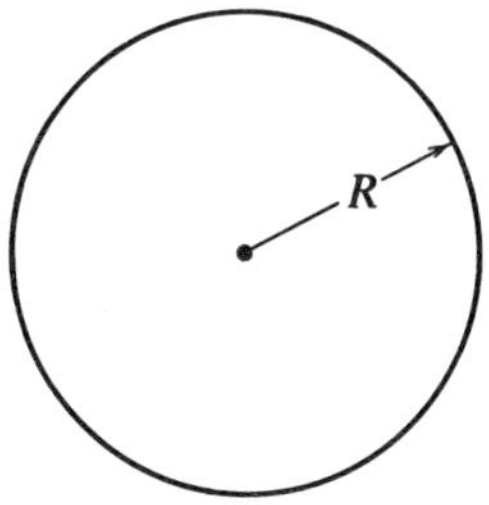

figure 16.8 circular membrane

The clamped circumference of the membrane means that u must satisfy the boundary condition $u(R, t) = 0$ for all $t \geqq 0$. It is customary to simplify the problem by considering only those solutions which are radially symmetric; that is, those which do not depend on θ. The mathematical model of the problem becomes

$$u_{tt} = c^2 \left(u_{rr} + \frac{1}{r} u_r \right)$$

where $u(R, t) = 0$, $u(r, 0) = f(r)$, $u_t(r, 0) = g(r)$, and where the last two conditions are initial conditions on the deflection and velocity. Assuming a product solution we let

$$u(r,t) = W(r)\,G(t).$$

Substituting into the given differential equation,

$$W\ddot{G} = c^2\left(W''G + \frac{W'G}{r}\right).$$

Dividing by WG,

$$\frac{\ddot{G}}{c^2 G} = \frac{1}{W}\left(W'' + \frac{1}{r}\,W'\right).$$

In the usual way, this gives two ordinary differential equations

$$\ddot{G} + c^2 k^2 G = 0$$

$$W'' + \frac{1}{r}\,W' + k^2 W = 0$$

with boundary-initial conditions

$$G(0) = f(r),\ G_t(0) = g(r),\ W(R) = 0.$$

The second of these differential equations is Bessel's modified equation and hence the solution is

$$W(r) = c_1 J_0(kr) + c_2 Y_0(kr)$$

where J_0 and Y_0 are Bessel functions of the first and second kind of order zero. Since the deflection of the membrane is always finite while Y_0 becomes infinite as r approaches 0, c_2 must be chosen to be zero. At the boundary, $W(R) = c_1 J_0(kR) = 0$, which means that k must be chosen so that kR is a zero of the Bessel function of order zero. Denoting the positive zeros of $J_0(x)$ by $\alpha_{10}, \alpha_{20}, \cdots, \alpha_{m0}, \cdots$ the allowable values of k are

$$k_m = \frac{\alpha_{m0}}{R}.$$

Corresponding to each of these values of k, the time equation has the solution

$$G_m(t) = a_m \cos ck_m t + b_m \sin ck_m t,$$

and hence, each of the functions

$$u_m(r,t) = (a_m \cos ck_m t + b_m \sin ck_m t)\,J_0(k_m r)$$

satisfies the two-dimensional wave equation (with the assumed symmetry property with respect to θ), and the boundary condition at the circumference. Further, any sum of these functions is also a solution since the boundary condition is homogeneous. Each u_m is called an "eigenfunction" of the problem; either k_m or ck_m is called an "eigenvalue."

In general, any one of the u_m, or any finite combination of them, would not reduce to $f(r)$ when $t = 0$. We form the infinite series

$$u(r, t) = \sum_{m=1}^{\infty} u_m(r, t).$$

Setting $t = 0$,

$$u(r, 0) = \sum_{m=1}^{\infty} u_m(r, 0)$$
$$= \sum_{m=1}^{\infty} a_m J_0(k_m r).$$

Therefore, the a_m must be chosen to be the Fourier-Bessel coefficients of the function f with respect to the orthogonal set $\{J_0(k_m r)\}_{m=1}^{\infty}$, that is,

$$a_m = \frac{2}{R^2 J_1^2(k_m R)} \int_0^R r f(r) J_0(k_m r)\, dr.$$

The coefficients b_m are obtained similarly. This example shows the strikingly similar roles played by the trigonometric functions and the Bessel functions in the solution of the wave equation. The trigonometric differential equation arises when rectangular cartesian coordinates are used. Bessel's equation arises naturally when cylindrical coordinates are used.

16.15 phase modulation of radio waves

The use of Bessel functions arises in a very natural way in transmission theory and information transfer. The power in a radio wave is carried in a relatively high-frequency *carrier* wave, with angular frequency ω_c. The information is transmitted by varying or *modulating* this wave at angular frequencies much lower than ω_c. If we represent the carrier wave by

$$V \sin(\omega_c t + \phi)$$

then V is called the *amplitude* and ϕ the *phase angle* of the carrier.

If the carrier is unmodulated, then the amplitude and phase angle are constants. If V is varied, the modulation is called *amplitude modulation*; if ϕ is varied, the modulation is called *phase modulation*.

While the "information" to be transmitted is usually far more complex, for the sake of elementary analysis let us restrict the discussion to the case where the modulation is a single low-frequency sine wave. The use of Bessel functions becomes necessary when describing phase modulation. Therefore, suppose

$$\phi = A \sin \omega_m t$$

where we are assuming that ω_m is an angular audio frequency, with $\omega_m \ll \omega_c$. Then,

$$v(t) = V \sin(\omega_c t + A \sin \omega_m t).$$

By a simple trigonometric identity,

$$v(t) = V[\sin \omega_c t \cos(A \sin \omega_m t) + \cos \omega_c t \sin(A \sin \omega_m t)].$$

In the course of doing Example 16.20, we obtained expressions for $\cos(x \sin\theta)$ and $\sin(x \sin\theta)$. Using these in the equation for $v(t)$,

$$\begin{aligned} v(t) &= V\Big[\sin \omega_c t \Big(J_0(A) \sin \omega_c t + 2 \sum_{n=1}^{\infty} J_{2n}(A) \cos 2n\omega_m t\Big) \\ &\qquad + 2 \cos \omega_c t \sum_{n=0}^{\infty} J_{2n+1}(A) \sin(2n+1)\,\omega_m t\Big] \\ &= V\big[J_0(A) \sin \omega_c t + J_1(A)\big(\sin(\omega_c + \omega_m)t - \sin(\omega_c - \omega_m)t\big) \\ &\qquad + J_2(A)\big(\sin(\omega_c + 2\omega_m)t + \sin(\omega_c - 2\omega_m)t\big) \\ &\qquad + J_3(A)\big(\sin(\omega_c + 3\omega_m)t - \sin(\omega_c - 3\omega_m)t\big) \\ &\qquad + \cdots \\ &\qquad + J_n(A)\big(\sin(\omega_c + n\omega_m)t + (-1)^n \sin(\omega_c - n\omega_m)t\big) \\ &\qquad + \cdots\big]. \end{aligned}$$

Hence, the phase modulated wave can be considered as a carrier wave of angular frequency ω_c and amplitude $J_0(A)$ along with "sidebands" consisting of sine waves with angular frequencies $(\omega_c \pm n\omega_m)$. The sidebands have amplitudes $J_n(A)$ which tend to zero as n becomes unbounded. In fact, A is usually small and hence the number of significant sidebands is small, determined of course by the values of $J_n(A)$. The effect of the modulation is to cause a change in the amplitude of the carrier considered as one component in the frequency spectrum. It is easy to show that there is no change in total power output of the transmitter as a result of the modulation even though there is a change in the distribution of power among the component frequencies.

exercises for sections 16.13–16.15

1. Find the Fourier-Bessel expansion of $f(x) = 1$ over the interval $[0, 3]$ in terms of Bessel functions of order 1.
2. Find the Fourier-Bessel expansion of $f(x) = x^2$ over the interval $[0, 1]$ in terms of Bessel functions of order 0.
3. Find the Fourier-Bessel expansion of $f(x) = 1 - x^2$ over the interval $[0, 1]$ in terms of Bessel functions of order 0.

4. Represent $f(x) = x$ on $[0, 2]$ by a Fourier-Bessel series involving Bessel functions of order 3.

5. Make a sketch of $J_0(\alpha_{10}x)$ and $J_0(\alpha_{20}x)$ where α_{10} and α_{20} are the first two zeros of the Bessel function of order 0.

6. Make a sketch of $J_1(\alpha_{11}x)$ and $J_1(\alpha_{21}x)$ where α_{11} and α_{21} are the first two nonzero zeros of the Bessel function of order one.

7. Before modulation, the power from a transmitter is given by $P = kV^2$ where k is a constant. After modulation, the power in a wave is

$$kV^2 \left[J_0^2(A) + 2 \sum_{n=1}^{\infty} J_n^2(A) \right].$$

Show that the power output is the same before and after modulation.

8. What are the zeros of the function $J_0(s_{30}x)$?

appendix a sequences, improper integrals, and series

a. 1 introduction

In this appendix we have grouped the significant definitions and theorems related to sequences, improper integrals, and infinite series that are needed in certain parts of this book. Some examples and discussion are included but the material presented here is intentionally sketchy. Some of the definitions and theorems are discussed in detail in most elementary courses. The other topics, particularly that of uniform convergence, are generally a part of a rigorous advanced calculus course.

a. 2 sequences of numbers

definition a. 1

A *sequence* is a function whose domain is the set of nonnegative integers. The functional values are called the *terms* of the sequence.

Conventional functional notation is seldom used for sequences, for instead of using $f(n)$ to designate the terms, letters with subscripts are used, such as a_n, b_n, V_n, α_n, etc. Also, we usually give the name "sequence" to a function whose domain is all integers, n, satisfying $n \geqq n_0$ where n_0 may be positive, negative, or 0, even though $n_0 = 0$ and $n_0 = 1$ are the most commonly used values. The complete sequence of terms is written either as $\{a_n\}$ or as

$$\{a_n\}_{n=n_0}^{\infty}$$

where n_0 indicates the integer at which the sequence begins.

Some examples of sequences are

$$\{n\}_{n=1}^{\infty} = \{1, 2, 3, 4, \cdots, n, \cdots\}$$
$$\{\cos n\pi\}_{n=0}^{\infty} = \{1, -1, 1, -1, \cdots, (-1)^n, \cdots\}.$$

Sequences are usually defined in one of the following three ways.

1. By an explicit rule in the form of an equation; for example,

$$\alpha_n = 2n\,; \quad \beta_n = 2^n\,; \quad \gamma_n = (-1)^n.$$

2. By an inductive or *recursive* formula which specifies each term of the sequence as a function of some previous ones; for example,

$$\alpha_1 = 2, \alpha_n = \alpha_{n-1} + 2\,; \quad \beta_1 = 2, \beta_n = 2\beta_{n-1}\,; \quad \gamma_1 = -1, \gamma_n = -\gamma_{n-1}.$$

3. By giving a few terms and expecting the rule to be inferred.

$$\{\alpha_n\} = \{2, 4, 6, 8, 10, \cdots\}\,;$$
$$\{\beta_n\} = \{2, 4, 8, 16, 32, \cdots\}\,;$$
$$\{\gamma_n\} = \{-1, 1, -1, 1, \cdots\}.$$

In inferring the terms of a sequence when only a few terms are given, we appeal to the good nature of the reader to understand the pattern.

The most important consideration for a sequence will usually be its behavior for large values of n.

definition a. 2

A sequence $\{a_n\}_{n=1}^{\infty}$ is said to *converge* to the real number L if for all $\varepsilon > 0$, there exists an integer, N, such that if $n > N$, then

$$|a_n - L| < \varepsilon.$$

If a sequence does not converge it is said to *diverge*.

Each of the three sequences above diverges, the first two because their terms become arbitrarily large and the third one since its terms do not remain within an arbitrarily small distance of one real number. If a sequence diverges because its terms become large, we write $\lim_{n\to\infty} \alpha_n = \infty$.

Usually the integer N which must be chosen is dependent on the value of ε.

example a.1: Show that the sequence $\{3n/(2n+1)\}_{n=1}^{\infty}$ converges to 3/2.

solution: Let N be the first integer bigger than $3/4\varepsilon - 1/2$.

$$|L - a_n| = \frac{3}{2} - \frac{3n}{2n+1} = \frac{6n+3-6n}{2(2n+1)} = \frac{3}{2(2n+1)}.$$

Let $n > N$. Then

$$|L - a_n| = \frac{3}{2(2n+1)} < \frac{3}{2(2N+1)} < \frac{3}{2\left(\dfrac{3}{2\varepsilon} - 1 + 1\right)} = \varepsilon$$

which shows the convergence. The estimate of N is obtained by assuming that the sequence *does* converge to 3/2 and determining N from that condition.

The idea of convergence is portrayed graphically as in Figure a.1. The sequence used is that of the previous example with the "arbitrary" ε chosen to be 1/8. The terms of the sequence are all within ε of the limit, 3/2, if N is bigger than or equal to the estimated value of 6 as given in the example.

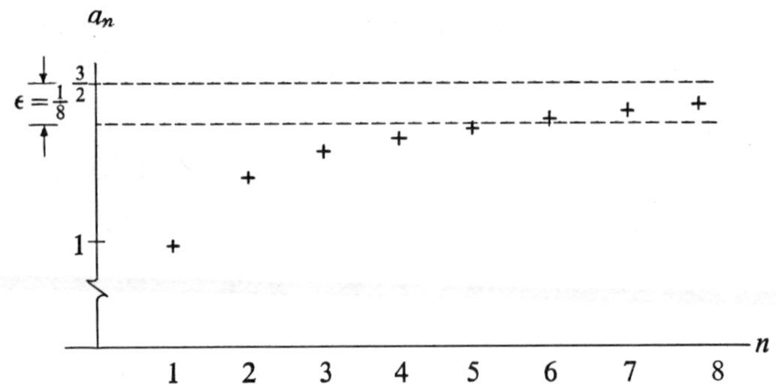

figure a. 1

The following theorem gives another condition for convergence of a sequence of real numbers. Sometimes it is used as the definition of convergence instead of Definition a.2. It allows consideration of the concept of convergence of a sequence without specific knowledge of the limit value.

theorem a. 1

The sequence $\{a_n\}_{n=1}^{\infty}$ converges if and only if for all $\varepsilon > 0$, there exists an integer N such that $|a_n - a_m| < \varepsilon$ whenever n and m are both greater than N.

The condition of Theorem a.1 is called the *fundamental* or *Cauchy criterion* for convergence.

definition a. 3

A sequence $\{a_n\}$ is said to be *monotone increasing* if $a_{n+1} \geqq a_n$ for all n, and is *monotone decreasing* if $a_{n+1} \leqq a_n$ for all n.

definition a. 4

A sequence $\{a_n\}$ is said to be *bounded above* if there exists a number M such that $a_n \leqq M$ for all n. A similar definition may be made for a sequence which is *bounded below.*

The notions of boundedness and monotonicity are usually extended to include those sequences which become bounded, or monotone, after finitely many terms of the sequence.

The following theorem combines the two previous definitions to give a criterion for convergence of monotone sequences.

theorem a. 2

If $\{a_n\}$ is monotone increasing and bounded above by M, then the sequence converges to a limit L which is $\leqq M$.

example a.2: Determine the convergence or divergence of the sequence

$$\left\{\sum_{k=n}^{\infty} \frac{1}{k}\right\}_{n=1}^{\infty}.$$

solution: We use Theorem a.2 by showing that the sequence is monotone (decreasing) and bounded (below).

$$\begin{aligned} a_n - a_{n+1} &= \sum_{k=n}^{2n} \frac{1}{k} - \sum_{k=n+1}^{2n+2} \frac{1}{k} \\ &= \frac{1}{n} - \frac{1}{2n+1} - \frac{1}{2n+2} \\ &= \frac{3n+2}{(2n+1)(n)(2n+2)} \end{aligned}$$

which is positive and thus the sequence is monotone decreasing. To show the sequence is bounded below, note that a_n is the sum of n terms,

$$a_n = \frac{1}{n} + \frac{1}{n+1} + \frac{1}{n+2} + \cdots + \frac{1}{2n-1} + \frac{1}{2n}.$$

The smallest of these terms is $1/2n$. Hence

$$a_n > \frac{1}{2n} + \frac{1}{2n} + \frac{1}{2n} + \cdots + \frac{1}{2n} = \frac{1}{2}.$$

which shows the terms are bounded below. The limit of the sequence therefore exists and is $\geqq 1/2$.

a. 3 sequences of functions

A sequence of functions $\{f_n\}_{n=1}^{\infty}$ is, in reality, many sequences of real numbers. For each x_0 on some interval on which each of the functions, f_n, is defined, the sequence of functions defines the sequence of numbers $\{f_n(x_0)\}_{n=1}^{\infty}$. If this sequence of real numbers converges, we say the sequence of functions is convergent at x_0. Those values of x for which the sequence converges comprise the domain of what is called the limit function, f_L, defined at each x by the equation

$$f_L(x) = \lim_{n\to\infty} f_n(x).$$

The domain of the limit function is also called the *domain of convergence* of the sequence, or, if the domain is an interval, the *interval of convergence.*

example a.3: Determine the limit function for the sequence of functions $\{f_n\}_{n=0}^{\infty}$ where $f_n(x) = x^n$. (See Figure a.2.)

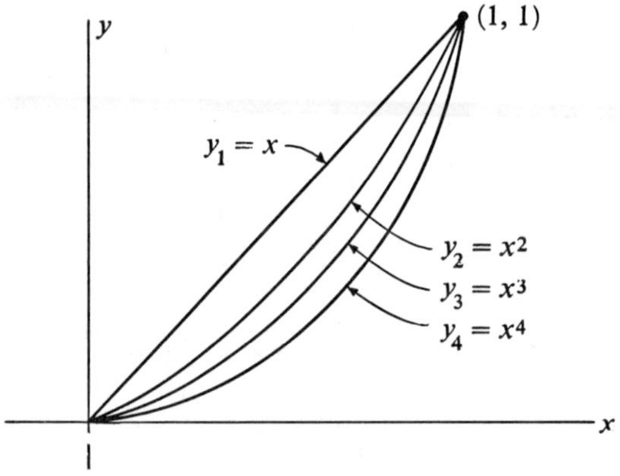

figure a. 2

solution: If $x > 1$, $\lim_{n\to\infty} x^n = \infty$; the limit function is undefined.

If $-1 < x < 1$, $\lim_{n\to\infty} x^n = 0$; from which $f_L(x) = 0$ on the interval.

If $x < -1$, $\lim_{n\to\infty} x^n$ is undefined; the limit function is undefined.

If $x = 1$, $\lim_{n\to\infty} 1^n = 1$, and hence $f_L(1) = 1$.

If $x = -1$, $\lim_{n\to\infty} (-1)^n$ is undefined.

In summary

$$f_L(x) = 0, \quad -1 < x < 1;$$
$$= 1, \quad x = 1;$$
$$\text{undefined for all other } x.$$

Note that each function of the sequence is continuous and differentiable on $[0, \infty)$ but the limit function only on $[0, 1)$.

example a.4: Determine the limit function for the sequence of functions $\{f_n\}_{n=1}^{\infty}$ defined by

$$f_n(x) = 0, \quad x \leqq 0$$
$$= n, \quad 0 < x < \frac{1}{n}$$
$$= 0, \quad \frac{1}{n} \leqq x.$$

(See Figure a.3.)

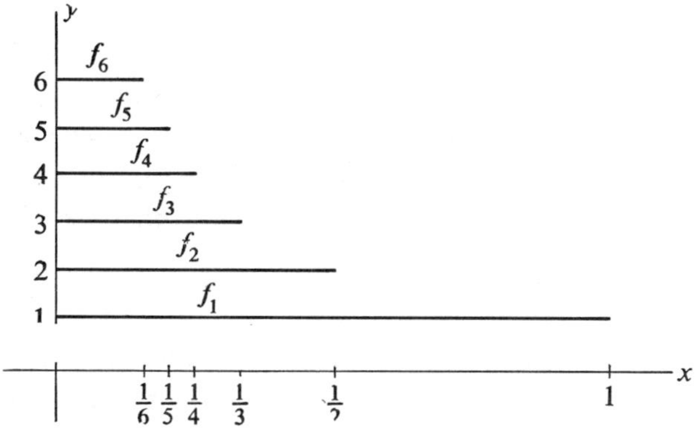

figure a. 3

solution: For each fixed x, $\lim_{n\to\infty} f_n(x) = 0$ so that the limit function is identically zero. Note that

$$1 = \lim_{n\to\infty} \int_0^{\infty} f_n(x)\,dx \neq \int_0^{\infty} \left[\lim_{n\to\infty} f_n(x)\right] dx = 0.$$

example a.5: Consider the sequence of functions $\{f_n\}_{n=1}^{\infty}$ defined on $[0, 1]$ by $f_n(x) = x^n/n$. Examine the sequence of derivatives and the derivative of the limit function.

solution: It is easy to show that $f_L(x) = 0$ for all x on $[0, 1]$ and hence $f_L'(x)$ is defined and equal to 0 throughout the interval.

Each f_n' is defined by the formula, $f_n'(x) = x^{n-1}$ and thus, $\lim_{n\to\infty} f_n'(x) = 0$ for x on $[0, 1)$ and 1 at $x = 1$. Note that

$$0 = f_L'(1) \neq \lim_{n\to\infty} f_n'(1) = 1.$$

How does the character of the elements of the sequence influence the limit function? Specifically, what requirements can be imposed on the sequence so that the limit function will be continuous, differentiable, or integrable over the interval of convergence? The three previous examples have been intended to show specific cases each of which contradicts at least one of the following properties:

1. $$\lim_{x\to x_0} \lim_{n\to\infty} f_n(x) = \lim_{n\to\infty} \lim_{x\to x_0} f_n(x).$$

2. $$\int_a^b \left[\lim_{n\to\infty} f_n(x)\right] dx = \lim_{n\to\infty} \int_a^b f_n(x)\, dx.$$

3. $$\left[\lim_{n\to\infty} f_n(x_0)\right]' = \lim_{n\to\infty} f_n'(x_0).$$

The importance and desirability of such properties is obvious enough. We define a type of convergence which will be a sufficient condition for the above three properties to hold. In the convergence process we demand a bit more than mere convergence of a sequence of numbers at each value of x.

The definition of convergence at x requires that for all $\varepsilon > 0$ there exists an N such that

$$|f_n(x) - f_L(x)| < \varepsilon$$

whenever $n > N$. Ordinarily, for a fixed value of ε, this value of N is different for each x; hence, it is dependent on x. However, if an N can be found which is independent of x, then the convergence is said to be *uniform* on the interval. To be more precise, we make the following definition.

definition a. 5

The sequence $\{f_n\}_{n=1}^{\infty}$ *converges uniformly* to f_L on (a, b) if for all $\varepsilon > 0$, there exists an N such that

$$|f_n(x) - f_L(x)| < \varepsilon$$

when $n > N$, for all x on (a, b).

Figure a.4 gives a geometric interpretation of the idea of uniform convergence. There must exist an N such that the graph of $f_n(x)$ lies within a 2ε strip of $f_L(x)$ if $n > N$.

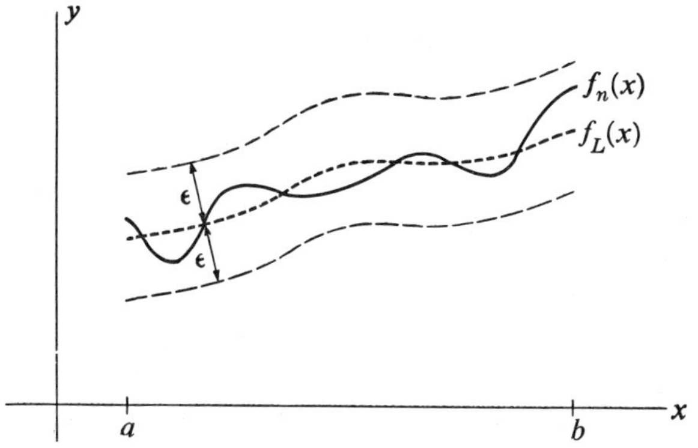

figure a. 4

When we say that a sequence of functions "converges on the interval (a, b)," we mean that at each point the sequence of functional values converges. Uniform convergence means more than this; it says that the number of terms required to get within any desired accuracy of the limit function can be made the same for the entire interval.

The following theorem is the Cauchy criterion for uniform convergence.

theorem a. 3

The sequence $\{f_n\}_{n=1}^{\infty}$ converges uniformly on (a, b) if and only if for all $\varepsilon > 0$ there exists an integer N such that

$$|f_n(x) - f_m(x)| < \varepsilon$$

for all x in (a, b) when n and m are $> N$.

The definition given for uniform convergence (and the statement of the Cauchy criterion) is for an open interval but there is no essential difference for intervals which are closed, or half-open and closed. The same central idea of being able to find one value of N which is independent of x is included in any definition.

example a.6: Show that the sequence of functions $\{f_n\}_{n=0}^{\infty}$, where

$$f_n(x) = x^n$$

does not converge uniformly on $[0, 1]$, but does converge uniformly on $[0, 1/2]$.

solution: The limit of this function was found in Example a.3 to be the zero function. To show that this sequence does not converge uniformly on [0, 1] we note that the nth element of the sequence f_n evaluated at $2^{-1/n}$ is $f_n(2^{-1/n}) = 1/2$. Therefore, the choice of N to make $|f_n(x) - f_L(x)|$ small is ultimately dependent on x and hence the convergence is nonuniform. If x is on [0, 1/2],

$$|f_n(x) - f_L(x)| = x^n < (\tfrac{1}{2})^n,$$

which can be made less than ε by choosing $n > (\log(1/\varepsilon))/\log 2$. Since this is independent of x, the convergence is uniform.

The following theorem will help to answer some of the questions posed earlier and will show why we often demand uniform convergence of a sequence of functions on an interval.

theorem a. 4

Consider a sequence of functions $\{f_n\}_{n=1}^{\infty}$, each of which is continuous on (a, b), that converges uniformly to f_L on (a, b). Then:

1. f_L is continuous on (a, b).
2. $$\int_a^b f_L(x)\,dx = \lim_{n\to\infty} \int_a^b f_n(x)\,dx.$$
3. If, further, the sequence $\{f_n'\}_{n=1}^{\infty}$ converges uniformly, then,
$$f_L'(x) = \lim_{n\to\infty} f_n'(x)$$
where x is any point on (a, b).

Properties 1, 2, and 3 in Theorem a.4 correspond exactly to the properties 1, 2, and 3 mentioned after Example a.5.

a. 4 improper integrals

The Riemann definite integral, $\int_a^b f(x)\,dx$, of a function is defined in elementary calculus with (at least) the following two requirements:

1. The interval of integration, (a, b), must be finite.
2. The function must be bounded on (a, b).

To distinguish this integral from what follows we will henceforth call it a *proper* integral, if it exists. There are times when it is convenient to abandon one or both of the conditions by defining a quantity called an *improper* integral.

definition a. 6

(a) If $\int_a^b f(x)\,dx$ is proper for any interval (a, b) then the quantity $\int_a^\infty f(x)\,dx$ is called an *improper integral of the first kind.*

(b) If $\int_a^b f(x)\,dx$ is proper for all open intervals (a, b) where $b < c$, but not for $b = c$, then $\int_a^c f(x)\,dx$ is called an *improper integral of the second kind.*

Definition a.6 gives a name to the symbols which have no meaning as Riemann proper integrals; the definition itself gives no meaning to the improper integrals. This we do by considering a limit of Riemann proper integrals.

definition a. 7

(a) The number, if it exists, given by

$$\lim_{t\to\infty} \int_a^t f(x)\,dx$$

is called the *value* of the improper integral

$$\int_a^\infty f(x)\,dx.$$

(b) The number, if it exists, given by

$$\lim_{t\to c^-}{}^{\dagger} \int_a^t f(x)\,dx$$

is called the *value* of the improper integral

$$\int_a^c f(x)\,dx.$$

In both cases, if the limit exists, the improper integral is said to *converge*; otherwise it *diverges.*

The next two theorems are statements of criteria for convergence of improper integrals of the first kind. The first is almost a restatement of the definition while the second is called Cauchy's criterion.

† We say $\lim_{x\to c^-} f(x) = L$ if for all $\varepsilon > 0$, there exists $\delta > 0$, such that $|f(x) - L| < \varepsilon$ if $c - x < \delta$. L is called the "left-hand limit" or the "limit from the left." A similar definition holds for right-hand limits. Note that $\lim_{x\to c} f(x)$ exists if and only if the right- and left-hand limits exist and are equal.

theorem a. 5

$\int_a^\infty f(x)\,dx$ converges if and only if for all $\varepsilon > 0$ there exists a number x_0 such that if $b > x_0$, then

$$\left|\int_b^\infty f(x)\,dx\right| < \varepsilon.$$

theorem a. 6

$\int_a^\infty f(x)\,dx$ converges if and only if for all $\varepsilon > 0$ there exists a number x_0 such that if x_1 and x_2 are both greater than x_0, then

$$\left|\int_{x_1}^{x_2} f(x)\,dx\right| < \varepsilon.$$

Definitions a.6 and a.7 are stated in terms of an integral which is improper because the upper limit of integration is infinite or the integrand is unbounded at the upper limit. Analogous definitions hold for those integrals which become improper because of an infinite lower limit or an unbounded discontinuity of the integrand at the lower limit.

The following example exhibits an entire class of improper integrals. This class will prove very useful in determining the convergence or divergence of many other improper integrals.

example a.7: Examine the integrals $\int_1^\infty x^p\,dx$ and $\int_0^1 x^p\,dx$ for convergence.

solution: (a) By definition

$$\int_1^\infty x^p\,dx = \lim_{t\to\infty}\int_1^t x^p\,dx$$

$$= \lim_{t\to\infty} \left.\frac{x^{p+1}}{p+1}\right|_1^t = \begin{cases} \dfrac{-1}{p+1}, & p < -1 \\ \infty, & p > -1 \end{cases}$$

$$= \lim_{t\to\infty} \ln x\Big|_1^t = \infty, p = -1.$$

Hence the integral converges for $p < -1$ and diverges for $p \geqq -1$.

(b) By definition $\int_0^1 x^p\,dx$ is proper (and equal to $1/(p+1)$) if $p \geqq 0$.

If $p < 0$,

$$\int_0^1 x^p\,dx = \lim_{t\to 0^+} \int_t^1 x^p\,dx$$

$$= \lim_{t\to 0^+} \frac{x^{p+1}}{p+1}\bigg|_t^1 = \begin{cases} \dfrac{1}{p+1}, & p > -1 \\ \infty, & p < -1 \end{cases}$$

$$= \lim_{t\to 0^+} \ln x\big|_t^1 = \infty, p = -1.$$

Hence the integral converges for $p > -1$ and diverges for $p \leqq -1$.

An integral may be called improper even if it is not strictly of the first or second kind. For example,

$$\int_0^\infty \frac{1}{x^2}\,dx$$

is improper because of the infinite upper limit (first kind) and also because $1/x^2$ becomes unbounded near the lower limit (second kind). An integral like this is analyzed by decomposing it into integrals which are improper for only one reason; for example,

$$\int_0^\infty \frac{1}{x^2}\,dx = \int_0^1 \frac{1}{x^2}\,dx + \int_1^\infty \frac{1}{x^2}\,dx.$$

Each of the improper integrals on the right must converge for the improper integral on the left to converge. Thus $\int_0^\infty dx/x^2$ diverges because $\int_0^1 dx/x^2$ diverges even though $\int_1^\infty dx/x^2$ converges.

Integrals are also called improper if the integrand has an unbounded discontinuity interior to the interval of integration. For example, the integrals

$$\int_{-1}^1 \frac{dx}{x^2} \quad \text{and} \quad \int_{-1}^1 \frac{dx}{\sqrt{|x|}}$$

both have unbounded discontinuities at the origin. We express each of the integrals in terms of integrals which are improper for only one reason,

$$\int_{-1}^1 \frac{dx}{x^2} = \int_{-1}^0 \frac{dx}{x^2} + \int_0^1 \frac{dx}{x^2} = \lim_{a\to 0^-} \int_{-1}^a \frac{dx}{x^2} + \lim_{b\to 0^+} \int_b^1 \frac{dx}{x^2}.$$

Since each of the last two limits does not exist, the improper integral is divergent.

In contrast,

$$\int_{-1}^{1} \frac{dx}{\sqrt{|x|}} = \lim_{a\to 0^-} \int_{-1}^{a} \frac{dx}{\sqrt{|x|}} + \lim_{b\to 0^+} \int_{b}^{1} \frac{dx}{\sqrt{|x|}} = 2 + 2 = 4,$$

which says that the improper integral is convergent to the number 4.

Sometimes a meaning different from that of Definition a.7 is assigned to improper integrals with unbounded discontinuities interior to the interval of integration or with both lower and upper infinite limits. This is done by allowing the limits to be taken in a dependent manner. The following definition will make this more precise.

***definition* a. 8**

(a) If $\int_{-r}^{r} f(x)\,dx$ is proper for all values of r, then the number, if it exists, given by

$$\lim_{r\to\infty} \int_{-r}^{r} f(x)\,dx$$

is called the *Cauchy principal value* of the improper integral $\int_{-\infty}^{\infty} f(x)\,dx$.

(b) Suppose f has an unbounded discontinuity for $x = c$, where $a < c < b$. Then the number, if it exists, given by

$$\lim_{\varepsilon\to 0} \left[\int_{a}^{c-\varepsilon} f(x)\,dx + \int_{c+\varepsilon}^{b} f(x)\,dx\right]$$

is called the *Cauchy principal value* (C.P.V.) of the improper integral

$$\int_{a}^{b} f(x)\,dx.$$

An improper integral may converge in the Cauchy sense but not in the ordinary sense; for example, the integral $\int_{-1}^{1} dx/x^2$ is divergent in the ordinary sense, but (as it is easy to show) has a Cauchy principal value of -2. However, if an improper integral converges, it converges to its C.P.V.

***theorem* a. 7**

If an improper integral converges in the sense of Definition a.7, it converges to its Cauchy principal value.

Evaluating the C.P.V. is often easier than computing the ordinary value and since the two values agree when the improper integral converges, it is useful (and sometimes sufficient) to know whether or not an improper integral converges. The question of convergence can thus be separated from the determination of its value. The following two theorems are the best available elementary tests for determination of convergence independently of actually using the definition.

theorem a. 8

Comparison Test. Let $0 \leqq f(x) \leqq g(x)$ for all $x \geqq a$.

(a) If $\int_a^\infty g(x)\,dx$ converges, then $\int_a^\infty f(x)\,dx$ converges.

(b) If $\int_a^\infty f(x)\,dx$ diverges, then $\int_a^\infty g(x)\,dx$ diverges.

The comparison test requires that you know some convergent and divergent integrals. The class of integrals $\int_1^\infty x^p\,dx$ is one of the more useful types. For example, the integrals

$$\int_1^\infty \frac{dx}{x^3+1} \quad \text{and} \quad \int_1^\infty \frac{|\sin x|}{x^3+1}\,dx$$

are both convergent by comparison to $\int_1^\infty x^{-3}\,dx$. Theorem a.8 can become rather difficult to use, since the inequality is sometimes hard to justify. The following test (sometimes called the *limit comparison test*) is much easier to apply.

theorem a. 9

Let $f(x) \geqq 0$ and $g(x) > 0$, and suppose

$$\lim_{x\to\infty} \frac{f(x)}{g(x)} = L.$$

(a) If L is finite and nonzero, then $\int_a^\infty f(x)\,dx$ and $\int_a^\infty g(x)\,dx$ both converge or both diverge.

(b) If $L = 0$ and $\int_a^\infty g(x)\,dx$ converges, then $\int_a^\infty f(x)\,dx$ converges.

(c) If $L = \infty$ and $\int_a^\infty g(x)\,dx$ diverges, then $\int_a^\infty f(x)\,dx$ diverges.

We might roughly summarize at least the first part of the limit comparison test by noting that if two functions are of the same order (that is, the limit of their ratio as $x \to \infty$ is a finite number), then improper integrals of these functions converge or diverge together. For example, an integral like

$$\int_2^\infty \frac{dx}{x^3 - 1}$$

is readily seen to converge since its integrand has the same order as $1/x^3$. Similarly,

$$\int_1^\infty \frac{dx}{x^2 + 2x + 8}$$

converges by using the limit comparison test with the function $1/x^2$.

The comparison test and the limit comparison test both apply to functions which are nonnegative. If the integrand function takes on negative values the question of convergence may become quite involved. For such improper integrals we are often content with proving the following type of convergence.

definition a. 9

The improper integral $\int_a^\infty f(x)\,dx$ is said to *converge absolutely* if the improper integral

$$\int_a^\infty |f(x)|\,dx$$

converges. If an improper integral converges, but not absolutely, it is said to *converge conditionally.*

theorem a. 10

If $\int_a^\infty f(x)\,dx$ converges absolutely, it converges.

a. 5 improper integrals with a parameter

definition a. 10

By an improper integral with a parameter, denoted by

$$\int_a^\infty f(x, u)\,dx$$

is meant the function of u defined by

$$F(u) = \lim_{t\to\infty} \int_a^t f(x, u)\,dx.$$

We say that the improper integral converges to the functional value, $F(u)$. The values of u for which the improper integral converges comprise the *domain* of F.

theorem a. 11

The improper integral $\int_a^\infty f(x, u)\, dx$ is said to converge to the functional value $F(u)$ if for all $\varepsilon > 0$ there exists an x_0 such that

$$\left| F(u) - \int_a^t f(x, u)\, dx \right| < \varepsilon$$

if $t > x_0$.

example a.8: Find the function to which the improper integral

$$\int_0^\infty e^{-ux}\, dx$$

converges.

solution:

$$\begin{aligned} F(u) &= \int_0^\infty e^{-ux}\, dx \\ &= \lim_{t\to\infty} \left(\frac{-e^{-ux}}{u} \right)\Bigg|_0^t \\ &= \lim_{t\to\infty} \left(-\frac{e^{-ut} - 1}{u} \right) \\ &= \frac{1}{u}, \qquad u > 0. \end{aligned}$$

Note in the previous example that the domain of the limit function is restricted to positive values of u, even though the function $1/u$ is defined also for negative u.

The determination of an elementary expression for the limit function may be impossible, unnecessary, and not of crucial importance. However, you will wish to know where the limit function is continuous and how to compute its derivative or integral. To make judgements of this kind we usually demand an additional property of the convergent improper integral.

definition a. 11

The improper integral $\int_a^\infty f(x, u)\, dx$ is said to *converge uniformly* to $F(u)$

on $[c, d]$ if for all $\varepsilon > 0$ there exists $x_0 > a$, such that

$$\left| F(u) - \int_a^t f(x, u)\, dx \right| < \varepsilon$$

for all u in $[c, d]$, when $t > x_0$.

example a.9: Show that $\int_0^\infty e^{-ux}\, dx$ converges uniformly to $1/u$ on every interval $[c, \infty)$ for which $c > 0$.

solution:

$$\left| F(u) - \int_a^t f(x, u)\, dx \right| = \left| \frac{1}{u} - \int_0^t e^{-ux}\, dx \right|$$

$$= \frac{e^{-ut}}{u}.$$

To establish uniform convergence, we must show that this value is less than an arbitrary ε, for $t > x_0$ where x_0 is a value to be specified. Since $c \leqq u$,

$$\frac{e^{-ut}}{u} \leqq \frac{e^{-ut}}{c}.$$

Choose x_0 such that $x_0 c > \ln 1/c\varepsilon$. Then

$$ut > ct > \ln \frac{1}{c\varepsilon}$$

and therefore

$$\frac{e^{-ut}}{u} < \frac{e^{-\ln(1/c\varepsilon)}}{c} = \varepsilon$$

which proves the uniform convergence.

The following theorem is called the Cauchy criterion for uniform convergence.

theorem a. 12

The improper integral $\int_a^\infty f(x, u)\, dx$ converges uniformly on $[c, d]$ if and only if there exists an x_0 such that

$$\left| \int_{x_1}^{x_2} f(x, u)\, dx \right| < \varepsilon$$

for all u in $[c, d]$ when x_1 and x_2 are $> x_0$.

The following theorem is the analog of Theorem a.4 for sequences.

theorem a. 13

Let f be continuous as a function of x and u for all $x \geqq a$ and all u in $[c, d]$. Suppose

$$\int_a^\infty f(x, u)\,dx$$

converges uniformly to $F(u)$ on $[c, d]$. Then:

(a) $F(u)$ is continuous on $[c, d]$.

(b) $$\int_c^d F(u)\,du = \int_a^\infty \int_c^d f(x, u)\,du\,dx.$$

(c) If, further,

$$\int_a^\infty \frac{\partial f(x, u)}{\partial u}\,dx$$

converges uniformly to $g(u)$, then $F'(u) = g(u)$ for all u in $[c, d]$; that is,

$$F'(u) = \int_a^\infty \frac{\partial f(x, u)}{\partial u}\,dx.$$

Therefore, on the interval of uniform convergence, a function defined by an improper integral is continuous, it may be integrated by interchanging the proper integration with the improper, and its derivative may be found by simply differentiating the integrand if the improper integral of partial derivatives converges uniformly.

Theorem a.13 shows the importance of uniform convergence. The next theorem gives a convenient test to show that an improper integral converges uniformly. With this test you can avoid appealing directly to the definition.

theorem a. 14

M Test. Let $M(x)$ be a function such that $|f(x, u)| \leqq M(x)$ for all u in $[c, d]$, and all x greater than some fixed value x_1. Suppose that $\int_a^\infty M(x)\,dx$ converges. Then $\int_a^\infty f(x, u)\,dx$ converges uniformly on $[c, d]$.

Using the M Test it is a simple matter to show that the integral $\int_0^\infty e^{-ux}\,dx$ converges uniformly on any interval $[c, \infty)$ for which $c > 0$. For the comparison function we choose $M(x) = e^{-cx}$. The improper integral

$\int_0^{\infty} e^{-cx}\,dx$ converges and on $[c, \infty)$, $e^{-ux} < M(x)$, so that the M test is applicable and uniform convergence is proven.

a. 6 infinite series of constants

definition a. 12

Let $\{a_n\}_{n=1}^{\infty}$ be a sequence. An expression of the form $\sum_{n=1}^{\infty} a_n$ is called an *infinite series.*

Note that Definition a.12 gives a name to the symbols; it gives no meaning. This we do by considering the *sequence of partial sums,* $\{s_n\}_{n=1}^{\infty}$ where $s_n = \sum_{k=1}^{n} a_k$.

definition a. 13

If the sequence of partial sums converges to a limit S, we say the infinite series $\sum_{n=1}^{\infty} a_n$ *converges* to the value S. If the limit of the sequence of partial sums does not exist, the series is said to *diverge*. The number S is called the *sum* of the infinite series.

Thus to determine if an infinite series converges, we first must form the sequence of partial sums. If this sequence converges, then the infinite series has meaning.

The next theorem is the Cauchy criterion for convergence of an infinite series.

theorem a. 15

$\sum_{n=1}^{\infty} a_n$ converges if and only if for all $\varepsilon > 0$ there exists N such that for all $n > N$ and all $m > 0$,

$$\left| \sum_{k=n+1}^{m+n} a_k \right| < \varepsilon .$$

To form the actual sequence of partial sums into terms independent of a summation (a so-called "closed" form) is not always possible or simple. The following example is a case when it *is* both possible and simple. We first give a name to the type of series being considered.

definition a. 14

A series of the form $\sum_{n=0}^{\infty} ar^n$ in which succeeding terms have a constant ratio is called a *geometric series.* The terms of the corresponding sequence of partial sums are called *geometric progressions.*

example a.10: Analyze the geometric series for convergence and divergence.

solution: A typical term of the sequence of partial sums is

$$s_n = a + ar + ar^2 + \cdots + ar^{n-1}.$$

Multiplying both sides of this equation by r we obtain

$$rs_n = ar + ar^2 + ar^3 + \cdots + ar^n.$$

Then the difference is found to be

$$s_n - rs_n = a - ar^{n-1}$$

from which

$$s_n = \frac{a(1 - r^{n-1})}{1 - r}.$$

If $-1 < r < 1$, $\lim_{n \to \infty} s_n$ exists and is equal to $a/(1-r)$. If $r > 1$, the limit does not exist and the geometric series diverges. Direct substitution into the initial expression for s_n shows that the series diverges for $r = 1$ and -1.

Most series cannot be analyzed in the same manner as in Example a.10 since there is usually no way of expressing the general term in convenient simplified form. To find a "closed" form expression normally depends on some kind of trick like that used in the previous example.

Fortunately it is not always important to know the value of the sum of an infinite series, but merely that such a sum exists. The question "Does the series converge?" may be considered independently from the question "To what does the series converge?" The second question is generally far more difficult to answer.

theorem a. 16

If $\sum_{n=1}^{\infty} a_n$ converges, then $\lim_{n \to \infty} a_n = 0$.

It is the contrapositive form of this theorem that yields a test for divergence: If the limit of the nth term of a series is nonzero, the series diverges.

The converse of Theorem a.16 is not true, for as we shall see, the series $\sum_{n=1}^{\infty} 1/n$ (the *harmonic* series) diverges, even though $\lim_{n\to\infty} 1/n = 0$.

theorem a. 17

The series $\sum_{n=1}^{\infty} a_n$ and $\sum_{n=m}^{\infty} a_n$ either both converge or both diverge.

Essentially, this theorem says that convergence or divergence is unaffected by the sum of finitely many terms.

The following test for convergence of an infinite series is analogous to Theorem a.2.

theorem a. 18

Let $a_n > 0$. The series $\sum_{n=1}^{\infty} a_n$ converges if and only if the sequence of partial sums is bounded above.

The next theorem describes the *integral test.*

theorem a. 19

Let $a_n > 0$ for all n. Suppose that for all x larger than some N there is a function $f(x)$ which is positive and nonincreasing with values $f(n) = a_n$. Then the series $\sum_{n=1}^{\infty} a_n$ and the improper integral $\int_N^{\infty} f(x)\,dx$ either both converge or both diverge.

Note that the value of the improper integral, if it converges, should not be considered as the value of the corresponding infinite series. While the integral test is necessary and sufficient, its use is limited to those series whose corresponding improper integral can be judged to converge or diverge.

example a.11: Determine the convergence properties of the "p-series,"

$$\sum_{n=1}^{\infty} n^p.$$

solution: In Example a.7 we have shown that the improper integral

$$\int_1^{\infty} x^p\,dx$$

converges for $p < -1$ and diverges for $p \geqq -1$. Hence, the corresponding properties for the series are exactly the same. Note that the series for $p = -1$, $\sum_{n=1}^{\infty} 1/n$, is called the *harmonic* series. (It diverges!)

The p-series and the geometric series are important, not only because they both appear quite naturally in the applications but also because they play an important role in conjunction with the tests described in the following two theorems.

theorem a. 20

Comparison Test. Let $0 \leqq a_n \leqq b_n$ for all n larger than some N. Then:

(a) If $\sum_{n=1}^{\infty} b_n$ converges, so does $\sum_{n=1}^{\infty} a_n$.

(b) If $\sum_{n=1}^{\infty} a_n$ diverges, so does $\sum_{n=1}^{\infty} b_n$.

The following theorem is much easier to use.

theorem a. 21

Limit Comparison Test. Let $a_n \geqq 0$ and $b_n > 0$, and suppose

$$\lim_{n\to\infty} \frac{a_n}{b_n} = L.$$

(a) If L is finite and nonzero, then the two series $\sum_{n=1}^{\infty} a_n$ and $\sum_{n=1}^{\infty} b_n$ either both converge or both diverge.

(b) If $L = 0$, and $\sum_{n=1}^{\infty} b_n$ converges, then $\sum_{n=1}^{\infty} a_n$ converges.

(c) If $L = \infty$ and $\sum_{n=1}^{\infty} b_n$ diverges, then $\sum_{n=1}^{\infty} a_n$ diverges.

Thus the limit comparison test says that if two series have terms of the same order, then they either converge or diverge together. Series of rational functions of n are particularly well suited to the limit comparison test with comparison being made to one of the p-series. For example, the series

$$\sum_{n=1}^{\infty} \frac{1}{2n+7} \quad \text{and} \quad \sum_{n=1}^{\infty} \frac{1}{2n-7}$$

both diverge in comparison (in the limit) to the known divergent series

$\sum_{n=1}^{\infty} 1/n$. The series

$$\sum_{n=1}^{\infty} \frac{1}{2n^2+7} \quad \text{and} \quad \sum_{n=1}^{\infty} \frac{1}{2n^2-7}$$

both converge by comparison to the known convergent series $\sum_{n=1}^{\infty} 1/n^2$.

The tests outlined in the following two theorems are probably the most popular since convergence is determined without reference to another series.

theorem a. 22

Ratio Test. If $\sum_{n=1}^{\infty} a_n$ is a series of nonnegative terms for which

$$\lim_{n\to\infty} \frac{a_{n+1}}{a_n} = R,$$

then:

(a) If $0 \leqq R < 1$, the series converges.
(b) If $R > 1$, the series diverges.
(c) If the limit of the ratio approaches 1 through values greater than 1, $\sum_{n=1}^{\infty} a_n$ diverges.

In other cases where $R = 1$, the test is inconclusive.

In passing we note that the ratio test is merely a formalization of the comparison test used with a geometric series.

theorem a. 23

Root Test. If $\sum_{n=1}^{\infty} a_n$ is a series of nonnegative terms for which

$$\lim_{n\to\infty} (a_n)^{1/n} = R,$$

then the conclusion is the same as in Theorem a.22.

Series of terms, some of which may be negative, are usually more difficult to analyze. In this case we often examine the series of absolute values.

***definition* a. 15**

The series $\sum_{n=1}^{\infty} a_n$ is said to *converge absolutely* if the series $\sum_{n=1}^{\infty} |a_n|$ converges. If a series converges, but not absolutely, it is said to *converge conditionally.*

theorem a. 24

If $\sum_{n=1}^{\infty} a_n$ converges absolutely, then it converges.

Theorem a.24 is frequently the key to testing series with mixed signs. The corresponding series of absolute values is tested; if this series converges so does the given series. If it diverges, the given series may yet converge conditionally.

definition a. 16

A series of terms in which the terms alternate in sign is called an *alternating* series.

theorem a. 25

Consider the alternating series $\sum_{n=1}^{\infty} (-1)^n a_n$ where a_n is nonnegative. If:

$$\text{(a)} \quad a_{n+1} \leqq a_n \quad \text{and} \quad \text{(b)} \quad \lim_{n\to\infty} a_n = 0,$$

then the series converges.

For example, the series

$$\sum_{n=1}^{\infty} \frac{(-1)^n}{n}$$

does not converge absolutely, but, by the alternating series test, it does converge conditionally.

a. 7 series of functions

definition a. 17

An infinite series of functions, $\sum_{n=1}^{\infty} u_n(x)$, is said to converge to a functional value $f_L(x)$ at x if for all $\varepsilon > 0$ there exists an $N(x)$ such that

$$|u_n(x) - f_L(x)| < \varepsilon, \text{ if } n > N(x).$$

To determine the domain of the limit function, any of the tests of Section a.6 may be used. Ordinarily, the ratio test is the most convenient to obtain an interval of absolute convergence.

example a.12: Determine the domain of convergence of the infinite series of functions

$$\sum_{n=0}^{\infty}\left(\frac{x}{x+1}\right)^n.$$

solution: Applying the ratio test we see that the domain of the limit function consists of those values of x for which

$$\left|\frac{x}{x+1}\right|<1.$$

This same inequality could be determined by noting that the given series is geometric with "common ratio" $x/(x+1)$. The solution set for this inequality is

$$\left\{x \,\middle|\, \frac{1}{x+1}>0\right\}\cap\left\{x \,\middle|\, \frac{2x+1}{x+1}>0\right\}=\{x \mid x>-\tfrac{1}{2}\}.$$

At the endpoint, $x=-1/2$, the series is $\sum_{n=0}^{\infty}(-1)^n$ which diverges. Therefore, the domain of convergence is

$$\{x \mid x>-\tfrac{1}{2}\}.$$

Finding the limit function in terms of other well-known elementary functions is generally impossible. In the above example, the fact that the series is geometric may be used to write the limit function as (see Example a.10)

$$f(x)=\frac{1}{1-\dfrac{x}{x+1}}=x+1, \qquad x>-\tfrac{1}{2}.$$

If a series of functions converges on an interval, the sum of finitely many of them may be considered to be an approximation to the limit function. Hence, the knowledge of the actual limit function may be relatively unimportant, since many evaluations are, in actuality, approximations anyway; for example, $\ln 2$, e^3, etc.

Even when you do not know an explicit expression for the limit function you will often wish to know information relating to continuity, differentiability, and integrability. We first define uniform continuity as it applies to an infinite series of functions.

definition a. 18

An infinite series of functions $\sum_{n=1}^{\infty} u_n(x)$ is said to converge uniformly to

$f_L(x)$ on (a, b) if there exists $N(\varepsilon)$ such that

$$\left|\sum_{n=1}^{M} u_n(x) - f_L(x)\right| < \varepsilon$$

for all x on the interval (a, b) if $M > N(\varepsilon)$.

theorem a. 26

An infinite series of functions $\sum_{n=0}^{\infty} u_n(x)$ converges uniformly on (a, b) if and only if there exists an $N(\varepsilon)$ such that $\sum_{k=m}^{n} u_k(x) < \varepsilon$, for all x in (a, b) when $m, n > N(\varepsilon)$.

theorem a. 27

Suppose a series of continuous functions $\sum_{n=1}^{\infty} u_n(x)$ converges uniformly to $f_L(x)$ on (a, b). Then:

(a) f_L is also continuous for all x on (a, b).

(b) $$\int_a^b f_L(x)\,dx = \int_a^b \left[\sum_{n=1}^{\infty} u_n(x)\right] dx = \sum_{n=1}^{\infty}\left[\int_a^b u_n(x)\,dx\right].$$

(c) If, further, $\sum_{n=0}^{\infty} u_n'(x)$ converges uniformly to $g(x)$, then $f_L'(x) = g(x)$ for all x in (a, b); that is,

$$\left[\sum_{n=0}^{\infty} u_n(x)\right]' = \sum_{n=0}^{\infty} u_n'(x).$$

The importance of knowing when the convergence of an infinite series of functions is uniform is highlighted by Theorem a.27. The next theorem describes a relatively easy test to establish uniform convergence without resorting to the definition. The test is essentially a comparison test since the terms of the series of functions are compared with a known series of constant terms. It is the analog of Theorem a.14, for improper integrals.

theorem a. 28

M Test for Series. Suppose $\sum_{n=1}^{\infty} M_n$ is a series of positive constants such that $|u_n(x)| \leqq M_n$ for all u in (c, d) and all n bigger than some fixed N. If $\sum_{n=1}^{\infty} M_n$ converges, then $\sum_{n=1}^{\infty} u_n(x)$ converges uniformly on (c, d).

Using the M test it is a simple matter to show, for example, that the series $\sum_{n=0}^{\infty} x^n$ converges uniformly on $[-1/2, 1/2]$. On the interval, $x^n \leqq (1/2)^n$, and since $\sum_{n=0}^{\infty} (1/2)^n$ converges, it follows that $\sum_{n=0}^{\infty} x^n$ converges uniformly. This type of argument may be extended to show that $\sum_{n=0}^{\infty} x^n$ converges uniformly on any interval $[-a, a]$ where $a < 1$.

example a.13: Examine the series

$$\sum_{n=0}^{\infty} \left(\frac{x}{x+1}\right)^n$$

for uniform convergence.

solution: From Example a.12 we know that this series converges for $x > -1/2$. Since it is a geometric series (that is, it has the same form as $\sum_{n=0}^{\infty} x^n$ except that $u = x/(x+1)$ replaces x), it converges uniformly on any interval $-a \leqq x/(x+1) \leqq a$ where $a < 1$. By graphing $x/(x+1)$, we see that the series is uniformly convergent on any interval $[c, d]$ where $-1/2 < c$ and $d < \infty$. (See Figure a.5.)

figure a. 5

example a.14: (a) Show that the function

$$f_L(x) = \sum_{n=1}^{\infty} \frac{\sin(2n-1)x}{(2n-1)(2n)}$$

is continuous at all x.

(b) Find $\int_0^{\pi/2} f_L(x)\,dx$.

(c) Can the series be differentiated term-by-term to obtain $f_L'(x)$?

solution: (a) To show that the series converges uniformly for all intervals, we use the M test with

$$M_n = \frac{1}{(2n-1)(2n)}.$$

Since the series

$$\sum_{n=1}^{\infty} M_n = \sum_{n=1}^{\infty} \frac{1}{(2n-1)(2n)}$$

is convergent and since

$$\left|\frac{\sin(2n-1)x}{(2n-1)(2n)}\right| \leqq M_n$$

for all x, it follows that the series converges uniformly on the entire x-axis.

(b) Since it is a series of continuous functions converging uniformly to the limit function, that limit function is continuous by Theorem a.27. Also,

$$\begin{aligned}\int_0^{\pi/2} f_L(x)\,dx &= \sum_{n=1}^{\infty} \left[\int_0^{\pi/2} \frac{\sin(2n-1)x}{(2n-1)(2n)}\,dx\right] \\ &= -\sum_{n=1}^{\infty} \frac{\cos(2n-1)x}{(2n-1)^2(2n)}\Bigg|_0^{\pi/2} \\ &= \sum_{n=1}^{\infty} \frac{1}{(2n-1)^2(2n)}.\end{aligned}$$

(c) Since the differentiated series

$$\sum_{n=1}^{\infty} \frac{\cos(2n-1)x}{2n}$$

does not converge for $x = 0$, it does not converge uniformly on any interval containing the origin. Thus f_L' cannot be obtained by term-by-term differentiation of the series.

a. 8 power series

For many elementary purposes the most important kind of series of variable terms is the type whose terms are multiplies of powers of x.

definition a. 19

A series of the form $\sum_{n=0}^{\infty} c_n(x-a)^n$, c_n's = constants, is called a *power series* with center at $x = a$.

We will state the theorems for the case $a = 0$. The more general power series is obtained from this one by a shift of the origin.

Power series have a very nice pattern of convergence which is described by the following theorem and determined by the ratio or root test.

theorem a. 29

Every power series $\sum_{n=0}^{\infty} c_n x^n$ behaves in one of the following three ways:

(a) It converges only for the value $x = 0$.
(b) It converges absolutely for all values of x.
(c) There exists a positive number R such that the series converges absolutely for $x < R$, and diverges for $x > R$.

The number R of Theorem a.29 is called the *radius of convergence*, and the interval $(-R, R)$ is called the *interval of convergence* of the power series. There is nothing in Theorem a.29 which describes the behavior at the endpoints of the interval of convergence.

example a.15: Find the interval of convergence of the power series

$$\sum_{n=0}^{\infty} \frac{x^n \sin n}{n^2}.$$

solution: A direct application of the ratio test leads to the limit

$$\lim_{n\to\infty} \left| \frac{x^n \sin n}{n^2} \frac{(n-1)^2}{\sin (n-1)} \right|$$

which does not exist for any value of x. However, notice that since $|\sin n| \leq 1$, we have that

$$\left| \frac{x^n \sin n}{n^2} \right| \leq \left| \frac{x^n}{n^2} \right|.$$

It is easy to show that the series

$$\sum_{n=1}^{\infty} \frac{x^n}{n^2}$$

converges for $[-1, 1]$, and consequently, by the comparison test, the given series is also convergent. Further, for $|x| > 1$,

$$\lim_{n \to \infty} \left| \frac{x^n \sin n}{n^2} \right| = \infty$$

and hence the series diverges for $|x| > 1$.

theorem a. 30

If the power series $\sum_{n=0}^{\infty} c_n x^n$ has a radius of convergence $R > 0$, then the series converges uniformly on the interval $[-r, r]$ where r is any number such that $0 < r < R$. If $R = \infty$, then r may be any positive number.

For example, since $\sum_{n=0}^{\infty} x^n$ converges to $1/(1-x)$ on $(-1, 1)$, it converges uniformly to that value on any interval $[-r, r]$ where $0 < r < 1$. Because of Theorem a.27, it follows that on $[-r, r]$,

$$\int_0^x \frac{dt}{1-t} = \sum_{n=0}^{\infty} \int_0^x t^n \, dt$$

or

$$-\log_e (1-x) = \sum_{n=0}^{\infty} \frac{x^{n+1}}{n+1}.$$

From that same theorem, part (c), it also follows that

$$\frac{1}{(1-x)^2} = \sum_{n=1}^{\infty} n x^{n-1}.$$

The coefficients in a convergent power series are related to the values of the derivative at the center of the series.

theorem a. 31

If the power series $\sum_{n=0}^{\infty} c_n x^n$ converges to a function $f(x)$ with nonzero radius of convergence, then

$$c_n = \frac{f^{(n)}(0)}{n!}$$

where $f^{(0)}(0)$ is understood to be $f(0)$.

a. 9 taylor's series

In the previous section we considered the nature of functions which are defined in terms of a power series. We now give a theorem which shows how to represent a function in a standard way.

theorem a. 32

If a function f and its first n derivatives are continuous in some neighborhood of a fixed point a, and if x is in this neighborhood, then $f(x)$ may be written

$$f(x)=\sum_{k=0}^{n-1}\frac{f^{(k)}(0)}{k!}(x-a)^k+R_n(x)$$

where

$$R_n(x)=\frac{(x-a)^n}{n!}f^{(n)}(\xi) \quad \text{where} \quad a<\xi<x.$$

The quantity $R_n(x)$ is called the *remainder* after n terms. Note that we can obtain a bound on the remainder,

$$|R_n(x)|\leqq \max|f^{(n)}(t)|\left|\frac{(x-a)^n}{n!}\right|$$

where the maximum of $f^{(n)}(t)$ is taken on the interval a to x.

The formula of Theorem a.32 is called *Taylor's representation* (or formula) of $f(x)$ with a remainder around $x=a$. The special form for $a=0$ is called Maclaurin's representation of $f(x)$ with a remainder.

If a function possesses derivatives of all orders at $x=a$, then the number of terms in Taylor's formula may be continued indefinitely. We arrive at an infinite power series, called *Taylor's series* for f in the neighborhood of $x=a$. Taylor's series will represent the function if the sequence of sums converges to $f(x)$, or equivalently, if $\lim_{n\to\infty} R_n(x)=0$.

It can be shown that each of the familiar elementary functions has a Taylor series which actually represents it on the interval of convergence of the series. Thus, the sequence of partial sums can be used to approximate functional values.

Three Maclaurin series should be known since they are so frequently used:

1. $$e^x=\sum_{n=0}^{\infty}\frac{x^n}{n!}=1+x+\frac{x^2}{2}+\frac{x^3}{6}+\cdots.$$

2. $$\sin x = \sum_{n=0}^{\infty} \frac{(-1)^n x^{2n+1}}{(2n+1)!} = x - \frac{x^3}{6} + \frac{x^5}{120} - \cdots.$$

3. $$\cos x = \sum_{n=0}^{\infty} \frac{(-1)^n x^{2n}}{(2n)!} = 1 - \frac{x^2}{2} + \frac{x^4}{24} - \cdots.$$

All three series converge and represent the function for all x.

There is also a Taylor's theorem and corresponding series for a function of two real variables. It is rather difficult to supply all the descriptive details in a short summary. Suffice it to say that under suitable conditions, a function, f, may be represented by an infinite series with center at (a, b) whose first few terms are

$$\begin{aligned} f(x, y) = f(a, b) &+ [f_x(a, b)](x - a) + [f_y(a, b)](y - b) \\ &+ \tfrac{1}{2}[f_{xx}(a, b)](x - a)^2 + \tfrac{1}{2}[f_{yy}(a, b)](y - b)^2 \\ &+ [f_{yx}(a, b)](x - a)(y - b) + \cdots. \end{aligned}$$

appendix b
proof of
stokes' theorem

We assume without proof that:

1. If C' is a smooth curve in the parameter plane, so is its image C on the surface S.
2. The positive unit normal with respect to the positive sense of integration is given by

$$\frac{\mathbf{r}_u \times \mathbf{r}_v}{|\mathbf{r}_u \times \mathbf{r}_v|} = \frac{\mathbf{i}\,(y_u z_v - z_u y_v) + \mathbf{j}\,(x_v z_u - x_u z_v) + \mathbf{k}\,(x_u y_v - x_v y_u)}{[(\mathbf{r}_u \cdot \mathbf{r}_u)\,(\mathbf{r}_v \cdot \mathbf{r}_v) - (\mathbf{r}_u \cdot \mathbf{r}_v)^2]^{1/2}}.$$

stokes' theorem

Let S be a smooth oriented surface

$$\mathbf{r}\,(u, v) = x\,(u, v)\,\mathbf{i} + y\,(u, v)\,\mathbf{j} + z\,(u, v)\,\mathbf{k}.$$

Let C be the boundary of S and suppose C is a piecewise smooth simple closed curve C, with the added hypothesis that $\mathbf{r}_{uv} = \mathbf{r}_{vu}$. Let $\mathbf{F}$ be continuously differentiable on S and suppose f, g, and h are the component functions of F. Then

$$\oint_C \mathbf{F} \cdot d\mathbf{r} = \iint_S (\nabla \times \mathbf{F}) \cdot \mathbf{N}\, dS$$

where $\mathbf{N}$ is the unit normal to S that points in a direction positive with respect to the line integration.

proof: Let α, β, and γ be the direction angles of $\mathbf{N}$. Then, expressed in rectangular coordinates, we must prove

$$\int_C f\,dx + g\,dy + h\,dz = \iint_S [(h_y - g_z)\cos\alpha + (f_z - h_x)\cos\beta + (g_x - f_y)\cos\gamma]\,dS.$$

The proof consists of considering the line integral one "piece" at a time. The proof for each piece is identical. Let S' and C' be the region and curve in the parameter plane that correspond to S and C. Then, transforming the integral to the parameter plane,

$$\int_C f\,dx = \int_{C'} f(x(u,v), y(u,v), z(u,v))\,(x_u\,du + x_v\,dv).$$

Applying Green's theorem in the plane to the integral on C',

$$= \iint_{S'} [(f\,x_v)_u - (f\,x_u)_v]\,du\,dv$$

$$= \iint_{S'} [f_u x_v + f\,x_{vu} - f\,x_{uv} - f_v x_u]\,du\,dv.$$

Since $\mathbf{r}_{uv} = \mathbf{r}_{vu}$, the two mixed partial derivatives are equal and hence, the integral is equal to

$$= \iint_{S'} (f_u x_v - f_v x_u)\,du\,dv.$$

We now expand f_u and f_v by the chain rule to obtain

$$\int_C f\,dx = \iint_{S'} [x_v(f_x x_u + f_y y_u + f_z z_u) - x_u(f_x x_v + f_y y_v + f_z z_v)]\,du\,dv$$

$$= \iint_{S'} [f_z(x_v z_u - x_u z_v) - f_y(x_u y_v - x_v y_u)]\,du\,dv$$

$$= \iint_{S'} \left[f_z \frac{x_v z_u - x_u z_v}{|\mathbf{r}_u \times \mathbf{r}_v|} - f_y \frac{x_u y_v - x_v y_u}{|\mathbf{r}_u \times \mathbf{r}_v|}\right] |r_u \times r_v|\,du\,dv$$

$$= \iint_{S'} [f_z \cos\beta - f_y \cos\gamma]\,|\mathbf{r}_u \times \mathbf{r}_v|\,du\,dv$$

$$= \iint_S [f_z \cos\beta - f_y \cos\gamma]\,dS$$

which proves the first part. Adding the pieces, the theorem is proved.

appendix c
table of mappings

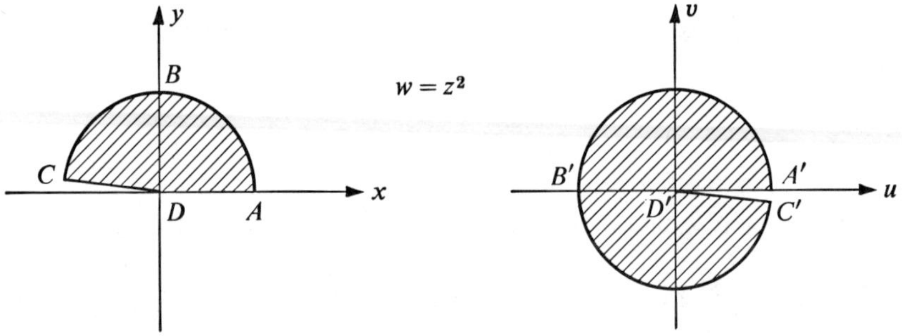

figure c. 1

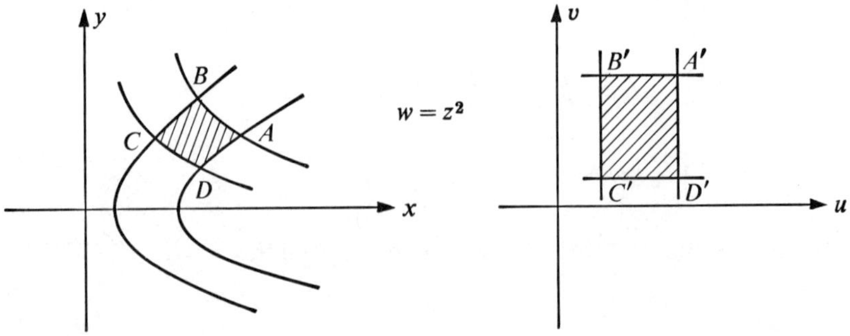

figure c. 2

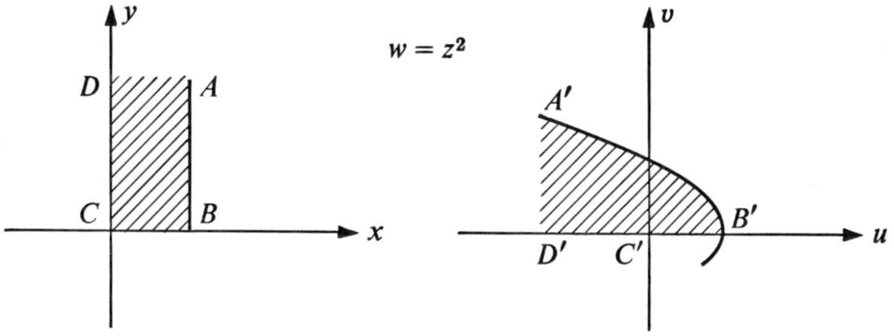

figure c. 3

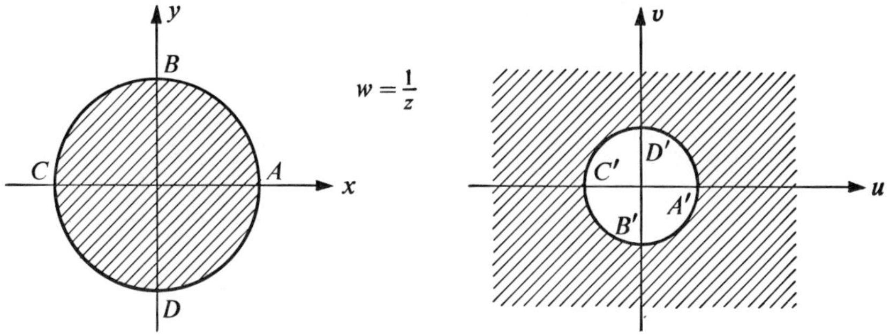

figure c. 4

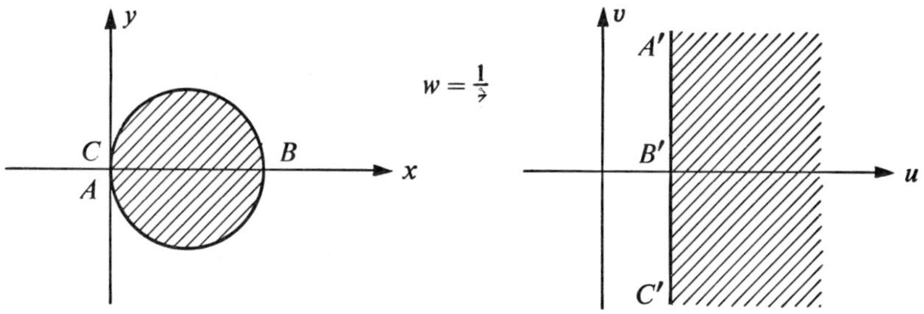

figure c. 5

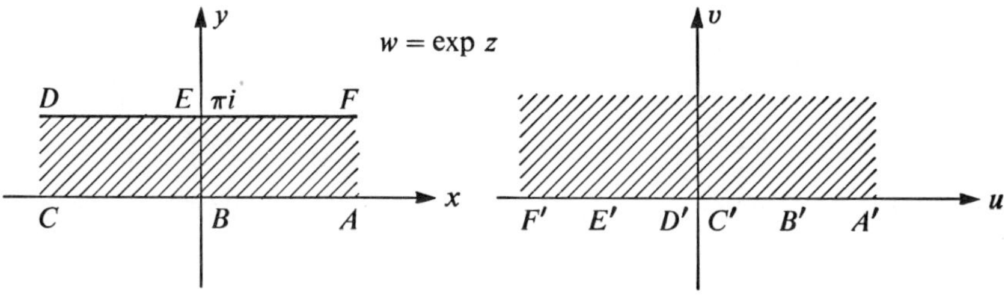

figure c. 6

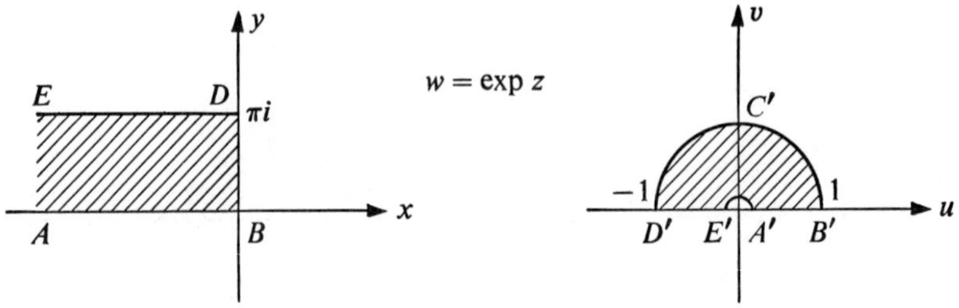

figure c. 7

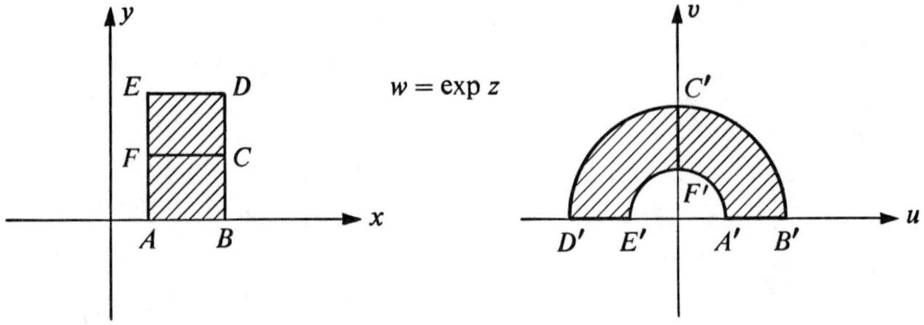

figure c. 8

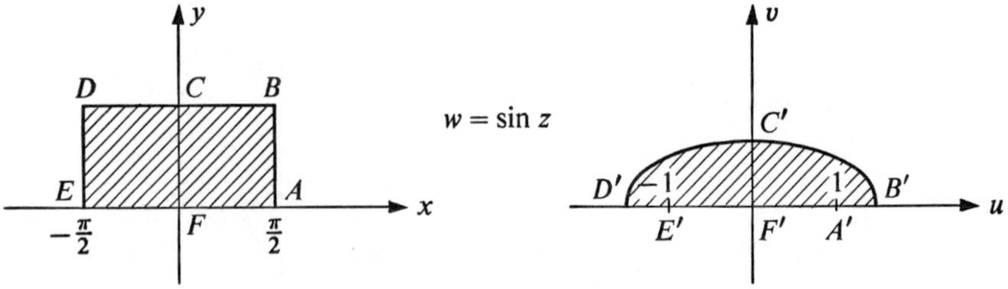

figure c. 9

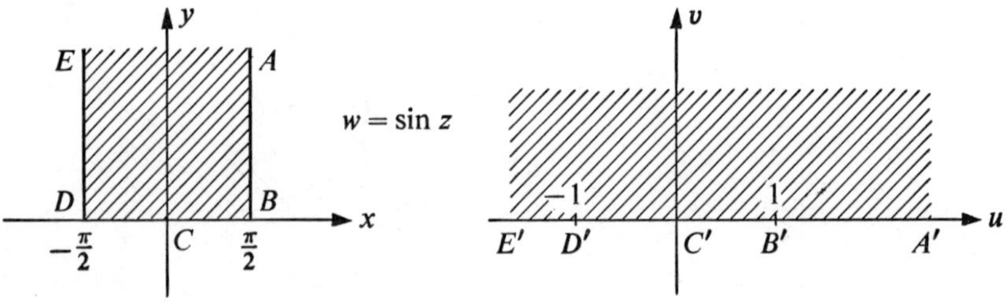

figure c. 10

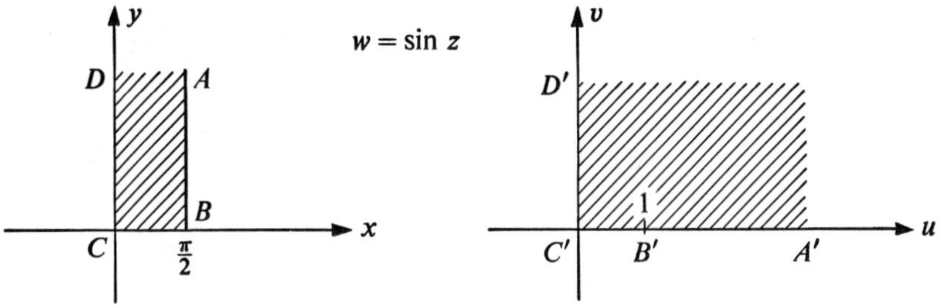

figure c. 11

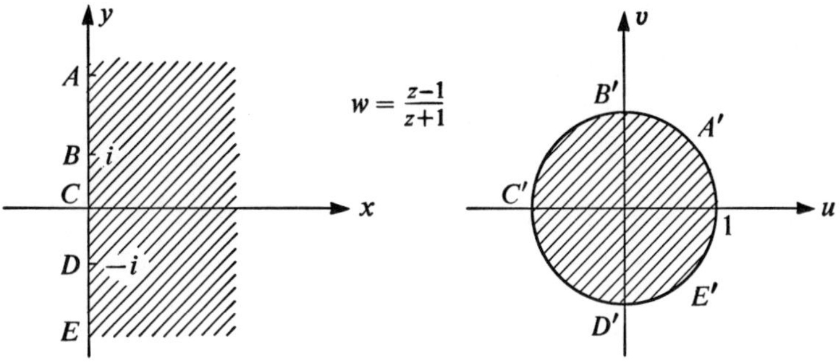

figure c. 12

appendix d table of values of J_0x and J_1x

x	$J_0(x)$	$J_1(x)$	x	$J_0(x)$	$J_1(x)$
0.0	1.00	0.00	2.0	0.22	0.58
0.1	1.00	0.05	2.1	0.17	0.57
0.2	0.99	0.10	2.2	0.11	0.56
0.3	0.98	0.15	2.3	0.06	0.54
0.4	0.96	0.20	2.4	0.00	0.52
0.5	0.94	0.24	2.5	−0.05	0.50
0.6	0.91	0.29	2.6	−0.10	0.47
0.7	0.88	0.33	2.7	−0.14	0.44
0.8	0.85	0.37	2.8	−0.19	0.41
0.9	0.81	0.41	2.9	−0.22	0.38
1.0	0.77	0.44	3.0	−0.26	0.34
1.1	0.72	0.47	3.1	−0.29	0.30
1.2	0.67	0.50	3.2	−0.32	0.26
1.3	0.62	0.52	3.3	−0.34	0.22
1.4	0.57	0.54	3.4	−0.36	0.18
1.5	0.51	0.56	3.5	−0.38	0.14
1.6	0.46	0.57	3.6	−0.39	0.10
1.7	0.40	0.58	3.7	−0.40	0.05
1.8	0.34	0.58	3.8	−0.40	0.01
1.9	0.28	0.58	3.9	−0.40	−0.03

x	$J_0(x)$	$J_1(x)$	x	$J_0(x)$	$J_1(x)$
4.0	−0.40	−0.07	8.0	0.17	0.23
4.1	−0.39	−0.10	8.1	0.15	0.25
4.2	−0.38	−0.14	8.2	0.12	0.26
4.3	−0.36	−0.17	8.3	0.10	0.27
4.4	−0.34	−0.20	8.4	0.07	0.27
4.5	−0.32	−0.23	8.5	0.04	0.27
4.6	−0.30	−0.26	8.6	0.01	0.27
4.7	−0.27	−0.28	8.7	−0.01	0.27
4.8	−0.24	−0.30	8.8	−0.04	0.26
4.9	−0.21	−0.31	8.9	−0.07	0.26
5.0	−0.18	−0.33	9.0	−0.09	0.25
5.1	−0.14	−0.34	9.1	−0.11	0.23
5.2	−0.11	−0.34	9.2	−0.14	0.22
5.3	−0.08	−0.35	9.3	−0.16	0.20
5.4	−0.04	−0.35	9.4	−0.18	0.18
5.5	−0.01	−0.34	9.5	−0.19	0.16
5.6	0.03	−0.33	9.6	−0.21	0.14
5.7	0.06	−0.32	9.7	−0.22	0.12
5.8	0.09	−0.31	9.8	−0.23	0.09
5.9	0.12	−0.30	9.9	−0.24	0.07
6.0	0.15	−0.28	10.0	−0.25	0.04
6.1	0.18	−0.26	10.1	−0.25	0.02
6.2	0.20	−0.23	10.2	−0.25	−0.01
6.3	0.22	−0.21	10.3	−0.25	−0.03
6.4	0.24	−0.18	10.4	−0.24	−0.06
6.5	0.26	−0.15	10.5	−0.24	−0.08
6.6	0.27	−0.13	10.6	−0.23	−0.10
6.7	0.29	−0.10	10.7	−0.22	−0.12
6.8	0.29	−0.07	10.8	−0.20	−0.14
6.9	0.30	−0.03	10.9	−0.19	−0.16
7.0	0.30	−0.00	11.0	−0.17	−0.18
7.1	0.30	0.03	11.1	−0.15	−0.19
7.2	0.30	0.05	11.2	−0.13	−0.20
7.3	0.29	0.08	11.3	−0.11	−0.21
7.4	0.28	0.11	11.4	−0.09	−0.22
7.5	0.27	0.14	11.5	−0.07	−0.23
7.6	0.25	0.16	11.6	−0.04	−0.23
7.7	0.23	0.18	11.7	−0.02	−0.23
7.8	0.22	0.20	11.8	0.00	−0.23
7.9	0.19	0.22	11.9	0.03	−0.23

x	$J_0(x)$	$J_1(x)$	x	$J_0(x)$	$J_1(x)$
12.0	0.05	−0.22	13.5	0.22	0.04
12.1	0.07	−0.22	13.6	0.21	0.06
12.2	0.09	−0.21	13.7	0.20	0.08
12.3	0.11	−0.19	13.8	0.19	0.10
12.4	0.13	−0.18	13.9	0.18	0.12
12.5	0.15	−0.17	14.0	0.17	0.13
12.6	0.16	−0.15	14.1	0.16	0.15
12.7	0.18	−0.13	14.2	0.14	0.16
12.8	0.19	−0.11	14.3	0.12	0.17
12.9	0.20	−0.09	14.4	0.11	0.19
13.0	0.21	−0.07	14.5	0.09	0.19
13.1	0.21	−0.05	14.6	0.07	0.20
13.2	0.22	−0.03	14.7	0.05	0.20
13.3	0.22	−0.01	14.8	0.03	0.21
13.4	0.22	0.02	14.9	0.01	0.21
			15.0	−0.01	0.21

appendix e
references

Agnew, R. P. *Differential Equations.* New York: McGraw-Hill Book Co., 1960.

Berg, P. W. and McGregor, J. L. *Elementary Partial Differential Equations.* San Francisco: Holden-Day, Inc., 1966.

Boas, M. L. *Mathematical Methods in the Physical Sciences.* New York: John Wiley & Sons, Inc., 1966.

Borisenko, A. I. and Tarapov, I. E. *Vector and Tensor Analysis with Applications.* Translated by R. A. Silverman. Englewood Cliffs, N.J.: Prentice-Hall, Inc., 1968.

Brand, L. *Vector Analysis.* New York: John Wiley & Sons, Inc., 1957.

Britton, J. R., Kriegh, R. B., and Rutland, L. W. *Calculus and Analytic Geometry.* San Francisco: W. H. Freeman & Co., 1966.

Churchill, R. V. *Complex Variables and Applications.* New York: McGraw-Hill Book Co., 1960.

Churchill, R. V. *Fourier Series and Boundary Value Problems.* New York: McGraw-Hill Book Co., 1941.

Crowder, H. K. and McCuskey, S. W. *Topics in Higher Analysis.* New York: Macmillan Co., 1964.

Esser, Martinus. *Differential Equations.* Philadelphia: W. B. Saunders Co., 1968.

Fisher, R. C. and Ziebur, A. D. *Calculus and Analytic Geometry.* Englewood Cliffs, N.J.: Prentice-Hall, Inc., 1965.

Golomb, M. and Shanks, M. E. *Elementary Ordinary Differential Equations.* New York: McGraw-Hill Book Co., 1965.

Kraus, J. D. *Electromagnetics.* New York: McGraw-Hill Book Co., 1953.

Kreider, D. L., Kuller, R. G., Ostberg, D. R., and Perkins, F. W. *An Introduction to Applied Linear Analysis.* Reading, Mass.: Addison-Wesley Publishing Co., Inc., 1966.

Kreyszig, E. *Advanced Engineering Mathematics.* New York: John Wiley & Sons, Inc., 1967.

Rainville, E. D. and Bedient, P. E. *Elementary Differential Equations.* New York: Macmillan Co., 1969.

Sokolnikoff, I. S. and Redheffer, R. M. *Mathematics of Physics and Modern Engineering.* New York: McGraw-Hill Book Co., 1966.

Spivak, M. *Calculus on Manifolds.* New York: W. A. Benjamin, 1965.

Taylor, A. E. *Advanced Calculus.* Boston: Ginn & Co., 1955.

Thomas, G. B. *Calculus and Analytic Geometry.* Reading, Mass.: Addison-Wesley Publishing Co., Inc., 1960.

Widder, D. V. *Advanced Calculus.* Englewood Cliffs, N.J.: Prentice-Hall, Inc., 1961.

Wylie, C. R. *Advanced Engineering Mathematics.* New York: McGraw-Hill Book Co., 1966.

answers to exercises

chapter one

exercises 1.1–1.6, p. 12

1. (a) $1/\sqrt{3}$. (b) $1/\sqrt{5}$. (c) $1/\sqrt{2}$. (d) 1. (e) $1/\sqrt{2}$.

2. (a) $2\sqrt{2}/\sqrt{3}$. (b) $4\sqrt{2}/\sqrt{5}$. (c) 1. (d) $\sqrt{2}$. (e) 1.

3. (a) Show that $\int_{-1}^{1} P_n(x)\, P_m(x)\, dx = 0$ for $m \neq n$. (b) No.

4. Show that $\int_{-\pi}^{\pi} \cos mx \cos nx\, dx = 0$ for $m \neq n$;

$\int_{-\pi}^{\pi} \sin mx \sin nx\, dx = 0$, for $m \neq n$; and that

$\int_{-\pi}^{\pi} \cos mx \sin nx\, dx = 0$.

5. When the vector is the zero vector.

7. $a_1 = \pm 1$, $b_1 = \pm 2\sqrt{3}$, $b_2 = \mp\sqrt{3}$, $c_1 = \mp 6\sqrt{5}$,
$c_2 = \pm 6\sqrt{5}$, $c_3 = \mp\sqrt{5}$.

8. (a) $\sqrt{31}/\sqrt{30}$. (b) $\left[\frac{\pi^3}{3} - \frac{3\pi}{2}\right]^{1/2}$. (c) $\left[\frac{\pi^3}{3} - \frac{5\pi}{3}\right]^{1/2}$.

9. (a) $\sqrt{8}$. (b) 0.

exercises 1.7, p. 18

2. (a) $3\mathbf{e}_1 - 4\mathbf{e}_2 + 2\mathbf{e}_3$. (b) $\frac{20}{11}\mathbf{e}_1 - \frac{5}{11}\mathbf{e}_2 - \frac{32}{11}\mathbf{e}_3$.

5. (a) $\left\{\frac{\mathbf{i}+\mathbf{j}}{\sqrt{2}}, \frac{\mathbf{i}-\mathbf{j}}{\sqrt{2}}, \mathbf{k}\right\}$.

(b) $\left\{\frac{3\mathbf{i}+2\mathbf{j}+\mathbf{k}}{\sqrt{14}}, \frac{\mathbf{i}+\mathbf{j}-5\mathbf{k}}{\sqrt{27}}, \frac{11\mathbf{i}-16\mathbf{j}-\mathbf{k}}{\sqrt{378}}\right\}$.

6. (a) $4\sqrt{2}\mathbf{e}_1 + 2\sqrt{2}\mathbf{e}_2 + 3\mathbf{e}_3$. (b) $\frac{25}{\sqrt{14}}\mathbf{e}_1 - \frac{7}{\sqrt{27}}\mathbf{e}_2 + \frac{31}{\sqrt{378}}\mathbf{e}_3$.

7. Determine a nontrivial linear combination equal to the zero vector.

8. Dimension is 1.

9. If $g = \sum_{n=1}^{N} c_n\phi_n$, then $(\phi_i, g) = \sum_{n=1}^{N} c_n(\phi_i, \phi_n) = c_i$.

11. $\{1, x, x^2\}, 3$.

exercises 1.8, p. 23

1. $\begin{pmatrix} 1 & -1 \\ 7 & 2 \end{pmatrix}$. **2.** $\begin{pmatrix} 7 & -12 \\ -1 & 4 \end{pmatrix}$. **3.** $\begin{pmatrix} -11 & 4 \\ 2 & 8 \end{pmatrix}\begin{pmatrix} 6 & 7 \\ 6 & -9 \end{pmatrix}$.

4. No. **5.** Basis elements may be chosen to be the matrices with 1 in the ij position and 0 elsewhere. The dimension of the space of $m \times n$ matrices is therefore mn.

6. No. Yes.

7. $\begin{pmatrix} 2 & 3 \\ 1 & 0 \\ 3 & 1 \end{pmatrix}$.

8. $\begin{pmatrix} 4 & 1 \\ 1 & 4 \end{pmatrix}, \begin{pmatrix} 0 & -7 \\ 7 & 0 \end{pmatrix}, \begin{pmatrix} 13 & 2 \\ 2 & 20 \end{pmatrix}$.

9. $\mathbf{B} = \begin{pmatrix} 1/8 & 3/16 \\ -1/4 & 1/8 \end{pmatrix}$, $\mathbf{BA} = \mathbf{I}$.

10. $\mathbf{AB} = (6)$, $\mathbf{BA} = \begin{pmatrix} 10 & -2 & 6 \\ 5 & -1 & 3 \\ -5 & 1 & -3 \end{pmatrix}$.

chapter two

exercises 2.1–2.3, p. 32

4. $y = ce^{-x}$. **5.** $y = c \csc x$. **6.** $(1+y)(1-x) = c(1+x)(1-y)$.

7. $y = \ln\left|\dfrac{-1}{e^x + c}\right|$. **8.** $i = ce^{-Rt/L}$.

9. $y = 3e^x$. **10.** $y = \dfrac{x^3}{3} + 2$.

11. $y = 7e^{3(x-1)}$. **12.** $y = 3e^{1+\cos x}$.

13. 6.97 min. **14.** 310.5 yrs.

15. 78.5 grams. **16.** $y =$ constant and $y = ce^{2x}$.

17. $y = -\dfrac{1}{2y(x+c)}$.

18. $y = \sqrt{2}\tan\sqrt{2}(x+c)$. **19.** $y^2 = 2(x+c)$. **20.** 11 A.M.

exercises 2.4–2.5, p. 41

1. $\ln y = \dfrac{x^2}{2y^2} + c$.

2. $\ln x = \dfrac{1}{2}\tan^{-1}\left(\dfrac{y}{x}\right) - \dfrac{1}{2}\ln\left(\dfrac{y^2 + x^2}{x^2}\right) + c$.

3. $x^2(x^2 - 2y^2) = c$.

4. $\tan^{-1}\left(\dfrac{y^2}{x^2}\right) - \dfrac{1}{4}\ln(x^4 + y^4) = c$.

5. $6y - 6x + 16\ln(3x + 6y + 5) = c$.

7. $\ln[2(x - \frac{4}{5})^2 + 2(x - \frac{4}{5})(y - \frac{4}{5}) + (y - \frac{4}{5})^2] - 4\tan^{-1}\dfrac{x + y - \frac{8}{5}}{x - \frac{4}{5}} = c$.

8. $2xy + x^2 - y^2 = c$. **10.** Exact, $x^2y + y^3/3 = c$.

11. Exact, $xy^2 - (x^3/3) = c$. **12.** Not exact.

13. Exact, $x^2 + xy - (y^2/2) = c$. **14.** Exact, $ye^x - x^2 = c$.

15. Exact, $x^2y + e^{xy} = c$. **17.** Only if $a = b$.

exercises 2.6, p. 49

1. I.F. $\dfrac{1}{x^2y}$, $\ln xy - \dfrac{y}{x} = c$.

2. I.F. $\dfrac{1}{x^{5/2}y^{3/2}}$, $3x^{-1/2}y^{-1/2} + x^{-3/2}y^{3/2} = c$.

3. No integrating factor of that form. $y/(1-y) = ce^{-1/x}$.

4. $\dfrac{1}{N}(M_y - N_x)$ is a function of x only.

5. $M_y - N_x = \dfrac{pN}{x} - \dfrac{qM}{y}$.

6. $x^3 = \frac{1}{3}[3cy - y^4]$.

7. $x^2 = 2y^2 - 2cy + c^2$.

8. $y = \dfrac{\sin^2 x}{3} + c \csc x$.

9. $y = ce^{-2x/3} - \dfrac{x}{2} + \dfrac{3}{4}$.

10. $y = -e^x \sec x + c \sec x$.

11. $y = \dfrac{x^2}{4} - \dfrac{x}{8} - \dfrac{1}{32} + ce^{4x}$.

12. $y = \frac{1}{2}(\sin x + \cos x) + ce^{-x}$.

13. $y = e^{-2x}(x + c)$.

14. (a) $i = \dfrac{(e^{-t} - e^{-Rt/L})}{R - L}$.

(b) $i = (R \sin t - L \cos t + Le^{-Rt/L})\dfrac{1}{R^2 + L^2}$.

(c) $i = \dfrac{L}{R^2}(e^{-Rt/L} - 1) + \dfrac{1}{R}t$.

15. $i = 1 - e^{-t}, 0 \leqq t \leqq 1$
$= 3 - t - (e + 1)e^{-t}, 1 \leqq t \leqq 2$
$= (e^2 - e - 1)e^{-t}, 2 \leqq t$.

16. $i = 2 - \dfrac{2}{5^{10}}(5 - t)^{10}, t \leqq 5$
$= 2, t \geqq 5$.

17. $i = 2$ for all t.

18. $i = 1;\ 0 \leqq t \leqq 1$
$= -1 + 2e^{-(t-1)};\ 1 \leqq t \leqq 2$.
$= (2e - e^2)e^{-t};\ 2 \leqq t$.

19. $\dfrac{1 - e}{1 + e}$.

20. $i(0) = \dfrac{1}{1 + e}$ amps.

exercises 2.7–2.8, p. 57

7. First line: $y = -2x - 2$; second line: $y = -6x + 6$; $y(4) = -18$.
8. First line: $y = 0$; second line: $y = x - 1$; third line: $y = 3x - 5$; $y(3) = 4$.
9. First line: $y = x + 1$; second line: $y = 4x - 5$; third line: $y = 9x - 20$; $y(4) = 16$.
10. All lines are $y = 3x - 4$, $y(5) = 11$.

12. $y_1 = -2 - 2x,\ y_2 = -2 - 2x - x^2,\ y_3 = -2 - 2x - x^2 - (x^3/3).$

13. $y_1 = \dfrac{x^2}{2},\ y_2 = \dfrac{x^2}{2} + \dfrac{x^3}{6},\ y_3 = \dfrac{x^2}{2} + \dfrac{x^3}{6} + \dfrac{x^4}{24}.$

14. $y_1 = \dfrac{x^2}{2} - 9x + 3,\ y_2 = -\dfrac{x^5}{20} + \dfrac{9x^4}{4} - 28x^3 + \dfrac{55}{2}x^2 - 9x + 3.$

15. $y_1 = y_2 = y_3 \doteq \dfrac{x^3}{3} - 4x + \dfrac{17}{3}.$

chapter three

exercises 3.1, p. 64

1. $x^2D^2 + xD - 1.$
2. $x^2D^2 + (2x-1)D.$
3. $x^2D^2 - D.$
4. $D^2 + 1 - x^2.$
5. $D^2 - 1 - x^2.$
6. $D^2 + D - 6.$
7. $x^2D^3 + (x + x^2)D^2 + (2x-1)D - 1 + x.$
9. $(2D-1)(D+2).$
10. $(D+1)(D-2)^2.$
11. $(D=2)(D+3)(D^2+4).$
12. $(D-3)^2.$
13. $2e^x \cos x.$
14. $2e^{-x}(\cos x - \sin x).$
15. $e^{2x}(38 + 39x + 9x^2).$
17. $y = e^{2x}(c_1x + c_2).$
18. $y = e^{3x}(c_0 + c_1x + c_2x^2 + c_3x^3).$
19. $y = e^{-x}(c_0 + c_1x + c_2x^2 + c_3x^3 + c_4x^4 + c_5x^5 + c_6x^6 + c_7x^7).$
20. $y = e^{3x}(c_1 + c_2x) + \frac{2}{27} + \frac{1}{9}x.$

exercises 3.2, p. 68

1. $y = c_1e^x + c_2e^{2x} + c_3e^{-4x}.$
2. $y = c_1e^{-x} + c_2e^{-3x}.$
3. $y = c_1e^{-2x} + c_2e^{2x}.$
4. $y = c_1 + c_2e^x.$
5. $y = c_1 + c_2e^x + c_3e^{7x}.$
6. $y = c_1x^2 + c_2x^{-2}.$
7. $y = c_1x^3 + c_2x^{-1}.$

exercises 3.3, p. 72

1. $\{\sin 3x, \cos 3x\}.$
2. $\{e^{3x}, e^{-3x}\}.$
3. $\{e^{-x}, e^{-4x}\}.$
4. $\{\sin x, \cos x\}.$

5. $\left\{e^{-x}, e^{+x/2}\sin\frac{\sqrt{3}}{2}x, e^{+x/2}\cos\frac{\sqrt{3}}{2}x\right\}$.

6. $\{e^{-x}, e^{2x}, e^{-x/2}, e^{3x/2}\}$.

7. $\{1, x, e^{-4x}, xe^{-4x}\}$.

8. $\{1, x, x^2, e^x, e^{-x}\}$.

9. $\{e^{2x}, e^{-x}, xe^{-x}\}$.

10. $\{e^x\cos\sqrt{5}x, e^x\sin\sqrt{5}x\}$.

11. $\{e^{-x}, \cos 2x, \sin 2x\}$.

12. $\{e^{-x}\cos 2x, e^{-x}\sin 2x, 1\}$.

13. $\{\cos x, \sin x, x\cos 2x, x\sin 2x, \cos 2x, \sin 2x\}$.

14. $\{e^{(7+3\sqrt{5})x/2}, e^{(7-3\sqrt{5})x/2}\}$.

15. $y=e^{3x}-xe^{3x}$.

16. $y=e^x$.

17. (a) $y=e^{2x}-xe^{2x}$. (b) $y=\frac{e^{2x}}{\varepsilon}[\varepsilon+1-e^{\varepsilon x}]$.

18. (a) $y=xe^{2x}$. (b) $y=\frac{e^{2x}}{\varepsilon}\sinh\varepsilon x \to xe^{2x}$ as $\varepsilon\to 0$.

(c) $y=\frac{1}{\varepsilon}e^{2x}\sin\varepsilon x \to xe^{2x}$ as $\varepsilon\to 0$.

19. $y=2e^{-x}$.

exercises 3.4–3.5, p. 76

2. $y=c_1+c_2e^x+x$.

3. $y=c_1\cos x+c_2\sin x+e^x$.

4. $y=c_1e^{2x}+c_2e^{-x}-3x+(3/2)-2xe^{-x}$.

5. $y=c_1e^{2x}+c_2e^{-2x}+c_3e^{-x}-xe^{-x}+x+(1/2)$.

7. $y=c_1e^{-x}+c_2+c_3x+(x^3/6)$.

8. $y=c_1e^x+c_2e^{-x}-x$.

9. $y=c_1e^x+c_2e^{-x}+\frac{1}{8}e^{3x}$.

10. $y=e^{-2x}(c_1x+c_2+\frac{1}{2}x^2)$.

11. $y=e^{-2x}(c_1x+c_2)+\frac{1}{4}x+c_3$.

exercises 3.6, p. 82

1. $y=e^x+c_1e^{2x}+c_2xe^{2x}$.

2. $y=\frac{1}{4}+e^x+c_1e^{2x}+c_2xe^{2x}$.

3. $y=\frac{1}{2}x^2e^{2x}+c_1e^{2x}+c_2xe^{2x}$.

4. $y=\frac{4}{25}\cos x+\frac{3}{25}\sin x+c_1e^{2x}+c_2xe^{2x}$.

5. $y=\frac{1}{6}x^3e^{2x}+\frac{1}{2}x^2e^{2x}+c_1e^{2x}+c_2xe^{2x}$.

6. $y=\frac{1}{6}x^3e^{2x}+c_1e^{2x}+c_2xe^{2x}$.

7. $y = -\frac{1}{3}\sin 2x + c_1 \cos x + c_2 \sin x.$

8. $y = -\frac{1}{4}x \cos 2x + c_1 \cos 2x + c_2 \sin 2x.$

9. $y = -\frac{1}{10}\cos 2x - \frac{1}{20}\sin 2x + c_1 + c_2 e^{-4x}.$

10. $y = \frac{4}{17}\cos 2x + \frac{1}{17}\sin 2x + c_1 e^{x}\cos 2x + c_2 e^{x}\sin 2x.$

11. $y = \frac{1}{4}xe^{x}\sin 2x + c_1 e^{x}\cos 2x + c_2 e^{x}\sin 2x.$

12. $y = \frac{1}{6}e^{2x} + \frac{1}{6}e^{-2x} + c_1 e^{-x} + c_2 e^{x} = \frac{1}{3}\cosh 2x + c_1 e^{-x} + c_2 e^{x}.$

13. $y = \frac{1}{4}xe^{x} - \frac{1}{4}xe^{-x} + c_1 e^{\ x} + c_2 e^{x} = \dfrac{x}{2}\sinh x + c_1 e^{-x} + c_2 e^{x}.$

14. $y = \frac{1}{4}x\cos x + \frac{1}{4}x^2 \sin x + c_1 \cos x + c_2 \sin x.$

15. $y = -xe^{x} + xe^{2x} + \frac{1}{6}e^{-x} + c_1 e^{x} + c_2 e^{2x}.$

16. $y = -2x - x^2 - \frac{1}{3}x^3 + c_1 + c_2 e^{x}.$

17. $y = -3x + c_1 + c_2 e^{x}.$ **18.** $y = \frac{8}{9}x + \frac{1}{6}x^2 + c_1 + c_2 e^{-3x}.$

19. $y = \frac{3}{4} + c_1 \cos 2x + c_2 \sin 2x.$ **20.** $y = \frac{3}{4}x + c_1 \cos 2x + c_2 \sin 2x.$

21. $y = \frac{7}{100}\cos 2x - \frac{1}{100}\sin 2x + c_1 e^{-x} + c_2 xe^{-x} + c_3 e^{2x}.$

22. $y = c_1 e^{x} + c_2 e^{-x} + c_3 \sin x + c_4 \cos x - \frac{1}{4}xe^{-x}.$

23. $y = c_1 e^{-2x} + e^{-x}(c_2 \cos 2x + c_3 \sin 2x) - \frac{1}{24}e^{x}.$

24. $y = c_1 e^{-x} + c_2 + c_3 x + \frac{1}{2}x^2.$

25. $y = c_1 e^{x} + e^{-x}(c_2 \cos x + c_3 \sin x) - \frac{1}{2}(x^2 + 1) - \frac{3}{5}\cos 2x - \frac{4}{5}\sin 2x.$

26. (a) $Ax + Bx^2.$ (b) $Ax + Bx^2.$

(c) $A\cos x + B\sin x + Cx^2 + Dx.$ (d) $Ax^2 e^{-x}.$

(e) $(Ax^2 + Bx^3)\,e^{-x}.$ (f) $Ax^2 e^{-x} + Be^{x}.$ (g) $Ae^{x}.$

(h) $Ax + Bx^2 + (Cx^2 + Dx^3)\,e^{-x}.$

exercises 3.7, p. 85

1. (a) $u_1 = 0,\ u_2 = (\frac{1}{2}x + \frac{5}{4})\,e^{-x}.$

(b) $u_1(x) = \frac{5}{8}e^{-x} + \frac{1}{4}xe^{-x};\ u_2(x) = \frac{5}{8}e^{-2x} + \frac{1}{4}xe^{-2x}.$

(c) $u_1(x) = (x+2)\,e^{-x};\ u_2(x) = -\frac{1}{2}xe^{-2x} - \frac{3}{4}e^{-2x}.$

3. $y = c_1 e^{-x} + c_2 e^{2x} + \frac{1}{3}xe^{2x}.$ **4.** $y = c_1 e^{-x} + c_2 xe^{-x} + xe^{-x}\ln x.$

5. $y = c_1 \cos 2x + c_2 \sin 2x - \frac{1}{4}\cos 2x \ln(\sec 2x + \tan 2x).$

6. $y = c_1 \cos 2x + c_2 \sin 2x - \frac{1}{2} + \frac{1}{4}\sin 2x \ln(\sec 2x + \tan 2x).$

7. $y = c_1 \cos 2x + c_2 \sin 2x + \frac{1}{12}\sin^4 2x + \frac{1}{4}\cos^2 2x - \frac{1}{12}\cos^4 2x$.
8. $y = c_1 e^x + c_2 e^{2x} - e^{2x}\cos(e^{-x})$.
9. $y = c_1 e^{2x} + c_2 x e^{2x} + x e^{2x}\ln x$.
10. $y = c_1 e^{-3x} + c_2 x e^{-3x} - \frac{1}{2}e^{-3x}\ln(x^2+1) + x e^{-3x}\tan^{-1} x$.
11. $y = c_1 e^{-x} + c_2 x e^{-x} + \frac{1}{4}x^2 e^{-x}(2\ln x - 3)$.
12. $y = c_1 e^{-x} + c_2 x e^{-x} + \dfrac{e^{-x}}{2x}$.
13. $y = c_1 x + \dfrac{c_2}{x} + \frac{1}{3}x^2 \ln x - \frac{4}{9}x^2$.
14. $y = c_1 x + c_2 e^x - \frac{1}{2}x e^{-x} + \frac{1}{4}e^{-x}$.
16. $y = c_1 x^{-1} + c_2 x^{-2}$.
17. $y = c_1 x + c_2 x^2 - 4x\ln x + \frac{1}{10}[\sin(\ln x) + 3\cos(\ln x)]$.
18. $y = c_1 x^3 + c_2 x^{-1} - \frac{1}{9}x^2(3\ln x + 2)$.
19. $y = c_1 \cos(3\ln x) + c_2 \sin(3\ln x)$.
20. $y = c_1 x + c_2 x^2 - \frac{3}{4} - \frac{1}{2}\ln x - x\ln x - \frac{1}{2}x(\ln x)^2$.
23. $y = c_1 e^x + c_2 e^{-x} - \frac{1}{2}\cos x$.
24. $y = c_1 e^{-x} + c_2 e^{-2x} + \frac{1}{6}e^x$.

exercises 3.8–3.10, p. 99

1. $i_t = \frac{225}{656}e^{-9t} - \frac{1025}{656}e^{-t}$; $i_{ss} = \frac{50}{41}\cos t + \frac{125}{82}\sin t$.
3. $i_t = e^{-5t/18}\left(-\dfrac{35}{37}\cos\dfrac{\sqrt{551}}{18}t + \dfrac{275}{37\sqrt{551}}\sin\dfrac{\sqrt{551}}{18}t\right)$;

   $i_{ss} = \dfrac{35}{37}\cos t + \dfrac{25}{37}\sin t$.
4. $i_t = e^{-t}(\sin t + 2\cos t)$; $i_{ss} = -2\cos t + \sin t$.
6. No effect.
7. $y = \frac{1}{8}\sin 2t - \frac{1}{4}t\cos 2t$.
9. At the positive values of $\tan^{-1} - \alpha$; $e^{2\pi\alpha}$.
10. $\sqrt{296/181}$.
11. 10 megacycles.
12. (a) $y = \frac{1}{2}\sin t - \frac{1}{2}t\cos t$.

    (b) $y = -\dfrac{4.75}{1.05}t + 5\sin 0.95t$.

    (c) $y = -5\cos t + e^{-0.1t}(0.5\sin t + 5\cos t)$.

13. $x(t)=\frac{5}{8}\cos t+\frac{3}{8},\ 0\leqq t\leqq\pi;\ x(t)=-\frac{1}{4},\ \pi\leqq t.$

14. $x(t)=0.75\cos 1.5t+0.5,\ 0\leqq t\leqq\frac{2}{3}\pi;\ x(t)=-0.25,\ \frac{2}{3}\pi\leqq t.$

15. $x(t)=\frac{11}{9}\cos\frac{3}{2}t+\frac{7}{9},\ 0\leqq t\leqq\frac{2}{3}\pi;\ x(t)=-\frac{4}{9},\ \frac{2}{3}\pi\leqq t.$

exercises *3.11, p. 103*

1. $y_1=\frac{1}{4}x^2+\frac{3}{2}x+c_1,\ y_2=\frac{3}{2}x-\frac{1}{4}x^2+c_2.$

2. $y_1=c_1e^x,\ y_2=c_2e^x.$

3. $y_1=c_1e^{\sqrt{2}x}+c_2e^{-\sqrt{2}x};\ y_2=c_1\cos\sqrt{2}x+c_2\sin\sqrt{2}x-x^2+\frac{3}{2}.$

4. $y_1=ce^{\left(\frac{x^5}{20}+\frac{2}{3}c_1x^3+c_1^2x\right)},\ y_2=\frac{1}{2}x^2+c_1.$

5. $y_1=\dfrac{1}{c_1(x+c_2)},\ y_2=-(x+c_2)^{-1}.$

6. $y_1=-\frac{1}{4}x^6-\frac{3}{2}x^5-c_1x^4-c_2x^3-c_3x^2-c_4x-c_5,\ y_2=\frac{1}{2}x^2+c_1.$

10. $y_1=c_1e^{3x}+c_2\cos 2x+c_3\sin 2x,$
$y_2=-1-5c_1e^{3x}+c_3\cos 2x-c_2\sin 2x.$

11. $y_1=c_1e^x-3c_2e^{-x}+(c_3-2c_4)\cos x+(2c_3+c_4)\sin x+4,$
$y_2=c_1e^x+c_2e^{-x}+c_3\cos x+c_4\sin x+1.$

12. $y_1=2c_1e^{-x}-2c_2e^{-5x}-\dfrac{23\sin 2x+14\cos 2x}{145},$

$y_2=c_1e^{-x}+c_2e^{-5x}+\dfrac{26\sin 2x-22\cos 2x}{145}.$

13. $y_1=\sin x+c_1x+c_2,\ y_2=-\frac{1}{6}x^3-\frac{1}{2}x^2+c_3x+c_4.$

14. $y_1=\sin x,\ y_2=c_1x+c_2.$ **15.** $y_1=\sin x,\ y_2=0.$

16. No solution exists.

chapter four

exercises *4.1–4.2, p. 110*

4. $\dfrac{2}{s^3}-\dfrac{2}{s^2}-\dfrac{5}{s},\ s>0.$ **5.** $\dfrac{s^2+2a^2}{s(s^2+4a^2)}.$ **6.** $\dfrac{1}{s}(1+2e^{-4s}).$

7. $\frac{1}{s^2}(1-e^{-2s})+\frac{2}{s}(e^{-2s}-2e^{-5s})$.

8. $\frac{e^{-\pi s}+1}{s^2+1}$. 9. $\frac{1}{s}(e^{-as}-e^{-bs})$. 10. $\frac{s-1}{s^2-2s+2}$.

exercises 4.3, p. 118

1. $\frac{s}{s^2+4}$ 2. $\frac{s-1}{s^2-2s+17}$. 3. $\frac{2}{(s+1)^3}$.

4. $\frac{s^2-4s+6}{(s-4)^4}$. 5. $\frac{1}{s-2}(1+2e^{-4s+8})$.

6. $\frac{1}{(s+1)^2}(1-e^{-2s-2})+\frac{2}{s+1}(e^{-2s-2}-2e^{-5s-5})$.

7. $\frac{3}{s^2+4s+13}$. 8. $\frac{1}{s^2+1}$. 9. $\frac{1}{s^2+1}-\frac{\sin 1}{s}$.

10. $\frac{2-s^2}{2s^3}$. 11. $\frac{2-(s-a)^2}{2(s-a)^3}$. 12. $\frac{1}{3}(1-e^{-3t})$.

13. $\frac{1}{9}(-1+\cosh 3t)$. 15. $\frac{1}{(1-e^{-\pi s})(s^2+1)}$.

16. $\frac{a}{(s^2+a^2)(1-e^{-\pi s/a})}$.

17. $\frac{1}{1-e^{-\pi s}}\left(\frac{-\pi^2 e^{-s\pi}}{s}-\frac{2\pi e^{-s\pi}}{s^2}-\frac{2e^{-s\pi}-2}{s^3}\right)$.

18. $\frac{(e^{-s}-2)e^{-s}+1}{(1-e^{-2s})s^2}$. 19. $\frac{2(e-{}^{Ls/2}-1)^2}{s(1-e^{-Ls})}$.

20. $\frac{2(e^{-s}-1)^2(e^{-s}+1)}{s(1-e^{-3s})}$.

exercises 4.4, p. 124

1. $e^{-t}-e^{-2t}$. 2. $\frac{1}{2}-e^{-t}+\frac{e^{-2t}}{2}$.

3. te^{-2t}. 4. $e^{-2t}\frac{\sinh\sqrt{2}t}{\sqrt{2}}$.

5. $e^{-2t}\cosh\sqrt{2}t - \sqrt{2}e^{-2t}\sinh\sqrt{2}t.$

6. $e^{-2t}(2t^2 - 4t + 1).$

7. $e^{-at}\cos bt - \dfrac{a}{b}e^{-at}\sin bt.$ **8.** $e^{-at}\sin bt.$

9. $\frac{1}{2}e^{-t} - 4e^{-2t} + \frac{9}{2}e^{-3t}.$ **10.** $e^{-2t}t^2(1 - \frac{1}{2}t).$

11. $-\frac{1}{9}e^{t} + \frac{1}{3}te^{t} + \frac{1}{9}e^{-2t}.$

12. $$-\frac{1}{64}e^{3t} + \frac{1}{24}te^{3t} + \frac{1}{64}e^{-3t/2}\left(\cos\frac{\sqrt{15}}{2}t - \frac{3}{\sqrt{15}}\sin\frac{\sqrt{15}}{2}t\right) + \frac{5}{48\sqrt{15}}e^{-3t/2}\sin\frac{\sqrt{15}}{2}t.$$

13. $e^{t}(3 + 2t) - 2e^{2t}.$ **14.** $7 - 3t - 7e^{-t}.$ **15.** $4t - \sin t.$

exercises 4.5, p. 128

1. $y = \frac{2}{3}e^{x} - \frac{2}{3}e^{-2x}.$ **2.** $y = \frac{1}{2}\sin 2x + \cos 2x.$

3. $y = \cosh 4x.$ **4.** $y = 2e^{-x/2}.$

5. $y = \frac{1}{4}(e^{x} - e^{-3x}).$ **6.** $y = \dfrac{11}{9}e^{3x} - \dfrac{2x}{3} + \dfrac{7}{9}.$

7. $y = e^{x}.$

8. $$y = -\frac{1}{8} + \frac{x}{4} + e^{-x}\left(\frac{17}{8}\cos\sqrt{3}x + \frac{23}{8\sqrt{3}}\sin\sqrt{3}x\right).$$

9. $y = -\frac{3}{4} + \frac{1}{2}x - \frac{1}{4}e^{-2x}.$

10. $y = \dfrac{5}{3}\sin x + \cos x - \dfrac{\sin 2x}{3}.$ **11.** $y = \cos ax.$

12. $y = \dfrac{1}{a}\sin ax.$ **13.** $i = 3e^{-t} - \frac{1}{2}e^{-2t} + t - \frac{3}{2}.$

14. $$i = \frac{22}{25}e^{-2t} + \frac{3}{5}te^{-2t} + \frac{3}{25}\cos t + \frac{4}{25}\sin t.$$

15. $$y = -\frac{73}{104}e^{-5t} + \frac{29}{8}e^{-t} + \frac{3}{26}\sin t + \frac{\cos t}{13}.$$

17. $y_1(t) = 0,\ y_2(t) = 1.$

18. $y_1(t) = \cos t + t - 1,\ y_2(t) = \sin t + t^2.$

19. $y_1(t) = \sin t + t^2,\ y_2(t) = \cos t - t^3.$

exercises 4.6, p. 135

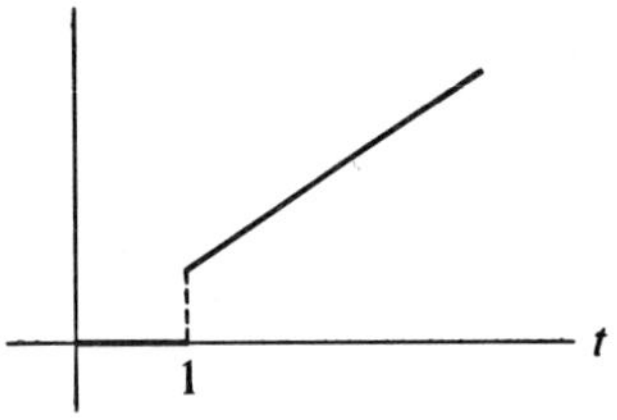

1.

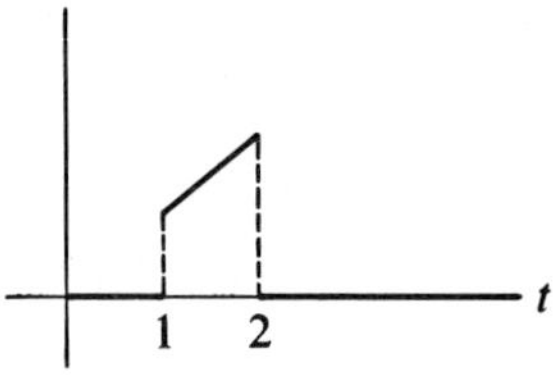

2.

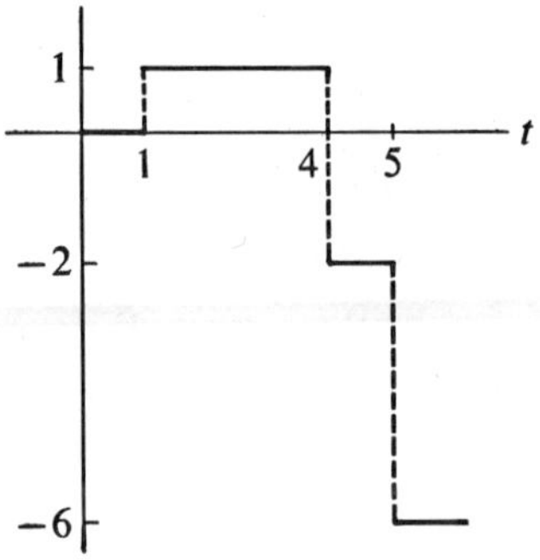

3.

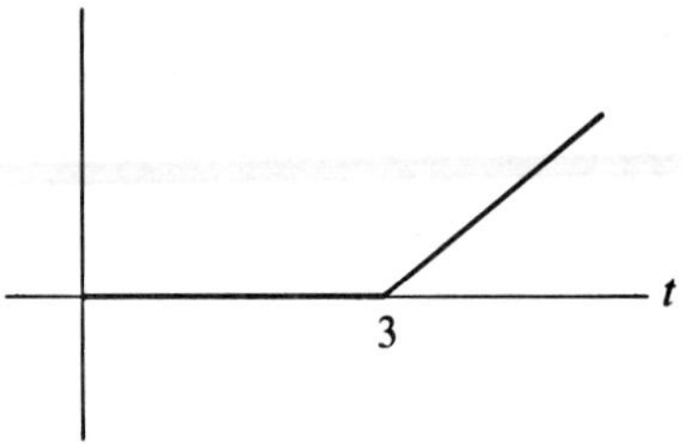

4.

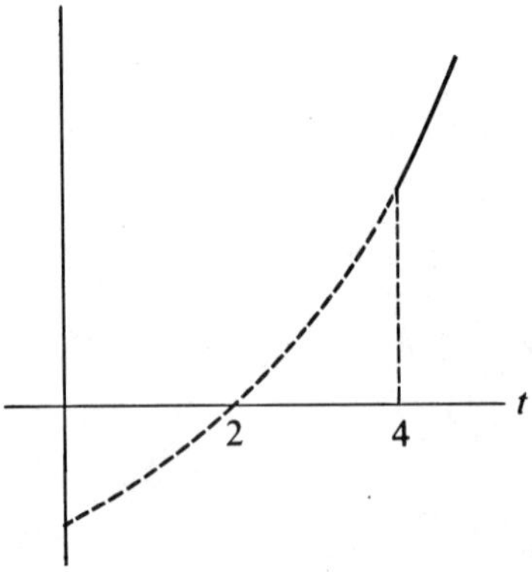

5.

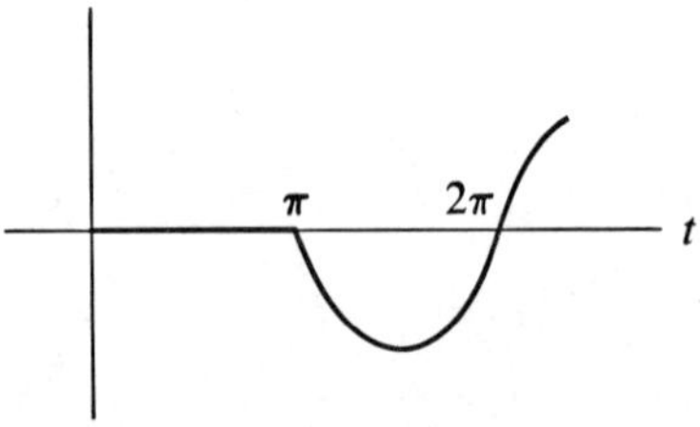

6.

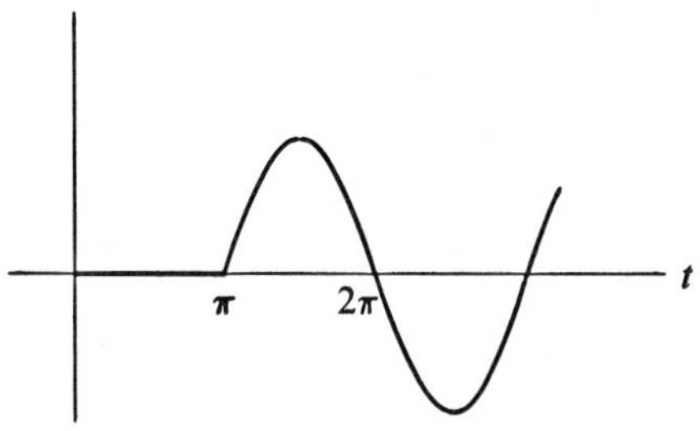

7.

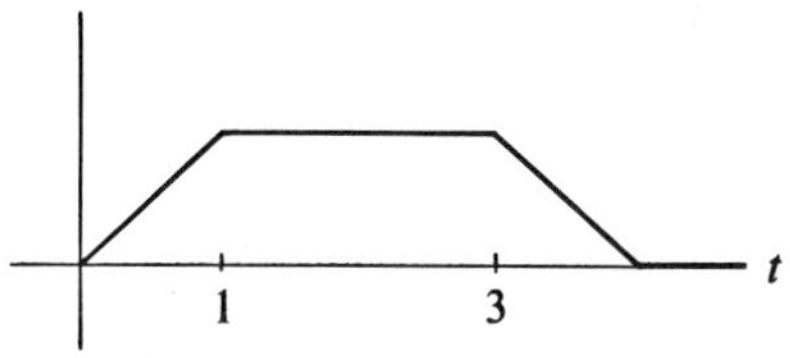

8.

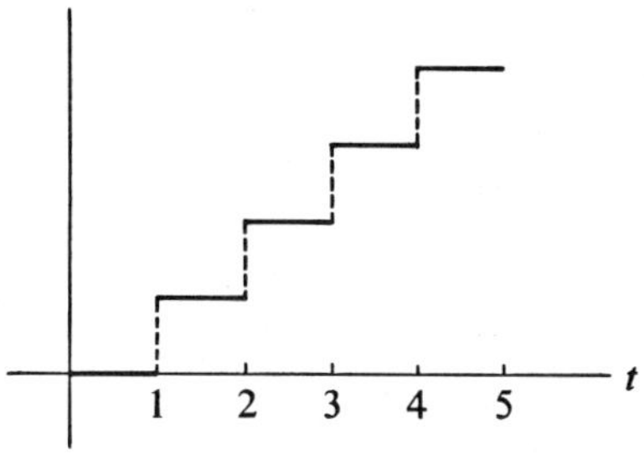

9.

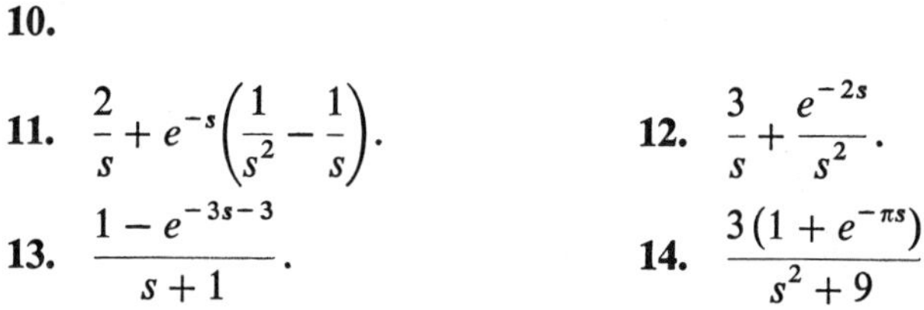

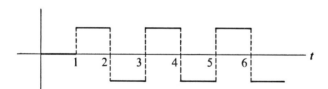

10.

11. $\dfrac{2}{s} + e^{-s}\left(\dfrac{1}{s^2} - \dfrac{1}{s}\right).$

12. $\dfrac{3}{s} + \dfrac{e^{-2s}}{s^2}.$

13. $\dfrac{1 - e^{-3s-3}}{s+1}.$

14. $\dfrac{3(1 + e^{-\pi s})}{s^2 + 9}.$

15. $\dfrac{2}{s^3} - e^{-3s}\left[\dfrac{2}{s^3} + \dfrac{6}{s^2}\right] - \dfrac{9e^{-5s}}{s}.$

16. $\dfrac{1}{s^2}(1 - e^{-s} - e^{-3s} + e^{-4s}).$

17. $\dfrac{1}{s^2}(1 - 2e^{-2s} + e^{-4s}).$

18. $\dfrac{1}{s^2} - \dfrac{1}{s}(e^{-s} + e^{-2s} + e^{-3s} + \cdots) = \dfrac{1}{s^2} - \dfrac{e^{-s}}{s(1 - e^{-s})}.$

19. $\frac{1}{s^2}(1-2e^{-s}+2e^{-3s}-2e^{-5s}+2e^{-7s}-\cdots)=\frac{1}{s^2}-\frac{2e^{-s}}{s^2(1+e^{-2s})}$.

20. $\frac{1}{s^2}-2\frac{e^{-s}}{s^2}+\frac{e^{-3s}}{s^2}-\frac{e^{-4s}}{s}$.

21. $(t-3)\,u_3$.

22. $\frac{1}{2}(t-4)^2u_4-\frac{1}{2}(t-1)^2u_1$.

23. $\frac{1}{2}\sin 2(t-1)u_1$.

24. $(e^{2(t-1)}-e^{(t-1)})u_1$.

25. $\cos 2(t-2)u_2$.

26. $e^{-2s}\left(\frac{1}{s^2}+\frac{2}{s}\right)$.

27. $e^{-3s}\left(\frac{2}{s^3}+\frac{6}{s^2}+\frac{9}{s}\right)$.

28. $-\frac{se^{-\pi s}}{s^2+1}$.

29. $e^{-2s}\left(\frac{2}{s^3}-\frac{2}{s^2}+\frac{1}{s}\right)$.

30. $\frac{e^{-3(s-1)}}{s-1}$.

31. $\frac{V_0}{R}(1-e^{-Rt/L})$.

32. $\frac{h}{R}(e^{-(t-a)/RC}u_a-e^{-(t-b)/RC}u_b)$.

exercises 4.7, p. 140

1. $\frac{2}{s^2(s^2+4)}$.

2. $\frac{s}{(s+1)(s^2+1)}$.

3. $1-\cos t$.

4. e^t-t-1.

5. $t+\frac{t^3}{6}$.

6. t.

7. $e^{-t}-e^{-2t}$.

8. $t^2+\frac{1}{12}t^4$.

9. $\frac{1}{6}e^{2t}-\frac{1}{6}e^{-t}\cos\sqrt{3}t+\frac{1}{2\sqrt{3}}\sin\sqrt{3}t$.

10. $t^2-\frac{1}{3}t^4$.

11. $t+\frac{t^2}{2}$.

12. $1+\frac{t^2}{2}$.

chapter five

exercises 5.1–5.3, p. 146

1. $y=1+x+x^2+\frac{x^3}{3}+\frac{x^4}{12}+\cdots=2e^x-x-1$.

2. $y = 1 + x + \frac{x^2}{2} + \frac{x^3}{6} + \frac{x^4}{24} + \cdots = e^x.$

3. $y = e^x.$

4. $y = 1 - 2x + 2x^2 - x^3 + \frac{1}{2}x^4 - \frac{1}{5}x^5 + \cdots = \frac{3}{4}e^{-2x} + \frac{1}{2}x^2 - \frac{1}{2}x + \frac{1}{4}.$

5. $y = 1 + x - \frac{7}{2}x^2 + \frac{25}{6}x^3 - \frac{79}{24}x^4 + \cdots = 2e^{-x} - e^{-3x}.$

6. $y = 2 + 4x + 8x^2 + 16x^3 + 32x^4 + \cdots = \frac{2}{1-2x}.$

7. $y = 2 - 2(x-1) + (x-1)^2 - \frac{(x-1)^4}{12} + \frac{1}{20}(x-1)^5 - \cdots.$

8. $y = 1 - (x-1) + \frac{1}{2}(x-1)^2 - \frac{1}{2}(x-1)^3 + \frac{3}{8}(x-1)^4 - \cdots.$

9. $y = 1 + (x-1) + \frac{1}{6}(x-1)^3 - \frac{1}{24}(x-1)^4 + \frac{1}{40}(x-1)^5 - \cdots.$

10. $y = 1 + ex + \frac{1}{2}e^2x^2 + \frac{1}{3}e^3x^3 + \frac{1}{4}e^4x^4 + \cdots + \frac{e^nx^n}{n} + \cdots = \ln\left(\frac{e}{1-ex}\right).$

11. $y = 1 + x^2 - x^3 + \frac{1}{2}x^4 - \frac{1}{6}x^5 + \frac{1}{24}x^6 - \cdots + (-1)^n \frac{x^n}{(n-2)!} + \cdots.$

12. $y = (1 + 2x^2 + 3x^4 + 4x^6 + \cdots + (n+1)x^{2n} + \cdots)$

$$+ \left(x + \tfrac{5}{3}x^3 + \tfrac{7}{3}x^5 + \cdots + \frac{2n+3}{3}x^{2n+1} + \cdots\right).$$

13. $y = 1 + x + 3x^2 + \frac{2}{3}x^3 + \frac{3}{8}x^4.$

14. $y = x - \frac{1}{3}x^3 + \Big(1 - \frac{3}{2}x^2 + \frac{1}{8}x^4 + \frac{1}{240}x^6 + \cdots$

$$+ \frac{3}{2^n n!(2n-1)(2n-3)}x^{2n} + \cdots\Big).$$

15. $y = \left(1 + \frac{3}{8}x^2 + \left(\frac{3x^2}{8}\right)^2\left(\frac{1}{2!}\right) + \cdots + \frac{1}{n!}\left(\frac{3x^2}{8}\right)^n + \cdots\right)$

$$+ \left(x + \frac{1}{4}x^3 + \cdots + \frac{n!}{(2n+1)!}\left(\frac{3}{2}\right)^n x^{2n+1} + \cdots\right).$$

exercises 5.4, p. 154

1. $y = a_0 \sum_{n=0}^{\infty} \frac{(-1)^n x^n}{n!}.$

2. $y = \sum_{n=0}^{\infty} \frac{x^n}{n!}\left[a_0(-1)^n + \frac{1}{2}\right].$

3. $y = \sum_{n=0}^{\infty} \frac{(-x)^n}{n!} [a_0 + x]$.

4. $y = \sum_{n=0}^{\infty} \frac{x^{2n}}{(2n)!} \left[a_0 - \frac{(-1)^n}{2}\right] + \sum_{n=0}^{\infty} \frac{x^{2n+1}}{(2n+1)!} \left[-a_0 + \frac{(-1)^n}{2}\right]$.

5. $y = \sum_{n=0}^{\infty} \frac{x^{2n}}{(2n)!} \left[a_0 + \frac{(-1)^{n+1}}{2}\right] + \sum_{n=0}^{\infty} \frac{x^{2n+1}}{(2n+1)!} \left[a_0 + \frac{(-1)^{n+1}}{2}\right]$.

6. $y = a_0 \sum_{n=0}^{\infty} \frac{(-1)^n}{2^n n!} x^{2n} + \sum_{n=0}^{\infty} \frac{(-1)^n 2^n n!}{(2n+1)!} x^{2n+1}$.

7. $y = a_0 \sum_{n=0}^{\infty} \frac{(-1)^n}{n! 3^n} x^{3n}$.

8. $y = a_0 \sum_{n=0}^{\infty} \frac{(-1)^n}{(2n)!} x^{2n} + a_1 \sum_{n=0}^{\infty} \frac{(-1)^n}{(2n+1)!} x^{2n+1}$.

9. $y = \sum_{n=0}^{\infty} a_{3n} x^{3n} + \sum_{n=0}^{\infty} a_{3n+1} x^{3n+1}$ where

$$a_{3n} = \frac{(-1)^n (3n-2)(3n-5)(3n-8)(3n-11)\cdots(10)(7)(4)(1)}{(3n)!} a_0$$

and $a_{3n+1} = \frac{(-1)^n (3n-1)(3n-4)(3n-7)\cdots(11)(8)(5)(2)}{(3n+1)!} a_1$.

10. $y = \sum_{n=0}^{\infty} a_{4n} x^{4n} + \sum_{n=0}^{\infty} a_{4n+1} x^{4n+1}$ where

$$a_{4n} = \frac{(-1)^n a_0}{(4n)(4n-1)(4n-4)(4n-5)\cdots(16)(15)(12)(11)(8)(7)(4)(3)}$$

and a_{4n+1}

$$= \frac{(-1)^n a_1}{(4n+1)(4n)(4n-3)(4n-4)\cdots(17)(16)(13)(12)(9)(8)(5)(4)}.$$

11. $y = \sum_{n=0}^{\infty} a_{5n} x^{5n} + \sum_{n=0}^{\infty} a_{5n+1} x^{5n+1}$ where

$$a_{5n} = \frac{(-1)^n a_0}{(5n)(5n-1)(5n-5)(5n-6)\cdots(15)(14)(10)(9)(5)(4)}$$ and

$$a_{5n+1} = \frac{(-1)^n a_1}{(5n+1)(5n)(5n-4)(5n-5)\cdots(16)(15)(11)(10)(6)(5)}.$$

12. $$y = a_0\left[1 - \frac{5}{2}x^2 + \frac{15}{8}x^4 - \frac{5}{16}x^6 - \sum_{n=4}^{\infty} \frac{15(2n-7)!}{2^{2n-4}n!(n-4)!}x^{2n}\right] + a_1(x - \tfrac{4}{3}x^3 + \tfrac{8}{15}x^5).$$

13. $$y = \sum_{n=0}^{\infty} \frac{x^{2n}}{(2n)!}[a_0(-1)^n + \tfrac{1}{2}] + \sum_{n=0}^{\infty} \frac{x^{2n+1}}{(2n+1)!}[a_1(-1)^n + \tfrac{1}{2}].$$

14. $$y = a_1 x + a_0\left[1 + \frac{1}{2}x^2 + \sum_{n=2}^{\infty} \frac{(2n-3)!}{2^{2n-2}n!(n-2)!(2n-1)}x^{2n}\right].$$

15. $$y = y_1 + \sum_{n=1}^{\infty} \frac{2^{2n-2}[(n-1)!]^2}{(2n+1)!}x^{2n+1}$$
where y_1 is the solution to Exercise 14.

16. $$y = a_0 \sum_{n=0}^{\infty} \frac{(-1)^{n+1}15}{(2n-1)(2n-3)(2n-5)}x^{2n} + a_1[x + 2x^3 + x^5].$$

17. $$y = a_0\left(1 + 2x^2 + \frac{x^4}{3} + a_1 \sum_{n=0}^{\infty} \frac{(-1)^n 3}{(2n+1)(2n-1)(2n-3)n!2^n}x^{2n+1}\right).$$

18. $$y = y_1 + \sum_{n=1}^{\infty} \frac{(-1)^n}{(2n+1)(2n-1)(2n-3)n!2^n}x^{2n+1},$$
where y_1 is the solution to Exercise 17.

19. $$y = a_0\left[1 + \frac{3}{2}x^2 + \sum_{n=2}^{\infty} (-1)^{n+1}\frac{(2n+1)(2n-3)!}{2^{2n-2}n!(n-2)!}x^{2n}\right] + a_1 x.$$

exercises 5.5–5.6, p. 158

1. R.S.P. at 0.
2. R.S.P. at 0, -1, I.S.P. at $+1$.
3. R.S.P. at 0, I.S.P. at 1.
4. I.S.P. at 0.
5. I.S.P. at 2.
6. $$y = a_0 \sum_{n=0}^{\infty} \frac{(-1)^n}{(2n)!}x^n + b_0 x^{1/2} \sum_{n=0}^{\infty} \frac{(-1)^n}{(2n+1)!}x^n.$$
7. $$y = a_0(1 + \tfrac{2}{3}x + \tfrac{1}{3}x^2) + b_0 \sum_{n=4}^{\infty} (n-3)x^n.$$

8. $y = a_0 x^{1/2} + b_0 x^{1/2} \sum_{n=1}^{\infty} \frac{x^n}{2^{n-1} n!}$.

9. $y = \frac{1}{2} a_0 \sum_{n=0}^{\infty} (-1)^n (n+1)(n+2) x^n$. 10. $y = a_0 (1 + x)$.

11. $y = a_0 \sum_{n=0}^{\infty} \frac{(-1)^n x^{2n+1}}{2^{2n} n! (n+1)!}$. 12. $y = a_0 \sum_{n=0}^{\infty} \frac{(-1)^n x^{2n+2}}{2^{2n-1} n! (n+2)!}$.

13. $y = a_0 x^{-1/2} \sum_{n=0}^{\infty} \frac{x^n}{2^n n!} + b_0 x +$

$$b_0 x \sum_{n=1}^{\infty} \frac{x^n}{(2n+3)(2n+1)(2n-1)\cdots(17)(15)(13)(11)(9)(7)(5)}.$$

14. $y = a_0 \sum_{n=0}^{\infty} \frac{(-1)^n x^{2n}}{2^{2n} (n!)^2}$. 15. $y = a_0 \sum_{n=0}^{\infty} \frac{(-1)^n x^{n+1}}{n! (n+1)!}$.

16. $y = b_0 x^{1/2} \sum_{n=0}^{\infty} \frac{x^n}{2^n n!}$.

17. $y = \sum_{n=0}^{\infty} a_n x^n$, where a_0 is arbitrary, $a_1 = 0$, and

$$a_n = \frac{1}{n(n+2)} [a_{n-2} - 2(n-1)(n-2) a_{n-1}], \quad n \geqq 2.$$

18. $y = x^{-1} \sum_{n=0}^{\infty} a_n x^n + x^{1/3} \sum_{n=0}^{\infty} b_n x^n$, where a_0 is arbitrary, $a_1 = a_0$, and

$$a_n = \frac{1}{n(3n-4)} [(n-2) a_{n-1} - 2a_{n-2}], \quad n \geqq 2;$$

b_0 is arbitrary, $b_1 = \frac{1}{21} b_0$, and

$$b_n = \frac{-1}{n(3n+4)} \left[\frac{3n-2}{3} b_{n-1} - 2b_{n-2} \right], \quad n \geqq 2.$$

19. $y = \sum_{n=0}^{\infty} a_n x^n + x^{3/2} \sum_{n=0}^{\infty} b_n x^n$, where a_0 is arbitrary, $a_1 = a_0$,

$$a_2 = -\tfrac{1}{2} a_0, \quad a_n = \frac{1}{n(2n-3)} [(n-3) a_{n-3} - a_{n-1}], \quad n \geqq 3;$$

and b_0 is arbitrary, $b_1 = -\frac{1}{5} b_0$, $b_2 = \frac{1}{70} b_0$,

$$b_n = \frac{1}{n(2n+3)} \left[\frac{2n-3}{2} b_{n-3} - b_{n-1} \right], \quad n \geqq 3.$$

20. $y = x^{-1} \sum_{n=0}^{\infty} a_n x^n + x^{1/3} \sum_{n=0}^{\infty} b_n x^n$, where a_0 is arbitrary, $a_1 = -a_0$,

$$a_n = \frac{1}{n(3n-4)}[a_{n-1} - a_{n-2}(n-3)(n-4)],\ n \geqq 2;$$

and b_0 is arbitrary, $b_1 = \frac{1}{7} b_0$,

$$b_n = \frac{1}{n(3n+4)}[b_{n-1} - b_{n-2}(n - \tfrac{8}{3})(n - \tfrac{5}{3})],\ n \geqq 2.$$

chapter six

exercises 6.1–6.2, p. 168

1. Expand and use the corresponding property for real numbers.
2. $\mathbf{A} \cdot \mathbf{B} = -10$.
3. $(a_1 b_2 - a_2 b_1)(\mathbf{e}_1 \times \mathbf{e}_2) + (a_1 b_3 - a_3 b_1)(\mathbf{e}_1 \times \mathbf{e}_3) + (a_2 b_3 - a_3 b_2)(\mathbf{e}_2 \times \mathbf{e}_3)$.
4. $2x + y - z = 6$.
5. 1 ft.lb.
6. $23/\sqrt{26}$.
7. $5x - 4y + z = 6$.
8. $-\frac{4}{3}(2\mathbf{i} - \mathbf{j} + \mathbf{k})$.
9. $A = 6$.
10. $\mathbf{M} = -4\mathbf{i} - \mathbf{j} + 7\mathbf{k}$.
11. 15.
12. $-31\mathbf{i} + 2\mathbf{j} - 29\mathbf{k}$.
13. -10.
14. Linearly independent.
15. Linearly dependent.
16. 0.
18. The value of $\mathbf{A} \cdot \mathbf{B} \times \mathbf{C}$ is unaffected by any cyclic permutation of the letters, **A, B, C**. A noncyclic permutation yields a result opposite in sign.

exercises 6.3–6.4, p. 173

2. $\mathbf{F}(t) = \sin t\mathbf{i} + 2\cos t\mathbf{j} + 3\mathbf{k}$, $\mathbf{F}(0) = 2\mathbf{j} + 3\mathbf{k}$.
3. $\mathbf{F}(u, v) = \cos u \sin v\mathbf{i} + \sin u \sin v\mathbf{j} + \cos v\mathbf{k}$, $\mathbf{F}(u + 2\pi, v) = \mathbf{F}(u, v + 2\pi) = \mathbf{F}(u, v)$.
4. $\mathbf{F}'(t) = \cos t\mathbf{i} - \sin 2t\mathbf{j} + 2t\mathbf{k}$.
5. $\mathbf{F}(t) = \left(\frac{t^2}{2} + c_1\right)\mathbf{i} + \left(\frac{t^3}{6} + c_2\right)\mathbf{j} + \left(\frac{t^4}{4} + c_3\right)\mathbf{k}$.
9. 2.
10. $\sqrt{(\mathbf{r}_u \cdot \mathbf{r}_u)(\mathbf{r}_v \cdot \mathbf{r}_v) - (\mathbf{r}_u \cdot \mathbf{r}_v)^2}$.

exercises 6.5–6.6, p. 183

1.

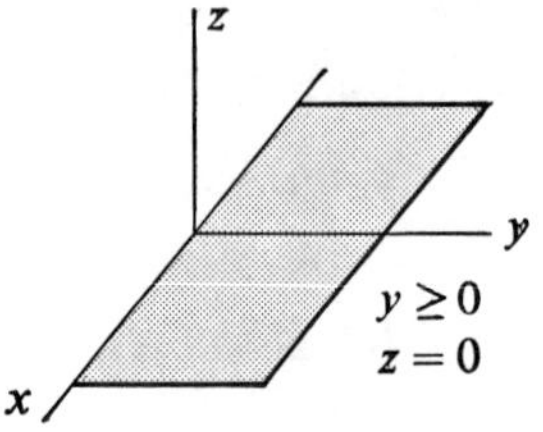

(a)

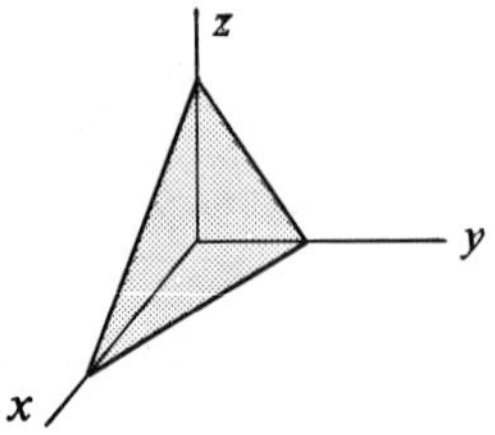

(b)

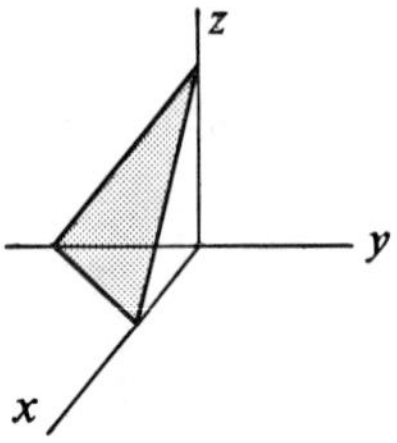

(c)

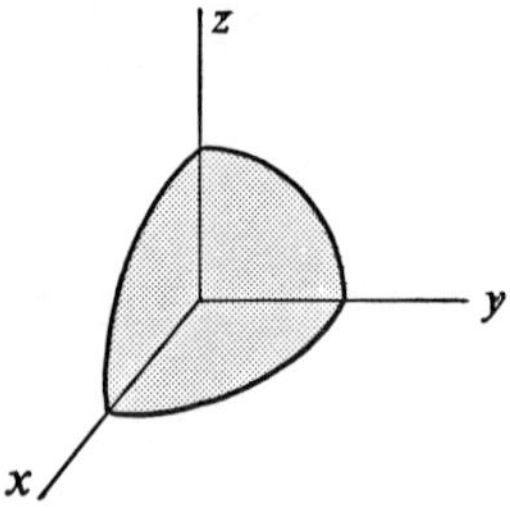

(d)

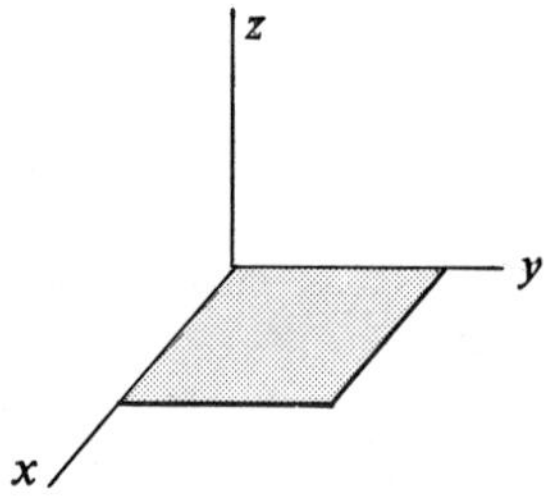

(e)

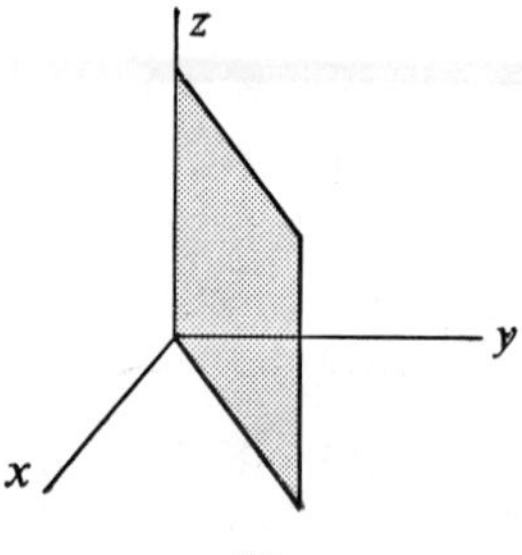

(f)

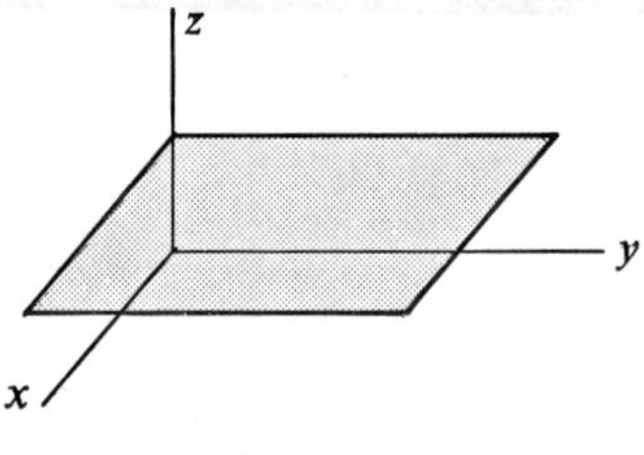

(g)

2.

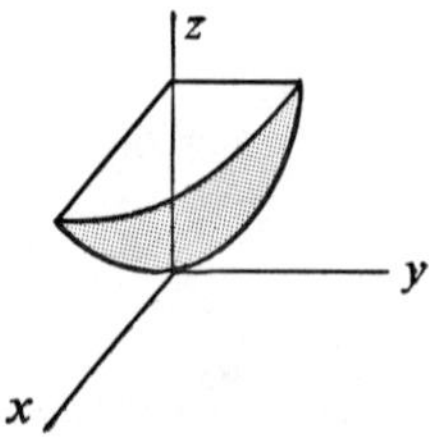

(a)

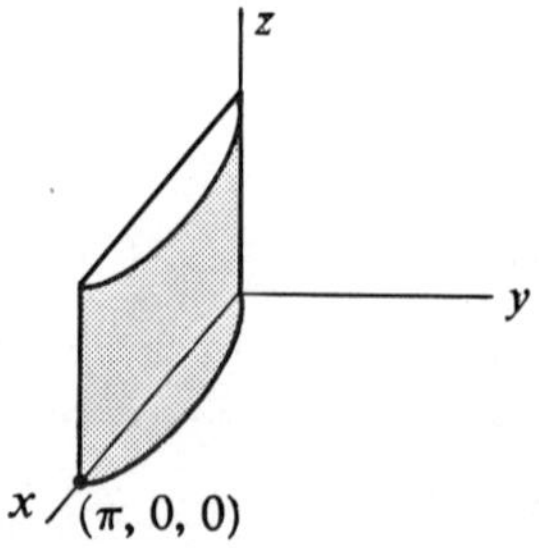

(b)

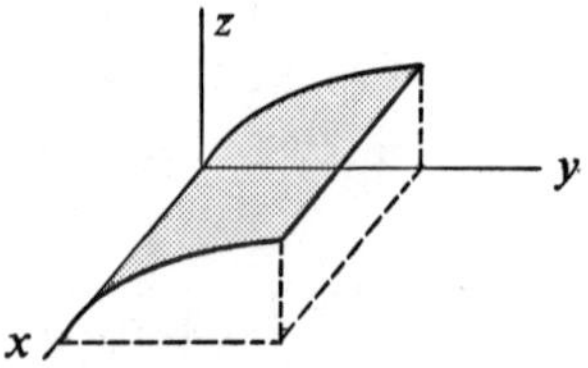

(c)

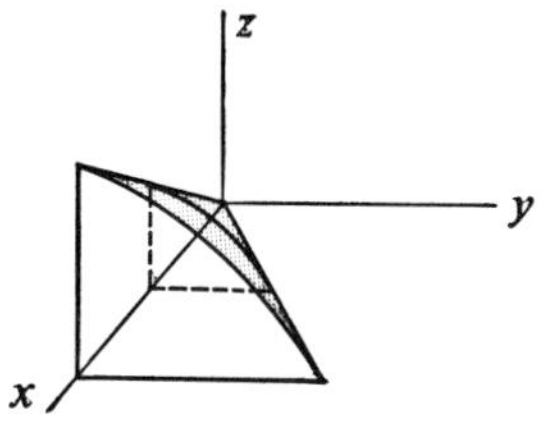

(d)

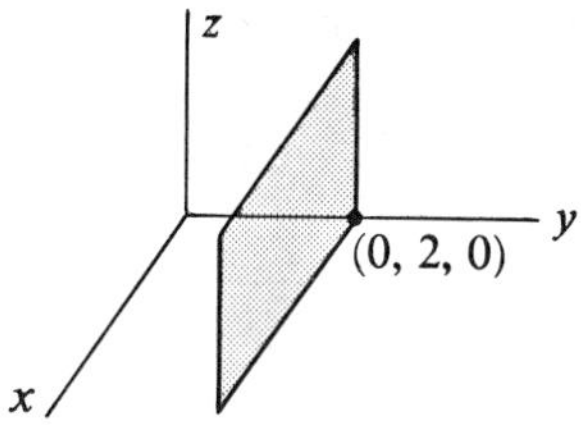

(e)

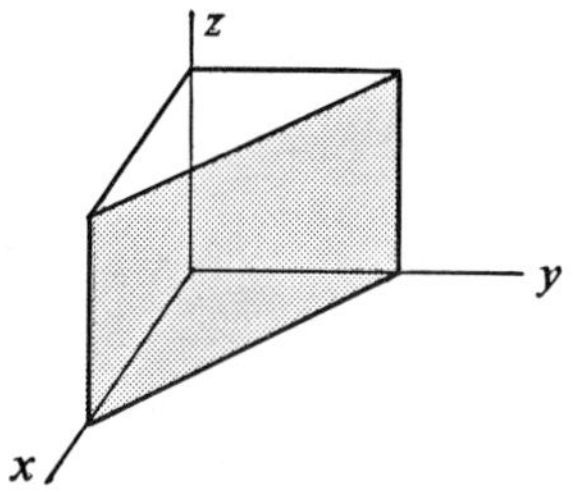

(f)

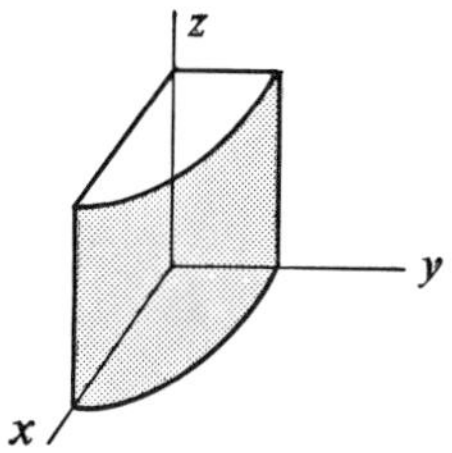

(g)

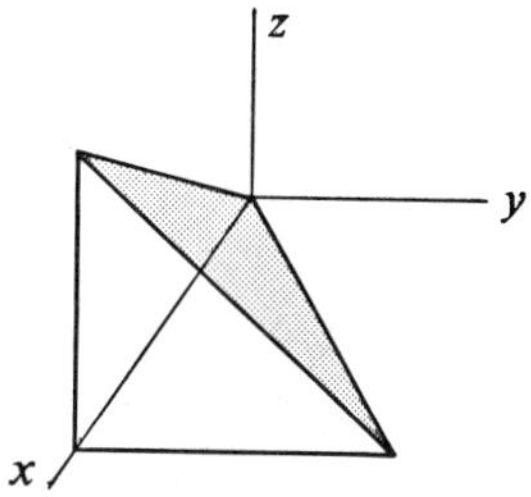

(h)

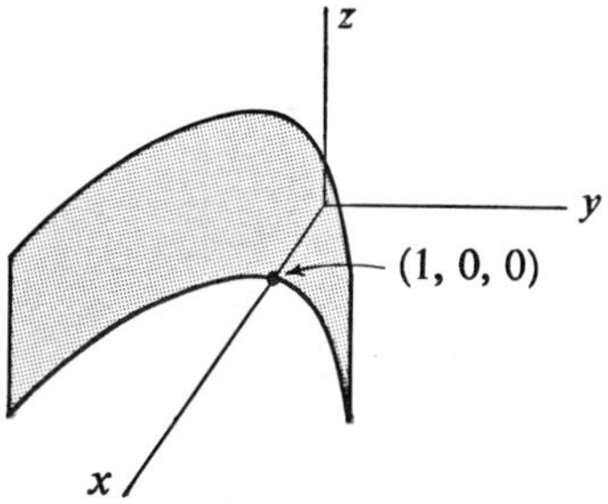

(i)

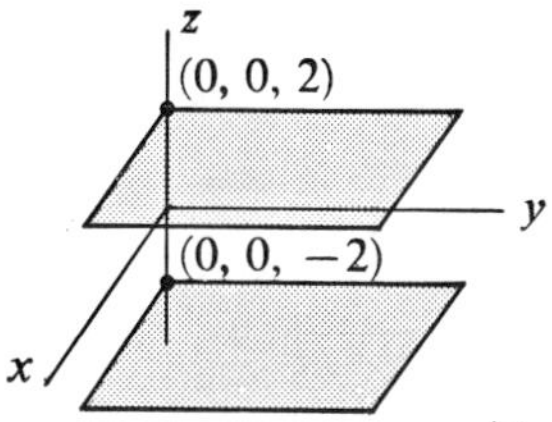

(j)

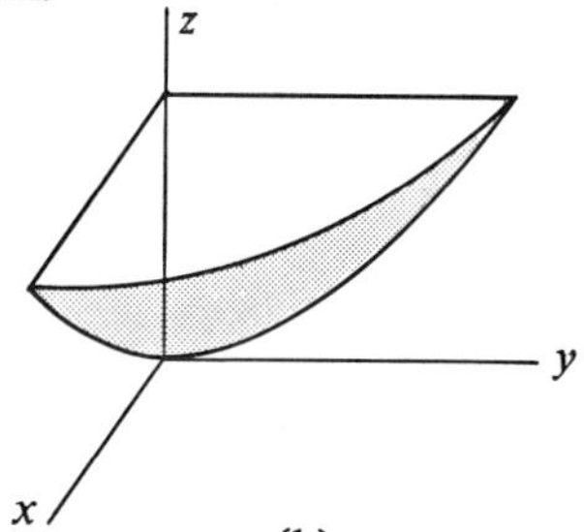

(k)

3.

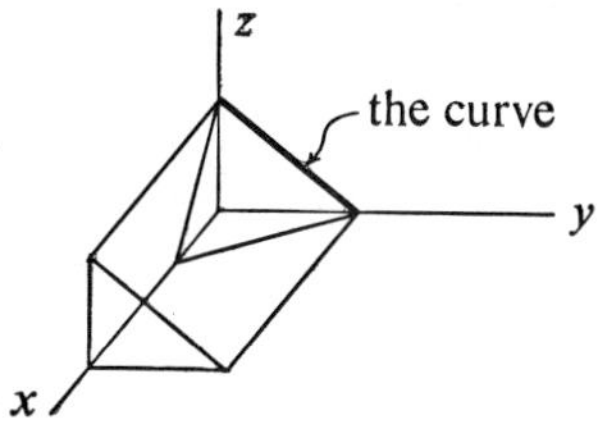

(a)

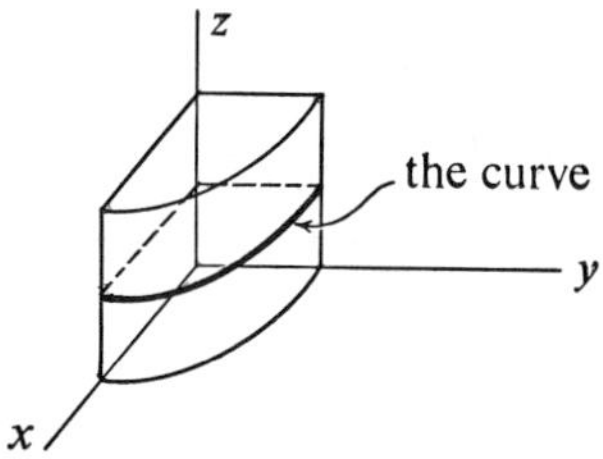

(b)

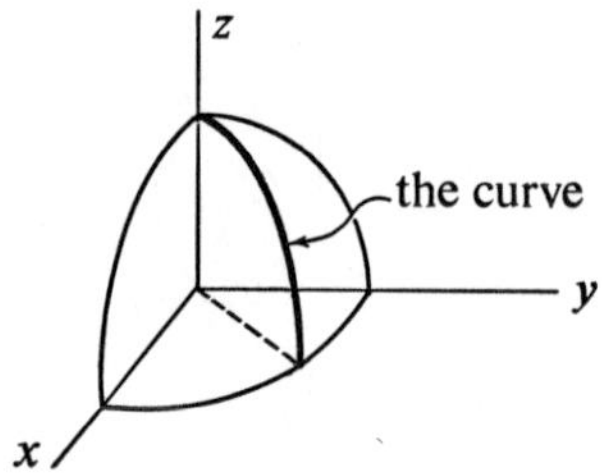

(c)

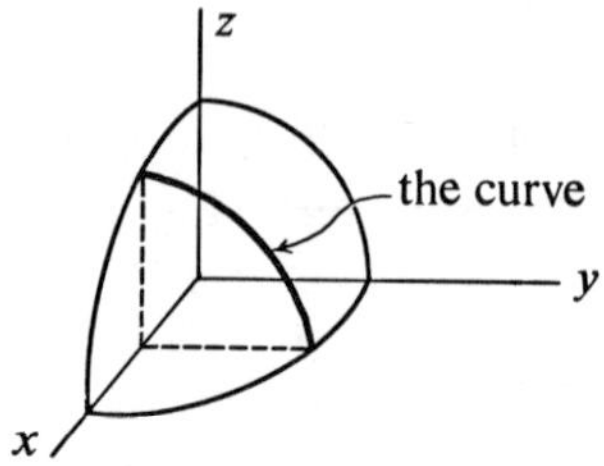

(d)

(e) No curve.

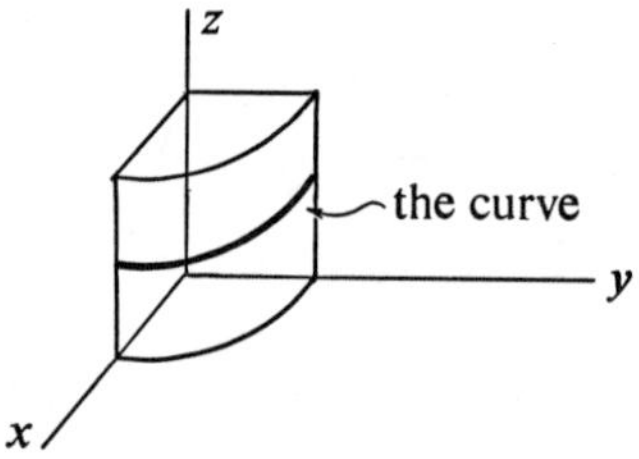

(f)

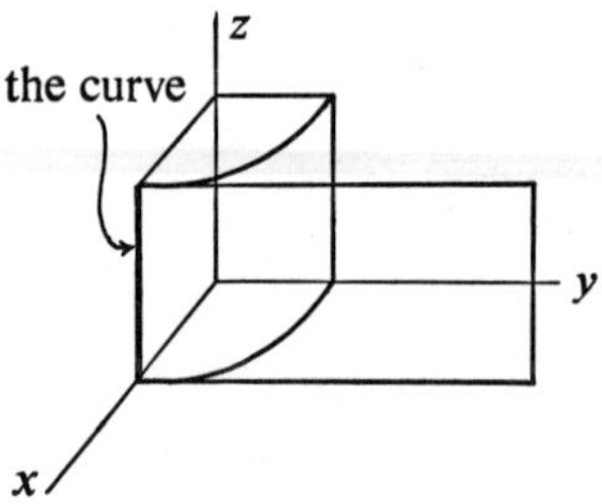

(g)

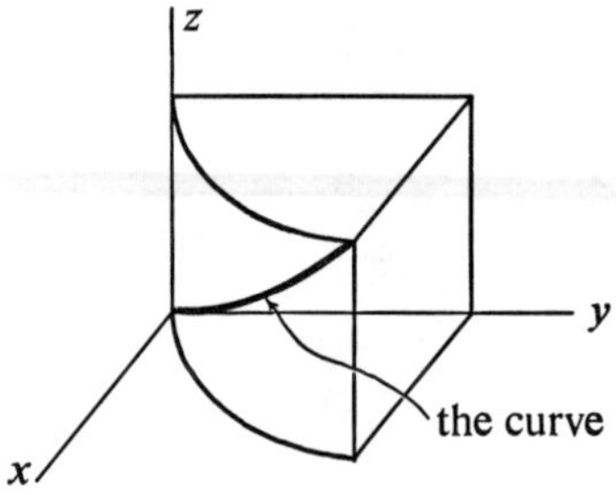

(h)

4.

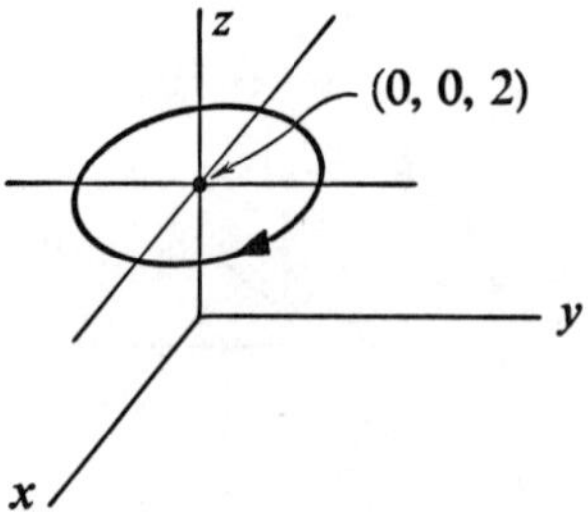

(a)

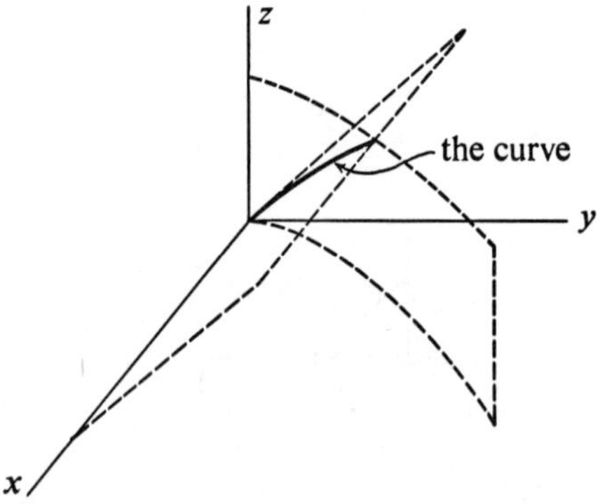

(b)

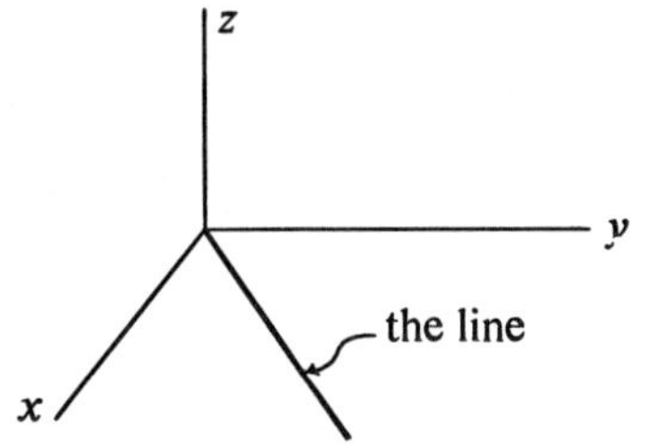

(c)

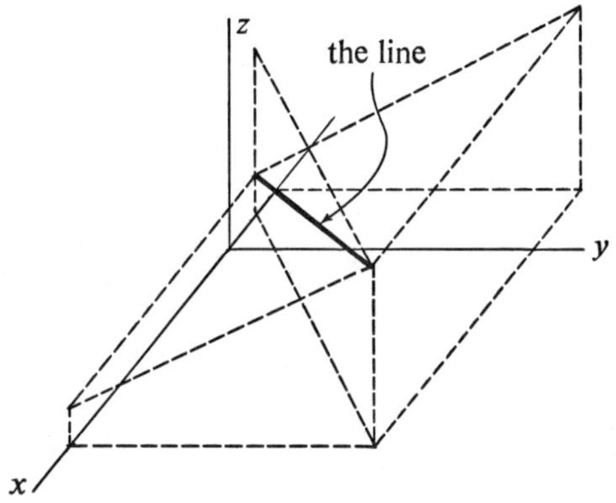

(d)

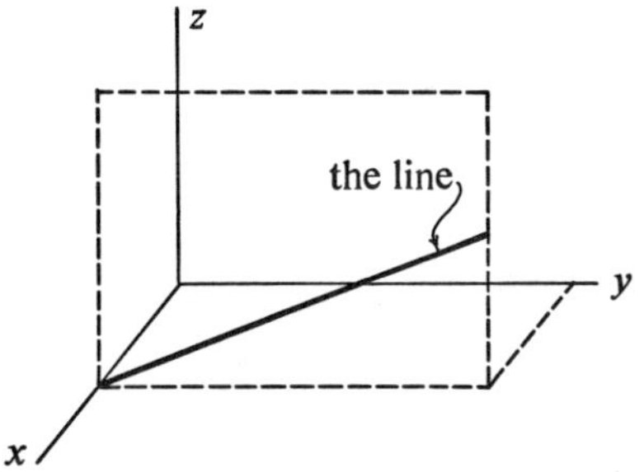

(e)

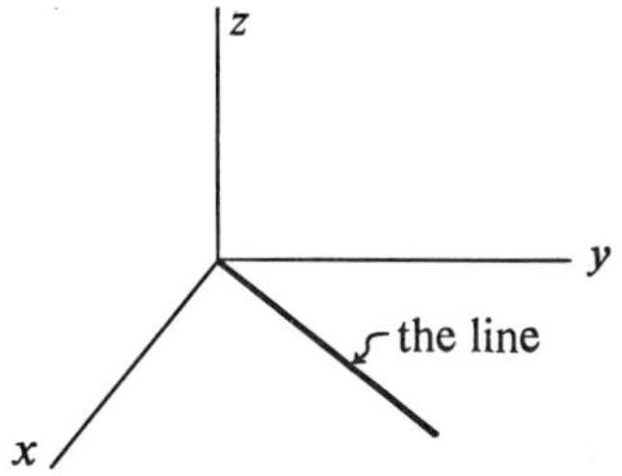

(f)

5. (a) $x = \sqrt{3}u \cos v,\ y = u \sin v,\ z = u^2.$
(b) $x = v,\ y = \sin v,\ z = u.$ (c) $x = u,\ y = v^2,\ z = v.$
(d) $x = u,\ y = u \sin v,\ z = u \cos v.$
(e) $x = u,\ y = 2,\ z = v.$ (f) $x = u,\ y = 2 - u,\ z = v.$
(g) $x = 2 \cos v,\ y = \sin v,\ z = u.$
(h) $x = u,\ y = v,\ z = u - v.$
(i) $x = \cosh v,\ y = \sinh v,\ z = u.$
(j) Two surfaces: $x = u,\ y = v,\ z = -2$; and $x = u,\ y = v,\ z = 2.$

6. (a) $x = 0,\ y = t,\ z = 1 - t.$ (b) $x = 4 \sin t,\ y = 4 \cos t,\ z = 2.$
(c) $x = \sqrt{2} \cos t,\ y = \sqrt{2} \cos t,\ z = 2 \sin t.$
(d) $x = 1,\ y = \sqrt{3} \cos t,\ z = \sqrt{3} \sin t.$
(e) No curve. (f) $x = 2 \cos t,\ y = 2 \sin t,\ z = 3.$
(g) $x = 2,\ y = 0,\ z = t.$ (h) $x = t,\ y = t^2,\ z = t^2.$

7. (a) $x^2+y^2=1,\ z=2.$ (b) $y=z,\ x=y^2.$
(c) $z=0,\ x=y,\ x\geqq 0.$ (d) $x-2y+1=0,\ z-3y-1=0.$
(e) $x=2,\ y=3z.$ (f) $y=2x,\ z=0.$

8. $\mathbf{R}(t)=(\mathbf{i}+2\mathbf{k})+(4\mathbf{i}+\mathbf{j}-\mathbf{k})\,t.$ **9.** $\mathbf{R}(t)=\mathbf{i}+(\mathbf{i}-2\mathbf{j})\,t.$

10. $\mathbf{R}(t)=a\cos t\mathbf{i}+b\sin t\mathbf{j}+\dfrac{ct}{2\pi}\mathbf{k}.$

11. By a constant vector in the direction of the line.

12. $\mathbf{R}(t)=(2-t)\,\mathbf{i}+(2-t)^2\,\mathbf{j},\ 0\leqq t\leqq 2.$

13. $\mathbf{R}(t)=(2\cos t)\,\mathbf{i}+(2+\sqrt{2}\sin t)\,\mathbf{j}+(2+\sqrt{2}\sin t)\,\mathbf{k}.$

14. $\mathbf{R}(u,v)=(1+4u-2v)\,\mathbf{i}+u\mathbf{j}+v\mathbf{k}.$

15. $\mathbf{R}(u,v)=(2\cos u\cos v)\,\mathbf{i}+2(1+\sin u\cos v)\,\mathbf{j}+(3+2\sin v)\,\mathbf{k}.$

16. $\mathbf{R}(u,v)=(2\cos u\cos v)\,\mathbf{i}+(\sin u\cos v)\,\mathbf{j}+(4\sin v)\,\mathbf{k}.$

17. $\mathbf{R}(u,v)=\sin u\cosh v\mathbf{i}+\tfrac{1}{2}\cos u\cosh v\mathbf{j}+\tfrac{1}{2}\sinh v\mathbf{k}.$

exercises 6.7–6.9, p. 196

1. $\mathbf{R}(s)=(4\mathbf{i}+\mathbf{j}+12\mathbf{k})\,s+(4\mathbf{i}+2\mathbf{j}+8\mathbf{k}).$

2. $\mathbf{T}=\dfrac{\mathbf{i}+2\mathbf{j}+3\mathbf{k}}{\sqrt{14}},\ \mathbf{N}=\dfrac{-11\mathbf{i}-8\mathbf{j}+9\mathbf{k}}{\sqrt{266}},\ \mathbf{B}=\dfrac{3\mathbf{i}-3\mathbf{j}+\mathbf{k}}{\sqrt{19}}.$

3. $\mathbf{T}=\dfrac{\mathbf{i}+\mathbf{k}}{\sqrt{2}},\ \mathbf{N}=-\mathbf{j},\ \mathbf{B}=\dfrac{\mathbf{i}-\mathbf{k}}{\sqrt{2}}.$

4. $\mathbf{R}(s)=(\mathbf{i}+\mathbf{k})\,s+(\mathbf{j}+2\pi\mathbf{k}).$

5.

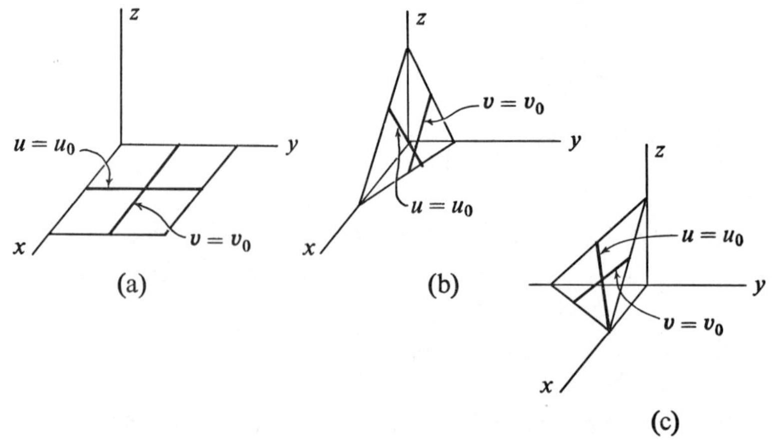

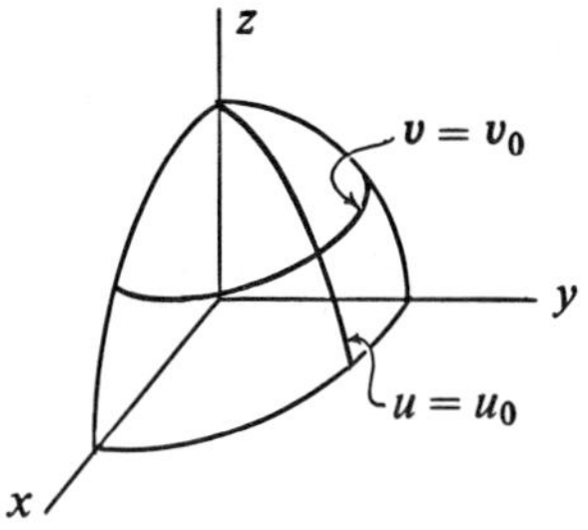

(d)

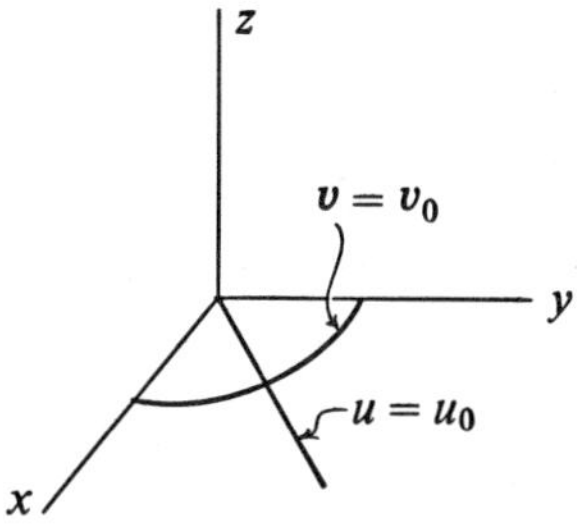

(e)

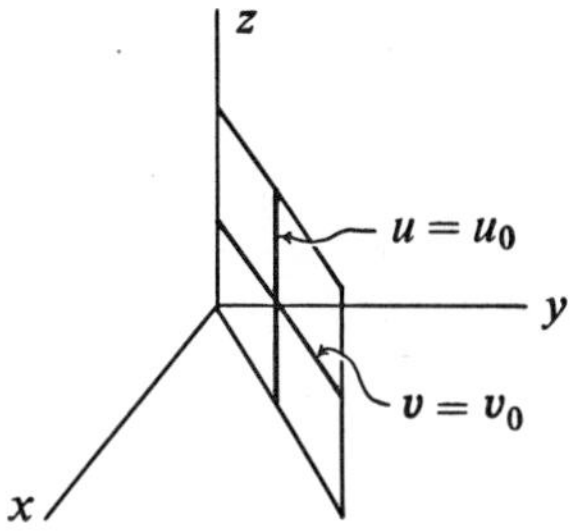

(f)

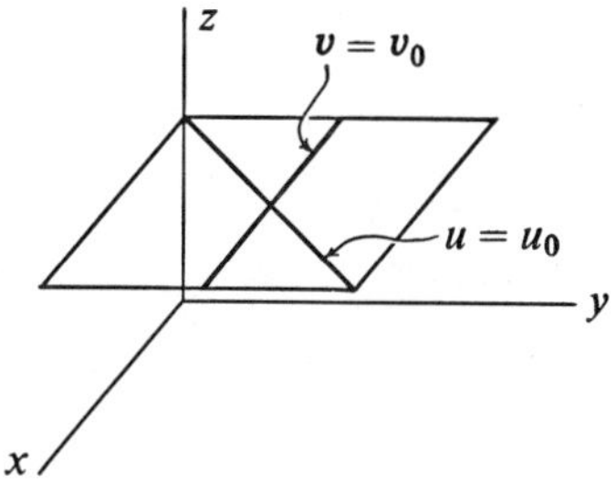

(g)

6. $\mathbf{i} + 3\mathbf{j} + \mathbf{k}$.

7. $\mathbf{i} - \mathbf{j} - \mathbf{k}$.

8. At $t = 2$: $\mathbf{v} = 4\mathbf{k}$, $\mathbf{a} = 2\pi^2\mathbf{i} - \pi^2\mathbf{j} + 2\mathbf{k}$, speed $= 4$, $\mathbf{a}_T = 2\mathbf{k}$, $\mathbf{a}_N = 2\pi^2\mathbf{i} - \pi^2\mathbf{j}$.

At $t = \frac{1}{4}$: $\mathbf{v} = \pi\mathbf{i} - \pi\dfrac{\sqrt{2}}{2}\mathbf{j} + \frac{1}{2}\mathbf{k}$, $\mathbf{a} = -\pi^2\dfrac{\sqrt{2}}{2}\mathbf{j} + 2\mathbf{k}$,

$$\text{speed} = \left(\pi^2 + \frac{\pi^2}{2} + \frac{1}{4}\right)^{1/2},$$

$$\mathbf{a}_T = \frac{(\pi^3 + 2)(2\pi\mathbf{i} - \pi\sqrt{2}\mathbf{j} + \mathbf{k})}{6\pi^2 + 1},$$

$$\mathbf{a}_N = -\frac{2\pi(\pi^3 + 2)}{6\pi^2 + 1}\mathbf{i} + \left(\frac{\pi^4\sqrt{2} + 2\pi\sqrt{2}}{6\pi^2 + 1} - \pi^2\frac{\sqrt{2}}{2}\right)\mathbf{j} + \left(2 - \frac{\pi^3 + 2}{6\pi^2 + 1}\right)\mathbf{k}.$$

9. $\mathbf{r}(t) = c_1 e^{c_2 e^t}\mathbf{i} + c_2 e^t\mathbf{j} + c_3 e^t\mathbf{k}$.

10. $\mathbf{r}(t) = \left(\dfrac{t^3}{3} + c_1\right)\mathbf{i} + \left(\dfrac{t^2}{2} + c_2\right)\mathbf{j} + \left(\dfrac{-t^4}{4} + c_3\right)\mathbf{k}$.

11. $\mathbf{r}(t) = (c_1\cos 2t + c_2\sin 2t)\mathbf{i} + (c_1\sin 2t - c_2\cos 2t)\mathbf{j} + c_3\mathbf{k}$.

chapter seven

exercises 7.1–7.3, p. 207

1.

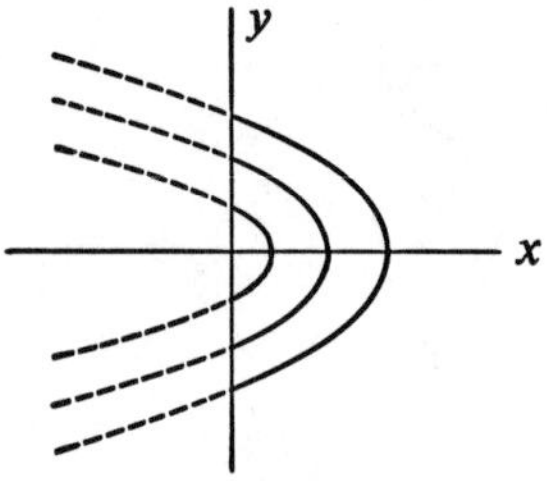

(a)

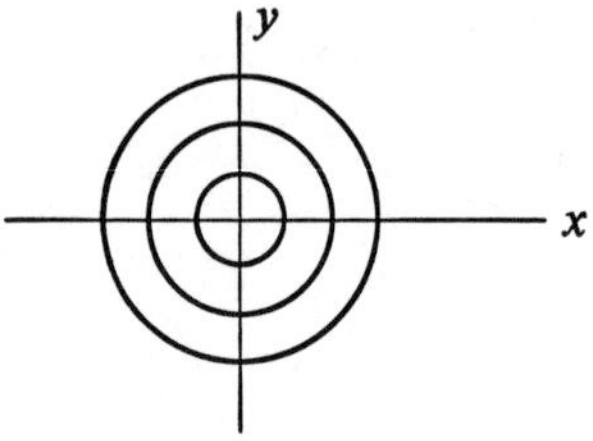

(b)

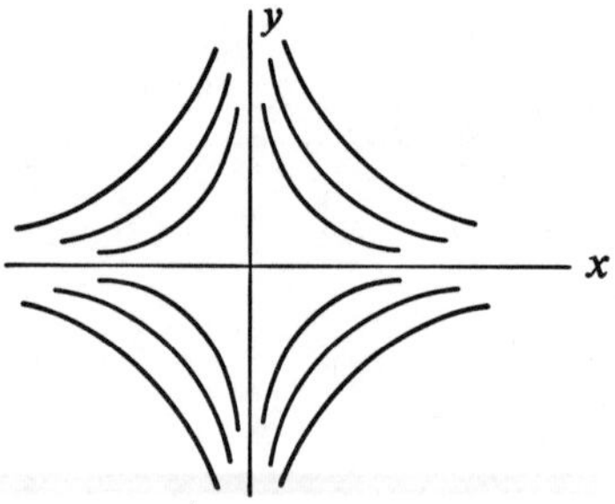

(c)

(d) All curves

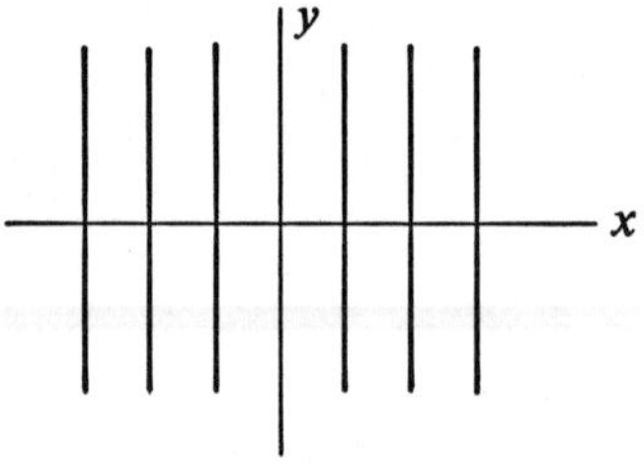

(e)

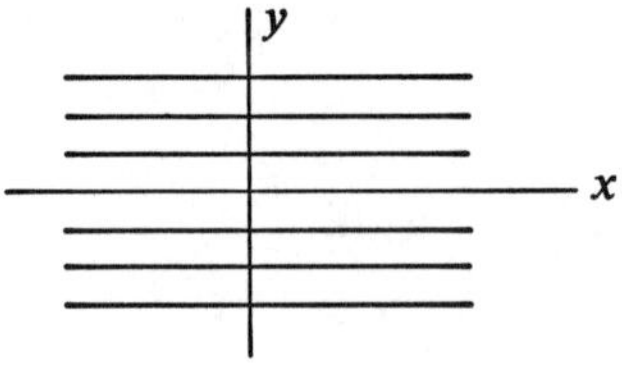

(f)

(g) Same as (e).
(h) Same as (b).
(i) Same as (b).

2.

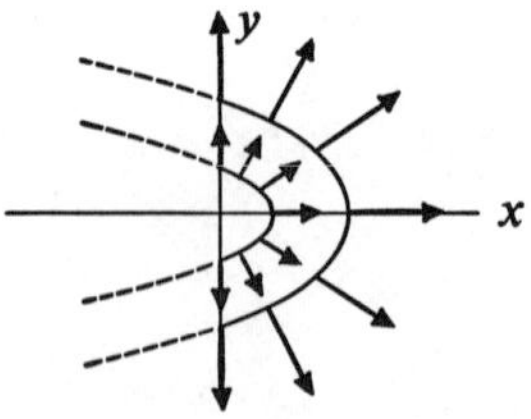

(a)

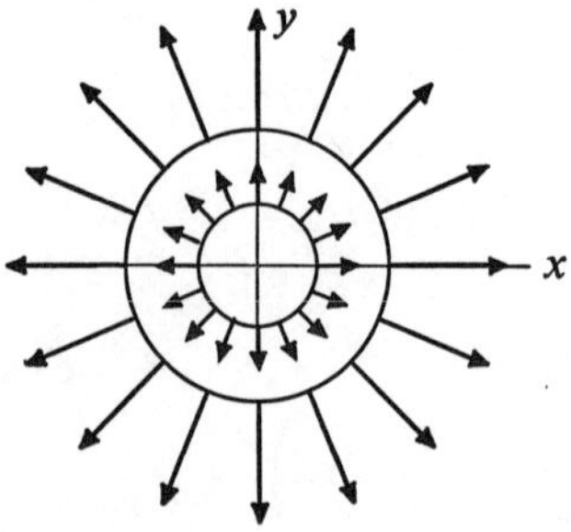

(b)

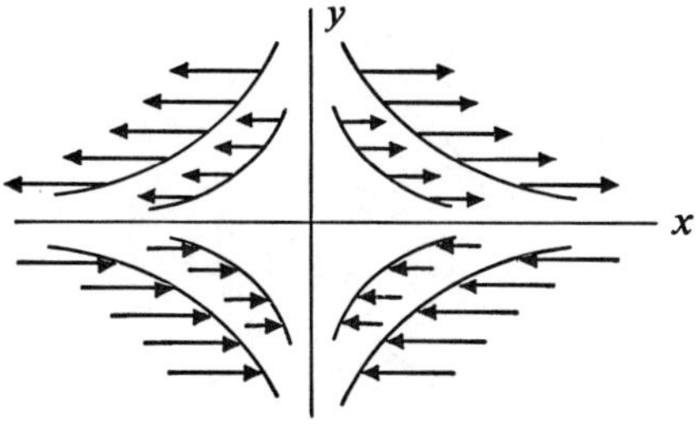
y
x

(c)

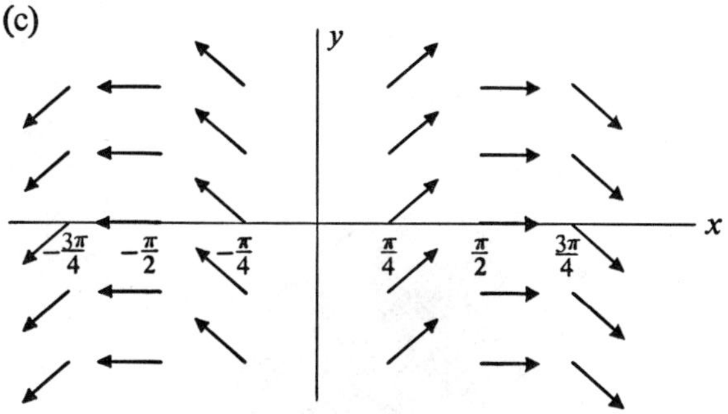
y
x
$-\frac{3\pi}{4}$
$-\frac{\pi}{2}$
$-\frac{\pi}{4}$
$\frac{\pi}{4}$
$\frac{\pi}{2}$
$\frac{3\pi}{4}$

(d)

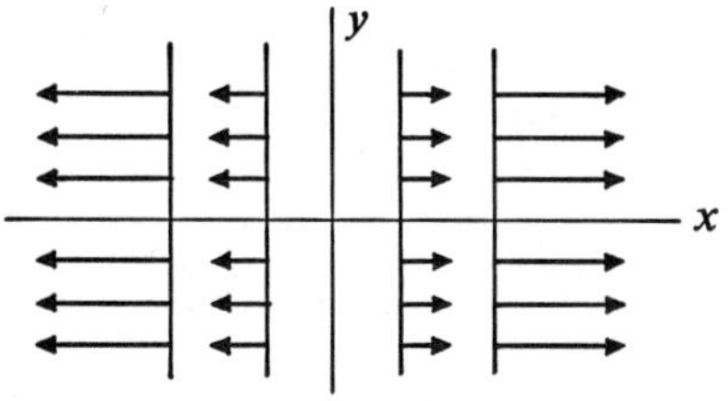
y
x

(e)

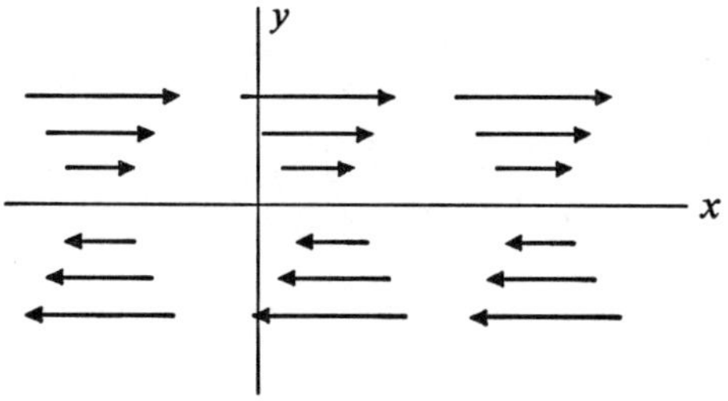
y
x

(f)

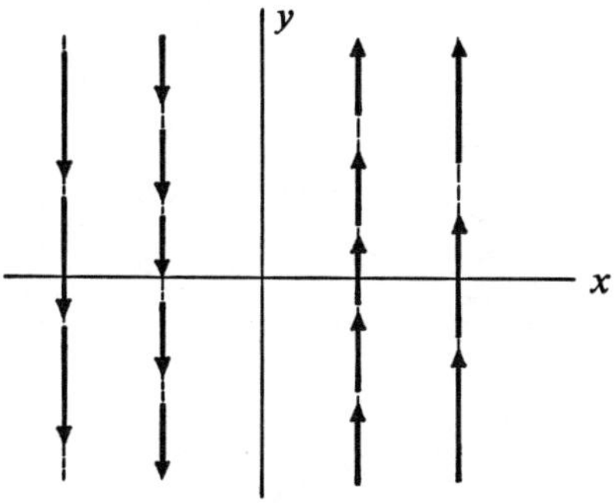
y
x

(g)

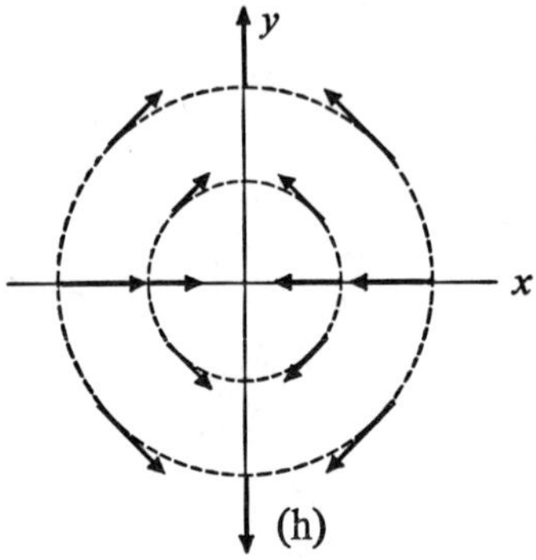

(h)

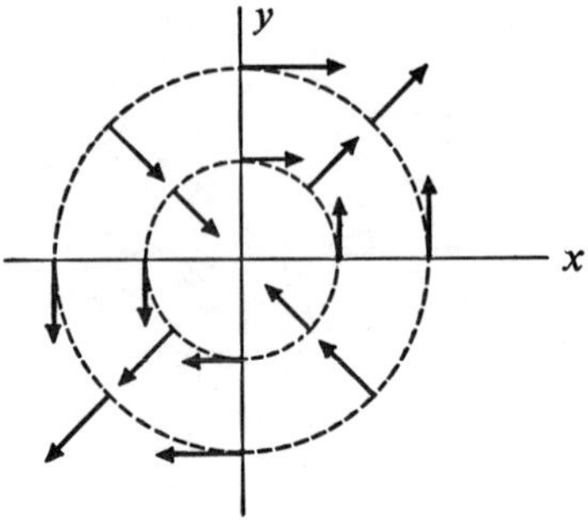

(i)

3.

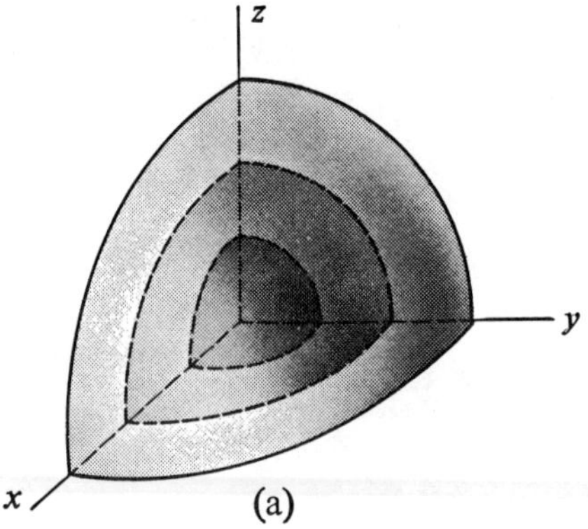

(a)

(b) Same as (a).

4.

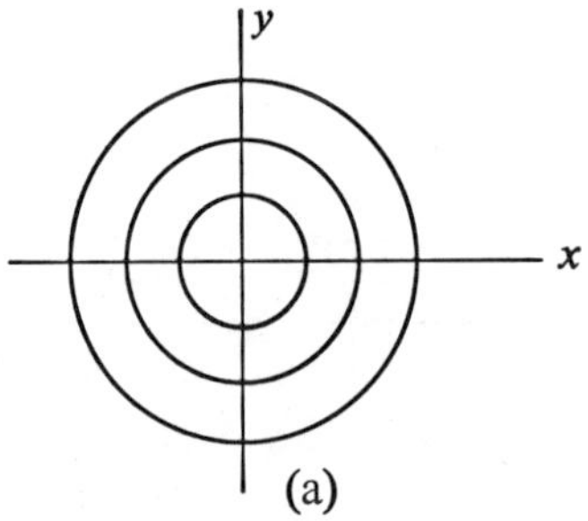

(a)

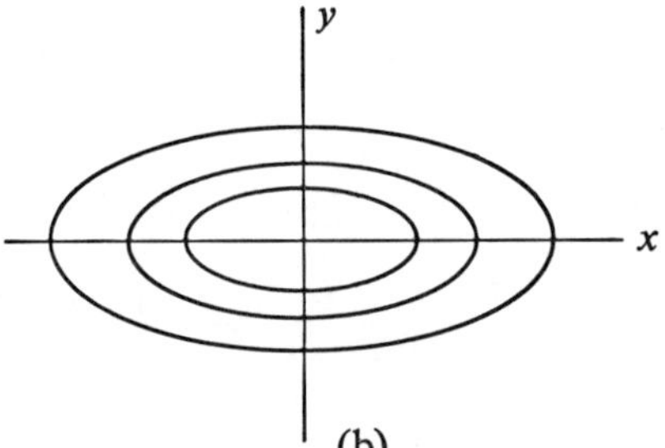

(b)

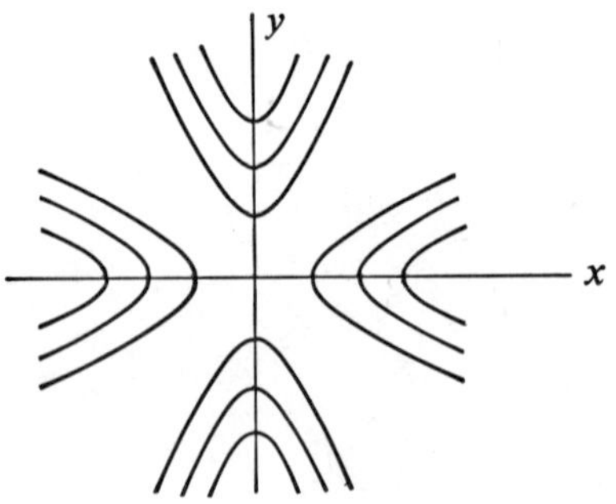

(c)

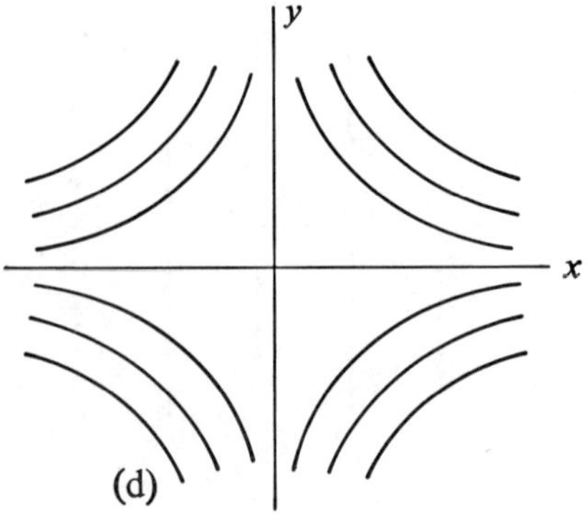

(d)

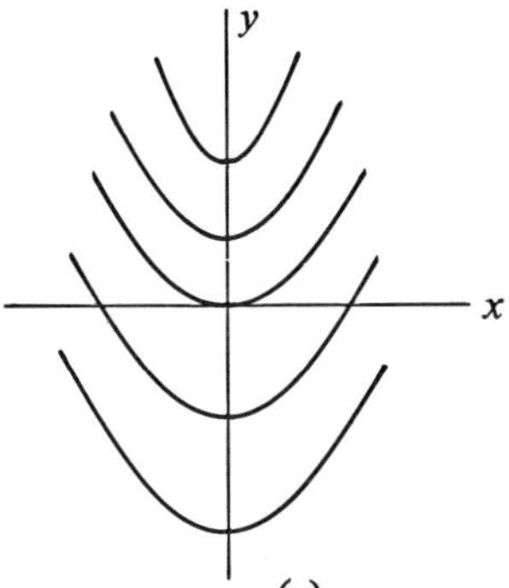

(e)

5.

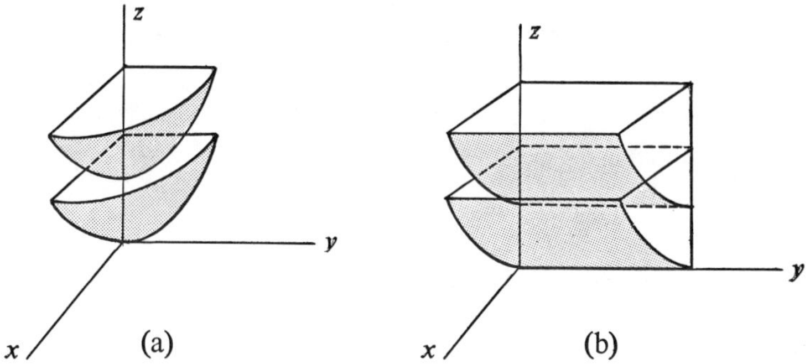

exercises 7.4, p. 214

1. (a) $yz\mathbf{i}+xz\mathbf{j}+xy\mathbf{k}$. (b) $2x\mathbf{i}+2y\mathbf{j}+2z\mathbf{k}$.
(c) $(\cos y - z\sin x)\,\mathbf{i}-(x\sin y-\sin z)\,\mathbf{j}+(y\cos z+\cos x)\,\mathbf{k}$.

2. $(6\mathbf{i}-\mathbf{j})/\sqrt{37}$. **3.** (a) -2. (b) 6, in direction $2\mathbf{i}+2\mathbf{j}+\mathbf{k}$.
(c) $(y+z)\,\mathbf{i}+(x+z)\,\mathbf{j}+(x+y)\,\mathbf{k}$.

4. (a) $f(x,y,z)=x+y+z$. (b) $f(x,y,z)=x^2+y^2+z^2$.
(c) $f(x,y,z)=(x^2+y^2)^{1/2}$.
(d) $f(x,y,z)=x^2yz+\cos yz-\frac{1}{4}\cos 2x^2$.
(e) $f(x,y,z)=x^2y+y^2z+z^2x$.

5. No. **6.** (a) $x\mathbf{i}+y\mathbf{j}+z\mathbf{k}$. (b) $yz\mathbf{i}+xz\mathbf{j}+xy\mathbf{k}$.
(c) $(\sin z-\sin x)\,\mathbf{i}+\cos y\mathbf{j}+x\cos z\mathbf{k}$.

7. $y(2x+z)\,\mathbf{i}+x(x+z)\,\mathbf{j}+xy\mathbf{k}$.

8. (a) $2x+2y-z=2$; $\mathbf{r}(t)=(\mathbf{i}+\mathbf{j}+2\mathbf{k})+t(2\mathbf{i}+2\mathbf{j}-\mathbf{k})$.
(b) $y=1$; $\mathbf{r}(t)=(\pi/2)\,\mathbf{i}+\mathbf{j}+t\mathbf{j}$.
(c) $x+y+z=6$; $\mathbf{r}(t)=(t+2)\,(\mathbf{i}+\mathbf{j}+\mathbf{k})$.

exercises 7.5, p. 219

1. (a) $yz + xz + xy$. (b) $6(x+z)$.
2. $\nabla \cdot \mathbf{F} = \nabla \cdot \phi\mathbf{G} = \phi(\nabla \cdot \mathbf{G}) + (\nabla\phi) \cdot \mathbf{G}$.
4. $e^x(\cos y\mathbf{i} - \sin y\mathbf{j})$. **5.** $38/\sqrt{14}$.
6. $\nabla \cdot \mathbf{v} = 1 \neq 0$. Sketch of vector field is shown below.

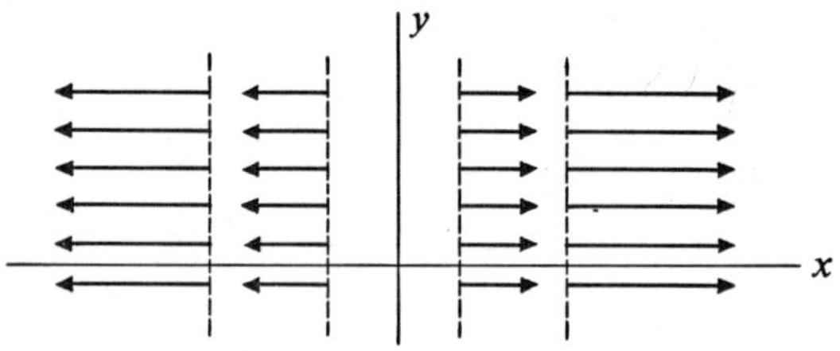

8. On the plane $x + y + z = 0$.

exercises 7.6, p. 224

1. (a) $f(x)\,\mathbf{i}$. (b) $f(y)\,\mathbf{i}$.
2. (a) $3x\mathbf{i} - 3y\mathbf{j}$. (b) $\mathbf{i} + \mathbf{j} + \mathbf{k}$. (c) 0. (d) 0.
3. No, a field may be both irrotational and solenoidal.
4. No, the curl of a scalar field is undefined. Yes, the divergence of a vector field is defined.
7. *Hint:* A central force field is described by $f(r)(x\mathbf{i} + y\mathbf{j} + z\mathbf{k})$.
8. (a)–(e) are solenoidal everywhere; (f) is solenoidal nowhere; (d), (e), and (f) are irrotational everywhere.
9. Unidirectional field must vary in the direction of the field to be nonsolenoidal. It must vary crosswise to the field to be nonirrotational.
10. Zero divergence.

chapter eight

exercises 8.1–8.3, p. 235

3. 219. **4.** 153. **5.** 153.
6. $153\sqrt{2}$. **7.** π. ($2\sqrt{2}$ is also acceptable.)
8. 0. **9.** $-\frac{25}{11}(\mathbf{i} - 3\mathbf{j} + \mathbf{k})$. **10.** $\sqrt{89}$, $6x - 2y - 7z = 7$.

exercises 8.4–8.5, p. 244

2. (a) 15. (b) 15. (c) 15.

3. (a) 111/4. (b) 9. (c) 267/10.

4. 64/35; $-8/3$. **5.** 0. **6.** $3\pi/8$; 0.

7. (a) -6π. (b) -2π. (c) 2π. (d) $2-2\cos 1$.

8. (a) $\sqrt{\pi/2}-7$. (b) $-34/3$.
(c) 0. (d) 2/3. (e) $3\sin 8-\sin 1$.

9. (a) 0. (b) 3. (c) 32.

10. (a) 0. (b) 4. (c) 310.

exercises 8.6–8.8, p. 257

3. 12π. **4.** $13\pi/3$. **5.** 4. **6.** 4π.

7. (a) 108. (b) 171/2. (c) 132. (d) 18π.

8. (a) 2. (b) 30. (c) $8\pi/3$. (d) $8\pi/3$.

9. $4\pi^2 ab$. **10.** $\dfrac{10^{3/2}-1}{54}$.

11. $\dfrac{5^{3/2}-1}{12}$. **12.** $\sqrt{2}$.

13. $\dfrac{19\sqrt{2}}{6}$. **14.** $\dfrac{5^{3/2}-1}{2}$.

exercises 8.9, p. 267

3. (a) 0. (b) -1. (c) $-4/3$. **4.** 0.

6. (a) 6. (b) 7/2. (c) $2e^2+e-3$.

7. (a) $3\pi/2$. (b) 0. **8.** 1.

9. **F** and **V** are orthogonal. **11.** 0. **12.** $50-2\pi$.

exercises 8.10, p. 274

3. (a) πa^2. (b) 3/2.

4. (a) $-9\pi/\sqrt{2}$. (b) 2π. (c) -1. (d) 0.

5. $\operatorname{div}\mathbf{F}=0$.

exercises 8.11, p. 280

1. $\bar{z}=1$. **2.** $\log_e 100$. **3.** 128/9.

4. (a) 256π. (b) 819.2π. (c) 0. (d) 0.

8. 18π. **9.** 1.

exercises 8.12, p. 282

1. 4. **2.** π. **3.** $\pi/2$. **4.** $2\pi+4\pi^2$.

5. $-1/3$. **6.** 18. **7.** $\pm 4\pi/\sqrt{2}$. **8.** 9.

9. $\pm\sqrt{2}\pi$. **10.** 0. **11.** 20,951/12.

12. $(\pi^2/4)+2$. **13.** -12π. **14.** 1/3. **15.** 1.

16. 4π. **17.** $\pi\sqrt{2}/16$. **18.** 3. **19.** $16\pi/3$.

20. 32π. **21.** 7/4. **22.** 0. **23.** $64\pi/3$.

24. 16π. **25.** 1/2. **26.** -8π. **27.** 8π.

28. 2. **29.** 0. **30.** 1/12. **31.** 8π.

chapter nine

exercises 9.1–9.3, p. 292

1. $A^{i1}B_1+A^{i2}B_2+A^{i3}B_3$, $i=1, 2, 3$.

2. $A^{i2}B_i$.

3. (a) $\begin{pmatrix} 22 & 15 & 14 \\ 4 & 6 & 26 \\ -2 & 24 & 5 \end{pmatrix}$. (b) $\begin{pmatrix} 6 & 12 & 11 \\ 30 & 12 & 44 \\ 6 & 30 & 59 \end{pmatrix}$.

5. $\bar{\mathbf{e}}^1=2\mathbf{i}+\mathbf{k}$, $\bar{\mathbf{e}}^2=\mathbf{i}+2\mathbf{j}+3\mathbf{k}$, $\bar{\mathbf{e}}^3=\mathbf{i}+\mathbf{j}+\mathbf{k}$;
$\bar{\mathbf{e}}_1=\frac{1}{3}\mathbf{i}-\frac{2}{3}\mathbf{j}+\frac{1}{3}\mathbf{k}$, $\bar{\mathbf{e}}_2=-\frac{1}{3}\mathbf{i}-\frac{1}{3}\mathbf{j}+\frac{2}{3}\mathbf{k}$, $\bar{\mathbf{e}}_3=\frac{2}{3}\mathbf{i}+\frac{5}{3}\mathbf{j}-\frac{4}{3}\mathbf{k}$.

6. $\bar{\mathbf{e}}^1=\mathbf{j}$, $\bar{\mathbf{e}}^2=-2\mathbf{i}+2\bar{x}^1\mathbf{j}$, $\bar{\mathbf{e}}^3=\mathbf{k}$; $\bar{\mathbf{e}}_1=\bar{x}^1\mathbf{i}+\mathbf{j}$, $\bar{\mathbf{e}}_2=-\frac{1}{2}\mathbf{i}$, $\bar{\mathbf{e}}_3=\mathbf{k}$.

7. $\bar{\mathbf{e}}^1 = \cos\bar{x}^2 \cos\bar{x}^3\mathbf{i} + \sin\bar{x}^2 \cos\bar{x}^3\mathbf{j} + \sin\bar{x}^3\mathbf{k}$;

$$\bar{\mathbf{e}}^2 = \frac{1}{\bar{x}^1 \cos\bar{x}^3}(-\sin\bar{x}^2\mathbf{i} + \cos\bar{x}^2\mathbf{j});$$

$$\bar{\mathbf{e}}^3 = -\frac{1}{\bar{x}^1}(\cos\bar{x}^2 \sin\bar{x}^3\mathbf{i} + \sin\bar{x}^2 \sin\bar{x}^3\mathbf{j} - \cos\bar{x}^3\mathbf{k});$$

$$\bar{\mathbf{e}}_1 = \cos\bar{x}^2 \cos\bar{x}^3\mathbf{i} + \sin\bar{x}^2 \cos\bar{x}^3\mathbf{j} + \sin\bar{x}^3\mathbf{k};$$

$$\bar{\mathbf{e}}_2 = -\bar{x}^1 \sin\bar{x}^2 \cos\bar{x}^3\mathbf{i} + \bar{x}^1 \cos\bar{x}^2 \cos\bar{x}^3\mathbf{j};$$

$$\bar{\mathbf{e}}_3 = -\bar{x}^1 \cos\bar{x}^2 \sin\bar{x}^3\mathbf{i} - \bar{x}^1 \sin\bar{x}^2 \sin\bar{x}^3\mathbf{j} + \bar{x}^1 \cos\bar{x}^3\mathbf{k}.$$

8. $\bar{\mathbf{e}}^1 = \dfrac{\sinh\bar{x}^1 \cos\bar{x}^2\mathbf{i} + \cosh\bar{x}^1 \sin\bar{x}^2\mathbf{j}}{\sinh^2\bar{x}^1 \cos^2\bar{x}^2 + \cosh^2\bar{x}^1 \sin^2\bar{x}^2}$;

$$\bar{\mathbf{e}}^2 = \frac{-\cosh\bar{x}^1 \sin\bar{x}^2\mathbf{i} + \sinh\bar{x}^1 \cos\bar{x}^2\mathbf{j}}{\sinh^2\bar{x}^1 \cos^2\bar{x}^2 + \cosh^2\bar{x}^1 \sin^2\bar{x}^2}; \bar{\mathbf{e}}^3 = \mathbf{k};$$

$$\bar{\mathbf{e}}_1 = \sinh\bar{x}^1 \cos\bar{x}^2\mathbf{i} + \cosh\bar{x}^1 \sin\bar{x}^2\mathbf{j};$$

$$\bar{\mathbf{e}}_2 = -\cosh\bar{x}^1 \sin\bar{x}^2\mathbf{i} + \sinh\bar{x}^2 \cos\bar{x}^2\mathbf{j}; \bar{\mathbf{e}}_3 = \mathbf{k}.$$

exercises 9.4, p. 296

3. $\mathbf{e}^1 = \mathbf{i} - \mathbf{j} - \mathbf{k}, \mathbf{e}^2 = -\frac{1}{2}\mathbf{i} + \frac{3}{2}\mathbf{j} + \mathbf{k}, \mathbf{e}^3 = \frac{1}{2}(\mathbf{i} - \mathbf{j})$.

4. $A_1 = 4, A_2 = 5, A_3 = 11; \quad A^1 = 8, A^2 = -11/2, A^3 = 9/2$.

6. The basis and the reciprocal set point in the same directions but with reciprocal lengths.

exercises 9.5, p. 299

1. $\mathbf{e}^1 = \dfrac{-3\mathbf{i} + 9\mathbf{j} + 21\mathbf{k}}{9}; \mathbf{e}^2 = \dfrac{\mathbf{i} + 3\mathbf{j} + 2\mathbf{k}}{9}; \mathbf{e}^3 = \dfrac{2\mathbf{i} - 3\mathbf{j} - 5\mathbf{k}}{9}$.

Covariantly: $\mathbf{A} = -6\mathbf{e}^1 + 34\mathbf{e}^2 + \mathbf{e}^3$.

Contravariantly: $\mathbf{A} = -\frac{120}{9}\mathbf{e}_1 + \frac{7}{9}\mathbf{e}_2 + \frac{32}{9}\mathbf{e}_3$.

2. $A^{1*} = \dfrac{-40\sqrt{3}}{3}, A^{2*} = \dfrac{7\sqrt{14}}{9}, A^{3*} = \dfrac{32\sqrt{38}}{9}$;

$$A_1^* = \frac{-6}{\sqrt{3}}, A_2^* = \frac{34}{\sqrt{14}}, A_3^* = \frac{1}{\sqrt{38}}.$$

3. $\mathbf{e}_1 = \frac{\sqrt{3}}{2}\mathbf{i} + \frac{1}{2}\mathbf{j}$, $\mathbf{e}_2 = -\mathbf{i} + \sqrt{3}\mathbf{j}$, $\mathbf{e}_3 = \mathbf{k}$;

$A^1 = 3\sqrt{3} + \frac{5}{2}$, $A^2 = -\frac{3}{2} + \frac{5}{4}\sqrt{3}$, $A^3 = -7$;

$\mathbf{e}^1 = \frac{\sqrt{3}}{2}\mathbf{i} + \frac{1}{2}\mathbf{j}$, $\mathbf{e}^2 = \frac{1}{4}(-\mathbf{i} + \sqrt{3}\mathbf{j})$, $\mathbf{e}^3 = \mathbf{k}$;

$A_1 = 3\sqrt{3} + \frac{5}{2}$, $A_2 = -6 + 5\sqrt{3}$, $A_3 = -7$.

5. $\mathbf{e}_1 = \frac{\sqrt{3}}{4}\mathbf{i} + \frac{1}{4}\mathbf{j} + \frac{\sqrt{3}}{2}\mathbf{k}$, $\mathbf{e}_2 = \frac{-3}{4}\mathbf{i} + \frac{3\sqrt{3}}{4}\mathbf{j}$, $\mathbf{e}_3 = \frac{-9}{4}\mathbf{i} - \frac{3\sqrt{3}}{4}\mathbf{j} + \frac{3\mathbf{k}}{2}$;

$\mathbf{e}^1 = \frac{\sqrt{3}}{4}\mathbf{i} + \frac{1}{4}\mathbf{j} + \frac{\sqrt{3}}{2}\mathbf{k}$, $\mathbf{e}^2 = -\frac{1}{3}\mathbf{i} + \frac{\sqrt{3}}{3}\mathbf{j}$, $\mathbf{e}^3 = -\frac{1}{4}\mathbf{i} - \frac{\sqrt{3}}{12}\mathbf{j} + \frac{1}{6}\mathbf{k}$;

$A^1 = \frac{7\sqrt{3} - 1}{4}$, $A^2 = -1 - \frac{\sqrt{3}}{3}$, $A^3 = \frac{-5 + \sqrt{3}}{12}$;

$A_1 = \frac{7\sqrt{3} - 1}{4}$, $A_2 = \frac{-9 - 3\sqrt{3}}{4}$, $A_3 = -\frac{15}{4} + \frac{3\sqrt{3}}{4}$.

exercises 9.6–9.7, p. 304

1. $g_{ij} = 0$, $i \neq j$.
2. $g_{11} = 1$, $g_{22} = (\bar{x}^1)^2 \cos^2 \bar{x}^3$, $g_{33} = (\bar{x}^1)^2$, $g_{ij} = 0$, $i \neq j$.
4. Each g_{ii} is the reciprocal of g^{ii}.
5. $A^1 = 2$, $A^2 = -1/9$, $A_3 = 7$.
6. $A_1 = 1$, $A_2 = 28$, $A_3 = -2$.
7. $\bar{g}_{11} = 3$, $\bar{g}_{12} = 1 = \bar{g}_{21}$, $\bar{g}_{13} = 1 = \bar{g}_{31}$, $\bar{g}_{23} = 5 = \bar{g}_{32}$, $\bar{g}_{22} = 3$, $\bar{g}_{33} = 13$.
8. $A_1 = -1$, $A_2 = 21$, $A_3 = 55$.

exercises 9.8, p. 308

1. $\bar{A}_1 = -4\sqrt{2}$, $\bar{A}_2 = -2$, $\bar{A}_3 = 7$.
3. $\bar{A}_{11} = 5/2$, $\bar{A}_{22} = -3$, $\bar{A}_{12} = -7\sqrt{2}/2$, $\bar{A}_{21} = 5\sqrt{2}/2$.
8. Contravariant tensor of rank 2.

exercises 9.9, p. 313

4. $\begin{pmatrix} 2 & -2 & 0 \\ -2 & 0 & 4 \\ -2 & 4 & 12 \end{pmatrix}$.

5. Tensor $S^{ij}V_i$ has contravariant components 5, − 5, − 15.
Tensor $S^{ij}V_j$ has contravariant components 5, − 5, − 15.

6. (a) 1. (b) 0. (c) 6.

exercises 9.10, p. 317

3. All are zero except $\{12, 2\} = \{21, 2\} = 1/r$ and $\{22, 1\} = -r$.

4. $\{12, 2\} = \{21, 2\} = 1/\rho$; $\{13, 3\} = \{31, 3\} = 1/\rho$;
$\{22, 1\} = -\rho\cos^2\phi$; $\{22, 3\} = \sin\phi\cos\phi$; $\{23, 2\} = \{32, 2\} = -\tan\phi$;
$\{33, 1\} - -\rho$.

5.
$$(F^i_{,j}) = \begin{pmatrix} 2r & -r^2\sin^2\theta & 2z \\ 2\sin^2\theta & r\sin 2\theta + \dfrac{r^2+z^2}{r} & 0 \\ z^2\cos\theta & -rz^2\sin\theta & 2rz\cos\theta \end{pmatrix}.$$

6. $(F_{i,j})$
$$= \begin{pmatrix} 2\rho\theta & \rho^2 - \dfrac{\theta^2\sin^2\phi}{\rho} & \theta\phi \\ -\dfrac{\theta^2\sin^2\theta}{\rho} & \begin{matrix} 2\theta\sin^2\phi + \rho^3\theta\cos^2\phi \\ -\theta\phi\rho\sin\phi\cos\phi \end{matrix} & \theta^2(\sin 2\phi + \tan\phi\sin^2\phi) \\ 0 & \rho\phi + \theta^2\sin^2\phi\tan\phi & \rho\theta + \rho^3\theta \end{pmatrix}.$$

8. Cylindrical coordinates: $\nabla\cdot\mathbf{F} = \dfrac{1}{r}\dfrac{\partial(rF^1)}{\partial r} + \dfrac{1}{r}\dfrac{\partial F^2}{\partial\theta} + \dfrac{\partial F^3}{\partial z}$ where

F^1, F^2, F^3 are contravariant components in cylindrical coordinates.

Spherical coordinates:

$$\nabla\cdot\mathbf{F} = \frac{1}{\rho^2}\frac{\partial}{\partial\rho}(\rho^2F^1) + \frac{1}{\rho\sin\phi}\frac{\partial}{\partial\phi}(F^2\sin\phi) + \frac{1}{\rho\sin\phi}\frac{\partial F^3}{\partial\theta} \text{ where}$$

F^1, F^2, and F^3 are contravariant components in spherical coordinates.

chapter ten

exercises 10.1–10.3, p. 326

5. No. For example, with this definition $f(x) = 1$ and $g(x) = -1$ would be "equal in the norm."

7. $I_{2^n+i} = \frac{1}{2^n}, 0 \leqq i < 2^n, n = 0, 1, 2, 3, \cdots$. **8.** (a) No. (b) No. (c) No. (d) Yes. (e) Yes. 4 must divide L.

exercises 10.4, p. 332

1. $\frac{\pi^2}{3} + 4 \sum_{n=1}^{\infty} \frac{(-1)^n}{n^2} \cos nx$. **3.** $\frac{1 + \cos 2x}{2}$.

4. $\frac{3}{2} + \sum_{n=0}^{\infty} \frac{2}{(2n+1)\pi} \sin (2n+1) x$.

5. $\frac{2}{\pi} + \frac{4}{\pi} \sum_{n=1}^{\infty} (-1)^{n+1} \frac{\cos 2nx}{4n^2 - 1}$. **6.** $\frac{2}{\pi} - \frac{4}{\pi} \sum_{n=1}^{\infty} \frac{\cos 2nx}{4n^2 - 1}$.

7. $\frac{1}{3} + \frac{4}{\pi^2} \sum_{n=1}^{\infty} \frac{(-1)^n}{n^2} \cos n\pi x$. **8.** $-\frac{4}{\pi} \sum_{n=0}^{\infty} \frac{\sin (2n+1) x}{2n+1}$.

9. $\frac{1}{2} + \frac{2}{\pi} \sum_{n=0}^{\infty} \frac{\sin (2n+1) \frac{\pi x}{2}}{2n+1}$. **10.** $\frac{8}{\pi} \sum_{n=1}^{\infty} (-1)^{n+1} \frac{n}{4n^2 - 1} \sin 2nx$.

exercises 10.5–10.6, p. 339

1. (a) Odd. (b) Even. (c) Neither. (d) Even. (e) Even. (f) Even. (g) Odd. (h) Odd. (i) Neither. (j) Neither. (k) Even. (l) Odd. (m) Even. (n) Odd.

3. 0. **4.** $\frac{4}{\pi} \sum_{n=0}^{\infty} \frac{\sin (2n+1) x}{2n+1}; \frac{2}{\pi} \sum_{n=0}^{\infty} \frac{\sin (2n+1) x}{2n+1}$.

5. $2 \sum_{n=1}^{\infty} \frac{(-1)^{n+1}}{n} \sin nx; \frac{\pi^2}{3} + 4 \sum_{n=1}^{\infty} \frac{(-1)^n}{n^2} \cos nx;$

$4 \sum_{n=1}^{\infty} \frac{(-1)^n}{n} \sin nx + \pi^2 + 12 \sum_{n=1}^{\infty} \frac{(-1)^n}{n^2} \cos nx$.

6. $5\cos 2x - 3 + \frac{8}{\pi}\sum_{n=0}^{\infty}(-1)^{n+1}\frac{n}{4n^2-1}\sin nx.$

8. (a) $\cos x$. (b) $\cos x$. (c) $\frac{2}{\pi} + \frac{4}{\pi}\sum_{n=1}^{\infty}\frac{(-1)^{n+1}}{(4n^2-1)}\cos 2nx.$

9. $\sum_{n=1}^{\infty}\frac{-2}{n}\sin nx.$

10. $-\frac{4}{\pi}\sum_{n=0}^{\infty}\frac{\cos(2n+1)x}{(2n+1)^2} + \frac{\pi}{2}.$

exercises 10.7, p. 345

1.

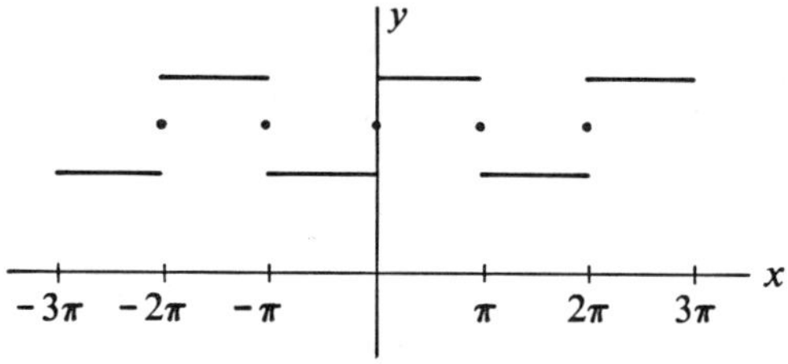

2.

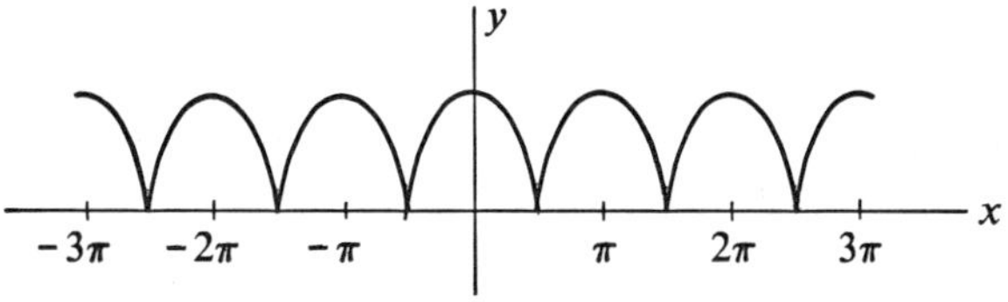

3.

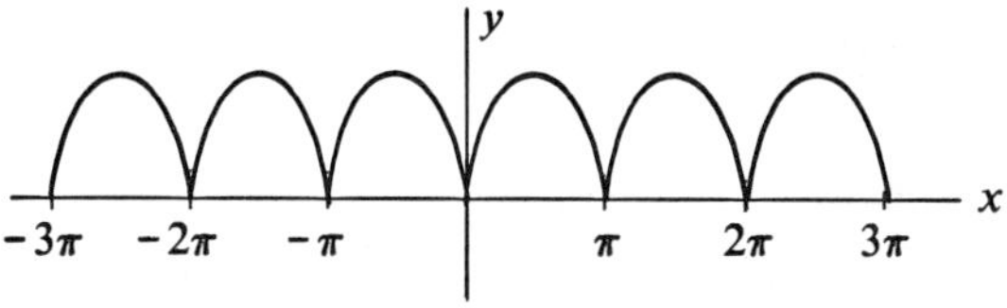

4.

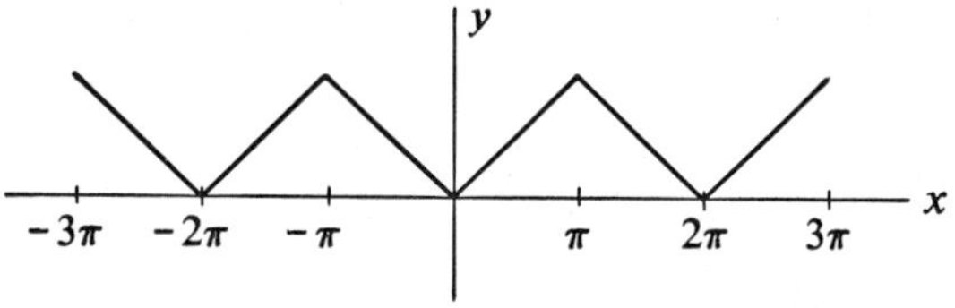

5.

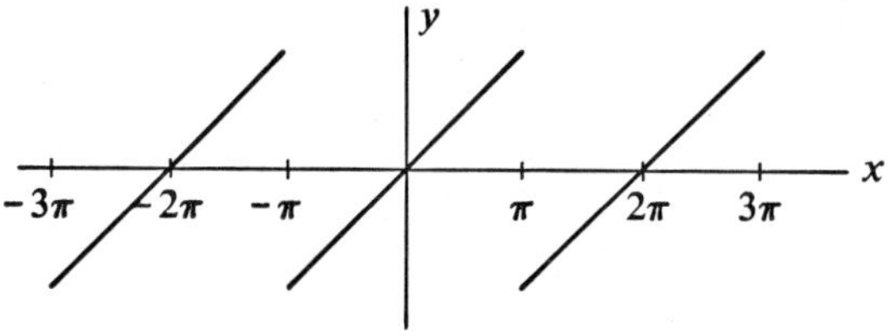

6.

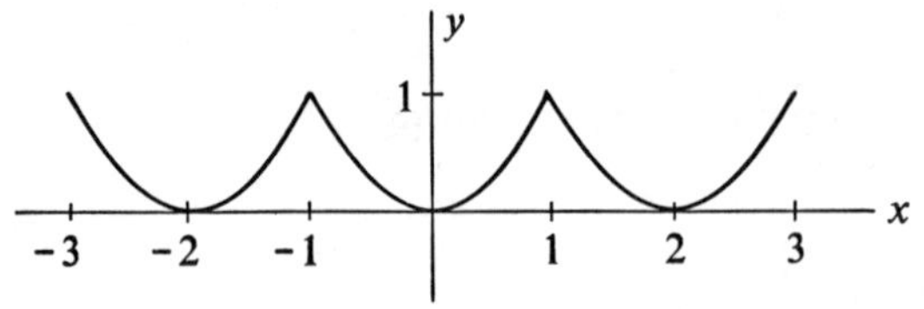

7.

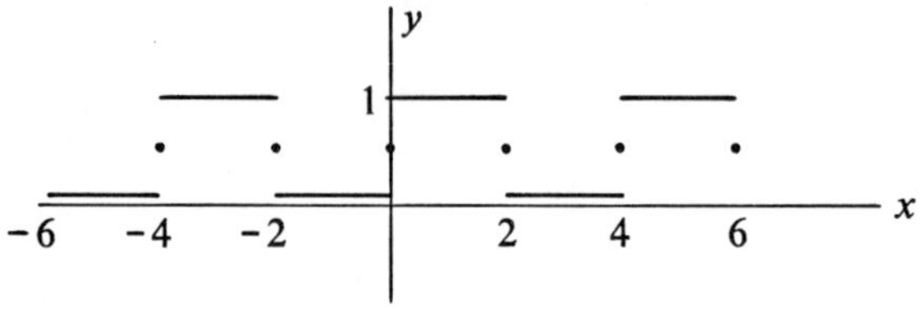

exercises 10.8–10.9, p. 352

1. (a) $f_0(x) = -x - 1, \ -2 < x < -1;$
$= x + 1, \quad -1 < x < 0.$
$f_e(x) = x + 1, \quad -2 < x < -1;$
$= -x - 1, \ -1 < x < 0.$

(b) $f_0(x) = -(x^2 - x), \ -2 < x < 0.$
$f_e(x) = x^2 - x, \quad -2 < x < 0.$

(c) $f_0(x) = \sin\frac{\pi x}{2}, \quad -2 < x < 0.$

$f_e(x) = -\sin\frac{\pi x}{2}, \quad -2 < x < 0.$

(d) $f_0(x) = -\cos\frac{\pi x}{2}, \ -2 < x < 0.$ $\quad f_e(x) = \cos\frac{\pi x}{2}, \ -2 < x < 0.$

(e) $f_0(x) = -e^{-x}, \ -2 < x < 0.$ $\quad f_e(x) = e^{-x}, \ -2 < x < 0.$

2. (a) Fourier sine series: $\sin x$.

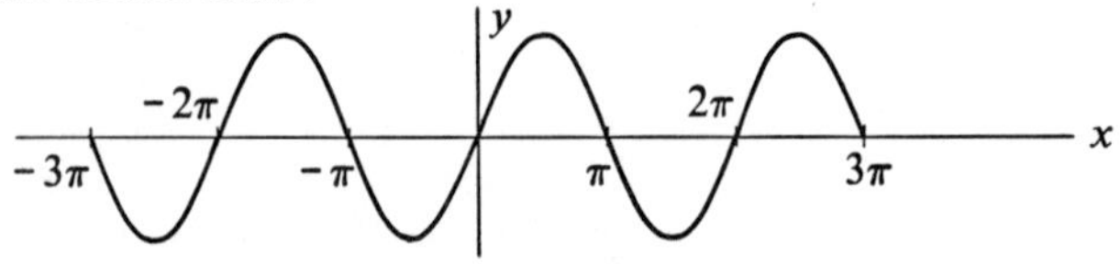

(b) Fourier cosine series: $\frac{2}{\pi} - \frac{4}{\pi}\sum_{n=1}^{\infty}\frac{\cos 2nx}{4n^2 - 1}.$

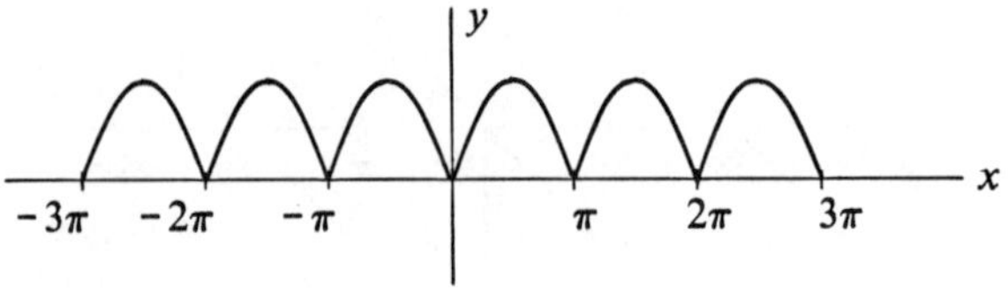

3. (a) Fourier sine series: $\frac{8}{\pi}\sum_{n=1}^{\infty}\frac{n}{4n^2-1}\sin 2nx.$

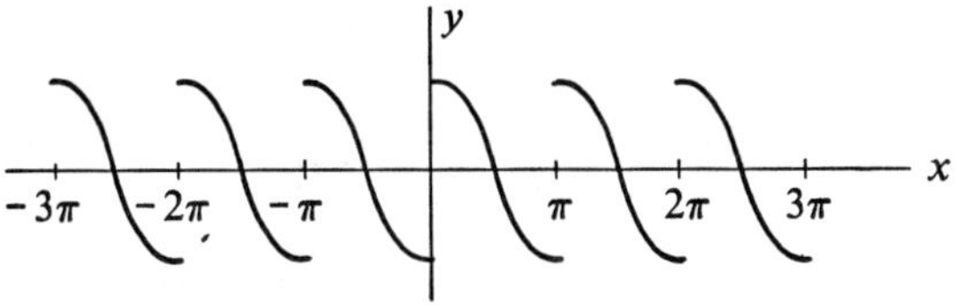

(b) Fourier cosine series: $\cos x.$

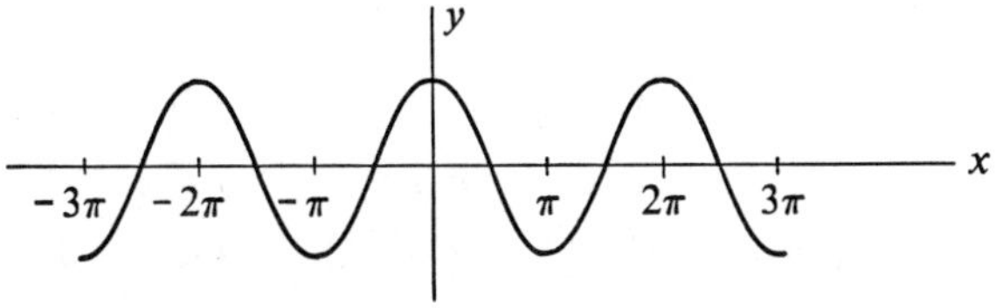

4. Cosine series: $\frac{1}{3}+\frac{4}{\pi^2}\sum_{n=1}^{\infty}\frac{(-1)^n}{n^2}\cos n\pi x.$

Sine series: $\sum_{n=1}^{\infty}\left(\frac{2(-1)^{n+1}}{n\pi}+\frac{4}{n^3\pi^3}[(-1)^n-1]\right)\sin n\pi x.$

5. (a) Cosine series. (b) Cosine series. (c) Cosine series.

exercises 10.10–10.11, p. 361

1. $8\left(\frac{1}{15}-\frac{1}{\pi^3}\right).$ **2.** $8/\pi^3.$ **3.** $\frac{\pi}{2}-\frac{4}{\pi}\cos x.$

4. $\frac{\pi^3}{6}-\frac{16}{\pi}.$ **5.** $12.$ **9.** $32/\pi^3.$

exercises 10.12, p. 371

1. (a) $F(\omega)=\frac{1}{\sqrt{2\pi}}\left[\frac{1}{1+\omega^2}-i\,\frac{\omega}{1+\omega^2}\right].$

(b) $F(\omega)=\frac{\sqrt{2}}{\sqrt{\pi}}\left[\frac{1}{1+\omega^2}\right].$

(c) $F(\omega)=\frac{1}{\omega\sqrt{2\pi}}[\sin\omega+i(\cos\omega-1)].$

2. (a) $B(\omega) = \frac{\sqrt{2}}{\sqrt{\pi}} \frac{\omega}{\omega^2 + 1}$; $A(\omega) = \frac{1}{\omega} B(\omega)$.

(b) Same as (a). (c) $B(\omega) = \frac{\sqrt{2}}{\sqrt{\pi}} \left[\frac{1 - \cos\omega}{\omega} \right]$; $A(\omega) = \frac{\sqrt{2}}{\sqrt{\pi}} \frac{\sin\omega}{\omega}$.

3. $f(x) = \frac{1}{\pi} \int_0^\infty \int_0^1 \cos\omega(x - t)\, dt\, d\omega$.

exercises 10.13, p. 375

3. Eigenfunctions 1, $\cos \pi x/L$, $\cos 2\pi x/L, \cdots$, $\cos n\pi x/L, \cdots$ corresponding to eigenvalues $0, 1, 4, \cdots, n^2, \cdots$.
4. Eigenfunctions $\sin[(2n+1)\,\pi x/2L]$, corresponding to eigenvalues $[(2n+1)\,\pi/2L]^2$.
5. $y'' + \lambda y = 0$, $y(0) = y(2L)$, $y'(0) = y'(2L)$.
6. $\|P_0\|^2 = 2$; $\|P_1\|^2 = 2/3$; $\|P_2\|^2 = 2/5$; $\|P_3\|^2 = 2/7$.
7. $\frac{1}{3}P_0(x) + \frac{2}{3}P_2(x)$.
8. $3(\sin 1 - \cos 1)P_1(x) + (-42\cos 1 + 7\sin 1)P_3(x)$.

chapter eleven

exercises 11.1–11.3, p. 383

2. (a) One possibility: $u_{xy} - u_{yx} = 0$. (b) $u_{xy} - u_{yx} = 0$.
5. (a) $u = c_1 f(x) + c_2$. (b) $u = g(y)\cos x + f(y)\sin x$.
   (c) $g(y) + f(y)\cos x + h(y)\sin x$.
   (d) $u = \frac{x^2 y}{2} + f(x) + x g(y) + h(y)$.
6. $u = \sin\frac{\pi x}{L} \cosh\frac{\pi y}{L}$.
7. $u = 3\sin(10\pi x)\cos(10\pi t)$.
10. $u = y\cosh x + \sin y \sinh x$.

exercises 11.4, p. 389

1. (a) $u = C_r e^{rx + \left(\frac{1-r}{2}\right) y}$. (b) $u = C_r e^{rx + \frac{1 \pm \sqrt{1+4r^2}}{2} y}$.

   (c) $u = C_{rp} e^{rx - (p+1+r)y + pz}$. (d) $u = C_r e^{rx + \frac{-r \pm \sqrt{r^2 - 4r}}{2} y}$.

3. (a) $u = f_1(x + \frac{1}{2}y) + f_2(x + \frac{1}{3}y)$.

(b) $u = f_1\left(x + \dfrac{i}{2}y\right) + f_2\left(x - \dfrac{i}{2}y\right)$.

(c) $u = f_1(x) + f_2(x + y)$. (d) $u = f_1\left(x + \dfrac{y}{5}\right) + f_2\left(x - \dfrac{y}{5}\right)$.

5. $u = C_r e^{rx + \frac{-r \pm ir\sqrt{3}}{2}y}$, not separable. **6.** (a) $u = C_\alpha e^{-\frac{\alpha}{x} - \frac{y^3}{\alpha}}$.

(b) $u = C_r e^{r(x-y)}$. (c) $u = \dfrac{\alpha^2}{y}(c_1 x + c_2 + c_3 \cosh \alpha x + c_4 \sinh \alpha x)$;

or $u = y^{-\alpha^2}(c_1 x + c_2 + c_3 \cos \alpha x + c_4 \sin \alpha x)$;
or $u = y^{-\alpha^2}(c_1 x^3 + c_2 x^2 + c_3 x + c_4)$.

(d) $u = C_r e^{rx + \frac{y}{r}}$. (e) $u = e^{\alpha \tan^{-1} y}(c_1 + c_2 e^{\alpha x})$.

(f) $u = C_{rs} e^{\frac{1}{2}\left(rx^2 + sy^2 + \frac{z^2}{rs}\right)}$. (g) $u = C_\alpha e^{\frac{x^2}{2} + \frac{y^2}{2} + \alpha x - \alpha y}$.

exercises 11.5–11.6, p. 396

2.

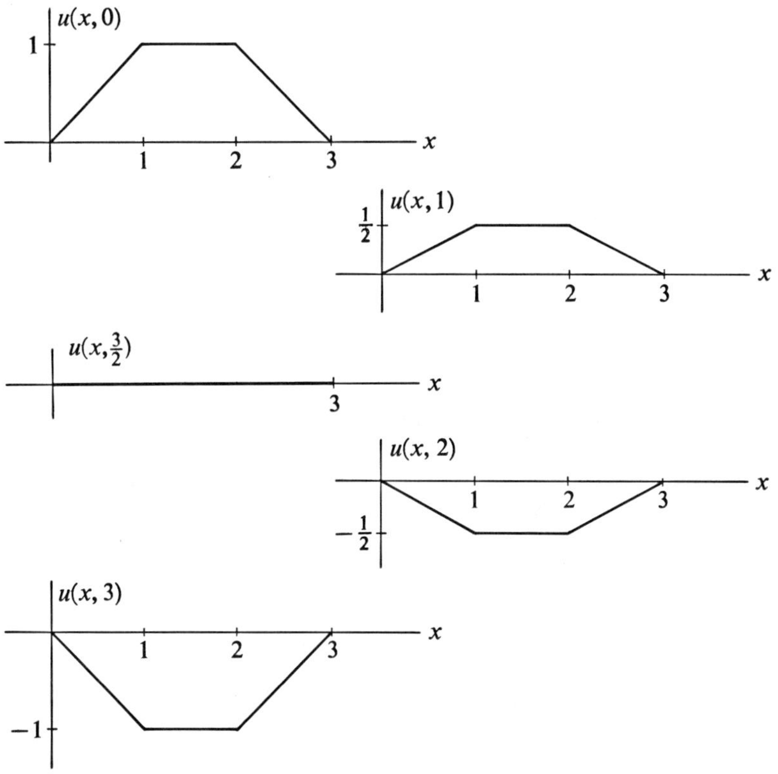

3.

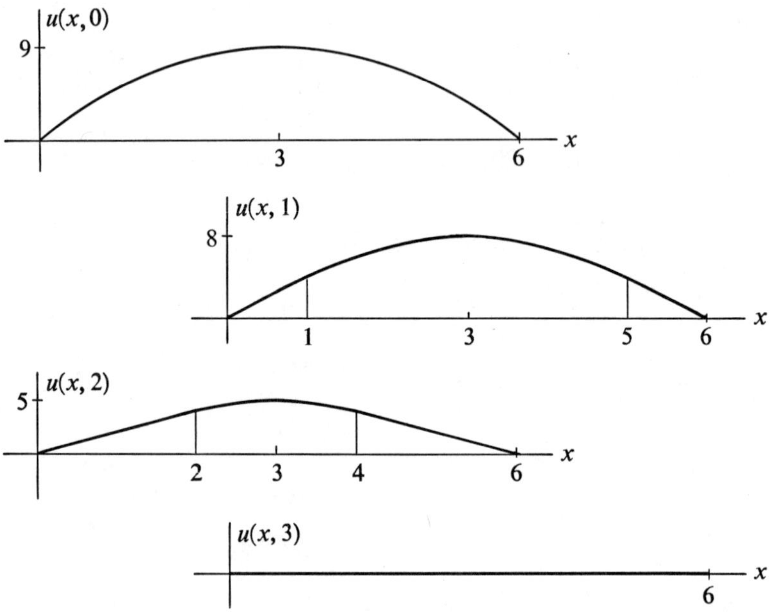
u(x,0)
9
3
6
x
u(x, 1)
8
1
3
5
6
x
u(x, 2)
5
2
3
4
6
x
u(x, 3)
6
x

4.

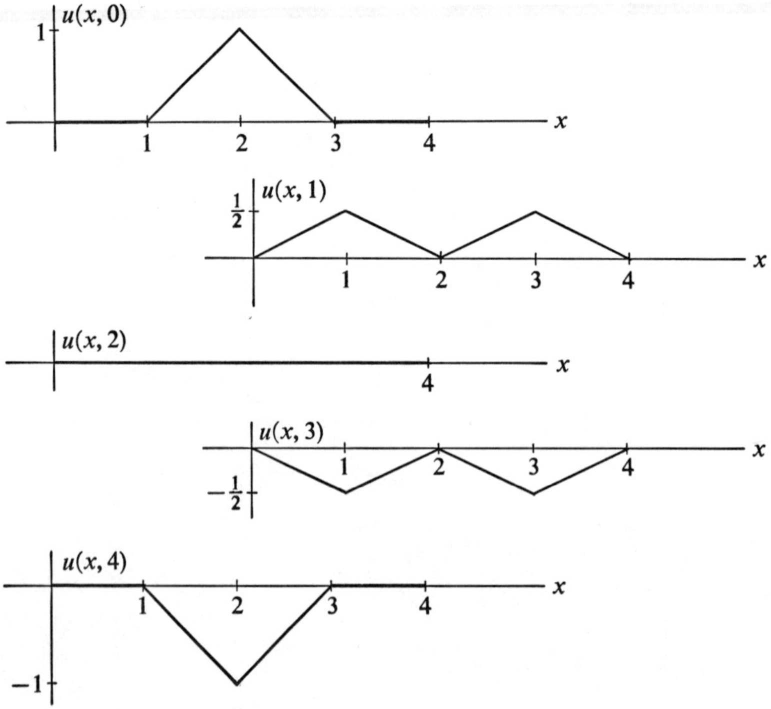
u(x,0)
1
1
2
3
4
x
u(x, 1)
1/2
1
2
3
4
x
u(x, 2)
4
x
u(x, 3)
1
2
3
4
x
−1/2
u(x, 4)
1
2
3
4
x
−1

6. $u(x,1) = \frac{x}{2}, 0 < x < 1,$
$\quad = \frac{1}{2}, 1 < x < 2,$
$\quad = -\frac{x}{2} + \frac{3}{2}, 2 < x < 3;$
$u(x, \frac{3}{2}) = 0; u(x,2) = -u(x,1);$
$u(x,3) = -u(x,0).$

7. $u(x,1) = 4x, 0 < x < 1,$
$\quad = -x^2 + 6x - 1, 1 < x < 5,$
$\quad = -4(x-6), 5 < x < 6;$
$u(x,2) = 2x, 0 < x < 2,$
$\quad = -x^2 + 6x - 4, 2 < x < 4,$
$\quad = -2x + 12, 4 < x < 6;$
$u(x,3) = 0;\ u(x,4) = -u(x,2);\ u(x,5) = -u(x,1);$
$u(x,6) = -u(x,0).$

8. $u(x,1) = \frac{1}{2}x, 0 < x < 1,$
$\quad = -\frac{1}{2}(x-2), 1 < x < 2,$
$\quad = \frac{1}{2}(x-2), 2 < x < 3,$
$\quad = -\frac{1}{2}(x-4), 3 < x < 4;$
$u(x,2) = 0;\ u(x,3) = -u(x,1);\ u(x,4) = -u(x,0).$

exercises 11.7, p. 400

1. $u(x,t) = \sin \pi x \cos c\pi t.$

2. $$u(x,t) = \sum_{n=0}^{\infty} \frac{8}{(2n+1)^3 \pi^3} \sin (2n+1) \pi x \cos c (2n+1) \pi t.$$

3. $u(x,t) = \sin 4\pi x \cos 4\pi ct - \sin 100\pi x \cos 100\pi ct.$

4. $$u(x,t) = \sum_{n=0}^{\infty} \left[\frac{8(-1)^n}{(2n+1)^2 \pi^2}\right] \sin \frac{(2n+1)\pi x}{2} \cos \frac{(2n+1)\pi ct}{2}.$$

5. $u(x,t) =$
$$\sum_{n=1}^{\infty} \frac{12}{(2n+1)^2 \pi^2} \sin \frac{(2n+1)\pi}{3} \sin \frac{(2n+1)\pi x}{3} \cos \frac{(2n+1)\pi ct}{3}.$$

exercises 11.8, p. 404

1. $$T(x,t) = \frac{8}{\pi^2} \sum_{n=0}^{\infty} \frac{(-1)^n}{(2n+1)^2} \sin \frac{(2n+1)\pi x}{2} e^{-[(2n+1)c\pi/2]^2 t}.$$

2. $$T(x,t) = \frac{32}{\pi^3} \sum_{n=0}^{\infty} \frac{1}{(2n+1)^3} \sin \frac{(2n+1)\,\pi x}{2}\, e^{-[(2n+1)\,c\pi/2]^2\, t}.$$

3. $$T_{ss} = 10 + \frac{80}{L} x.$$

4. $$T(x,t) = 20 + \frac{50}{L} x + \frac{1}{\pi} \sum_{n=1}^{\infty} \left[\frac{(-1)^{n+1}\, 80 + 20}{n} \right] \sin \frac{n\pi x}{L}\, e^{-(cn\pi/L)^2\, t}.$$

chapter twelve

exercises 12.1–12.3, p. 413

1. (a) $32 + 47i$. (b) $-i$. (c) $5(1 - i\sqrt{3})$.
(d) $8i$. (e) $8i$. (f) $3 + 4i$.
(g) $(5 + 2i)/29$. (h) $x^2 - y^2$. (i) $x^2 + y^2$.
(j) $1/10$.

2. (a) $2 \operatorname{cis}(-\pi/3)$. (b) $2 \operatorname{cis}(\pi/2)$. (c) $5 \operatorname{cis} 0$.

3. Rotation by 90° ccw.

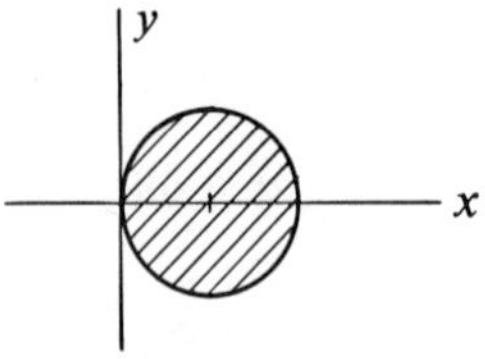

6. (a)

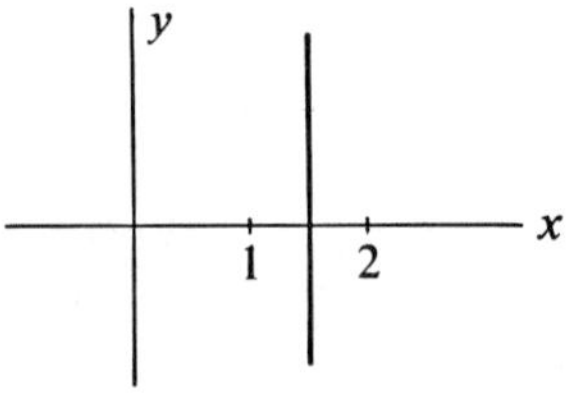

(b)

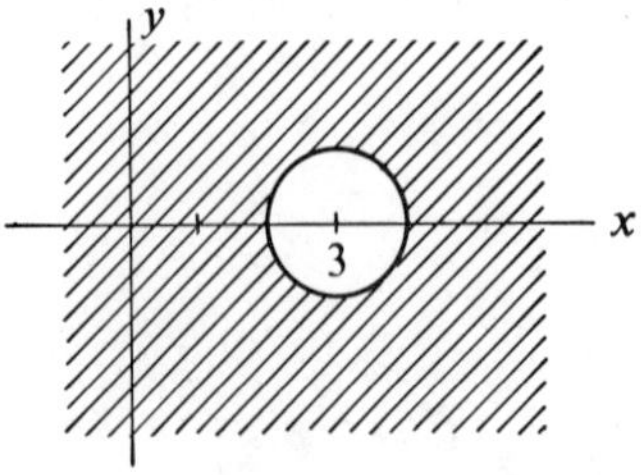

(c)

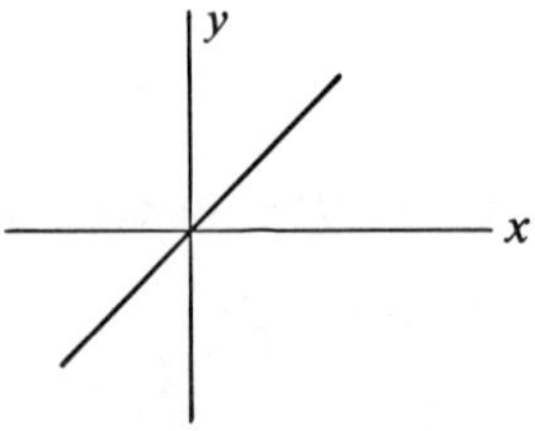

(d)

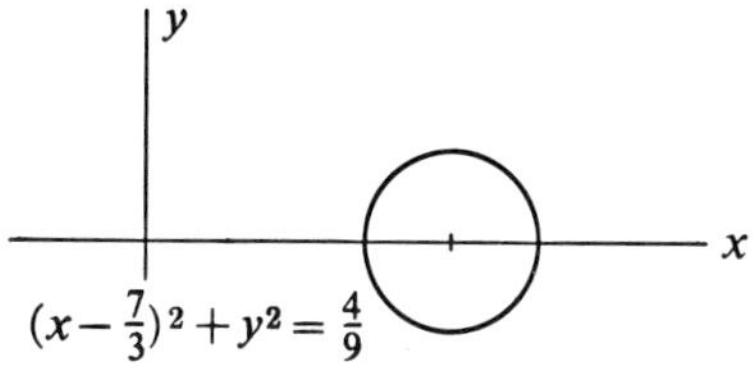

(e)

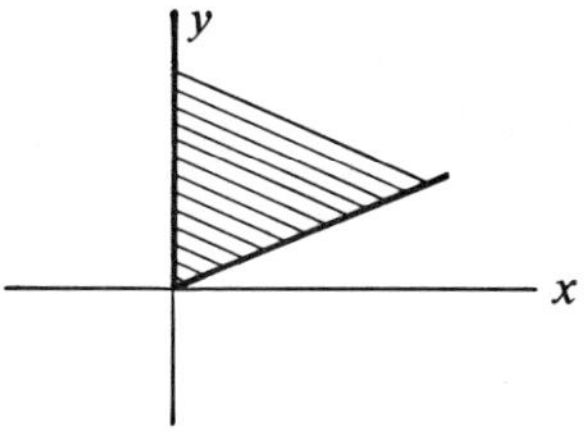

(f)

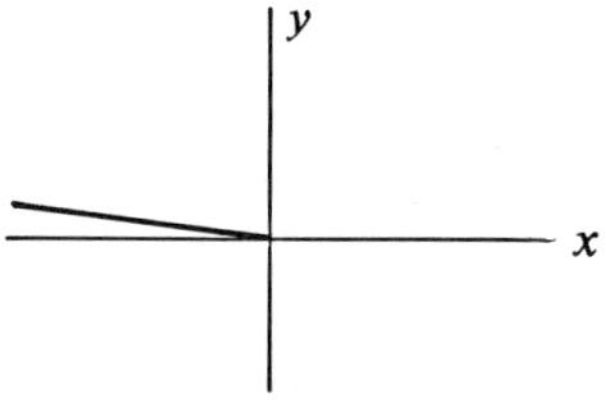

(g)

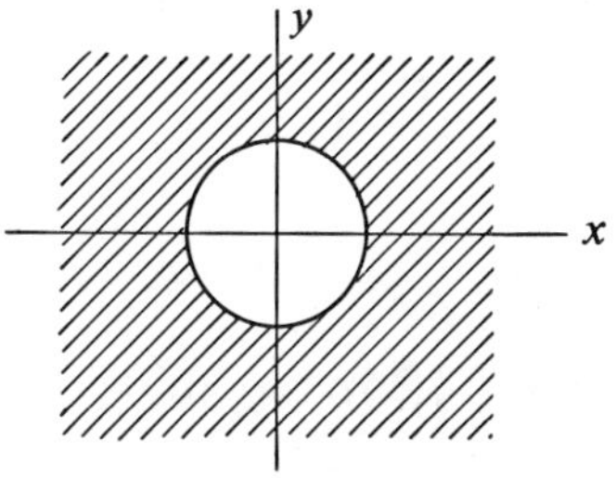

(h)

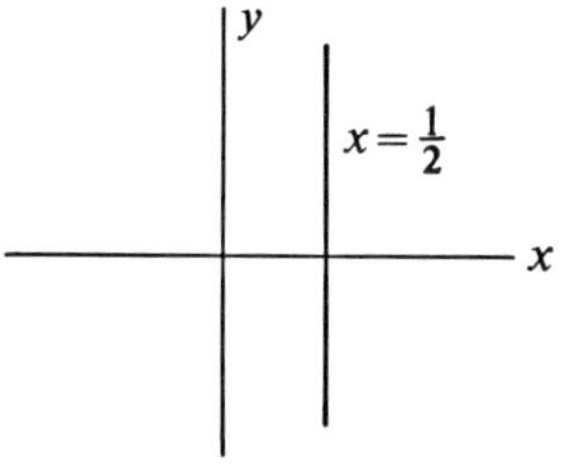

(i)

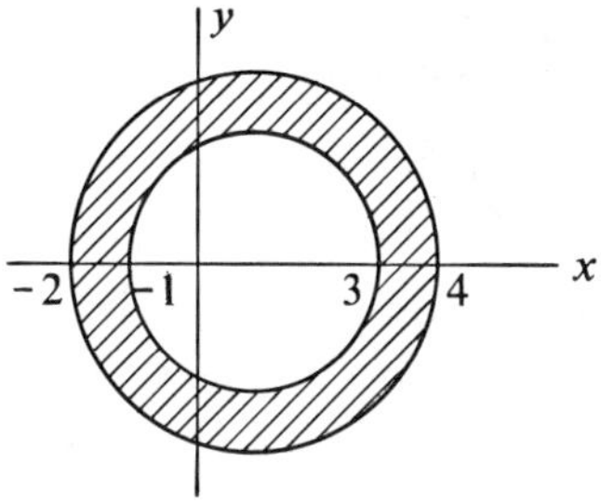

(j)

7. (a) $v = u - 1$. (b) $\begin{Bmatrix} u = x^2 - x^4 \\ v = 2x^3 \end{Bmatrix}$. (c) $|w| < 4$.

(d) $|w| < 1$; $0 < \operatorname{Arg} w < \pi$. (e) $(u - \frac{1}{2})^2 + v^2 < \frac{1}{4}$.

8. (a) $u = v$ in the first quadrant.

(b) Image of left side is the segment $u = 0$, $0 \leqq v \leqq \sinh \pi/2$. Image of bottom side is $v = 0$, $0 \leqq u \leqq 1$. Image of right side is $v = 0$, $1 \leqq u \leqq \cosh \pi/2$. Image of top is part of ellipse

$$\frac{u^2}{\cosh^2 \pi/2} + \frac{v^2}{\sinh^2 \pi/2} = 1.$$

(c) First quadrant portion of the annulus is $1 \leqq u^2 + v^2 \leqq e^{\pi}$.

(d) $|w| > \frac{1}{2}$. (e) $u < \log_e 3$, $0 < v < \pi/3$.

12. (a) $z\bar{z}$. (b) $i\bar{z}$. (c) z^3.

13. (a) $u = x^2 - y^2 + x, v = y - 2xy$.
(b) $u = x^2 + y^2 + 5x, v = 5y$.
(c) $u = \dfrac{x^3 + x^2 + x + 1 - 3xy^2}{(x^2 - y^2 + 1)^2 + 4x^2y^2}, v = \dfrac{3x^2y - y^3 + 2xy + y}{(x^2 - y^2 + 1)^2 + 4x^2y^2}$.

exercises 12.4, p. 421

4. (a) $3z^2 + 1$. (b) $\dfrac{-2z^5 - 6z^4 - 2z - 2}{(z^4 - 1)^2}$.
(c) $\dfrac{-1}{z^2}$. (d) $\dfrac{ad - bc}{(cz + d)^2}$.
(e) $18(15z^{14} + 30z^9)(z^{15} + 3z^{10} + 17)^{17}$.

5. (a) $\cos x \cosh y - i \sin x \sinh y$. (b) $1/z$.
(c) $-\sin x \cosh y - i \cos x \sinh y$.

6. (a)–(e) Analytic nowhere.
(f) Analytic everywhere except at $z = i$.
(g) Analytic nowhere.

9. (a) $e^x \cos y + ie^x \sin y$.
(b) z^3. (c) $-iz$. (d) $-iz^2/2$.
(e) $\operatorname{Arg} z - i \log_e|z|$.
(f) $\log_e|z| + i \operatorname{Arg} z$.

11. (a) $v = y$. (b) $u = -y$.
(c) $v = -e^x \cos y$. (d) $u = -\operatorname{Arg} z$.
(e) $v = -\dfrac{y}{x^2 + y^2}$. (f) $v = \cos x \sinh y$.
(g) $v = \dfrac{y^2 - x^2}{2} + (y - x)$.

13. Coefficients of x^2 and y^2 are equal in magnitude but of opposite sign.

16. (a) Lines parallel to the coordinate axes.
(b) Confocal ellipses and hyperbolas, foci at $(-1, 0)$ and $(1, 0)$.
(c) Concentric circles, center at the origin, and rays emanating from the origin.

exercises 12.5–12.6, p. 430

1. (a) $1 \operatorname{cis} \pi/4, 1 \operatorname{cis} 3\pi/4, 1 \operatorname{cis} 5\pi/4, 1 \operatorname{cis} 7\pi/4$.

(b) $2 \operatorname{cis} \pi/6$, $2 \operatorname{cis} 5\pi/6$, $2 \operatorname{cis} 3\pi/2$.
(c) $2 \operatorname{cis} \pi/5$, $2 \operatorname{cis} 3\pi/5$, $2 \operatorname{cis} \pi$, $2 \operatorname{cis} 7\pi/5$, $2 \operatorname{cis} 9\pi/5$.
(d) $2 \operatorname{cis}(-\pi/4)$, $2 \operatorname{cis} 3\pi/4$.

2. (a) $4 \operatorname{cis} \pi/4$, $4 \operatorname{cis} 3\pi/4$, $4 \operatorname{cis} 5\pi/4$, $4 \operatorname{cis} 7\pi/4$.
(b) $z = 3 \pm \sqrt{3}i$.
(c) $1 \operatorname{cis} 0$, $2 \operatorname{cis} \pi/4$, $\operatorname{cis} \pi/2$, $2 \operatorname{cis} 3\pi/4$, $\operatorname{cis} \pi$, $2 \operatorname{cis} 5\pi/4$, $\operatorname{cis} 3\pi/2$, $2 \operatorname{cis} 7\pi/4$.

3. One way: $f_1(z) = |z|^{1/2} \operatorname{cis} \theta/2$, $0 \leqq \theta < 2\pi$, $f_2(z) = |z|^{1/2} \operatorname{cis} \theta/2$, $2\pi \leqq \theta < 4\pi$. Branch cut is along positive real axis. To show discontinuity, examine functional values on either side of the branch cut.

4. $f_1(z) = |z|^{1/5} \operatorname{cis} \theta/5$, $-3\pi/2 < \theta \leqq \pi/2$, etc. This yields a 5-sheeted Riemann surface.

6. (a) $u^2 + v^2 = e^4$. (b) $v = u(\tan 2)$.
(c) $u^2 + v^2 = e^{2 \tan^{-1}(v/u)}$. (d) $0 < |w| \leqq 1$.

7. (a) $z = \log_e 2 \pm i(2n+1)\pi$.
(b) No solution. (c) $z = i(4n+1)\pi/2$. (d) $z = 2n\pi i$.

8. $\operatorname{Re} z < 0$.

exercises 12.7–12.9, p. 437

2. *Hint:* Recall that the square of the focal distance is the difference of the squares of the semimajor and semiminor axes.

3. Typical image.

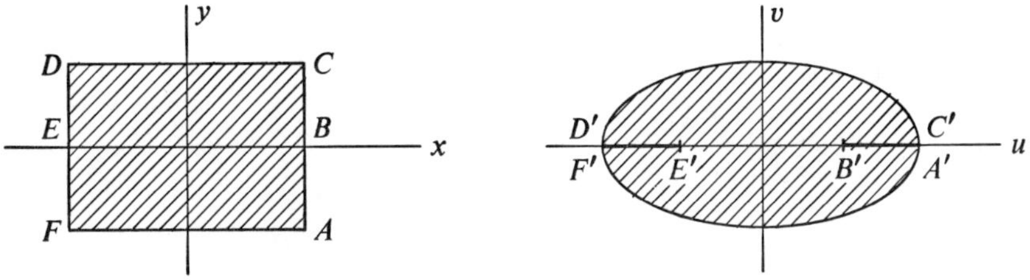

4. (a) $z_n = (4n+1)\dfrac{\pi}{2} + i \cosh^{-1} 2$.

(b) $z_n = \pm \dfrac{\pi}{3} \pm 2n\pi$.

(c) $z_n = (4n+3)\dfrac{\pi}{2} + i \sinh^{-1} 1$; $(4n+1)\dfrac{\pi}{2} + i \sinh^{-1}(-1)$.

(d) $z_n = (4n+1)\dfrac{\pi}{2} \pm 2i$.

5. The line $y = 0$ and the lines $x = (2n+1)\frac{\pi}{2}$.

8. $(\sin^2 x \cosh^2 y + \cos^2 x \sinh^2 y)^{1/2}$,
$(\cos^2 x \cosh^2 y + \sin^2 x \sinh^2 y)^{1/2}$.

10. (a) $2n\pi i$. (b) $i(2n\pi + 1)$.
(c) $i\left(2n\pi + \frac{\pi}{2}\right)$.

11. (a) $e^{-1} \operatorname{cis} 2n\pi$. (b) $\operatorname{cis}(1 + 2n\pi)$. (c) $\operatorname{cis} 2n\pi$.

12. (a) z.
(b) $ze^{2n\pi(\operatorname{Arg} z) - 4mn\pi^2} \operatorname{cis} 2m\pi \log_e |z|$.
(c) $z + 2n\pi i$. (d) $z + 2n\pi i$.

13. (a) $e^{\frac{\pi}{4} + 2n\pi} \operatorname{cis}(-\log_e \sqrt{2})$. (b) $ie^{\frac{\pi}{2} + 2n\pi}$.
(c) $e^{-2n\pi}$. (d) $\operatorname{cis}(2\sqrt{3}n\pi)$.

chapter thirteen

exercises 13.1–13.3, p. 444

1. (a) $u = \dfrac{t}{t^4 - t^2 + 1}$, $v = \dfrac{1 - t^2}{t^4 - t^2 + 1}$.
(b) $\left(u + \frac{1}{24}\right)^2 + v^2 = \left(\frac{5}{24}\right)^2$.
(c) $\{w \mid |w - \frac{1}{2}| < \frac{1}{2}\} \cap \{w \mid \operatorname{Im} w < 0\}$.
(d) $\{w \mid |w + i| > 1\} \cap \{w \mid \operatorname{Im} w < 0\}$.
(e) $|w| = 1$. (f) $\operatorname{Re} w = \frac{1}{2}$.

2. (a) No rotation. (b) Rotation by π. (c) No rotation.

3. (a) $v = \text{constant}$, $u = \text{constant}$. (b) Positive imaginary axis.
(c) Negative real axis. (d) $-\frac{\pi}{3} < \operatorname{Arg} w < \frac{\pi}{6}$; $1 < |w| < 16$.
(e) The line $u = v$ in the first quadrant.
(f) $\{-3 < u < -1,\ v > 0\}$.
(g) $u = \sin t \cosh t$, $v = \cos t \sinh t$.
(h) $u = e^x \cos x$, $v = e^x \sin x$.

4. (a) $z \neq 0,\ z \neq \pm 1$. (b) $z \neq \pm 1,\ \pm i;\ \ z \neq 0$.

(c) $z \neq 0$. (d) $z \neq -\dfrac{b}{2a}$.

(e) Everywhere. (f) $z \neq (2n+1)\dfrac{\pi}{2}$.

(g) $z \neq n\pi$. (h) $z \neq 0$.

5. (a) $\dfrac{\pi}{4}$. (b) $\dfrac{\pi}{2}$.

(c) π. (d) π.

7. (a) $w = -1$. (b) $z \neq \pm 1$. (c) Yes.

8. (a) $u = 2t^3 - 6t^2 - 3t;\ v = 2t^3 - 3t - 1$.

(b) $u = -26t^3 + 54t^2 - 27t;\ v = -9(2t^3 - 8t^2 + 9t - 3)$.

9. (a) No. (b) No. (c) No. (d) Yes.

10. (a) $z(t) = t + i\left(\dfrac{1-2t}{3}\right)$.

(b) $z(t) = 2\cos t + i(\sin t)$.

(c) $z(t) = (1 + 2\cos t) + i(-1 + 2\sin t)$.

(d) $z(t) = \cosh t + i(\sinh t)$.

11. (a) $u = \frac{1}{9}(5t^2 + 4t - 1),\ v = \frac{2}{3}t(1-2t)$.

(b) $u = 4\cos^2 t - \sin^2 t;\ v = 2\sin 2t$.

(c) $u = 4(\cos t + \sin t) + 4\cos 2t;$
$v = 2(-1 + 2\sin t - 2\cos t + 2\sin 2t)$.

(d) $u = 1,\ v = \sinh 2t$.

12. (a) $z(t) = (t+2) + 4i(t+1)$.

(b) $z(t) = (t+1) + ie\left(\dfrac{t}{3} + 1\right)$.

(c) $z(t) = t + 2i$.

exercises 13.4–13.5, p. 454

1. (a) $w = \dfrac{3z + 2i}{iz + 6}$. (b) $w = -i\,\dfrac{z-1}{z+1}$.

(c) $w = -z + 1 + i$. (d) $w = \dfrac{2z+1}{z-1}$. (e) $w = z$.

2. (a) $1 \pm \sqrt{2}$. (b) $\dfrac{3 \pm \sqrt{5}}{2}$. (c) All points.

3. $w = z$.

4. (a) $\dfrac{w - w_1}{w - w_2}\dfrac{w_3 - w_2}{w_3 - w_1} = \dfrac{z - z_1}{z_3 - z_1}$.

(b) $\dfrac{w - w_1}{w_3 - w_1} = \dfrac{z - z_1}{z - z_2}\dfrac{z_3 - z_2}{z_3 - z_1}$.

(c) $\dfrac{w - w_1}{w_3 - w_1} = \dfrac{z - z_1}{z_3 - z_1}$.

5. (a) $w = \dfrac{az + 1}{z + a}$. (b) $w = \dfrac{az - 1}{z + a}$.

6. $\left(u - \dfrac{5}{6}\right)^2 + v^2 = \dfrac{1}{36}$.

7. $w_1 = 2z,\ w_2 = w_1 + 1,\ w_3 = \dfrac{1}{w_2},\ w_4 = -\tfrac{7}{2}w_3,\ w = \tfrac{5}{2} + w_4$.

9. $w = -\dfrac{2z + 1}{z}$.

10. $w = i\,\dfrac{z(i - 1) + 2}{z(1 - i) + 2i}$.

11. $w = \dfrac{iz + 1}{z + i}$.

12. $w = -\dfrac{(z + 1 - i)^4 - i}{(z + 1 - i)^4 + i}$.

13. $w = \left(\dfrac{iz + 1}{z + i}\right)^4$.

14. $w_1 = \dfrac{az}{d},\ w_2 = w = w_1 + \dfrac{b}{d}$.

exercises 13.7, p. 461

1. $v = \dfrac{100}{\pi}\tan^{-1}\left[\dfrac{2ay}{x^2 + y^2 - a^2}\right]$.

2. $v = \dfrac{100}{\pi}\tan^{-1}\left[\dfrac{+\,10xy}{-(x^2 + y^2)^2 + 3(x^2 - y^2) + 4}\right]$.

3. $v = 100\left[1 - \dfrac{2}{\pi}\tan^{-1}\dfrac{y}{x}\right] = \dfrac{200}{\pi}\tan^{-1}\dfrac{x}{y}$.

4. $T = \dfrac{200}{\pi}\tan^{-1}\left[\dfrac{\cos\dfrac{\pi y}{2a}}{\sinh\dfrac{\pi x}{2a}}\right]$.

5. $T = \dfrac{100}{\pi} \tan^{-1}\left[\dfrac{x^2 + y^2 - 1}{2y}\right].$

6. $T(x, y) = \dfrac{200}{\pi} \sin^{-1} \dfrac{1}{10} [\sqrt{(y + 5)^2 + x^2} - \sqrt{(y - 5)^2 + x^2}].$

7. $T(x, y) = \dfrac{100}{\pi} \tan^{-1}\left[\dfrac{2y_1}{1 - x_1^2 - y_1^2}\right]$, where $\begin{cases} x_1 = \dfrac{x}{2}\left(1 + \dfrac{1}{x^2 + y^2}\right) \\ y_1 = \dfrac{y}{2}\left(1 - \dfrac{1}{x^2 + y^2}\right) \end{cases}.$

chapter fourteen

exercises 14.1–14.2, p. 469

1. (a) $S_n = a(z_1 - a) + z_1(z_2 - z_1) + \cdots + z_{n-1}(b - z_{n-1}).$
(b) $S_n^* = z_1(z_1 - a) + z_2(z_2 - z_1) + \cdots + b(b - z_{n-1}).$
(c) $\dfrac{S_n + S_n^*}{2} = \dfrac{b^2 - a^2}{2}.$

2. (a) 0. (b) 8.
3. (a) $6 + 5i$. (b) $\sqrt{61}$.

4. (a) $\frac{128}{15} + i\frac{57}{2}$. (b) $\frac{28}{3}[1 + 3i]$.

5. $\dfrac{\pi e}{2}$.
6. (a) $2\pi i$. (b) $-4\pi i$.

7. $\dfrac{\pi}{2} + \frac{1}{2}\log_e 50$.

8. (a) 0. (b) $\dfrac{i\pi}{8}$. (c) $\frac{3}{2} + i\frac{28}{3}$. (d) $49\pi i$. (e) $-2\pi i$.

9. (a) 2. (b) $\pi^2 e^{2\pi}$. (c) 1. (d) $\log_e\sqrt{2} + \dfrac{\pi}{4}$. (e) 1.

10. (a) $\dfrac{1}{\pi}(-1 + \cosh 3\pi)$. (b) $\dfrac{2\cosh\pi}{\pi}$.

11. $2\pi i$.
12. $2\pi i$.

exercises 14.3, p. 475

1. $1 - \cos 1 \cosh 1 + i \sin 1 \sinh 1$.
2. πi.
3. $\frac{2}{3}(-1 - i)$.
4. $\frac{2}{3}(1 + i)$.

5. $\frac{511}{3}$. **6.** $\frac{2}{3}(-1+i)$.

8. (a) 0. (b) $-\cos 1 + \sin 1$.

(c) $\log_e \frac{1}{2} + i\left[\tan^{-1}\frac{1}{2} - \tan^{-1}\frac{1}{3} - \frac{\pi}{4}\right]$.

(d) $\log_e \sqrt{10} + i \tan^{-1}\frac{1}{3}$. (e) $\frac{2 \sinh 6\pi}{3}(\sin 3 - i \cos 3)$.

(f) $\frac{63}{4} - 6i$. (g) $\frac{1}{2}\tan^{-1}\frac{i}{2} = \frac{i}{4}\ln 3$.

exercises 14.4–14.7, p. 486

1. (a) $4\pi i$. (b) $2\pi i$. (c) 0. (d) 0. **2.** When $m \geqq 0$. **3.** 0.

4. (a) Does not exist. (b) $-2\pi i$. (c) $-2\pi i$.
(d) Does not exist. (e) 0. (f) 0.

5. $2\pi i$. **6.** (a) Does not exist. (b) 0.

7. (a) 0. (b) -2π. (c) 2π. (d) -2π. (e) 0.

8. (a) $2\pi i$. (b) 0. (c) $2\pi i$. (d) 0. (e) $\pi/4$.
(f) 0. (g) 0. (h) $14\pi i$. (i) $\pi i/3$. (j) $-\pi i/3$.

9. For example, let $f(x) = x^2 \sin 1/x$, $x \neq 0$, and 0 for $x = 0$. Impossible for complex functions.

chapter fifteen

exercises 15.1–15.4, p. 496

1. Diverges. **2.** Converges to 0. **3.** Converges to 0.

4. Diverges. **5.** Diverges. **6.** Converges to 0.

7. Converges to 2/3. **8.** Diverges.

9. Converges to 2/3. **10.** Converges to 0. **11.** Converges.

12. Converges. **13.** Diverges. **14.** Diverges. **15.** Converges.

16. Diverges. **17.** Converges. **18.** Diverges. **19.** Converges.

20. Converges for $p < -1$, diverges for $p \geqq -1$.

21. Converges for $\operatorname{Im} z = 0$. **22.** Converges for $\operatorname{Re} z < 0$.

23. $|z + \frac{15}{8}| > \frac{5}{8}$.

24. $|z - \frac{2}{15}| < \frac{8}{15}$.

25. (a) $|z| < 1$. (b) $|z| < \frac{1}{3}$.

26. $|z + \frac{10}{3}| < \frac{5}{3}$.

27. $|z - a| < 1$.

28. $|z - a| < 1$.

29. $|z| < 2$.

30. $|z - 4| < 3$.

exercises 15.5, p. 503

1. $\sum_{n=0}^{\infty} (-1)^n \frac{z^{2n}}{(2n)!}$, converges for all z.

2. $\sum_{n=0}^{\infty} (-1)^n \frac{z^{2n+1}}{(2n+1)!}$, all z.

3. $\sum_{n=0}^{\infty} (-1)^n \frac{z^n}{n!}$, all z.

4. $\sum_{n=0}^{\infty} (-1)^n (z-1)^n$, $|z-1| < 1$.

5. $\sum_{n=0}^{\infty} \frac{(z-i)^n}{(1-i)^{n+1}}$, $|z-i| < \sqrt{2}$.

6. $\sum_{n=0}^{\infty} (-1)^n \frac{z^{2n}}{(2n+1)!}$, all z.

7. $\sum_{n=0}^{\infty} (-1)^n (n+1)(z-1)^n$, $|z-1| < 1$.

8. $\sum_{n=0}^{\infty} z^{3n}$, $|z| < 1$.

9. $\sum_{n=0}^{\infty} (-1)^n \frac{z^{4n+2}}{(2n+1)!}$, all z.

10. $\sum_{n=0}^{\infty} \frac{z^{2n}}{n!}$, all z.

11. $(z-1)^2 + 2(z-1) + 1$, all z.

12. $\frac{1}{7} \sum_{n=0}^{\infty} (-1)^n \left[\frac{3(z-2)}{7}\right]^n$, $|z-2| < \frac{7}{3}$.

13. $(z+1) - 2$, all z.

14. $z^2 - 8z + 2$, all z.

15. $(z-1)^2 - 6(z-1) - 5$, all z.

16. $\frac{1}{i+4} \sum_{n=0}^{\infty} (-1)^n \left(\frac{2}{i+4}\right)^n (z-2)^n$, $|z-2| < \frac{\sqrt{17}}{2}$.

17. $-\frac{1}{4} \sum_{n=0}^{\infty} \left[1 + \frac{(-1)^n}{3^{n+1}}\right] (z-1)^n$, $|z-1| < 1$.

18. $\frac{i}{4}\sum_{n=0}^{\infty}(-1)^n\left[\frac{1}{(2i+1)^{n+1}}-\frac{1}{(1-2i)^{n+1}}\right](z-1)^n, |z-1|<\sqrt{5}.$

19. $\frac{1}{6}\sum_{n=0}^{\infty}(-1)^n(n+1)(n+2)(n+3)z^n, |z|<1.$

20. $\sum_{n=0}^{\infty}\left(1-\frac{1}{2^{n+1}}\right)(-1)^n(z+1)^n, |z+1|<1.$

21. $-\sum_{n=0}^{\infty}z^{4n}, |z|<1.$

22. $e(1-z+3z^2\cdots).$

23. $\cos 1+z\sin 1-\frac{z^2\cos 1}{2}-\frac{z^3\sin 1}{6}\cdots.$

24. $1+z+\frac{1}{2}z^2-\frac{1}{8}z^4\cdots.$

25. $1+\frac{z}{2}-\frac{z^2}{8}+\frac{z^3}{16}-\frac{5z^4}{128}\cdots.$

exercises 15.6, p. 509

1. $\sum_{n=0}^{\infty}\frac{z^{3-n}}{n!}, 0<|z|.$

2. $\frac{i}{2}\sum_{n=0}^{\infty}\left[\frac{(-1)^n}{(i+1)^{n+1}}+\frac{1}{(i-1)^{n+1}}\right](z-1)^n, 0\leqq|z-1|<\sqrt{2};$

$\frac{i}{2}\sum_{n=0}^{\infty}[(-1)^n(i+1)^n-(i-1)^n]z^{-n-1}, \sqrt{2}<|z-1|.$

3. $-\sum_{n=0}^{\infty}z^{4n}, 0\leqq|z|<1, \sum_{n=0}^{\infty}z^{-4(n+1)}, 1<|z|.$

4. $\sum_{n=0}^{\infty}(-1)^n(n+1)(z-1)^n, 0\leqq|z-1|<1;$

$\sum_{n=0}^{\infty}(-1)^n(z-1)^{-n-z}(n+1), 1<|z-1|.$

5. $\frac{1}{2}\sum_{n=0}^{\infty}(-1)^n(n+1)(n+2)(z-1)^n, 0\leqq|z-1|<1;$

$\frac{1}{2}\sum_{n=0}^{\infty}(-1)^n(n+1)(n+2)(z-1)^{-n-3}, |z-1|>1.$

6. $\frac{1}{6}\sum_{n=0}^{\infty}(-1)^n(n+1)(n+2)(n+3)(z-1)^n, 0 \leqq |z-1| < 1;$

$\frac{1}{6}\sum_{n=0}^{\infty}(-1)^n(n+1)(n+2)(n+3)(z-1)^{-n-4}, 1 < |z-1|.$

7. $\sum_{n=0}^{\infty}(-1)^n z^{2n-2}, 0 < |z| < 1; \sum_{n=0}^{\infty}(-1)^n z^{-2n-4}, 1 < |z|.$

8. $\sum_{n=0}^{\infty}\frac{e(z-1)^{n-2}}{n!}, |z-1| > 0.$

9. $-\sum_{n=0}^{\infty}\left(\frac{z+1}{2}\right)^n + \frac{1}{z+1}, 0 < |z+1| < 1 \text{ and } 1 < |z+1| < 2;$

$\sum_{n=0}^{\infty}\left(\frac{2}{z+1}\right)^{n+1} + \frac{1}{z+1}, 2 < |z+1|.$

10. $\frac{1}{z} - \frac{z}{3} - \frac{1}{45}z^3 \cdots, 0 < |z| < \pi.$

11. (a) $-\sum_{n=0}^{\infty} z^{2n}, 0 \leqq |z| < 1; \sum_{n=0}^{\infty} z^{-2n-2}, 1 < |z|.$

(b) $\sum_{n=0}^{\infty}\frac{(z-1)^{n-1}(-1)^n}{2^{n+1}}, 0 < |z-1| < 2;$

$\sum_{n=0}^{\infty}\frac{2^n(-1)^n}{(z-1)^{n+2}}, |z-1| > 2.$

(c) $-\sum_{n=0}^{\infty}\frac{(z+1)^{n-1}}{2^{n+1}}, 0 < |z+1| < 2;$

$\sum_{n=0}^{\infty}\frac{2^n}{(z+1)^{n+2}}, |z+1| > 2.$

12. (a) $\frac{1}{z}, |z| > 0.$

(b) $\sum_{n=0}^{\infty}(-1)^n(z-1)^n, 0 \leqq |z-1| < 1;$

$\sum_{n=0}^{\infty}(-1)^n(z-1)^{-n-1}, 1 < |z-1|.$

(c) $\sum_{n=0}^{\infty} (-1)^n \frac{(z-2)^n}{2^{n+1}}, 0 \leqq |z-2| < 2;$

$\sum_{n=0}^{\infty} (-1)^n \frac{2^n}{(z-2)^{n+1}}, 2 < |z-2|.$

13. (a) $\frac{i}{z} + \frac{1}{i+1} \sum_{n=0}^{\infty} \left[\frac{1}{i^n} + (-1)^n\right] z^n, 0 < |z| < 1;$

$\frac{i}{z} + \frac{1}{i+1} \sum_{n=0}^{\infty} [-i^{n+1} + (-1)^n]\, z^{-n-1}, 1 < |z|.$

(b) $\sum_{n=0}^{\infty} (-1)^n (z-i)^n [i^{-n} + (1+i)^{-n-2}] - \frac{i}{i+1} \cdot \frac{1}{z-i},$

$0 < |z-i| < 1;$

$\sum_{n=0}^{\infty} (-1)^n \left[\frac{i^{n+1}}{(z-i)^{n+1}} + \frac{(z-i)^n}{(1+i)^{n+2}}\right] - \frac{i}{i+1} \cdot \frac{1}{z-i},$

$1 < |z-i| < \sqrt{2};$

$\sum_{n=0}^{\infty} \frac{(-1)^n}{(z-i)^{n+1}} [i^{n+1} + (1+i)^{n-1}] - \frac{i}{i+1} \cdot \frac{1}{z-i}, \sqrt{2} < |z-i|.$

(c) $\sum_{n=0}^{\infty} (-1)^n (z-1)^n \left[\frac{1}{2^{n+1}} \cdot \frac{1}{1+i} + i - \frac{i}{(1+i)(1-i)^{n+1}}\right],$

$0 \leqq |z-1| < 1;$

$$\sum_{n=0}^{\infty} (-1)^n (z-1)^n \left[\frac{1}{2^{n+1}} \cdot \frac{1}{1+i} - \frac{i}{(1+i)(1-i)^{n+1}}\right] + \frac{(-1)^n i}{(z-1)^{n+1}}, 1 < |z-1| < \sqrt{2};$$

$$\sum_{n=0}^{\infty} (-1)^n (z-1)^n \left[\frac{1}{2^{n+1}} \cdot \frac{1}{1+i}\right] + \frac{(-1)^n}{(z-1)^{n+1}} \left[i - \frac{(1-i)^n i}{1+i}\right], \sqrt{2} < |z-1| < 2;$$

$$\sum_{n=0}^{\infty} \frac{(-1)^n}{(z-1)^{n+1}} \left[i - \frac{i(1-i)^n}{1+i} + \frac{2^n}{1+i}\right], 2 < |z-1|.$$

exercises 15.7, p. 512

1. Simple zeros at ± 1 and $\pm i$. **2.** Zero of order 4 at 0.

3. Simple zeros at $\pm n\pi$.

4. Simple pole at $z = 1$, simple zeros at $z = \pm i$.

5. Double zeros at $\pm n\pi$, double poles at $\pm (2n+1)\frac{\pi}{2}$.

6. Essential singularity at 0, double zeros at $z = \pm \frac{1}{n\pi}$.

7. Simple zeros at $z = \pm 2n\pi i$.

8. Zero of order 4 at $z = 6$, simple pole at 0.

9. Simple zero at $z = 1$, simple pole at $z = \pm i$.

10. Simple zero at -5. **11.** Simple zeros at $\pm i$.

12. Simple zero at -1, double pole at 0.

13. Essential singularity at 0. **14.** Simple zeros at $\pm (2n+1)\frac{\pi}{2}$.

15. Simple zeros at $\pm i$, simple pole at 0. **16.** Simple zero at 0.

17. Removable singularity at $z = 0$, simple zeros at $\pm n\pi$, $n \neq 0$.

18. Simple pole at $z = 0$, simple zeros at $z = \pm n\pi$, $n \neq 0$.

19. Simple pole at $z = -1$, double pole at $z = 2$, triple pole at $z = 1$.

20. Removable singularity at $z = 0$, simple zeros at $z = \sqrt{\pm n\pi}$.

exercises 15.8, p. 516

1. e at $z = 1$. **2.** $-i \sinh 2$, $z = -2i$.

3. -1 at $(2n+1)\frac{\pi}{2}$. **4.** $\frac{1}{120}$ at $z = 0$.

5. $(-1)^n$ at $(2n+1)\frac{\pi}{2}$. **6.** 1 at $z = 0$.

7. $\frac{\sqrt{2}}{8}(-1 \mp i)$ at $z = e^{\pm \pi i/4}$, $\frac{\sqrt{2}}{8}(1 \mp i)$ at $z = e^{\pm 3\pi i/4}$.

8. $\dfrac{1}{7\operatorname{cis}\dfrac{9\pi}{7}}$ at $\operatorname{cis}\dfrac{3\pi}{14}$; $\dfrac{1}{7\operatorname{cis}3\pi}$ at $\operatorname{cis}\dfrac{\pi}{2}$; $\dfrac{1}{7\operatorname{cis}\dfrac{5\pi}{7}}$ at $\operatorname{cis}\dfrac{11\pi}{14}$;

$\dfrac{1}{7\operatorname{cis}\dfrac{3\pi}{7}}$ at $\operatorname{cis}\dfrac{15\pi}{14}$; $\dfrac{1}{7\operatorname{cis}\dfrac{\pi}{7}}$ at $\operatorname{cis}\dfrac{19\pi}{14}$; $\dfrac{1}{7\operatorname{cis}\left(-\dfrac{\pi}{7}\right)}$ at $\operatorname{cis}\dfrac{23\pi}{14}$;

$\dfrac{1}{7\operatorname{cis}\left(-\dfrac{3\pi}{7}\right)}$ at $\operatorname{cis}\dfrac{27\pi}{14}$.

9. $\frac{3}{2}$ at $z=0$, $-\frac{5}{2}$ at $z=-2$, 1 at $z=-1$.

10. 0 at $z=0$. **11.** πi. **12.** $\dfrac{\pi i}{2}$. **13.** $2\pi i$. **14.** 0.

15. $2\pi i$. **16.** 0. **17.** 0.

exercises 15.9, p. 518

1. $-2\pi i$. **2.** 0. **3.** $\pi i(1-e^{-2})$. **4.** 0. **5.** $2\pi i$.

6. $\dfrac{56\pi i}{25}$. **7.** $4\pi i$. **8.** 0. **9.** $-4\pi i$. **10.** 0.

11. (a) 0. (b) 0. (c) πi. (d) $-\pi i$.

12. (a) $-8\pi i$. (b) $-\pi i$. (c) Same as (b). (d) Does not exist.

13. (a) $6\pi i$. (b) 0. (c) Undefined. (d) 0. (e) $\dfrac{8\pi i}{3}$.

14. (a) $-8\pi i$. (b) $208\pi i$. (c) Same as (b).
(d) Same as (b). (e) $216\pi i$.

exercises 15.10, p. 530

1. $\dfrac{\pi}{\sqrt{2}}$.

2. $\dfrac{2\pi}{\sqrt{a^2-1}}$.

3. $\dfrac{2\pi}{\sqrt{1-a^2}}$.

4. $\dfrac{\pi}{4}$.

5. $\dfrac{3\pi}{8}$.

6. $\dfrac{\pi}{2\sqrt{2}}$.

7. $\dfrac{\pi}{2\sqrt{a^2+a}}$.

8. $\dfrac{2\pi}{1-m^2}$.

9. $\dfrac{\pi}{\sqrt{2}}$.

10. 0.

11. $\dfrac{\pi}{2}$.

12. $\dfrac{\pi}{4}\left[\dfrac{\sqrt{13}-3}{2}\right]^{1/2}$.

13. $\dfrac{5\pi}{96}$.

14. $\dfrac{10\pi e^{-9}}{54}$.

15. $\dfrac{\pi}{e}$.

16. 0.

17. $\dfrac{\pi\sqrt{2}}{6}$.

18. $s\pi$.

19. $\dfrac{\pi}{2\cosh\dfrac{m\pi}{2}}$.

20. $\dfrac{\pi}{8}$.

chapter sixteen

exercises 16.1–16.3, p. 538

1. $-\Gamma\left(\dfrac{4}{3}\right)$.

2. $\Gamma(p+1)$.

3. $\Gamma\left(\dfrac{7}{4}\right)$.

4. $\frac{2}{125}$.

5. $\dfrac{(-1)^n}{n!}$.

9. All real x.

exercises 16.4–16.5, p. 545

3. $\frac{1}{2}B(\frac{1}{2},\frac{5}{2})=\dfrac{3\pi}{16}=-0.59$.

4. $\sqrt{\pi}\,\dfrac{\Gamma(1/4)}{4\Gamma(3/4)}$.

7. π.

9. 1/120.

10. $\frac{1}{2}B(\frac{1}{2},\frac{1}{4})=2.62$.

12. $\frac{4}{3}B(\frac{1}{3},\frac{4}{3})$.

13. $4B(\frac{1}{2},\frac{1}{4})$.

16. (a) 0.882.

exercises 16.6–16.8, p. 551

4. $y = c_1 J_{1/2}(x) + c_2 J_{-1/2}(x)$. **5.** $y = c_1 I_3(x) + c_2 K_3(x)$.

6. $y = c_1 J_{1/2}\left(\frac{x}{2}\right) + c_2 J_{-1/2}\left(\frac{x}{2}\right)$.

exercises 16.9–16.12, p. 561

10. $-2J_1(x) + \int J_0(x)\,dx + c$. **11.** $-xJ_0(x) + \int J_0(x)\,dx + c$.

12. $-J_0(x) - 2J_2(x) + c = -J_2(x) - \frac{2}{x} J_1(x) + c$.

13. $x^2 J_1(x) + xJ_0(x) - \int J_0(x)\,dx + c$.

14. $x^3 J_1(x) - 2x^2 J_2(x) + c$.

exercises 16.13–16.15, p. 570

1. $f(x) \sim \sum_{j=1}^{\infty} A_j J_1(s_{j1}x)$ where the $3s_{j1}$ are the zeros of $J_1(x)$ and

$$A_j = \frac{2}{9J_2^2(3s_{j1})} \frac{1}{s_{j1}^2}\left[-3s_{j1}J_0(3s_{j1}) + \int_0^{3s_{j1}} J_0(x)\,dx\right].$$

2. $f(x) \sim \sum_{j=1}^{\infty} A_j J_0(s_{j0}x)$ where the s_{j0} are the zeros of $J_0(x)$ and

$$A_j = \frac{2}{s_{j0}J_1(s_{j0})}\left[1 - \frac{2J_2(s_{j0})}{s_{j0}J_1(s_{j0})}\right].$$

3. $f(x) \sim \sum_{j=1}^{\infty} A_j J_0(s_{j0}x)$ where the s_{j0} are the zeros of $J_0(x)$ and

$$A_j = \frac{4J_2(s_{j0})}{s_{j0}^2 J_1^2(s_{j0})}.$$

4. $f(x) \sim \sum_{j=1}^{\infty} A_j J_3(s_{j3}x)$ where the $2s_{j3}$ are the zeros of $J_3(x)$ and

$$A_j = \frac{1}{2J_4(2s_{j3})\, s_{j3}^3}\left[-4s_{j3}^2 J_2(2s_{j3}) - 8s_{j3}J_1(2s_{j3}) - 8J_0(2s_{j3}) + 1\right].$$

Index

Absolute convergence
of improper integrals, 586; of series, 492, 594
Abstract vector space, 3
Acceleration, 193
Addition
of arrows, 2; of complex numbers, 405; of matrices, 20; of operators, 61; of tensors, 309
Additive inverse of an arrow, 1, 3
Admissible functions, 60
Alternating series, 595
Amplitude, 92
Analytic function
definition of, 420; derivative of, 484
Angle preserving, 442
Angular frequency, 89, 95
Angular rotation vector, 201
Annihilator, 60, 77
Approximating polynomial, 142
Approximation
by straight lines, 51; in the mean, 353; Picard, 55; pointwise, 359; to Bessel functions, 555
Arc length function, 183
Arc of a curve, 182
Area
directed, 166; surface, 246
Argument, 407
Arrow, 1
Associated tensor, 312
Auxiliary conditions
homogeneous, 382; nonhomogeneous, 403
Auxiliary equation, 68

Basis
definition of, 15; for solution space, 66; local, 290; reciprocal, 288, 293
Basis set for $PC(a, b)$, 15, 325
Beats, 93
Behavior at infinity, 439
Bessel equation, 153, 547
Bessel functions
definition of, 548; modified, 550; zeros of, 558
Bessel inequality, 358
Bessel integral form, 557
Bessel modified equation, 550
Beta function, 539
Bilinear functions, 446
Binormal, to a curve, 187
Boundary value problem, 27, 379, 456
Bounded domain, 255
Bounded sequence, 575
Bounded set, 410
Branch functions, 424

Cartesian ordered triples, 161
Cartesian three-dimensional vector space, 4
Cauchy criterion
for improper integrals, 582; for infinite series, 590; for sequences, 574; for uniform convergence, 588
Cauchy-Euler equation, 86
Cauchy-Goursat Theorem, 476
Cauchy inequality, 485
Cauchy integral formula, 481
Cauchy Principal Value, 521, 584
Cauchy-Riemann conditions, 418
Central force field, 201
Centripetal acceleration, 194
Characteristic function, 130
Christoffel symbols, 315
Circle of convergence, 495
Circulation, 233
Circulation density vector, 234
Closed line integral, 230, 241
Closed set, 410
Closed surface, 255
Closed surface integral, 255, 277
Column of a matrix, 19
Column vector, 21
Combination, linear, 13, 353
Comparison test
for improper integrals, 585; for series, 492, 593
Complete set, 378
Complete solutions, 27
Complete trigonometric set, 325
Complex conjugate, 409
Complex derivative, 415
Complex field, 405, 439
Complex function, 411, 440
Complex integration, 463, 472
Complex logarithm, 434
Complex mapping, 411
Complex plane, 406
Complex sequence, 489
Complex series, 491
Complex trigonometric function, 431
Component functions, 200
Components
contravariant, 290; covariant, 296; of a tensor, 305; of a vector, 16
Condition for incompressibility, 217
Conditional convergence
of a series, 594; of improper integrals, 586

Conformal mapping, 442
Conjugate, complex, 409
Conjugate harmonic function, 421
Connected, multiply, 263
Connected set, 410
Connected, simply, 262, 273, 476
Conservation of mass, 217
Conservative field, 212, 242, 273, 281
Constant level surface, 175
Constant magnitude curves, 204
Constant magnitude surfaces, 204
Continuity equation, 217
Continuous vector function, 172
Contour integration, 522
Contraction, 311
Contravariant components, 290
Contravariant tensor, 306
Contravariant Transformation Law, 300
Convergence
circle of, 495; conditional, 586; domain of, 493, 576, 595; interval of, 576, 600; in the mean, 323; of a Fourier series, 340; of a sequence, 573; of an improper integral, 581; of an infinite series, 590; pointwise, 323; radius of, 495; uniform, 578, 587, 597
Convolution, 137
Coordinate curves, 190
Coordinate system
curvilinear, 292; oblique, 287
Cosine series, 345
Covariant components, 296
Covariant differentiation, 313
Covariant tensor, 306, 307
Covariant Transformation Law, 300
Critical damping, 90
Cross product, 164
Cumulative distribution function, 545
Curl, 220, 234
Current
steady state, 90; transient, 88
Curves
constant magnitude, 204; definition of, 177; in the complex plane, 440; normal vector to, 187; of constant slope, 53; orientation of, 181; piecewise smooth, 181; plane, 181; projecting cylinders of, 178; rectifiable, 183; simple, 183; skew, 181; smooth, 181; twisted, 181; vector function of, 180
Curvilinear coordinate system, 292

D'Alembert's method, 387, 393
Damping factor
in electric circuit, 89; in mechanical system, 95
Decay rate, 29
Deformation of path, 241, 264, 480
Derivative
covariant, 314; directional, 207; normal, 265; of a complex function, 415; of a vector function, 172; of Fourier series, 344; of an analytic function, 484
Differentiable vector function, 172
Differential, total, 37
Differential equations
functions defined by, 546; homogeneous, 59, 68; linear, 26, 59; method of infinite series, 141; method of Laplace transform in, 125; ordinary, 25; partial, 377; separable, 27, 386
Differentiation, covariant, 313
Dimension
of a vector space, 16; of solution space, 66
Directed area, 166
Direction field, 53
Directional derivative, 207
Dirichlet conditions, 340, 367
Dirichlet problem, 380, 456
Divergence
definition of, 215; of a sequence, 573; of an improper integral, 581; of an infinite series, 590; Theorem, 275
Domain
bounded, 255; definition of, 241, 410; multiply connected, 263; of convergence, 493, 576, 595; simply connected, 262, 476; standard, 260, 276
Dominated by a function, 108
Dot product, 8, 162
Driving function, 59
Dyad, 311

Eigenfunctions, 373, 568
Eigenvalue, 373
Electric circuits, 87
Electric field, 201
Electric potential, 199
Electrostatic potential, 459
Equality
in the mean, 322; in the norm, 322; of operators, 60; Parseval's, 358; pointwise, 322
Equation
Cauchy-Euler, 86; differential, 25, 546; heat, 278, 400; homogeneous, 59, 68; integral, 137; Laplace, 380, 456; linear, 26, 42, 59; parametric, 169; reduced, 59; system of, 100, 127; telegraph, 401; wave, 388
Equipotential surfaces, 203
Error function, 543
Error in the mean, 353
Essential singularity, 511

Euclidean geometric vectors, 161
Euclidean vector space, 3
Even extension, 345
Exact differential equation, 38
Expansion
  Fourier, 319; of functions, 319; of operators, 61; orthogonal, 319
Exponential
  function, 427; integral, 545; order, 109; shift, 63
Extended complex number system, 439
Extension
  even, 345; odd, 345; periodic, 341, 350

Factorial, 533
Field
  central, 201; complex, 405, 439; direction, 53; gravitational, 201; inverse square law, 205; irrotational, 221, 244, 274; magnetic, 217; scalar, 199; unidirectional, 206, 223; vector, 200
First Shifting Theorem, 110
Fixed points, 449
Force field, central, 201
Forms of solution, 144
Fourier-Bessel series, 565
Fourier coefficients, 327
Fourier cosine integral, 369
Fourier cosine series, 347
Fourier cosine transform, 369
Fourier expansion, 319
Fourier integrals, 362, 524
Fourier linear combination, 353
Fourier method, 17
Fourier sine integral, 369
Fourier sine series, 348
Fourier sine transform, 369
Fourier series
  definition of, 325; derivative of, 344
Fourier transform, 369
Frequency
  angular, 89, 95; fundamental, 399
Fresnel integral, 545
Frobenius series, 156, 547
Function
  admissible, 60; analytic, 420; arc length, 183; Bessel, 548; beta, 539; bilinear, 446; branch, 424; characteristic, 130; complex, 411, 440; defined by differential equation, 546; defined by improper integral, 531; defined by indefinite integral, 542; driving, 59; generating, 556; harmonic, 218, 280, 421; input, 59; inverse, 434; Neumann, 549; piecewise continuous, 5; pulse, 130; sine integral, 545; "turn on," 129; unit step, 130, 370; vector, 172; Weber, 549
Fundamental frequency, 399
Fundamental region of exp z, 429
Fundamental Theorem of Algebra, 486
Fundamental Theorem of Complex Integration, 472

Gamma function
  definition of, 532; extended, 537; Table of, 534
General solution, 27, 378
Generating function, 556
Geometric series, 492, 591
Geometric vector, 1
Geometric vector addition, 2
Gibbs phenomenon, 360
Gradient vector field, 210
Gravitational field, 201
Gravitational potential, 199
Green's Theorem, 259

Half-life, 29
Harmonic functions, 218, 280, 421
Harmonic series, 592
Heat equation, 278, 400
Homogeneous auxiliary conditions, 382
Homogeneous equation, 59, 68

Identity mapping, 446
Imaginary part of z, 406
Imaginary period, 429
Impedance, 91
Improper integrals
  definition of, 580; use in defining functions, 531; with a parameter, 531, 586
Incompressible, 218
Indefinite integral
  of Fourier series, 344; use in defining functions, 542
Independence
  linear, 14; of path, 240, 244, 471; of surface, 254, 270
Inequality
  Bessel, 358; Cauchy, 485; triangle, 11, 408
Infinite series
  methods in differential equations, 141; of constants, 590; sum of, 590
Initial condition, 379
Initial value problem, 27
Inner product
  definition of, 7; integral, 9, 321; of two tensors, 312; weighted, 321
Input function, 59
Integral
  Fresnel, 545; improper, 531, 580; indefinite, 542; line, 227, 463; surface, 250
Integral equations, 137

Integral inner product, 9
Integral powers of z, 423
Integral test, 592
Integrating factor, 42, 44
Integration
complex, 463; contour, 522; of rational functions, 521
Interval of convergence, 576, 600
Invariance of scalar fields, 199
Inverse bilinear transformation, 446
Inverse function, 434
Inverse Laplace transform, 119
Inverse square law field, 205
Inversion, 447
Irrotational field, 221, 244, 274
Isolated zero, 510
Isothermal surface, 203
Iterative method, 55

Laguerre polynomials, 326
Laplace's equation, 380, 456
Laplace transform
definition of, 106; existence of, 109; general properties, 115; inverse, 119; linearity property, 111; of derivative, 112; of indefinite integral, 114; of periodic function, 116; short table of, 108; use in solving differential equations, 124
Laplacian, 218
Laurent series, 504, 556
Legendre polynomials, 13, 151, 374
Length of arc, 183
Level curves, 203
Level surfaces, 203
Limit Comparison Test
for improper integrals, 585; for series, 593
Limit of a vector function, 171
Line integral
closed, 230, 241; definition of, 227, 463; in the complex plane, 463; of a vector field, 232
Linear combination, 13, 353
Linear differential equation, 26, 59
Linear equations
of the first order, 42; with constant coefficients, 26, 68; nonhomogeneous, 59, 72
Linear independence
definition of, 14; of arrows and ordered triples, 167
Linearity property, 8
Liouville's Theorem, 485
Local basis, 290
Logarithm, complex, 434
Low frequency transmission line, 401

"M" Test
for improper integrals, 589; for series, 597
Magnetic field, 217
Magnification factor, 445
Mapping
complex, 411; conformal, 442; identity, 446; of upper half-plane, 452
Matrices, 19
Mean convergence, 323
Mean square error, 353
Mechanical systems, 94
Method of
Frobenius, 156; successive antidifferentiations, 74; undetermined coefficients, 76, 147
Mixed tensor, 307
Modified Bessel functions, 550
Modulation, 569
Modulus, 407
Monotone sequence, 574
Morera's Theorem, 485
Multiple points, 182
Multiplication
of matrices, 21; scalar, 2, 20
Multiply connected domain, 263

Natural angular frequency
electrical, 89; mechanical, 95
Neumann function, 549
Newton's Law of Cooling, 31
Nonhomogeneous linear equation, 59, 72
Norm
of a partition, 464; of a vector, 9
Normal derivative, 265
Normal to surface, 212
Normal vector to a curve, 187
Normed vector space, 7

Oblique coordinate system, 287
Odd extension, 345
Open set, 410
Operator addition, 61
Operators
expansion of, 61; polynomial, 61
Order
of coefficients, 349; of a differential equation, 26; of a pole, 511
Ordered n-tuples, 5
Ordered triples, 3, 161
Ordinary differential equations, 25
Ordinary points, 155
Orientable surface, 193
Orientation of curves, 181
Orthogonal
definition of, 10; curvilinear coordinate system, 292; expansion, 319
Orthonormal, 11
Outer product of tensors, 310

Overdamping, 89
Overtones, 399

Parameter line, 169
Parameter plane, 170
Parametric equations, 169
Parseval's equality, 358
Partial derivative of a vector function, 172
Partial differential equation, 377
Particular solution
to nonhomogeneous equation, 73; to ordinary differential equation, 27; to partial differential equation, 378
Partition
norm of, 228, 464; of a curve, 228, 463
Period, imaginary, 429
Periodic extension, 341, 350
Phase modulation, 569
Phase shift, 92
Picard approximations, 55
Piecewise continuous function, 5
Piecewise smooth curves, 181
Plane curve, 181
Point at infinity, 439
Pointwise approximation, 359
Pointwise convergence, 323
Pointwise equality, 322
Polar form, 407
Pole, 511
Polynomial operator, 61
Polynomials
Legendre, 13, 151, 374; Tchebichef, 326
Positive definite, 8
Potential difference, 244
Potential
electric, 199, 459; gravitational, 199; scalar, 199; vector, 224, 271, 280
Power function, 435
Power of an operator, 60
Power series, 494, 498, 600
Principal part, 511
Principal Value, Cauchy, 521, 584
Principal values of arg z, 408
Principle of deformation of path, 264, 480
Principle of superposition, 74
Product
dot, 8, 162; inner, 7, 312, 321; outer, 310; scalar triple, 166; vector triple, 167
Product method, 385
Product of operators, 61
Product solution, 397
Projecting cylinders, 178
Projection, scalar, 163
Pulse function, 130

Radius of convergence
definition of, 495; of a power series, 600
Rank, 305
Ratio test, 492, 594
Rational function, integration of, 521
RC circuit, 126
Reactance, 91
Real part of z, 406
Reciprocal basis, 288, 293
Rectifiable curve, 183
Rectified sine wave, 331
Reduced equation, 59
Reduction of order, 85
Reflection, 447
Removable singularity, 511
Residue, 513
Residue Theorem, 517
Resonance
electrical, 93; mechanical, 96
Response, 46
Riemann surface, 427
RL circuit, 45, 134
RLC circuit, 87
Root Test, 493, 594
Rotation, 447
Rotation vector, angular, 201
Row of a matrix, 19
Row vectors, 21

Scalar field
definition of, 199; invariance of, 199; variation of, 207
Scalar multiplication, 2, 20
Scalar projection, 163
Scalar triple product, 166
Scale factors, 304
Second Shifting Theorem, 133
Separable differential equation, 27, 386
Sequence
bounded, 575; complex, 489; monotone, 485; of functions, 576; of numbers, 572; of partial sums, 590
Series
alternating, 595; complex, 491; geometric, 492, 591; harmonic, 592; infinite, 141, 590; Laurent, 504, 556; of functions, 595; power, 494, 498, 600; Taylor, 498, 602
Set
bounded, 410; closed, 410; open, 410
Shifting Theorem
First, 110; Second, 133
Simple curve, 183
Simply connected, 262, 273, 476
Sine integral function, 545
Sine series, 345
Sine wave, rectified, 331
Singularity
definition of, 420; essential, 511; polar, 511; removable, 511

Singular point
irregular, 155; of a differential equation, 155; regular, 155
Skew curve, 181
Sliding block, 97
Smooth curve, 181
Smooth surface, 193
Snowplow problem, 30
Solenoidal, 218
Solution
general, 378; particular, 27, 378; product, 397; set, 27; steady state, 403
Solution space
definition of, 66; dimension of, 66
Speed, 194
Standard domain, 260, 276
Steady state current, 90
Steady state solution, 403
Stokes Theorem, 268, 604
Sturm-Liouville problem, 372
Subspaces of $PC[-L, L]$, 333
Substitution, 33
Sum
of an infinite series, 590; of tensors, 309
Superposition, 74, 381
Surface
closed, 255; constant level, 175, 204; definition of, 174; equipotential, 203; isothermal, 203; normal to, 212; orientable, 193; Riemann, 427; smooth, 193
Surface area, 246
Surface integral, 250, 255
Symmetric property, 8
System of equations, 100, 127

Table of
Bessel functions, 610; Erf x, 544; Laplace transforms, 105; Mappings, 606; Values of gamma function, 534; Zeros of J_1 and J_2, 560
Tangent vector, 185, 441
Taylor series, 142, 498, 602
Tchebichef polynomials, 326
Telegraph equations, 401
Temperature distribution, 456, 461
Tensor
addition, 309; associated, 312; contravariant, 296; covariant, 306; mixed, 307
Tone, 399
Total differential, 37
Transform
Fourier, 369; Laplace, 106
Transformation, bilinear, 446
Transformation law
contravariant, 300; covariant, 300
Transient current, 88
Transient distribution, 404
Translation, 446
Transmission line
differential equation of, 391; low frequency, 401
Transpose of a matrix, 23
Traveling waves, 389
Triangle inequality, 11, 408
Trigonometric function, complex, 431
Triples, ordered, 3, 161
Trivial linear combination, 14
"Turn on" function, 129, 130
Twisted curve, 181

Underdamping, 90
Unidirectional field, 206, 223
Uniform convergence, 578, 587, 597
Unit step function, 130, 370

Value at infinity, 439
Variation of parameters, 82
Variation of scalar field, 207
Vector
column, 21; derivative, 172; Euclidean, 3, 161; geometric, 1; row, 21; tangent, 185, 441
Vector components, 16
Vector equation of line, 180
Vector field
conservative, 212, 242, 273, 281; definition of, 200; gradient, 210
Vector function
of one real variable, 169; of two real variables, 170; representing a curve, 180; representing a line, 180; representing a surface, 176
Vector moment, 166
Vector potential, 224, 271, 280
Vector space
abstract, 3; dimension of normed, 7
Vector triple product, 167
Velocity, 193
Vibrating membrane, 567
Vibrating string problem, 390

Wave, traveling, 389
Wave equation
D'Alembert's method, 388, 393; derivation of, 390; product solution, 397
Weber function, 549
Weighted inner product, 321
Work, 164
Wronskian, 18, 66, 84

Zero
isolated, 510; of a function, 510
Zero arrow, 2
Zeros of the Bessel functions, 558